THE
NEUROSPORA
COMPENDIUM
Chromosomal Loci

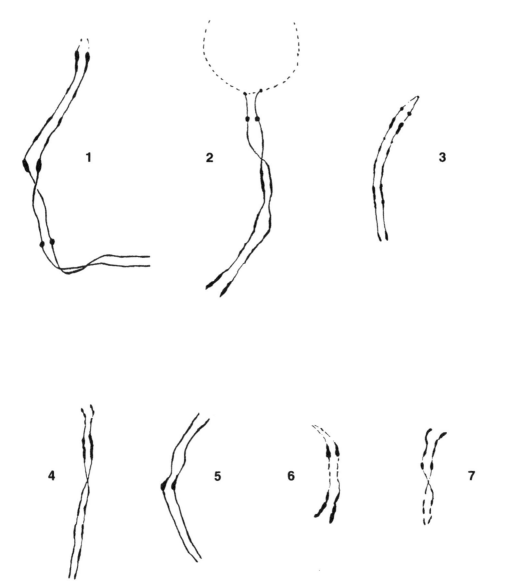

FRONTISPIECE Barbara McClintock's unpublished sketches of the seven *Neurospora crassa* chromosomes. Pachytene bivalents stained with aceto-orcein. Courtesy of E. G. Barry.

THE NEUROSPORA COMPENDIUM

Chromosomal Loci

David D. Perkins
Department of Biological Sciences
Stanford University
Stanford, California

Alan Radford
School of Biology
Leeds University
Leeds, United Kingdom

Matthew S. Sachs
Department of Biochemistry,
and Molecular Biology
Oregon Graduate Institute of
Science and Technology
Beaverton, Oregon

ACADEMIC PRESS

A Harcourt Science and Technology Company

SAN DIEGO   SAN FRANCISCO   NEW YORK   BOSTON   LONDON   SYDNEY   TOKYO

Cover photograph: Rodlets on the surface of wild-type *Neurospora crassa* conidia. The hydrophobic rodlets promote aerial dissemination by keeping the conidia dry and powdery. Rodlets consist of a secreted hydrophobin, which is specified by the gene *eas* (*easily wettable*). Scanning EM photograph of freeze-etched material. Reprinted with permission from *Nature* [ref. (142)]. Copyright 1978 Macmillan Magazines Ltd.

This book is printed on acid-free paper.

Academic Press
A Harcourt Science and Technology Company
525 B Street, Suite 1900, San Diego, California 92101-4495, USA
http://www.academicpress.com

Academic Press
Harcourt Place, 32 Jamestown Road, London NW1 7BY, UK
http://www.academicpress.com

Library of Congress Catalog Card Number: 00-102573

International Standard Book Number: 0-12-550751-8

Printed and bound by CPI Group (UK) Ltd, Croydon, CR0 4YY
Transferred to Digital Print 2011

Contents

PREFACE   vii
ACKNOWLEDGMENTS   ix

I   Introduction   1
Resources   3

II   Information on Individual Loci   7
Explanatory Foreword   7
Synonymous Gene Symbols   8
Alphabetical Entries for Genes and Other Loci   12

Appendix 1   *Neurospora* Genetic Nomenclature   199
Appendix 2   Genetic Maps and Mapped Loci   207
Appendix 3   Data for RFLP Mapping   231
Appendix 4   The *Neurospora crassa* Mitochondrial
             Genome   241
Appendix 5   Expressed Sequences from Different Stages of
             the *Neurospora* Life Cycle: Putative
             Identification of cDNAs   243
References   261

Preface

Those who work with *Neurospora* have chosen the organism for a variety of reasons. B. O. Dodge, who described the sexual phase in 1927 and who pioneered analysis of the *Neurospora* life cycle, was so impressed by its advantages for genetic work that he tried to persuade T. H. Morgan and his *Drosophila* group to adopt the organism. Although not persuaded, Morgan encouraged a graduate student, Carl Lindegren, to explore the potentialities of *Neurospora*. Ten years later, knowledge of Dodge's and Lindegren's work led Beadle and Tatum to adopt *Neurospora* in their search for nutritional mutants, with the result that *Neurospora* gained wide recognition and became the fungal counterpart of *Drosophila*. Their 1941 paper "Control of Biochemical Reactions in *Neurospora*" broke down the barriers that had separated biochemistry and genetics. The *Neurospora* work went on to establish the similarity of genetic mechanisms in fungi to those in *Drosophila*, maize, and other "higher" eukaryotes. It also inspired genetic studies of other microorganisms — *Escherichia coli*, *Chlamydomonas*, *Aspergillus*, *Sordaria*, *Ustilago*, *Saccharomyces*, *Schizosaccharomyces*, *Coprinus*, *Podospora*, *Ascobolus*, *Schizophyllum*, *Ophiostoma*, *Cochliobolus*, *Magnaporthe*, and others.

Neurospora is not only phylogenetically distinct from *Saccharomyces*, but also more complex in both structure and life cycle. More than half the expressed genes identified thus far in *Neurospora crassa* have no recognizable homolog in the yeast genome. Although *E. coli* and *Saccharomyces* have become the workhorses of genetics and molecular biology, *Neurospora* remains the preferred model for filamentous fungi and the organism of choice for numerous problems that cannot be investigated effectively using bacteria or yeast.

A combination of features make *Neurospora* ideal for studying circadian rhythms, vegetative incompatibility, mitochondria and mitochondrial plasmids, and nuclear trafficking in the common cytoplasm of a heterokaryon. Meiosis and ascospore differentiation occur without cell division within a single giant cell, the ascus, where the morphology and precisely programmed movements of chromosomes, nuclei, and organelles can be observed effectively with the light microscope. Natural populations are readily sampled.

In the six decades since Beadle and Tatum, *Neurospora* workers have accumulated a wealth of biological and genetic information. Correspondingly rich resources have become available for research. Genetically characterized wild-type and mutant strains and strains from worldwide natural populations can be obtained from a stock center. Genomic and cDNA libraries, clone banks, and individual cloned genes are available. Physical mapping and genome sequencing are in progress.

In 1982, information on all known chromosomal genes was brought together by Perkins, Radford, Newmeyer, and Björkman for publication in *Microbiological Reviews* in a summary paper popularly known as the "compendium." The present volume is its successor.

David Perkins
Alan Radford
Matthew Sachs

Acknowledgments

We are deeply grateful to the many colleagues who provided unpublished information or examined parts of the text and helped improve its accuracy and completeness. Unpublished sources are acknowledged by citation in the References section. Those who reviewed entries in their areas of specialization are too numerous to name. Mary Anne Nelson was especially helpful in providing new EST and RFLP information from the University of New Mexico Genome Project. We thank Dorothy Newmeyer for continued encouragement and advice. Previously unpublished diagrams or photographs have been contributed by Barry Bowman (Fig. 66), David Catcheside (Fig. 57), Marta Goodrich-Tanrikulu (Figs. 17 and 18), George Marzluf (Fig. 50), Tom Schmidhauser (Fig. 7), and Matthew Springer (Figs. 14, 23, 28, 29, 35, 58, and 60). Photographic prints of previously published figures were made available by Ross Beever, Ramesh Maheshwari, N. B. Raju, and Matthew Springer.

Work in our laboratories was supported by Grants AI-01462 and K6-GM-4899 from the National Institutes of Health and MCB-9728675 from the National Science Foundation to D.D.P., Grants GM-47498 from the National Institutes of Health and MCB-9630910 from the National Science Foundation to M.S.S., and grants from the UK Biotechnical and Biological Sciences Research Council to A.R.

Chapter I

Introduction

This volume, which succeeds the 1982 compendium entitled "Chromosomal loci of *Neurospora crassa*" (1596), brings together data on all of the nuclear genes that have been identified and characterized in the 75 years since the sexual phase of *Neurospora* was discovered and the genus was named. Coverage includes publications through September 1999. The number of mapped loci has doubled since 1982. Newly added genes are described here, and new information is provided on the genes that were included previously. We use "chromosomal loci" to include not only protein-encoding genes but also rearrangement breakpoints, centromeres, telomeres, and genes that encode functional RNAs. Linkage maps have been revised and augmented. Known sites of gene action are shown in diagrams of biosynthetic and catabolic pathways. The text is concerned primarily with the organization and function of each locus and only secondarily with allelic variation or the properties that distinguish specific alleles. Rearrangement breakpoints are not given separate entries, but all mapped chromosome rearrangements are described individually in ref. (1578). Here, as in the 1982 compendium, "chromosomal" in the title refers to the nuclear chromosome complement. An appendix is devoted to the mitochondrial genome.

Investigation of the *Neurospora* genome began in the mid-1920s, when C. L. Shear and B. O. Dodge described the life cycle (Fig. 1), named the genus, and showed that the two mating types of *Neurospora crassa* segregate so as to give 1:1 ratios in randomly isolated ascospores and 4:4 ratios in individual unordered asci (1899). Dodge found that morphological variants also showed Mendelian segregation. In the 1930s, Carl Lindegren used morphological mutants, centromeres, and mating type to construct the first genetic maps, which consisted of six loci in linkage group I (1200) and four in linkage group II (1203). Markers suitable for mapping became abundant in the 1940s, when mutagens were used effectively to obtain auxotrophs and an array of other mutant types and techniques were devised for mutant enrichment. By 1949, six linkage groups had been established (907) and the seventh was soon added (1713). From eight loci mapped in

1937, the number had increased to about 50 in 1949 (907), 75 in 1954 (103), and 500 in 1982 (1596). Now, at the millennium, the number of mapped loci exceeds 1000 (Appendix 2).

Physical knowledge of the genome increased in parallel with the growth of genetic knowledge. Barbara McClintock showed in 1945 that the seven tiny *Neurospora* chromosomes can be identified individually using light microscopy (Frontispiece; Fig. 2). She and J. R. Singleton went on to describe chromosome morphology and behavior during meiosis and mitosis in the ascus (1307, 1918) (Fig. 3). Genetic evidence of chromosome rearrangements was confirmed cytologically (907, 1307, 1918). Genetically mapped rearrangements soon enabled linkage groups to be assigned to each of the seven cytologically distinguishable chromosomes (1580). The discovery of insertional and quasiterminal rearrangements made it possible to map genes by duplication coverage [see ref. (1566)]. The complete meiotic karyotype was reconstructed in three dimensions using the synaptonemal complex with its associated recombination nodules (740). Electrophoretic separation of whole-chromosome DNAs provided estimates of the DNA content of individual chromosomes and yielded a value of 43 megabases for the entire haploid genome (1501, 1505)—about three times that of yeast and roughly one-third that of *Drosophila* or *Arabidopsis*. Physical mapping (54) and genome sequencing are now in progress.

When the first *Neurospora* compendium was prepared in 1982, a gene could not be recognized until a phenotypically recognizable variant had been obtained, either as a mutant in the laboratory or as a novel allele from the wild. Molecular genetics has changed all that. Genes can now be recognized on the basis of nucleotide sequence and mapped using restriction fragment length polymorphisms, all before any phenotype is known. Numerous genes in the present compilation have been identified using the cloned wild-type allele. To determine the phenotype of such a gene, a null allele or a mutant allele must be obtained by gene replacement or by repeat-induced point mutation (RIP). Genes without known mutant phenotypes may soon

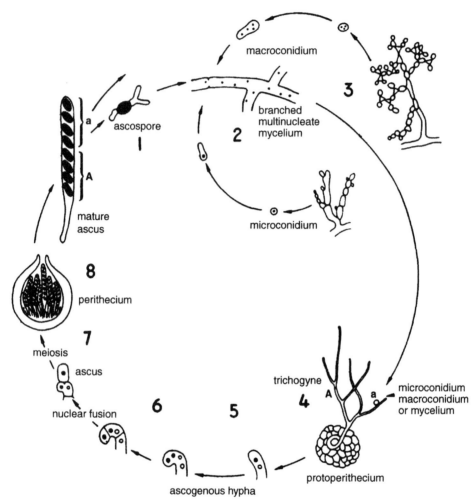

FIGURE 1　The *Neurospora crassa* life cycle. Stages 1–4, from ascospore germination to fertilization, constitute the vegetative phase, and stages 4–8, from fertilization through ascospore maturation, constitute the sexual phase. Nuclear fusion and meiotic prophase occur 4–5 days after fertilization at 25°C. Ascospores are shot from the perithecium from the ninth day. (1) The black ascospores, approximately 17×26 μm, can be isolated manually from an agar surface without the use of micromanipulation apparatus. Ascospore dormancy is broken by heat or chemicals. A germinating ascospore is shown, with hyphae growing from both ends. (2) The mycelium consists of branched, threadlike hyphae made up of multinucleate cells separated by perforate cross-walls through which nuclei and cytoplasm can pass readily. Linear growth of hyphae can exceed 5 mm/h at 37°C. (3) Vegetative spores are of two developmentally distinct types: powdery orange conidia 6–8 μm in diameter, mostly multinucleate, and smaller microconidia, which are uninucleate. (4) Protoperithecia are formed by coiling of hyphae around ascogonial cells. Specialized hyphae, the trichogynes, are attracted to cells of the opposite mating type, from which they pick up nuclei and transport them to the ascogonium within the protoperithecium. Fertilizing nuclei may originate from macroconidia, microconidia, or mycelium. Usually only one fertilizing nuclear type contributes to the contents of a perithecium, but exceptions can occur and are revealed when the fertilizing parent is heterokaryotic (1431). (5) Upon fertilization, the haploid *A* and *a* nuclei do not fuse but proliferate in heterokaryotic ascogenous hyphae. (6) The final conjugate division before ascus formation occurs in a binucleate hook-shaped structure, the crozier. Nuclei of opposite mating type fuse, and the diploid zygote nucleus immediately enters meiosis. The fusion nucleus is the only diploid nucleus in the life cycle. Until ascospore walls are laid down following a postmeiotic mitosis, the ascus remains a single, undivided cell within which the two parental genomes are present in a 1:1 ratio in a common cytoplasm. (7) Details of meiosis and ascus development are diagrammed in Fig. 3 and documented photographically in ref. (1676). Asci within a perithecium do not develop synchronously. Numerous asci are initiated successively from the same ascogenous hypha. Mature eight-spored asci measure about 20×200 μm. (8) The perithecial wall is maternal in origin. Each mature perithecium, 400–600 μm in diameter, develops a beak that terminates in an osteole through which ascospores are forcibly shot in groups of eight, each comprising the contents of an ascus. Adapted from ref. (645).

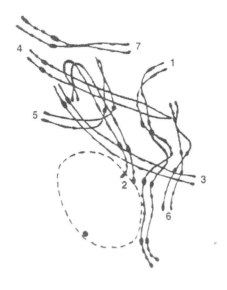

FIGURE 2 The *Neurospora crassa* chromosome complement in a pachytene nucleus. Following the convention used by McClintock, the chromosomes are numbered in order of their measured length at pachytene. Chromosome 5 has been shifted to the right in the drawing to avoid a confusing overlap with parts of 2 and 4. The overlap in the actual nucleus was not a problem because the arms were in different focal planes. The chromosome 1 bivalent is homozygous for inversion *H4250*. Orcein staining, × 4000. Photograph by E. G. Barry from ref. (1580).

outnumber those for which the mutant phenotype has been determined.

Accounts of the knowledge obtained using *Neurospora*, and evaluations of the significance of *Neurospora* for genetics and molecular biology are given in refs. (459)–(461), (893a), (894), and (1575). The development of *Neurospora* cytogenetics and chromosome cytology is outlined in refs. (461), (1573), (1575), (1578), and (1580). Reference (1669) provides an overview of the *Neurospora* genome, with an emphasis on molecular aspects.

RESOURCES

Stocks and Clones

The preferred source of genetically defined cultures is the Fungal Genetics Stock Center (FGSC), Department of Microbiology, University of Kansas Medical Center, Kansas City, KA 66160-7420. The FGSC specializes in *Neurospora* stocks for research. Over 7500 *Neurospora* stocks are maintained in the main collection, including key mutant alleles at most of the known gene loci. Every effort is made to ensure authenticity and, where possible, to provide stocks with defined genetic backgrounds and free of modifiers and extraneous genes. In addition, FGSC maintains strains from specialized collections, including >4000 isolates from wild populations. Stock lists are

available on line, and a catalog of the main collection is published biennially as a supplement to *Fungal Genetics Newsletter*. Cultures are listed under the following headings: Single Mutants, Multiple Mutants, Nonchromosomal Mutants, Wild-Type and Wild-Collected Strains (these represent all the known species), Chromosome Rearrangements, and Reference Strains, Testers, and Stocks for Special Purposes. The last category includes linkage testers, standard wild types and mating-type testers, diagnostic testers for determining species, heterokaryon-incompatibility testers, strains for mutant enrichment and replication, progeny from crosses for RFLP mapping, strains for obtaining protoplasts or spheroplasts, strains for disruption of essential genes, strains for directed transformation, strains for assaying aneuploidy, strains that produce only microconidia, special strains for teaching, strains for photobiology and circadian rhythms, Spore-killer testers, transport mutants, and strains for studying mutations in the *ad-3* region.

FGSC also has more than 15,000 clones, including EST banks, individual cloned genes, and a growing number of libraries, including seven genomic libraries, nine cDNA libraries, and three two-hybrid libraries.

The FGSC web site provides links to other culture collections that include *Neurospora* in their holdings. The American Type Culture Collection maintains approximately 200 *Neurospora* stocks, most of which are redundant with those in FGSC.

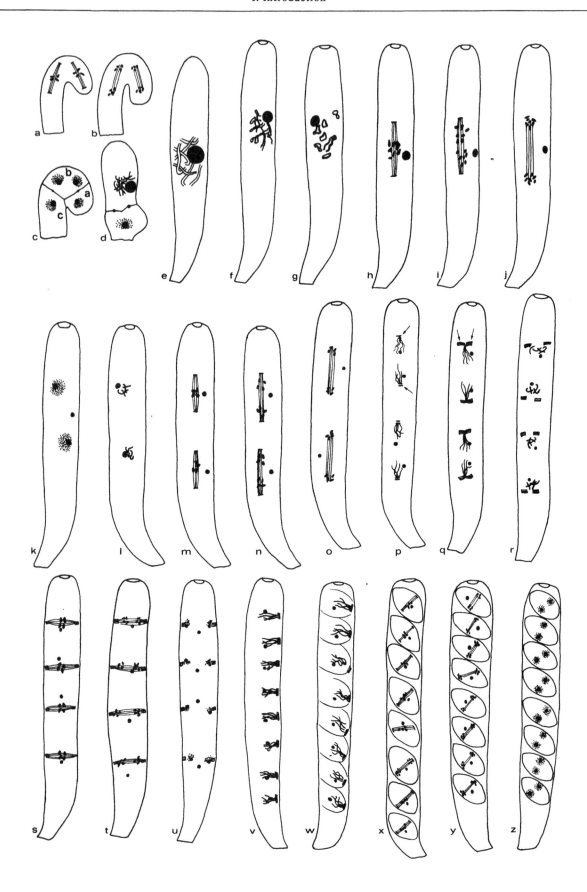

Sources of Information

The Internet

(1) *http://www.fgsc.net*   (Fungal   Genetics   Stock Center, Kansas City, KS)
This site provides:

- The FGSC Catalog, including *Neurospora* wild types, mutants, strains from natural populations, and strains for testing, reference, and other special purposes. The *Neurospora* holdings are predominantly *N. crassa*, but other *Neurospora* species are included, as well as some representatives of related genera.
- Lists of available cloned genes and gene libraries.
- *Fungal Genetics Newsletter* (full text for recent years). Includes new acquisitions by FGSC, recent publications, and current genetic maps.
- Cumulative RFLP map data.
- A bibliography of publications on *Neurospora* from 1843 to the present. This is a nearly complete listing that combines the 1965 bibliography of Bachmann and Strickland (81) with annual bibliographies from all of the *Neurospora Newsletters* and *Fungal Genetics Newsletters* issued in subsequent years.
- The complete 1982 *Neurospora* compendium from *Microbiological Reviews* 46: 426–570.
- Abstracts of Neurospora Conferences and Fungal Genetics Conferences.
- Forms for on-line ordering from FGSC and forms for depositing strains and plasmids.
- Other features, including descriptions of methods and provisions for discussion and the posting of news. Users of this compendium are encouraged to send corrections and additions to FGSC for posting at their web site.

Information in the FGSC web site is available in formats both for searching and for browsing. Links are provided to other useful sites. A European mirror of the FGSC site will be found at the U.K. Human Genome Mapping Project Resource Centre (HGMP) *http://www.hgmp.mrc.ac.uk/research/fgsc/index.html.*

(2) *http://www.unm.edu/~ngp/*   (University of New Mexico Neurospora Genome Project)
This site provides an annotated proteome (549a) with searchable databases of EST sequences for genes expressed at different stages in the *Neurospora* life cycle. (See Appendix 5 for the January 2000 version.) The project also provides and maintains an RFLP mapping program to facilitate placement of new genes using cumulative data for 18 ordered tetrads from a highly polymorphic cross (1452). The RFLP tabulation can be found at the FGSC web site. (See Appendix 3 for the January 2000 version.)

(3) *http://www.genome.ou.edu/*   (University of Oklahoma Advanced Center for Genome Technology)
EST sequences are listed for *Neurospora* genes expressed at dawn and dusk in the circadian cycle.

(4) *http://websvr.mips.biochem.mpg.de/*   (Munich Information Centre for Protein Sequences)

(a) *http://websvr.mips.biochem.mpg.de/cgi-bin/proj/medgen/homecrassa*
Includes information linking *Neurospora* genes to sequences, with literature citations.

(b) *http://websvr.mips.biochem.mpg.de/proj/medgen/mitop/*
Consolidates information on both nuclear- and mitochondrial-encoded genes and their proteins. Files are given for *N. crassa* and four other species. Interrelated sections include "Gene Catalogues," "Protein catalogues," "Homologies," and "Pathways and Metabolism."

(c) *http://www.mips.biochem.mpg.de/proj/neurospora/*
Neurospora genomic DNA sequences for linkage groups II and V.

(5) *http://fungus.genetics.uga.edu*   (Fungal Genome Resource, University of Georgia)
This site provides data on the physical mapping of the *Neurospora* genome.

FIGURE 3 Ascus development, meiosis, and ascospore formation in *Neurospora crassa*. Successive drawings are not to the same scale. (a)–(d) are enlarged several times relative to later stages. Mature ascospores are at least 2-fold larger than ascospores that have just been formed. For clarity, fewer than seven chromosomes are shown in most nuclei. The nuclear envelope (not shown) remains intact during each division. (a) Conjugate metaphase in the crozier. (b) Conjugate telophase. (c) Three-celled crozier. The binucleate penultimate cell (b) grows into an ascus as in (d). The apical cell (a) fuses with the stalk cell (c). (d) Young ascus after nuclear fusion. (e) Pachytene ascus with extended (paired) chromosomes and enlarged nucleolus (see Fig. 2). (f) Diffuse diplotene. The ascus pore forms at this stage. (g) Diakinesis. (h) Metaphase I. (i) Anaphase I. Chromosome separation is asynchronous. (j) Telophase I. (k) Interphase I. (l) Prophase II. (m) Metaphase II. (n) Anaphase II. (o) Telophase II. (p) Early interphase II. Spindle plaques have just formed (arrows). (q) Late interphase II with enlarged double spindle plaques (arrows). (r) Prophase III. Spindle plaques have separated. (s) Metaphase III. Spindles orient across the ascus. (t) Anaphase III. (u) Telophase III. (v) Interphase III. All eight nuclei have lined up in single file. (w) Ascospore delimitation. (x) Metaphase IV (one ascospore is at anaphase IV). (y) Telophase IV. (z) Binucleate ascospores. From ref. (1676), with permission from Urban & Fischer Verlag.

Published Sources of Information

The most generally useful publication will be Rowland Davis's comprehensive book, "Neurospora" (461). Other useful publications are the book, "Fungal Genetics," by Fincham, Day, and Radford (645) (basic background information); and reviews by Davis and de Serres (462) (basic methods and media) and Perkins and Barry (1580) (cytogenetics).

Also useful are the *Fungal Genetics Newsletter* and the 1982 compendium from *Microbiological Reviews* (1596). (These are also available at the FGSC web site.) The 1965

bibliography of Bachmann and Strickland (81) is notable for its meticulously prepared subject index.

Meetings

A Neurospora Conference is held at 2-year intervals, alternating with the more inclusive biennial Fungal Genetics Conference. An elected Neurospora Policy Committee is responsible for coordination and planning of the Neurospora Conference. Committee members are identified in the current *Fungal Genetics Newsletter.*

Chapter II

Information on Individual Loci

EXPLANATORY FOREWORD

In this section, information is gathered on each established nuclear gene and on chromosome features such as centromeres, tips, and the nucleolus organizer. Entries are arranged alphabetically by symbol. Some categories of mutants are prefaced by a generic entry that concerns the entire group (e.g., *al, arg, het, mat, mus, rec*). Synonyms are cross-referenced to the preferred symbol. Abbreviations are defined in Table 1. Synonymous gene symbols are listed in Table 2.

Within each entry for a gene locus, the name follows the symbol. The linkage group is then indicated and the arm is specified, if known. This is followed by linkage data relating the gene to other loci. If the gene has been cloned and sequenced, accession numbers are given for the sequence. The pSV50 cosmid library number or other reference number of clones containing the sequence may also be given. The phenotype is described in a subsequent paragraph, including such attributes as enzyme deficiency (including EC number), dominance, interaction with other genes, fertility, and stability. The described phenotype is that of mutant strains unless stated otherwise. Scoring information, technical applications, and alternate names may be given.

At the time the previous compendium was prepared, a gene could not be recognized or mapped until a phenotypically recognizable mutant or variant had been discovered. In contrast, genes can now be recognized and defined solely on the basis of DNA sequence, without knowledge of the phenotype. Some of the genes discovered in this way have no obvious null phenotype. For many others, the null phenotype is unknown. The gene name may then be based on the inferred gene product or the known function of orthologues or homologues in other organisms. Making the name explicit in such a situation would ideally require addition of the suffix "-like" (for example, *ras*-like), but for brevity the suffix is usually omitted, just as the suffix "-requirement" is omitted, although it is implied, when auxotrophic mutants are named. Similarly, qualifying adjectives have often been omitted. When the function of a gene is inferred from sequence homology, "structural gene" usually implies "putative structural gene." Likewise, "encodes" is used in place of "inferred to encode." A gene symbol without a superscript is used either to designate a locus or to symbolize a mutant allele at that locus. Which usage applies in a particular instance should be obvious from the context.

In the paragraph on linkage, recombination values may be given that originated from more than one cross. Differences between these values may be due to sampling error (sample size is usually less than 100). The differences may also reflect differences in *trans*-acting *rec* genes, which are known to regulate recombination in specific local regions. Values given as percentages or as numerical fractions indicate the frequency of reciprocal recombination per meiotic product, unless it is indicated explicitly that the value is for reciprocal crossing-over events per tetrad. (The number of crossing-over events per tetrad is double that of crossovers per random meiotic product for the same interval.)

When gene order is based on duplication coverage or on three-point crosses, the sequence of loci is often known with confidence even though interval lengths are unknown or imprecisely known. Where linkage information is given in the format, "Locus *c* is between *a, b* and *d, e*," the word "and" separates markers to the left of *c* from markers to the right.

RFLP data from two widely used crosses are tabulated in ref. (1452) and are reproduced here as Appendix 3. These come either from 18 tetrads (the first cross) or from 18 random isolates (the second cross). Data from these crosses have been extremely valuable for assigning cloned sequences to a linkage group and for indicating their approximate location. The two RFLP-marked crosses were not intended to provide information for precise mapping. Because of the small numbers, estimates of distance are imprecise and gene order is often uncertain.

Results from the first RFLP-scored cross are presented here as numbers of asci of the three types, PD (parental ditype), NPD (nonparental ditype), and T (tetratype), using

TABLE 1
Abbreviations[a]

3-AT	3-Amino-1,2,4-triazole (aminotriazole)	NO	Nucleolus organizer
bp	Base pair(s)	NOR	Nucleolus organizer region
BT	Beadle and Tatum wild type or origin (1471)	NPD	Nonparental ditype tetrad
cAMP	Cyclic adenosine monophosphate (cyclic AMP)	nt	Nucleotide(s) or nucleotide pair(s)
cDNA	Complementary DNA	OR	Oak Ridge wild type (1471) or origin
DEAE	Diethylaminoethyl (-cellulose)	ORF	Open reading frame
EC	Enzyme classification number	PABA	p-Aminobenzoic acid
EDTA	Ethylenediaminetetraacetic acid	PCR	Polymerase chain reaction
Em	Emerson wild type (1471)	PD	Parental ditype tetrad
EM	Electron microscope	PIR	Protein sequence database at the Protein Information Resource (PIR)
EMBL	Sequence database at European Bioinformatics Institute (EBI)	RAPD	Random amplified polymorphic DNA (2214)
EMBO	European Molecular Biology Organization	rDNA	DNA sequence(s) specifying ribosomal RNA
EMS	Ethylmethane sulfonate	RFLP	Restriction fragment length polymorphism (1339, 1340)
EST	Expressed sequence tag		
FGSC	Fungal Genetics Stock Center	RIP	Repeat-induced point mutation (1886)
FPA	p-Fluorophenylalanine	RL	Rockefeller–Lindegren wild type (1471)
G6PD	Glucose-6-phosphate dehydrogenase	rRNA	Ribosomal RNA
GenBank	Sequence database at National Center for Biotechnology Information (NCBI)	SAM	S-Adenosylmethionine
		SEM	Scanning electron microscope
HMG	High-mobility group (domain)	SHAM	Salicylhydroxamic acid
ITS	Internal transcribed spacer	SPB	Spindle pole body
kb	Kilobase pairs (duplex DNA), kilobases (single-stranded)	SSR	Simple sequence repeat
		Swissprot	Protein sequence database at European Bioinformatics Institute (EBI)
kD	Kilodalton		
MMS	Methylmethane sulfonate	T	Tetratype tetrad
MT	4-Methyltryptophan	Tn	Transposon
NADH	Nicotinamide adenine dinucleotide	UNM	University of New Mexico
MNNG	N-Methyl-N'-nitro-N-nitrosoguanidine	URF	Unassigned reading frame
NG	N-Methyl-N'-nitro-N-nitrosoguanidine	YAC	Yeast artificial chromosome

[a]Additional abbreviations are given in the legends of Figs. 5, 11, 18, and 40 and a footnote to Table 3.

TABLE 2
Synonymous Gene Symbols[a]

Replaced symbol	Symbol now in use	Replaced symbol	Symbol now in use
ac	ace	amy(SF26)	exo-1
ace-6	suc	ANT-1[b]	azs; has
acon-1	fl	arg-7	arg-4
acp	aac	arg-8	pro-3
acp[i]	acu-16	arg-9	pro-4
act	cyh	arg(CD-15)	cpc-1
adg	arg-11	arg(CD-55)	cpc-1
age-1-3	so	arg(RU1)	am
age-3	al-1	arg(RU3)	arg-13
alx-1[b]	azs; has	arg[R]	pmb (?)
amr	nit-2	arom	aro
ami	sor-4	arp1	ro-4

Replaced symbol	Symbol now in use	Replaced symbol	Symbol now in use
asc(DL95)	*mei-1* (?)	*fas*	*cel*
asc(DL243)	*mei-1* (?)	*fdu-1*	*ud-1*
asco	*lys-5*	*flm-1*	*os-1*
asp	*asn*	*flm-2*	*os-4*
aspt	*asp*	*frq-5*	*prd-1*
atp-9	*oli*	*G*	*gul*
aur	*al-1*	*gla*	*sor-4*
bas	*pmb* (?)	*glm*	*gln-1*
bat	*pmb*	*glp-3*	*ff-1*
Ben	*Bml*	*gluc-2*	*gluc-1*
bis	*pk*	*gly*	*glp*
bli-7	*eas*	*gpi-2*	*gpi-1*
bm-1	*pmb*	*grg-1*	*ccg-1*
c	*het-c*	*ham*	*so*
c	*col-3*	*hist*	*his*
c	*cot-4*	*hs*	*hom*
c	*cy*	*hsp83*	*hsp80*
can	*cnr*	*hspe*	*hsp80*
ccg-2	*eas*	*hsps-1, hsps*	*hsp70*
ccg-7	*gpd-1*	*hsps-2*	*grp78*
ccg-12	*cmt*	*i*	*en(am)-1*
cla-1	*frq⁷*	*i*	*het-i*
col-3	*bn*	*inos*	*inl*
col-7	*rg-1*	*iv*	*ilv*
col-13	*vel* (?)	*kex*	*scp*
col-l4	*sc*	*lni-1*	*cpc-1*
col-15	*vel* (?)	*lni-2*	*lacc*
col-le	*le-1*	*lox*	*lao*
col(C-L2b)	*mel-1*	*lysᴿ*	*su(mtr)-1* (?)
cox-5	*cya-4*	*m*	*pe*
crp-1	*cyh-2*	*nac*	*met-6*
cys(oxD1)	*cys-15*	*matᵈ*	*rug*
cyt-3	*cyt-4*	*mbic*	*Bml*
cyt-U-10	*cyt-4*	*mc-1*	*nuo19.3*
cyt-12	*cyc-1*	*me*	*met*
cyt-U-14	*cyt-4*	*mel-2*	*bal*
cyt-U-19	*cyt-19*	*mep-3*	*mep-1*
cyt-20ᶜ	*un-3(55701)*	*mep(3)*	*mep-1*
cyt-U-28	*cyt-25*	*met-4*	*cys-10*
cyt(289-56)ᶜ	*un-3(55701)*	*meth*	*met*
cyt(297-24)	*cyt-21*	*mfA-1*	*ccg-4*
d	*het-d*	*mig*	*tre*
dgr-3	*sor-4*	*mik-2*	*nrc-1*
dgr-4	*sor-3*	*mms*	*mus*
e	*het-e*	*mo-3*	*sk*
en-am	*en(am)-2*	*mo(M111)*	*at*
er	*rg-1*	*mo(KH160)*	*shg*
f	*su([mi-1])-1*	*mo(NM213t)*	*bld*

Continues

TABLE 2 *Continued*

Replaced symbol	Symbol now in use	Replaced symbol	Symbol now in use
mo(P1163)	*dr*	*rec-5*	*rec-2*
mo(P2402t)	*un-20*	*rec-w*	*rec-2*
mo(R2441)	*cwl-1*	*rec-x*	*rec-3*
mod-5	*cpc-1* (?)	*rec-z*	*rec-1*
moe-1	*sk* (?)	*ro-5*	*ro-4*
mom	*tom*	*ro-8*	*ro-4*
morph	*mo*	*ro-9*	*da* (?)
mpp-2	*pep*	*ros*	*al-3*
MS5	*nmr*	*Rsp*	*R*
mt	*mat*	*s*	*arg-12^S*
mt	*mtr*	*t*	*scot* (?)
mts	*cpc-1*	*sdv-10*	*asd-1*
mus(SA60)	*mei-2*	*sdv-15*	*asd-3*
nac	*cr-1*	*smco-2*	*sc*
NC-ras	*ras-1*	*snz*	*pdx-1*
neu	*mtr*	*sor-A*	*sor-1*
nik-1	*os-1*	*sor-B*	*sor-2*
nit-5	*nit-4*	*sor^r-14*	*sor-5*
nuh-4	*uvs-3*	*sor^r-15*	*sor-4* (?)
nuh-4	*mus-9*	*sor^r-17*	*sor-3*
nuo9.6	*acp*	*sor^r-19*	*sor-5*
orn-1	*arg-5*	*sor(DS)*	*sor-4*
orn-2	*arg-6*	*sor(T9)*	*sor-4*
orn-3	*arg-4*	*spco-1*	*col-4*
ovc	*cut* (?)	*spco-2*	*wa*
pab-3	*pab-2*	*spco-3*	*spco-7*
pcon	*nuc-2*	*spco-13*	*spco-7* (?)
pdc-1	*cfp*	*spco-13*	*moe-2* (?)
pdx-2	*pdx-1*	*su-2, etc.*	*su(trp-3^{td2})-2, etc.*
Ph-mod-D	*sor-4*	*su^m*	*su(pe)*
phe-3	*phe-2*	*su-B*	*su(bal)*
phen	*phe*	*su-C*	*su(col-2)*
pmn	*mtr*	*su(pro-3)*	*arg-6*
Pm-N	*mtr*	*sup(...)*	*su(...)*
pph-2	*cna-1*	*sup-1, -3, etc.*	*su([mi-1])-1, -3, etc.*
ppz-1	*pzl-1*	*sw*	*per-1*
prl-1	*oli*	*t*	*scot*
prol	*pro*	*t(289-4)*	*cyt-22*
prt	*pts*	*td*	*trp-3*
ps15-1	*ndk-1*	*thi-lo*	*thi-4*
psp	*ndk-1*	*thr-1*	*ile-1*
put-1	*spe-1*	*tr^u*	*uc-5*
pyr-5^e	*pyr-1 pyr-3*	*try*	*trp*
Q, q	*nic-1*	*tryp*	*trp*
ras-2	*smco-7*	*tub-2*	*Bml*
rco-3	*dgr-3* (?)	*tyr-3*	*tyr-1*
Rsp	*R*	*tyr-s*	*tys*
rec-4	*rec-2*	*tyr(NM160)*	*phe-1*

Continues

Replaced symbol	Symbol now in use	Replaced symbol	Symbol now in use
un(STL6)	*fls*	*un(83106)*	*un-6*
un(b39)	*un-5*	*uve-1*	*phr*
un(44409)	*un-1*	*vac-5*	*htl*
un(46006)	*un-2*	*van*	*pho-4*
un(55701)	*un-3*	*ylo-3*	*fl*
un(66204)	*un-4*	*ylo-4*	*al-1*

[a] The replaced symbols (and corresponding names) have been placed in synonymy for a variety of reasons, mainly to correct violations of priority, to honor priority when allelism was shown with an already named gene, or to avoid redundancy when a proposed symbol was found to already be in use. A new name or symbol may have been adopted when the original designation was shown to be incorrect or misleading (e.g., *met-4*). Symbols originally used for several auxotrophs were changed to conform to the standard symbols adopted by biochemists (e.g., *hs* to *hom*, *tryp* to *trp*). For nomenclature conventions, see Appendix 1. Symbols and synonyms for mitochondrial genes are given in Appendix 4.

[b] Used for the double-mutant strain *azs; has*.

[c] Strain *55701* may be composed of two loci, one of which is *cyt-20*. See the entry for *un-3*.

[d] The morphological mutant formerly called *mat* was renamed (745) with the object of freeing the symbol *mat* to designate the mating-type locus, which had previously been called *mt*. The change brought *Neurospora* into conformity with usage in other fungi.

[e] A strain called *pyr-5* was found to be a *pyr-1 pyr-3* double mutant.

the format: T, NPD/total. (The total number is PD+NPD+T.) Because each crossover event in an ascus recombines only two of the four chromatids, the frequency of crossover meiotic products (on which map distance is based) is half the tetratype frequency when there are no NPDs. NPDs are produced when four-strand double exchanges occur in an interval between markers. If NPDs are present, the best estimate of crossover frequency in the interval is obtained by multiplying the number of NPDs by 6 and adding this to the number of T's before dividing by 2. (The multiplication compensates for cryptic multiple crossovers that result in T or PD asci rather than NPD.) A significant excess of PD asci over NPD asci constitutes proof of linkage [see refs. (645) and (1547)].

Results from the second RFLP cross, which involves random ascospore isolates, are given as a numerical fraction, crossover products/total.

In deriving recombination estimates from the RFLP data of both crosses, we have omitted from the calculation any NPD tetrads (cross 1) or double-crossover progeny (cross 2) in intervals that contain few or no single crossovers. In making the omission, we assume that the apparent double crossovers are spurious, having resulted from experimental or recording errors or from gene conversion, rather than from bona fide reciprocal crossing over. Such errors or conversions appear to be abundant in the existing RFLP data.

Rearrangement breakpoints define loci that have often been crucial for determining gene order. Rearrangements have not been given separate entries in the main section, however, but breakpoints that are well-defined are included in lists of mapped loci and are shown on the genetic maps (Appendix 2). Most of the breakpoints listed are from insertional or quasiterminal rearrangements, which have been widely used to establish gene order and to map

chromosome tips and centromeres. Rearrangements with breakpoints that are located less precisely are not included in the lists in Appendix 2 but are described in Appendix 4, which provides full information on all characterized rearrangements, whether mapped precisely or mapped only to linkage group.

Rearrangement symbols have been abbreviated in the locus entries, lists, and maps. For example, *In(IR;IL)-OY323* is shortened to *In(OY323)*. Multiple breakpoints in the same chromosome are distinguished using superscripts, for example, *In(OY323)^L* and *In(OY323)^R*. Rearrangement symbols are written in full in refs. (700) and (1578).

With a few exceptions, entries are not provided for sequences of unknown function ("anonymous" loci), even if they have been RFLP-mapped, nor are these loci listed with the maps in Appendix 2. RFLP linkage data are given in Appendix 3. Gene loci recognized on the basis of EST sequences are given in Appendix 5; these are named and included in the entries only if strong inference exists for homology with genes of known function in other organisms.

The symbol *Tel* designates a locus determined by RFLP mapping of a telomeric DNA sequence (1810). *Tip* is used to designate a linkage-group terminus inferred to lie beyond the most distal genetic marker. A *tip* may be marked by the breakpoint of a quasiterminal chromosome rearrangement.

In addition to gene loci, rearrangement breakpoints, and chromosome landmarks, entries are included that describe transposable elements, both active (*Tad*) and degenerate (e.g., *Pogo*, *Punt*, *Guest*). Entries also are included for genes from other organisms that have been integrated into the *Neurospora* chromosomes. Examples are *amdS*, *ble*, *hph*, and *tk*.

Allele numbers have been given in the entries for mutant genes only when there is some question about allelism, when two alleles differ significantly in phenotype, or when a mutant has been referred to previously only by allele number. Allele numbers of mutant strains can be found in FGSC listings or in the cited references. Reference (103) provides a table relating allele numbers to the locus symbols that were first assigned in 1954. When a particular mutant gene has not been definitely assigned to a locus, it may be symbolized temporarily by an appropriate base symbol followed by the allele number in parentheses, as *met(T70)*, *mo(D301)*, or *un(74E)*.

For map location and linkage analysis, we have cited mainly the most definitive data sources that establish the location of a gene relative to its immediate neighbors. Earlier references that established linkage originally or that are less precise have usually been omitted. Although they represent tremendous effort, the linkage data in these publications now are of mainly historic interest. The older data may be found in the 1982 compendium (1596) or in refs. (103), (321), (322), (426), (907), (1255), (1369), (1383), (1548), (1585), (1587), (1592), (1593), (1603), (1604), (1928), (1986), and (2014). Original sources of RFLP data are not identified by Nelson *et al.* (1447), nor do we do so. The source usually is one of the publications referenced elsewhere in the same entry.

All of the important references cannot be cited in the entries, especially for loci that have been studied extensively. References for the cloning and sequencing of genes are usually not cited because they can be obtained on-line from the sequence databases. The publications that are cited should lead the reader to other significant literature. Theses and abstracts have been cited only if they are known to contain pertinent information that has not been documented adequately in a published reference. Unpublished sources are cited by number and are identified as such in the References section, where they are listed alphabetically by name of the contributor and are interspersed with the cited publications.

ALPHABETICAL ENTRIES FOR GENES AND OTHER LOCI

A, a

These one-letter symbols are used in most contexts as abbreviations for *mat A* and *mat a*, designating idiomorphs at the mating-type locus in linkage group I.

aab-1: am α-binding protein-1

IIIR. Linked to *trp-1* and *con-7* by RFLP (370).
Cloned and sequenced: EMBL/GenBank AF026550.
Specifies subunit of a nuclear heteromeric complex that binds to a CCAAT box in the upstream activating sequence of *am*. AAB-1 is similar to *Saccharomyces cerevisiae* HAP5. Disruption of *aab-1* by RIP not only

cuts in half the level of glutamate dehydrogenase but also affects growth and development adversely, resulting in short aerial hyphae and reduced conidiation. This indicates that other genes in addition to *am* are subject to regulation by *aab-1* (370).

aac: ADP/ATP carrier protein

Unmapped.
Cloned and sequenced: Swissprot ADT_NEUCR, EMBL/GenBank X00363, PIR A03182, XWNC, GenBank NCADPATP.
Encodes ADP/ATP carrier protein (translocase), adenine nucleotide translocator, exchanging ADP and ATP across the mitochondrial inner membrane (50). Called *acp*.

aaf: acetylaminofluorine requirement

Unmapped. The data are said to be consistent with one gene.
Complex phenotype. The requirement is satisfied by 2-acetylaminofluorine, certain azo dyes, or certain single amino acids. Cold sensitive. Found among progeny of a *rib-1* strain that had become tolerant to 2-acetylaminofluorine (2080).

aag-1: accelerated acetate growth

II. Linked to *arg-5* (33%), *acu-5* (26%) (367).
Growth is faster than that of wild type on acetate as the sole carbon source and slower on sucrose as the sole carbon source. Derepressed, with significant constitutive levels of acetyl-CoA synthetase and glyoxylate cycle enzymes on sucrose; semidominant (367).

aap-2: amino acid permease-2

IR. Linked to *al-2* (0T/18 asci) (1447).
Cloned and sequenced: EMBL/GenBank AF053231.
Encodes a polypeptide predicted to have 12 membrane-spanning regions. The predicted AAP-2 polypeptide has significant sequence similarity to GABA transport proteins in other organisms; its function in *Neurospora crassa* is unknown (1270).

ace: acetate

Acetate mutants *ace-1* through *ace-7* are auxotrophs that grow on 0.3% sodium acetate, as do *suc* mutants (which often grow better on acetate than on succinate). A carbon source also is needed. Most *ace* mutants grow better when the carbon source is maltose rather than sucrose (1122). Acetate mutants, except *ace-1* and *ace-5*, can grow on various Tweens as the sole carbon source (232). The mutants differ in their ability to grow on complex media. Unlinked genes *ace-2*, *-3*, and *-4* are involved with the pyruvate dehydrogenase complex

(1496), a complex enzyme that in yeast is formed from polypeptides encoded by three different genes, with the activities pyruvate dehydrogenase (E1) (EC 1.2.4.1), dihydrolipoamide acetyltransferase (E2) (EC 2.3.1.12), and lipoamide dehydrogenase (E3) (EC 1.8.1.4). A separate set of acetate mutations called *ac-1, -2, -3, -4,* and *-5* (2011) was lost before being mapped or tested for allelism with the mutants now called *ace-1* through *ace-7.*

ace-1: acetate-1

IIR. Between *un-20* (15%) and *eas* (1%), *fl* (11%) (103, 1546).

Requires acetate. Poor growth on complex complete medium. Grows well on acetate (0.1%) aided by ethanol (0.5%) [E. L. Tatum and L. Garnjobst, cited in refs. (103) and (1582)]. Ascospore maturation and germination are slow. Germination is best on sucrose minimal medium with yeast extract and ethanol (714). Not the same as *ac-1* of ref. (2011), which was lost. Called *ac.*

ace-2: acetate-2

IIIR. Between *pro-1* (1%; 9%; 36 kb) and *com* (5%), *ad-4* (4%; 7%), *met-8* (448, 1122, 1546).

Cloned and sequenced: Swissprot ODP2_NEUCR, EMBL/GenBank J04432, EMBL NCPNUC, GenBank NEURPNUC; pSV50 clones 1:6G, 11:1B, 18:12G; pCRD129, pCRD130 (448).

Requires acetate. Will not use succinate or ethanol (556). Encodes the dihydrolipoamide acetyltransferase (EC 2.3.1.12) component (E2) of pyruvate dehydrogenase complex (EC 1.2.4.1) (1496). Good growth on complex complete medium (1122). Not the same as *ac-2* of ref. (2011).

ace-3: acetate-3

IR. Between *lys-3* (≤1%; 1%), *In(OY323)^R* and *nic-1* (<1%). Included in duplications from *In(OY323)×In(NM176)* (9, 115, 1122, 1971).

Cloned/partially sequenced candidates: Either GenBank AI399057 (EST W17H2) or AI397651 (EST SC5E2), which, show homology to the α- and β-subunits of the pyruvate dehydrogenase E1 component, respectively (1448).

Requires acetate. Lacks pyruvate dehydrogenase complex activity (EC 1.2.4.1) (1496). Grows poorly on complex complete medium (1122). Conidiation is best at 25°C, not 34°C (1546). Not the same as *ac 3* of (2011).

ace-4: acetate-4

IVL. Between *cys-10* (19%; 27%) and *fi* (10%) (1122).

Cloned/partially sequenced candidates: Either GenBank AI399057 (EST W17H2) or AI397651 (EST SC5E2), which show homology to the α- and β-subunits of the pyruvate dehydrogenase E1 component, respectively (1448).

Requires acetate. Grows on complex complete medium (1122). Lacks pyruvate dehydrogenase complex activity (EC 1.2.4.1) (1496). Lipoate acetyltransferase fails to aggregate to form the core of the pyruvate dehydrogenase complex. As a result, there is high activity of the free components pyruvate dehydrogenase and lipoamide reductase (1496). Not the same as *ac-4* of ref. (2011).

ace-5: acetate-5

VR. Between *gul-1* (<1%) and *ure-1* (<1%) (1121, 1122).

Requires acetate. Grows poorly on complex complete medium (1122). Not the same as *ac-5* of ref. (2011).

ace-6: acetate-6

Name not used because *suc* has priority. See *suc.*

ace-7: acetate-7

IR. Between *nic-2* (4%; 7%) and *cr-1* (1%; 3%) (1122).

Requires acetate. Structural gene for glucose-6-phosphate dehydrogenase (G6PDH) (EC 1.1.1.49); the enzyme in two revertants is qualitatively different from wild-type G6PDH. Pyruvate dehydrogenase and pyruvate carboxylase activities are normal (1122). Grows well on complex complete medium. Unable to use xylose as a carbon source, resembling *suc* mutants and differing from the wild type and all other *ace* mutants in this respect.

ace-8: acetate-8

VIIL. Between *thi-3* (1%), *T(T54M50)* and *qa-3* (2% or 3%) (1120, 1578).

Structural gene for pyruvate kinase (EC 2.7.1.40) (1123). Growth and conidiation are best on acetate plus ethanol or ethanol plus L-alanine.

ace-9: acetate-9

IIR. Between *nuc-2* (2%) and *arg-12* (3%), *aro-1* (10%) (1788, 1789).

Requires acetate. Very weak activity of the pyruvate dehydrogenase complex. Grows well on complex medium (1788, 1789).

acon-1: aconidiate-1

Allelic with *fl.*

acon-2: aconidiate-2

IIIR. Linked to *vel* (6%), *tyr-1* (9%; 14%), probably between them (1293, 1582).

Macroconidiation is defective. Allele RS91 is heat-sensitive, with macroconidiation blocked early in the developmental pathway at 34°C. No minor constriction budding. Scanning EM photograph (1958). Some conidia are formed at 25°C, but growth is subnormal (1293). Homozygous fertile (1582).

acon-3: aconidiate-3

IVL. Between *cys-10* (1%; 6%) and *cut* (33%) (1582). A report of linkage in VIL was not confirmed.

Macroconidiation is blocked (1293). No major constriction budding. Scanning EM photograph (1958). Some conidia have been observed low in slants at 25°C. Female sterile (1582).

acp-1: acyl carrier protein

Unmapped.

Cloned and sequenced: Swissprot ACPM_NEUCR, EMBL/GenBank X83578, NCMTACP, PIR S00491, S17647 (1825).

Encodes acyl carrier protein precursor (ACP), NADH-oxidoreductase 9.6-kDa subunit (EC 1.6.5.3, EC 1.6.99.3). Mitochondrial acyl carrier protein (240) is part of complex I (1774, 1825), where it has a role in both lipid metabolism and complex formation [1824, 1825, 2149; see *nuo* in ref. (2299)]. Called *nuo9.6* (1843). The symbol *acp* was used temporarily for genes specifying ADP/ATP carrier protein (*aac*) and inducible acetate permease (*acu-16*).

acp^i: inducible acetate permease

Changed to *acu-16*.

acr-1: acriflavine resistant-1

IL. Linked to *mat* (8%; 12%) (931).

Shows low-level resistance to acriflavine (2 µg/ml on agar medium) (931). Scoring may be difficult with some tester strains because differences in sensitivity probably exist between different laboratory strains, with wild-type STA4 being less sensitive to acriflavine than is wild-type Pa, which was used in the original study (1546).

acr-2: acriflavine resistant-2

III. Between *acr-7* and *T(T54M140b)*, *thi-4* (0/286), *sc* (3%; 6%), *spg* (1%; 11%) (931, 1546, 1578, 1592). *acr-2* and *trp-1* (IIIR) cosegregated at the second division in 1 of 13 asci (914), which would favor a right-arm location for *acr-2*. However, *acr-2* is in IIIL if the *T(OY339)* breakpoint is left of *Cen-III* (1546).

Cloned and sequenced: EMBL/GenBank D45893, PIR S72537, S78458 for the main open reading frame, plus the two short upstream open reading frames PIR S72535 and S72536; cosmid pDC107 (27).

Mutant alleles obtained by selecting for resistance to acriflavine (927, 928) are resistant to 3-AT (927) but not to acridine orange, acridine yellow, or malachite green (27). They also are not cross-resistant to cycloheximide and oligomycin, unlike pleiotropic drug-resistant (*PDR*) mutants of *Saccharomyces*. An excellent stable marker, fully fertile. Small inocula should be used to avoid false-positive tests (280). Acriflavine is used at 50 µg/ml in minimal agar medium (1592) (higher concentrations may be used) or 3-AT at 0.5 mg/ml, both added before autoclaving. Resistance is dominant (27, 931). For cloning, a library therefore was constructed using resistant mutant KH2, in which an asparagine residue is changed to lysine. Disruption of *acr-2*^+ by RIP results in greater sensitivity than that of wild type. Resistance thus is due to a functional change in the ACR-2 protein rather than to loss of function of the protein (27). The gene includes two small upstream open reading frames, which are probably regulatory (27).

acr-3: acriflavine resistant-3

IL. Between *un-16* (1%; 5%) and *suc* (1%; 5%), *In(H4250)*^L. Probably right of *ta* (1582, 1592).

Resistant to acriflavine and to malachite green, but not to 3-AT, on agar medium (931). Reported not resistant to malachite green in liquid medium or to acridine orange, acridine yellow, or 3-AT (27). Resistance is dominant in duplications from *In(H4250)* and *T(39311)* (1546) and in heterokaryon tests (931). Scoring on agar is clear when uniform inocula of appropriate size are used, but false-negative or false-positive scoring may result if test inocula are too small or too large. Tests should be read at 2 and 4 days (34°C) to detect possible delayed expression of resistance. On minimal agar medium at 34°C, acriflavine is used at 10 µg/ml (1592) and malachite green at 2 µg/ml (931).

acr-4: acriflavine resistant-4

IL. Linked to *mat*, *acr-3* (5%) (932).

Resistant to acriflavine (50 µg/ml) when *acr-4* is combined with the morphological mutation *shg* (932). Also resistant to acridine orange and acridine yellow, but not to 3-AT or malachite green (27).

acr-5: acriflavine resistant-5

IIR. Between *arg-5* (6%) and *pe* (9%) (1593).

Resistant to acriflavine (50 µg/ml) when *acr-5* allele KH16 is combined with morphological mutation *mo(KH161)* (932). Allele JLC74, however, is readily scorable in strains of wild-type morphology (1546). Not resistant to 3-AT, acridine orange, acridine yellow, or malachite green (27).

acr-6: acriflavine resistant-6

IIIR. Linked to *shg* (0/368) (932).

Resistant to acriflavine (50 μg / ml). The *shg* strain in which *acr-6* originated is acriflavine-sensitive (932). Also resistant to acridine orange and acridine yellow, but not to 3-AT or malachite green (27).

acr-7: acriflavine resistant-7

IIIL. Between *r(Sk-2)-I* (7%), *T(NM183)* (5%) and *acr-2* (12%), *thi-4* (7%), *sc* (12%; 14%) (280, 1582, 2121). A report of VI linkage in ref. (1603) is incorrect.

Resistant to acriflavine (50 μg / ml). Not resistant to 3-AT, malachite green, acridine orange, or acridine yellow (27, 1582). Several *acr-7* strains have become female-infertile after vegetative transfer (1582). In the presence of *cum*, scoring of *acr-7* requires very small inocula that are best obtained by floating conidia down onto agar medium without touching the surface (2121).

act: actidione resistant

Changed to *cyh*.

act: actin

VR. Linked near *inl* (2092).

Cloned and sequenced: Swissprot ACT_NEUCR, EMBL/GenBank U78026.

Structural gene for actin. Filamentous actin is localized primarily to hyphal tips. Transcript levels decrease upon induction of conidiation (2092). The symbol *act* was once used for mutants resistant to cycloheximide (then called actidione). These were then renamed *cyh: cycloheximide-resistant*. The symbol *act* is retained here for the *actin* gene. Doing so violates priority, but is less likely to confuse than would be the creation of a new symbol for *actin*. For an actin-related gene, see *arp3*.

acu: acetate utilization

Unable to use acetate as a carbon source. When ammonium acetate (3 mg/ml) is the sole carbon source, the wild type shows sparse but positive growth, in contrast to clear blanks for *acu* mutants. Mutants are selectable by inositol-less death on acetate medium. *acu-1*, *acu-5*, *acu-6*, and *acu-7* do not behave as respiratory mutants in tetrazolium-overlay tests on acetate medium (593). Positions of mutants in the metabolic pathway are shown in Fig. 4.

acu-1: acetate utilization-1

VR. Right of *asn* (21%) (651).

Unable to use acetate as a carbon source (651, 652). Selected by inositol-less death on acetate medium.

acu-2: acetate utilization-2

IVR. Between *leu-2* (11%) and *pan-1* (6%) (651).

Unable to use acetate as a carbon source. Reduced level of oxoglutarate dehydrogenase (EC 2.3.1.61). Poor recovery from ascospores (651, 652).

acu-3: acetate utilization-3

VR. Between *inl* (7%) and *asn* (20%) (1149).

Cloned and sequenced: Swissprot ACEA_NEUCR, PIR S26858, S22057, EMBL/GenBank X62697, GenBank NCACU3; EST NM3F4.

Structural gene for isocitrate lyase (EC 4.1.3.1). Unable to use acetate as a carbon source. Some revertants produce a temperature-sensitive enzyme (1149) (Fig. 4).

acu-4: acetate utilization-4

IR. Right of *arg-1* (5/29) (651).

Unable to use acetate as a carbon source (651, 652).

acu-5: acetate utilization-5

IIR. Right of *arg-5* (6%), *aro-3* (7%) (651).

Cloned and sequenced: Swissprot ACSA_NEUCR, PIR S09244, SYNCAA, EMBL / GenBank X16989, Z47725, GenBank NCACU5, NCACCOASY.

Unable to use acetate as a carbon source. Structural gene for acetyl coenzyme A synthetase (EC 6.2.1.1) (acetyl-CoA ligase, acyl-activating enzyme) (652). See Figs. 4 and 17.

acu-6: acetate utilization-6

VIL. Left of *cys-1* (3%) (143, 651).

Unable to use acetate as a carbon source (651). Structural gene for phosphoenolpyruvate carboxykinase (EC 4.1.1.31) (143, 652) (Fig. 4). Strains with some complementing alleles possess protein that is electrophoretically similar to the enzyme. A temperature-sensitive partial revertant enzyme maps at the original locus (143). Interallelic complementation (651).

acu-7: acetate utilization-7

IIIR. Linked to *dow* (0/72) (1582).

Unable to use acetate as a carbon source. Recovery from ascospores is poor (~25%) (651). Reduced level of oxoglutarate dehydrogenase (EC 2.3.1.61) (652) (Fig. 4).

acu-8: acetate utilization-8

IIR. Between *trp-3* (8%) and *un-15* (3%) (1263).

Cloned and sequenced: Swissprot ACU8_NEUCR, PIR A36316, EMBL / GenBank M31521, GenBank NEUACU8, EST NC4G6.

Structural gene for acetate permease (1263) (Fig. 4).

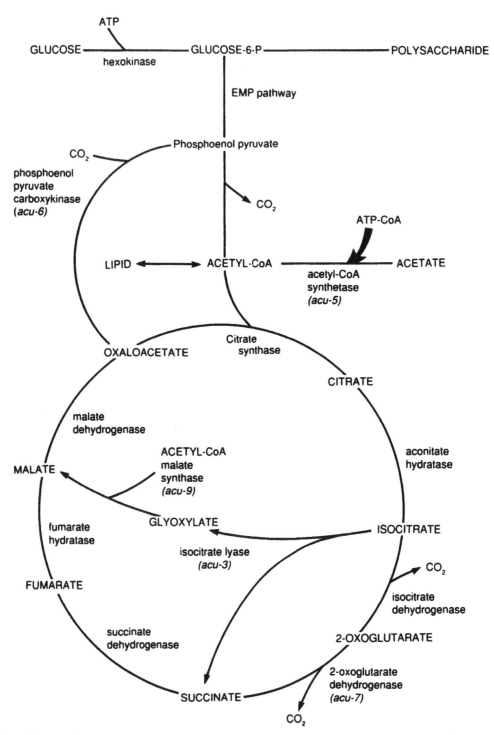

FIGURE 4 The tricarboxylic acid cycle and the anaplerotic glyoxylate cycles showing *acu* mutants affected in acetate utilization. Not shown is *acu-8*, specifying acetate permease. From ref. (406), with permission from Wiley-VCH Verlag.

acu-9: acetate utilization-9

VII. Linked to *nic-3* (29%) (405).
Cloned and sequenced: Swissprot MASY_NEUCR, PIR S17774, EMBL/GenBank X56627, GenBank NCACU9.
Structural gene for malate synthase (glyoxysomal) (EC 4.1.3.2) (1784) (Fig. 4). Mutant obtained by RIP (405).

acu-10: acetate utilization-10

Unmapped.
Resistant to fluoroacetate (1512).

acu-11: acetate utilization-11

VII. Linked to *arg-10* (21%) (1512).
Resistant to fluoroacetate (1512).

acu-12: acetate utilization-12

IIR. Linked to *trp-3* (24%) (1512).
Resistant to fluoroacetate (1512).

acu-13: acetate utilization-13

IIR. Linked to *trp-3* (23%) (1512).
Resistant to fluoroacetate. No growth on ethanol (1512).

acu-14: acetate utilization-14

VI? Perhaps loosely linked to *ylo-1* (404).
Does not complement any other *acu* mutant. There is no clear enzyme defect, but the levels of acetyl-CoA synthase and isocitrate lyase are affected (367).

acu-15: acetate utilization-15

Unmapped.
Cloned and sequenced: Swissprot AC15_NEUCR, EMBL/GenBank Y11565, GenBank NC11565.
Encodes acetate metabolism transcriptional activator; a zinc-finger protein that binds upstream of *acu* genes (178). Mutants isolated through RIP show a reduction in growth on acetate. Unlike the *facB* mutants of *Aspergillus*, they are not appreciably more resistant to fluoroacetate than is the wild type. When transferred to acetate growth medium, they do, however, show weak induction of acetyl-CoA synthetase and poor induction of the glyoxylate enzymes (404).

acu-16: acetate utilization-16

Unmapped.
Not allelic with other *acu* mutations. Described as lacking an inducible acetate transport system (1696). More recently found to take up acetate into the mycelium but with no apparent glyoxalate cycle flux (406, 2082). Originally called *acp^i*: *inducible acetate permease*, but renamed because characterization of its function has been revised and because the symbol *acp* is used for a gene specifying acyl carrier protein (240).

ad: adenine

For the purine biosynthetic pathway, see Fig. 5. *ad-5* catalyzes two sequential steps in the pathway, and, by analogy with yeast, the gene specifies a single polypeptide with two enzymatic activities. *ad-1* appears to be an isogene. A similar situation pertains in yeast, with *ade_{a16}* and *ade_{17}* being the isogenes for these two steps. *ad-4* carries out two lyase reactions on different but structurally related substrates at different steps in the pathway. *ad-2* has until now been regarded as specifying the fifth step in the adenine pathway, but its homolog in yeast also specifies the second step. Growth of mutants in terminal (post-AICAR) steps of the adenine biosynthetic pathway is aided by histidine, which has a sparing effect on *ad-1, ad-4,* and *ad-8* (295, 1314). Some *ad-5* alleles are aided by histidine, others inhibited (295). Mutants of *ad-3B, ad-4, ad-8,* and *ad-9* have been used to study the effect of histidine on purine pool utilization (1543). Mutants at the *ad-3A* and *ad-3B* loci accumulate purple pigment when adenine is limiting. Smaller amounts of the pigment may be seen in other post-AIR genes such as *ad-4* and *ad-5* (997, 1360). The pigment may consist of polymerized AIR (1543). Mutants affecting earlier biosynthetic steps are epistatic to *ad-3* and later mutants with respect to purple pigment production (997). *ad-3B* (and presumably also *ad-3A*) cultures accumulate spontaneous mutants at other *ad* loci; these prevent pigment production and improve the growth rate of the double mutant (1368). Mutants of *ad-3B, ad-4, ad-8,* and *ad-9* have been used to study the effect of histidine on purine pool utilization (1543). For regulation of purine catabolism, see the review in ref. (1281). For regulation of purine biosynthesis, see refs. (767) and (1545). For interrelation of purine, histidine, and tryptophan pathways, see ref. (1543). Indole may strongly inhibit adenine mutants (1159). Adenine mutants at the various loci were originally assigned to complementation groups designated by capital letters (997). Relation of most of these groups to steps of the biosynthetic pathway are given in Fig. 10 of ref. (249).

ad-1: adenine-1

VIL. Between *ylo-1* (6%) and *Cen-VI* (1% or 2%), *T(AR209), glp-4* (0; 2%), *rib-1* (3%; 5%) (1986, 2160).
Uses adenine or hypoxanthine (1360, 1620). Accumulates AICAR (160, 1764) and SAICAR (160). By analogy with

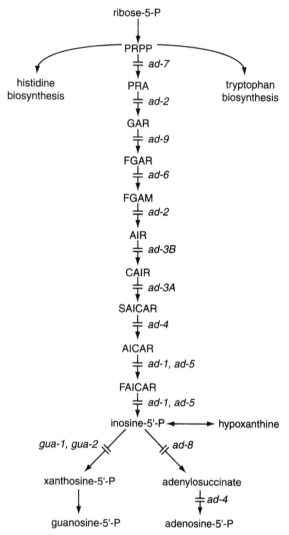

FIGURE 5 The purine biosynthetic pathway and sites of action of *ad* and *gua* genes (160, 249, 650, 736, 767, 964, 996). Abbreviations: PRPP, 5-phosphoribosyl pyrophosphate; PRA, 5-phosphoribosylamine; GAR, 5'-phosphoribosylglycinamide; FGAR, γ N-5'-phosphoribosylformylglycinamide; FGAM, 5'-phosphoribosylformylglycinamidine; AIR, 5'-phosphoribosyl-5-aminoimidazole; CAIR, 5'-phosphoribosyl-5-aminoimidazole-4-carboxylate; SAICAR, 5'-phosphoribosyl-5-aminoimidazole-4-N-succinocarboxamide; AICAR, 5'-phosphoribosyl-5-aminoimidazole-4-carboxamide; FAICAR, 5'-phosphoribosyl-5-formamidoimidazole-4-carboxamide. From ref. (1596), with permission from the American Society for Microbiology.

yeast, a putative isogene that, with *ad-5*, specifies AICAR transformylase (EC 2.1.2.3) and IMP cyclohydrolase (EC 3.5.4.10) (Fig. 5). Used to study purine transport (1765). Ascospores are white and inviable in homozygous *ad-1* × *ad-1* crosses, and the *ad-1* ascospores may be white in heterozygous crosses (1546). Called complementation group M.

ad-2: adenine-2

IIIR. Between *thi-2* (1%) and *trp-1* (1%; 7%) (18, 426).

Requires adenine or hypoxanthine (1360). The mutant is defective in phosphoribosylglycinamide cycloligase (AIR synthetase) (EC 6.3.3.1) (249), which catalyzes the fifth step in purine biosynthesis (Fig. 5). The yeast homolog is bifunctional, also specifying the phosphoribosylamine–glycine ligase (EC 6.3.4.13), the second step in the pathway, for which no gene has been identified in *Neurospora*. Allele 70004(t) is heat-sensitive (34 vs 25°C) (1360) and osmotic-remediable (1273). Called complementation group H.

ad-3A: adenine-3A

IR. Between *ure-4*, *his-3* (1% or 2%) and *ad-3B* (<1%) (156, 492).

Cloned and partially sequenced: EMBL/Genbank AI416404; EST NC2G7.

Requires adenine or hypoxanthine (1360). Blocked in the interconversion of CAIR+aspartate to SAICAR (SAICAR synthase, EC 6.3.2.6) (650) (Fig. 5). Produces purple pigment, permitting direct visual selection. Pigment is secreted with low adenine supplement (e.g., 0.1 mM), not with high (2 mM) (503, 1360, 1542). Fertility is reduced in crosses between alleles (771). There is no interallelic complementation (486, 1896). Either forward mutation (504) or reverse mutation (1500) can be measured precisely. Production of purple pigment permits visual detection of mutants (650). *ad-3A* mutants N23 and N24 have been used extensively as mutagen testers: N23 reverts with agents that cause base-pair substitutions, and N24 reverts with agents that cause frame shifts (1500). Pigment production has been used to assess the effect of histidine and tryptophan on purine nucleotide synthesis (1543). *SK(ad-3A)* is a mutation at or near *ad-3A* and may be a cryptic *ad-3A* allele. It does not require adenine for growth. In crosses of *SK(ad-3A)* × *ad-3A*, the *ad-3A* progeny die. Possibly *SK(ad-3A)* fails to make enough adenine to support their growth (516). Translocations *T(Y155M64) ad-3A* (493, 1582) and *T(Y112M15) ad-3A* (774) each have one breakpoint inseparable from *ad-3A*. The letter A in the locus symbol does not refer to mating type; it was appended because *ad-3A* mutants were originally called complementation group A (487).

ad-3A and *ad-3B* are two genetically and functionally distinct loci separated by a short but functionally complex region of unknown but essential function (492, 771). They have been used to develop an *ad-3A* and *ad-3B* specific-locus assay system for quantitative genetic and molecular studies of mutation [see introduction to ref. (497)]. The marked *ad-3A*+*ad-3B* heterokaryons used in the assay system shelter not only mutations in single genes but also recessive lethal multilocus deletions, enabling these to be characterized. The system has been used in massive studies with various physical and chemical agents (495–497, 499, 500, 505, 506, 827). Characterized mutations that were induced by various agents are available from FGSC.

ad-3B: adenine-3B

IR. Between ad-3A (<1%) and T(P7442)^L, nic-2 (3%) (492).

Uses adenine or hypoxanthine (1360). Blocked in the interconversion of AIR to CAIR (AIR carboxylase, EC 4.1.1.21) (650) (Fig. 5). Produces purple pigment, permitting direct visual selection. Pigment is secreted with low adenine supplement (e.g., 0.1 mM), not with high (2 mM) (503, 1360, 1542). Pigment production has been used to assess the effect of histidine and tryptophan on purine nucleotide synthesis (1543). Reduced inter-allelic fertility (487, 771). Complementation maps (489, 501). Relation of mutagens to complementation patterns (490). Mutants with nonpolarized complementation patterns on the right side of the complementation map grow on minimal medium if supplied with CO_2; other mutants do not respond to CO_2 (491). Used in conjunction with ad-3A for studies of mutagenesis (see ad-3A). Rearrangement T(I→III)Y112M4i ad-3B, which has a breakpoint inseparable from ad-3B, was the first insertional translocation reported in fungi (485). Allele 7-017-0137 shows "fixed instability," mutating to an unstable prototrophic allele (97). Alleles 2-17-126 and 12-21-28 and numerous others are supersuppressible amber mutants (772, 1461, 1860). Called complementation group B.

ad-4: adenine-4

IIIR. Between met-8 (1%; 4%), com (0%; 5%) and col-16, leu-1 (1%; 3%) (426, 1591).

Cloned and partially sequenced: EMBL/GenBank AI416404; EST NC2G7.

Requires adenine. Cannot use hypoxanthine or inosine (1314). Growth on adenine (at least of allele 44206) is improved by the addition of histidine, and improved still more by the addition of histidine plus methionine (1314). Structural gene for adenylosuccinase (EC 4.3.2.2), which controls two reactions in adenine synthesis, namely, SAICAR to AICAR in the common part of the purine pathway and adenylosuccinate to adenosine 5'-P in the adenine-specific branch (249, 736, 2240) (Fig. 5). Accumulates a slight amount of purple pigment when adenine is limiting (1360). Used for the first demonstration of complementation between alleles in vivo (736) (simultaneous with independent demonstration in am) and in vitro (2239). Enzyme in revertants

(2240). Used to study purine transport (1544). Alleles 44206 and 44415 are heat-sensitive (34 vs 25°C) (907, 1360) and osmotic-remediable at 30°C (1273); the enzyme synthesized at 30°C by heat-sensitive strains has altered properties (1273). Originally called complementation group F.

ad-5: adenine-5

IL. Between phe-1 and arg-1 (1%) (914, 1592).

Uses adenine or hypoxanthine (1360). Structural gene for a bifunctional enzyme encoding AICAR transformylase (EC 2.1.2.3) and inosine-5'-monophosphate (IMP) cyclohydrolase (EC 3.5.4.10). Accumulates AICAR (160, 1764) and SAICAR (160) (Fig. 5). By analogy with yeast, a putative isogene of ad-1. Some alleles are stimulated by histidine and may not grow on hypoxanthine unless histidine is present; others may be inhibited by histidine (295, 736). Produces some purple pigment, but less than ad-3A and ad-3B (997). Evidence of ref. (249), apparently enzymatic, suggests that some ad-5 mutants lack both the AICAR formyltransferase and IMP cyclohydrolase activities of the bifunctional polypeptide, but others lack only one or the other. Allele Y112M192, for example, is blocked only at the formyltransferase step (1763, 1764). Called complementation group J.

ad-6: adenine-6

IVR. Between ilv-3 (9%) and pan-1 (1% or 2%), chol-1 (1%), cot-1 (2%; 6%) (1124, 1255).

Blocked in phosphoribosylformylglycinamidine synthase (FGAM synthase, EC 6.3.5.3) (249) (Fig. 5). Uses adenine or hypoxanthine (1360). Inhibited by caffeine in the presence of adenine (2279). Called complementation group I.

ad-7: adenine-7

VR. Between cot-2 (4%) and T(EB4)^R, ro-4 (4%), pab-2 (8%) (321, 322, 1578).

Cloned: Orbach-Sachs clones X22H08, X24H05.

Uses adenine or hypoxanthine (1360). Lacks phosphoribosylpyrophosphate amidotransferase (EC 2.4.2.14), the first enzyme in de novo purine biosynthesis (996) (Fig. 5). Ascospores from homozygous ad-7 × ad-7 crosses are white (allele Y175M256). Allele P73B171(t) is temperature-sensitive.

ad-8: adenine-8

VIL. Between het-8 (12%), ser-6 (15%) and cpl-1 (7%; 11%), aro-6 (8%; 12%) (822, 963, 1425, 1582).

Requires adenine; cannot use hypoxanthine (997). Lacks adenylosuccinate synthase (IMP-aspartate ligase, EC

6.3.4.4) (964) (Fig. 5). Has little hypoxanthine uptake and low hypoxanthine phosphoribosyltransferase; both of these effects are partly counteracted in the *ad-1; ad-8* double mutant (1765). Low hypoxanthine phosphoribosyltransferase is also found in *mep(3)* and *mep(10)*. Used to study purine transport (1544, 1765). Fine-structure mapping and intralocus complementation (963–965). Called complementation group E.

ad-9: adenine-9

IR. Between *met-6* (9%; 16%), *tre* (7%) and $T(T54M94)^M$, $In(OY348)^R$, *nit-1* (3%; 15%) (1578, 1592).
Cloned: pSV50 clone 6:3G (1819).
Uses adenine or hypoxanthine (997). Lacks phosphoribosylglycinamide formyltransferase (GAR transformylase, EC 2.1.2.2). Controls conversion of GAR to FGAR (249) (Fig. 5). Called complementation group D.

adg: adenine–arginine

Allelic with *arg-11*.

adh: adherent

VIIL. Between *cyt-7* (9%) and *nic-3* (4%; 11%). Linked to *do* (0/53), *spco-4* (4%) (1582, 1592).

Morphology is abnormal. Conidia are not powdery and do not shake loose. Makes arthroconidia but few blastoconidia. Morphologically distinct from *do* and *spco-4*. Complements *spco-4* (1582, 1592) (Fig. 6).

ads: adenine sensitive

IV. Linked to *col-4* (966).
Growth is completely inhibited at 35°C by 10 μM adenine; high concentrations are inhibitory at 25°C. Growth is poor on minimal medium at 35°C, compared to 25°C. Inhibition is not relieved by vitamins, amino acids, guanine, guanosine, or guanylic acid; there is no growth response to guanosine in the absence of adenine (962, 966).

aga: arginase

VIIR. Between *wc-1* (2%) and *arg-10* (24%; 27%) (463, 1380). Near *un-10* (1264).
Cloned and sequenced: Swissprot ARGI_NEUCR, EMBL/GenBank L20687, EMBL NCAGA, GenBank NEUAGA.
Structural gene for arginase (EC 3.5.3.1) (463, 1380) (Figs. 10–12). Unable to form ornithine from arginine; arginine is thus unable to satisfy the proline requirement of *pro-3* in the *pro-3, aga* double mutant. The proto-

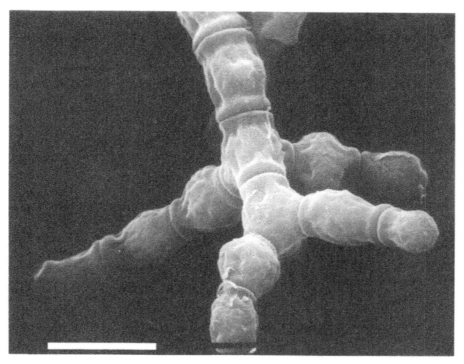

FIGURE 6 The mutant *adh: adherent*. Macroconidia of this type, called arthroconidia ("joint" conidia), differ in mode of formation from blastoconidia ("bud" conidia), shown in Fig. 31B. Bar length = 10 μm. Scanning EM photograph by M. L. Springer. From ref. (1958), with permission from Cold Spring Harbor Laboratory Press.

trophic single mutant develops a polyamine requirement in presence of arginine. This is due to feedback inhibition of ornithine biosynthesis by arginine, combined with a catabolic block in ornithine formation from arginine (458, 463). Two forms of arginase polypeptide (41- and 36-kDa) are produced (200). These differ at their N-termini and arise as a consequence of two different translation initiation codons being used in each of two differentially expressed mRNAs transcribed from alternative promoters (1264). The *aga* mutant has been used to study polyamine synthesis (466) and to obtain mutants defective in siderophore uptake (359); see *sit*. The triple mutant *arg-12; ota; aga* was used to look at deoxyhypusine and hypusine modification of a 21-kDa polypeptide (2261) that is the precursor of elongation factor eIF5A (see *eif5A*). Sideramine production is completely blocked in the absence of ornithine in the triple mutant *arg-5; ota; aga*, which is used to study iron transport (2227, 2229) and iron storage (1296, 1297).

age: aging of conidia

Used by K. D. Munkres for mutations showing reduced conidial longevity in the light. Not expressed in the dark or, with vitamin E or reduced glutathione, in the light. Deficient in enzymes that destroy free radicals and peroxides. Scored by plating efficiency after incubation of mature slant cultures at 30°C, 100% relative humidity, in continuous white fluorescent light (24 J m^{-2}). Most *age* mutants can also be scored by a defect in conidiophore development on Vogel's sorbose–sucrose plates. The *age* mutants were initially selected as spontaneous variants from the f$_1$ of Oak Ridge wild types. Variants with increased conidial longevity also can be selected. Longer life span is correlated with slower growth (1394, 1395).

age-1: aging of conidia-1

IR. Mutations designated *age-1* were interpreted as being at a complex of recombinable sites distal to *al-1*, with individual sites being symbolized 1.1, 1.2, 1.3, etc. Reversion or misscoring may possibly have led non-mutant progeny of intercrosses to be misinterpreted as crossovers (1394).

The short-lived phenotype of *age-1* mutants is associated with abnormal morphology. Conidiophores are absent, and conidia are formed on the agar surface. The various *age-1* mutations are stated to be dominant to wild type and qualitatively indistinguishable from one another in longevity, morphology, and enzymatic defects. Mutations designated 1.3 are apparently alleles of *so* (1394).

age-2: aging of conidia-2

VIR. Right of *ws-1* (8%) (1394).

Conidiophores are absent, and conidia are formed on the agar surface. Dominant to wild type (1394).

age-3: aging of conidia-3

An *al-1* allele [also called *ylo-4* (1394)].

agg: accelerated growth on galactose

Poorly defined genetically. Inheritance is not simple, with nuclear and perhaps cytoplasmic components (125). Best considered as designating a mutant strain, not a gene locus.

Differs from wild type in being able to use lactose as a sole carbon source in agitated cultures and in producing four times the wild-type amount of growth in standing cultures. Used to establish the existence of two β-galactosidases in *Neurospora* (127) and to study responses of the two enzymes to inducers (126). Called L5.

agr-1: altered glucose repression-1

Unmapped.

Glucoamylase and other hydrolases that are normally repressed by glucose are synthesized and secreted constitutively (1913).

al: albino

Carotenoid synthesis is impaired in mutants designated *albino* (Fig. 7). Orange macroconidia and mycelia are characteristic of wild-type strains in all of the conidiating *Neurospora* species. (Some *Neurospora discreta* and *Neurospora intermedia* strains are yellowish.) Carotenoids also are synthesized in microconidia (2120) and in perithecia and protoperithecia (1569), where their presence is normally masked but is revealed when the *per* mutation deletes black pigment. Different mutant *albino* alleles result in strains that are white, yellow, pink, or purple, that have only traces of color, or in which only peripheral conidia become pigmented (1570). An unstable constitutive variant has been described (1139). Carotenoids are also affected by genes designated *ylo*, by modifiers of intensity such as *int* and *vvd*, and by genes affecting induction by light (*wc, ccb, lis, blr*). Rapid development of carotenoids is induced by blue light (479, 837, 1378, 1760, 2295). Control by light is at the level of transcription initiation. For references and a review of photobiology, see ref. (1141). The blue light photoreceptor is still intact in the triple mutant *al-1;al-2;al-3* (1760). Carotenoid production is maximized if dark-grown cultures are placed at ∼5°C immediately after exposure to inducing light (835). Albino mutants can be scored in submerged colonies if platings are grown at 25°C in the dark and transferred to 5°C under light (319, 933). Carotenoid synthesis is also developmentally regulated. Wild-type macroconidia slowly become pigmented in the dark, and conidia slowly become pigmented in the mutants *wc-1* and *wc-2*, which

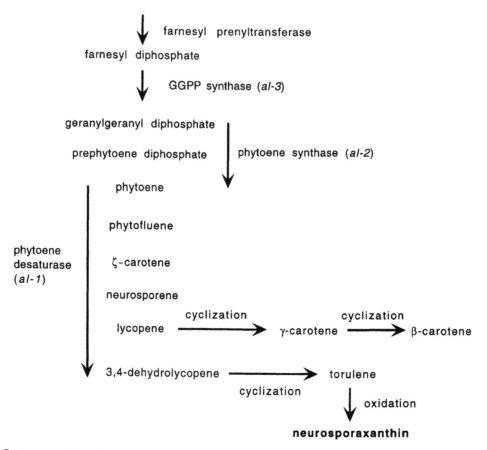

FIGURE 7 Carotenoid biosynthesis in *Neurospora*, showing positions of *albino* loci in the pathway. Original figure from T. J. Schmidhauser. A dual function is strongly suggested for *al-2* by its homology to a gene in *Xanthophyllomyces* that encodes a bifunctional protein involved both in geranylgeranyl diphosphate and in cyclization of lycopene to β-carotene and torulene (2146a).

are blind to photoinduction. For developmental regulation and photoregulation during conidiation, see refs. (1188) and (1189). For background and biochemical methodology, see ref. (1703). Albino mutants have been used to examine the role of carotenoids in photoprotection (1907, 2084). Epimutations of *albino* genes have been useful in studies of transgene-induced gene silencing ("quelling") (390, 392, 1737). The most commonly recovered *albino* mutations are at *al-1* and *al-2* (700, 1845), which were once thought to be contiguous but are now known to be separated by other loci unrelated to carotenoid biosynthesis (1546, 1554). Cultures of *osmotic* mutants and of *eas* may become intensely orange; this can be ascribed to physical effects as cells adhere, rather than to changes in carotenoid production. Mutations that delete or modify the black pigmentation of perithecia and ascospores are not called *albino* but have been given other names, such as *per*, *ws*, and *bs*.

al-1: albino-1

IR. Between *hom* (1%) and *cnr* (1%) (1554, 1578, 1580). On a common plasmid with *hom* (1816).

Cloned and sequenced: Swissprot CRTI_NEUCR, EMBL/ GenBank M57465, PIR A35919, GenBank NEUAL1A, EMBL NCAL1A; pSV50 cosmid 3:11H, Orbach-Sachs clone G15C10.

Structural gene for phytoene dehydrogenase (phytoene desaturase) (EC 1.3.-.-) (1816), which catalyzes five steps in carotenoid synthesis (Fig. 7). Alleles differ widely in phenotype, ranging from white (e.g., allele 4637) and "aurescent" (pigment in peripheral conidia and conidiophores, allele 34508) to yellow mycelia and conidia (alleles ALS4, RES-25); see ref. (1570). Alleles ALSl4, RES-6, 34508, and RES-25 contain large amounts of phytoene (>99% of total neutral carotenoids) (748, 2018). Allele RWT-ylo accumulates ζ-carotene and smaller amounts of neurosporene, suggesting a leaky block between them (2084). A mutation called *ylo-4* (also *age-3*) proved to be a pigmenting *al-1* allele (1582). Early fine-structure mapping (933, 2020) needs to be reevaluated because the flanking marker *arg-6* was positioned incorrectly (1554). For complementation tests, see refs. (933), (2018), and (2019). Used in studies of posttranscriptional gene silencing (quelling) (392, 1737). Translocation *T(4637)*, which is inseparable from *al-1*, was the first *albino* mutant and one of the first chromosome rearrangements to be identified and studied in *Neurospora* (1307). For visualization of pachytene chromosomes and the synaptonemal complex in crosses heterozygous for *T(4637)*, see refs. (1578). *T(4637)* was used to show that transcription of *al-1* proceeds toward the centromere.

al-2: albino-2

IR. Between *cyh-1* (8%; 13%), *T(STL76)* and *Tp(T54M94)^R*, *arg-6* (1%), *hom*, *al-1*. Near *os-5* (<1%) (1554, 1559, 1578, 1580, 1603, 1605).

Cloned and sequenced; Swissprot PSY_NEUCR, EMBL/ GenBank L27652, PIR A53583, GenBank NEUAL2X, EMBL NCAL2X; Orbach-Sachs clones X14D06, X24C01, X25F09, G2G08.

Encodes phytoene synthase (EC 2.5.1.-) (1817), a particulate enzyme (838) (Fig. 7). Regulated developmentally and by light. Phenotypes of different alleles range from white and pale rose-white for l5300 and Y254M165 (2020) to purple for MN58a (319). For complementation, see refs. (933) and (2019). Fine-structure mapping (933, 2020) needs to be reevaluated because the flanking marker *arg-6* was located incorrectly (1554).

al-3: albino-3

VR. Between *pho-2* and *inl* (1%; 1T/18 asci) (747, 1447, 1582, 2185).

Cloned and sequenced: Swissprot GGPP_NEUCR, EMBL/ GenBank U20940, X53979, PIR S15662; pSV50 clones 23:1A, 23:1A, Orbach-Sachs clone X17:A07 (1451).

Structural gene for the trifunctional enzyme encoding geranylgeranyl pyrophosphate synthase (EC 2.5.1.1), geranyltransferase (EC 2.5.1.10), and farnesyltransferase (EC 2.5.1.29) (2161) (Fig. 7). Encodes two overlapping transcripts, AL-3(m), expressed in mycelia following light induction, and AL-3(c), induced developmentally and by light, and subject to circadian control (58). AL 3(c) is photoregulated only early in conidiation, unlike transcripts of other *albino* genes (1188). Allele Y234M470 (*al-3^ros*), formerly called *rosy* (105), becomes partially pigmented but is readily distinguished from wild type. *ylo-1* can be scored in *al-3^ros* (Y234M470) (1582). Other alleles (e.g., RP100) (2185) are white with a trace of pink pigment. Null mutations are probably lethal [N. Romano, unpub-

lished, cited in ref. (1737)], as might be expected if *al-3^+* function is required for viability. Placement of *al-3* early in the biosynthetic pathway is consistent with its having another function that is essential. Compounds affecting protein kinase C activity alter the response of *al-3* to light, implicating protein kinase C in *al-3* regulation (59). Among progeny of strains duplicated for *al-3^+*, the only surviving mutants are those that have undergone mild RIP (95).

alc-1: allantoicase-1

IIR. Linked to *pe* (10%), probably on the opposite side of *pe* from *xdh-1* (24%) (1712). Linked very close to *arg-12* by RFLP (1277).

Cloned and sequenced: Swissprot ALC_NEUCR, EMBL/ GenBank J02927, PIR A35829, GenBank NEUALCA, EMBL NCALCA.

Defective in purine catabolism. Unable to use allantoic acid as the sole nitrogen source. Structural gene for allantoicase (EC 3.5.3.4) (1712) (Fig. 8); for regulation, see Fig. 50. *alc-1* expression requires induction, and this requires a functional *nit-2* product, with binding of

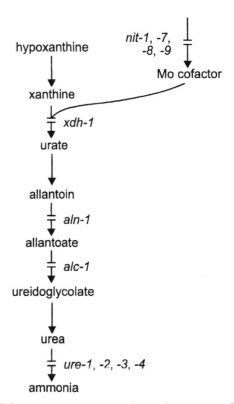

FIGURE 8 The purine catabolic pathway, showing sites of action of the *xdh*, *aln*, *alc*, and *ure* genes (862, 1087, 1712). The molybdenum cofactor is essential to both xanthine dehydrogenase and nitrate reductase (2102). From ref. (1596), with permission from the American Society for Microbiology.

NIT2 to two GATA elements in the *alc-1* promoter (1151).

alcoy (Acronym for al-1;cot-1;ylo-1)

Genotype: *T(IR;IIR)4637 al-1; T(IVR;VR)R2355, cot-1; T(IIIR;VI)1, ylo-1* (1592).

A linkage tester strain containing three unlinked reciprocal translocations, each tagged with a visibly scorable marker, marking linkage groups I–VI (Fig. 9). Linkage of a gene to *al-1*, *cot-1*, or *ylo-1* in a cross to *alcoy* allows assignment to linkage group by a single follow-up cross. A majority of new point mutants are linked to one of the *alcoy* markers (1592). An improved version, *alcoy;csp-2*, carries the VII marker *csp-2* in addition to the three original markers (1572, 1584). *alcoy* strains have been used cytologically in studies of the synaptonemal complex and recombination nodules (740, 1231).

aln-1: allantoinase-1

VII (1712).

Defective in purine catabolism. Unable to use allantoin or any purine intermediate prior to it as the sole nitrogen source. Lacks allantoinase (EC 3.5.2.5) (1712) (Fig. 8); for regulation, see Fig. 50.

Alpha-tubulin

See *tba*.

alx-1: alternate oxidase-1

Used for a double-mutant strain called ANT-1.

alx-2: alternate oxidase-2

Unmapped.

Lacks inducible cyanide-insensitive respiration. Cannot grow on antimycin A. Complements the double-mutant strain called *alx-1* or ANT-1 (591).

am: amination deficient

VR. Between *sp* (4%; 8%, 450 kb), *ure-2* (2%) and *gul-1* (<1%), *ace-5*(<1%), *ure-1* (1%; 150 kb), *his-1* (3%) (217, 219, 220, 256, 1086, 1124, 1941).

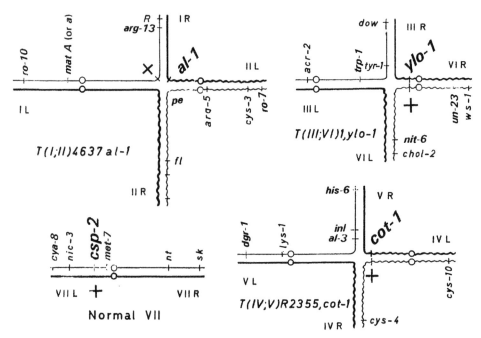

FIGURE 9 Linkage relations in a cross of the linkage tester *alcoy; csp-2 × Normal sequence*. Chromosomes are shown paired as in meiotic prophase I. The *alcoy; csp-2* chromosomes are drawn as heavy lines, and those from the standard-sequence parent are shown as thin lines. Segments from odd-numbered linkage groups are shown as straight lines and even-numbered as wavy lines. The only mutant markers present in the *alcoy* parent are *al-1, cot-1, ylo-1*, and *csp-2*. These are shown in large letters. *al-1* is inseparable from the *T(I; II)4637* breakpoint in linkage group I, *cot-1* is a few units proximal from the *T(IV; V)R2355* breakpoint in IV, and *ylo-1* is a few units from the *T(III; VI)1* breakpoint in V (arbitrarily shown as centromere proximal). Also shown are the locations of markers present in the standard-sequence follow-up testers, a few key markers for orientation, and some useful distal markers that are far removed from the breakpoints. The latter are used only when no linkage to the *alcoy* markers is detected. Interval lengths and relative distances are not to scale, but gene order is as shown. Figure from D. D. Perkins.

Cloned and sequenced: Swissprot DHE4_NEUCR, EMBL/ GenBank K01409, PIR A00381, DENCEN, GenBank NEUAMG, EMBL NCAMG, EST NM2B4; pSV50 clones 12:2A, 15:1B, 17:2A, 32:3H, Orbach-Sachs clones X1C05, X5F02, X6B01, G7G02, G9A10, G9D12.

Structural gene for NADP-specific glutamate dehydrogenase (EC 1.4.1.4) (639) (Fig. 49). Requires a source of α-amino nitrogen for growth; alanine is a good supplement (1940). Readily scorable at 25°C; leaky at 34°C (100). Leaky growth and adaptation on minimal medium are prevented by 0.02 M glycine (1536, 1537) or by en(am)-1, en(am)-2, or nit-2. The am mutants show abnormal regulation of NADH-glutamate dehydrogenase and are synergistic with nit-2 in this effect (442). Some am alleles (e.g., RU1) suppress the pyrimidine requirement caused by pyr-3 (CPS⁻ACT⁺) mutations (2212). Used for the first demonstration of complementation between alleles in vivo (646) (simultaneous with independent demonstration in ad-4). In vitro complementation (640). Used for studies of the complementation mechanism (386, 387, 2191) and for fine-structure mapping (634, 641). Used to study the mechanism of recombination (218, 221), including the control of intralocus recombination by rec-3 (1939–1941), the relation of gene conversion and crossing over (219), and the orientation of the fine-structure intragenic map relative to flanking markers (217). Used to study the mechanism of RIP (636) and to generate new alleles by RIP (643). Instrumental in the discovery of the Tad retrotransposon (1065). Used to study the colinearity of gene and gene product, internal suppressors (226, 635, 888), and the action of amber supersuppressors (1859, 1860). Used as a site for targeting homologous integration of transforming DNA (1343), for transposon targeting (277), and for identifying upstream regulatory regions (371, 670, 671). The functional defects in several mutant enzymes with single amino acid replacements have been defined. The product of am¹ fails to bind NADPH (2191); those of am², am³, am¹⁹, am¹³⁰, and am¹³¹ are stabilized in an inactive conformational form (66, 386, 639, 1063, 2023) and all are complementable by am¹. Allele am¹⁴ is osmotically reparable and thought to have unstable quaternary structure (635). Used in a study showing glutamine to have a role as co-repressor of uricase synthesis (2184). Used to study nitrogen assimilation and metabolism (939) and nitrogen metabolite repression (354, 564). Efficient procedure for selecting new am mutants (1058). am mutants suppress the p-phenylalanine resistance of mtr (1058) and fpr-6 (1335). Spectrum of UV- and nitrous acid-induced mutants (1066). Used to study the effect of mutants on intron processing (1054) and the effect of mutations to rare codons (those with A in the third position) in reducing translation efficiency (1053). Allele am¹⁷ has a chain-terminating amber codon at residue 313 of GDH,

based on amino acid replacements in revertants and ssu1 (1861). Allele 6 is a frameshift with insertion in Ser5 codon (1912). Allele 126 is highly unstable (1062). Allele 132 is a deletion (2245). In(UK2-y), T(UK9-18), T(mpr13-1), and T(mpr15-2) are inseparable from am (1578).

amdS: acetamide utilization

Introduced from Aspergillus nidulans.

Encodes an acetamidase that confers the ability to use acetamide as the sole nitrogen source. This Aspergillus gene has been used in preference to the Neurospora gene Bml as a dominant selectable marker for transformation (2257). Transformants containing a single copy of amdS are not inactivated by RIP and, therefore, can be recovered from crosses. In contrast, when the essential Neurospora gene Bml is used as a selectable marker, transformed genes cannot be readily recovered from crosses because the functional Bml⁺ allele is inactivated by RIP. See also ble.

amr: ammonium regulation

Allelic with nit-2.

amt: aminotransferase

VR. Between crp-4 (8T/18 asci) and Tel-VR (2T/18 asci) (1452).

Cloned and sequenced: cDNA clones NM8H12, NP2A7.

Encodes putative amidinotransferase (EC 2.1.4.1) (875a, 1408).

amy-1: amylase-1

Allelic with sor-4.

amy(SF26): amylase(SF26)

Allelic with exo-1.

amyc: amycelial

IL. Between ad-5 and Cen-I (914).

Recessive. On sucrose media, amyc forms dotlike granular colonies of irregular budding vesicular elements. On permissive media, made either with acetate (plus α-ketoglutarate, succinate, malate, or certain amino acids) or with amino acids as C and N sources, it forms hyphae and macroconidia but apparently is still colonial (532, 2117). cAMP induces conidiation even on sucrose (532). Conidia show multipolar germination (2118). Photographs (1510, 2117). Low oxygen consumption and depressed amino acid pools (532), abnormal mitochondria (1509), surface glycopeptide (532), wall composition (403), recovery of antigenic arc representing the isozyme of malate dehydrogenase associated with conidiation (1539). Ultrastructure (1509, 2117, 2119). Used extensively in a balanced heterokaryon to evaluate nuclear distribution (71) and to detect lethal recessive mutations (70), anticipating the method of Stadler (1980). Techniques are described in ref. (72).

an: anaerobic
Unmapped.

Reported to be facultatively anaerobic, showing weak growth on enriched medium in the absence of oxygen. Prototrophic and indistinguishable from wild type under aerobic conditions. Not glucose-repressed. Anaerobic cultures aconidial, with reduced cytochrome oxidase and malate dehydrogenase activities, mitochondrial changes, and production of ethanol. Obtained by filtration enrichment with recycling. The symbol An$^+$ was used to specify the mutant phenotype, An$^-$ the wild phenotype (922, 923). Strains deposited in FGSC as *anaerobic* mutants have been tested unsuccessfully for their ability to grow under strictly anaerobic conditions (162).

ANT-1: antimycin sensitive
Not a locus designation. The symbol ANT-1 was used (595) to designate the double-mutant strain *azs; has*, which was also called *alx-l*. See *azs* and *has*.

anx14: annexin 14
Unmapped.
Cloned and sequenced: GenBank AF036871.
Encodes annexin XIV (225). Related to annexins in *Dictyostelium* and in animals. (No annexins are found in *Saccharomyces cerevisiae*.)

aod-1: alternate oxidase deficient-1
IV. Linked to *Cen-IV* (0T/17 asci) (1447). Left of *trp-4* (23%) (164, 2033).
Cloned and sequenced: Swissprot AOX_NEUCR, EMBL/GenBank L46869, PIR S65752, GenBank NEUALTO; pSV50 clone 23:7F.
Structural gene for the alternate terminal oxidase that is used instead of cytochrome *c* oxidase when cytochrome-mediated, cyanide-sensitive respiration is inhibited (1192). AOD-1 is identified immunologically as M_r 36,500 and 37,000 polypeptide species present in high concentrations under inducing conditions (1135). Alternate oxidase activity can be stimulated by some, but not all, nucleotides (1344, 1524, 2142). Strains carrying *aod-1* plus either [*mi-1*] or [*mi-3*] are viable (165, 2033).

Called *aod-B* (165). Three recessive mutant alleles, originally called *aod-1*, *aod-2*, and *aod-3*, later were called *aod-1.1*, *aod-1.2*, and *aod-1.3* (2033).

aod-2: alternate oxidase-2
II. Linked to *arg-5* (7%), *thr-3* (16%), *trp-3* (36%) (164, 2033).
The alternate oxidase system fails to be induced when cytochrome-mediated, cyanide-sensitive respiration is inhibited (164). The *aod-2* defect appears to eliminate the increase in the level of *aod-1* transcript observed under these conditions (1192).

Aph-1: Aphidicolin resistance-1
Unmapped. Mutant alleles show Mendelian segregation (1354).
Resistant to aphidicolin; an inhibitor of DNA polymerase *a* at 200–800 µg/ml. Resistance is dominant. With mutant strain E2-4-1, enzyme activity *in vitro* is resistant to aphidicolin inhibition. With mutant C-3, the pH optimum for enzyme activity is changed. Mutations were obtained by plating nitrosoguanidine-treated s*lime* in medium containing 20 µg/ml aphidicolin. Stocks of Aph^R; *slime* have been kept as heterokaryons on minimal agar medium (1354).

apu: accumulation of purines
Unmapped. Not linked to *ad-7* (VR) or *mat*, *aza-1* (IL).
Excretes purines. Obtained among prototrophic revertants of *ad-7* (which blocks the first step in *de novo* purine synthesis). Secretion was assayed by cross-feeding on plates seeded with *ad-3A* conidia. Purine secretion by *apu* occurs later than with *aza-1* allele 67-12 (4 days vs 36 hr, 25°C), and colony size is larger (996).

ar-2: (mating type) A/a related
Allelic with *eat-2*.

arg: arginine
For details of the arginine biosynthetic pathway, see Fig. 10. The most comprehensive references on arginine are (457) and (458). Arginine mutants have been used extensively for studies of compartmentation (436, 471, 475) and for studies of control of flux through the arginine pathway (468) (also see *ota* entry). Crossing is inhibited by high arginine levels; 0.1 or 0.2 mg/ml in crossing medium is satisfactory. Lysine and arginine show competitive inhibition, and all arginine auxotrophs are inhibited by lysine. Crosses or strains involving both requirements usually can be handled by adjusting the ratio; 0.8 mg/ml L-arginine-HCl:1.6 mg/ml

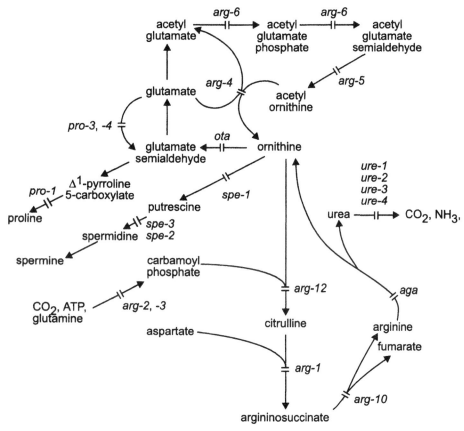

FIGURE 10 The biosynthetic pathways for arginine, proline, polyamines, and associated intermediary metabolism, showing sites of gene action in biosynthesis and arginine catabolism (457, 465, 862, 1087, 1310, 1379, 1380, 1967, 2165, 2291). Carbamoyl phosphate for pyrimidine synthesis is made as a separate pool by a distinct enzyme (see Fig. 53). Interchange between the two pools occurs only in certain mutant combinations. From ref. (1596), with permission from the American Society for Microbiology. For details of the polyamine pathway, see Fig. 62.

L-lysine-HCl is recommended for crosses (1970). Leaky arginine mutants (e.g., *arg-2*, *arg-3*, *arg-13*) are less leaky on nitrate medium (457) or canavanine plus lysine (1716). Leakiness of germinating ascospores of *arg-1* and *arg-3* is prevented by 0.05 mg of lysine/ml with no canavanine (1476). Lysine resistance is conferred on *arg-1* by a probable transport mutant *lys^R*. Some *arg* genes were originally called *cit* or *orn*. For degradative or related steps in arginine metabolism, see *aga, ota, spe,* and *ure.*

Regulation: Arginine biosynthesis and catabolism are controlled in a major way by compartmentation (455, 458) (Figs. 11, 12). Acetyl glutamate synthase and acetyl glutamate kinase are feedback-regulated by arginine (458). With one exception, the enzymes of arginine biosynthesis are not repressed below levels seen in minimal medium when arginine is added to cultures. The exception is carbamoyl phosphate synthetase A (CPS-A), the *arg-2*-encoded small subunit of which is repressed 5-fold. Upon limiting cultures for arginine, all biosynthetic enzymes except one increase concomitantly by about 3- to 5-fold; the CPS-A small subunit increases by as much as 20-fold (458, 677). These "derepressions" can also be brought about by starvation for other amino acids, such as tryptophan, lysine, and histidine; they require the normal product of the *cpc-1* locus (458, 1768). See *cpc-1*. The catabolic enzymes, arginase and ornithine aminotransferase, are present without induction and are elevated by only 2- to 5-fold in response to nitrogen limitation or the addition of arginine (but not ornithine) to the medium. These enzymes are not affected by mutations at the *nit-2* and/or *cpc-1* loci.

arg-1: arginine-1

IL. Between *ad-5* (1%) and *eth-1* (<1%) (1462, 1592).
Cloned: pSV50 clones 9:11F, 30:6G (1818).
Uses arginine, but not its precursors (1968). Lacks arginosuccinate synthetase (EC 6.3.4.5) (1463) (Figs. 10 and 11). Interallelic crosses produce perithecia, but

CYTOSOL

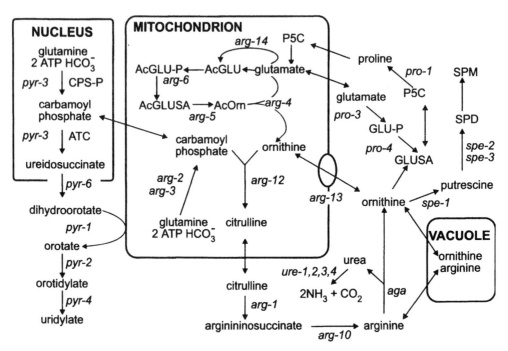

FIGURE 11 A schematic view of the organization of arginine, pyrimidine, proline, and polyamine metabolism in *N. crassa*. Boxes enclose cellular organelles to delineate the reactions carried out within different cellular compartments. The names of genes are shown next to the metabolic steps executed by their products. The product of *arg-13* is positioned on the mitochondrial membrane, where it is predicted to act as a carrier. The genes *pro-3* and *pro-4* originally were called *arg-8* and *arg-9*. Abbreviations: ATC, aspartate carbamoyltransferase; CPS-P, pyrimidine-specific carbamoyl-P synthetase; Ac, acetyl; GLU, glutamate; GLU-P, glutamyl phosphate; GLUSA, glutamate semialdehyde; ORN, ornithine; P5C, pyrroline-5-carboxylate; SPD, spermidine; SPM, spermine. Adapted from refs. (1213) and (458), with permission from the Genetics Society of America and the American Society for Microbiology, respectively.

most ascospores are white and inviable (1462). Leaky *arg-1* mutants are frequent among mutants selected as citrulline-resistant variants of *arg-12ˢ*; *pyr-3*; most of these show interallelic complementation, and many are transport-deficient (2089). *arg-1* does not grow well on some complex complete media unless extra arginine is added.

arg-2: arginine-2

IVR. Between *col-4* (1%; 2%) and *T(S4342)ᴸ*, *pyr-3* (1%; 3%) (196, 1369, 1372, 1578, 1580, 1596, 1716, 1928).

Cloned and sequenced: Swissprot CARA_NEUCR, Y3KD_NEUCR, EMBL/GenBank J05512, EMBL NCPSAS, GenBank NEUPSAS; pSV50 clones 6:7C, 11:7E, 20:5B, 24:8D.

Uses citrulline or arginine (1968). Structural gene for the small subunit of arginine-specific carbamoyl phosphate synthetase (EC 6.3.5.5), a two-subunit enzyme (471). The small subunit enables the enzyme to use glutamine as a nitrogen donor (458, 468). (The large subunit is

specified by *arg-3*; Figs. 10 and 11.) For regulation and compartmentation, see ref. (458). Arginine-specific negative translational regulation requires an evolutionarily conserved upstream open reading frame in the mRNA transcript that causes ribosome stalling (2187). The arginine–citrulline requirement can be suppressed by *pyr-3* (CPS⁺ACT⁻) mutations (455, 1716); see *pyr-3*. Leakiness is prevented by canavanine; lysine overcomes the side effects of canavanine (1716). At least some alleles can grow on minimal medium in 30% CO₂ (230). Leakiness is decreased if CO₂ is removed or if uridine is added (1721). Translocation *T(MEP24)* is inseparable from *arg-2* (451).

arg-3: arginine-3

IL. Between *eth-1* (1%) and *csp-1* (1%), *T(39311)ᴿ* (1592, 1882).

Cloned and partially sequenced: GenBank AI392110, EST NC3A12, pSV50 clones 9:12E, 10:6C, 13:6A, 15:5E, 26:11B, 28:6G, 28:12D; Orbach-Sachs clones X1F03,

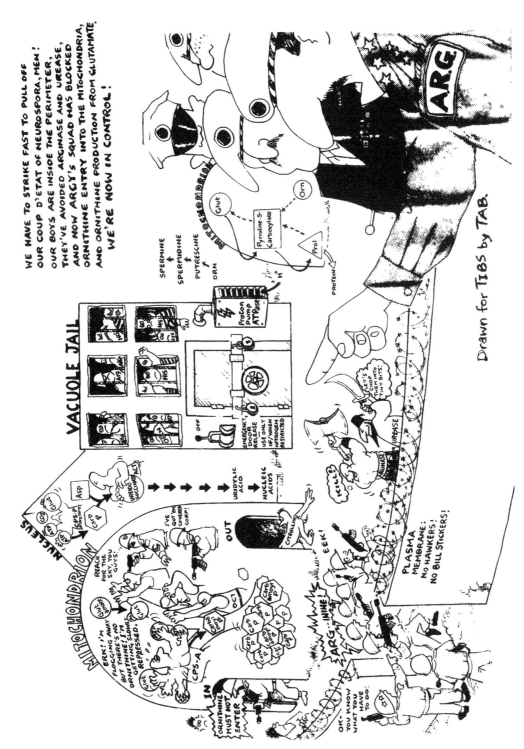

FIGURE 12 An alternative view of compartmentation involving arginine and related metabolites. Reprinted from ref. (475) [Davis, R. H., and R. L. Weiss (1988). Novel mechanisms controlling arginine metabolism in *Neurospora*. *Trends Biochem. Sci.* 13: 101–104.] Copyright 1988, with permission from Elsevier Science.

X8:D04, X9C06, X10H06, X17B01, X17D07, X22H02, G1A11, G3E09, G4G05, G6F06, G9C04, G13C04.

Uses citrulline or arginine (1968). Structural gene for the large component of arginine-specific carbamoyl phosphate synthetase (EC 6.3.5.5), a two-subunit enzyme (458). This component can form carbamoyl phosphate *in vitro*, using ammonia as a nitrogen donor (458). (The small component is specified by *arg-2*; Figs. 10 and 11). For regulation and compartmentation, see ref. (458). The arginine–citrulline requirement can be suppressed by *pyr-3* (CPS$^+$ACT$^-$) (455, 1312); see *pyr-3*. Allele 30300 can grow on minimal medium in 40% CO_2 (230, 363). Translocation MEP35 is inseparable from *arg-3*.

arg-4: arginine-4

VR. Between *sp* (1%; 11%) and *inl* (2%; 4%) (1585, 1895).

Uses ornithine, citrulline, or arginine (1968). Lacks acetylornithine-glutamate transacetylase (EC 2.3.1.35) (acetylornithine acetyltransferase) (514, 2166) (Figs. 10 and 11). Weakly suppresses CPS$^-$ATC$^+$ *pyr-3* mutants (1313). Alleles 21502 and 34105 (later called *arg-4* and *arg-7*) were originally thought to be genetically distinct because they complemented (1963), but an intercross produced no recombinants (1313) and both lack the same enzyme (514, 2166).

arg-5: arginine-5

IIR. Between *bal* (1%; 9%), *T(ALS176)* and *aro-3*, *pe* (6%; 18%) (800, 1546, 1548, 1578).

Structural gene for acetylornithine transaminase (EC 2.6.1.11) (1379) (Figs. 10 and 11). Uses ornithine, citrulline, or arginine (1968). Sideramine production is completely blocked in the absence of ornithine in the triple-mutant strain *ota; arg-5; aga*, which has been used to study iron transport (2227, 2229) and iron storage (1296, 1297). Called *orn-1*.

arg-6: arginine-6

IR. Between *al-2* (1% or 2%), *T(T54M94)^R* and *hom* (1%), *al-1* (1%; 4%) (1554, 1578, 1580).

Cloned and sequenced: Swissprot AR56_NEUCR, EMBL/GenBank L27746, PIR A53429, EMBL NCPPOA, GenBank NEUPPOA, EST NC1G8; Orbach-Sachs clones X3A12, X13G12, X17D10.

Uses ornithine, citrulline, or arginine (1968). A bifunctional gene specifying arginine-sensitive acetylglutamate kinase (EC 2.7.2.8) (457) and *N*-acetylglutamyl phosphate reductase (EC 1.2.1.38) (Figs. 10 and 11). The polyprotein precursor has an N-terminal mitochondrial targeting sequence followed by the kinase domain, a connecting region, and the reductase domain. Both the N-terminal-targeting sequence and the connecting region are cleaved by the mitochondrial-processing peptidase and -processing-enhancing protein to yield two mature enzymes in the mitochondrion. Processing of the polyprotein precursor has been reconstituted *in vitro* (1535). L-Methionine may inhibit (1546). Can be selected as a suppressor of *pro-3* by virtue of the feedback insensitivity of ornithine synthesis (2204). For supersuppressible amber alleles, see ref. (474). Called *orn-2*.

arg-7: arginine-7

Allelic with *arg-4*. Also called *orn-3*.

arg-8: arginine-8

Changed to *pro-3*.

arg-9: arginine-9

Changed to *pro-4*.

arg-10: arginine-10

VIIR. Between *dr* (12%), *T(5936)*, *arg-11* (1% or 2%) and *nt* (1%; 12%) (1462, 1548, 1578).

Uses arginine, but not its precursors (1462). Lacks argininosuccinate lyase (EC 4.3.2.1) (644) (Fig. 10). Accumulates arginosuccinate on limiting arginine (644). Viable ascospores from interallelic crosses are rare but the viable ones are often *arg$^+$*, whereas most *arg$^-$* ascospores remain colorless and inviable (1462). All *arg-10* alleles tested showed spasmodic growth in growth tubes at low arginine concentrations (738). *arg-10* does not grow well on some complete media unless extra arginine is added.

arg-11: arginine-11

VIIR. Between *dr*, *T(5936)* and *arg-10* (1% or 2%) (1548, 1578).

Requires arginine or citrulline, plus low levels of a purine and a pyrimidine (556, 1465, 1964). Inhibited by guanidine, sarcosine, and serocyamine (1963). Complements *arg-10* fully in heterokaryons (1465). The relation of this mutant to arginine biosynthesis or metabolism is obscure. Growth requirements vary markedly with CO_2 concentration and inoculum size. At 0% CO_2 or with small inocula, the requirement for all three supplements is absolute; with increasing CO_2 concentration or inoculum size, pyrimidine and then purine can be omitted. At 30% CO_2, all three supplements can be omitted (230, 364, 1465). Growth rate and morphology are highly variable among progeny from *arg-11* × wild type (1465). Grows spasmodically in growth tubes (738). Allele 44601 was formerly called *un(44601)* (907) and *adg* (556).

arg-12: arginine-12

IIR. Between *pe* (1%; 5%) and *T(NM177)^R*, *aro-1* (<1%) (731, 1580, 2242).

Cloned: pSV50 clone 16:9H (699).

Structural gene for ornithine carbamoyl transferase (OCT; EC 2.1.3.3) (452, 473, 2242). Uses citrulline or arginine (Figs. 10 and 11). The leaky allele *arg-12^S*, discovered as a suppressor of a *pyr-3* mutant and initially called *s* (908), reduces OCT activity by over 98% without imposing any arginine requirement. *arg-12^S* suppresses the pyrimidine requirement of *pyr-3* alleles that lack only pyrimidine-specific carbamoyl phosphate synthetase (CPS). This is because *arg-12^S* accumulates arginine-specific carbamoyl phosphate, which can then be used for pyrimidine synthesis (455) (see *pyr-3*). Nonleaky *arg-12* alleles cannot cause such suppression because the exogenous arginine that is required for growth results in repression of the arginine-specific CPS (455). Arginine mutants at all other arginine biosynthetic loci can be obtained efficiently as tight double mutants, using *arg-12^S* as the starting material (457). The double mutants *pro-4*; *arg-12^S* and *pro-3*; *arg-12^S* are prototrophic, due to diversion of ornithine to the proline pathway. The double mutant *arg-5*; *arg-12^S* cannot use exogenous ornithine (453, 465).

arg-13: arginine-13

IR. Between *os-1* (1%) and *so* (2%; 12%) (1592).

Cloned and sequenced: Swissprot AR13_NEUCR, EMBL/ GenBank L36378, EMBL NCARG13A, GenBank NEUARG13A; pSV50 clones 7:12B, 21:10D, 24:10H.

Structural gene for a nuclear-encoded protein that transports ornithine across the mitochondrial inner membrane (1213) (Fig. 11). Homologous with the human mitochondrial ornithine transporter responsible for hyperornithinaemia–hyperammonaemia–homocitrullinuria syndrome (274). Responds well to arginine or citrulline, but poorly to ornithine (457, 1313). Acts as a suppressor of the pyrimidine requirement of CPS⁻ACT⁺ mutations of *pyr-3* (1313). Leaky on minimal medium; scoring is cleared by the addition of lysine. Interallelic crosses are sterile. Called *arg(RU3)*.

arg-14: arginine-14

IVR. Between *T(S4342)^L*, *arg-2* (1%) and *T(NM152)^L*, *pyr-3* (1%) (457).

Cloned and sequenced: EMBL/GenBank L35484, EMBL NCARG1A, GenBank NEUARG1A; pSV50 clones 7:12B, 21:10D, 24:10H.

Structural gene for *N*-acetylglutamate synthase (EC 1.4.1.13). Uses arginine, citrulline, or ornithine (Fig. 11). Unlike the genes specifying other arginine biosynthetic enzymes, expression of *arg-14* appears relatively constant over the course of the asexual life cycle (2287). Point mutants were selected as tight double mutants using *arg-12^S* (457). Allele S1229 is inseparable from translocation *T(S1229)* (111, 112, 1580).

arg(CD-15)

Changed to *cpc-1*.

arg(CD-55)

Changed to *cpc-1*.

arg(RU1)

Allelic with *am*.

arg^R: arginine resistant

Allelic with *pmb*.

aro: aromatic

Used for genes concerned with the biosynthesis of aromatic amino acids and *p*-aminobenzoic acid (PABA). See Fig. 13 for pathway and sites of gene action. Excepting *aro-6*, -7, and -8, the genes designated *aro* are auxotrophs able to grow on a mixture of PABA, tyrosine, tryptophan, and phenylalanine. The first step in the pathway is catalyzed by three isozymes subject to feedback by different end products of the branched pathway and specified by different widely separated genes (*aro-6*, -7, -8). The second, third, fourth, fifth, and sixth steps are specified by a cluster gene that produces a single transcript (728, 2205), which produces a pentafunctional polypeptide (389, 2205). (The cluster gene is here called *aro-1*. Domains specifying the five functions were named *aro-1*, *aro-2*, *aro-4*, *aro-5*, and *aro-9* before it was realized that they were not separate genes.) The final step prior to branching is specified by a unifunctional gene (*aro-3*), which is at a separate chromosomal locus from the *aro* cluster gene although linked to it at a distance. The third and fourth steps are paralleled by similar reactions in the quinate catabolic pathway (Figs. 13 and 54). Thus, the *aro-9* enzyme ARO-9 can be replaced by QA-2, and under appropriate conditions, ARO-1 can be replaced by the QA-3. Supplement levels: 40–80 mg/l each tyrosine, tryptophan, and phenylalanine; 0.25 mg/l PABA (344, 800). Also called *arom*.

aro-1: aromatic-1 cluster gene

IIR. Between *arg-12* (<1%), *T(NM177)^R* and *ff-1* (4%; 6%) (1102, 1578, 1580, 2044). Intracluster map shows the order of domains to be 1, 9, 5, 4, 2 (728, 1726).

Cloned (315); pSV50 clone 31:9G.

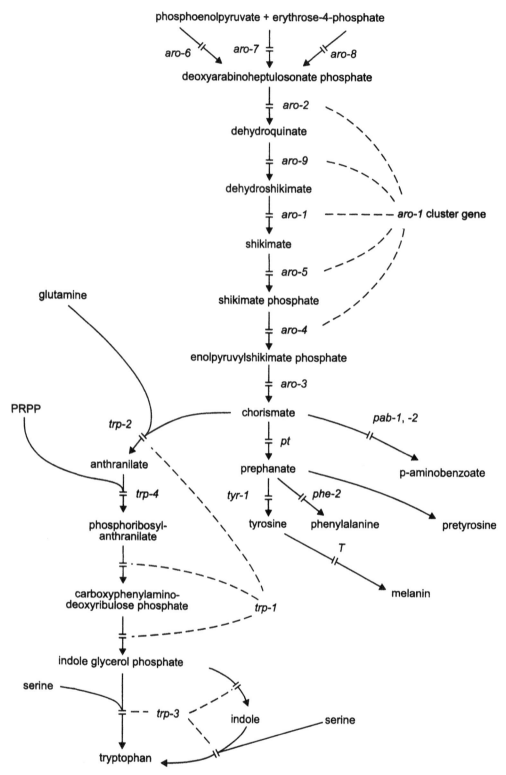

FIGURE 13 The biosynthetic pathways of the aromatic amino acids, showing sites of action of the *aro, trp, pt, phe, tyr,* and *T* genes (193, 305, 396, 523, 601, 728, 822, 896, 999, 1038, 2227, 2264). The conversion of chorismate to *p*-aminobenzoate has not been demonstrated in *Neurospora*. In the conversion of tyrosine to melanin, the later steps are nonenzymatic. Products of the *trp-1* and *trp-2* genes form an enzyme aggregate with three properties: a given *trp-1* mutant may block one or more of the reactions; *aro-9*$^+$ activity (biosynthetic dehydroquinase) can be replaced by the product of *qa-2*$^+$, the equivalent gene in the catabolic pathway (1726); and pretyrosine accumulates when other pathways are blocked (see *phe-2*). PRPP: 5-phosphoribosyl pyrophosphate. From ref. (1596), with permission from the American Society for Microbiology.

Structural gene for the aromatic biosynthetic pathway leading to tryptophan, tyrosine, phenylalanine, and PABA (Fig. 13). A multifunctional "cluster gene" (705) specifying five enzymes (389, 705, 731, 2205). Clustering of functions was discovered (800). Reviews are in refs. (728) and (1325). The order of regions that specify the five functions is still conveniently represented by the symbols established when it was thought that five separate genes were involved: *(arg-12)* ... *aro-1 aro-9 aro-5 aro-4 aro-2* ... *(ace-1)*. These symbols actually represent five domains of the pentafunctional polypeptide, specifying products that catalyze five different steps in the pathway. The native enzyme is a dimer of the pentafunctional chains (389, 705, 1235). We now designate the entire cluster gene as a single locus, *aro-1*. Mutants exist that block individual steps; also polar mutants exist, symbolized *aro(p)*, that eliminate more than one function. There are some discrepancies between genetic mapping and mapping by polarity [reviewed in ref. (1325)]. Genetic data indicate that transcription begins at the *aro-2* end (303, 728). *aro-1*, *-2*, and *-9* can use shikimate (0.3 mg/ml) as an alternative to the mixture of four aromatics shown for *aro-1* (2057). Mutants that block different individual steps complement each other in heterokaryons (731). Complementation between alleles that block the same single step has been detected only for *aro-2* and *aro-1* (297, 731). Polar mutants are divided into six classes (A–F) based chiefly on their complementation behavior (731); types D–F are semicolonial with yellowish-orange conidia (294). Single-function *aro-9* mutants were obtained by selecting in *qa-1*, which is noninducible for catabolic dehydroquinase activity (1726). Translocation *T(C161)aro* (called *arom-2*) is inseparable from the *aro-1* cluster and lacks several activities (800). Noncomplementing alleles M26, M1039, Ml065, M1108, M1162, M1172, and Y306M54 (abbreviated M54) are suppressible by amber suppressors (*ssu*, "supersuppressors") (294, 303, 1862). Details regarding individual components are given in the following paragraphs.

aro-1 domain: Specifies dehydroshikimate reductase (EC 2.5.1.19) (731, 800). Accumulates dehydroshikimate, which induces dehydroshikimate dehydrase in the catabolic pathway (800). Suppressed by *qa-4*, which lacks dehydroshikimate dehydrase; this allows induction of the catabolic enzyme quinate (shikimate) dehydrogenase, which substitutes for the biosynthetic enzyme dehydroshikimate reductase (305). A lag in the growth of *aro-1* on shikimate occurs with sucrose or glucose as the carbon source. This is overcome by substituting 1% glutamate for the sugar (942).

aro-2 domain: Specifies dehydroquinate synthetase (EC 4.6.1.3) (731). *aro-2* point mutants should not be confused with strain C161, called *arom-2* (800), which lacks several activities in the *aro-1* cluster, including *aro-*

2 function (800), and is inseparable from translocation *T(C161)* (1580).

aro-4 domain: Specifies 3-enolpyruvate shikimic acid-5-phosphate synthetase (EC 4.2.1.10) (731).

aro-5 domain: Specifies shikimate kinase (EC1.1.1.25) (731).

aro-9 domain: Cloned. Active site sequence PIR S14749; pSV50 clone 31:9G. Encodes biosynthetic dehydroquinase (EC 2.7.1.71) (1726). Requires shikimic acid or a mixture of four aromatic products when a *qa* mutant is present that eliminates catabolic dehydroquinase. *aro-9*; *qa*+ grows on minimal medium without supplement. Single-function *aro-9* mutants were obtained first by selecting in *qa-1*, a regulatory mutant that lacks catabolic dehydroquinase activity (1726) (Fig. 54). Conversely, *aro-9* is used to select *qa-1* and *qa-2* mutants (1724, 1726). Homozygous crosses produce white ascospores.

aro-3: aromatic-3

IIR. Between *arg-5* (1%; 3%) and *T(NM177)^L*, *nuc-2* (347, 800, 1333). Not part of the *aro-1* cluster gene.

Cloned and sequenced (872): Swissprot AROC_NEUCR, EMBL/Genbank U25818, PIR S11236.

Structural gene for chorismate synthase (EC 4.6.1.4) (704, 731) (Fig. 13). *aro-3* is an unusual example of a chorismate synthase gene because it specifies a bifunctional polypeptide with both chorismate synthetase and flavin reductase activities (872). Requires a mixture of PABA, tyrosine, tryptophan, and phenylalanine for growth. Shows interallelic complementation (731). Leaky, giving hazy growth on minimal medium at 4 days, 34°C; tests should be scored promptly (1546).

aro-4: aromatic-4

Part of the *aro-1* cluster gene. See *aro-4* domain under *aro-1*.

aro-5: aromatic-5

Part of the *aro-1* cluster gene. See *aro-5* domain under *aro-1*.

aro-6: aromatic-6

VIL. Between *ad-8* (8%; 12%), *cpl-1* (5%) and *lys-5* (3%) (822, 1815).

Structural gene for 3-deoxy-D-arabino-heptulosonic acid-7-phosphate synthase [DAHP synthase (Tyr), EC 4.1.2.15], one of the three isozymes inhibitable by tyrosine, phenylalanine, and tryptophan, respectively (Fig. 13). Grows on minimal medium except when both tryptophan and phenylalanine are present to inhibit the alternate synthases (822). Both activity-negative and allosteric-inhibition-negative alleles have been found (821). An *aro-6 ad-8; aro-7 aro-8* stock is available (FGSC 4492).

aro-7: aromatic-7

I. Between *arg-1* (4%) and *his-3* (1% or 2%) (822).

Structural gene for DAHP synthase (Phe) (EC 4.1.2.15), one of the three isozymes inhibitable by tyrosine, phenylalanine, and tryptophan, respectively (Fig. 13). Grows on minimal medium except when both tyrosine and tryptophan are present (822). Both activity-negative and allosteric-inhibition-negative alleles have been found (821). An *aro-7 ad-3 aro-8; aro-6* stock is available (FGSC 4492).

aro-8: aromatic-8

IR. Between *so* (7%; 11%) and *R* (4%) (822, 2129).

Purified and partially sequenced: Swissprot AROF_NEUCR; PIR P80576 (2178).

Structural gene for DAHP synthase (Trp) (EC 4.2.1.15), one of three isozymes inhibitable by tyrosine, phenylalanine, and tryptophan, respectively (Fig. 13). Grows on minimal medium except when both phenylalanine and tyrosine are present (822). Both activity-negative and allosteric-inhibition-negative alleles have been found (821). *aro-7 aro-8 nic-1; aro-6* stocks are available (FGSC 4489, 4490).

aro-9: aromatic-9

Part of the *aro-1* cluster gene. See *aro-9* domain under *aro-1*.

aro(p): aromatic (polar)

Symbol used for polar mutants in the *aro-1* cluster gene. See *aro-1*.

arom: aromatic

Symbol changed to *aro*.

arp1: actin-related protein 1

Allelic with *ro-4*.

arp3: actin-related protein 3

IIR. Linked near *arg-12* (2092).

Cloned and sequenced: Swissprot ARP3_NEUCR, EMBL/GenBank U79737; Orbach-Sachs clone X7F05.

Encodes actin-related protein 3. Transcript levels decrease upon the induction of conidiation and then increase slightly during conidial differentiation (2092).

ars-1: aryl sulfatase-1

VIIL. Between *thi-3* (2%; 5%) and *qa, Cen-VII, met-7* (1%), *ile-1* (1%) (340, 1328, 1418).

Cloned and sequenced: EMBL/GenBank U89492; Orbach-Sachs clone X23G09.

Structural gene for aryl sulfatase (EC 3.1.6.1) (1329); see Fig. 24. Scored by color reaction with *p*-nitrophenyl sulfate. Because the enzyme is repressed in the wild type by traces of inorganic sulfate and other compounds present in all normal agars, screening on plates is carried out in an *eth-1^r^ cys-11* background, in which *ars*⁺ colonies have detectable, derepressed activity (1328). Scoring in crosses does not require this special background if the germinated spores are grown with cysteic acid (1 m*M*) as the sole sulfur source (MgCl₂ replacing MgSO₄). Scoring method (1418). Reversion (1330). Mutants lacking aryl sulfatase were first isolated by screening in *Neurospora crassa* (1328) and were later shown to be allelic with a "natural" aryl sulfataseless gene, introgressed into *N. crassa* from one isolate of *Neurospora tetrasperma*, and with a gene that produces an electrophoretic variant enzyme in another natural isolate of *N. tetrasperma* (1329). Positively activated by *cys-3* (1517); regulation reviewed (1283) (Fig. 24). Integration of *ars-1* DNA at transgenic sites can eliminate the need for positive activation of expression (2148).

asc: ascus development

This symbol has been used for recessive mutants that affect ascus (or ascospore) development (518, 519). Some are barren; others result in much ascospore abortion. Only those *asc* mutations that have been mapped are listed here. Five remaining unmapped recessive mutants complement each other and *mei-1*. The *asc* mutants described by A. M. DeLange were shown not to be mutagen sensitive (775). See also *asd*, and *mei*.

asc(DL95): ascus development (DL95)

IVR.

Perhaps a *mei-1* allele.

asc(DL243): ascus development (DL243)

IVR.

Perhaps a *mei-1* allele.

asc(DL879): ascus development (DL879)

II. Linked to *arg-5* (3%) (518).

Recessive. Meiosis is impaired in homozygous crosses. Seventy percent of ascospores abort and the total ascospore number is reduced, as are pachytene pairing and recombination (519). Viable ascospores are usually disomic for one or more linkage groups. Not tested for allelism with *mus-7*, which is homozygous-barren and maps in the same region.

Asc(KH2A83): Ascus development (KH2A83)

VR. Left of *al-3*; probably right of *his-1* (701, 844).

Ascospore abortion is 10–95% in heterozygous crosses and 90–95% in homozygous crosses. The frequency of aneuploids is increased. Sensitive to UV, but not to MMS or histidine (701).

asco: ascospore maturation

Used for a *lys-5* allele with an autonomously expressed ascospore-ripening defect (1986).

Ascospore Color Mutants

Mutant genes with autonomous effects on ascospore pigmentation include *asco, bs, per, ws, cys-3, lys-5,* and *pan-2.* For applications, see refs. (597), (999), (1138), (1431), (1618), and (1985). *cys-3* may be best for demonstrating segregation patterns in asci (1546, 1676) (Fig. 25). Except for *bs* and *per*, ascospores that lack wild-type pigment fail to germinate. Brown *bs* ascospores are fully viable. The unpigmented ascospores of *per* germinate spontaneously without heat shock and are killed by exposure to 60°C.

asd-1: ascus development-1

VIL. Between *nuo19.3* (5T/18 asci) and *cmt* (2 or 3T/18 asci), *Bml* (5 or 6T/18 asci) (1447).

Cloned and sequenced: EMBL/GenBank U70861. On a common cosmid (pSV50 12.7B) with *asd-3*.

ASD-1 is homologous with *Aspergillus* rhamnogalacturonase B, which cleaves pectin (1449). Identified as a gene expressed preferentially under conditions that induce the sexual cycle. Essential for sexual development, but vegetative growth is normal. Expression is very low in *mat A^{m44}*. In crosses homozygous for RIP-induced *asd-1* mutants (1449, 1450), perithecial beaks are short, asci are normal in size and number, the two meiotic divisions are normal, but following the third division the distribution of nuclei is highly abnormal, ascospore delimitation fails, and no ascospores are produced (1450, 1673). Called *sdv-10*.

asd-2: ascus development-2

VIIL. Linked to *nic-3* (15%), *ars-1* (36%) (1450).

Sexual development is blocked in homozygous mutant crosses. Perithecia are small, without beaks. No asci are formed, or only a few very short asci (1450, 1673). The mutation was probably caused by ectopic integration of nonhomologous transforming DNA into *asd-2$^+$* (1450).

asd-3: ascus development-3

VIL. Between *nuo19.3* (5T/18 asci) and *cmt* (2 or 3T/18 asci), *Bml* (5 or 6T/18 asci) (1447).

Cloned and sequenced: EMBL/GenBank AF136235. On the same cosmid (pSV50 12.7B) as *asd-1*.

ASD-3 is homologous to sugar-transporter proteins (1319). Identified as a gene expressed preferentially under conditions that induce the sexual cycle. RIP-mediated disruption of *asd-3$^+$* yielded strains resembling *asd-1* mutants, with aberrant sexual development but normal vegetative growth. In homozygous mutant crosses, early sexual development is normal and many asci are formed, but ascospores are never delineated. Originally called *sdv-15*.

Asd-4: Ascus development-4

IVR. Near *met-5* (1277).

Cloned and sequenced (626a, 1277).

Encodes a GATA-type zinc finger DNA-binding protein unlike previously known GATA factors NIT-2, WC-1, WC-2, and SRE. Prevents development of asci. No detectable vegetative phenotype. Dominant. Obtained by RIP. The effect is reversed by introduction of the wild-type gene Asd-4$^+$. Called BF99 (626a, 1277).

Asm-1: Ascospore maturation-1

VR. Linked to *al-3* (0T/18 asci) (49, 1447).

Cloned and sequenced: EMBL/GenBank U51117; Orbach-Sachs clones X17A07, X19C05.

A key regulatory gene involved in sexual development. Contains a conserved DNA-binding domain (49). The abundant product is nucleus-localized. The deletion mutant has a complex phenotype with vegetative defects recessive and sexual-phase defects ascus-dominant. Matted mycelium, stunted aerial hyphae, slow conidial germination, no protoperithecia. In crosses heterozygous for a deletion of Asm-1 (Asm-1$^+$ × ΔAsm), all eight ascospores fail to mature in a large majority of asci. This is due to failure of transvection. Ascospore maturation depends on close meiotic pairing of Asm$^+$ with its homologous allele (48). For other genes with ascus-dominant effects, see *Ban, eat-2, Fsp, Pad, pk^D, Prf* and *R*.

asn: asparagine

VR. Between *inv* (4%; 9%) and *pyr-6* (6%), *pl* (1%; 9%) (321, 322, 1383, 2014).

Requires asparagine for growth; does not respond to aspartic acid (2048). Lacks asparagine synthetase (EC

6.3.5.4). Complementation between alleles (1240). *asn* auxotrophs appear dominant in heterokaryons when forced with *ad*, but recessive when forced with *nic*. May be inhibited by histidine (1546). Temperature-sensitive alleles T51M147 and T51M158 were originally classed as *un* mutants because growth is inhibited on organic complete medium. Symbol changed from *asp* (1579).

asp: aspartate

VR. Between *at* (0%; 3%) and *per-1* (16%; 26%) (1582, 1585, 1604).

Growth is aided by aspartate or glutamate. Also some response to homoserine or leucine (556). Grows adaptively on minimal medium; adapting more rapidly at 25°C than at 36°C. Inhibited by alanine; 0.5 mg/ml alanine in test medium aids scoring. Symbol changed from *aspt*. The symbol *asp* was formerly used for asparagine. In 1973, new symbols were adopted to conform to bacterial usage for amino acid auxotrophs (1579), with *asn* used for asparagine and *asp* for aspartate.

aspt: aspartate

Symbol changed to *asp*.

at: attenuated

V. Linked to *Cen-V* (0T/57 asci) (1602). Between *lys-1* (2%; 10%), *cyt-9* (2%, 5%) and *asp* (0%; 3%) (1582, 1604).

Conidia are formed in small flecks or granular clumps on an agar surface, especially in a crescent at the top of a slant (1604) (Fig. 14). Growth and pigmentation are slower than wild type, but the surface of a slant is covered. A good marker most easily scored on minimal medium, where conidiation is less profuse than on complete medium. Allele NM221t is expressed better at 34°C than at 25°C. Called *morph(M111)* (1604).

atp: ATPase

F-type ATPases consist of two major components, the catalytic core (F_1) and the membrane proton channel (F_0). F_1 is composed of five polypeptide subunits, α, β, γ, δ, and ε, which are all nuclear-encoded. F_0 has three subunits—a, b, and c; a and b are encoded in the mitochondrion and c in the nucleus. For a review, see ref. (208).

atp-1: ATPase-1

I or V. Determined by blotting chromosomal DNA on CHEF gels.

Cloned and sequenced: Swissprot ATPA_NEUCR, EMBL/ Genbank M84191, PIR JC1111, EMBL NCATPA, GenBank NEUATPA; EST SC1H4.

Structural gene for the ATP synthase α-subunit of the catalytic core F_1, mitochondrial precursor (EC 3.6.1.34). This subunit contains a nucleotide-binding site, but the bound nucleotide is not hydrolyzed during the catalytic cycle (212).

FIGURE 14 The mutant *at: attenuated*, showing the formation of conidia in aggregates. Clumping of conidia can be seen macroscopically. Bar length 0.1 mm. Scanning EM photograph from M. L. Springer.

atp-2: ATPase-2

II. Linked to *Cen-II*, *arg-5* (0T/18 asci) (1447).

Cloned and sequenced: Swissprot ATPB_NEUCR, EMBL/ GenBank M84192, PIR JC1112, EMBL NCATPB, GenBank NEUATPB, NCF1ATPB; EST NP6B1.

Structural gene for the ATP synthase β-subunit of the catalytic core F_1, mitochondrial precursor (EC 3.6.1.34). The site of ATP synthesis is largely formed by this polypeptide (with a smaller contribution from the α-subunit) (212).

atp-9: ATPase-9

Used for *oli*.

atr-1: aminotriazole resistant-1

IL. Between *suc* (0/39), *In(H4250)^L* and *T(39311)^R* (1582, 1603).

Resistant to 0.5 mg/ml 3-AT in solid medium (added before autoclaving). Not resistant to acriflavine. Abnormal vegetative morphology. Female sterile. Resistance is recessive in heterozygous duplications from *T(39311)* (1582). Histidine in test media neutralizes the toxicity of 3-AT to wild type [3-AT affects histidine biosynthesis (1768)]. Resistance to 3-AT also is shown by *acr-2* (931), *leu-1*, and *leu-2* (1048, 1582).

aur: aurescent

The name originally used for *al-1* allele 34508, which becomes visibly pigmented only in mature peripheral conidia and conidiophores. See *al-1*.

aza: azapurine resistant

Wild type is resistant to azapurines. A *cpc-1* strain, which is sensitive, was used for selecting azapurine-resistant mutants. Three resistance loci are known. *aza-1* and *aza-2* mutants have been lost, but their described characteristics and map locations should enable recurrences to be recognized. They were selected using aza-adenine, but at least one allele at each locus also showed resistance to azaguanine. *aza-3* was selected for resistance to azaguanine using a different procedure, and resistance to aza-adenine has not been determined.

aza-1: azapurine resistant-1

IL. Left of *mat* (23%) (995).

Resistant to purine analogues 8-aza-adenine and 2-6-diaminopurine. (One allele of four was also resistant to 8-azaguanine.) Obtained in *cpc-1* (*mts*), which is inhibited by the analogs. Selected and scored using 1 mg/ml 8-aza-adenine in medium. Resistance is recessive in a heterokaryon. At least one allele secretes purines. called aza-adenine-resistant (995). Strains lost (991).

aza-2: azapurine resistant-2

IL. Linked to *mat* (2%), *aza-1* (39%) (995).

Resistant to purine analogues 8-aza-adenine and 8-azaguanine. Sensitive to 2-6-diaminopurine. Obtained and scored as described for *aza-1*. Resistance is recessive in a heterokaryon. Called aza-adenine-resistant (995). Strains lost (991).

aza-3: azapurine resistant-3

III. Linked to *trp-1* (14%) (884).

Resistant to purine analogues 8-azaguanine and 6-mercaptopurine. Obtained upon limiting adenine in an adenine auxotroph, which is inhibited by the analogs. Selected and scored using *ad-3A ad-3B; ad-2* on medium with 2 μg of adenine sulfate and 200 μg/ml azaguanine. Relative resistance of *aza-3^R* and *aza-3^S* to 8-aza-adenine not determined. Resistance is recessive in a heterokaryon. Hypoxanthine can be used as the sole purine supply (883, 884). Called azaguanine-resistant.

azs: azide sensitive

Unmapped. Unlinked to *has* (595).

Cannot produce the inducible azide-sensitive respiratory pathway when grown in the presence of chloramphenicol (595). Obtained from the double-mutant strain ANT-1 (antimycin-sensitive), which was also the source of *has* (588, 591, 595). The double mutant *azs; has* has been used to obtain oligomycin-resistant mutants (596) and mutants deficient for succinic dehydrogenase (590).

B^m: mauve

IL. Left of *nit-2* (30%) (739).

Colonies are mauve on special dye medium where the wild type is blue (739).

bal: balloon

II. Between *T(AR179)* (hence of *thr-2, thr-3*) and *T(ALS176)*, *arg-5* (1%; 7%) (1578, 1580, 1582, 1585). Probably left of *Cen-II* (714, 800).

Forms a smooth, slow-growing hemispherical colony (1548). Fully female-fertile, which is uncommon for colonial mutants having such restricted growth. Recessive in heterokaryons. Glucose-6-phosphate dehydrogenase deficiency (1853, 1854, 1857). (*col-2* and *fr* are also deficient in G6PD.) Reduced levels of NADPH (233), linolenate (241), and peptides in the cell wall (2249). Morphology is subject to alteration by modifiers that are commonly present in laboratory stocks, resulting in spreading growth and conidiation. See *su(bal)*. Photograph (235, 1851, 1853). Allele C-1405 called *mel-2* (1409, 1585).

Ban: Banana

IL. Left of *mat* (14%), Probably left of *leu-3* (1684).

Each ascus delimits a single giant ascospore enclosing all four meiotic products and their mitotic derivatives (Fig. 15). The ascus trait is dominant and almost completely penetrant. The mature giant ascospores are capable of germination and usually give rise to mixed cultures. The giant ascospore in *Ban* is formed not by hardening of the ascus wall, as in *Iasc*, but by hardening of the ascus membrane within a normal wall, which it may not fill completely. Vegetative defects are recessive. Morphology is abnormal, with short aerial hyphae and no protoperithecia (1684). In older perithecia, the prefusion nuclei in the croziers revert to mitosis, which is synchronized and favorable for cytological observation (Fig. 16). (Croziers in *Prf* show similar behavior.) *Ban* has been used to rescue segregants that would not survive as homokaryotic haploids (1674). *Ban* arose in a burst of spontaneous mutants in a cross involving *mei-3*, which is also located in IL (1469). Some *Ban* stocks may contain *mei-3* or a *mei-3* partial revertant that was recovered from the same cross.

bas: basic amino acid transport

Probably allelic with *pmb*. Also called *bas.*[a]

bat: basic amino acid transport

Allelic with *pmb* (513, 1776).

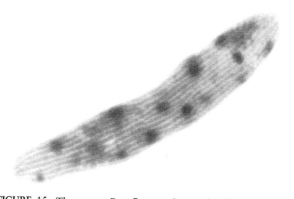

FIGURE 15 The mutant *Ban: Banana*. A maturing giant ascospore from *Ban⁺ × Ban*, showing striations throughout its length. Sixteen nuclei can be seen before the ascus wall becomes opaque. Photograph by N. B. Raju. From ref. (1684).

bd: band

IVR. Between *met-5* (9%) and *nit-3* (19%) (1185, 1794).

Because dense bands of conidia are produced on appropriate solid medium at intervals of about 24 hr (1792, 1794), the mutant has been used extensively to study circadian rhythms (239, 560, 621, 1382, 1794). *bd* has no effect on the underlying clock mechanism, but allows the visible expression of rhythm (622). Growth rate is about 70% that of the wild type (232). Conidiation is enhanced, even on slants (232). CO_2 inhibits conidiation and, thus, inhibits banding; *bd* is much less sensitive than wild type to this effect of CO_2 (1792). Biotin starvation leads to a phenocopy in wild type and to increased persistence of banding in *bd* (2206). Originally identified in a *bd; inv* strain called "timex" (1791). *bd* alone is sufficient to cause banding (1794). Used to study conidiation under nonstarvation conditions (1820). Used in a study of morphological differentiation patterns such as concentric rings and radial zonations (528). Expression is affected by changing the concentrations of agar, sugar, and salts. Conidial scatter is eliminated in the double mutant *bd; csp* (239). Conveniently scored by conidial banding on agar in long tubes or large plates at 25°C in constant dark or in a dark–light cycle, but not in constant light (1791).

ben: benomyl resistant

Symbol changed to *Bml*.

Beta-tubulin

See *Bml*.

bis: biscuit

Changed to *pk*. See p. 270 of ref. (1592).

bld: bald

IVR. Right of *pdx-1* (4%) (1592).

Aerial hyphae at the top of a slant are devoid of conidia. More extreme at 25°C than at 37°C, where powdery conidia are formed in a zone below the top. Called *mo(NM213t)* and incorrectly stated to be wild type at 34°C (1592).

ble: bleomycin resistant

Introduced from transposon *Tn5*.

This gene from *Tn5*, which confers resistance to bleomycin and phleomycin, has been placed under control of the *Neurospora am* promoter and used as a dominant selectable marker for *Neurospora* transformation (74). See also *amdS*.

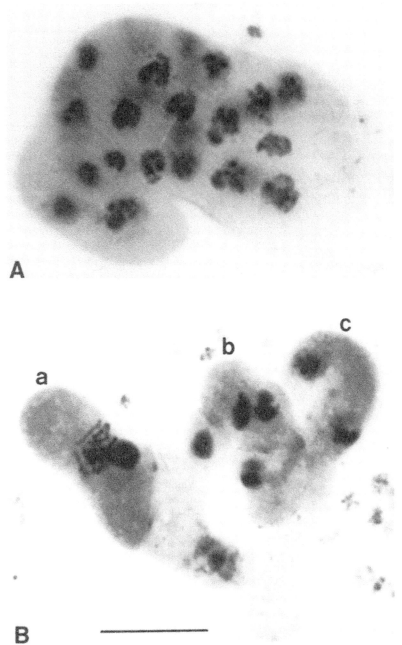

FIGURE 16 (A) multinucleate crozier cyst from *Ban$^+$* × *Ban*. Rather than fusing, the haploid pronuclei have reverted to dividing by mitosis in rounds that are synchronous. (B) Normal crozier development in the wild type, for comparison. (a) A young ascus following fusion of the haploid parental nuclei. (b) A three-celled crozier with a binucleate penultimate cell, which will give rise to an ascus. (c) A crozier with the two nuclei in conjugate division. ×1500. Stained with haemotoxylin. Photographs by N. B. Raju, from ref. (1684).

bli-3: blue light induced-3

IVR. Right of *trp-4* (1T/15 asci). Linked to *Fsr-51* (0T/15 asci), *Fsr-13*, *vma-8* (0T/18 asci) (1447).

Cloned and sequenced: Swissprot BLI3_NEUCR, EMBL/ GenBank X81318, GenBank NCBLI3.

Increased *bli-3* mRNA level is detected after 2-min exposure of a dark-grown culture to blue light. Obtained as a cDNA clone by differential screening, using probes from illuminated mycelia and dark controls (1946).

bli-4: blue light induced-4

IIR. Linked to *Fsr-34* (0T/18 asci), between *eas* (1T/18 asci) and *Fsr-17* (5T, 1NDP/18 asci), *leu-6* (9T, 1NPD/18 asci) (1447).

Cloned and sequenced: Swissprot BLI4_NEUCR, EMBL/ GenBank X89499, GenBank NCBLI4GEN.

Encodes a blue-light-induced, short-chain alcohol dehydrogenase (1530). An increased *bli-4* mRNA level is detected after 2-min exposure of a dark-grown culture to blue light.

bli-7: blue light induced-7

Allelic with *eas*.

bli-13: blue light induced-13

Unmapped.

Also strongly induced by continuous white light (1946).

Bml: Benomyl resistant

VIL. Between *cys-2* (2%), *cpc-1* and *ad-1*, *Cen-VI*. Linked to *ylo-1* (2% or 3%) and *un-13* (1/64), perhaps to the right (197, 1582, 1815).

Cloned and sequenced: Swissprot TBB_NEUCR, EMBL/ GenBank M13630, PIR A25377, GenBank NEUTUBB; EST NC3D10; pSV50 clone 24:11A; Orbach-Sachs clones X15H04, X18D12, G12C10, G14C11, G15C03, G15E01.

Encodes the β-subunit of tubulin. Mutation causes semi-dominant resistance to benomyl (197), a benzimidazole fungicide that affects hyphal tip extension and polarity. Mutations showing different degrees of resistance to benomyl all map to the *Bml* locus (2095). Typically, 0.5–1 µg/ml benomyl inhibits the wild type but not the mutant. Benomyl causes multiple germ-tube emergence from germinating macroconidia (268) and affects the level of *Bml* mRNA during germination (882). A Phe-to-Tyr change at position 167 confers resistance in the original mutant, 511 (1503). A Glu-to-Gly change at position 198 confers both resistance to the related benzimidazole fungicide carbendazim and sensitivity to diethofencarb, an *N*-phenylcarbamate fungicide to which the wild type is normally resistant (696). Additional amino acid changes can result in simultaneous resistance to both fungicides (694, 695). *Bml* has been used as a dominant selectable marker to introduce unselected genes by transformation (1503, 2167). Its use has the disadvantage, however, that introduced genes cannot be recovered efficiently in the progeny when these transformants are crossed, presumably because duplicate β-tubulin-encoding sequences in the transformants are subject to inactivation by RIP, causing the progeny to be inviable. Consequently, *hph* has seen wider use as a dominant selectable marker for transformation (1502, 1504, 1977). *Bml* has been called *Ben* (197); *mbic* (931), and *tub-2* (1503).

bn: button

VII. Linked to *met-7* (0/93), *wc-1* (1%) (1585). Right of *T(T54M50)*, hence, of *thi-3* (2%) (1548, 1578).

Nonconidiating, restricted colonial growth (102, 1548). Altered 6-phosphogluconate dehydrogenase (EC 1.1.1.44) (1852, 1854). (6PGD is also affected by *col-10*.) Levels of NADPH (233) and linolenic acid (241) are reduced. Used to study apical growth and polarization (2116). Germination may be better on minimal than on complex complete medium (1546). Allelic with *col-3* and complements *col-2* in heterokaryons (1565).

bs-1: brown ascospore-1

IR. Linked to *un-1* (9%), probably to the right (1603).

Ascospores are brown rather than black at maturity. Viability is not impaired. Expressed autonomously, allowing visual scoring in heterozygous asci (1603). Used to study factors affecting second-division segregation frequencies (1138). Translocation *T(NM139) bs* has a similar phenotype but is not allelic with *bs-1*, although one breakpoint is in IR proximal to *al-2* (1580).

c

Used in early publications as a symbol for *het-c*, *cot-4*, or *cy*

ca-1: calcium transport-1

I. Linked to *cys-9* (10%), *un-1* (10%) (693).

Optimal growth is obtained in liquid medium containing low calcium (693). Alleles 19 and 31 grow optimally with 10 µM calcium in the medium (408, 693) and are defective in vacuolar calcium transport (408). Allele 101, which grows optimally at 100 µM calcium, results in a shortened circadian rhythm (693). The effects of calcium on growth are not observed on solid medium.

ca-3: calcium transport-3

I. Linked to *ca-1* (5%), *cys-9*, *un-1* (10%) (693).

Optimal growth is obtained in liquid medium containing 10 µM calcium. The effect of calcium on growth is not observed on solid medium (693).

ca(23-2): calcium transport(23-2)

Unmapped.

A long lag phase is seen in medium containing low calcium Phase of the circadian rhythm is shifted by temperature but not by light (1491).

caf-1: caffeine resistant

VL. Between *dgr-1* (8%), *T(AR30)* (19%) and *mus-16* (4%), *T(AR33)* (5%; 12%), *lys-1* (4%; 14%) (926, 1578, 1582, 1598, 1602).

Resistant to caffeine (927). Resistance is dominant in duplications from *T(AR33)* (1598). Scoring is clear at 25°C and poor at 34°C, on slants with 2.5 mg/ml caffeine in minimal medium without sorbose (1322). Also readily scorable using conidial suspensions spotted on plates containing 2 mg/ml caffeine in minimal sorbose medium (1582).

can: canavanine resistant

Changed to *cnr*.

car: carbohydrate

IVL. Linked to *cys-10* (1%) (815).

Affects carbohydrate transport. Altered morphological rhythm is associated with a deficiency in the low-affinity glucose transport system. On glucose, *car* produces dense and sparse mycelium in cycles (period about 50 hr). On acetate, *car* is insensitive to the light–dark cycle and has a normal conidiation cycle with period about 24 hr. The cycle is not circadian. Periodicity is affected by the composition of the medium rather than by time. The *car* mutant originated from a cross between *pat* and *acu-7*, but *acu-7* is not necessary for expression of the phenotype. Called *Lpcar: long period carbohydrate* (815). The symbol *car* has also been used for some carotenoid mutants, at least one of which is an *al-2* allele (2019), and for a carbohydrate mutant (2007) that is not available for testing.

cat: catalase

Expression of the three structural genes is regulated differentially during development and in response to heat shock and treatment with oxidants. Products of the three genes differ in electrophoretic mobility and molecular weight (366, 1917).

cat-1: catalase-1

IIIR. Linked to *T(1)*, *ylo-1* (1/26), *trp-1* (17%) (1447, 1917).

Cloned and partially sequenced: EMBL/GenBank AI392552; EST SC3A3.

Encodes major catalase enzyme (EC 1.11.1.6) predominant in growing mycelium (366). Used to study the effects of singlet oxygen on CAT-1 heme (1216).

cat-2: catalase-2

VIIR. Probably linked to *for*, *frq* (≤10T+0NPD/18 asci), *cox-8* (≤6T+1NPD/18 asci) (1447, 1917).

Cloned and partially sequenced: GenBank AA901787; EST NM 10 E2.

Structural gene for minor catalase enzyme (EC 1.11.1.6) predominant in mycelium after heat shock (366, 1917).

cat-3: catalase-3

IIIR. Between *Fsr-45* (2T, 1NPD/18 asci) and *Tel-IIIR* (5T/18 asci) (1447).

Cloned and partially sequenced: EST EMBL/GenBank AA901970; EST NM6H12.

Structural gene for minor catalase enzyme (EC 1.11.1.6) predominant in mature conidia (366).

cax: calcium exchanger

VIL. Linked to *Tel-IVL* (3T/18 asci) (1447).

Cloned and sequenced: EMBL/GenBank AF053229; Orbach-Sachs clones X5D07, G3A09, G23H08.

Encodes a calcium transport protein in the vacuolar membrane. The predicted polypeptide contains 11 membrane-spanning helices. *cax* mutants grow normally (1272).

cbh-1: cellobiohydrolase-1

Unmapped.

Cloned and sequenced: Swissprot GUX1_NEUCR, EMBL/GenBank X77778, GenBank NCCBH1, PIR S42093.

Structural gene for the major 1,4-β-cellobiohydrolase (exo-glucanase, exocellobiohydrolase) (EC 3.2.1.91), a component of the cellulase complex.

ccb-1: constitutive carotenoid biosynthesis-1

Unmapped.

Mycelia in dark-grown liquid cultures become deep orange, but the mRNA levels of light-regulated genes do not increase (1206, 1208). Carotenoid synthesis after 2 days of growth in the dark resembles that of light-induced wild type. mRNA levels of carotenoid biosynthetic genes are increased following exposure to blue light, but those of developmentally regulated genes expressed during conidiation (*con-8*, *con-10*, and *eas*) are not. Conidia are formed in liquid medium (1208). Recessive. Obtained by selection using an *al-3-promoter::mtr* fusion strain in which *mtr*-mediated uptake of tryptophan and *p*-fluorophenylalanine depends on light induction (286).

ccb-2: constitutive carotenoid biosynthesis-2

Unmapped.

Mycelia in dark-grown liquid cultures become deep orange, but the mRNA levels of light-regulated genes do not increase (1206, 1208). Poor growth and conidiation. Hyphae are short, with chains of separated cells. mRNA level for the developmentally regulated gene *eas* is increased in response to blue light, but mRNAs for carotenoid genes show little or no increase. Attempts to cross or to form heterokaryons have been unsuccessful (1208). Origin same as for *ccb-1*.

ccg-1: clock-controlled gene-1

VR. Between *Cen-V* (4T/18 asci) and *ilv-2* (4T/13 asci) (1447).

Cloned and sequenced: Swissprot GRG1_NEUCR; EMBL/ GenBank X14801, EMBL NCMORCCG, GenBank NEUMORCCG; pSV50 clone 29:6C.

Encodes a highly abundant transcript that peaks early in the circadian morning (147, 1226). Regulated by development and light (61, 1206); light-regulation normally requires *wc-1* and *wc-2*, but the *cot-1* mutation suppresses the requirement for *wc-2* (61). A *ccg-1::tyrosinase* reporter gene has been used to screen for mutants with altered expression of conidiation-related genes (1093). Allelic with *grg-1*, which was isolated on the basis of regulation by glucose (575, 1315, 2055, 2186).

ccg-2: clock-controlled gene-2

Allelic with *eas*.

ccg-4: clock-controlled gene-4

IL. Between *mat* (3T/18 asci; 0/18) and *Cen-I* (2T/18 asci) (1447).

Cloned and sequenced: EMBL/GenBank AF088909, U46085, X81318; EST NM9C1.

The transcript level peaks early in the circadian morning and is regulated by both light and developmental signals (147). The *ccg-4* sequence shows similarity to a mating-type-specific pheromone precursor gene from *Cryphonectria parasitica* and *Magnaporthe grisea*, including multiple repeats of a putative pheromone sequence bordered by KEX2 processing sites. Consistent with mating-type-specific pheromone activity, the expression of *ccg-4* is restricted to cells of mating type A (189). See *eat-1, eat-2, mfa-1,* and *scp* for other genes that are expressed differentially in the two mating types.

ccg-6: clock-controlled gene-6

VR. Linked to *al-3, inl* (0 or 1T/18 asci) (1447).

Cloned and sequenced: Swissprot CCG6_NEUCR, EMBL/ GenBank U46086, AF088908; EST NC5G12.

RNA level peaks late in the circadian night and is regulated by both light and developmental signals (147).

ccg-7: clock-controlled gene-7

Allelic with *gpd-1*.

ccg-8: clock-controlled gene-8

V. Linked to *lys-1, Cen-V* (0T/18 asci) (147, 1447).

Cloned and sequenced: Swissprot CCG8_NEUCR; EMBL/ GenBank U46087, AF088907.

The mRNA level peaks late in the circadian night but is not regulated by either light or developmental signals (144, 147).

ccg-9: clock-controlled gene-9

VIIL. Linked to *nic-3* (2T/18 asci) (1447).

Cloned and sequenced: EMBL/GenBank U46088, AF088906.

Transcript peaks late in the circadian night and is regulated by light and developmental signals (147). Sequence comparisons suggest that *ccg-9* encodes trehalose synthase (1909).

ccg-12: clock-controlled gene-12

Allelic with *cmt*.

cdr-1: cadmium resistant-1

II. Left of *arg-12* (35%) (1182).

Resistant to 10–20 mM cadmium chloride on solid medium, but similar to wild type when grown immersed in liquid medium. Not cross-resistant to zinc, copper, or cobalt. Not osmotic-sensitive. Distinctive morphology (1182). Stock lost (1277).

cdr-2: cadmium resistant-2

VIIL. Between *nic-3* and *met-7* (1182).

Similar to *cdr-1* in resistance, but morphologically distinct (1182). Stock lost (1277).

cdr-3: cadmium resistant-3

Unmapped. Not in II or VII (1182).

Present in *Nurospora intermedia* strain P3 (FGSC 1768), which originated from a volcanic area of hot springs and vents in Unzen, Japan. Similar to *cdr-1*, but morphologically distinct (1182).

cel-1: chain elongation-1

IVR. Linked to *pan-1* (1%), *cot-1* (0/17) (1585).

Impaired synthesis of saturated fatty acids attributed to defective fatty acid chain elongation (871). Defective fatty acid synthase complex (603) (Fig. 17). Activity of the α-subunit of fatty acid synthase is reduced to 1.2% of wild type. Growth is normal with palmitate as supplement (239). Tween 40 provides a convenient supplement. The requirement is leaky at 22°C (603), but not at 34°C (1585). Sensitivity to oligomycin is increased [L. R. Forman and S. Brody, cited in ref. (536)]. Used to change fatty acid composition (237) and make the circadian clock sensitive to fatty acids and temperature (239, 1294). Effects on the circadian clock are reviewed in ref. (1130). Temperature compensation of the clock is lacking in *cel-1* (1295). Used to study membrane lipid-phase transitions and electrical properties (679, 680). Used to incorporate photolabile azido fatty acid probes for membrane studies (342). Palmitate is used more efficiently in *cel-1* than in wild type for the synthesis of unsaturated fatty acids (754). Called *ol: oleic* (1585) and *fas: fatty acid synthesis* (603). The initial report of oleic utilization (1585) was incorrect (871), probably because of impurities.

cel-2: chain elongation-2

IVR. Linked to *cot-1* (1%) (755), *bd* (232).

Impaired synthesis of saturated fatty acids (755) (Fig. 17). The β-subunit of fatty acid synthase is defective. Activity is lower than in *cel-1*, making *cel-2* potentially more useful for enriching membranes in supplemental fatty acids. Revertible. Fertility is reduced. Growth is normal on palmitate (232). Palmitate is used more efficiently in *cel-2* than in wild type for the synthesis of unsaturated fatty acids (1997).

cell-1: cellobiase/cellulase

Unmapped. Segregates as a single gene, independent of *gluc-1*.

Both cellulase and cellobiase are produced constitutively. Levels of aryl-β-glucosidase are not affected. Recessive to wild type in heterokaryons (1420). Isolated using *gluc-1* and selecting for high activity in destroying the β-glucoside esculin (582). (Cellulase, cellobiase, and aryl-β-glucosidase normally are induced simultaneously by cellobiose.)

cem-1: condensing enzyme with mitochondrial function

Unmapped

Cloned and sequenced: EMBL/GenBank AF021234.

A nuclear gene that encodes a mitochondrial protein, 3-oxyacyl (acyl carrier protein) synthase (EC 2.3.1.41) (250).

Cen: Centromere

Knowledge of the physical structure and organization of the centromeres is based on the only sequenced centromere region, *Cen-VII* (275, 340). The *Neurospora* centromere is structurally more similar to that of *Drosophila* than to that of *Saccharomyces*. If centromeres of the other *Neurospora* chromosomes are comparable in size and content, centromeric repeats may make up as much as 7% of the *Neurospora* genome. A defective *copia*-like transposable element *Tcen* is associated exclusively with all seven centromere regions, which also contain other relic transposons and many defective *Tad* elements (275).

Several methods are available for the genetic mapping of centromeres relative to flanking gene loci: (i) tetrad

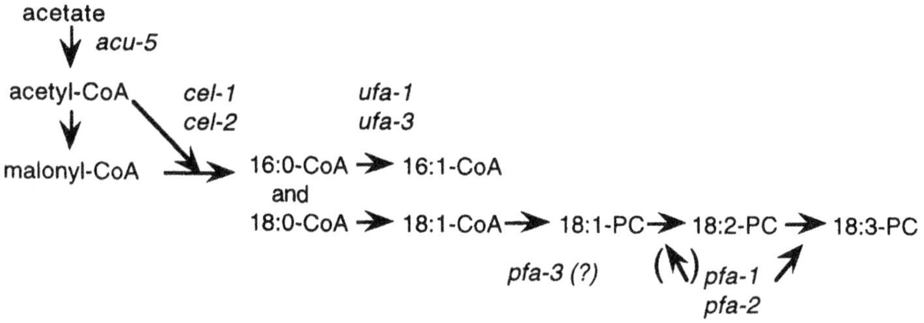

FIGURE 17 The biosynthetic pathway for the major fatty acids of *Neurospora*, showing the sites of action for the *cel*, *ufa*, and *pfa* genes. Abbreviations for the fatty acids are as follows: 16:0, palmitate; 16:1, palmitoleate; 18:0, stearate; 18:1, oleate; 18:2, linoleate; 18:3, linolenate. The fatty acids generally are found esterified to complex lipids such as phosphatidylcholine (PC). The genes *ufa-2* and *ufa-4* have pleiotropic effects and may affect auxiliary electron transport reactions. The genes *pfa-4* and *pfa-5* appear to affect the pathway of biosynthesis of 18:3 indirectly, as do *bal*, *fr*, and *col-2*. Original figure from Marta Goodrich-Tanrikulu.

analysis with physically ordered asci (1199); (ii) analysis of unordered asci in crosses heterozygous for Spore killer (1602) or other known centromere markers; (iii) duplication coverage of a flanking gene in a segment carrying gene loci known to be located in a given arm (1566); and (iv) cotranslocation to another chromosome arm of a flanking gene together with other loci whose arm is known. The third method is the least laborious when appropriate rearrangements are available. Linkage group arms were first defined as left (L) and right (R) by ref. (103), using *mat, pe, ser-l, pdx-l, ilv-l, rib-l,* and *nt* as reference markers.

Cen-I: Centromere-I

I. Between *arg-3* (2%), *T(39311)*^{R} and *sn* (1/271 second-division segregation), *os-4*, *T(AR173)*^{L}, *his-2* (733, 1578, 1580, 1605, 1963).

The amount of crossing over per unit of physical distance is much lower near *Cen-I* than in more distal regions of I or in the overall genome (1299, 1740). Recombination between *arg-3* and *his-2* varies from 3% to 18%, depending on *rec* genotype (338).

Cen-II: Centromere-II

II. Between *T(AR179)*^{R} and *T(ALS176)*, *arg-5* (0T/18 asci) (1447, 1578, 1580) (*bal* also lies between these translocations). Ordered asci indicate that *Cen-II* is probably between *bal* and *arg-5* (714, 800).

Cen-III: Centromere-III

III. Between *acr-7* (2T/18 asci) and *thi-4* (1T/18 asci), *sc* (4T/230 asci). Linked to *crp-2* (0T/18 asci) (103, 1447). Order relative to *acr-2* is uncertain.

Cen-IV: Centromere-IV

IV. Right of *ace-4, cut, fi* by inference from recombination distances. Left of *psi* [D. R. Stadler, A. M. Towe, and M. Loo, cited in ref. (1220)], *T(ALS159)* (1580).

Cen-V: Centromere-V

V. Between *lys-1* (5%), *cyt-9* (2%), *T(MB67)*^{R} and *asp*. No recombination with *at* (0/57 asci) (1578, 1602).

Cen-VI: Centromere-VI

VI. Between *ad-1* (l% or 2%) and *T(AR209), rib-1* (l%; 4%), *pan-2* (2%) (296, 1580, 1986). *glp-4* is also between *ad-1* and *rib-1* (2160).

Cen-VII: Centromere-VII

VII. Between *ars-1, qa* cluster and *met-7* (1%) (340). Cloned and sequenced: EMBL/GenBank L23168, AF079510, GenBank NEURSCENTR.

The 450-kb region that contains *Cen-VII* is recombination-deficient and A+T-rich relative to flanking regions. The region contains repetitive sequences that hybridize to sequences that map at or near the centromeres of all other linkage groups. The repeated sequences are divergent, with differences that are predominantly GC→AT transitions, suggesting that they have been subjected to RIP.

Recombination frequencies relative to physical length decline dramatically in this region, which comprises 11% of the linkage group VII DNA (340). Reduction extends into the arms, gradually diminishing in the 200-bp interval *Cen-VII–wc-1* [G. Macino, cited in ref. (340)]. Sequencing of a 16-kb portion revealed a tightly nested cluster of simple sequence repeats and degenerate retrotransposon-like elements—*Tcen*, which is *copia*-like and is found exclusively in centromere regions, *Tgl1*, identified as *gypsy*-like by sequence resemblance to the Ty3-2 transposon of *Saccharomyces*, and *Tgl2*, another immobile putative *gypsy*-like element. Multiple degenerate fragments of the Line-like *Tad* element also are present. The defective transposons all contain transition mutations typical of RIP. Multiple clusters of similar content are present elsewhere in the *Cen-VII* region (275).

cfp: cellular filament polypeptide

VIIR. Between *for* (1 or 2T/18 asci) and *T(5936)* (810, 1447). In a cosmid adjoining that carrying *ras-3* (1267).

Cloned and sequenced: Swissprot DCPY_NEUCR, EMBL/GenBank L09125, U56927, PIR JN0782, EMBL NCCFPX, GenBank NEUCFP; EST NC3D2; Orbach-Sachs cosmid G15G6.

Structural gene for pyruvate decarboxylase (EC 4.1.1.1), which forms 8- to 10-nm cytoplasmic filaments (1739). These filaments are abnormally sized and shaped in strains mutant at the *sn* locus (34, 40, 1739). Identified independently and called *pdc-1* (1267).

cfs(OY305): caffeine sensitive(OY305)

IR. Between *mat* and *al-2* (2279).

Growth is inhibited by caffeine (0.2 mg/ml) and stimulated by adenine. Morphologically abnormal. Not sensitive to caffeine in the presence of adenine. Growth is slow on minimal medium. Probably UV-sensitive by spot test. Not tested for allelism with *ad-3A, ad-3B,* or *ad-9* (2279).

cfs(OY306): caffeine sensitive(OY306)

IR. Near *al-2*, probably to the right (2279).

Growth is inhibited by caffeine (0.2 mg/ml) and by adenine. Morphologically abnormal. Grows slowly on minimal medium and is not stimulated on complete medium. Complements *cfs(OY305), cfs(OY307)* (2279).

cfs(OY307): caffeine sensitive(OY307)

IR. Between *mat* and *al-2*. Near *cfs(OY305)* (2279).

Growth is inhibited by caffeine (0.2 mg/ml) and by adenine. Morphologically abnormal. Grows slowly on minimal medium and is not stimulated on complete medium. Complements *cfs(OY305)*, *cfs(OY306)* (2279).

chol: choline

There are four identified auxotrophic genes that respond to choline. The catalytic activity of the first is phosphatidylethanolamine N-methyltransferase (EC 2.1.1.17). The product of the second gene, methylene-fatty-acyl-phospholipid synthase (EC 2.1.1.16), catalyzes the second and third methylations in the pathway of phosphatidylcholine biosynthesis from phosphatidylethanolamine (Fig. 18). Two of the *chol* genes are known to be involved in the synthesis of phosphatidylcholine.

chol-1: choline-1

IVR. Linked to *ad-6* (1%), probably to the right (1255).

Requires choline (905). Also uses mono- or dimethylaminoethanol (892) (Fig. 18). Deficient in S-adenosylmethionine phosphatidylethanolamine N-methyl transferase (EC 2.1.1.17) (430, 1806), the enzyme catalyzing the first step in the phosphatidylcholine (lecithin) biosyn-

thetic pathway. Morphology on limiting choline is abnormal, colonial (430). Colonies from single conidia on minimal agar medium resemble inhibited *mat-A/mat-a* duplications, with swollen hyphae and dark pigment in the presence of phenylalanine+tyrosine (1546). Phospholipid composition is abnormal on limited concentrations of supplement (934). Best scored late on minimal medium, where the mutant grows slightly and then stops (1546). Used to study the inhibition of cytochrome-mediated respiration and of conidiation when lecithin is depleted by choline starvation (1006, 1007). Used to demonstrate that mitochondria reproduce by division (1233). The *chol-1* mutation lengthens the period of circadian rhythm of conidiation. In the *chol-1; frq* double mutant, the period is lengthened to over 50 hr on low choline. In combination with a null *frq* mutation, *chol-1* remains rhythmic on low choline (1131).

chol-2: choline-2

VIL. Between *Tip-VIL* and *nit-6* (6%; 8%) (1582, 1585).

Requires choline (895). Also uses di- but not monomethylaminoethanol (892) (Fig. 18). Deficient in methylene-fatty-acyl-phospholipid synthase (EC 2.1.1.16), catalyzing steps 2 and 3 in the phosphatidylcholine (lecithin) biosynthetic pathway. The only allele, 47904t, is leaky on minimal medium at 22°C, but not at 34°C (934). Phospholipid composition is abnormal on limiting

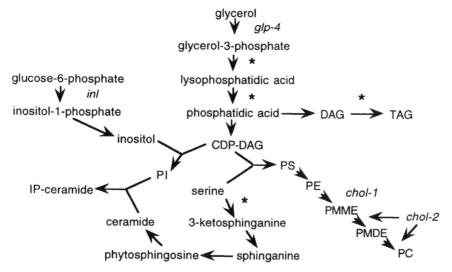

FIGURE 18   The biosynthetic pathway for the major fatty acid-derived lipids. Sites where fatty acids are incorporated are marked with asterisks. The major phospholipids are phosphatidylserine (PS), phosphatidylethanolamine (PE), phosphatidylcholine (PC), and phosphatidylinositol (PI). The sites of action of the *inl* and *chol* genes are shown. Abbreviations: DAG, diacylglycerol; TAG, triacylglycerol; IP-ceramide, inositolphosphorylceramide. The PC intermediates phosphatidylmonomethylethanolamine (PMME) and phosphatidyldimethylethanolamine (PMDE) do not accumulate in wild type. Minor reactions occur, such as the phospholipase-mediated conversion of PC to DAG and the conversion of DAG to PC, which occurs in *chol* mutants supplemented with choline. For simplicity, these are not illustrated. Sites of action of the *chol-3*, *chol-4*, and *un-17* genes are not known. Original figure from Marta Goodrich-Tanrikulu.

choline (934). Growth is colonial on limiting supplement at 34°C and on minimal at 25°C. Resistant to aminopterin.

chol-3: choline-3

VR. Between *at* (18%) and *al-3* (19%; 47%) (1582, 1593). Requires choline, but growth is subnormal even on optimal supplement (1593). The mutant was obtained and the requirement identified in the E. L. Tatum laboratory.

chol-4: choline-4

IV. Linked to *cot-1* (32%), *chol-1* (3 prototrophs among 24 intercross progeny) (1582, 1593). Responds to choline, but growth is subnormal even on optimal supplement (1593). The mutant was obtained and the requirement identified in the E. L. Tatum laboratory (1593).

chr: chrono

VI. Between *chol-2* (10%) and *pan-2* (712). Altered period of circadian conidiation rhythm. The one known allele is incompletely dominant, with a 23.5-hr period at 25°C in *csp*⁺ background. Temperature compensation is good above 30°C (711, 713).

chs-1: chitin synthase-1

V. Between *rDNA* (5/18), *con-4a* (1/18) and *am* (4/18), *chs-3* (5/18) (1447). Cloned and sequenced: Swissprot CHS3_NEUCR, EMBL/GenBank M73437, PIR E38192, EMBL NCCHT, GenBank NEUCHT. Structural gene for chitin synthase 1 (EC 2.4.1.16) (chitin-UDP acetylglucosaminyl transferase 1). The RIP-induced mutant is slow-growing and morphologically abnormal (2269).

chs-2: chitin synthase-2

IVR. Linked to *nit-3* (6/18), *Fsr-4* (0/17), *uvs-2* (0/18) (1447). Cloned and sequenced: Swissprot CHS2_NEUCR, EMBL/GenBank X77782, PIR B45189, EMBL NCCHITSYN, GenBank NCCHITSYN. Structural gene for chitin synthase 2 (EC 2.4.1.16) (chitin-UDP acetylglucosaminyl transferase 2). RIP-induced mutant has no phenotype other than increased sensitivity to edifenphos (543).

chs-3: chitin synthase-3

VR. Linked to *am* (1 or 2/18), *inl* (1 or 2/18) (1447). Cloned and sequenced: Swissprot CHS3_NEUCR, EMBL M73437, GenBank AF127086, PIR A41638.

Structural gene for CHS-3 (EC 2.4.1.16) (chitin-UDP acetylglucosaminyl transferase 3), a class I chitin synthase, which is essential (175, 203).

chs-4: chitin synthase-4

I. Linked to *leu-4*, *mat*, *arg-3* (0/18) (1447). Cloned and sequenced: Swissprot CHS4_NEUCR, EMBL/GenBank U25097, EMBL NC25097, PIR S61886. Encodes the structural gene for chitin synthase 4 (EC 2.4.1.16) (chitin-UDP acetylglucosaminyl transferase 4) (542). RIP-induced mutant is abnormal only on sorbose (542).

chs-5: chitin synthase-5

IVR. Between *arg-14* (2/18) and *cot-1* (1/18) (1447). Cloned and partially sequenced (2267): GenBank A 1397637; EST SC5D3. Structural gene for chitin synthase 5 (EC 2.4.1.16) (chitin-UDP acetylglucosaminyl transferase 5) (2267).

ci-1: cycloheximide inducible-1

Unmapped. Cloned and sequenced: EMBL/GenBank X15033, GenBank NCCI1, PIR S09643. Encodes a 70-kDa polypeptide that is a member of the cytochrome P-450 family (69). Highly induced by the partial inhibition of protein synthesis resulting from low doses of cycloheximide or the depletion of polyamines; not induced by heat shock or starvation (24, 805).

cia35: chaperone protein 35

Unmapped. Cloned and sequenced: EMBL/GenBank AJ001726, GenBank NCCIA35. Encodes a chaperone protein involved in the assembly of mitochondrial complex I (1105).

cia80: chaperone protein 80

Unmapped. Cloned and sequenced: EMBL/GenBank AJ001712, NCCIA80 (1105). Encodes a chaperone protein involved in the assembly of mitochondrial complex I (1105).

cit-1: citrate synthase-1

IIR. Linked to *preg* (0T/18 asci) (1447). Cloned and sequenced: Swissprot CISY_NEUCR, EMBL/GenBank M84187, PIR S41563, EMBL NCCIT1A, GenBank NEUCIT1A; EST NM6F5. Structural gene for citrate synthase, mitochondrial isozyme (EC 4.1.3.7) (627).

cit-2: citrate synthase-2

VR. Between *inl* (1T/18 asci) and *cya-2* (1T/18 asci) (1447). Structural gene for citrate synthase, microbody isozyme (EC 4.1.3.7) (627).

cl: clock

VR. Right of *pk* (1% or 2%) (569, 1965).

Shows noncircadian periodicity. Spreading flat colonial morphology, forming dense bands. The mycelium becomes increasingly dense until growth ceases in all but a few hyphae; these reinitiate the cycle. Grows 1 cm/day. Band size and period are modifiable. Photographs (569, 2026). Asci from *cl* × *wild type* are normal. Crosses homozygous for *cl* give flaccid asci with unordered ascospores, similar to *pk* × *pk*. So do *cl* × *pk*, suggesting allelism (1965). *pk (bis)* sometimes forms growth bands (569). However, substantial crossing-over frequencies and the recovery of *pk; cl* double mutants (569) indicate that *pk* and *cl* are at separate loci. Increased activity of L-glutamine:D-fructose 6-phosphate amidotransferase was found in crude extracts of *cl* and morphological mutants at five other loci (1758).

cla-1: clock affecting-1

Identical to *frq*[7] (397).

cmc: carboxymuconate cyclase

Unmapped.

Cloned and sequenced: Swissprot CCCM_NEUCR, EMBL/GenBank L27538, PIR A55525, GenBank NEUCCMUCY.

Encodes carboxy-*cis,cis*-muconate cyclase (EC 5.5.1.5) (3-carboxy-*cis,cis*-muconate lactonizing enzyme) (CMLE) (1306).

cmd: calmodulin

VR. Between *inl* (3T/18 asci), *cya-2* (2T/18 asci) and *tom70* (1 or 2T/17 asci). Linked to *hsp80*, *hsp83* (0T/18 asci) (1447).

Cloned and sequenced: Swissprot CALM_NEUCR, EMBL/GenBank L02963, PIR S58703, S28188, S32452, GenBank NEUCLMDLN; EST SP4B1.

Encodes the Ca^{2+}-binding protein that is a key regulator of many cellular processes. Purification (1507) and properties (422). Stimulation of adenylate cyclase (1710); cyclic nucleotide phosphodiesterase (1508, 1898, 2070); protein kinase (2136, 2137). Binds to proteins associated with the cytoskeleton (284, 1506). *Neurospora crassa* calmodulin activates the calmodulin-dependent protein phosphatase catalytic subunit of calcineurin, which is encoded by *cna-1* (874, 1652). It can be ubiquitinated by a Ca^{2+}-dependent mechanism (2303).

Calmodulin antagonists affect the circadian rhythm (1432, 2067) and light regulation (1775). Mutants with hypersensitive responses to the calmodulin antagonist chlorpromazine have been isolated [*cpz-1* and *cpz-2* (2031)]; these mutations affect light regulation and circadian rhythms. Calmodulin antagonists may also affect chitin synthase activity (2024).

cmt: copper-metallothionein

VIL. Between *asd-1* (2T/18 asci) and *Bml*, *Cen-VI* (3T/18 asci; 1/18) (1388, 1447).

Cloned and sequenced: Swissprot MT_NEUCR, EMBL/GenBank X03009, GenBank NCCMTG, PIR A24641, SMNC; Orbach-Sachs clones X16H10, G4H01, G9G12, G10C10.

Structural gene for copper metallothionein (1387, 1388). RNA level peaks late in the circadian night, but is not induced by light or developmental signals (147). Copper-inducible promoter (1812); metal binding of the polypeptide (151, 602, 1164); purification (1168); use as a heavy-metal bioabsorbent (1538). Strongly resembles a *Colletotrichum gloeosporioides* polypeptide predicted to be expressed during appressorium formation (943). Allelic with *ccg-12*.

cna: calcineurin A

VR. Linked near *cyh-2* (1652).

Cloned and sequenced: Swissprot P2B_NEUCR, EMBL/GenBank M73032, GenBank NEUCAM.

Encodes calcineurin A, the Ca^{2+}- and calmodulin-regulated protein kinase catalytic (α) subunit of calcineurin (874, 1652). The subunit encoded by *cna* interacts functionally with the mammalian catalytic subunit (2135). The *Neurospora* catalytic subunit is partially myristoylated when coexpressed with the mammalian regulatory subunit in insect cells (614). Reduction in *cna* expression is imposed by antisense RNA expression. Reduction in calcineurin function is imposed by the addition of cyclosporin A or FK506, which increases hyphal branching and changes hyphal morphology, ultimately causing growth arrest (1652). For a review of fungal protein kinases, see ref. (535).

cnb: calcineurin B

Unmapped.

Cloned and sequenced: EMBL/GenBank Y12814, AF034089, GenBank NCCALCINB.

Encodes the regulatory (β) subunit of calcineurin. An insertional mutation at *cnb* results in swollen and septate hyphae during vegetative growth, indicating that calcineurin may modulate the conidiation pathway (1092).

cni-1: cyanide insensitive-1

Unmapped. Not in V (592).

Respiration in the first 24 hr of growth is not sensitive to cyanide or antimycin. Initially insensitive to salicylhydroxamic acid (SHAM), but SHAM and cyanide together are inhibitory (592).

Cyanide-insensitive respiration and cytochrome c level decrease markedly in late log phase and stationary phase, whereas cytochrome aa_3 increases rapidly (1073). By electron microscopy, mitochondria appear defective in early log phase but later resemble wild type (1072). Behaves as a cold-sensitive ribosome assembly mutant. Small subunits are not assembled at low temperatures; the respiratory differences between early and late growth noted previously were found at 30°C, and normal aa_3 production occurred throughout growth at 37°C (1051). Electron-spin resonance data (591). Selected by failure to reduce tetrazolium in overlay after inositol-death enrichment (593).

cnr: canavanine resistant

IR. Between $T(4637)al\text{-}1$ and $In(OY323)^R$. Therefore, between hom (1%), $al\text{-}1$ and $lys\text{-}3$, $nic\text{-}1$ (11%) (1546, 2064).

Resistance to canavanine (904, 2064, 2065) is due to a constitutive enzyme that cleaves L-canavanine to hydroxyguanidine plus a compound that can be converted to L-homoserine (1218). Resistance, therefore, is dominant. Resistance is altered secondarily by modifiers that affect the rate of uptake (1218). Many laboratory strains are resistant, but a few are sensitive (1549). Best scored on 0.2 mg/ml L-canavanine H_2SO_4, autoclaved in medium, at 2 days and 34°C (1546, 1585). Sensitive strains have been used to select resistant mutants defective in basic amino acid transport (1736, 1782, 2233). Called can, but pmb rather than cnr is the *Neurospora* counterpart of CAN in *Saccharomyces cerevisiae*.

cog: recognition

IR. Between $his\text{-}3$ (1%; 3%) and $ad\text{-}3A$ (2%; 7%) (45, 216, 336). Located about 3 kb from the $his\text{-}3$ coding sequence (2275), at the opposite end from the $his\text{-}3$ promoter (216).

The *cis*-acting recombinator responsible for increasing recombination within $his\text{-}3$ (45) and crossing over between $his\text{-}3$ and $ad\text{-}3$ (336), but only in the absence of $rec\text{-}2^+$. Highly polymorphic alleles have been described. cog^{La} and cog^{S79a}, which are probably identical, mediate high recombination, which is dominant. Low recombination is mediated by $cog^{LA'}$, cog^{EA}, and cog^{ST74A} (2276). Generic designations are cog^+ for high-recombination alleles and cog for low (2275). The effect of a specific cog allele is limited to the chromatid that contains that allele, as evidenced by crosses

heterozygous for translocation $TM429$, which has a breakpoint between mutant sites within the $his\text{-}3$ locus (336). Recombinators with similar rec specificity are presumably present in other regions under $rec\text{-}2$ control, and cog-like recombinators with differing specificity are presumably located in regions controlled by other rec genes. See rec (Fig. 57).

coil: coiled

IVR. Between $arg\text{-}2$ (2%) and $leu\text{-}2$ (12%) (141).

Growth of hyphae from ascospores or conidia is initially spiraled in a clockwise coil on the surface of agar or on a membrane placed on agar (Fig. 19). Scorable microscopically in young cultures under low magnification (141). Counterclockwise coiling has been reported for $rco\text{-}1$ (2256).

col: colonial

The name *colonial* has been used primarily for mutants with restricted mycelial growth that is self-limiting on agar medium. The radial growth of each colony does usually not exceed a few millimeters. Colonial mutants vary widely in texture, density, conidiation, pigment, and fertility. Mutants of the series $col\text{-}5$ to $col\text{-}l7$, described in ref. (719), were in some instances assigned new locus names without having been tested for allelism with already-named morphological mutants having similar map locations. Some were subsequently shown to be

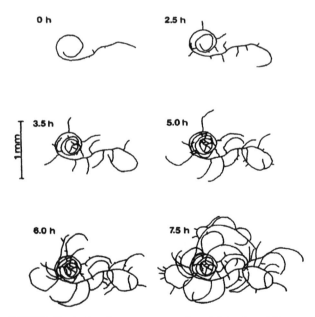

FIGURE 19 A time-lapse sequence of *coil-1* hyphae following ascospore germination, on agar medium overlaid with cellophane to prevent hyphae from penetrating the medium. From ref. (141).

recurrences (e.g., *col-3* of *bn*, *col-7* of *rg-1*, *col-14* of *sc*); others may be recurrences but have never been tested (possible examples are *col-5* and *col-8* with *col-1* or *cot-1*; *col-13* and *col-15* with *vel*). Two colonial mutants with nongerminating ascospores are symbolized *le-1*, *le-2*. For reviews covering morphological mutants and morphogenesis, see refs. (235), (1281), (1353), (1848), (1851), and (2117). See ref. (384) for growth rates and hyphal diameters of numerous colonial mutants.

col-1: colonial-1

IVR. Linked to *pan-1* (0/47 asci), *cot-1* (3%) (102, 708).

No macroconidiation. Growth is cyclic at moderate temperatures and steady at 38°C. The double mutant *col-1; cot-1* grows better than *col-1* at 24°C and better than *cot-1* at 38°C (708). *pe; col-1*, which forms microconidia (102), was used in early mutation studies, but *col-4; pe fl* was found to be better (727). The *pe; col-1* strain forms microconidia at 25°C, but macroconidia at 35°C. Cell-wall analysis; photograph (507). Not tested for allelism with *col-5*, *col-8*.

col-2: colonial-2

VII. Linked to *met-7*, *met-9* (1%), probably to the left (1585, 1592). Right of *T(T54M50)* (1578).

Colonial morphology (102). Photographs (235, 1851, 1853). Altered NADP-specific glucose-6-phosphate dehydrogenase (EC 1.1.1.49) (242, 1854). (*bal* and *fr* are also deficient in G6PD.) Reduced NADPH level (233). Reduced linolenic level (241). Accumulates much neutral lipid (1484). Pyridine nucleotide levels (234). Suppressor *su(col-2)* increases the linear growth rate and influences the electrofocusing pattern of *col-2* G6PD (1853). Homozygous *col-2 × col-2* crosses fail to mature; the immature asci frequently are nonlinear and occasionally show dichotomization (1965). Hyphae swell with age to 20-μm diameter (384).

col-3: colonial-3

Allelic with *bn*.

col-4: colonial-4

IVR. Between *met-1* (4%), *mus-26* (3%) and *arg-2* (1% or 2%) (1369, 1715, 1928).

Spreading colonial morphology, forming dense balls of conidia high in a slant (103). Scorable microscopically after ascospore germination (Fig. 20). Cell-wall-autolyzing enzyme (1247). Reduced amount of cell-wall peptides (2249). The mutant allele is dominant in heterozygous duplications from *T(S1229)* (110). Used in combination with *pe fl* to produce microconidiating

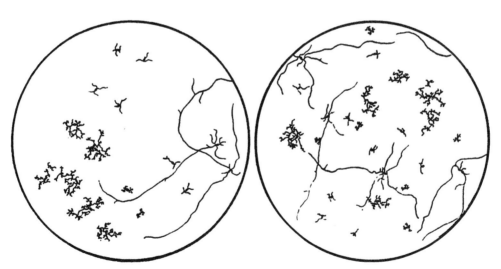

FIGURE 20 A camera lucida drawing of germinating ascospores, showing gene segregation in a cross of *col-4 × pyr-2*. Ascospores were placed on minimal agar medium in petri dishes, heat-activated at 60°C for 30 min, and incubated at 25°C for 18 hr. The *pyr-2* mutant grows sufficiently on minimal medium to permit the identification of *col-4 pyr-2* double mutants. The plate on the left shows segregations in two asci with ascospore pairs isolated in linear order. In the first ascus, genotypes were *col⁺ pyr*, *col⁺ pyr*, *col pyr⁺*, and *col pyr⁺* (from top to bottom). Those in the second ascus were *col⁺ pyr⁺*, *col⁺ pyr*, *col pyr*, *col pyr⁺*. The other plate shows the result of plating ascospores from the same cross at random. Genetic analyses can be made directly using either procedure or by transferring germinated ascospores to appropriately supplemented culture tubes for further testing. The random method allows much larger numbers to be observed conveniently. Drawing by M. B. Mitchell. From ref. (2177), with permission.

colonial growth suitable for reversion experiments (727). Called *spco-1* (719), *c* (727).

col-5: colonial-5

IVR. Linked to *cot-1* (1%), probably to the right (1604).

Dense, nonconidiating, poorly pigmenting, slow-spreading, colonial morphology (719, 1604). Cell-wall analysis and photograph (507). Reverts readily (719). Not tested for allelism with *col-l*, which it resembles and which maps in the same region. Called *col(B28)* (1604).

col-6: colonial-6

IV. Linked to *Cen-IV* (0/28 asci), *pan-1* (21%) (719). Colonial morphology. Slow ascospore germination (719).

col-7: colonial-7

Allelic with *rg-1*.

col-8: colonial-8

IVR. Linked to *pan-1* (4%; 13%) (719).

A fluff of hyphae may form at the top of a slant. Not tested for allelism with *col-l* or *col-5* (719). Peptides in cell wall are reduced in amount (2249).

col-9: colonial-9

VR. Between *inl* (16%) and *asn* (5%) (1383).

Forms small, slow-growing colonies. Reverts readily (719).

col-10: colonial-l0

IIL. Left of *cys-3* (14%), at or near *pi*. A single wild-type progeny (1/81) from *col-10* × *pi* may have been a revertant (719, 1592).

Slow-growing, dense, compact morphology, with no conidia (719). Altered 6-phosphogluconate dehydrogenase (1852, 1854). (*col-3* also affects 6PGD.) Crosses of *col-l0* (R2438) × *pi* (B101) resemble R2438 × R2438 and B101 × B101 in producing abnormal asci with flaccid walls and unordered ascospores, suggesting allelism (1965). However, R2438 and B101 are distinctly different in morphology. B101 has not been tested for 6PGD. As a marker, *pi* (B101) is preferable to R2438 because of growth rate, stability, and ease of handling; *ro-7* in the same region is preferable to both (1582).

col-12: colonial-12

I. Linked to *mat* (17%; 22%) (719).

Growth from ascospores is slow (719).

col-13: colonial-13

Probably allelic with *vel*.

col-14: colonial-14

Allelic with *sc* (1546).

col-15: colonial-15

Probably allelic with *vel*.

col-16: colonial-16

IIIR. Between *acr-2* (6%), *pro-1* (10%), *ad-4* and *leu-1* (1%) (1546).

Balls of powdery conidia are formed at top of slant on glycerol complete medium. Female fertile. A good marker, preferable to *com*, which grows more slowly and does not conidiate (1582). Complements *mo-4* and *spco-15* (719).

col-17: colonial-17

VII. Linked to *nt* (14%), *spco-5* (6%) (719).

Very slow-growing (719).

col-18: colonial-18

VIR. Between *un-23* (3/28) and *T(OY320)* (1582, 1589).

The original mutant allele was obtained in progeny of *Dp(VIR → IIIR)OY329* × *Normal*, presumably as a result of RIP (1589).

col-le: colonial lethal

Changed to *le-1*.

com: compact

IIIR. Between *ace-2* (5%) and *ad-4* (<1%; 5%) (1122, 1587).

Forms small, slow-growing colonies (507, 1548). Grows better on complete medium than on minimal medium (1582). Cell-wall analysis. Photograph (507).

Complex I

The NADH:ubiquinone respiratory complex of the inner mitochondrial membrane, with subunits specified both by nuclear genes and by mitochondrial genes. See *nuo*, *ndh*. See Appendix 4.

con: conidiation

The symbol *con* has been used to designate developmentally controlled genes that were identified because their mRNA levels are high during the development of macroconidia (158, 1773, 1956, 1958). *con* mRNA levels can also increase during microconidia and ascospore formation (1959). The *con* genes that have been analyzed appear to be subject to complex regulation by environmental and developmental cues; regulatory genes including *rco-1*, *rco-3*, *wc-1*, and *wc-2* have roles in their expression. None of the *con* genes for which null mutants have been obtained appears to be essential for conidiation.

The symbol *con* was also previously proposed for hypothetical "control of recombination" sites postulated to confer specificity of response to a *rec* gene product on genes that are subject to *rec*-mediated recombination control (335). However, no genetic variants are known in the postulated sequences, and there has been no direct demonstration that the sequences conferring specificity of response are in separate *cis*-acting control genes as distinct from recombinators such as *cog*. Specificity may be encoded in the recombinator itself. In contrast, *con* has been widely used as the symbol for genes that are transcribed preferentially during conidiation, and that usage is adopted here.

con-1: conidiation-1

IIIL. Left of *Cen-III*, *crp-2* (6T/18 asci) (1447).
Encodes conidiation-specific protein 1 (158).

con-2: conidiation-2

V. Between *rDNA* (2/18; 4T/18 asci) and *con-4a* (2/18), *lys-1* (8T/18 asci) (1447).
Cloned (158).
Encodes conidiation-specific protein 2 (158).

con-3: conidiation-3

VI. Linked to *Bml* (0/18) (1447).
Cloned (158).
Encodes conidiation-specific protein 3 (158).

con-4: conidiation-4

V. Between *rDNA* (4/18), *con-2* (2/18) and *chs-1* (1 or 2/18), *am* (4/18) (1447).
Cloned (158).
Encodes conidiation-specific protein 4 (158).

con-5: conidiation-5

IVR. Linked to *cot-1* (0/18) (158, 1447).
Cloned (158)
Encodes conidiation-specific protein 5 (158). Induced by blue light (1145).

con-6: conidiation-6

II. Left of *arg-12* (2 or 3/18) (1447).
Cloned and sequenced: Swissprot CON6_NEUCR, EMBL/Genbank L26036, EMBL NCCON6X, GenBank NEUCON6X.
Encodes conidiation-specific protein 6 (158, 2209). The CON-6 protein is present at high levels in free conidia (2209). Putative disruption by RIP caused no obvious conidiation or germination defect (2209). Regulatory studies (84, 1144, 2256).

con-7: conidiation-7

IIIR. Between *ace-2* (4T/18 asci), *cyt-8* (3T/18 asci) and *ad-2*, *trp-1* (0/17; 1T/18 asci) (1447).
Cloned (158).
Encodes conidiation-specific protein 7 (158).

con-8: conidiation-8

IR. Linked to *phr* (0/18), which maps right of *os-1* (15%) in conventional crosses (1908). Linked to *nic-1*, *Fsr-15* (4/18) (1447). The IL location between *nit-2* (2/17) and *mat* (2/17), shown in ref. (1447), appears to be incorrect (1908).
Cloned and sequenced: Swissprot CON8_NEUCR, EMBL/GenBank X07040, GenBank NCCON8, PIR S02210.
Encodes conidiation-specific protein 8 (158, 1731). Regulatory studies (1731).

con-9: conidiation-9

IVR. Linked to *cot-1* (0/18) (158, 1447).
Cloned (158).
Encodes conidiation-specific protein 9 (158), which is expressed at high levels in late-phase vegetative growth and newly formed conidia (1773).

con-10: conidiation-10

IVR. Linked to *cot-1* (0/18) (158, 1447). Adjacent to *con-13* (1730).
Cloned and sequenced: Swissprot CONX_NEUCR, EMBL/GenBank M20005, PIR A31849, EMBL NCCON10A, GenBank NEUCON10A. In the same cDNA clone (pCon-10a) with *con-13*.
Encodes conidiation-specific protein 10 (158, 1730), which has sequence similarities to stress-induced proteins in other organisms (1154). CON-10 polypeptide is localized to the conidiophores where it is seen distributed in the cytoplasm (1957). A RIP mutant in which *con-10* and *con-13* were disrupted showed no obvious conidiation or germination phenotype (except for not producing these polypeptides) (1957). Fusions of the 5'-region to *hph* enabled selection for mutants affecting transcript

levels of *con-10* and other conidiation genes during vegetative growth; mutants isolated in this way include *rco-1* and *rco-3* (1241, 1242, 2256). Regulatory studies (84, 1144, 1242, 2256). Promoter elements have been defined that are important for regulated expression of *con-10* (409, 1153, 1154).

con-11: conidiation-11

VIR. Right of *Cen-VI, Fsr-50* (6T/17 asci). Linked to *trp-2* (0/18) (1447).

Cloned (158) and sequenced (184).

Encodes conidiation-specific protein 11 (158), which is induced late in conidiation close to the time of appearance of CON-10 and free conidia. The protein has a predicted size of 51 kDa and an apparent size of approximately 100 kDa. The protein is N-glycosylated and may have other posttranslational modifications. CON-11 can be localized to the periplasmic space by cell fractionation. Normal vegetative transcription is very low, but is slightly derepressed in double-mutant strains with a RIP-inactivated *rco-1*. Although *con-11* is not induced by stress during vegetative growth, the predicted protein sequence is very similar to that of *rds1* of *S. pombe*, a stress-response gene of unknown function (184).

con-13: conidiation-13

IVR. Adjacent to *con-10* (1730), which is linked to *cot-1* (158, 1447).

Cloned and sequenced: Swissprot COND_NEUCR, EMBL/ GenBank M35120, PIR JH0363, EMBL NCCON13, GenBank NEUCCON13. In the same cDNA clone (pCon-10a) with *con-10*.

Encodes conidiation-specific protein 13 (814, 1730). CON-13 polypeptide is localized to the conidiophore (1957). A RIP mutant in which *con-13* and *con-10* were disrupted showed no obvious defects in conidiation or germination phenotype (except for not producing these polypeptides) (1957).

cor-1: cobalt resistant-1

IIIR. Linked to *dow* (1%, probably to the right) (2221).

Resistant to inhibition by cobalt (20-fold more resistant than control) and nickel (4-fold more resistant than control), but not by zinc or copper. Cobalt uptake is partially blocked (1374). A cobalt-binding glycoprotein is overproduced (1780).

cot-1: colonial temperature sensitive-1

IVR. Between *pan-1* (2%) and *his-4* (1%; 6%) (1369, 1585, 1592).

Cloned and sequenced: Swissprot COT1_NEUCR, EMBL/ GenBank X97657, PIR S22711, GenBank NCCOT1.

Encodes a serine/threonine protein kinase (EC 2.7.1.-). The *cot-1* gene is homologous to protein kinase genes *udk1* of *Ustilago*, TB3 of *Colletotrichum*, *orb6* of *Schizosaccharomyces*, the warts tumor suppressor gene of *Drosophila*, and the myotonic dystrophy kinase gene of humans (570). Two sense transcripts are encoded that differ in length. The deduced transcripts differ in their regulatory domains. The ratio of the transcripts is regulated by blue light and depends on functional products of *wc-1* and *wc-2* (1142). Null mutations obtained following disruption of the locus by RIP are not remediable, retaining a tight colonial phenotype at all temperatures (2268). The temperature sensitivity reflected in the locus name is due to a posttranslational defect of the original allele, C102t (2268). The following description applies to C102t unless otherwise indicated. Growth is colonial and extremely restricted at 34°C, with increased hyphal branching, but growth, morphology, and fertility are completely normal at 25°C and below (Fig. 21). Linear growth is maximum at 24°C (708), becoming colonial at 32°C. The mutation is a single-base-pair change in the coding domain. Only one of the two COT-1 isoforms is detectable at restrictive temperatures, and the activity of Ser/Thr kinase is reduced whereas that of types 1 and 2B phosphatase (calcineurin) is increased (760). Colonies from ascospores or conidia are viable at restrictive temperatures and continue to grow slowly with dense branching, but they do not conidiate. They quickly resume normal growth when shifted to a permissive temperature (1369, 2079). Recessive in duplications (1580); apparent dominance in heterokaryons (708) may have resulted from a shift in nuclear ratios. Used in studies of septation and branching (1213), growth-inhibiting mucopolysaccharide (1718, 1719), sulfate transport (1280). Cell-wall analysis (708). Growth is stimulated by lysine or arginine (0.1 mM) on glucose medium at high temperatures (1210). Because of its high viability and tightly restricted growth at restrictive temperatures and its normality at 25°C, *cot-1* C102t has had valuable technical applications. For example, crosses homozygous for C102t have been used in combination with sorbose for experiments with *rec* genes, where high-density ascospore platings are required for precise quantitative analysis of intralocus recombination (332, 1940, 2083). In another application, when C102t is shifted up after initial growth at the permissive low temperature, linear growth is arrested and hyphae assume a "bottle brush" appearance with small side branches (398, 1368) (Fig. 21). This has been used to select ultraviolet-sensitive (UV-sensitive) mutants by subsurface survival on UV-irradiated plates containing *p*-aminobenzoic acid, which absorbs lethal wavelengths (317, 1839). C102t conidia or ascospores from homo-

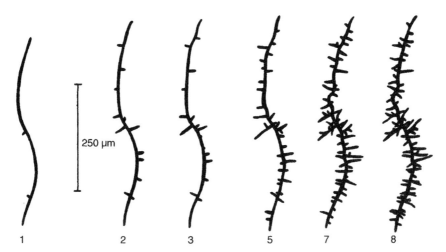

FIGURE 21   Branching and arrest of linear growth of *cot-1* allele C102 following the transfer of a mycelium from 25 to 37°C. Hours following the transfer are given below each drawing. From ref. (398), with permission from the British Mycological Society.

zygous C102t crosses have been used for replication in a protocol involving transfer by filter paper (1210). For suppressors of *cot-1*, see *gul, ro*. Most suppressors of *cot-1* are mutations at various *ropy* loci, which inactivate cytoplasmic dynein or dynactin (248, 1634).

cot-2: colonial temperature sensitive-2

VR. Between *pk* (8%), *ser-2* (5%), *T(EB4)^L* and *ad-7* (4%) (321, 1546, 1578, 1603). Recombines with *inv* (5%) (598).

Colonies are small but fully viable at 34°C. Growth and morphology are nearly, but not completely, normal at 25°C (719). Invertase is altered; it is not clear whether *cot-2* is the structural gene for a second subunit or whether *cot-2* affects invertase structure indirectly, perhaps by altering the carbohydrate moiety (598). See *inv*. Cell-wall peptides are reduced in amount (2249). Ascospores are round in homozygous *cot-2* × *cot-2* crosses, but are shaped normally in heterozygous crosses (117). Some *cot-2* strains carry *mei-3*, which was discovered in the original *cot-2* strain (1469); most strains used by Eggerding *et al.* (598) are free of *mei-3*, however, and *mei-3* cannot be responsible for the effects on invertase (1476).

cot-3: colonial temperature sensitive-3

IV. Left of *pan-1* (16%; 25%). Probably right of *arg-2* (719, 1582).

Forms small, fully viable colonies at 34°C. Morphology is normal at 25°C, but the growth rate is reduced (719, 2107, 2109). Septae behind hyphal tips are not plugged (2109).

cot-4: colonial temperature sensitive-4

VR. Between *rol-3* (5%) and *inl* (10%) (1383).

Forms small colonies at 34°C. At 25°C, growth is spreading (719) and the morphology resembles that of *cum, sn*, and perhaps *sp*, with late-forming blooms of conidia on aerial hyphae. Female fertility is good in heterozygous crosses, but no perithecia are formed in crosses homozygous for *cot-4* (1582).

cot-5: colonial temperature sensitive-5

III. Between *T(P2869)^R*, *T(B18)^R* (1578) and *het-c* (3%), based on recombination in crosses *cot-5 het-C pyr-4 thr-2* × *het-c* and *pi het-C pyr-4 thr-2* × *T(NM149) het-c* (924) [these genotypes were given incorrectly or incompletely in ref. (924)].

Little or no growth at 34°C; colonial at 30°C. Colonies form short, aerial hyphae at 25°C (719). Female-sterile. Morphologically distinct from *fs-1* (1582).

cox: cytochrome oxidase

Cytochrome *c* oxidase (EC 1.9.3.1) (shortened to *cytochrome oxidase* in gene names) is a hetero-oligomeric enzyme located at the inner mitochondrial membrane. Three of its subunits (subunits I, II, and III) are encoded mitochondrially. The remaining subunits are nuclear-encoded.

cox-4: cytochrome oxidase-4

III. Linked to *acr-2* (1 or 2/15) (445), *cyt-8* (1766) by RFLP.

Cloned and partially sequenced: Swissprot COX4_NEUCR, EMBL/GenBank M12116, GenBank NEUCOX4.

Encodes cytochrome *c* oxidase subunit IV (1770). A synthetic peptide corresponding to part of the mitochondrial import-sequence causes increased permeability to small solutes when added to rat liver mitochondria (1942). RFLP data indicate that *cox-4* is close to *cyt-8*; whether they are allelic has not been established.

cox-5: cytochrome oxidase-5

Allelic with *cya-4*.

cox-6: cytochrome oxidase-6

V. Near *inl* (445).

Cloned and sequenced: Swissprot COX6_NEUCR, EMBL/GenBank M12118, GenBank NEUCOX6.

Encodes cytochrome oxidase subunit VI. Possible allelism with *cya-2* not tested.

cox-8: cytochrome oxidase-8

VIIR. Linked to *cit* (2T + 1NPD/15 asci), *un-2* (2/17), *lacc* (3/17) (1447).

Cloned (1770).

Cytochrome *c* oxidase subunit VIII; similar to *Saccharomyces cerevisiae* cytochrome oxidase subunit VIIa (1767).

cpc-1: cross pathway control-1

VIL Between *ylo-1* (1%; 3%; 143 kb), *un-13* (105 kb) and *Cen-VI*. Inseparable from $T(IBj5)^R$ (457, 1578).

Cloned and sequenced: Swissprot CPC1_NEUCR, EMBL/GenBank J03262, PIR A30208, EMBL NCCPC, GenBank NEUCPC; pSV50 clone 22:11A; Orbach-Sachs clones G2C08, G8C01, G8C03, G15E08, G16H02, G18D02.

Encodes a bZIP transcriptional activator similar to *Saccharomyces cerevisiae* GCN4 (1525). Simultaneously affects both basal levels and the ability to derepress enzymes in arginine and other amino acid biosynthetic pathways (1768). The CPC-1 polypeptide binds to sequences related to 5'-ATGACTCAT-3' (577); these sequences are present in the 5'-regions of genes controlled by CPC-1. *cpc-1* mutations interfere with the cross-pathway activation of amino acid biosynthetic enzymes in response to amino acid starvation or imbalance (1768). *cpc-1* mutants are sensitive to 5-MT, 3-AT, FPA, and some nucleoside analogues (326, 1080); also sensitive to high concentrations of some amino acids (121). *cpc-1* mutants were originally isolated as arginine auxotrophs by selection in the mutant *arg-12ˢ*. The *cpc-1; arg-12⁺* single mutant is prototrophic (120, 457). Growth is delayed after ascospore germination; the delay is not alleviated by arginine or precursors. Scorable by delayed growth (120, 457). Sequence changes determined for inactivating

mutations CD-15 and IBj-5 (1526). Independently isolated as *mts-1* (327, 1080). Independently isolated as *lni-1*, a regulator of *lacc* (2043, 2296). *cpc-1* mutations are epistatic to the *lah-1* mutation, which affects *lacc* expression (833, 2041), and the *lah-1* mutation appears to affect *cpc-1* expression (833). *T(IBj5)* and *T(MN9)* are inseparable from *cpc-1* (1578).

cpc-2: cross pathway control-2

VIIR. Linked to *arg-10* (18%). Right of *T(5936)* (1099).

Cloned and sequenced: EMBL/GenBank X81875, GenBank NCCPC2, PIR S57839; EST NC4D7.

Encodes a protein entirely composed of WD repeat segments (1386). Simultaneously affects both basal levels and the ability to derepress enzymes in arginine and other amino acid biosynthetic pathways (1386). Sensitive to *p*-fluorophenylalanine and 3-AT (1386). Female-sterile (1099).

cpc-3: cross pathway control-3

VR. Linked to *cyh-2* (0 or 1/16) (1452, 1795).

Cloned and sequenced: EMBL/Genbank X91867, GenBank NCCPC3.

Encodes the homolog of *Saccharomyes cerevisiae* Gcn2p. The predicted CPC-3 polypeptide contains an eIF2 kinase domain and a histidyl-tRNA synthetase-like domain. *Neurospora* strains in which the *cpc-3* gene function was disrupted by gene replacement did not activate the transcription of amino acid biosynthetic genes in response to amino acid starvation. Whereas the level of *cpc-1* transcript does not appear to be affected in the *cpc-3* mutant, the level of CPC-1 polypeptide does not show a large increase in response to amino acid starvation (1795). Thus, *cpc-3* has a posttranscriptional effect on *cpc-1* expression, presumably at the level of translation, in analogy to the effect of functional GCN2 product on GCN4 translation in *Saccharomyces* (876, 1768).

cpd-1: cyclic phosphodiesterase-1

IVR. Right of *pyr-2* (19%) (851).

Encodes orthophosphate-regulated cyclic phosphodiesterase (EC 3.4.1.-). Isolated by filtration enrichment from a *pho-2* progenitor as cAMP could not be used as the sole phosphorus source (849, 851). Short aerial hyphae on solid medium. Rhythmic conidiation on solid medium resembles that of *bd*.

cpd-2: cyclic phosphodiesterase-2

IL. Right of *arg-1* (6%) (851).

Impaired activity of orthophosphate-regulated cyclic phosphodiesterase (EC 3.4.1.-). Reduced adenylate cyclase activity. Isolated by filtration enrichment from a *pho-2* progenitor as cAMP could not be used as the sole phosphorus source (849, 851). Aerial hyphae are short on solid medium; rhythmic conidiation on solid medium resembles that of *bd*.

cpd-3: cyclic phosphodiesterase-3

IL. Right of *arg-1* (11%) (852).

Impaired activity of orthophosphate-regulated cyclic phosphodiesterase (EC 3.4.1.-). Reduced adenylate cyclase activity. Isolated by filtration enrichment from a *pho-2* progenitor as cAMP could not be used as the sole phosphorus source (852). Hyphal morphology resembles that of *cr-1*.

cpd-4: cyclic phosphodiesterase-4

IIR. Right of *arg-12* (7%) (852).

Altered activity of orthophosphate-regulated cyclic phosphodiesterase. Reduced adenylate cyclase activity (852). Isolated by filtration enrichment from a *pho-2* progenitor as AMP could not be used as the sole phosphorus source. Does not form aerial hyphae or conidia in low-phosphate medium.

cpk: cAMP-dependent protein kinase

IIIR. Linked to *un-17* (5%) (1403).

Mutation results in a cAMP-dependent protein kinase activity that is independent of cAMP addition (1403). The function of CPK is unknown at present (1632).

cpl-1: chloramphenicol sensitive-1

VIL. Between *ad-8* (7%; 11%) and *aro-6* (5%), *lys-5* (6%), *un-25* (2%, 4%) (346, 1582, 1815).

The mutant is sensitive to chloramphenicol (<0.5 mg/ml added to autoclaved medium) and antimycin A (1 μg/ml); wild type is resistant to 4 mg/ml chloramphenicol. Protein synthesis is not grossly altered. Cyanide-insensitive and azide-insensitive respiratory systems are still present. The cytochrome spectrum is normal on minimal medium. Obtained by inositol death enrichment and replica plating, using a strain of genotype *inl; trp-3; sn cr-1* (346, 349). Scorable on 0.5 mg/ml chloramphenicol, autoclaved in the medium (1582).

cpt: carpet

IIR. Between *arg-5* (3%) and *T(NM177)^L*, *pe* (6%) (1580, 1585).

Flat, slow-growing mycelium. Aerial hyphae are short, and no conidia are formed unless the culture becomes desiccated (1958). Produces microconidia (793), but much less abundantly than the double mutants *fl; dn* and *pe fl* (1582). Homozygous fertile.

cpz-1: chlorpromazine sensitive-1

Unmapped.

Mycelial growth is hypersensitive to the calmodulin antagonists chlorpromazine, trifluoperazine, imipramine, and W5 (2031). The effects of chlorpromazine on circadian rhythm in wild type are abrogated in the mutant (2031).

cpz-2: chlorpromazine sensitive-2

Unmapped.

Mycelial growth is hypersensitive to the calmodulin antagonists chlorpromazine, trifluoperazine, imipramine, and W5 (2031). The effects of chlorpromazine on circadian rhythm in wild type are abrogated in the mutant (2031). Growth is temperature-sensitive (1035). Addition of spermidine to the medium or transformation with the *spe-3* gene encoding spermidine synthase eliminates the temperature-sensitive phenotype of *cpz-2* (1035) and causes circadian rhythm in the mutant to be chlorpromazine-sensitive; growth of mycelia in the mutant remains chlorpromazine-sensitive under these conditions (1035).

cr-1: crisp-1

IR. Between *nic-2* (4%; 7%), *ace-7* (1%; 3%) and *cys-9* (3%), *un-1* (5%) (1414, 1592). Included in duplications from *T(4540)*, which do not include *cr-2* or *cr-3* (1578). One of the first mutants to be mapped in *Neurospora* (1200).

Cloned and sequenced: Swissprot CYAA_NEUCR, EMBL/GenBank D00909, GenBank NEUNAC; EST NP2C2; pSV50 clones 1:10G, 4:7C.

Structural gene for adenylate cyclase (EC 4.6.1.1) (2077). The mutant has little or no endogenous adenosine 3',5'-phosphate (cyclic AMP; cAMP) (1524, 2076). Slow mycelial growth. Conidiates rapidly and profusely, close to the surface of agar. Produces very short conidiophores, bearing conidia in tight clusters (1200, 1201). Photographs (1005, 1256). Excellent as a marker. Scorability and viability are good. Morphology may vary on different media. Strains carrying various alleles differ in growth habit. (For example, B123 strains are flat and restricted whereas allele L strains are spreading.) Some *cr-1* alleles will conidiate in submerged shake culture (1146, 1633). Mutant B123 can be replicated using a needle replicator (1256). *cr-1* ascospores may require longer to mature than *cr^+*. Crosses homozygous for allele B123 exude intact linear asci (1256). Frequently occurring modifier mutations can alter morphology and the ability of *cr-1* to use glycerol (720, 1778). For a suppressor of *cr-1*, see *hah*.

Used to determine what functions are controlled by cAMP (1524). The abnormal morphology is partially corrected by exogenous cAMP (1742, 1743, 2076, 2077). Guanosine 3′,5′-phosphate also stimulates mycelial elongation (1743). Phosphodiesterase inhibitors do not counteract the morphological effect of *cr-1* (1743). Other anomalies are also normalized by exogenous cAMP: overproduction and excretion of NAD(P) glyco-hydrolase (1005); altered induction and localization of β-glucosidase (1779); inability to use glycerol and certain other carbon sources (1147, 2078); limited secretion of extracellular alkaline protease (1798); and ability to use alanine as a gluconeogenic substrate (1460). Cyclic nucleotide levels differ in mycelia and conidia (1742, 1743). The mutant is constitutively thermotolerant (434). Used to study cAMP-binding protein (2104), cell-wall polymer (1916). Called *nac* (1089). Allele CE4-11-67 called *con* (1408, 1409).

cr-1 suppresses the apolar growth of *mcb* at 37°C. *mcb* therefore can be used to select new *crisp-1* mutations (247). The double mutants *sn cr-1* and *cr-1 rg-1* form small conidiating colonies suitable for replica plating (349, 1256, 1553, 1830, 1979). These have been used to obtain UV-sensitive mutants (1256, 1830) and to measure recessive lethal mutation (Fig. 22) (1980). The triple mutant *sn cr-1; csp-2* can be overlayered without displacing conidia (1454, photograph, 1457).

cr-2: crisp-2

IR. Between *cr-3* (11%), *T(NM103)* and *al-2* (18%); hence, right of *thi-1* (18%) (720, 1582).

Conidiation is delayed. Fine, pale-pigmented conidia are then produced in clumps over the agar surface (720). Cell-wall peptides are reduced in amount (2249). Cyclic AMP levels are normal. Overproduces and excretes NAD(P) glycohydrolase; this is not cured by exogenous cAMP (1005).

cr-3: crisp-3

IR. Between *cr-1* (13%), *T(4540)^R* and *cr-2* (11%) (720, 1582).

Conidiation is delayed. Fine, pale conidia are then produced uniformly over the agar surface (720). The amount of cell wall peptides is reduced (2249). Cyclic AMP levels are normal. Overproduces and excretes NAD(P) glyco-hydrolase; this is not cured by exogenous cAMP (1005).

cr-4: crisp-4

IV. Linked to *pdx-1* (6%), *cot-1* (22%) (1593)
Resembles *cr-1* in appearance (1147).

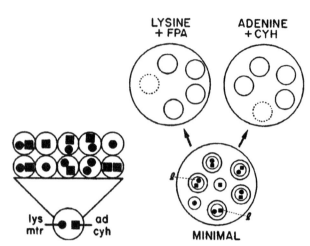

FIGURE 22 The method developed by D. R. Stadler for use of *sn; cr-1* strains in quantitative studies of recessive lethal mutation. Mutations are detected using a heterokaryon of constitution (*sn cr-1; mtr; lys-5 + sn cr-1 cyh-1; ad-2*). *ad-2* and *lys-5* are complementing recessive auxotrophic markers, and *mtr* and *cyh-1* are recessive mutations conferring resistance to fluorophenylalanine and cycloheximide. *sn cr-1* conidia grow into small conidiating colonies suitable for replica plating. (Left) A single conidium containing one nucleus of each genetic type. On minimal medium it grows into a colony that produces many conidia of each of three classes, heterokaryons and the two types of homokaryons. (Right) On minimal medium, the conidia (inner circles) that contain both nuclear types grow into colonies (outer circles) that produce conidia able to grow on replica plates of both test media, unless a recessive lethal mutation has arisen in one of the ancestral nuclei. Reprinted from ref. (1980) [Stadler, D., and H. Macleod (1984). A dose–rate effect in UV mutagenesis in *Neurospora. Mutat. Res.* **127**: 39–47.] Copyright 1984, with permission from Elsevier Science.

cr-5: crisp-5

III. Linked to *ad-4* (19%) (974, 975).
Resembles *cr-1* in appearance (974, 975).

cre-1: carbon catabolite regulation

IVR. Right of *trp-4* (5 or 6T, 1NPD/17 asci) (1447).
Cloned and sequenced: EMBL / Genbank AF055464.
Homologous to the *Aspergillus nidulans* carbon catabolite repressor *creA* and able to bind to CREA target sequences in *Aspergillus* when expressed heterologously. Used to examine whether the *Neurospora* gene acts similarly to *creA* by testing whether CRE-1 and binding sites for CRE-1 are important for transcriptional efficiency and glucose regulation of *crp-2* and *Fsr40* promoters. Disruption of CRE-1 binding had no effect on transcription rate during steady-state growth or carbon shifts, indicating that transcriptional control of ribosomal genes by carbon source is not mediated by *cre-1* (480).

crib-1: cold-sensitive ribosome biosynthesis-1

IV. Linked to *met-1* (6%) (1814).

Defective ribosome biosynthesis below 20°C; attributed to
a defect in ribosomal ribonucleic acid (rRNA) processing
(1756). Grows at 6% of the wild-type rate at 10°C and
79% at 25°C. 37S cytosolic ribosomal subunits are
underaccumulated, and relatively little stable 17S rRNA
is produced at low temperatures. Not a conditional
lethal mutation (1754, 1814). Conditionally defective in
the expression of *S*-adenosylmethionine synthetase
activity (1759).

crib(PJ31562): cold-sensitive ribosome biosynthesis (PJ31562)

IVR. Linked to *crib-1* (3%) (1753).

Defective in the biosynthesis of cytosolic ribosomes at
10°C, but normal at 25°C. Grows at 16% of the wild-
type rate at 10°C and 90% at 25°C. Underaccumulates
17S rRNA and, hence, 37S ribosomal subunits. Shows
partial complementation in forced heterokaryons with
crib-1 (1753).

crp-1: cytoplasmic ribosomal protein-1

Allelic with *cyh-2*.

crp-2: cytoplasmic ribosomal protein-2

III. Linked to *Cen-III* (0T/18 asci) (1447).

Cloned and sequenced: Swissprot RS14_NEUCR, EMBL/
GenBank X53734, GenBank NCCRP2, PIR S11667.

Encodes 40S ribosomal protein S14. Used in studies of
promoter elements (435).

crp-3: cytoplasmic ribosomal protein-3

IVR. Right of *trp-4* (3T, 1NPD/18 asci) (1447).

Cloned and sequenced: Swissprot RS17_NEUCR, EMBL/
GenBank M63879, PIR S34441, EMBL NCCRP3,
GenBank NEUCRP3.

Encodes 40S ribosomal protein S17 (1905).

crp-4: cytoplasmic ribosomal protein-4

VR. Right of *inl* (9T/18 asci), *tom70* (4T/18 asci), *Fsr-20*
(1T/18 asci). Linked to *sdv-6* (0T/18 asci) (1447).

Cloned and sequenced: EMBL/GenBank AF054907.

Encodes the large subunit of a ribosomal protein homologous to yeast L1 and rat L5, which bind to 5S rRNA
(481).

crp-5: cytoplasmic ribosomal protein-5

Unmapped.

Cloned and sequenced: Swissprot RS26_NEUCR, EMBL/
GenBank X55637, PIR S13082, R4NC26, GenBank
NEUCRP5A, NCRPSSU, M95597.

Encodes 40S ribosomal protein S26E (2188).

crp-6: cytoplasmic ribosomal protein-6

Known only as part of the *ubi::crp-6* fusion gene, which
encodes ribosomal protein S27a.

crp-7: cytoplasmic ribosomal protein-7

IR. Linked to *con-8* (0/18), *Tel-IR* (1T/18 asci). Adjoins *phr*
(1115a).

Cloned and sequenced: EMBL/GenBank AB015207.

Encodes a ribosomal protein homologous to *Saccharomyes
cerevisiae* YS25 (1115a).

crp-12: cytoplasmic ribosomal protein-12

Called *rps*.

crp(S7): cytoplasmic ribosomal proteinS7

IIIR. Adjoins *ro-2* (2154).

Cloned and sequenced: EMBL/GenBank U73847.

Encodes a small-subunit ribosomal protein homologous to
the rat ribosomal protein S7 (2154).

csh: cushion

IR. Between *thi-1* (12%; 20%) and *ad-9* (5%) (1592).

Restricted colonial growth (1585).

csp: conidial separation

In addition to the two genes so named, other mutations are
known to result in defective separation of macroconidia.
Insertional translocation *T(MD2) csp* and the duplication progeny it generates are phenotypically similar to
csp-1 and *csp-2*, suggesting that there may be yet another
csp locus between *arg-5* and *pe* in IIR (1578). The *os-8*
mutant and the unmapped histidine kinase *nrc* genes
also are defective in conidial separation.

csp-1: conidial separation-1

IL. Between *arg-3* (1%) and *T(39311)*[R] (1582, 1882).

Conidia fail to separate and become airborne (1882) (Fig.
23). Recessive. Cultures in agar slants are readily scored
by a "tap test." In water, conidia are freed at one-tenth
the wild-type concentration (1882). Carotenoids tend to
be yellowish in young cultures (1582). Used in connection with *bd* for the study of circadian rhythms (239).
Useful in student laboratories to avoid contamination
(1876). Used in combination with *csp-2* in a teaching
exercise to demonstrate complementation (1880).

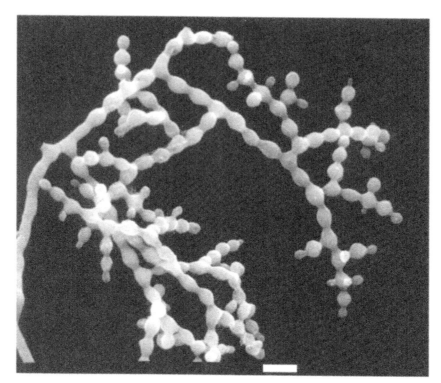

FIGURE 23   The mutant *csp-1*: *conidial separation-1*. Bar length-10 μm. Scanning EM photograph from M. L Springer.

csp-2: conidial separation-2

VIIL. Linked to *thi-3* (<1%), probably to the right. Left of *T(T54M50)*, *ace-8* (1582, 1882).

Conidia fail to separate and become airborne. Cultures on agar are readily scored by a tap test. Resembles *csp-1*. Conidia are freed in water suspension, but only long after induction of aerial growth and at 1/100 the concentration obtained using wild type. A *csp-1; csp-2* double mutant releases no detectable free conidia under the same conditions (1882). Most *csp-2* alleles complement *csp-1* in forced heterokaryons to form the wild-type number of free conidia (1882), but *csp-2* allele UCLA102 does not (1878). Conidiating colonies of the *csp-2; sn cr-1* strain on replica plates can be overlayered without dislodging conidia (1454); photograph (1457).

csr-1: cyclosporin resistant-1

IR. Right of *his-2* (20%) and *cyh-1* (1%; 3%) (358, 2111).

Cloned and sequenced: Swissprot CYPH_NEUCR, EMBL/GenBank J03963, X17692, PIR A30809, EMBL NCCYCLO, GenBank NEUCYCLO, NCCYCGEN; EST WO1C2.

Encodes peptidyl-prolyl *cis–trans* isomerase precursor (EC 5.2.1.8) (PPIase) (cyclophilin) (rotamase) (cyclosporin-A-binding protein) (CPH). Encodes both cytoplasmic and mitochondrial cyclophilins, which differ at their N-termini and which appear to be translated from alternative transcripts differing in their 5′-leader regions (2112). The absence of cyclophilin, or failure to bind it, confers resistance to cyclosporin. Resistance is recessive. Catalyzes protein folding (1826). The mitochondrial enzyme is part of the mitochondrial protein import pathway and interacts with chaperones HSP70 and HSP60 (1702); its function can be discriminated from that of chaperonin (2170). Called *cyp*.

cum: cumulus

IIIL. Between *Tip-IIIL* and *Sk-2^K* (2%), *Sk-3^K* (2%), *r(Sk-2)-1* (4%), *acr-7* (5%; 18%), *acr-2* (18%) (280, 1582).

Initially grows as a colony, then spreads and sends up blooms of aerial hyphae that conidiate profusely at the shallow end on agar slants. Good female fertility. Similar in morphology to *sn*, *sp*, and *cot-4* at 25°C (1582). Colonies are tighter at higher temperature (1322). Very small inocula (floated conidia) should be used when scoring *acr-7* in *cum* isolates (2121).

cut: cut

IVL. Between *cys-10* (28%; 37%) and *fi* (4%; 10%); therefore, linked to *ace-4* (1559, 1582).

The osmotic-sensitive phenotype and the morphology resemble those of mutants named *osmotic*. Scorable either by failure to grow on medium containing 4% NaCl or by morphology. The morphology approaches normal at high humidity (1116). Conidia do not shake loose when a slant is tapped. The IR linkage shown by the original allele, HK53, was due to an unrecognized I;IV translocation *T(HK53)*, which is inseparable from *cut* (1580, 1592). An allelic point mutation, LLMI, shows linkage in IVL and segregates independently of I markers (1559). *ovc* is probably an allele (94).

cwl: cross wall

Defective in septum formation in mycelia and ascogenous cells. Vegetative hyphae tend to bleed, forming an exudate on the agar surface and in lens-shaped pockets beneath. Slow-growing, aconidial. Stocks are conveniently kept as heterokaryons with a helper strain such as *a^m1* (1564). The phenotypically normal heterokaryon forms perithecia when it is used as maternal parent. In crosses homozygous for *cwl*, paraphasoid cells are septate, whereas ascogenous cells contain no cross walls (1680). Helper nuclei from the maternal parent thus contribute to paraphasoidal cells, but not to ascogenous hyphae.

cwl-1: cross wall-1

IIR. Between *arg-5* (3%; 6%) and *arg-12* (12%) (1582).

Hyphal septa are largely absent. Subject to alteration by modifiers that restore septa and increase the growth rate, but the original mutant gene can be extracted by crossing. Recessive (719, 825, 1582). Called *mo(R2441)*.

cwl-2: cross wall-2

IIR. Between *arg-12* (10%), *un-20* (4%) and *fl* (9%) (1546).

Hyphal septa are largely absent. Recessive in heterokaryons, but the cross-wall phenotype appears erratically. Mutant allele 1098 originated in *flm-1* (1546).

cy: curly

IL. Linked to *arg-1* (0/34), *ad-5* (1/54), probably to the left (1582).

Young hyphae are seen to be curled when examined under low magnification on agar or on the walls of culture tubes. Growth rate and gross morphology are normal (1366). Conidia fail to shake loose in a tap test, as with *csp* mutants (1955). Called *c*.

cya-1: cytochrome a-1

IL. Linked to *mat* (6%, shown to the left) (170).

Deficient in cytochrome *aa₃*. Cannot reduce tetrazolium. Very slow growth. Female sterile (170). Possibly not really a *cya* mutant (162).

cya-2: cytochrome a-2

VR. Linked to *al-3* (3%, indicated to the right) (170). Linked to *inl* (one apparent NPD, 0T/18 asci), *cmd*, *hsp80*, *hsp83* (2T/18 asci) (1447).

Deficient in cytochrome *aa₃*. Cannot reduce tetrazolium. Slow growth. Female-sterile (170). Possible allelism with *cox-6* not tested.

cya-3: cytochrome a-3

VIL. Between *chol-2* (10%) and *cyt-2* (10%) (170).

Deficient in cytochrome *aa₃*. Cannot reduce tetrazolium. Slow growth (170). The *aa₃* deficiency is suppressed by antimycin A (167). Spectrum (167).

cya-4: cytochrome a-4

IIR. Between *arg-5* (2/39; 1T/18 asci) and *preg* (1T+one double exchange/18 asci). Linked to *Fsr-55* (0T/18 asci) (1339, 1447, 1769).

Cloned and sequenced: Swissprot COX5_NEUCR, EMBL/GenBank E91991, M12117, M24389, PIR A30134, B25629, OTNCV, GenBank NEUCOX5, NEUCOXV; Orbach-Sachs clones X13G05, G22A10.

Structural gene for cytochrome *c* oxidase subunit V (1769). Deficient in cytochrome *aa₃*. Cannot reduce tetrazolium. Growth is slow. Spectrum (167, 170). Called *cox-5*.

cya-5: cytochrome a-5

IVR. Right of *pan-1* (2%) (1437).

Cloned and sequenced: EMBL/GenBank AF002169, EMBL NCAF2169.

Homologous to yeast PET309, which regulates the expression of mitochondrially encoded COXI (388). Deficient in cytochrome *aa₃*. Subunit 1 polypeptide of cytochrome *c* oxidase is absent by immunological criteria and by analysis of mitochondrial translation products. Growth is slow. Recovery from crosses is poor (10%). Selected as a tetrazolium nonreducer (173, 1437, 1438). The circadian period is shortened (236, 560). Called *cya-U-34*.

cya-6: cytochrome a-6

IVR. Right of *pan-1* (2%) (1437).

Deficient in cytochrome *aa₃*. Alleles called *cya-6-2* and *cya-6-35* are heat-sensitive. Selected at 41°C by slow growth on salicylhydroxamic acid and resistance to tetrazolium. At least five subunits of cytochrome *c* oxidase are present at 41°C by immunological criteria, but they are not associated. Complements *cya-5* (1437).

cya-7: cytochrome a-7

VR. Right of *Cen-V*. An earlier report of linkage in III was incorrect (162).

Deficient in cytochrome aa_3. Allele cya-7^{13} is heat-sensitive. Selected at 41°C by slow growth on salicylhydroxamic acid and resistance to tetrazolium. At least five subunits of cytochrome c oxidase are present at 41°C by immunological criteria, but they are not associated (1437). Called cya-7-13.

cya-8: cytochrome a-8

VIIL. Between *Tip-VIIL* and *T(ALS179)* (5%), *cyt-7* (7%), *adh* (19%), *nic-3* (39%) (1593)

Deficient in cytochrome aa_3; phenotypically similar to [*mi-3*] (162). Very slow, sparse, transparent growth with no or little conidiation. Severity of the defect is variable with mutants of independent origin. Many germinants from ascospores fail to survive, but they can be rescued readily in a heterokaryon with a^{ml} *ad-3B cyh-1*. New *cya-8* mutations are found in ~20% of the progeny of *eas(UCLA191)* (1546).

cya-9: cytochrome a-9

IVR. Left of *trp-4* (9%) (162).

Deficient in cytochrome aa_3. Mitochondrial-encoded subunit 1 of aa_3 is of a larger apparent size than in wild type; mitochondrial-encoded subunit 3 is less abundant than in wild type (173).

cyb-1: cytochrome b-1

VR. Between *al-3* (24%) and *his-6* (10%) (170).

Deficient in cytochrome b. Cannot reduce tetrazolium. Slow growth. Spectrum (167, 170). Suppresses the aa_3 deficiency of *cyt-2* and the mitochondrial mutant [*mi-3*] (167).

cyb-2: cytochrome b-2

Unmapped. A report of VI linkage (170) may be incorrect (162).

Deficient in cytochrome b; erratic deficiency of aa_3 (162). Cannot reduce tetrazolium. Very slow growth. Reduced female fertility. Spectrum (167, 170) Shortened circadian period (238, 244, 560).

cyb-3: cytochrome b-3

IIL. Between *Tip-IIL* and *ro-3* (9%) (1546).

Deficient in cytochrome b. Cannot reduce tetrazolium. Heat-sensitive. Grows slowly, with mutant phenotype at 38 or 39°C, but nearly normal mycelial growth rate at 25 and 34°C (2207). Growth from germinated ascospores is delayed at 34°C (1546). The circadian period is shortened (560).

cyc-1: cytochrome c-1

IIR. Between *thr-3* (35%; 38%) and *trp-3* (15%; 18%) (170, 2015).

Cloned and sequenced: Swissprot CYC_NEUCR, EMBL/GenBank X05506, NCCYTCR, NCCYC1R, PIR A00041, CCNC.

Structural gene for cytochrome c. Growth rate is reduced; female sterile (170). Mutations called *cyc-1-1* and *cyt-12* are allelic (0/1000 recombination; no complementation) (201). Strains carrying the allele called *cyt-12* (or *cyc-1-12* or *cyt-12-1*) are partially deficient for cytochrome aa_3, whereas *cyc-1-1* is not. The cytochrome aa_3 level in mitochondria is temperature-sensitive (2015). Import of apocytochrome c into mitochondria is blocked by an mRNA splicing deficiency in *cyc-1-1* (201). Although the gene was named *cyt-12* when first mapped, it has been called *cyc-1* in recent publications (201, 2015). *cyc-1* is the preferred name because it is more explicit and more accurate.

cyh-1: cycloheximide resistant-1

IR. Between *nit-1* (6%) and *T(STL76)*, *al-2* (8%; 13%) (929, 1554, 1580).

Resistant to cycloheximide (929, 1458). Resistance is recessive in duplications (2122). Dominance reported in forced heterokaryons (929, 1458) may have been due to skewed nuclear ratios (2122). Protein synthesis on ribosomes of the mutant *cyh-1* proceeds in the presence of cycloheximide in a cell-free system (1643). Readily scored on slants with 10 μg/ml cycloheximide autoclaved in the medium. Excellent as a marker and valuable for selecting somatic recombinants or deletions in heterozygous duplications (1458, 2123). Used as a component of heterokaryons for testing spontaneous mutations in mutator strains (1010). Used to show that the cycloheximide-induced phase shift of the circadian clock involves protein synthesis (1435). Cycloheximide induces changes in the pattern of synthesized polypeptides in wild type, but not in the mutant (1609). Called *act-1: actidione resistant-1*.

cyh-2: cycloheximide resistant-2

VR. Between *lys-2* (<1%) and *leu-5* (1% or 2%), *sp* (2%; 9%) (929, 1582, 1603).

Cloned and sequenced: Swissprot RL2A_NEUCR, EMBL/GenBank X06320, PIR S01744, R6NC7A, EMBL NCCRP1G, GenBank NCCR; Orbach-Sachs clones X23C08, G16H11.

Encodes 60S ribosomal protein L27A, homologous to yeast CYH2 (1096). Resistant to cycloheximide (929, 1458). Protein synthesis on mutant ribosomes proceeds in the presence of cycloheximide in a cell-free system (1643). Excellent as a marker, readily scored on slants with 10 μg/ml cycloheximide autoclaved in the medium or with 1 μg/ml added after autoclaving. Resistance in heterokaryons has been reported to be dominant (929, 1234) or recessive (1840); dominance may depend on nuclear ratios or on the medium. A modifier of resistance has been isolated (1745). Used in mutagenicity test systems (1234). Used to show that the cycloheximide-induced phase shift of the circadian clock involves protein synthesis (1435). The double mutant *cyh-1; cyh-2* grows slowly and is much more insensitive to cycloheximide than is either single mutant (929). Called *act-1, crp-1*.

cyh-3: cycloheximide resistant-3

Unmapped. Unlinked to *cyh-2*. Stated to be distinct from *cyh-1* (2169).

Resistant to 100 μg/ml cycloheximide. The double mutant *cyh-2; cyh-3* is morphologically abnormal, resistant to more than 2400 μg/ml (2169). The one known allele, CH96, was called *act-5* (2168), *act-3* (2169).

cyp: cyclophilin

See *csr-1*.

cys-1: cysteine-1

VIL. Between *cys-2* (1%; 3%) and *ylo-1* (8%) (1414, 1986).

Uses sulfite, thiosulfate, cysteine, or methionine. The original isolate (allele 84605) also had a partial requirement for tyrosine and showed high tyrosinase activity at 25°C, but not at 35°C (903, 1414, 1901). These properties reverted, whereas the cysteine requirement is stable (903). Used in studies of intra- and interlocus recombination (1414, 1987, 1989, 1994–1996).

cys-2: cysteine-2

VIL. Between *un-4* (4%), *T(T39M777)*, *T(Y18329)* and *cys-1* (1%; 3%) (1414, 1578, 1986).

Uses cysteine or methionine. Strains carrying *cys-2* alleles are heterogeneous in response to thiosulfate, but do not use sulfite (1414). Lacks sulfite reductase, as do *cys-4* and *cys-10* (1160) (Fig. 45). No interallelic complementation. Used in studies of intra- and interlocus recombination (see *cys-1*).

cys-3: cysteine-3

IIL. Between *ro-7*, *pi* (4%) and *T(AR18)^L*, *pyr-4* (18%; 21%) (1414, 1580, 1592).

Cloned and sequenced: Swissprot CYS3_NEUCR, EMBL/Genbank M26008, PIR A30225, A37066, EMBL NCCYS3, GenBank NEUCYS3.

Encodes a nucleus-localized, bZIP DNA-binding protein that regulates genes of sulfur uptake and metabolism (e.g., sulfate permeases, aryl sulfatase, choline sulfatase) (539, 1279, 1287, 1329, 1519) (Fig. 24). Uses sulfite, cysteine, or methionine; little or no response to thiosulfate (1279, 1414). Grows well on methionine. Resistant to chromate (1279). Consensus DNA-binding site (1191). Studied extensively (1281). Functional analysis (414, 686, 689, 1518). Conserved in *Neurospora intermedia* and in *Neurospora crassa* Oak Ridge and Mauriceville wild types (1287). Expression from a heterologous promoter in *N. crassa* causes the expression of sulfur-controlled genes (1517). Expression positively autoregulated; also regulated by sulfur catabolite repression and by *scon-1* and *scon-2* (1287). Mutation of the F-box of the regulator *scon-2* eliminates *cys-3* expression (1111). Transcriptional and posttranscriptional regulation of *cys-3* RNA levels (1287). Regulation of CYS-3 polypeptide half-life by sulfur (2054). *cys-3* ascospores darken slowly or not at all, even when a *cys-3* strain is the fertilizing parent and when a strain carrying the heat-sensitive allele NM27t is crossed at 25°C, the permissive temperature for growth. *cys-3* can be used effectively as an autonomous ascospore color mutant for demonstrating segregation patterns in asci (1584) (Fig. 25). Addtion of methionine to crossing medium promotes darkening but fails to give good recovery of *cys-3* progeny. Recovery of a few percent *cys-3* progeny is possible in well-aged crosses (1414, 1582).

cys-4: cysteine-4

IVR. Between *rug* (10%), *T(NM152)^R* and *uvs-2* (5%) (1414, 1578, 1580, 1591, 1993).

Uses cysteine; slight response to thiosulfate (1414). Poor growth on methionine. Lacks sulfite reductase, as do *cys-2* and *cys-10* (1160) (Fig. 45).

cys-5: cysteine-5

IL. Between *leu-4* and *ser-3* (0.1%) (1592, 2198). Linked to *cys-11* (<1%) but probably at a distinct locus, with the order *leu-3 cys-5 cys-11 mat* in a cross showing no negative interference (1415).

Uses sulfite, thiosulfate, cysteine, or methionine (1414). Lacks 3'-phosphoadenosine-5'-phosphosulfate reductase (EC 1.8.99.4) [F.-J. Leinweber, cited in refs. (1414) and (1415)] (Fig. 45). Enzymatically distinct from *cys-11* (adenosine 5'-triphosphate sulfurylase), which it complements in heterokaryons [F.-J. Leinweber, cited in refs. (1414) and 1415)]. Leaky, but not so as to interfere with scoring. Ascospores may be oozed from perithecial beaks rather than being forcibly expelled. For good recovery of *cys-5* progeny, crossing media should be supplemented even when the protoperithecial parent

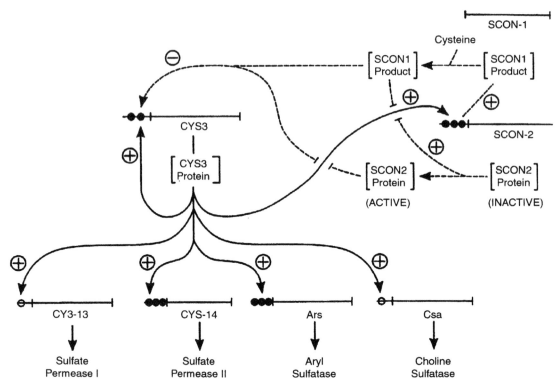

FIGURE 24   A speculative model of the sulfur regulatory circuit of *N. crassa*. The positive-acting *cys-3* gene is itself expressed only when the cells become limited for sulfur. CYS-3 is a *trans*-acting DNA-binding protein that shows autogenous control, increasing its own expression, and that also turns on *cys-14, ars,* and other structural genes of the circuit. CYS-3-binding sites have been identified in the 5'-promoter regions of *ars, cys-14, cys-3,* and *scon-2*. The sulfur-controlled genes *scon-1* and *scon-2* prevent CYS-3 function during conditions of sulfur catabolite repression. The *scon-1* product is postulated to be the factor that recognizes the metabolic repressor, cysteine. The SCON-2 β-transducin protein alone does not inhibit *cys-3* function, but apparently requires modification by or interaction with the unknown product of *scon-1*. An active form of SCON-2 or a SCON-1–SCON-2 complex is postulated to inhibit the transcription of *cys-3* directly (as diagrammed) and/or by binding to the CYS-3 protein to prevent its positive autogenous control. Note that CYS-3 positively controls *scon-2*, which negatively controls *cys-3*, thus constituting a complex set of feedback regulatory loops. Interactions that have been demonstrated experimentally are shown with solid lines, whereas speculative interactions are shown with dashed lines. Solid circles represent known CYS-3-binding sites, and open circles represent presumed CYS-3-binding sites. The gene specifying choline sulfatase has not been identified. From ref. (1287), with permission from the *Annual Review of Microbiology*, Volume 51, © 1997, by Annual Reviews (www.annualreviews.org).

is *cys*⁺; otherwise *cys-5* ascospores may fail to blacken. *cys(NM86)* and *cys(85518)*, which initially were listed as *cys-5* alleles, are now designated *cys-11*.

cys-6, -7, -8: cysteine-6, -7, -8

Lost. Identity with other loci was never established (1414, 1619).

cys-9: cysteine-9

IR. Between *cr-1* (3%) and *T(4540)ᴿ, thi-1* (13%) (1414, 1578).

Cloned and sequenced: Swissprot TRXB_NEUCR, EMBL/GenBank D45049, EMBL NCCYS9, GenBank NEUCYS9.

Structural gene for thioredoxin reductase (EC 1.6.4.5) (1499). Circadian rhythm is affected in the mutant

(1499). Uses sulfite, thiosulfate, cysteine, or methionine. Somewhat leaky (1414).

cys-10: cysteine-10

IVL. Between *Tip-IVL, T(AR33)* and *acon-3* (1%; 6%), *ace-4* (19%; 33%), *cut* (28%; 37%). Linked to *uvs-3* (3%; 7%) (1122, 1414, 1578, 1596).

Uses cysteine, cystathionine, homocysteine, or methionine, with a slight response to thiosulfate (893, 1160, 1414). Lacks sulfite reductase, as do *cys-2* and *cys-4* mutants (1160) (Fig. 45). Käfer (1008) has obtained good growth on thiosulfate. Growth is better on casein hydrolysate than on methionine (1546). *cys-10 chol-1* double mutants grow better on methionine alone than on methionine plus choline (1414). Name changed from *met-4*; see ref. (1414).

FIGURE 25 Maturing asci from a cross heterozygous for *cys-3*, which has pleiotropic effects on ascospore maturation and cysteine biosynthesis. Mature asci show four black and four white spores. The white ascospores received the *cys-3* mutation. Asci with all eight spores unpigmented are immature. One ascus at top center and two asci at upper left show first-division segregation. The remaining mature asci show second-division segregation patterns resulting from crossing over between *cys-3* and the centromere. The length of mature ascospores is ~26 μm. Photograph by N. B. Raju. From ref. (1676), with permission from Urban & Fischer Verlag.

cys-11: cysteine-11

IL. Linked to *cys-5* (<1%). Between *leu-3* (8%) and *mat* (5%). Probably a locus distinct from *cys-5*, with the order *leu-3 cys-5 cys-11 mat* in crosses showing no negative interference (1414, 1415).

Uses sulfite, thiosulfate, cysteine, or methionine (1414). Affects adenosine 5'-triphosphate sulfurylase (EC 2.7.7.4) [1278; F.-J. Leinweber, cited in refs. (1414) and (1415)] (Fig. 45). Enzymatically distinct from *cys-5* (3'-phosphoadenosine-5'-phosphosulfate reductase), which it complements in heterokaryons [F.-J. Leinweber, cited in refs. (1414) and (1415)]. Mutant *cys(85518)* also lacks adenosine 5'-triphosphate sulfurylase (1278) and thus is allelic with *cys-11* rather than with *cys-5*, in harmony with recombination data (1415). Called *cys(NM86)* (1414).

cys-12: cysteine-12

IR. Linked to *al-1* or *al-2* (0/76). Right of *ad-9* (12%) (1415).

Uses cysteine or methionine (1415). No information on what precursors are used.

cys-13: cysteine-13

IR. Right of *his-3* (2%), *tyr-2*, *T(T112M4i) ad-3B* (1279, 1546, 1578).

Resistant to chromate. No demonstrable requirement. Deficient in sulfur permease I (conidial type) (1278, 1279) (Fig. 24). Scored on minimal agar medium with

25 mM chromate and 0.25 mM methionine after 3 days or longer at 34°C. (*cys-3* mutants are also chromate-resistant.)

cys-14: cysteine-14

IV. Linked to *cot-1* (21%) (1279).

Cloned and sequenced: Swissprot CY14_NEUCR, EMBL/GenBank M59167, PIR A37956, GenBank NEUSPII; pSV50 clone 16:2B.

Structural gene for sulfate permease II (mycelial type) (1046, 1278, 1279) (Fig. 24). CYS-14 is a membrane protein with 12 predicted membrane-spanning domains (1046) and was the first member of an evolutionarily conserved transporter protein superfamily to be found (1287, 1783). Localized to the plasma membrane (984). CYS-3 regulatory elements in the promoter have been defined (1193). mRNA half-life measured (1046). The single mutant is deficient in sulfate transport in the mycelial stage, but is sensitive to chromate. Resistant to selenate. The double mutant *cys-14; cys-13* cannot transport inorganic sulfate, but grows on methionine; both single mutants are prototrophic (1279).

cys-15: cysteine-15

IVR. Between *pdx-1* (0T/55 asci), *T(S1229)^L* and *arg-14*, *met-1* (3%) (111, 1490, 1580).

Uses sulfite, thiosulfate, cysteine, or methionine, but is unable to use sulfate (1414, 1490). Only one allele is known, with the requirement not separated from a deficiency of D-amino acid oxidase (1490); a closely linked coincident lesion is thought to be responsible. (See *oxD* for D-amino acid oxidase mutants that have no cysteine requirement.) Called *cys(oxD1)*.

cys(oxD1)

Changed to *cys-15*.

Cytochrome

Cytochrome-deficient Mendelian mutations have been subdivided into four main classes: *cyt*, deficiency of more than one cytochrome; *cya*, deficiency of cytochrome aa_3; *cyb*, deficiency of cytochrome *b*; and *cyc*, deficiency of cytochrome *c* (170, 1631). Apart from diagnostic spectra, they are characterized by slow growth (4–7 days for conidiation of mutants vs 3 days for the wild type) and by the inability to reduce tetrazolium (170). These properties have been used in screening new mutants. *cni* also affects cytochrome spectra (592). The relation of *tet* (tetrazolium-resistant nonreducer) to cytochrome mutations is not known. Cytochrome defects that result from mutations of the mitochondrial genome (e.g., [*mi-1*], [*mi-3*]) are not considered here except as they interact with chromosomal genes such as *su([mi-1])*; see Appendix 4.

cyt-1: cytochrome-1

IL. Between *leu-3* (5%; 8%) (1128) and *T(OY321), leu-4* (1599).

Cloned and sequenced: Swissprot CY1_NEUCR, EMBL/GenBank X05235, NCCYTC1R, PIR A27187.

Structural gene for cytochrome *c* (202). Deficient in cytochromes aa_3 and *b*. Very slow growth. Female sterile (170, 1371). Growth is slower on organic complete medium than on minimal, presumably owing to inhibition by yeast extract (1371, 1592). Mutant *cyt(U-9)* is linked and may be an allele of C115, the original *cyt-1* mutation (162, 170).

cyt-2: cytochrome-2

VIL. Between *cya-3* (10%) and *lys-5* (6%) (170, 1371, 1986)

Cloned and sequenced: Swissprot CCHL_NEUCR, EMBL/GenBank J05075, PIR A34365, EMBL NCCYC8, GenBank NEUCYC8.

Structural gene for cytochrome *c* heme lyase (EC 4.4.1.17) (551). Completely deficient in cytochromes aa_3 and *c*. Very slow growth. Female sterile. Complements *cyb-2* (170, 1371). Spectrum (170). Cytochrome aa_3 deficiency is suppressed by *cyb-1* and by antimycin A (167).

cyt-3: cytochrome-3

Allelic with *cyt-4*.

cyt-4: cytochrome-4

IR. Between the breakpoints of *T(AR173)*; hence, between *Cen-I, os-4, sn,* and *nuc-1, lys-4* (1578).

Cloned and sequenced: Swissprot CYT4_NEUCR, EMBL/GenBank M80735, PIR A38227, GenBank NEUPROPHOS.

Encodes a mitochondrial protein homologous to cell-cycle protein phosphatase (2115). The mutant is deficient in mitochondrial RNA processing (544) and hence, deficient in cytochromes aa_3 and *b*. Grows slowly. The circadian period is altered (560). The phenotype is partially suppressed by the electron transport inhibitor antimycin (166). The *cyt-4-7* mutant and alleles 5, 10, and 14 listed next are all defective in the splicing of large mitochondrial rRNA (162, 166). *cyt-4* complements mutants at *cyt-18* and *cyt-19*, genes that are also required for the splicing of large mitochondrial rRNA (166). Five alleles are cold-sensitive. Mutants called *cyt-3-5*, *cyt-U-10*, and *cyt-U-14* (170) proved to be *cyt-4* alleles and were designated *cyt-4-5, cyt-4-10,* and *cyt-4-14*, respectively (162).

cyt-5: cytochrome-5

IVR. Left of *trp-4* (9%). Linked to *cyt-19* (<1%; 5%; 10%) (170). Near *sdv-11, sdv-12* (162).

Cloned and sequenced: Swissprot RPOM_NEUCR, EMBL/GenBank L25087, EMBL NCCYT5MIT, GenBank NEUCYT5MIT.

Deficient in cytochromes aa_3 and *b*. Slow growth. Not defective in the splicing of mitochondrial rRNA, unlike *cyt-19*, which is closely linked but complements *cyt-5* (166, 170). A clone that includes the sequence for mitochondrial RNA polymerase rescues the mutant phenotype (369). Allele *cyt-5-4* was called *cyt-4*.

cyt-6: cytochrome-6

VII. Linked to *wc-1* (2%). Indicated to the left (170).

Deficient in cytochromes aa_3 and *b*. Grows slowly. Not tested for allelism with *slo-2* (170).

cyt-7: cytochrome-7

VIIL. Between *T(ALS179)* and *adh (9%), nic-3* (18%) (1546).

Deficient in cytochromes aa_3 and *b*. Slow growth. Possibly allelic with *su([mi-1])-5* (170). Allele 20 is cold-sensitive (1546).

cyt-8: cytochrome-8

IIIR. Linked near *ad-4* (170). Between *ace-2* (1 or 2T/18 asci) and *ad-2* (4T/18 asci), *trp-1* (3/18) (1447).

Deficient in cytochromes aa_3 and *b*. Scored by slow growth on glycerol at 15 or 18°C. Possibly allelic with *su([mi-*

1])-1 (170). RFLP data indicate that *cyt-8* is close to *cox-4*; whether they are allelic has not been established.

cyt-9: cytochrome-9

VL. Between *lys-1* (5%) and *T(MB67)ᴿ*, *Cen-V* (4T/57asci), *at* (2T/57 asci; 5%) (170, 1602).
Deficient in cytochromes *aa₃* and *b*. Slow growth (170).

cyt-12: cytochrome-12

Changed to *cyc-1*.

cyt-15: cytochrome-15

VIR. Between *pan-2* (4%) and *trp-2* (3%) [H. Bertrand, unpublished, cited in ref. (2151)].
Mitochondria of the mutant are deficient in cytochromes *aa₃* and *b*, whereas cytochrome *c* is present in higher than normal amounts. Involved in the assembly of small subunits of mitochondrial ribosomes [H. Bertrand, unpublished, cited in ref. (2151)].

cyt-18: cytochrome-18

IR. Linked to *al-2* (10%) *nic-1* (1%; 5%) (399, 1262).
Cloned and sequenced: Swissprot SYYM_NEUCR, EMBL/ GenBank M17118, PIR A27158, SYNCYT, GenBank NEUTYRSM.
Encodes tyrosyl-tRNA synthetase (EC 6.1.1.1) (26). Heat-sensitive; grows slowly. Stimulated by the addition of tyrosine. Deficient in cytochromes *aa₃* and *b* at 37°C (1631). Mitochondrial protein synthesis and the assembly of small mitochondrial subunits also are abnormal (401). A novel large RNA precursor (35S) is found at restrictive temperatures (401). The intervening sequence of large (25S) mitochondrial rRNA is not excised. Used for direct measurement of tyrosyl-tRNA synthetase participation in RNA splicing (26, 285). Alleles 289-67 and 299-9 differ in the speed of turn-off of RNA processing when the temperature is shifted (1262). Complements *cyt-4* and *cyt-19*, which are also required for splicing large mitochondrial rRNA (166).

cyt-19: cytochrome-19

IVR. Between *T(S4342)ᴸ*, *arg-14* and *T(NM152)ᴸ*, *pyr-3* (5%). Linked to *cyt-5* (<1%; 5%; 10%) (166, 170, 1377).
Deficient in cytochromes *aa₃* and *b*. Slow growth. Required for splicing of mitochondrial large rRNA. Mutant is defective in intron splicing from mitochondrial transcripts (402). Complements *cyt-5*. Complements *cyt-4* and *cyt-18*, which are also required for splicing (166). Called *cyt-U-19* (170).

cyt-20: cytochrome-20

IL. Linked to *mat* (0/151) (399).
Cloned and sequenced: Swissprot SYV_NEUCR, EMBL/ GenBank M64703, PIR A41251, GenBank NEUXXX.
Encodes cytoplasmic and mitochondrial valyl-tRNA synthetase; possibly has other functions (1104). Deficient in the small subunit of mitochondrial ribosomes, but contains normal ratios of 19S to 25S rRNA in whole mitochondria (401). Either the *cyt-10* mutation is allelic with *un-3*, or, if *un-3* strain 55701 is complex, *cyt-10* is a component. The *cyt-10* component is responsible for the temperature sensitivity of strain 55702, which is also ethionine-resistant. The transport deficiency responsible for ethionine resistance is not temperature-sensitive. *cyt-20* has also been called *cyt(289-56)* (1104). See *un-3*.

cyt-21: cytochrome-21

IL. Between *ro-10* (1%) and *nit-2* (5/18), *Fsr-12* (9T/18 asci). Linked to *mus-40*, *pzl-1* (1/18) (976, 1447). An earlier report of linkage in II (399, 1596) was incorrect.
Cloned and sequenced: Swissprot RT24_NEUCR, EMBL/ GenBank J03533, X06360, PIR A29927, EMBL NCCYT21, GenBank NEUCYT21.
Encodes mitochondrial ribosomal protein S-24 (1107). Allele 297.24 (1631), called *cyt-21-1*, inactivates an RNA splice site; activation of a cryptic splice site causes an in-frame deletion in the polypeptide (1108). At 25°C, deficient in cytochromes *aa₃* and *b*. Defective in mitochondrial ribosome assembly. Deficient in the small subunit of mitochondrial ribosomes. The 19S rRNA is degraded rapidly. The effect is more severe at higher temperatures. No growth at 40°C. The phenotype resembles that of the mitochondrial mutant [*mi-1*] (401), but the deficiency is not suppressed by *su([mi-1])-4* or *su([mi-1])-5* (399). Called *cyt(297.24)*.

cyt-22: cytochrome-22

IIIL. Linked to *cum* (6%), *acr-7* (19%) (1134). Left of *con-1* (4/15), *cyt-8* (5/18) (1447).
Encodes a mitochondrial ribosomal protein. Also called *cyt(289-4)* (1103).

cyt-25: cytochrome-25

VI? Linked to *pan-2* (32%) (162)?. Does not complement *cyt-15* in heterokaryons (2151).
Cytochromes *aa₃* and *b* are deficient, whereas cytochrome *c* is present in higher than normal amounts. Deficient in large subunits of mitochondrial ribosomes, whereas small subunits appear to be in excess. Female-sterile with no functional protoperithecia, but fertile as male parent (2151). Called *cyt-U-28*.

cyt(289-56)

Changed to *cyt-20*.

cyt(297-24)

Changed to *cyt-21*.

cyt-U-28: cytochrome-U-28

Changed to *cyt-25*.

d: heterokaryon incompatibility-d

Used in early publications as a symbol for *het-d*.

da: dapple

IIL. Right of *T(NM149)*. Linked to *thr-3* (3%), probably to the right (1578, 1585, 1603).

Produces flecks of conidia on agar surface (1585). Conidia separate very easily (1958).

Dab-1: Dead and buried-1 (Transposable Element)

Cloned and sequenced: EMBL/GenBank Y14976; cosmid G14:3A.

A degenerate retrotransposon-like element of the *gypsy* class present in multiple copies in the genome. Lacks long terminal repeats and shows earmarks of RIP (177).

del: delicate

VIR. Between *rib-1* (12%), *pan-2* (6%) and *os-9*, *T(OY329)*, *trp-2* (0; 13%) (1603).

Growth is less profuse than wild type, with prolific hyphal branching flat on the agar and fine aerial growth at the top of a slant. Good scorability (1548).

Deletion

The synonym *Df: Deficiency* is preferred. A single-gene deletion can be treated as an allele.

des: δ-subunit ATP synthase

Unmapped.

Cloned and partially sequenced: EMBL/GenBank AI398451; EST W13F5.

Encodes the δ-subunit of ATP synthase (1864).

Df(. . .): Deficiency

A segmental rearrangement having two or more contiguous genes missing. *Deficiency* is preferred to the synonymous term *Deletion*.

dgr-1: deoxyglucose resistant-1

VL. Between *rDNA*, *T(UK2-33)*, *In(UK2-y)^L* and *T(AR30)*, *caf-1* (7%) (35, 1578).

Resistant to inhibition by 2-deoxy-D-glucose. Grows more slowly than wild type on standard medium but growth is initially faster than that of wild type on media with mono- or disaccharides plus deoxyglucose. Conidiation is precocious. The greatest differential growth in deoxyglucose is obtained with 0.1% cellobiose, trehalose, lactose, fructose, or galactose (35, 578, 1593). For convenient scoring, see ref. (1642).

dgr-2: deoxyglucose resistant-2

I. Right of *mat* (15%), *arg-1* (2%) (35). Linked to *csp-1* (~1%) (573).

Resistant to inhibition by 2-deoxy-D-glucose and sorbose (35). Completely or partially resistant to carbon catabolite repression of glucoamylase, invertase, and glucose transport system II. Thermostability of invertase is altered. Lacks glucose transport system I.

dgr-3: deoxyglucose resistant-3

Allelic with *sor-4*.

dgr-4: deoxyglucose resistant-4

IR. Linked to *al-2* (1%), probably to the left (35).

Resistant to inhibition by 2-deoxy-D-glucose and sorbose (35). Completely or partially resistant to carbon catabolite repression of glucoamylase, invertase, and glucose transport system II. Thermostability of invertase is altered. Glucose transport system I is enhanced.

dim: defective in methylation

The designation *dim* refers to *trans*-acting mutants showing reduction in or absence of DNA methylation at loci that are normally methylated (e.g., the ψ_{63} and ζ–η regions in Oak Ridge strains at the *Fsr-63* and *Fsr-33* loci). The designation does not refer to mutants showing reduced methylation when starved for methionine, such as *eth-1* (at semipermissive temperatures) (660), and *met* and *cys* mutants starved for methionine (1732). It should be noted that DNA methylation in *Neurospora* is somewhat sensitive to growth conditions, and methylation levels can be increased, even in wild-type strains, by supplementation with methionine (662, 1885).

dim-1: defective in methylation-1

III. Linked to *mus-20*, *trp-1* (1885).

DNA methylation level is reduced by ~40% overall and multiple sites are affected, although to different extents.

Mutant alleles may be partially dominant (660). Female sterility of *dim-1* is complementable. Obtained by screening mutants that decrease the sensitivity of *mus-20* to 5-azacytidine (662).

dim-2: defective in methylation-2

VIIR. Between *wc-1* and *un-10, arg-10* (1094).

Specifies DNA methyltransferase (EC 2.1.1.-). Mutations abolish all detectable DNA methylation in vegetative cells but have no visible phenotype and do not affect fertility. Detected by screening for reduced methylation of the *Bam*HI site in the 5S RNA pseudogene, *Fsr63* (660, 661). RIP is not reduced by *dim-2* (1885). *dim-2* was used to demonstrate that quelling does not depend on DNA methylation (390), that DNA methylation indirectly prevents transcription elongation in *Neurospora* (1746), and that reversion to *Am*+ of *Tad::am* strains depends on DNA methylation (276).

dim-3: defective in methylation-3

VR. Between *am* and *inl* (861).

Methylation is reduced at all of several methylated loci that were examined. Origin same as for *dim-2* (660).

dim-4: defective in methylation-4

IIR. Linked to *arg-12, aro-1, ure-3* (812).

Reduces RIP frequency, unlike *dim-2* (1885). Obtained in a visual screen for reduced RIP, using a *pan-2* duplication (812).

Dip-1: Diploid ascospores-1

IIIR. Between *ro-2* (7%) and *phe-2* (1%) (1322).

The mutant is vegetatively indistinguishable from wild type. Crosses heterozygous for *Dip-1* typically give about two-thirds abnormal asci, most commonly with four ascospores but not uncommonly with eight. Other asci have three spores, or two, or one giant spore. The ascospores in asci with fewer than four spores are often misshapen and usually do not germinate. Surprisingly, *Dip* × *Dip* crosses give almost entirely eight-spored asci. Hence, *Dip* is not simply ascus-dominant in the usual sense. Spores from eight-spored asci are always haploid, and the *Dip* segregants can only be identified by progeny testing. Individual spores from four-spored asci generally give rise to cultures that are self-fertile because they are heterozygous or heterokaryotic for *mat*. When a four-spored ascus is heterozygous for multiple markers, a germinated ascospore gives rise to a culture in which the diploidy quickly breaks down, resulting in a complex heterokaryon. Many kinds of mitotic segregants can be isolated from the conidia of such a culture. Cytologically, the nuclei giving rise to ascospores are clearly of higher ploidy than haploid, but the exact number of

chromosomes cannot be distinguished (1683). Crosses to *Dip* can shelter sexually lethal mutant genes such as *Asm-1* and enable them to be recovered from crosses (1322).

dir: dirty

IR. Right of *pa* (37%) (1200).

Conidia are few and misshapen. A yellowish exudate may be produced (1201). Photograph (1200). One of the first markers to be mapped. Possibly *os-1*. Stock lost.

dn: dingy

IVR. Linked to *rug* (1%). Right of *pyr-2* (4%) (1369, 1585).

Produces gray patches of microconidia in addition to macroconidia (1369). On glycerol complete medium, orange conidia are abundant at top of slant, with "dingy" growth on the flat surface below. Growth is slower than normal. The double mutant *fl; dn* produces abundant microconidia, similar to *pe fl*. Microconidia from *fl; dn* were found to be less viable, however (865). *fl; dn* is fully fertile in homozygous crosses (1584). For this reason, it may be preferred to *pe fl* as a microconidiating strain.

DNA Repair Mutants

See *mei-2, mei-3, mus, nuh, phr, upr, uvs*. Reviewed (946, 1838).

do: doily

VIIL. Linked to *spco-4* (<1/400) (1592). Left of *nic-3* (1%; 3%) (587, 1585).

Growth is restricted, colonial, and at 4% of the wild-type rate [(587); D. R. Stadler, cited in ref. (1585)]. Cell-wall galactosamine is 0.5% of the wild-type level, UDPGalNAc content is 3%, and the specific activity of UDPGlcNAc-4-epimerase in cell extracts is 20%. Partial back-mutations can differentially affect cell-wall and alcohol-soluble galactosamines, indicating pleiotropy (587). Cell-wall peptides are reduced in amount and have an altered DEAE elution profile (2249).

dot: dot

IR. Linked to *ad-9* (0/44), right of *thi-1* (2%) (1592).

Growth is colonial (1585) and more restricted on glycerol complete medium than on minimal. A possible maternal effect is seen in *dot*+ progeny from heterozygous crosses (1546).

dow: downy

IIIR. Between *ty-1* (21%), *un-17* (23%), *T(AR17)^L* and *T(AR17)^R*, *erg-3* (10%, 14%). Linked to *acu-7* (0/72), *ro-11* (0/63) (1578, 1582, 1589, 1592).

Conidia are formed on a soft, matty mycelium that covers the agar in a slant (1592). An excellent marker with good viability, fertility, and scorability.

Dp(. . .): Duplication

A rearrangement having two or more genes duplicated, either in tandem or ectopically. Strains duplicated for specific segments can be obtained as a major class of progeny from crosses heterozygous for insertional or quasiterminal rearrangements or from intercrosses between rearrangements having breakpoints appropriately positioned in the same two chromosome arms or in the same arm. For theory and applications, see ref. (1566). For information on individual duplications, see ref. (1578). See figure entitled "Duplication coverage."

DUPLICATION COVERAGE Segments of the *N. crassa* genome that can be obtained as viable nontandem duplications in progeny of crosses heterozygous for insertional or quasiterminal chromosome rearrangements. Numbers to the right of each dashed line identify the rearrangement used to obtain duplications for the indicated segment. Genes that bracket breakpoints are shown, together with a few key markers for orientation, but most markers have been omitted. The figure does not show segments that can be obtained as duplications from overlapping rearrangements. The duplications shown here, and the rearrangements that produce them, have been used for determining dominance and dosage in studying regulation (1333), genetic mapping (1566, 1605), relating physical maps to genetic maps (1935), obtaining genetically defined truncated chromosomes (91a), studying chromosome instability (1468, 1469, 1835, 1938), identifying and manipulating vegetative incompatibility genes (1421, 1560, 1797), and obtaining mutations by RIP (1589). Reproduced from ref. (1578).

Mapping by duplication coverage was pioneered by Dorothy Newmeyer in a paper (1475) that described the first *Neurospora* inversion, showed that one of its breakpoints is genetically terminal, and described the partial-diploid (duplication) progeny that result from crossing over when the inversion in heterozygous. The inversion was used in crosses to determine whether linked markers were located inside or outside the duplicated segment. The method was extended by using other duplication-generating rearrangements (1592), including both insertional and terminal translocations. Growth of duplication progeny was shown by Newmeyer to be greatly inhibited when they were heterozygous for alleles at an included vegetative incompatibility locus. Duplications thus provided a practical way for identifying individual *het* genes. Use of the method is illustrated in refs. (1560) and (1566).

dr: drift

VIIR. Between *for* (3%) and *T(5936), arg-10* (12%) (1578, 1580, 1604).

Conidia are formed in dense masses at the top of a slant. Growth elsewhere is flat on the agar surface. Good scorability (1604). Called *mo(P1163)*.

dys-1: deoxyhypusine synthase-1

Unmapped.

Cloned and sequenced: Swissprot DHYS_NEUCR, EMBL/ Genbank U22400.

Specifies the enzyme eukaryotic initiation factor 5A-deoxyhypusine synthase (EC 1.1.1.249) (2051). Enzyme purification (2052). See *eif5A*.

e: heterokaryon incompatible-e

Used in early publications as a symbol for *het-e*.

eas: easily wettable

IIR. Between *ace-1* (1%) and *fl* (9%) (1546).

Cloned and sequenced: Swissprot RODL_NEUCR, EMBL/ GenBank X62170, X67339, PIR A46222, GenBank NCBLI7DNA.

Specifies a self-assembling protein that forms a hydrophobic layer of rodlets on the surface of conidia. This remarkable rodlet hydrophobin is soluble only in 100% trifluoroacetic acid (2071). In the *eas* mutant, conidia are devoid of rodlets (142) and, hence, easily wettable (1875) in contrast to the hydrophobic wild type (Fig. 26). *eas* superficially resembles *csp* mutants in that conidia do not readily become airborne, but it differs from *csp* in that conidia do not remain joined in the proconidial chains (1875, 1958). It is somewhat sensitive to high osmotic pressure, and aging cultures may appear brighter orange than wild type. Conidiating cultures are readily scored by tapping an inverted slant, transferring conidia to water, or adding a drop of water

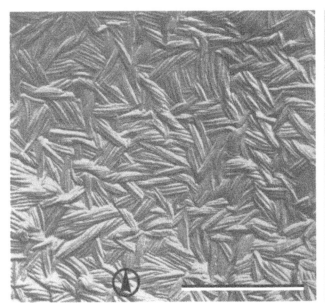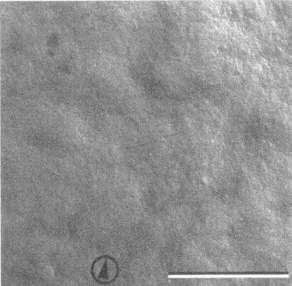

FIGURE 26   Replicas of the surface of macroconidia of wild-type (left) and the mutant *easily wettable* (right), showing the absence of rodlets in *eas*. Bar length = 0.5 µm. Scanning EM photograph. Reprinted with permission from *Nature* [ref. (142)]. Copyright 1978 Macmillan Magazines Ltd.

to the culture. Transcription of *eas* is inducible by blue light and is under circadian control (145, 586, 1143, 1227), peaking late at night (147). Clock, light, and developmental regulation are controlled by distinct *cis*-acting upstream elements (145). Induction of *eas* by light does not require an intact circadian oscillator (60). Transcription of *eas* is normal in the mutants *acon-2* and *acon-3*, but not in *fl* (60). *eas* allele UCLA191 is inseparable from a rearrangement that blocks recombination distal to the *eas* locus; this led originally to incorrect mapping of *eas* at the IIR tip. Allele KH5.9 is a translocation (1578). Location of *eas* right of *ace-1* is based on crosses using RIP-derived point mutant JD105 (1546). Because conidia do not become airborne, double mutants have been constructed that combine *eas* with markers used for teaching (700, 1790). Genes originally identified as clock-controlled gene *ccg-2* and blue-light-inducible gene *bli-7* proved to be *eas* alleles (145, 586, 1019, 1143).

eat: encodes anonymous transcript

Six genes closely flanking the mating-type idiomorphs were defined on the basis of transcripts, with the order *eat-6 eat-5 mat eat-1 eat-2 eat-3 eat-4*. Those right of *mat* are located in the polymorphic region called "variable" in ref. (1690). Regulation differences and the phenotype of a mutant inactivated by RIP indicate that genes in this region are concerned with mating and sexual development. Transcripts of *eat-1* and *eat-2* are unique in showing mating-type-specific differences in size and expression (1691).

eat-1: encodes anonymous transcript-1

IL. Right of *mat*. Between *mat* and *eat-2* (1691).
Cloned and sequenced: EMBL/GenBank AF037226.
Transcript size and expression are different for the two mating types. Attempts to identify alleles inactivated by RIP have been unsuccessful. The alleles are perhaps pseudogenes, reflecting the accumulation of recessive mutations resulting from close linkage to *mat* (1691). For other genes expressed differentially in the two mating types, see *ccg-4*, *eat-2*, *mfa-1*, and *scp*.

eat-2: encodes anonymous transcript-2

IL. Right of *mat*. Between *eat-1* and *eat-3* (1691).
Cloned and partially sequenced: EMBL/GenBank AF037227.
Transcript size and expression are different for the two mating types. The *eat-2* allele in *mat A* contains a tandemly repeated 119-bp segment that is present only once in the *eat-2* allele of *mat a*. The sequenced region is >95% similar in *mat A* and *mat a*. A portion of a 909-bp ORF corresponding to the first 58 amino acids of the putative product shows 95% identity to a domain present in *pma-1* (*plasma membrane ATPase-1*). [This homology was shown first by S. M. Mandala (1925).] A mutant *eat-2* allele, obtained by RIP in a *mat A* strain, is abnormal in both vegetative and sexual phases. Growth is very slow but is better on crossing medium (with no ammonium nitrogen) than on Vogel's minimal growth medium. No conidia are formed. Perithecia are unbeaked and devoid of ascospores when the mutant is crossed as either male or female. The mutant vegetative phenotype is recessive. The mutant sexual phenotype appears to be dominant. Alternatively, normal fertility may be dependent on the presence of functional *eat-2* sequences from both mating types. A few ascospores can be obtained from crosses when the male parent carries the mutant *eat-2* allele in a heterokaryon with the *a^{m1} ad-3B cyh-1* inactive-mating-type helper strain. These ascospores can germinate, but the germlings are inviable (1691). For other genes expressed differentially in the two mating types, see *ccg-4*, *eat-1*, *mfa-1*, and *scp*.

eat-3: encodes anonymous transcript-3

IL. Right of *mat*. Between *eat-2* and *eat-4* (1691).
Cloned.
Encodes as anonymous transcript (1691).

eat-4: encodes anonymous transcript-4

IL. Right of *mat*. Between *eat-3* and the right end of the variable flanking region (1691).
Cloned.
Encodes an anonymous transcript (1691).

eat-5: encodes anonymous transcript-5

IL. Left of *mat*. Between *eat-6* and *mat* (1691).
Coloned.
Encodes an anonymous transcript (1691).

eat-6: encodes anonymous transcript-6

IL. Left of *mat*. Between *un-3* and *eat-5* (1691).
Coloned.
Encodes an anonymous transcript (1691).

edr-1: edeine resistant-1

VI. Linked to *ad-1*, *pan-2* (<1%) (2069).
Resistant to 200 mg/ml edeine. Only a fraction of *edr* conidia grow. Recessive. Called *ed^r-1* (2069).

edr-2: edeine resistant-2

VIL. Left of *ad-1* (19%) (2069).
Resistant to 200 mg/ml edeine. Only a fraction of *edr* conidia grow. Recessive (2069). In intact cells, edeine

inhibits the synthesis of protein, DNA, and RNA of the wild type but not of the mutant. *In vitro*, edeine inhibits protein synthesis equally for both mutant and wild type. Hence, uptake is probably impaired in the mutant (2172). Used to examine the effect of edeine on crossing over and intragenic recombination (2068). Called *ed^r-2, ed^r-29*.

eif5A: eukaryotic initiation factor 5A

Unmapped.

Cloned and sequenced: Swissprot IF5A_NEUCR, EMBL/GenBank U02638, PIR S55278, EMBL NCIF5A, GenBank NCEIF5A.

Encodes eukaryotic initiation factor 5A (eIF-5A homolog) (2050), which is a substrate for deoxyhypusine synthase, the product of *dys-1* (2050, 2051).

en(am)-1: enhancer-1 of am

VR. Between *am* (8%) and *inl* (1%). Linked to *gln* (1%) (256, 642, 647).

In *en(am)-1 am* double mutants, *en(am)-1* blocks the adaptation of *am* on minimal medium that is devoid of amino nitrogen (647) (Fig. 49). The double mutants are inhibited by ammonium and grow adequately only when glutamate is the sole nitrogen source. The *en(am)-1* single mutant grows well on minimal medium, but it is unable to use, as the sole nitrogen source, proline (256), methionine, alanine, isoleucine, valine, urocanate, hypoxanthine, uridine, urea, or bovine serum albumin (350). Relative resistance to *p*-fluorophenylalanine or ethionine and complete resistance to 0.02 *M* glycine cosegregate with *en(am)-1* (642). Glutamate synthase (GOGAT) is normal (566). The single mutant is scored using proline as the sole N source (256) or (better) by resistance to 0.2 m*M* *p*-fluorophenylalanine (642). Called *I* (566).

en(am)-2: enhancer-2 of am

IIR. Linked near *pe* (1906).

en(am)-2 counteracts the leakiness of *am* on minimal medium. The *en(am)-2; am* double mutant grows well on L-alanine (25 mM), 0.5% casein hydrolysate (1906), or glutamate (5 mM) (566). The *en(am)-2* single mutant grows normally on minimal medium. *en(am)-2* lacks glutamate synthase (GOGAT) (Fig. 50). The double mutant *en(am)-2; am* lacks both NADP-glutamate dehydrogenase (1738) and glutamate synthase activities (566). Frequent revertants of *en(am)-2;am* on suboptimal medium are attributed to back-mutation at *am* (1906). Called *en-am*.

En(pdx): Enhancer of pdx-1 pigment

IL. Linked to *mat* (5%), probably to the left (1274).

The *En(pdx); pdx-1* double mutant excretes yellow pigment when grown on Vogel's medium (2163) or Westergaard and Mitchell's medium (2208) supplemented with suboptimal pyridoxine plus ammonium sulfate at 5 mg/ml, but not at 1 mg/ml. Pigment is not produced by either single mutant (1274). The addition of a nonlimiting concentration of pyridoxine inhibits production of the pigment on both media. Production of the pigment also is inhibited in heterokaryons between complementing *pdx* alleles. *En* is dominant over *En^+* in heterokaryons between noncomplementing *pdx* alleles (1662).

er: erupt

Allelic with *rg-1*.

erg: ergosterol

Ergosterol mutants have been detected by resistance to nystatin and other polyene antibiotics. The known mutants are female-sterile. Intercrosses for allelism tests can be made, however, by using a heterokaryon as the female parent (787, 1385). Biosynthesis is illustrated in Fig. 27.

erg-1: ergosterol-1

VR. Between *pk* (2%) and *asn* (9%) (787).

Cloned and sequenced: EMBL/GenBank U59671, Swissprot ERG2_NEUCR; Orbach-Sachs clone G17F12.

Lacks fecosterol Δ^8, Δ^7-isomerase [5-ergosta-8,24(241)-dien-3-ol isomerase] (787). Defective in the conversion of fecosterol [5-ergosta-8,24(241)-dien-3-ol] to episterol (Fig. 27). Ergosterol is deficient in the cell membrane, conferring strong resistance to nystatin and other polyene antibiotics (785, 787). Infertile as female. Slow growth, reduced conidiation. The *nys^r* mutants of ref. (1385) are blocked in the same reaction, but have not been tested for allelism with *erg-1*.

erg-2: ergosterol-2

VR. Left of *inl* (6%) (787).

Ergosterol is deficient in the cell membrane, thereby conferring slight resistance to nystatin and other polyene antibiotics (784, 785). Lacks 24(28)-dehydroergosterol hydrogenase or reductase (EC 1.-.-.-), which mediates the terminal step of ergosterol synthesis (787) (Fig. 27). Poorly fertile as female. Good growth and conidiation.

erg-3: ergosterol-3

IIIR. Between *T(AR17)R, dow* (10%, 14%) and *Tip-IIIR* (1582).

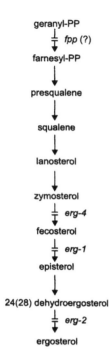

geranyl-PP
↕ *fpp* (?)
farnesyl-PP
↓
presqualene
↓
squalene
↓
lanosterol
↓
zymosterol
↕ *erg-4*
fecosterol
↕ *erg-1*
episterol
↓
24(28) dehydroergosterol
↕ *erg-2*
ergosterol

FIGURE 27   The probable pathway of sterol biosynthesis, showing sites of gene action (787, 1385). From ref. (1596), with permission from the American Society for Microbiology.

Cloned and sequenced: Swissprot ERG3_NEUCR, EMBL/ GenBank X77955, GenBank NCERG3, PIR S44170; Orbach-Sachs clone G18A10.
Encodes sterol Δ14,15-reductase (EC 1.-.-.-) (1533). Ergosterol is deficient in the cell membrane. Resistance to nystatin and other polyene antibiotics is increased slightly (784, 785). Female-sterile, but tiny protoperithecia are formed. Growth is slow, conidiation is reduced, and the production of aerial hyphae is uneven. Mutants are resistant to the steroidal glycoside α-tomatine and sensitive to the phytoalexins pisatin and biochanin A (1532, 1894). The *erg-3* null mutant is complemented by sequences encoding the transmembrane domain of human lamin B receptor, indicating defective sterol C-14 reductase activity (1647).

erg-4: ergosterol-4

IR. Linked to *al-1* (10%) (1582).
Ergosterol is deficient in the cell membrane, conferring a slight resistance to nystatin. Lacks C-24 (zymosterol)-methyl transferase (EC 2.1.1.41). Accumulates zymosterol (Fig. 27). Infertile as female (787). Growth is slow, colonial at 34°C, spreading at 25°C (1582).

erp38: ERP38 protein

Unmapped.

Cloned and sequenced: Swissprot ER38_NEUCR, EMBL/ GenBank Y07562, GenBank NCERP38.
Encodes a putative protein disulfide isomerase ERP38 precursor (EC 5.3.4.1), a stress-inducible member of the PDI family (986).

esr: enhancer of spermidine requirement

Mutations that increase the need for spermidine. The trait is scored by combining the putative *esr* mutation with a leaky *spe* mutation and testing strains on polyamine inhibitors dicyclohexylamine or α-difluoromethyl-ornithine (DFMO). Because arginine imposes a spermidine requirement on strains carrying *aga*, these can also be used. Ascospores and (usually) conidia of *esr* mutants germinate and then pause, often for more than 1 day, before continuing to grow. The known *esr* mutations, originally called BMH (464), were named and given locus numbers by ref. (451).

esr-1: enhancer of spermidine requirement-1

VR. Between *pyr-6* and *inl* (7%) (464).
Increases the need for spermidine (464).

esr-2: enhancer of spermidine requirement-2

I. Linked to *mat* (10%) (464).
Increases the need for spermidine (464). A similar mutant, *esr(BMH8)*, is linked to *mat* at 3% and may be allelic.

esr-3: enhancer of spermidine requirement-3

Unmapped. Not linked to *mat* (1L) or *inl* (VR) (464).
Increases the need for spermidine (464).

eth-1: ethionine resistant-1

IL. Between *arg-1* (<1%) and *arg-3* (<1%) (1336, 1592).
Cloned and sequenced: Swissprot METK_NEUCR, EMBL/ GenBank U21547, PIR S65800.
Structural gene for S-adenosylmethionine synthetase (EC 2.5.1.6) (981, 1045, 1298) (Fig. 45). Heat-sensitive, not growing at 37°C (1336). Resistant to ethionine at 24°C (1336). Resistance attributed to the overproduction of methionine (1028, 1030) and/or reduced production of the toxic S-adenosylethionine (98, 981). In *eth-1*⁺/*eth-1*⁻ partial diploids, ethionine resistance is dominant but temperature sensitivity is recessive (98). Deficient in DNA methylation at semipermissive growth temperatures (660). Levels of several enzymes that are normally repressible by methionine are not repressed by methionine in *eth-1ʳ* even at growth-permissive temperatures (260, 1324, 1874). Both heat sensitivity and ethionine resistance are reparable by high osmotic pressure (1324).

For high-resolution analysis of this chromosome region, see refs. (1299) and (1740).

eth-2: ethionine resistant-2

Provisional name for a mutational lesion present, together with *cyt-20*, in the *un-3* mutant 55701. See *un-3*.

exo-1: exoamylase-1

IL. Linked to *mat* (7%), *ad-3A* (22%). Stated to be left of *mat* (615).

Shows hyperproduction of β-amylase, α-amylase, glucoamylase, invertase (β-fructofuranosidase), pectinases, and (to a lesser extent) trehalase (432, 433, 764, 765, 1641, 1954, 2002). Enzymes are secreted abundantly upon depletion of an exogenous carbon source (765). A polysaccharide is also released (615). Conidial enzyme levels are increased sevenfold. Amino sugar content of the cell wall is altered (764, 765). The initial allele was called SF26, *exo$_a$-1* (615, 765). A probable second allele was found in *inl* 89601 *a* (372, 615, 2002). With allele SF26, high amylase and high invertase cosegregated in 91 isolates (764). With alleles from the 89601 strain in a mixed-background cross, high amylase and high invertase each acted as if due to a single major gene with many modifiers; high amylase and high invertase usually cosegregated and were not correlated with alkaline phosphatase levels (2002). The relation of *exo-1* to gene VI-178, which reverses the repression of invertase and trehalase production by mannose (1323), is not known. For a linked gene defective in glycoamylase, see *sor-4*, the T9 allele of which was called *gla* in ref. (106) and *amy* in ref. (615). *exo-1* was not tested for allelism with *sor-4* T9, but is stated to map on the opposite side of *mat*. Because *inl* 89601 *a* has been used to obtain mutants by inositol-less death enrichment, *exo-1* may be present but unrecognized in some laboratory stocks.

f: fast

Changed to *su([mi-1])-f*.

fas: fatty acid synthesis

Changed to *cel*.

fba: fructose bisphosphate aldolase

Unmapped.
Cloned and sequenced: Swissprot ALF_NEUCR, EMBL/ GenBank L42380, GenBank NEUF16B; EST NC4F12.
Encodes fructose 1,6-bisphosphate aldolase (EC 4.1.2.13) (2259).

fdh: formate dehydrogenase

VR. Contiguous with *leu-5* (380).
Cloned and sequenced: Swissprot FDH_NEUCR, EMBL/ GenBank L13964, PIR A47117, EMBL NCFDHA, GenBank NEUFDHA; EST NM1D6.
Structural gene for formate dehydrogenase (EC 1.2.1.2). Under developmental control. Expressed during conidiation and early germination (380).

fdu-1: fluorodeoxyuridine resistant-1

Allelic with *ud-1*.

fdu-2: fluorodeoxyuridine resistant-2

IVR. Right of *cys-4* (2%) (265).
Resistant to 5-fluorodeoxyuridine, 5-fluorouracil, and 5-fluorouridine. Resistance is partially dominant in heterokaryons. Involved in the regulation of *pyr-3*, *udk*, and *ud-1* (267). Scored by spotting a conidial suspension on medium containing 4×10^{-5} M 5-fluorodeoxyuridine, filter-sterilized (885).

fes-1: iron–sulfur subunit-1

Unmapped.
Cloned and sequenced: Swissprot UCRI_NEUCR, EMBL/ Genbank X02472, GenBank NCCYRFES.
Encodes the iron–sulfur subunit of ubiquinol–cytochrome *c* reductase (EC 1.10.2.2) (841).

ff: female fertility

Infertile as female, but fully fertile as male (fertilizing) parent. Besides those listed, 32 more *ff* mutants were obtained (1000). These are not listed separately because they are unmapped and lack locus numbers. These comprise 28 complementation groups. They have been characterized with respect to the position of the block in perithecial development, dominance, effects on vegetative growth, supplementability, and independence of mating type (1000). These mutants are available from FGSC. A different symbol, *fs* (*female sterile*), has been used by other workers for mutants having the same phenotype. See *fs* for a listing of additional mutants, including some that have other names and that show impaired female fertility as part of a pleiotropic syndrome.

ff-1: female fertility-1

IIR. Between *aro-1* (5%) and *un-20* (4%) (2044).
Female sterile, with no protoperithecia (2044, 2045). Glycerol utilization is enhanced. The carbon source affects vegetative expression (2098). Aconidial on liquid

glycerol medium. Conidiation is good, but aerial hyphae are reduced and surface growth is heavy on sucrose minimal slants. Growth on glycerol reduces levels of both pyruvate dehydrogenase and dihydrolipoyl transacetylase (418). Conveniently scored by failure to produce protoperithecia on small slants of synthetic cross medium (7 days, 25°C). Allele T30 was originally recognized as female-sterile and called *ff-1* (2044, 2045). Allele JC744 was first characterized by glycerol utilization and called *glp-3* (418).

ff-2: female fertility-2

Unmapped.
No protoperithecia. Normal vegetative morphology (880).

ff-3: female fertility-3

IR. Right of *os-1* (3%) (368).
Defective in protoperithecial production. Abnormal morphology. Found in strain T22, which also contains *ty-3* and *ty-4*. Not allelic to *ty-3* or *T* (368, 881). Not tested for allelism with *so*.

ff-5: female fertility-5

IIIR. Between *pro-1* (2%) and *met-8* (1%) (2044).
Produces sterile brown protoperithecia without trichogynes. Darkens medium. Vegetatively normal. Closely resembles *ff-6*. Not allelic with *ty-1* or *ty-2* (2044).

ff-6: female fertility-6

IIIR. Linked near *ty-1* (880).
Produces many large black protoperithecia, but no perithecia are formed after fertilization. Black pigment is excreted into the medium (880).

fi: fissure

IVL. Between *ace-4* (10%), *cut* (4%; 10%) and *pyr-1* (10%; 15%), *pdx-1* (31%; 44%) (1122, 1582).
Produces exudate in fissures formed under the agar and on the surface. Variable expression. Best scored on minimal synthetic cross medium (2208) at 34°C, pH 6. Some isolates difficult to score (1122, 1585).

fkr-1: FK506 resistant-1

VR. Between *leu-5* (4%) and *ure-2* (13%) (119).
Resistant to inhibition by the immunosuppressant FK506 (123). Phenotypically normal in other respects, including sensitivity to cyclosporin A. Immunodetectable FK506-binding protein is present in the resistant mutant (123). Incompletely dominant to wild type in heterokaryons.

fkr-2: FK506 resistant-2

VR. Between *inl* (5%) and *cot-2* (3%) (123).
Cloned and sequenced: Swissprot FKBP_NEUCR, EMBL/ GenBank X55743, PIR S11090, GenBank NSFKBPA; EST NC4A7.
Resistant to inhibition by the immunosuppressant FK506. Phenotypically normal in other respects, including sensitivity to cyclosporin A. Lacks immunodetectable FK506-binding protein (FKBP), which functions as a peptidyl-prolyl *cis–trans* isomerase (EC 5.2.1.8). Incompletely dominant to wild type in heterokaryons (123).

fl: fluffy

IIR. Between *ace-1*, *eas* (9%) and *mus-23* (1%, 5%), *trp-3* (3%) (85, 1592). In a common cosmid with *mus-23* (85).
Cloned and sequenced: EMBL/GenBank AF022648; Orbach-Sachs clones X24A11, G15G05.
Encodes a developmentally regulated transcription factor (C6 zinc cluster protein) required for normal conidial morphogenesis, resulting in mycelia devoid of macroconidia (Fig. 28). Mutant alleles have been sequenced (85). Development of macroconidia is blocked in null mutants (1199, 1958). Because *fluffy* strains are aconidiate and highly fertile (1202), they are used routinely as female parents in tests for mating type and for chromosome rearrangements (1558, 1595, 1606). Produces few microconidia when dry, but when wetted produces sufficient microconidia to have been used in early irradiation and mutation studies (1204, 1787). Large numbers of microconidia can be obtained under certain conditions (1744, 2120, 2219). Double mutants *pe fl* (102, 1389) and *fl; dn* (1563) produce abundant microconidia; *fl; dn* (but not *pe fl*) is fertile when homozygous. Photograph of microconidial formation (1510); see also ref. (1744). SEM photograph (1958). Nuclear numbers in microconidia (102, 133, 902). Mutant allele *fl^Y*, which has highly conservative missense mutations (85), is exceptional in producing macroconidia, and these are yellow (1594). Cell-wall analysis (403). Immunoelectrophoretic pattern (1539). Paradoxical high alcoholic glycolysis on nitrate medium (159). Deficiency of isocitrate lyase on acetate medium (2117). Used to study the regulation of *al-1*, *al-2*, and *al-3* (1188). When *fl* strains of opposite mating type are inoculated separately on crossing medium in plates, a double line of perithecia forms where they meet, similar to the barrage in *Podospora* (776, 780) (Fig. 37). *fl* ascospores show high spontaneous germination (1546, 1673, 2201). Allele C-1835 was called *acon* (1409, 1585).

fld: fluffyoid

IVR. Left of *his-5* (2%) (1928). Near *arg-14* (2255).
Resembles *fl* in appearance, usually forming no macroconidia (1585) or only minor constrictions (1958).

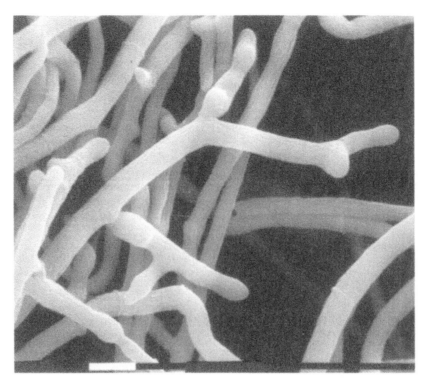

FIGURE 28   The mutant *fl*: *fluffy*. Bar length = 10 μm. Scanning EM photograph from M. L. Springer.

Capable of producing conidia when dessicated or starved, however (573) (Fig. 29).

flm-1: flame-1

An *os-1* allele.

flm-2: flame-2

An *os-4* allele.

fls: fluffyish

IR. Between *nit-1* (5%; 19%) and *al-1* (6%; 19%) [P. St. Lawrence, cited in refs. (1546) and (1548)].

Cultures are initially aconidiate, resembling *fl*, but are capable of conidiating later. Vegetative growth is slow (1546, 1958). Shows a suboptimal growth response to methionine (1548). Called *un(STL6)*.

fmf-1: female and male fertility-1

I. Between *mat* (2%; 15%) and *cr-1* (2%). Linked to *arg-1* (1003).

When *fmf-1* is present in either parent, male or female, perithecial development is blocked 15 hr after fertilization, prior to meiosis. Perithecia attain only 40% normal

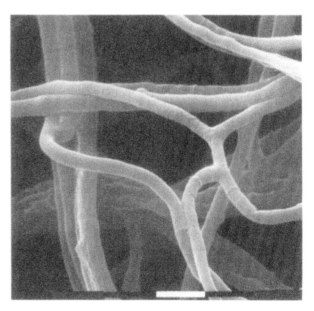

FIGURE 29   The mutant *fld*: *fluffyoid*. Bar length = 10 μm. Scanning EM photograph from M. L. Springer.

diameter. Recessive in heterokaryons. Can be crossed as one component of a heterokaryon, either as female or as male. Female fertility is also restored in mixed-mating-type (*fmf-1 A + fmf⁺ a*) heterokaryons that are homokaryotic for *tol* (1003). Called PBJ6 (1002).

for: formate

VIIR. Between *wc-1* (6%) and *frq* (2%), *oli* (5% or 6%), *dr* (3%), *un-10* (6%) (1094, 1228, 1585, 1604)

Cloned and sequenced: Swissprot GLYC_NEUCR, EMBL/GenBank, M81918, PIR A42241, GenBank NEUSERHMT; EST NC5A6; pSV50 clone 31:5E.

Structural gene for cytosolic (but not mitochondrial) serine hydroxymethyltransferase (EC 2.1.2.1) (260, 412, 1308) (Fig. 45). Expression is subject to cross-pathway control by *cpc-1* (1308). The mutant grows well on formate (0.3 mg/ml, autoclavable) or formaldehyde and has increased formyl THF synthetase, methylene THF dehydrogenase, isocitrate lyase, and glyoxalate aminotransferase activities (411, 412). Growth is aided slightly if glycine, histidine, or choline is added to formate (843). The *for* mutant also responds to supplements that either spare or supply methylene tetrahydrofolate. Will grow on a mixture of methionine and adenine and suboptimally on adenine alone (843). In auxanograms, *for* responds weakly to adenine or ascorbate alone, but strongly to adenine plus either histidine, tryptophan, or ascorbate [for possible rationale, see ref. (1472)]. Serine hydroxymethyltransferase purification (1100). Used to examine folate storage in dormant conidia (1101) and pathways of one-carbon metabolism (989). Folic acid auxotrophs have not been found in *Neurospora*, presumably because folate is not transported into the cell.

fox-2: fatty acid oxidation-2

Unmapped.

Cloned and sequenced: EMBL/GenBank X80052, NCFOX2, PIR S54786; EST NM4H8.

Structural gene for a multifunctional β-oxidation protein involved in fatty acid degradation (664).

fpp: farnesyl pyrophosphate synthetase

Unmapped.

Cloned and sequenced: Swissprot FPPS_NEUCR, EMBL/GenBank X96944, GenBank NCFPPSGEN, PIR S71434, S71436.

Encodes farnesyl pyrophosphate synthetase (EC 2.5.1.10). Involved in membrane sterol synthesis. Not induced by light (891).

fpr: fluorophenylalanine resistant

The mapped *fpr* mutants were selected as FPA-resistant. Many mutants isolated by resistance to FPA were found to be bradytrophs (very leaky auxotrophs); most of these required one of many amino acids. The resistance, which disappears when the required growth factor is added, may arise from *trans*-inhibition of amino acid transport or by induction of the cross-pathway control response

(1059, 1768). It is not established whether the mapped *fpr* mutants are bradytrophs.

fpr-1: fluorophenylalanine resistant-1

VR. Linked to *cyh-2* (<1%) (1067, 2230).

Resistant to *p*-fluorophenylalanine and 4-methyltryptophan. Isolated as resistant to FPA in the presence of *su(mtr)-1*. Resistance is recessive in a heterokaryon. Used in a mutagenicity test system (1234). Suppressed by several *lys* and *arg* auxotrophs, which apparently give the double mutant greater sensitivity to FPA with no increase in uptake (1067). Scored on solid medium containing 10 μg/ml FPA or 60 μg/ml MT (1057).

fpr-3: fluorophenylalanine resistant-3

IIIR. Linked to *trp-1* (<1%), *thi-2* (5%) (1057).

Resistant to FPA but not to 4-methyltryptophan. Not resistant in the presence of indole. Amino acid uptake is normal through both transport systems I and II, as defined in ref. (1057). Isolated in *su(mtr)*. Scored on solid medium containing 10 μg/ml FPA (1057), added before autoclaving.

fpr-4: fluorophenylalanine resistant-4

VR. Right of *inl* (11%) (1057).

Resistant to *p*-fluorophenylalanine and 4-methyltryptophan. Isolated in *su(mtr)*. Not tested for amino acid uptake. Scored on solid medium containing 10 μg/ml FPA or 60 μg/ml MT (1057).

fpr-5: fluorophenylalanine resistant-5

I. Left of *al-2* (25%) (1057).

Resistant to *p*-fluorophenylalanine but not to 4-methyltryptophan. Isolated in wild type. Not tested for amino acid uptake. Scored on solid medium containing 10 μg/ml FPA (1057).

fpr-6: fluorophenylalanine resistant-6

VIR. Between *pan-2* and *trp-2* (477).

Resistant to *p*-fluorophenylalanine. The only allele, UM300, was found as a variant unable to take up arginine to satisfy the requirement of *arg* mutants. This blockage is manifest mainly when ammonium is in the medium. Uptake of many other metabolites (amino acids, uridine, sugars) is also affected. The primary defect is unknown (451, 477). FPA resistance is suppressed by *am* (1335). Not tested for allelism with *cpc-1* or *mod-5*, which both map in the same area and cause increased rather than decreased uptake. Called UM300, *fpr(UM300)*.

fr: frost

IL. Between *ro-10* (18%) and *T(OY330)^L*, *un-5* (6%) (1578, 1582).

Cloned and sequenced: EMBL/GenBank AB021703; Orbach-Sachs clone X18G08 (1946a).

Delicate multiple hyphal branching on an agar surface (719) (Fig. 30). Aerial growth is coral-like, with no blastoconidia (1548, 1955) (Fig. 31). Deficient in G6PD (EC 1.1.1.49) (as are *col-2* and *bal*) (1854, 1857). Addition of Ca²⁺ almost completely corrects the abnormal branching (533). Partially deficient in linolenic acid (241). Morphology is partially corrected by exogenous linolenic acid (1743, 1847). Cell-wall peptides are reduced in amount (2249). Low adenylate cyclase activity and low cyclic AMP (1847, 1855). Used to determine what functions are controlled by cAMP (1524). Unlike *cr-1*, the morphology of *fr* is not corrected by exogenous cyclic nucleotides (1743, 1856). Morphology was reported to be corrected by theophylline (1847), but in another study (1743) no correction by phosphodiesterase inhibitors was seen. Cell-wall analysis; photograph (235, 507, 1851). Recessive in duplications (1580). Female sterile.

frq: frequency

VIIR. Between *wc-1*, *for* (3%) and *oli* (<2%), *un-10* (1094, 1228, 1309).

Cloned and sequenced: Swissprot PER_NEUCR, PIR S04653, EMBL/GenBank U17073; pSV50 clones 6:5A, 8:3B, 26:2A, 30:1A (55).

A series of *frq* alleles result in altered periods in the circadian rhythm cycle of conidiation. For reviews, see refs. (144), (148), (558)–(560), (562), (617), (618), (620), (1130), (1224), and (1225) (Fig. 33). *frq* mutants that are incompletely dominant and that change period length are the following: *frq¹*, 16.5 hr; *frq²*, 19.0 hr; *frq³*, 24.0 hr; *frq⁴*, 19.0 hr; *frq⁶*, 19.0 hr; *frq⁷*, 29.0 hr, and *frq⁸*, 29.0 hr (561). The recessive allele *frq⁹* shows erratic periodicity; strains carrying this allele have been critical for understanding the gene's function (1309, 1320, 1321). Scoring is accomplished by zonation in growth tubes or plates using strains that contain *bd* and preferably *csp-1* or *csp-2* (618, 1130) (Fig. 32). The presence of *csp* shortened the period length by about 1 hr in one strain tested (529). Temperature compensation and interactions with other loci have been described (617, 711–713). FRQ appears to function in transcription (561), and nuclear localization is essential for function (1236). The levels of *frq* transcript and FRQ polypeptide show circadian cycles, and this cycling is central to establishing the circadian rhythm (57, 561). Multiple forms of the gene product result from alternative translation initiation sites and time-specific phosphorylation (709, 1214, 1434). The mechanism of temperature resetting has been analyzed (1215).

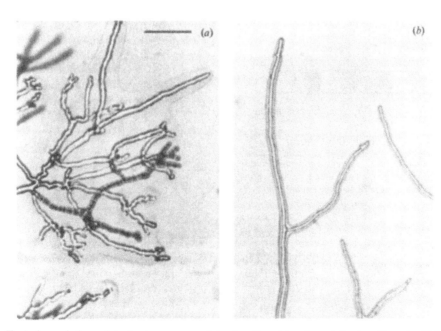

FIGURE 30 Typical morphology of the *frost* mutant on minimal medium (a) and on minimal medium plus 500 mM Ca²⁺ (b). The calcium supplement effectively restores wild-type morphology. Bar length = 50 μm. From ref. (533), with permission from the Society for General Microbiology.

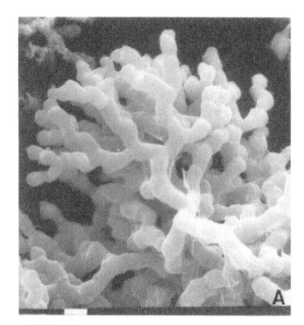 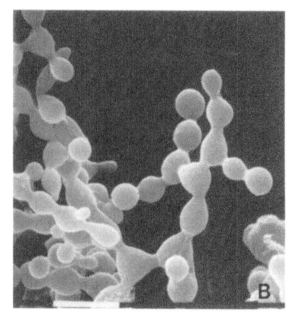

FIGURE 31 (A) Aerial hyphae of the mutant *fr: frost.* (B) Conidiating wild type, for comparison, showing the formation of typical blastoconidia. Bar length = 10 μm. Scanning EM photographs by M. L. Springer. From refs. (1957) and (1958), with permission from Cold Spring Harbor Laboratory Press.

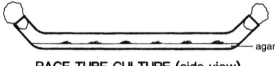

RACE TUBE CULTURE (side view)

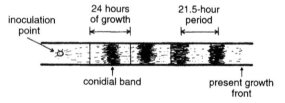

RACE TUBE CULTURE (top view)

FIGURE 32 Diagram of a race-tube culture of the *band* mutant, expressing a circadian rhythm of conidiation. Dark areas are bands of dense conidiation, which occur at subjective dawn in the cycle. Upper panel: Side view of a culture that has completed growth to the end of the tube. Raised areas represent regions of conidiation at circadian intervals. Lower panel: Top view of a culture that has grown as far as the present growth front. The position of the growth front (vertical lines on the tube) is marked at 24-hr intervals, in red safelight. Because the linear growth is nearly constant, the position of the center of the conidial band relative to the growth-front marks can be used to determine the time of formation of each conidial band. The period length then is given by the number of hours between the centers of successive bands. Adapted from ref. (618), [Feldman, J. F. (1983). *BioScience* 33: 426–431], with permission, © 1983 American Institute of Biological Sciences.

Until *frq* was cloned and sequenced, the possibility remained that alleles with different period lengths represented a cluster of different genes. Once allelism was established, the original symbols *frq-1, frq-2,* etc. were changed to frq^1, frq^2, etc. To avoid possible confusion, the symbol *frq* is reserved for this locus (621). However, mutations at numerous loci other than *frq* have been shown to affect the establishment of the rhythm [e.g., *wc-1* and *wc-2*; (431)] or the period length (e.g., *chr* and *prd*). Also, period length and/or temperature compensation are affected by mutations at loci initially recognized on the basis of quite different characteristics, including *cel, chol-1, cys-9, fas,* and *phe-1.* For a more inclusive list of such loci, see references (558), (561), (1130), and (1133). A mutant called *cla* proved to be frq^7 (397).

frq-5: frequency-5

Changed to *prd-1.*

fs: female sterile

Infertile as female but fully fertile as the male (fertilizing) parent. No or few functional perithecia are produced. *ff (female fertility)* and *pp (protoperithecia)* have been used by other workers to name mutants having a similar phenotype. Female fertility may also be impaired or absent in certain mutants that were named for other traits, e.g., *cyt-1, -2, erg, fr, glp-3, gul-3, -4, leu-1, R, ro, sk, so, ssu, ty-1, -2, var-1.* Many female-sterile mutants

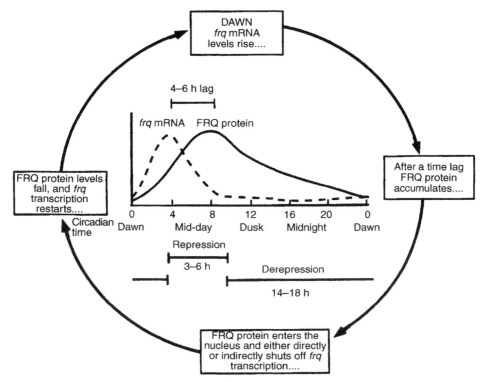

FIGURE 33 The *Neurospora frq* feedback loop. Circadian time (CT) is a formalism used to compare the properties of circadian rhythms from organisms or strains with different endogenous periods, whereby the period is divided into 24 equal parts with each part defined as 1 circadian hour. By convention, CT0 represents subjective dawn and CT12 represents subjective dusk. From ref. (144), where it was adapted from ref. (148) [Bell-Pederson, D., N. Garceau, and J. J. Loros (1996). Circadian rhythms in fungi. *J. Genet.* 75: 387–401.], with permission. Copyright Indian Academy of Sciences, Bangladore.

affect vegetative morphology or growth rate. Tests show that different *female sterile* mutants are blocked at different points in perithecial development (1000). Female sterility appears to have no genetic or functional relationship to mating type (1000). Crosses homozygous for any of the listed *fs* genes can be made and progeny can be obtained by using a heterokaryon of marked *fs* and *fs+* strains as the female parent (1423, 1424); the same is true for most of the mutants called *ff* (1000). Numerous additional female-sterile mutants have been isolated that have not been mapped and/or tested for allelism with the mutants listed here (176, 518, 920, 1000).

fs-1: female sterile-1

I or II. Linked to *T(4637) al-1*. Unlinked to *mat* (1423).

Perithecia are absent or infrequent when the mutant is used as female. Fertile as male. Vegetative growth is somewhat stringy, slightly slower, and paler than wild type. Recessive in heterokaryons. Complements *fs-2, -3, -4, -5, -6, -n* (1423, 1424). Shown in crosses to be nonallelic with *fs-2, -3, -5, -n* (1423).

fs-2: female sterile-2

II. Probably linked to *fs-1* (1423), *cot-5* (14/48) (1582).

No perithecia are formed when the mutant is used as female. Fertile as male. Morphology is abnormal, somewhat colonial. Grows at 25°C, but not 34°C. Recessive in heterokaryons. Complements *fs-1, -3, -4, -5, -6, -n* (1424)

fs-3: female sterile-3

IL. Left of *mat* (16%) (1423).

No perithecia are formed when the mutant is used as female. Fertile as male. Vegetative growth is slightly slower and paler than wild type. Recessive in heterokaryons. Complements *fs-1, -2, -4, -5, -6, -n* (1423, 1424).

fs-4: female sterile-4

I? Linked to *mat* (22%) (1423), but may be inseparable from a chromosome rearrangement (1424).

No perithecia are formed when the mutant is used as female. Fertile as male. Complements *fs-1, -2, -3, -5, -6, -n.*

Vegetative growth is slightly slower and paler than wild type. Recessive in heterokaryons (1423, 1424).

fs-5: female sterile-5

I or II. Probable loose linkage with *fs-1* (1423).

Perithecia are absent or rare when *fs-5* is used as female. Some strains produce occasional perithecia and ascospores. Fertile as male. Grows slowly, with aerial growth near the agar surface. Cultures turn brown with age (1423). Complements *fs-1, -2, -3, -4, -6, -n* (1424).

fs-6: female sterile-6

I or II. Linked to *T(4637) al-1*. Unlinked to *mat* (1423).

No perithecia when *fs-6* is used as female. Fertile as male. Vegetative growth is slightly slower and paler than wild type. Recessive in heterokaryons. Complements *fs-1, -2, -3, -4, -5, -n* (1423, 1424).

fs-n: female sterile-n

I. Linked to *mat* (35%; 45%) and to *T(4637) al-1* (1423).

No perithecia when *fs-n* is used as female. Fertile as male. Vegetative growth is slightly slower and paler than wild type. Recessive in heterokaryons. Three complex segregations in 51 asci suggest two closely linked genes. If so, they do not complement each other, although they complement *fs-1, -2, -3, -4, -5, -6* (1423, 1424).

Fsp-1: Four spore-1

IIR. Right of *pe* (4%) (1546).

Some of the asci contain four large ascospores rather than the normal eight (Fig. 34). In these asci, ascospores are formed at the four-nucleate stage following meiosis II. The mutant allele is dominant with variable penetrance, depending on genetic background. Ascospores from four-spored asci produce homokaryotic cultures. Some large, heterokaryotic ascospores are found in rare three-spored or two-spored asci. One postmeiotic mitosis is omitted in the four- and three-spored asci, and two divisions are omitted in the two-spored asci. Vegetative morphology is normal (1675). Expression is temperature-sensitive. Asci are four-spored at 25°C and above (1677). Heterokaryotic large ascospores in three-spored asci have been used in the study of *Spore killer* to rescue segregants that would not survive as homokaryotic haploids (1674). The developmental basis of the *Four-spore* mutants of *Neurospora crassa* is distinct from that of the normally four-spored species *Neurospora tetrasperma* [for a dominant eight-spored mutant of *N. tetrasperma*, see ref. (272)].

Fsp-2: Four spore-2

IR. Right of *nic-2* (8%) (1677).

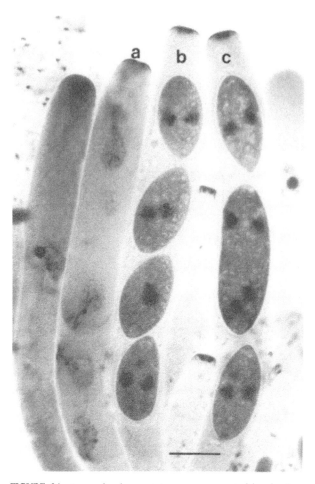

FIGURE 34 Ascus development in a cross parented by the *Four spore* mutant *Fsp-1*. (a) Interphase II. The four nuclei with their duplicate spindle–pole–body plaques have failed to undergo mitosis but have lined up on one side of the ascus and initiated ascospore delimitation. (b) An ascus showing four binucleate spores after a single mitosis. No additional mitoses occur after the ascospores become mature. (c) The middle large spore in this three-spore ascus enclosed two of the four meiotic products. Hemotoxylin staining. Bar length = 10 μm. Photographs by N. B. Raju, from ref. (1675).

In crosses heterozygous for *Fsp-2*, nearly all asci are four-spored at 16°C, eight-spored at 25°C (518). Crosses homozygous for *Fsp-2* make four-spored asci at 16 and 20°C, but a mixture of four- and eight-spored asci at 25°C. The developmental basis is similar to that of *Fsp-1*. No asci contain more than four ascospores at any temperature when both *Fsp-1* and *Fsp-2* are present in a cross, either homozygous or heterozygous (1677). The double mutant has been used to reduce significantly the effort required for tetrad analysis (1602, 1888).

Fsr: 5S rRNA

Almost all of the approximately 100 genes that specify 5S rRNA are widely dispersed in the genome (672, 1893).

Sequences flanking the individual 120-bp 5S regions are mostly divergent, but they share an upstream consensus sequence that is required for transcription (1889). Many 5S genes have been partially or completely sequenced, and over 20 have been mapped using RFLPs in their flanking sequences as markers. Most of the 5S genes are of the same sequence type, α, but other isotypes are found, with as many as 15 differences in the 120-bp 5S region (1334, 1339). The variant 5S rRNAs are present in the ribosomes (1892). The symbol *Fsr* is used both for genes and for pseudogenes.

Fsr-1: 5S rRNA-1

IR. Linked to *al-2* (0T/18 asci; 0/18), *un-7, met-6, cr-1* (0/18) (1447). Adjacent to *ad-9* (573).
Cloned and partially sequenced (1339).
Specifies α-type 5S rRNA (1339).

Fsr-3: 5S rRNA-3

IIR. Between *preg* (5T + 2 double crossovers/18 asci), *vma-2* (1T/18 asci) and *eas* (1T/18 asci), *trp-3* (16+1 double crossover/41), *Fsr-34, Fsr-21* (4/41) (1339, 1447).
Cloned and partially sequenced (1339).
Specifies α-type 5S rRNA (1339).

Fsr-4: 5S rRNA-4

IVR. Linked to *Tel-IVR* (1T/18 asci), *chs-2, uvs-2* (0/17) (1447). Near *Tel-IVR* by inclusion in duplications from quasiterminal translocation *T(ALS159)* and by linkage to *T(T54M50)* and *T(OY337)* (1/23) (1339).
Cloned and sequenced: EMBL/GenBank J01247, GenBank NEURG5P.
A 5S rRNA pseudogene (ψ4); apparently not transcribed (1887).

Fsr-9: 5S rRNA-9

V. Linked to *Cen-V, lys-1* (1T/18 asci), *Fsr-16* (0T/14 asci; 0/18) (1447).
Cloned and sequenced (1339).
Specifies β-type 5S rRNA (1339).

Fsr-12: 5S rRNA-12

IL. Linked to *nit-2* (0 or 1/16), *vma-4* (0T/18 asci). Left of *mat* (4T/18 asci) (1447).
Cloned and sequenced: EMBL/GenBank K02478, GenBank NEURGAD.
Specifies δ-type 5S rRNA. [Possibly a pseudogene (1339, 1893).]

Fsr-13: 5S rRNA-13

IVR. Right of *trp-4* (0T/18 asci). Linked to *vma-8* (0T/18 asci) (1447). Between *pyr-1* and *cot-1* (1339).

Cloned and sequenced: EMBL/GenBank K02477, GenBank NEURGABP.
Specifies β'-type 5S rRNA (1339, 1893).

Fsr-16: 5S rRNA-16

V. Linked to *Cen-V, lys-1* (1T/14 asci), *Fsr-9* (0T/14 asci; 0/18) (1447).
Cloned and sequenced: EMBL/GenBank K02470, GenBank NEURGAAB.
Specifies α-type 5S rRNA (1339, 1893).

Fsr-17: 5S rRNA-17

IIR. Between *eas* (6T, 1NPD/18 asci), *trp-3* (1/16) and *leu-6* (0/16; 9T, 1NPD/18 asci) (1447).
Cloned and sequenced: EMBL/GenBank K02947, K02948, GenBank NEURGAAH, NEURGAAI.
Specifies α'-type 5S rRNA (1339).

Fsr-18: 5S rRNA-18

IR. Linked to *nic-1* (0/18), *arg-13* (2T/18 asci) (1447).
Cloned and partially sequenced (1339).
Specifies α-type 5S rRNA (1339).

Fsr-20: 5S rRNA-20

VR. Right of *ro-4* (1/18) (1447).
Cloned and sequenced: EMBL/GenBank K02479, GenBank NEURGAG.
Specifies γ-type 5S rRNA (1339, 1893).

Fsr-21: 5S rRNA-21

IIR. Between *arg-12* (1/17) and *trp-3* (1/17; 4/41) in the order *preg Fsr-3 Fsr-21 trp-3*. Linked to *Fsr-34* (0/41; 0/18) (1339, 1447).
Cloned and sequenced: EMBL/GenBank NEURGAAE, K02473, also NEURGAAJ, K02949 (OR strains), and NEURGAAK, K02950 (BT strains).
Specifies α-type 5S rRNA (1339, 1893). An allelic difference is present between strains of Beadle–Tatum (BT) and Oak Ridge (OR) ancestry.

Fsr-30: 5S rRNA-30

II.. Linked to *mat* (0T/18 asci) (1447).
Cloned and partially sequenced (1339).
Specifies 5S rRNA of uncertain type (1339).

Fsr-32: 5S rRNA-32

IIL. Between *pyr-4* (10/41), *Fsr-52* (8/41; 4T/18 asci) and *Cen-II* (1T/18 asci), *arg-5* (2/41) (1339, 1447).

Cloned and sequenced: EMBL/GenBank NEURGAAF, K02474, also NEURGAAL, K02951 (OR strains), and NEURGAAM, K02952 (BT strains).

Specifies α-type 5S rRNA (1339, 1893). An allelic difference is present between strains of Beadle–Tatum (BT) and of Oak Ridge (OR) ancestry.

Fsr-33: 5S rRNA-33

IR. Linked to lys-4 (0 or 1T/18 asci). Right of his-2 (1 or 2T/18 asci) (1447).

Cloned and sequenced: EMBL/GenBank NEURGATH, M29278 (θ), NEURGAZE, M11398 (ζ–η).

In Oak Ridge and related laboratory wild types, the Fsr-33 locus is occupied by a duplication that includes two heavily methylated pseudogenes (types ζ and η) that differ by numerous transition mutations (1886, 1890). This region was the first in Neurospora to be recognized as methylated. In numerous wild-collected strains, in wild type Abbot 4A, and in other Neurospora species, the locus contains a single Fsr gene of type θ (766, 1891). Sequence comparisons indicate that ζ and η originated by tandem duplication of an 0.8-kb segment that included θ and that this was followed by ~270 GC-to-AT mutations that are characteristic of RIP (766).

Fsr-34: 5S rRNA-34

IIR. Between eas (1T/18 asci) and trp-3 (1/18; 4/41). Linked to bli-4 (0T/18 asci), Fsr-21 (0/59) (1339, 1447).

Cloned and partially sequenced (1339).

Specifies α-type 5S rRNA (1339).

Fsr-38: 5S rRNA-38

Unmapped.

Cloned and sequenced: EMBL/GenBank K02475, GenBank NEURGAAG.

Specifies α-type 5S rRNA (1893).

Fsr-45: 5S rRNA-45

IIIR. Between ro-2 (3/18) and cat-3 (2T, 1NPD/18 asci), Tel-IIIR (3T, 1NPD/18 asci) (1447).

Cloned and sequenced: EMBL/GenBank K02471, GenBank NEURGAAC.

Specifies α-type 5S rRNA (1339, 1893).

Fsr-50: 5S rRNA-50

VI. Linked to Bml (0/18), Cen-VI, pan-2 (0T/18 asci) (1447). Between ylo-1 and trp-2 (1339).

Cloned and sequenced (1339).

Specifies α-type 5S rRNA (1339).

Fsr-51: 5S rRNA-51

IVR. Between arg-14 (2/18), trp-4 (2T/18 asci) and cot-1 (1/18). Linked to bli-3 (0T/15 asci) (1447).

Cloned and sequenced: EMBL/GenBank K02476, GenBank NEURGAB.

Specifies β-type 5S rRNA (1339, 1893).

Fsr-52: 5S rRNA-52

IIL. Between pyr-4 (2/41), Fsr-32 (3/18; 8/41; 4T/18 asci), Fsr-54 (8/41) and Cen-II (5T/18 asci), arg-5 (10/41) (1339, 1447).

Cloned and sequenced: EMBL/GenBank K02469, GenBank NEURGAAA.

Specifies α-type 5S rRNA (1339, 1893).

Fsr-54: 5S rRNA-54

IIL. Between pyr-4 (2/41), Fsr-52 (8/41), Fsr-32 (3/18; 0/41) and arg-5 (10/41) (1339, 1447).

Cloned and partially sequenced: EMBL/GenBank K02472, GenBank NEURGAAD.

Specifies α-type 5S rRNA (1339, 1893).

Fsr-55: 5S rRNA-55

IIR. Between Cen-II, arg-5 (1T/18 asci; 2/39), con-6 (1/18) and preg (4/39; 1T+1 double crossover/18 asci). Linked to cya-4 (0T/18 asci) (1339, 1447).

Cloned and partially sequenced (1339).

Specifies α-type 5S rRNA (1339).

Fsr-62: 5S rRNA-62

IVR. Linked to Fsr-63 (0/18), mtr (0/18), arg-14 (1/18) (1447).

Cloned and partially sequenced (1339).

Specifies β-type 5S rRNA (1339).

Fsr-63: 5S rRNA-63

IVR. Between Cen-IV (2T/15 asci) and pyr-1 (1T/18 asci), mtr (0/18), arg-14 (1/18). Linked to Fsr-62 (0/18) (1447).

Cloned and sequenced (1339).

In Abbott 12a and Abbott 4A strains, the locus is occupied by an α-type 5S rRNA gene. In Oak Ridge strains, however, the 5S rRNA gene is interrupted by a 1.9-kb transposon, Punt. The copy of Punt in Fsr-63 and the surrounding 5S rDNA sequences are methylated and have the earmarks of RIP. This pseudogene is called ψ_{63} (1266).

fz: fuzzy

Unmapped.

Morphology is abnormal. Identified as one component of the multiple mutant combination that results in the cell-wall-less "slime" phenotype (607).

G: Gulliver

Changed to *gul: gulliver*.

Gamma-tubulin

See *tbg*.

gap: gap

IL. Between *mat* (6%) and *Cen-I* (4%) (1200).
Conidia are formed in a few scattered clusters on long hyphae otherwise devoid of conidia. Photograph (1200). (Stock lost.)

gch-1: GTP cyclohydrolase-1

Unmapped.
Cloned and sequenced: Swissprot GCH1_NEUCR, EMBL/GenBank Z49758, GenBank NCGTPCHI.
Encodes GTP cyclohydrolase I (EC 3.5.4.16) (GTP-CH-I) (1253).

gdh-1: glutamate dehydrogenase-1

VIL. Linked to *tom22* (0T/16 asci), *nuo19.3* (9T/18 asci) (1447). Apparently not linked in III as originally reported (1027).
Cloned and sequenced: EMBL/GenBank L20497, PIR S66039, EMBL NCNADSPEC, GenBank NEUNADSPEC.
Structural gene for NAD-specific glutamate dehydrogenase (EC 1.4.1.2) (1027) (Fig. 49).

gla-1: glucoamylase-1

Unmapped.
Cloned and sequenced: Swissprot AMYG_NEUCR, EMBL/GenBank X67291, GenBank NCGLA1, PIR S36364, S13710, S13711; EST NM6B8.
Structural gene for the major extracellular glucoamylase (EC 3.2.1.3) by homology and by RIP knock-out. Mutants have reduced starch halos and reduced growth on starch as the sole carbon source (2005). Residual glucoamylase in the null mutant may be due to *gla-2*.

gla-2: glucoamylase-2

IL. Linked to *mat* (3/68), *sor-4* (≤ 10 kb), *his-3* (3/65) (576).
Cloned and partially sequenced: Orbach-Sachs clone X10E5 (576).

Structural gene for extracellular glucoamylase (EC 3.2.1.3) by homology. Little sequence homology to *gla-1*. Homologous to a glucoamylase from *Schwanniomyces occidentalis* (576).

glm: glutamine

Symbol changed to *gln*.

gln-1: glutamine-1

VR. Linked to *inl* (2%, probably to the right) (1707).
Requires glutamine (1707). Probably a glutamine synthetase structural gene (447, 1376, 1781) (Figs. 49 and 50). Sensitive to chlorate on both ammonium and glutamate; resistant to chlorate on glutamine (565). Inactivation by oxidation (8); alterations in activity during development are associated with hyperoxidation (2097). NADPH-nitrate reductase, NAD(P)H-nitrite reductase, and uricase are freed from repression by ammonium or glutamate but not by glutamine in the *gln-1ᵃ* mutant (564, 567, 1648, 2184). A *gln-1ᵇ* allele, defective in the β-subunit, is more derepressed than mutant R1015 (*gln-1ᵃ*), which is defective in the α-subunit (565, 1650). Glutamine synthetase can be found in an octameric form composed of either β-subunits exclusively or β-subunits plus α-subunits and in a tetrameric form consisting exclusively of α-subunits (1376). The α- and β-subunits are encoded by different mRNAs (1140). The enzyme is allosterically inhibited by alanine, glycine or serine, and a β-subunit mutant sensitive to glycine has been isolated (873). For interaction with *am*, see ref. (939). Dominant in heterokaryons (446). Regulated by *nit-2* and *nmr-1* (269). The *gln-1bR8* strain has been used to obtain a mutant with altered L-amino acid oxidase regulation (271). Called *glm* (1707).

gln-2: glutamine-2

Unmapped. Unlinked to *gln-1*.
Affects glutamine synthetase α-polypeptide (270). Transferase activity drastically is reduced and that of glutamine synthetase is slightly reduced. Resistant to α-methyl-DL-methionine-*SR*-sulfoximine, an inhibitor of glutamine synthetase activity (270).

glp: glycerol phosphate

Mutants with altered ability to use glycerol as a carbon source. For pathways of glycerol utilization in various organisms, see the diagram in ref. (2160) or (2098). Scored in slants on minimal synthetic cross medium (2208) with 2% glycerol vs 2% sucrose as the carbon source (2160). The poor growth of wild type on glycerol medium is markedly improved by adding 0.5%

L-asparagine and 100 µg/ml ascorbic acid (361). Symbol changed from *gly*.

glp-1: glycerol phosphate-1

IR. Probably between *ad-9* (2%) and *nit-1* (11%) (890, 1478).

Unable to use glycerol as the sole carbon source (1478). Can use dihydroxyacetone or glyceraldehyde (525). Probably regulatory. Deficient in inducible glycerol kinase under normal conditions (890, 1481); wild-type levels of normal enzyme are induced by cold or by deoxyribose with some, but not with all, alleles (525, 890). Glycerol transport is normal (525). Fine-structure map (526). Scored on slants of minimal synthetic cross medium (2208) with 2% glycerol vs 2% sucrose as the carbon source (2160). Poor growth of wild type on glycerol medium is markedly improved by adding 0.5% L-asparagine and 100 µg/ml ascorbic acid (361). Called *gly, gly-u*.

glp-2: glycerol phosphate-2

IIR. Between *T(ALS176)*, *arg-5* (8%) and *T(NM177)^L*, *pe* (7%). Linked to *aro-3* (3%) (527, 1578).

Unable to use glycerol, dihydroxyacetone, or glyceraldehyde as the sole carbon source (525, 527). Lacks both mitochondrial and cytosolic flavin-linked glycerol-3-phosphate dehydrogenase (EC 1.1.99.5) (527). Scored on slants of minimal synthetic cross medium (2208) with 2% glycerol vs 2% sucrose as the carbon source (2160). Poor growth of wild type on glycerol medium is markedly improved by adding 0.5% L-asparagine and 100 µg/ml ascorbic acid (361). Three independent isolates are *ropy*-like in vegetative morphology but are female-fertile, unlike most *ropy* mutants (419, 527). A report of complementation groups at this locus (420) is an error (417). Fine-structure map (526). Called *gly-2*.

glp-3: glycerol phosphate-3

Allelic with *ff-1*.

glp-4: glycerol phosphate-4

VI. Between *ylo-1* (1%; 6%), *ad-1* (0; 2%) and *rib-1* (3%; 4%), *pan-2* (4%; 6%) (2160).

Unable to use glycerol as the sole carbon source (2160). Uses dihydroxyacetone or glyceraldehyde (525). Lacks both inducible and constitutive glycerol kinase (EC 2.7.1.30) (2160), but there is some doubt that these are two different enzymes [ref. (417), based on ref. (526)]. Scored on slants of minimal synthetic cross medium (2208) with 2% glycerol vs 2% sucrose as the carbon source (2160). The poor growth of wild type on glycerol medium is improved markedly by adding 0.5%

L-asparagine and 100 µg/ml ascorbic acid (361). A revertant has altered kinetic properties (526). Allele G660 originated in *Neurospora tetrasperma* and was introgressed into *Neurospora crassa* (155, 2160). Fine-structure map (526). See Fig. 18.

glp-5: glycerol phosphate-5

I. Left of *cr-1* (15%) (2160).

Unable to use glycerol as the sole carbon source. Lacks glyceraldehyde kinase (2160), but the significance of this is uncertain because of the findings of ref. (2098). Scored on slants of minimal synthetic cross medium (2208) with 2% glycerol vs 2% sucrose as the carbon source (2160). Poor growth of wild type on glycerol medium is improved markedly by adding 0.5% L-asparagine and 100 µg/ml ascorbic acid (361). Allele M1051 originated in *Neurospora tetrasperma* and was introgressed into *Neurospora crassa* (155, 2160).

glp-6: glycerol phosphate-6

V. Left of *inl* (30%) (1654).

Deficient in NAD-linked glycerol-3-phosphate dehydrogenase (EC 1.1.1.8). Called 42-94 (914, 1654).

glt: glycyl-leucyl-tyrosine resistant

Unmapped.

Unable to transport oligopeptides necessary to support the growth of specified amino acid auxotrophs (2236). Has only 10% of wild-type uptake rate (2237). (The oligopeptide uptake system transports tri-, tetra-, and pentapeptides, but not di- or higher than pentapeptides.) However, whereas a *leu-2 glt^R* strain is unable to use oligopeptides as a source of leucine when ammonium nitrate is provided as a nitrogen source, the double mutant but not the *leu-2* single mutant can use small peptides as a nitrogen source, apparently because of an induced extracellular peptidohydrolytic activity (2238). Originally obtained using *tys* by selecting mutants resistant to glycyl-L-leucyl-L-tyrosine but still sensitive to tyrosine (2236). See ref. (2232) for a review of peptide uptake.

gluc-1: β-glucosidase-1

IIIR. Linked to *dow* (10%) (579).

Activity of the thermostable aryl-β-glucosidase is reduced to 10% of wild type (582) in one allele and to <1% in a second-step mutant then called *gluc-2*, which showed 0/200 recombination with the original mutant and is probably allelic (581). Low activity is dominant in heterokaryons (1246). Selected by *p*-nitrophenylgluco-side-staining reaction (580). Scored either by breakdown of the β-glucoside esculin (0.01%) measured by fluorescence at pH 5.5 (582) or by precipitation of ferric ammonium citrate (0.1%) by esculetin (2 days, 25°C) (579).

gluc-2: β-glucosidase-2

Probably allelic with *gluc-1*.

gly: glycerol utilization

Symbol changed to *glp*.

gna-1: guanine nucleotide-α-1

IIIR. Linked to *ro-2* (0/17) (2130).

Cloned and sequenced: Swissprot GBA1_NEUCR, EMBL/ GenBank L11452, U56090, EMBL NCGPROTAA, GenBank NEUGPROTAA; Orbach-Sachs clones X9F04, G9C03.

Encodes guanine nucleotide-binding protein α-1-subunit (2130). The predicted amino acid sequence shows significant homology to the G_i family found in higher organisms. The null mutant shows female infertility, osmotic sensitivity, reduced hyphal growth rate, and defective macroconidiation (972), has decreased intracellular cAMP levels, and is more resistant to some environmental stresses than wild type (973, 2260). Mutants that are predicted to maintain an activated state of GNA-1 have increased intracellular cAMP levels, overproduce aerial hyphae, and are more sensitive to environmental stresses than wild type (2260). Null-mutant extracts have reduced Mg^{2+}-dependent adenylyl cyclase and cAMP phosphodiesterase activity. Mg^{2+}-dependent adenylyl cyclase activity in wild-type extracts can be inhibited by using antibody against GNA-1 (973).

gna-2: guanine nucleotide-α-2

VR. Linked to *inl* (0/18) (2130).

Cloned and sequenced: Swissprot GBA2_NEUCR, EMBL/ GenBank L11453, EMBL AF004846, Genbank NEUGPROTAB; EST NP3A7; Orbach-Sachs clones X23E01, X24E12.

Encodes guanine nucleotide-binding protein α-2-subunit (2130). Expressed in both vegetative and sexual stages. The null mutant has no obvious phenotype. Mutants containing predicted GTPase-deficient (activated) alleles of *gna-2* have increased aerial hyphae and reduced conidial germination, but normal female fertility. The null mutant has normal levels of Mg^{2+}-dependent adenylyl cyclase activity but reduced levels of cAMP phosphodiesterase activity (973). *gna-1, gna-2* double mutants are more impaired than *gna-1* strains in osmotic sensitivity and female fertility (80).

gna-3: guanine nucleotide-α-3

Probably IV, near *uvs-2* (199)

Cloned and partially sequenced.

Encodes guanine nucleotide-binding protein α-3-subunit (199).

gnb-1: guanine nucleotide-β-1

IIIR. Linked to *con-7*, *trp-1* (0/18) (199).

Cloned and partially sequenced.

Encodes guanine nucleotide-binding protein β-subunit. The null mutant is female-sterile and has reduced levels of GNA-1 protein under some growth conditions (199).

gpd-1: glyceraldehyde-3-phosphate dehydrogenase-1

IIR. Linked to *arg-12* (0/18) (1777), *vma-2* (1T/18 asci). Between *nuo78*, *preg*, *cit-1* (5T/18 asci) and *eas* (3T/18 asci) (1447).

Cloned and sequenced: Swissprot G3P_NEUCR, EMBL/ GenBank U56397, U67457, U56379; EST NC2H9.

Structural gene for glyceraldehyde-3-phosphate dehydrogenase (EC 1.2.1.12) (1777, 1910). RNA peaks late in the circadian night and enzyme activity peaks in late morning; gene regulation other than circadian control has not been found (575, 1910). Used in a study of phylogenetic relationships among homothallic and heterothallic species of *Neurospora* and *Sordaria* (1639). Also called *ccg-7*.

gpi-1: glucose phosphate isomerase-1

IVR. Linked to *ad-6* (10%) (1401).

Lacks glucose phosphate isomerase (phosphohexoisomerase) (EC 5.3.1.9). Grows slowly and colonially on glucose or sucrose. Unable to use fructose, but growth on glucose is stimulated by added fructose. Growth is enhanced in double mutants with either *sor-4*(T9) or *pp*. Allele T66M37 was originally called *gpi-2* (1401).

gpi-2: glucose phosphate isomerase-2

Used for *gpi-1* allele T66M37.

gran: granular

VR. Between *pab-2* (1%; 8%) and *his-6* (8%; 27%) (1582, 1592). Linked to *pl* (0/75).

Delicate granular conidiation, with conidia adherent rather than powdery (1585). Exaggerated major and minor constrictions (1958) (Fig. 35). Sparsely branched hyphae (719). Morphologically distinct from *pl*. Cell-wall peptides are reduced in amount (2249).

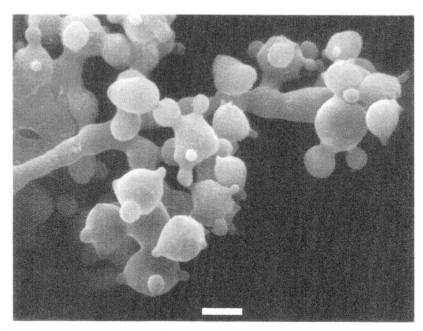

FIGURE 35   The mutant *gran: granular*, showing late budding. Bar length = 10 μm. Scanning EM photograph from M. L. Springer.

grey: grey

IVR. Linked to *cot-1* (4%) (932).
Produces gray conidia (microconidia?) in the presence of *cr-1* (932).

grg-1: glucose-repressible gene-1

Allelic with *ccg-1*.

grn: guanyl-specific RNase

Unmapped.
Sequenced: Swissprot RNN1_NEUCR, PIR NCNCT1.
Specifies guanyl-specific ribonuclease N1 (EC 3.1.27.3) (2038).

grp78: glucose-regulated protein 78

I. Linked to *mat*, *Fsr-30* (0T/18 asci) (1447).
Cloned and sequenced: EMBL/GenBank Y09011, GenBank HCHSP70GR.
Encodes a glucose-regulated and heat-shock-inducible protein of the *hsp70* family, homologous with *Saccharomyces KAR2/GRP78*. The mRNA level is low in dormant conidia and then increases at germination, low in young aerial hyphae and then increases (811). Called *hsps-2*.

gs: gamma sensitive

The symbol *gs* was used for a series of mutants sensitive to γ-radiation (1316). These were not mapped or preserved. They are listed in ref. (1596) but are not included here.

gs-1: glucan synthase-1

Allelic with *cot-2*.

gsp: giant spore

IL. Left of *mat* (10%) (1966).
Some asci contain a single giant ascospore and others have two very large ascospores or four double-size ascospores, whereas some asci contain eight ascospores. Proportions of these types vary on different crossing media. The giant ascospores have multiple germination pores. The mutant ascus phenotype is recessive. Vegetative morphology is normal, but growth is weak (1148, 1966).

gsy-1: glycogen synthase-1

Unmapped.
Cloned and sequenced: EMBL/GenBank AF056080.
Encodes glycogen synthase (EC 2.4.1.11, UDP-glucose-glycogen glucosyltransferase) (2075).

gtp-1: guanine triphosphate binding-1

Unmapped.

Cloned: Orbach-Sachs clone G11G09 (1267).

The predicted protein shows similarity to the product of *S. pombe gtp1* and vertebrate DRG ("developmentally regulated gene") protein. This class of proteins is found in organisms from bacteria (*obg* in *Bacillus subtilis*, an essential gene possibly involved in regulating DNA replication) to fungi (*gtp1* in *S. pombe*, with no obvious mutant phenotype) to animals (*DRG*, induced in developing brain tissue). Contains a GTP-binding domain shown to be critical to function.

gua: guanine

The guanosine-specific branch of the purine biosynthetic pathway diverges with the conversion of inosine 5′-monophosphate to xanthosine 5′-monophosphate. In yeast there are at least four genes encoding four distinct isozymes for this function, IMP dehydrogenase (EC 1.1.1.205). In *Neurospora*, *gua-2* definitely is blocked in this enzyme, and *gua-1* may specify an inducible isozyme. No mutant is known yet for the subsequent step, conversion of xanthosine-5′-P to guanosine-5′-P by guanosine-5′-monophosphate synthase (EC 6.3.5.2). See Figs. 5 and 62.

gua-1: guanine-1

I. Linked to *arg-3* (8%) (2278), probably between *his-2* (3%) and *cr-1* (3%) (1582).

Requires guanine. Deficient in IMP dehydrogenase (EC 1.1.1.205) (10% of wild type in allele OY301) (767) (Fig. 5). Inhibited (competitively) by adenine and by complex complete medium. Adapts phenotypically after several days and grows up on minimal or complete medium, but retains the requirement on subculture. Adenine prevents or decreases adaptation. Guanosine is preferred to guanine as a supplement because of greater solubility. Best scored at 2 and 3 days on slants of minimal + 1 mg/ml adenine vs minimal + 0.2 mg/ml guanosine (2278).

gua-2: guanine-2

IVR. Unlinked to *gua-1* (25% prototrophs in intercross) (768). Linked to *cot-1* (5%) (767).

Requires guanine. No IMP dehydrogenase activity (EC 1.1.1.205) (Fig. 5). Slow growth, poor conidiation. Does not adapt to grow on minimal medium (767, 768)

Guest (Transposable Element)

Cloned and sequenced: Orbach/Sachs cosmid G36F (2274).

A DNA transposable element discovered as a 98-bp fragment in wild-type ST74A, in the region between *his-3* and *cog*. The element is flanked by a 3-bp duplication of target sequence and has terminal inverted repeats (TIRs) similar to those of non-retrotransposon-type elements in other organisms. Multiple copies of sequences that are similar, but not identical, to the TIR are present elsewhere in the genome (2274, 2275).

gul: gulliver

This name was given to suppressors of *cot-1* that give large colonies at restrictive temperatures, where unsuppressed *cot-1* forms tiny colonies (1714). Scorable in the presence of *cot-1* at 34°C, 2 days after the transfer of small inocula to solid medium. Of 36 independent *gul* mutants, 25 were *gul-1* alleles (2079).

gul-1: gulliver-1

VR. Between *am-1* (<0.01%; <1%) and *ace-5* (<1%) (1121, 1941).

Cloned (217).

Modifies the colony size of *cot-1* at restrictive temperatures (1714, 2079). *cot-1; gul-1* colonies exceed 20 mm in diameter after 60 hr at 33°C, compared to 1 mm for *cot-1; gul-1*+ (2079). Female fertile, with viable ascospores. Dominant in heterokaryons (217). Recombination within *gul-1* is unaffected by *rec-3*, which acts on the nearby *am-1* locus (1941). Called *G* (1714).

gul-2: gulliver-2

Unmapped.

Modifier of the colony size of *cot-1* at restrictive temperatures. Phenotype similar to *gul-1* (2079).

gul-3: gulliver-3

IVR. Linked to *cot-1* (10%), *pyr-2* (7%) (2079).

Modifier of the colony size of *cot-1* at restrictive temperatures. Female sterile. *gul-3* ascospores are black but inviable. *gul*− progeny can be obtained, however, from *gul*+/*gul*− pseudo-wild disomic ascospores. Reported unable to make heterokaryons (2079).

gul-4: gulliver-4

VII. Linked to *nic-3* (17%) (2079).

Modifier of the colony size of *cot-1* at restrictive temperatures. Resembles *gul-3* (2079).

gul-5: gulliver-5

VI. Linked to *trp-2* (10%) (2079).

Modifies the colony size of *cot-1* at restrictive temperatures. Female fertile. *gul-5* ascospores are black, but inviable (2079).

gul-6: gulliver-6

Unmapped. Unlinked to *cot-1* (IVR), *inl* (VR), *nic-3* (VIIL), *gul-5* (VI), or *gul-2* (2079).

Modifier of the colony size of *cot-1* at restrictive temperatures. Said to resemble *gul-3* (2079). However, ascospore ripening and recovery from ascospores have been found to be good (1582).

hah: high aerial hyphae

VR. Linked to *inl* (13%) (1404).

Obtained as a suppressor of the colonial growth and low cAMP level of *cr-1*. High aerial hyphae are formed on agar medium. The double mutant *cr-1; hah* conidiates, but the single *hah* mutant does not (1399, 1404). See also ref. (1020).

hak-1: high-affinity potassium transporter

Unmapped.

Cloned and sequenced: EMBL/GenBank AJ009759, GenBank NCR9759.

Encodes a K^+–H^+ symporter homologous to the HAK potassium transporter of *Schwanniomyces occidentalis*. The potassium transporter is inducible when cultures are K^+-starved (842). See also *trk-1*.

ham: hyphal anastomosis

Allelic with *so* (809a).

has: hydroxamic acid sensitive

Unmapped. Unlinked to *azs* (595).

Lacks the salicylhydroxamic acid (SHAM)-sensitive respiratory pathway. Cannot produce the hydroxamate-sensitive respiratory pathway when grown in the presence of chloramphenicol. Grows slowly in the presence of antimycin A (595). The double mutant *has; azs* is unable to grow in the presence of antimycin A, whereas wild type and *has+; azs* grow well. The *has; azs* strain (called *ANT-1*: antimycin-sensitive) was used to obtain mutants resistant to oligomycin (596) and mutants deficient in succinate dehydrogenase (590).

hbs: homebase

II. Linked to *Fsr54* (0 or 1/18). Left of *con-6* (1 or 2/18), *arg-12* (3 or 4/18) (1447).

An anonymous DNA fragment discovered when it was inserted as a duplicate copy in the *am* locus (strain 77s15) (1055, 1060).

hda-1: histone deacetylase-1

Unmapped.

Cloned and sequenced (545).

The sequence resembles *HDA1* in *Saccharomyces* (545).

hda-2: histone deacetylase-2

IR. Linked to *arg-13* (2T/18 asci) (545).

Cloned and sequenced (545).

The sequence resembles *HOS2* in *Saccharomyces* and *phd1* in *S. pombe* (545).

hda-3: histone deacetylase-3

Unmapped.

Cloned and sequenced (545).

The sequence resembles *RPD3* in *Saccharomyces* (545).

helper-1

Used for the strain $a^{m1}ad$-3B *cyh-1* (FGSC No. 4564), which is able to form vigorous wild-type heterokaryons with OR-compatible mutant strains of either mating type (1564). Because *mat a* allele a^{m1} is inactive, the *helper-1* component of such a heterokaryon is a passive partner when the heterokaryon is used as parent in a cross. Only nuclei of the active partner participate in karyogamy, meiosis, and the production of sexual progeny. The *helper-1* strain has proved useful for crossing otherwise infertile or poorly fertile strains (1565), for rescuing and sheltering inviable or unstable genotypes (114, 1576), and for determining whether strains of undetermined mating type are OR-heterokaryon-compatible.

het: heterokaryon (vegetative) incompatibility

If two strains carry different alleles at one or more *het* loci, they are unable to form stable heterokaryons (715, 716). Incompatibility due to *het* genes is strictly vegetative; fertility of crosses is not reduced when parent strains are vegetatively incompatible. Vegetative incompatibility may be manifested in several ways: (a) Failure to form stable heterokaryons (715, 889), best seen using complementing auxotrophic or other forcing markers. (b) Cell death following the fusion of unlike hyphae (721) or after the microinjection of cytoplasm or extracts into unlike strains (2226). The hyphal segments involved are sealed off and die; incompatible nuclei do not migrate through septal pores into adjoining cells. Microinjection implicates proteins in the killing reaction (2213, 2226). (c) Abnormal growth, morphology, pigmentation, and cell death in colonies of meiotically generated partial diploids that are heterozygous for one or more *het* genes (1421, 1475, 1560) (Fig. 36). Duplication-generating chromosome rearrangements enable individual *het* genes to be identified, mapped, and characterized one at a time in an otherwise haploid

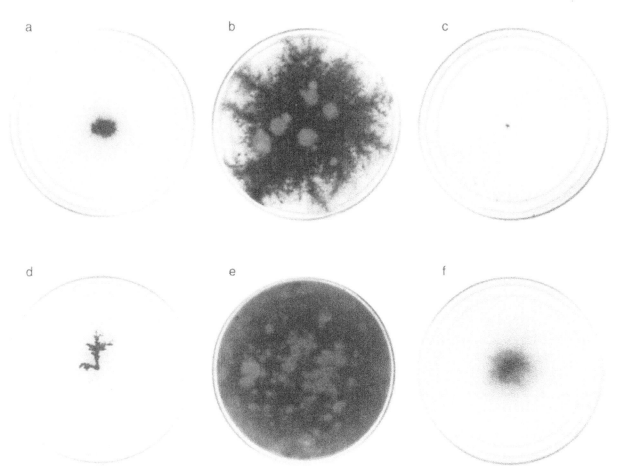

FIGURE 36 Abnormal growth of duplication (partial diploid) strains that are heterozygous at different *het* loci. Differences in morphology and severity of inhibition are apparent. (a) *mat A/mat a* from a cross heterozygous for *In(H4250)*. (b) *het-C/het-c* from *T(NM149) het-C × het-c*. (c) *het-E/het-e* from *T(T54M50) het-e × het-E*. (d) *het-8^{OR}/het-8^{AD}* from *T(T39M777) het-8^{OR} × het-8^{AD}*. (e) *het-9^{OR}/het-9^{PA}* from *T(AR209) het-9^{OR} × het-9^{PA}*. (f) *het-10^{OR}/het-10^{AD}* from *T(5936) het-10^{OR} × het-10^{AD}*. Strains were inoculated onto minimal agar medium supplemented with phenylalanine + tyrosine, without which brown pigment does not form. Pale-colored areas on the brown-pigmented surface in (b) and (e) are patches of noninhibited growth resulting from breakdown of the duplication. From ref. (1421), with permission from the Genetics Society of America. For additional photographs of *het*-incompatible heterozygous partial diploids, see refs. (1475) (*mat*) and (979) (*het-c, het-6*).

background using the abnormality of heterozygous duplication progeny as a criterion (1560, 1578). Cloning of *het* genes has also made use of the abnormality of heterozygous transformants (1346, 1934). (d) Occurrence of a barrage reaction following the confrontation of vegetatively incompatible strains (776, 780, 1597). When the interacting strains are of opposite mating type, *het* incompatibility is manifested by the appearance of two rows of perithecia separated by a clear zone within which killing occurs (Fig. 37). The barrage is seen most clearly when strains are used that do not form macroconidia.

het-c, -d, -e, and *-i* were detected in laboratory strains, using heterokaryon tests (715, 1630, 2225). Discovery of

het-5 through *het-10* was based on the abnormality of duplication progeny when a series of duplication-generating rearrangements were crossed with strains from nature (1421). *het* genes are polymorphic in natural populations of *Neurospora crassa* (1425). The mating-type idiomorphs *A* and *a* also act as *het* genes in *N. crassa* (134, 721, 1475, 1629). The *mat A/mat a* vegetative incompatibility reaction depends on the presence of a functional *tol$^+$* allele (977, 1466). Genetic differences at other loci can affect the vigor, stability, and speed of formation of heterokaryons (51, 488, 980). Dominant suppressors of the incompatibility reaction have been reported that affect one or more *het* loci (51). Rockefeller–Lindegren (RL) wild types are *het-C, het-D,*

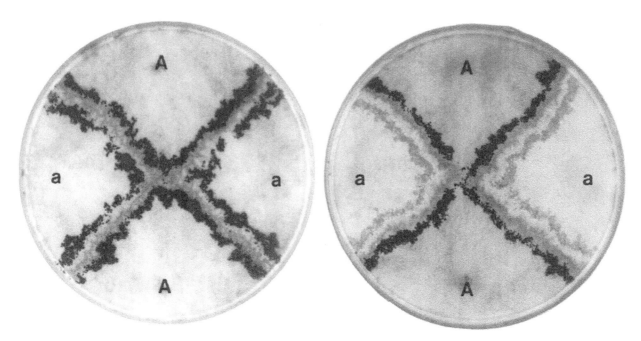

FIGURE 37   Barrage formation as a manifestation of vegetative (heterokaryon) incompatibility. Strains of opposite mating type, *mat A* and *mat a*, were inoculated to crossing medium in alternate quadrants of the petri dish at the positions marked. When unlike strains come together, a clear zone of inhibition is formed and lines of perithecia develop on each side of this barrage. Each strain acts as the maternal parent of perithecia on its own side of the barrage, and the strain on the opposite side acts as the fertilizing parent. This is shown in the plate to the right, where the *mat a* parent carried *per* allele PBJ1. The *perithecial color* mutant blocks formation of black pigment in the maternally generated perithecial wall. The nonblack perithecia, which become orange as carotenoids develop, were fully as numerous as the black perithecia opposite them (1569). Parents in these tests all carried a mutant *fluffy* allele, which improves visibility by eliminating macroconidia. Photographs from D. D. Perkins.

het-E (2225), and *het-I* (2222). St. Lawrence 74A and Oak Ridge (OR) wild types are *het-C, het-d, het-e* (2225), and *het-i* (1630, 2222). By definition, OR strains are *het*^{OR} for *het* loci other than these four (1421), e.g., *het-5*^{OR}. Tester strains are available for identifying and scoring the known *het* genes (700, 1588). The Wilson–Garnjobst testers for *het-c, -d,* and *-e* are complicated by differences at another locus or loci affecting heterokaryon formation (980, 2223). For map locations of *het* genes and the rearrangements used for testing, see Fig. 1 in ref. (1588). The inactive-mating-type helper strain $a^{m1}ad$-3B cyh-1 (779) is useful for determining whether a strain is *het*-compatible with OR strains (1564). *het* gene differences affect the transmission of mitochondrial plasmids between strains, both vegetatively (509) and in sexual interactions (508). Vegetative incompatibility is reviewed in refs. (742), (1176), and (1568).

het-c: heterokaryon incompatibility-c

IIL. Between *cot-5* (3%) and *Pad-1, pyr-4* (1%), *ro-3*. In a common cosmid with *Pad-1*. Included in duplications from *T(NM149)* and *T(AR179)*, but not from *T(P2869)* (924, 1109).

Cloned and sequenced: EMBL/GenBank L77234, EMBL NCHETC, GenBank NEUHETC; Orbach-Sachs clone G22H5.

Specifies a polypeptide that contains a hydrophobic sequence, a leucine-rich domain, and a glycine-rich domain. Allele specificity is due to a highly variable domain of about 40 amino acid residues (1797). Stable heterokaryons are not formed by strains with *het-c* alleles that differ in specificity (715, 716). Heterozygous duplications show inhibited "brown flat" morphology, spreading to cover a slant but not conidiating (1552, 1560) (Figs. 36 and 37). Two alleles found in laboratory strains were originally called *het-C* and *het-c* (715). Following the discovery and demonstration of a third allele, *het-c*^{PA}, the symbol *het-C* was changed to *het-c*^{OR} and the symbol *het-c* was changed to *het-c*^{EM} (1797). Three specificity types have been recognized and shown to be polymorphic in nature. These are designated *het-c*^{OR}, *het-c*^{GR}, and *het-c*^{PA}. (*het-c*^{EM} is functionally identical to *het-c*^{GR}.) Polymorphisms in other Neurospora species suggest that the known *het-c* alleles originated prior to divergence of the species (1797). Differences at *het-c* are more effective than those at *het-d, het-e,* or *mat* in preventing the transmission of mitochondrial plasmids between strains with different allele specificity (509). Called *c*.

het-d: heterokaryon incompatibility-d

IIR. Right of *fl* (25%) (715). Included in duplications from *T(ALS176)* (1562) and *T(OY337)* (1578).

Stable heterokaryons are not formed by strains *het-D+ het-d* (715, 716); *het-D/het-d* duplications show inhibited spreading growth on slants, with fine subsurface hyphae and no conidia. These are distinguishable from *het-C/het-c* duplications, which have coarser texture (1562).

het-e: heterokaryon incompatibility-e

VIIL. Between *cya-8*, *T(ALS179)* and *nic-3* (28%) (1578, 2225). Included in duplications from *T(T54M50)*.

The killing reaction following the fusion of *het-E* and *het-e* mycelia is more rapid and severe, and growth inhibition in *het-E/het-e* duplication strains is more severe than for the vegetative incompatibility mediated by different alleles at *het-c* or *mat* (1560, 2225) (Fig. 36).

het-i: heterokaryon incompatibility-i

I or II. Linked to *T(IR;IIR)4637 al-1* (1630). Second-division segregation in five of eight asci (2223).

Recognized by cessation of growth of forced heterokaryons. In heterokaryons between the strains used by Pittenger (1630), *het-i* nuclei were eliminated and *het-I* nuclei were retained if the initial frequency of *het-I* exceeded 30%. When more than 70% of nuclei were *het-i*, growth continued without a change of ratio. In heterokaryons between the RL strains used by Calligan and Wilson, nuclei called *Hi* are eliminated from *HI + hi* heterokaryons regardless of the starting ratio (2223). The genes *HI* and *hi* ("heterokaryon instability") from RL strains are believed to be alleles of *het-I* and *het-i*, which initially were called *I* and *i*.

het-5: heterokaryon incompatibility-5

IR. Between *R*, *T(MD2)^L* and *T(MD2)^R*, *T(NM169d)*, *un-18* (978, 1578).

Vegetative incompatibility, initially recognized by inhibited duplications, was confirmed using heterokaryon tests (1421, 1425). Can be scored in duplication progeny from crosses that are heterozygous for either *T(MD2)* or *T(NM103)* (1588).

het-6: heterokaryon incompatibility-6

IIL. Between *T(AR18)^L*, *hsp70*, *un-24* (0/222; 14 kb), and *T(AR18)^R*; hence, between *cys-3* and *T(P1869)*, *cot-5* (1421, 1935).

Cloned and sequenced: GenBank AF206700 (OR), AF208542 (PA); Orbach-Sachs clones G8G1, X14C1.

Duplications heterozygous for *het-6* are severely inhibited. Escape from inhibition occurs by deletion of a segment carrying one of the *het* alleles from the partial diploid produced by *T(AR18)* (1938). Vegetative incompatibility was recognized by the production of inhibited duplication progeny when strains from nature were crossed by *T(AR18)*, *T(P2869)*, or *T(NM149)* (1421, 1578). Confirmed by heterokaryon tests (978). Alleles at *het-6* and *un-24* are in linkage disequilibrium, with no coupling-phase recombinant in strains from natural populations (1351).

het-7: heterokaryon incompatibility-7

IIIR. Between *T(D305)^L*, *T(UK8-18)^R* and *T(AR17)^L*. Hence, right of *un-17*, *nit-7* and left of *dow* (1578). Linked to *cyt-19* (5%; 9%) (1377). Included in duplications from *T(D305)* (1421).

Vegetative incompatibility was recognized by the production of inhibited duplication progeny in crosses of wild strains by *T(D305)*. No heterokaryon tests.

het-8: heterokaryon incompatibility-8

VIL. Between *ser-6* (8%) and *ad-8* (3%; 12%), *T(T39M777)* (1172, 1421, 1425).

Vegetative incompatibility initially was recognized by the production of inhibited duplications from crosses of wild strains by *T(T39M777)* (Fig. 36) and was confirmed using heterokaryon tests (1425). Crosses of *T(T39M777)* with isolates from nature indicate that there are either three alleles (*het-8^OR*, *het-8^HO*, and *het-8^PA*) or another polymorphic *het* locus closely linked to *het-8* (924).

het-9: heterokaryon incompatibility-9

VIR. Between *T(AR209)* and *T(OY329)^L*; hence, between *Cen-VI* and *trp-2* (1578).

Vegetative incompatibility was recognized by the production of inhibited duplications from crosses of wild strains by *T(AR209)*. No heterokaryon test. Photograph of heterozygous duplication colony (1421).

het-10: heterokaryon incompatibility-10

VIIR. Between *dr*, *T(5936)* and *Tip-VIIR*. Included in duplications from *T(5936)* (1421, 1578).

Recognized by the production of inhibited duplications in crosses of wild strains by *T(5936)* (1421). No heterokaryon test (Fig. 36).

het-11: heterokaryon incompatibility-11

Unmapped.

Identified and cloned using duplications of random sequences obtained by ectopic integration of cosmids from a library (1345, 1346).

het-12: heterokaryon incompatibility-12

V (1492).

Identified by the slow growth of transformants obtained using a linkage group V cosmid. Thought at first to be a suppressor of het-c-mediated vegetative incompatibility and then reinterpreted as containing an independent functional het gene (1492, 1493). Published information is insufficient to establish the validity of the postulated het-12 locus.

hgu-4: histidylglycine uptake-4

VR. Between cyh-2 (7%) and ure-2 (10%) (2230).

Unable to use L-histidylglycine to support his-6 (2235). Reduced by approximately 50% in its ability to transport most of the amino acids tested. Resistant to many amino acid analogs (2230).

hH1: histone H1

VIIL. Linked to ccg-9, pho-4 (658).
Cloned and sequenced.
Encodes histone H1 (658).

hH2A: histone H2A

VII. Linked to Cen-VII (1T/18 asci), hH2B (on a common plasmid) (2032).
Cloned and partially sequenced: EMBL/GenBank AI392462; EST SC3B9.
Encodes histone H2A (2032).

hH2B: histone H2B

VII. Linked to Cen-VII and hH2A (on a common plasmid) (2032).
Cloned and sequenced: Swissprot H2B_NEUCR, PIR PN0142.
Encodes histone H2B (1032).

hH3: histone H3

IIR. Adjoins hH4-1 (2248). On a common cosmid with hH4-1 and aro-9 (861). Between Fsr-55 and Fsr-3 (1447).
Cloned and sequenced: Swissprot H3_NEUCR, EMBL/GenBank X01612, GenBank NCHISH3, PIR S07350; EST NP4A11.
Encodes histone H3 (2248).

hH4-1: histone H4-1

IIR. Adjoins hH3 (2248). On a common cosmid with hH3 and aro-9 (861). Between Fsr-55 and Fsr-3 (1447).

Cloned and sequenced: Swissprot H4_NEUCR, EMBL/GenBank X01611, GenBank NCHISH4.
Encodes histone H4-1 (2248).

hH4-2: histone h4-2

IIIR. Linked to trp-1 (0T/18 asci) (861).
Cloned and sequenced.
Encodes histone H4-2.

his: histidine

The biosynthetic pathway is shown in Fig. 38. Most histidine auxotrophs are inhibited by complex media or by certain combinations of amino acids with which histidine does not effectively compete for permeases of the basic, neutral, and general amino acid transport systems; see Fig. 47. A histidine mutant can grow on minimal medium plus histidine in the presence of either a basic amino acid or a competing neutral amino acid, but not in the presence of both (809, 1243, 1291). Histidine mutants were not obtained in early mutant hunts where complex media were used, but were recovered on histidine-supplemented minimal medium (809, 1159). General studies (329, 809, 2196). Enzymes of histidine biosynthesis are derepressed coordinately with those of tryptophan, arginine, lysine (292), and other amino acids [reviewed in ref. (1768)]. See cpc-1. Called hist.

his-1: histidine-1

VR. Between sp (11%), am (3%), ure-1 (1%) and pho-2 (3%), al-3, inl (1%; 10%) (218, 747, 809, 1086, 1122, 2014).

Requires histidine (809). Accumulates imidazole glycerol phosphate. Lacks imidazole glycerol phosphate dehydrase (EC 4.2.1.19) (42, 43) (Fig. 38). Expression appears positively regulated by α-isopropylmalate, a leucine biosynthetic intermediate that functions with leu-3 in leucine biosynthetic gene regulation (1048). Intralocus complementation (329). Recombination between his-1 alleles is controlled by rec-1 (314, 990, 2083). Called C84.

his-2: histidine-2

IR. Between un-2 (<1%), T(AR190) and T(AR173)^R, nuc-1 (1%) (314, 1331, 1580).
Cloned: pSV50 clones 10:6F, 6:11E, Orbach-Sachs clones X7H07, X20D08, X21G12, G13B04, G13H11, G19B11.
Requires histidine (809). Affects ATP phosphoribosylpyrophosphate pyrophosphorylase (EC 2.4.2.17) (23) (Fig. 38). Intralocus complementation (329). Recombination between his-2 alleles is controlled by rec-3 (337) and is not affected by rec-1 (314). Called C94.

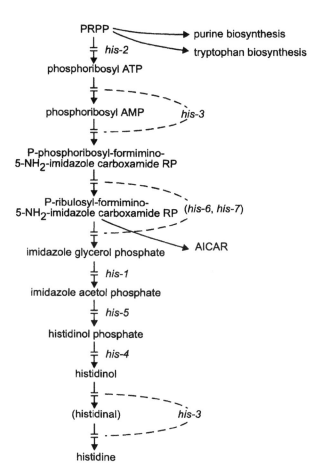

FIGURE 38 The biosynthetic pathway of histidine, showing the sites of gene action (23, 42, 43, 329, 1350, 2196). Abbreviations: ATP, adenosine 5'-triphosphate; AMP, adenosine 5'-monophosphate; PRPP, 5-phosphoribosyl pyrophosphate. AICAR (5'-phosphoribosyl-5-aminoimidazole-4-carboxamide ribosyl phosphate) is an intermediate in purine synthesis. See Fig. 5. For relations between histidine and purine synthesis, see ref. (1543). Modified from ref. (1596), with permission from the American Society for Microbiology.

his-3: histidine-3

IR. Between *met-10*, *lys-4* (1%) (1022, 1322) and *cog* (1%; 3%) (314, 338), *ure-4* (1%) (156), *ad-3A* (1%; 3%) (492).

Cloned and sequenced: EMBL/GenBank AF045455, AF045456, M27531, EMBL NCHIS3, GenBank NEUHIS3; EST W08H2; pSV50 clones include 6:4E, 6:11C, 12:5C; Orbach-Sachs clone G3F06.

Requires histidine (809). Complex multidomain structural gene encoding histidinol dehydrogenase (EC 1.1.1.23), phosphoribosyl-ATP-pyrophosphohydrolase (EC 3.6.1.31), and phosphoribosyl-AMP cyclohydrolase (EC 3.5.4.19) (23, 1350) (Fig. 38). Mutants produce cross-reacting material (428). Supersuppressible amber alleles identified (303). Individual alleles may lack only the early reaction(s), or only histidinol dehydrogenase, or both. Mutants that lack only histidinol dehydrogenase accumulate histidinol (23, 329, 2196). Used to study intralocus complementation and recombination (22, 23, 45, 314, 329, 331, 336, 2194, 2195, 2197). Intralocus recombination is regulated by *cog* and *rec-2* (45, 336) and is not affected by *rec-1* (314). Translocation *T(TM429)*, with one breakpoint in *his-3*, has been used to show that *cog* is *cis*-acting (336). Fine-structure interallelic complementation map (1511). Correlation of noncomplementing mutants with *his-3* restriction site differences (557). Efficient integration of gene constructs by gene replacements at the *his-3* locus (47, 1265, 1771). Differences in sequence between the Lindegren and St. Lawrence strains (328). Called C140, T1710 (=C1710).

his-4: histidine-4

IVR. Between *cot-1* (1%; 4%) and *met-5* (4%) (809, 1585).

Requires histidine (809). Lacks histidinol phosphate phosphatase (EC 3.1.3.15) (42, 43). Accumulates L-histidinol phosphate (Fig. 38). Allele P143h is heat-sensitive (wild type at 25°C), whereas C141 is not heat-sensitive (1591); both are leaky (1412).

his-5: histidine-5

IVR. Between *pyr-3* (1%) and *tol*, *trp-4* (3–7%) (329, 1466, 1911, 1928).

Requires histidine. Accumulates imidazole acetol phosphate and some imidazole glycerol phosphate. Evidently lacks imidazole acetol phosphate transaminase (EC 2.6.1.9) (23, 329, 2196) (Fig. 38). Intralocus complementation (329). Recombination control (314).

his-6: histidine-6

VR. Between *pyr-6* (6%; 18%), *un-9* (6%), *oak* (6%) and *T(NM149)* (0/499), *Tel-VR* (30 kb) (1592, 1593, 1603, 1809).

Cloned (1809).

Requires histidine. Blocked prior to IGP (329, 2196), presumably in either EC 5.3.1.16 or EC 2.4.2.-, which catalyze the only two steps not definitively genetically characterized (Fig. 38). No intralocus complementation (95 alleles) (329). Intralocus recombination (314). Used in the first cloning of a *Neurospora* telomere (1809).

his-7: histidine-7

IIIR. Between *leu-1* (8%; 20%) and *thi-2* (1% or 2%) (329, 426, 1582, 2044, 2196).

Cloned and partially sequenced: EMBL/GenBank AI398871; EST W09G5.

Requires histidine. Blocked prior to IGP, presumably in either EC 5.3.1.16 or EC 2.4.2.-, as the only two steps not definitively characterized genetically (329, 2196); likely encodes a bifunctional amidotransferase/cyclase, from EST (Fig. 38). Intragenic recombination (314).

hist: histidine

Symbol changed to *his*.

Histidine Sensitivity

Many mutagen-sensitive mutants are also sensitive to inhibition by histidine (see *mus*, *uvs*, *mei*, and Table 3) (946, 1467, 1830). Certain strains that were first identified as mutant on the basis of histidine sensitivity are not sensitive to the few mutagens tested [see, for example, ref. (520) and *hss-1*]. The mechanism underlying histidine sensitivity is not understood.

hlp-1: histidinol permeability-1

VIIR. Between *sfo* (1%; 9%) and *hlp-2* (8%; 25%), *nt* (28%; 37%) (879).

Enables a *his-3* allele to use L-histidinol. Proposed to be due to increased uptake through basic L-amino acid transport system III [as defined in ref. (1522)]. The *hlp-1* mutation confers increased sensitivity of *lys* and *arg* mutants to inhibition by arginine and lysine, respectively (879).

hlp-2: histidinol permeability-2

VIIR. Between *sfo* (3%; 7%), *hlp-1* (8%; 25%) and *nt* (29%) (879).

Enables a *his-3* allele to grow on L-histidinol. Growth of *his-3; hlp-2* on histidine is inhibited by methionine, isoleucine, valine, and asparagine (879).

hom: homoserine

IR. Between *Tp(T54M94)^R*, *arg-6* (1%) and *al-1* (<1%), *al-2* (2%; 7%), *cnr* (1%). Between breakpoints of *T(STL76)* and *T(4637)*; hence, left of *al-1* (1548, 1578, 1585). Located on a common cosmid with *al-1* (1816).

Cloned and partially sequenced: EMBL/GenBank AI392256; EST SC2C6, pSV50 clone 3:11H.

Uses homoserine, or methionine + threonine (2066). Affects aspartate semialdehyde dehydrogenase (EC 1.2.1.11) (987) (Fig. 45). Inhibited on complex complete medium and by methionine and other amino acids (2066); supplemented minimal medium should therefore be used. Symbol changed from *hs*.

hph: hygromycin phosphotransferase

Introduced from *Escherichia coli*.

Neurospora crassa is sensitive to growth inhibition by hygromycin B. The bacterial gene *hph* encodes a phosphotransferase that confers resistance to the fungus (1977), where it has had numerous applications. The *hph* gene has been incorporated into cosmid vectors as a selectable marker for use in transformation (1502, 1504). Selection for *hph* expression has been used to obtain site-specific integration events (277). *hph* has been used to study gene silencing (955, 1529). As a reporter, *hph* has been used to obtain both *cis*- and *trans*-acting mutations that affect gene regulation by asking for increased *hph* expression when expression is normally repressed (677, 1241). Hygromycin phosphotransferase enzyme activity can be measured in *N. crassa* extracts (677, 678).

hpp: housekeeping phosphate permease

Between *al-2* and *nic-1* (1327).

Selected as a suppressor of the double mutant *pho-4; pho-5*, which lacks both of the two high-affinity permeases and is unable to grow at low phosphate concentration and high pH. On ordinary Vogel's (high phosphate) medium, the triple mutant *hpp; pho-4; pho-5* accumulates brownish pigment and does not grow as well as the *pho-4; pho-5* double mutant (1327).

hs: homoserine

Symbol changed to *hom*.

hsp30: heat shock protein 30

I. Linked to *mei-3*, *Cen-I*, *un-2*, *his-2* (0T/18 asci) (1447).

Cloned and sequenced: Swissprot HS30_NEUCR, EMBL/GenBank M55672, J05601, EMBL NCHSP30, GenBank NEUHSP30, PIR A38360.

Encodes a 30-kDa heat-shock protein. Mutant obtained by RIP shows reduced viability at high temperature and altered location of a 22-kDa protein in mitochondria (1637). Lacks a small α-crystallin-related heat-shock protein, shows poor survival during heat shock on a nutrient medium with restricted glucose, and accumulates unphosphorylated glucose at high temperature. Hexokinase is reduced by more than 35% in the mutant relative to wild type. HSP30⁺ may protect hexokinase from thermal inactivation (1635).

hsp70: heat shock protein 70

IIL. Between *T(AR18)^L* and *un-24*, *het-6*, *T(AR18)^R* (1935).

Cloned and sequenced: Swissprot HS70_NEUCR, EMBL/GenBank U10443, GenBank AAA82183.

Encodes a major stress-inducible 70-kDa heat-shock protein (1026), which reaches its highest level in late aerial hyphae (811). Homolog of *Escherichia coli dnaK*. A molecular chaperone involved in the translocation motor for the import of proteins through the mitochondrial membrane into the matrix (2300) (Fig. 64). Called *hsps*, *hsps-1*. For another member of the *hsp70* family, see *grp78*.

hsp80: heat shock protein 80

VR. Between *inl* (3T/18 asci), *cya-2* (2T/18 asci) and *tom70* (1 or 2T/17 asci). Linked to *cmd*, *hsp83* (0T/18 asci) (1447).

Cloned and partially sequenced: EMBL/GenBank AI398579, PIR A56820; EST W01G6.

Encodes an 80-kDa heat-shock protein (1747); a cytosolic molecular chaperone of the eukaryotic stress 90 family (674). Hybridizes with *Saccharomyes cerevisiae hsp83* probe (1636). Called *hspe-1*, *hsp83*.

hsp83: heat shock protein 83

Allelic with *hsp80*.

hsp88: heat shock protein 88

Unmapped.

Cloned and sequenced: GenBank AF069523, AAC23862.

Encodes an 88-kDa heat-shock protein that interacts with HSP30. Homolog of the mammalian Hsp110 family (1638).

hsp98: heat shock protein 98

Unmapped.

Partially sequenced: Swissprot HS98_NEUCR, EMBL P31540, PIR S28174, Genbank AAB24758.

Encodes a 98-kDa heat-shock protein (2143) showing homology with HSP104 of *Saccharomyces*. HSP98 is enriched in the microsomal fraction of heat-shocked cells.

hspe-1: heat shock protein 80

Symbol changed to *hsp80*.

hspp-1: heat shock peroxidase

VI. Linked to *Bml*, *Cen-VI*, *pan-2* (0T/18 asci) (1447).

Encodes a heat-shock-inducible peroxidase (1025).

hsps: heat shock protein 70

Symbol changed to *hsp70*.

hss-1: histidine sensitive-1

IVR. Linked to *cot-1* (19%), *cys-4* (2%) (1593).

Sensitive to histidine (0.5 mg/ml) (1593), but not to UV (317), MMS, or γ rays (1828).

htb-1: high-temperature banding-1

IVR. Linked to *bd* (27%), near *T(R2355)*, *cot-1* (1433).

Allows clear expression of circadian conidiation rhythm at 36°C (1433).

htb-2: high-temperature banding-2

Unmapped. Unlinked to *bd* (1433).

Allows clear expression of circadian conidiation rhythm at 36°C (1433).

htl: hyphal tip lysis

IR. Right of *ad-3* (1%; 10%) (204). Linked to *nuo12.3* (3/18 asci), *cys-9* (1T/18 asci) (1447).

Cloned: Orbach-Sachs clone G8H12 (204).

Encodes an abundant 19-kDa protein of unknown function. Mutants generated by RIP show lysis of hyphal tips when grown on solid medium. Patches of bright orange-red pigment are accumulated. Hyphal tips below the agar surface produce large balloons or lens-shaped structures. Growth on high concentrations of NaCl is indistinguishable from that of wild type (204). Called *vac-5* (1447).

i: heterokaryon incompatibility-i

Used in early papers as the symbol for *het-i*. The symbol *i* was also used for *en(am)-1* and for an unmapped intensifier of carotenoid pigment (1904).

Iasc: Indurated ascus

VR. Right of *Cen-V* (1968).

The wall of an indurated ascus hardens and darkens, making the entire ascus resemble a giant ascospore. Pores are formed and striations appear. The spore can be induced to germinate. Dominant with variable expression. Some heterozygous asci are normal. The ascus phenotype resembles that described for indurated asci of *Neurospora tetrasperma* (546). Vegetative growth is weak (1966). Indurated asci can be produced by biotin deficiency in wild-type strains of *N. tetrasperma* (725, 1680). The *Neurospora crassa* strain deposited in FGSC as *Iasc* (FGSC No. 3424) has been reported not to produce indurated asci, but the crosses were made on medium supplemented with biotin. Whether biotin concentration affects expression in this strain has not been determined. (In contrast to *Iasc*, the single giant ascospore in the *Banana* mutant is formed by hardening of the ascus membrane within a normal wall rather than by hardening of the ascus wall itself.)

ile-1: isoleucine-1

VII. Between *ars* (1%) and *wc-1* (3%). Probably right of *met-7* (<1–2%) (1328, 1582, 1585).

Uses isoleucine, α-amino-*n*-butyric acid, threonine (2064), or canavanine (129). Affects threonine dehydratase (EC 4.2.1.16) (1056, 1059) (synonym; threonine deaminase; 1056) (Fig. 39). Leaky on minimal medium; treacherous to score with large inocula. Tests should be read early (24 hr if at 34°C). Moderate inhibition by methionine (2064). May be screened as tiny germlings from ascospores germinated on minimal medium (1328). Name changed from *thr-1* (1056).

ilv: isoleucine + valine

Mutants designated *ilv* require both amino acids. The three *ilv* loci specify enzymes that catalyze corresponding steps in the parallel biosynthetic pathways of isoleucine and valine (Fig 39). These enzymes are located in the mitochondria (157, 1161), and they may indirectly affect electron transport (157). The enzymes may be in an aggregate; see review in ref. (456). A ratio of 20–30% isoleucine to 80–70% valine is optimal for growth (194). At least some *ilv* mutant strains are inhibited by norleucine, norvaline, phenylalanine (194), or tryptophan (1172).

Regulation: Enzyme production in response to end-product-derived signals depends on *leu-3*[+] product and α-isopropylmalate. In *leu-3*[+], threonine deaminase production is repressed as a function of available isoleucine, acetohydroxy acid synthetase as a function of valine, and isomeroreductase and dihydroxy acid dehydratase as a function of isoleucine and leucine. In the absence of effective *leu-3* product and/or α-IPM, enzyme production is repressed even under severe end-product limitation (1497). Called *iv*.

ilv-1: isoleucine + valine-1

VR. Between *per-1* (4%) and *ilv-2* (<1%; 9%), *lys-2* (4%; 7%) (10, 21, 919). Crosses with *ilv-2* gave prototroph

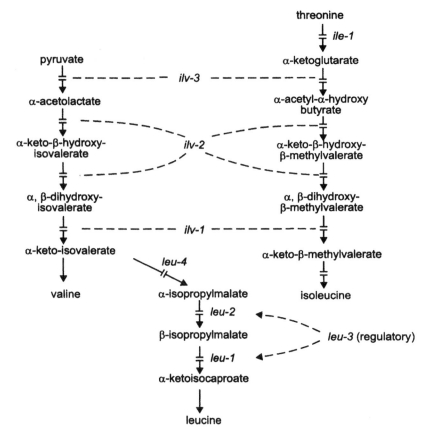

FIGURE 39 The biosynthetic pathways of isoleucine, valine, and leucine, showing sites of gene action (37, 290, 798, 1056, 1419, 1670, 2174). Isoleucine and valine are synthesized along parallel pathways catalyzed by common enzymes. The leucine precursors α- and β-isopropylmalate formerly were called β-hydroxy-β-carboxyisocaproate and α-hydroxy-β-carboxyisocaproate. From ref. (1596), with permission from the American Society for Microbiology.

frequencies that were variable and sometimes unusually high (<1% to 9%) (1069).

Requires both isoleucine and valine or the corresponding keto acids (1069, 2138, 2174). Affects dihydroxy acid dehydratase (EC 4.2.1.9) (37, 1419) (Fig. 39). Most alleles are leaky (1069). Leucine has a sparing effect on the valine requirement (194). Called *iv-1*; groups 2 and 3.

ilv-2: isoleucine + valine-2

VR. Between *ilv-1* (<1%; 9%) and *lys-2* (1069, 1919).

Cloned and sequenced: Swissprot ILV5_NEUCR, EMBL/ GenBank M84189, EMBL NCILV2A, GenBank NEUILV2A; pSV50 clones 30:5C, 30:10F.

Structural gene for α-keto-β-hydroxylacyl reductoisomerase (EC 1.1.1.86) (1919, 2174). Requires both isoleucine and valine (Fig. 39). Known alleles are not leaky (1069). Allele T313 is heat-sensitive (1069). The ILV-2 protein is present at high levels in the cell (1070). Called *iv-2*; group 1.

ilv-3: isoleucine + valine-3

IVR. Linked to *met-2* (0/129) (1582). Between *leu-2* (4%) and *ad-6* (9%) (985, 1124).

Cloned: pSV50 clone2:4C.

Requires both isoleucine and valine. Accumulates pyruvate. Very low acetohydroxy acid synthetase activity (EC 4.3.1.18) (290) (Fig. 39). Markedly inhibited by methionine (1582). The level of *ilv-3* and *leu-1* mRNAs increases in an *ilv-3* mutant starved for branched-chain amino acids (985). Alleles recombine and complement (1124, 2174). Called *iv-3*; group 4.

In(. . .): Inversion

A rearrangement in which two or more contiguous genes are inverted relative to the normal sequence. For information on published individual inversions, see ref. (1578).

In(OY323): Inversion (IL;IR) (OY323)

A pericentric inversion with breakpoints between *nit-2* and *leu-3* in IL and between *lys-3* and *ace-3* in IR (115). Reference (1578) is incorrect in stating that the right breakpoint is left of *lys-3*.

inl: inositol

VR. Between *pho-2* (3%; 4%), *al-3* (1%; 1T/18 asci) and *pab-1* (1%; 10%) (673, 747, 1447, 2014).

Cloned: pSV50 clones include 13:10B, 17:9D, 18:6C.

Lacks myoinositol-1-phosphate synthase (glucocycloaldalase, EC 5.5.1.4). Growth is colonial on low levels of inositol (698). Tends to extrude dark pigment into the medium when grown on suboptimal inositol. Composition of phospholipids and cell walls is abnormal on limiting inositol (698, 829, 830, 934) (Fig. 18). Inhibited by hexachlorocyclohexane (697, 877, 1827). Conidia are subject to death by unbalanced growth on minimal medium (2003, 2010), a property exploited for mutant enrichment ("inositolless death") (1181, 1292) because double mutants are at a selective advantage. Heat-sensitive allele 8320, which grows at one-third the wild-type rate on race tubes, is especially useful for mutant enrichment (1631, 2022). Used in the first experiments reporting the transformation of *Neurospora* (1356, 1358, 2245). Used to study transport systems for glucose (1792) and sulfate (1280) . Used for studying induced reversion (732). Used for studying the mechanism of inositol-less death (1292, 1391), the mutagenicity of ferrous ions, and regulation of mitochondrial membrane fluidity; reviewed in ref. (1391). Spontaneous reversion rates (727). Heat-sensitive allele 83201 shows slow semicolonial growth in liquid minimal at 25°C (1280), but looks normal on slants (1546). *opi* acts as a partial suppressor of allele 89601, which contains cross-reacting material (2305). *su(inl)* is an allele-specific partial suppressor of another *inl* allele (735). A mutant *exo-1* allele is present in *inl* (89601) stock FGSC 498 and may, therefore, be present in stocks of mutants derived from it by inositol-less death (372, 615, 2002). Allele 46802 is nonrevertable and inseparable from *T(46802)* (727, 1578). Called *inos*.

inos: inositol

Symbol changed to *inl*.

int: intense

IVR. Linked to *pan-1* (0/50) (1592).

Brighter orange than wild type, perhaps because of morphology rather than carotenoid content (1592).

inv: invertase

VR. Between *ro-4* (5%; 8%), *pab-2* (3%), *mus-11* (4%) and *asn* (4%; 9%) (1582, 1794, 1828).

Cloned: Plasmid pNC2 (293).

Structural gene for invertase (EC 3.2.1.26). Unable to use sucrose as the carbon source. Grows well on glucose or fructose and fairly well on casamino acids or yeast extract. Uninducible by normal inducers (1793). Mutants can be identified efficiently by a procedure using o-tolidine, glucose oxidase, and horseradish peroxidase to assay colonies *in situ* by halo formation (1150). Invertase also is affected by *cot-2*.

ipa: it pokes along

IL. Between *mat* (20%) and *arg-1* (1%) (110, 1932).

Hyphae from germinating ascospores or conidia grow out for long distances without branching. Cultures are one day late growing up (110). Modifies *pro-3*. The double mutant *pro-3; ipa* does not respond to arginine and does not grow as well as *pro-3;ipa⁺* on proline, citrulline, or ornithine. As a single mutant, *ipa* grows on minimal medium at half the wild-type rate. Arginine uptake is normal; *arg-2; ipa* or *arg-5; ipa* can grow on arginine. Inhibition studies suggest that *ipa* may be unable to shunt exogenous arginine into the proline pathway (1932).

ipm-1: isopropylmalate-1

Unmapped. Unlinked to *ipm-2* or *leu-4* (1708).

Altered isopropylmalate (IPM) permeability. The mutant is able to use β-isopropylmalate to support the growth of *leu-4* and for the induction of β-IPM isomerase and β-IPM dehydrogenase in contrast to *ipm⁺*, which is unable to take up this intermediate (1708, 1709).

ipm-2: isopropylmalate-2

Unmapped. Unlinked to *ipm-1* or *leu-4* (1708).

Altered isopropylmalate permeability. Improves β-IMP uptake of *ipm-1* mutants in supporting the growth of *leu-4* (1708, 1709). The *ipm-2* single mutant is relatively ineffective in promoting β-IPM permeability.

iv: isoleucine + valine

Symbol changed to *ilv*.

kex-1: kex-like-1

Changed to *scp*.

kin-1: kinesin-1

Unmapped.

Cloned and sequenced: Swissprot KINH_NEUCR, EMBL/GenBank L47106, EMBL NCNKIN, GenBank NEUKIN.

Specifies a plus-end-directed motor molecule functionally homologous to the kinesin heavy chain of animals. The motor domain at the amino-terminal end shows homology with those of conventional kinesins, and there is a short region of homology near the C-terminus (1999). The central stalk domain is unique in sequence, however. A null mutant obtained by RIP shows severely altered morphology and branching, reduced linear growth, abnormal nuclear distribution in mycelia, and increased nuclear number in conidia. Motility of organelles, examined using video microscopy, is unaffected. The Spitzenkörper is difficult to detect. The defect is thought to be in the transport of secretory vesicles to sites of cell-wall biosynthesis (1870). Apical movement of vesicular organelles to the Spitzenkörper is defective, and protein secretion into the medium is deficient. Used to examine the kinesin–dynein (*kin; ro-1*) double mutant, which is defective in oppositely directed microtubule motors, and to compare it with the single mutants (1871). Used to examine the role of the motor domain in determining the direction of movement in molecular chimeras of the *Neurospora* kinesin with an oppositely oriented kinesin homolog, NCD (*nonclaret disjunctional dominant* in *Drosophila*) (870). Growth rate of the null mutant was used to assay the functional efficiency of mutant kinesins from various sources (1068). For a review of fungal molecular motors, see ref. (2258).

krev-1: Krev-1-like

IL. Linked to *nit-2*, *mus-18* (0/18) (969, 970, 1447). A report of linkage to *ser-1* and *pro-1* in IIIR apparently was in error (1446).

Cloned and sequenced: EMBL/GenBank AB000281.

Homolog of mammalian *Krev-1*, a member of the RAS superfamily. Null mutants obtained by RIP are fertile and vegetatively normal. Overexpression of *krev* mutants inhibits perithecial development (969).

kyn-1: kynureninase

VII. Linked to *nic-3* (30%), *wc-1* (20%) (1180).

Partially defective in the induction of the inducible kynureninase I (EC 3.7.1.3) isozyme by kynurenine, indole, or tryptophan but has normal levels of the constitutive kynureninase II isozyme. Possibly regulatory. Scored and selected by a low level of anthranilate accumulation on medium supplemented with a high level of tryptophan; this results in lowered UV fluorescence compared to wild type (1813). The inducible kynureninase is inactivated by incubation with L-alanine or L-ornithine. The inactivated enzyme is resolved to the apoenzyme by dialysis and reactivated by incubation with pyridoxamine 5′-phosphate plus pyruvate or with pyridoxal 5′-phosphate (2049).

Lactose Utilization

Lactose is a poor carbon source for *Neurospora* (127). The ability to use lactose has a multigenic basis. In early studies, strains were identified that showed impaired ability to grow on lactose. Properties of β-D-galactosidase (lactase) were unaltered; levels were normal when grown on sucrose, but depressed on lactose (1136, 1137). These strains differed from wild type at several loci, each with a small and additive effect on lactose utilization; e.g., three component genes from "*lac*" strain 31389 × wild type were shown to be unlinked and were designated *n-lac-1*, *pow(n-lac-2)* (powdery conidia), and *floc(n-lac-3)* (flocculent morphology). These genes are not specific for lactose utilization, however; they also showed an altered adaptation response to other carbon sources. No major gene was identified that qualifies to be designated *lac* (665). The failure to find a single-gene mutant can be ascribed to the fact that *Neurospora* has two β-galactosidases (127, 1179).

lacZ: lactose Z

From *Escherichia coli*.

The *E. coli* structural gene for β-galactosidase is a reliable reporter gene for quantitative measures of gene expression in *Neurospora crassa* (574, 1771). For examples, see refs. (80), (146), (409), (1237), (1238), (1517), and (2209). Enzyme activity from the introduced *lacZ* is manyfold higher than that of the endogenous *Neurospora* β-galactosidases (1771).

lacc: laccase

VIIL. Between *nic-3* (14%) and *thi-3* (7%) (2296).

Cloned and sequenced: Swissprot LAC1_NEUCR, LAC2_NEUCR, EMBL/GenBank M18333, M18334, EMBL NCLCCA, NCLCCB, GenBank NEULCCA, NEULCCB.

Laccase structural gene (EC 1.10.3.2) (benzenediol:oxygen oxidoreductase, urishiol oxidase), extracellular. Alleles from wild-type strains TS and OR sequenced (726). The laccase polypeptide is processed posttranslationally at both its N- and C-termini (726). *lacc* mutants do not show a detectable phenotype except that laccase is not produced (2296). Gene expression is induced by inhibitors of RNA or protein synthesis or by certain aromatic molecules (684, 1209). The secreted enzyme forms a blue halo of oxidized *o*-tolidine around colonies in test plates. Screening for mutants that overproduced laccase led to the identification of *lah-1* (2041). Screening for additional mutants that did not allow the overexpression of laccase in the *lah-1* background established that functional *cpc-1* was required for *lacc* induction (833, 2042, 2296). Called *lni-2*.

lah-1: laccase halo-1

IL. Between *nit-2* (2%) and *leu-3* (5%) (2041).

Laccase is derepressed. Hypersensitive to cycloheximide (2041). Acts as a negative regulator of *cpc-1* (833). Identified by the development of a blue halo when *o*-tolidine is added to colonies of *exo-1* grown on xylose plus sorbose (2041). Used to obtain a mutant in *lacc*, the laccase structural gene (2296). Called *halo-1* (832).

lao: L-amino acid oxidase

IIIR. Linked to *un-17* (5%) (511).

Cloned and sequenced: Swissprot OXLA_NEUCR, EMBL J05621, PIR A38314 (1480).

Structural gene for L-amino acid oxidase (EC 1.4.3.2) (1480). Under general nitrogen metabolite regulation (1914) (Fig. 50). Also called *lox*.

le-1: lethal-1

IVR. Linked to *pan-1* (1% or 2%) (719). Right of *cot-1* (14%) (1405, 1408).

Ascospores containing *le-1* are black, but they fail to germinate unless given special treatment (719). Autonomously expressed. Growth is colonial and aconidiate, with dense granular aerial mycelium turning brown with age (1405, 1408). Photographs (719, 1408). Cell-wall peptides are reduced in amount (2249). Alleles B55 and S4355 of ref. (719) were presumed to be allelic with similar mutants called *col-le-1* (CM3) and *col-le-2* in ref. (1405) and (1408), but direct tests were not made.

le-2: lethal-2

VIIL. Linked to *met-7* (7%). Indicated to the left (719).

Ascospores containing *le-2* are black, but they fail to germinate except for a few that are recovered after aging. Autonomously expressed. Compact colonial growth (719).

leu: leucine

For the biosynthetic pathway, see Fig. 39. Leucine mutants have been used extensively for studies of regulation (799, 1281, 1640). When grown on Difco Bacto Agar–sorbose medium, leucine auxotrophs acquire suppressors that are leaky auxotrophs blocked at various steps in sulfur metabolism. Apparently, these blocks allow more efficient use of the traces of leucine in the agar (793, 797). Most aliphatic and aromatic amino acids can inhibit the growth of leucine mutants at appropriate concentrations, probably as a result of competition for a common uptake system (793).

leu-1: leucine-1

IIIR. Between *ad-4* (1%; 5%), *col-6* (1%) and *his-7* (8%) (426, 985, 1122, 1591, 2044).

Cloned and sequenced: Swissprot LEU3_NEUCR, EMBL/GenBank U01061, EMBL NC061; EST NP2D4.

Requires leucine (1705, 1706). Structural gene for α-isopropylmalate dehydrogenase (EC 1.1.1.85) (798) (Fig. 39). Accumulates α-isopropylmalate (α-IPM) and β-isopropylmalate (799). Synthesis of enzyme also requires the function of regulatory gene *leu-3*$^+$ and the presence of α-IPM, which acts as an inducer (799). Resistant to aminotriazole (1048). Female sterile (1424).

Used to study reversion and competition in heterokary-ons (1762). The 5'-leader of the mRNA precursor contains an intron that does not interrupt a coding region (1190). Allele 33757 contains a 1-bp insertion predicted to result in a frameshifted, prematurely terminated polypeptide (1190).

leu-2: leucine-2

IVR. Between trp-4 (2%) and ilv-3 (4%) (1124, 1928). Cloned (1047).

Requires leucine (1705, 1706). Structural gene for iso-propylmalate isomerase (EC 4.2.1.33) (804, 1709) (Fig. 39). Altered heat inactivation of hybrid enzymes (804). Structural differences of hybrid enzymes (1709). Accumulates α-IPM (799). Synthesis of enzyme also requires the function of regulatory gene leu-3$^+$ and the presence of α-IPM, which acts as an inducer (799). Induction measured at the transcript level is very rapid (1276). Resistant to 3-aminotriazole (1048). Alleles show intralocus complementation (796). Allele 37501 is heat-sensitive (30 vs 20°C), leaky at 25°C (1704).

leu-3: leucine-3

IL. Between nit-2 (12%; 18%), lah-1, In(OY323)^L and T(OY322)^L, cyt-1 (5%; 8%), T(OY321) (115, 1578, 1582).

A regulatory gene. The mutant requires leucine (1705). Prevents the synthesis of α-isopropylmalate isomerase and β-isopropylmalate dehydrogenase and prevents full de-repression of α-isopropylmalate synthase. Is also involved in the regulation of isoleucine and valine synthesis (799, 1497, 1640) (Fig. 39). The LEU-3 regulator, which from the rapid rate of LEU-3-dependent leu-2 induction appears to be present irrespective of inducing conditions (1276), may also have more global regulatory functions (52). Neurospora crassa LEU-3 might be a Zn(II)2Cys6 binuclear cluster DNA-binding protein, as is its functional yeast homolog, LEU3 (2094). However, the DNA-binding sequences used by the Saccharomyces cerevisiae protein are not found in the promoter region that is upstream of the Neurospora leu-1 mRNA transcript (1190), which is regulated by LEU-3. The original mutant allele, 47313, is leaky, but some other alleles, e.g., R156, are not.

leu-4: leucine-4

IL. Between cyt-1, T(OY321) and cys-5 (≪1%) (801, 1578, 2198).

Cloned: pSV50 clone 9:7A.

Requires leucine. Structural gene for α-isopropylmalate synthase (EC 4.1.3.12) (798, 799, 804) (Fig. 39). Feedback-negative mutants (798, 799). Hybrid synthe-tases with altered properties (804). Complementation between alleles (804, 2198).

leu-5: leucine-5

VR. Between cyh-2 (1%) and sp (3%; 9%) (378, 1582, 1603, 1651).

Cloned and sequenced: Swissprot SYLM_NEUCR, EMBL/GenBank M30472, PIR A33474, SYNCLM, EMBL NCLEURS, GenBank NEUMTLEURS.

Structural gene for mitochondrial leucyl-tRNA synthetase (EC 6.1.1.4) (378, 1651). Mutant 45208 has a partial leucine requirement at low temperatures, a tighter leucine requirement at 34°C, and stops growth at 37–39°C, regardless of leucine supplementation (556, 1651). The 45208 allele contains a Thr to Pro mutation at amino acid 135, which is presumed to result in the phenotypic defect (378). Mutations mapping in the leu-5 region appeared to affect either cytoplasmic leucyl-tRNA synthetase or mitochondrial leucyl-tRNA synthe-tase separately or both simultaneously (138, 803), and leu-5 was proposed to be a gene complex consisting of structural genes for both enzymes. This proved to be incorrect; the cytoplasmic leucyl-tRNA synthetase is specified instead by leu-6 (153). In mitochondrial mutant [cni-3], mitochondrial tRNA synthetase is increased greatly whereas the cytoplasmic enzyme is unchanged (802). Allele 45208 is somewhat unstable (803). It causes alterations in unrelated enzymes, apparently via mistranslation (1183, 1651), and it was used to study the hypothesis that senescence is due to faulty protein synthesis (1183). Assembly of glycerol kinase and GPDH into the inner mitochondrial membrane is not impaired in leu-5 (421). Recovery from ascospores is poor at 34°C (on complex complete medium); ascospores are best germinated at 25°C.

leu-6: leucine-6

IIR. Between trp-3 (2/16), Fsr17 (0/16; 5T/18 asci) and Tel-IIR (1T, 1NPD/16 asci) (153, 378, 1447).

Cloned and sequenced: Swissprot SYLC_NEUCR, EMBL/GenBank, M30473, X13021, PIR A33475, S04532, SYNCLC, EMBL NCLEUR01, NCLEURSC, GenBank NEULEURSC; EST SM1D7.

Structural gene for cytoplasmic leucyl-tRNA synthetase (EC 6.1.1.4) (153, 379, 1113). The reported effect of leu-5 mutations to alter the K_m of the cytoplasmic leucyl-tRNA synthetase in crude extracts (138) was not reproduced with the purified cytoplasmic enzyme (1113), although the stability of the enzyme in extracts was decreased when extracts were prepared from cells incubated at nonpermissive temperatures. Subject to cross-pathway control through cpc-1. Histidine starvation induces new transcription start sites (379).

lgd: laggard

VIL. Linked to chol-2 (7/32). Left of T(OY350) (1589).

Growth is delayed greatly following ascospore germination. Morphologically wild type when grown. Obtained in progeny of $Dp(VIL \rightarrow IR)OY350 \times Normal$, presumably by RIP (1589).

Lipid Biosynthesis
See Fig. 18.

lis-1: light-insensitive-l
IR. Between *ad-3* (6%) and *al-1* (16%) (1514).

Circadian conidiation is not suppressed in constant light (1514). The growth rate is normal. Photoinduced genesis of carotenoids and phase shifting of periodic conidiation are not altered (1711). Recessive in heterokaryons.

lis-2: light insensitive-2
VI. Between *chol-2* (11%) and *trp-2* (26%) (1515).

Resembles *lis-1*, but the growth rate is reduced somewhat (1515, 1711).

lis-3: light insensitive-3
VR. Right of *inl* (4%) (1514).

Resembles *lis-1*, but the growth rate is reduced markedly (1514, 1711).

lni-1: laccase noninducible-1
Allelic with *cpc-1*.

lni-2: laccase noninducible-2
Changed to *lacc-1*.

lox: L-amino acid oxidase
Symbol changed to *lao*.

lp: lump
II. Right of *thr-3* (10%). Linked to *bal* (25%) (1585, 1603).

Restricted colonial growth. Grows more rapidly than *bal* and forms aerial hyphae (1585).

lpl: lysophospholipase
IR. Right of *his-3* and *cog*, which it adjoins (2276).

Cloned and sequenced: EMBL/GenBank AF045574, AF045575.

Encodes lysophospholipase (EC 3.1.1.5). Shows homology to *Saccharomyces PBL1* and *SPO1*. Alleles differ in wild types St. Lawrence 74A and Lindegren 25a (2276).

lys: lysine
Lysine biosynthesis is by the α-aminoadipate pathway in *Neurospora* and other higher fungi (2163) (Fig. 40). Enzymes of lysine biosynthesis are subject to cross-pathway control (1768); see *cpc*. All lysine auxotrophs are inhibited competitively by arginine (548). (See the *arg* entry for a medium that allows growth while providing both lysine and arginine requirements.) Resistance to arginine is conferred on *lys-1* by a presumed transport mutant arg^R. Complex interactions between *lys*, *pyr*, and *arg* mutants have been described (910).

lys-1: lysine-1
VL. Between *caf-1* (4%; 14%) and *cyt-9* (5%), *at* (1%; 20%) (926, 1582, 1598).

Uses lysine, α-aminoadipic acid, or ε-hydroxynorleucine (α-amino-ε-hydroxycaproic acid) (751, 2113). Accumulates homocitrate on limiting lysine concentrations (887) (Fig. 40). Fine structure; complementation between alleles (15).

lys-2: lysine-2
VR. Between *ilv-1*, *ilv-2* (4%; 7%) and *cyh-2* (<1%), *leu-5* (9%) (10, 1603, 1651).

Requires lysine. Will not use ε-hydroxynorleucine (751). Probably blocked in α-amino acid reductase (EC 1.2.1.31), converting aminoadipic semialdehyde to saccharopine (2113) (Fig. 40).

lys-3: lysine-3
IR. Between *al-1* (9%) and $In(OY323)^R$, *ace-3*, *nic-1* (9, 1971). Not included in duplications from $T(OY323) \times In(NM176)$; hence, left of *ace-3* (115, 1578).

Cloned by complementation (1815) and sequenced: GenBank 142777.

Requires lysine. Partial response to glutaric acid (1478). Lacks homocitrate synthase activity (EC 4.1.3.21) (887, 1478) (Fig. 40). Some alleles, such as 37402, are autonomous ascospore lethals or semilethals, producing mostly immature white spores [photograph in ref. (1986)]. Viability of the ascospores carrying the mutant allele is improved by long incubation (14). With other alleles, such as DS6-85, ascospores blacken and germinate normally. Malate and citrate are accumulated on limiting lysine concentrations (887). Four complementation groups have been identified (19). Allele 37402 was called *asco*.

lys-4: lysine-4
IR. Between *nuc-1* (1%), *met-10* and *his-3* (1%) (492, 1022, 1331).

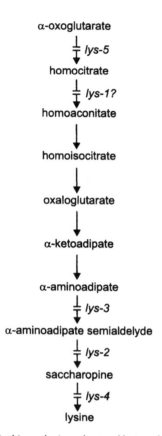

FIGURE 40 The biosynthetic pathway of lysine, showing sites or probable sites of gene action (246, 751, 887, 933, 1479, 2113). α-Amino-ε-hydroxycaproic acid can be converted to α-aminoadipate semialdehyde (2292), but apparently it is not an intermediate. Modified from ref. (1596), with permission from the American Society for Microbiology.

Requires lysine (751). Lacks saccharopine-cleaving enzyme activity, saccharopine dehydrogenase (L-lysine-forming) (EC 1.5.1.7) (2113) (Fig. 40). Complementation between alleles (20). Uses 0.5 mg/ml lysine.

lys-5: lysine-5

VIL. Between *cyt-2* (6%), *aro-6* (3%), and *un-4* (2%; on a common cosmid) (1596, 1815, 1986).
Cloned by complementation (1815).
Requires lysine. Partial response to glutaric acid (1478). Lacks homocitrate synthase activity (EC 4.1.3.21) (887, 1478) (Fig. 40). Some alleles, such as 37402, are autonomous ascospore lethals or semilethals, producing mostly immature white spores [photograph in ref. (1986)]. Viability of the ascospores carrying the mutant allele is improved by long incubation (14). With other alleles, such as DS6-85, ascospores blacken and germinate normally. Malate and citrate are accumulated on limiting lysine concentrations (887). Four complementa-

tion groups have been identified (19). Allele 37402 was called *asco*.

lys^R: lysine resistant

IR. Between *his-3* and *nic-2* (878, 1083).
Growth of *arg-1 lys^R* is resistant to normal inhibition by L-lysine. Proposed to be due to basic amino acid transport system (1083). May be allelic with *su(mtr)-1* (1082).

m: microconidial

Used for *pe*.

ma-1: malate utilization-1

Unmapped. Probably in a left arm, any of III–VII.
Unable to use malate as the carbon source when the tricarboxylic acid cycle is blocked by a *suc* (pyruvate carboxylase) mutation. Altered mitochondrial malate dehydrogenase. Scorable only in *suc* (1393, 1396, 1398).

ma-2: malate utilization-2

IIR. Between *un-20* and *ace-1* (210).
Unable to use malate as the carbon source when the tricarboxylic acid cycle is blocked by a *suc* (pyruvate carboxylase) mutation. Altered mitochondrial malate dehydrogenase. Scorable only in *suc* (1393, 1396, 1398).

mac: methionine–adenine–cystine

Used for *met-6* allele 65108. See *met-6*.

mad-1: MADS box-1

I. By correlation with cosmids assigned to linkage groups (573).
Cloned: Orbach-Sachs clone G11C02 (1268).
Cloned by PCR with degenerate primers encoding MADS box conserved domains. The predicted protein is virtually identical to the product of *Saccharomyces MCM1* (1268). Proteins including MADS box domains are DNA-binding proteins present in plants, animals, and fungi. The Mcm1 protein is a regulator of mating-type specific gene expression in *Saccharomyces* (1268).

mak-1: mitogen-activated protein kinase-1

Unmapped.
Cloned: Orbach-Sachs clone X13B11 (1267).
Cloned by PCR with degenerate primers encoding MAPK (mitogen-activated protein kinase)-conserved domains (1267). The predicted protein product shows similarity

to the product of *Saccharomyces MPK1* (also known as *SLT2*) (1267).

mak-2: mitogen-activated protein kinase-2

VII. By correlation with cosmids assigned to linkage groups (573).

Cloned: Orbach-Sachs clone G13A03 (1267).

Cloned by PCR with degenerate primers encoding MAPK (mitogen-activated protein kinase)-conserved domains (1267). The predicted protein shows similarity to the products of *Saccharomyces FUS3* and *KSS1*, which are involved in mating pheromone response in budding yeast (1267).

mak-3: mitogen-activated protein kinase-3

Unmapped.

Cloned and partially sequenced: EMBL/GenBank AI398492, AI416420; EST W08A7, W08A8.

The predicted protein resembles the product of *HOG1* of *Saccharomyces*, which is involved in cell-wall regulation and response to osmotic stress. Also resembles HOG1 of *Zygosaccharomyces rouxii* (1267).

mat: mating type

IL. Between *un-3* (0.1% or less) and *un-16* (<1%) (917, 1470, 1546). Between *eat-5* and *eat-1* (1691).

Genes at the *mat* locus are master regulators of mating, postfertilization development, and nuclear identity. Strains must be of opposite mating type, *mat A* and *mat a*, for a complex of events to occur that are associated with sexual reproduction and morphogenesis. These include the attraction of trichogynes to cells of opposite mating type (82, 182), pickup and transport of nuclei to the ascogonium, growth and development of perithecia, proliferation of heterokaryotic ascogenous hyphae, conjugate nuclear divisions in precrozier and crozier cells, and karyogamy. Vegetative cultures deleted for the *mat* genes are viable, grow normally, and form protoperithecia (629).

In *Neurospora crassa*, strains of opposite mating type are vegetatively incompatible. *mat A* + *mat a* combinations are unable to form stable heterokaryons (134, 721, 1629, 1786). Vegetative fusion is usually followed by cell death (721), but some *A* + *a* heterokaryons grow slowly (517, 779, 794). Heterozygous *A/a* duplications are highly abnormal, with inhibited growth, spider-like morphology, and darkening of agar (1475, 1561). Incompatibility in heterokaryons or in heterozygous duplications is relieved by the spontaneous deletion of either allele (517, 1468). Vegetative incompatibility is not expressed during the sexual phase. An active allele at the unlinked *tol* locus is necessary for the *mat*-mediated incompatibility reaction to occur in either heterokaryons or duplications. *tol* does not affect sexual compatibility or mating, however (1466). The sexual and vegetative manifestations of *mat* have been resolved mutationally, but not by genetic recombination (1470). Mutants selected by loss of vegetative incompatibility usually lose both sexual and vegetative functions simultaneously, and both functions usually are restored simultaneously in revertants selected for restoration of fertility (one null mutant gives atypical revertants) (517, 777, 779).

The *mat* locus is occupied by nonhomologous genes in the two mating types. *mat A* and *mat a* therefore are called idiomorphs rather than alleles (1332). Both idiomorphs have been cloned and sequenced (see following entries). The *A* idiomorph contains three open reading frames, the *a* idiomorph only one (Fig. 41). *mat A-1* and *mat a-1* appear to be essential for both mating and sexual development, whereas *mat A-2* and *mat A-3* increase the efficiency of sexual development but are not essential for

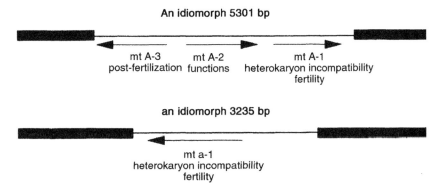

FIGURE 41 Structural and functional regions of the *mating type* idiomorphs of *N. crassa*. The idiomorphic DNA sequences are indicated by lines, and the conserved flanking regions are indicated by boxes. Arrows indicate the direction of transcription. From ref. (744) [Glass, N. L., and M. A. Nelson (1994). Mating-type genes in mycelial ascomycetes. In "The Mycota" (R. Brambl and G. A. Marzluf, eds.), Vol. I, pp. 295–306, Fig. 1.], with permission. Copyright 1994 Springer-Verlag.

ascospore production (629). When the idiomorph in a *mat A* strain is replaced by the *a* idiomorph, the strain becomes fully functional as *mat a* (356). All target genes essential for development as *mat a* therefore are present in strains of the opposite mating type.

Classical genetics has failed to reveal the presence of any genes related to mating functions in regions flanking *mat*, but molecular analysis has revealed regions contiguous to *mat* that are involved in mating and sexual development. A flanking region right of *mat* is divergent in sequence between species and often between mating types of the same species (1337, 1690). This region contains *eat* genes that appear to be involved in mating. *eat-1* is transcribed in *mat A* strains but not in *mat a*, in which a smaller transcript is produced. An allele of *eat-2* that has been inactivated by RIP is recessive vegetatively (slow growing, aconidiate), but is dominant (sterile) in the sexual phase (1691). Genes linked to mating type encode putative pheromone precursors that are expressed specifically in *mat A* (*ccg-4*) or in *mat a* (*mfa-1*); see also *scp*.

In the early literature, *A* was called + (plus) or *A*, whereas *a* was called − (minus) or *B* [e.g. ref. (547)]. The locus symbol was changed from *mt* to *mat* in 1997 to conform with usage in most other fungi (745). For reviews of mating type, including homologies and comparisons with other species and genera, see refs. (137), (407), (744), (1098), (1639), and (1976). Reference (407) is concerned especially with sexual development.

mat A: mating type A

IL.

Cloned and sequenced: sequences for component genes follow; Orbach-Sachs clones X1B02, X5A06, X5E06, X5E07, X11D06.

Symbol for the *A* idiomorph, with three open reading frames, *mat A-1*, *mat A-2*, and *mat A-3*, which are read divergently (630, 631) (Fig. 41). The abbreviated symbol

A or *mat A* is preferred for use in contexts where the *A-1*, *A-2*, and *A-3* constitution is irrelevant.

mat A-1: mating type A-1

IL. Right of *mat A-2*.

Cloned and sequenced: Swissprot MATA_NEUCR, EMBL/ GenBank M33876, PIR S65582, EMBL NCAMTR, GenBank NEUAMTR.

Encodes a DNA-binding protein similar to *Podospora* FMR1 in mating function (53). Mutational analysis (1796) (Fig. 42). The MAT A-1 polypeptide has a short but decisive stretch of homology with MAT1p of *Saccharomyces*. Inactivation by RIP shows *mat A-1* to be essential for mating identity, mating functions that follow fertilization, and *mat*-associated vegetative incompatibility (743, 1796). Mutant *mat A-1^{m99}*, which is heterokaryon-compatible and fertile with *mat a* when used as a female but not as a male, is due to a base-pair substitution that results in a truncated protein. Mutant *A-1^{m13}*, which is heterokaryon-compatible and produces perithecia but no ascospores with *mat a*, is a frameshift (1796). Other mutants lose both the mating and the vegetative incompatibility functions simultaneously (777) as a result of frameshifts in *mat A-1* (741).

mat A-2: mating type A-2

IL. Between *mat A-1* and *mat A-3*.

Cloned and sequenced: Swissprot MATC_NEUCR, EMBL/ GenBank M33876, PIR S65583, EMBL NCAMTR, GenBank NEUAMTR.

Encodes a putative DNA-binding protein similar to *Podospora* SMR1 in mating function (631). Strains mutant in *mat A-2* are normal in mating and fertility but are unaffected in vegetative incompatibility (629, 743). Strains mutant in both *A-2* and *A-3* are able to mate normally but are impaired for postmating steps in ascospore genesis, suggesting that MAT A-2 and MAT A-3 are redundant (629).

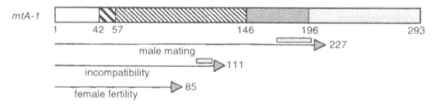

FIGURE 42 Summary of the proposed regional specialization in the *mat A-1* polypeptide of *N. crassa*, which is symbolized by a rectangle. The finely hatched rectangle encompasses a region of similarity to FMR1 of *Podospora*, MAT-1 of *Cochliobolus*, MATα1of *Kluyveromyces*, and MATα1p of *Saccharomyces*. The coarsely hatched area (positions 42–57) corresponds to the region of greatest similarity to MATα1p of *Saccharomyces*. The dark-shaded area represents the region of similarity unique to MAT A-1 and FMR1 of *Podospora*. The lightly shaded area represents the acidic C-terminus. The arrows represent the regions found to be minimally sufficient for the functions noted. The shaded segments above the arrows represent regions found to be required for the activity noted. From ref. (1796) [Saupe, S., L. Stenberg, K. T. Shiu, A. J. Griffiths, and N. L. Glass (1996). The molecular nature of mutations in the *mt A-1* gene of the *Neurospora crassa A* idiomorph and their relation to mating-type function. *Mol. Gen. Genet.* 250: 115–122, Fig. 5.], with permission. Copyright 1996 Springer-Verlag.

mat A-3: mating type A-3

IL. Left of *mat A-2*.

Cloned and sequenced: Swissprot MATD_NEUCR, EMBL/ GenBank M33876, PIR S65584, EMBL NCAMTR, Genbank NEUAMTR.

Encodes an HMG box DNA-binding protein similar to *Podospora* SMR2 in mating function (631, 1615). MAT A-3 contains a transcription activation domain (83). Not concerned with vegetative incompatibility. Strains mutant in *A-3* are normal in mating and fertility. Strains mutant in both *A-2* and *A-3* also are able to mate but are impaired for postmating steps in ascospore genesis, suggesting that MAT A-2 and MAT A-3 are redundant (629).

mat a: mating type a

IL.

Symbol for the *a* idiomorph, containing a single open reading frame (356). The abbreviated symbols *a* and *mat a* are preferred in most contexts.

mat a-1: mating type a-1

IL.

Cloned and sequenced: Swissprot MATB_NEUCR, EMBL/ GenBank M54787, Genbank NEUMTA1A, Orbach-Sachs clones X13E01, X22F11 (1978).

Encodes an HMG box DNA-binding protein, presumed to be a transcriptional regulator. Similar to *Podospora* FPR1 in mating function (53). The HMB box region is sufficient for DNA binding but not for mating-type activity. DNA-binding activity does not appear to be necessary for vegetative incompatibility (1614), but does appear to be necessary for mating activity. Changes within the carboxy-terminal region of the polypeptide eliminate vegetative incompatibility function without affecting mating-type function. The *mat a* gene is essential for mating identity, mating functions after fertilization, and *mat*-associated vegetative incompatibility (1614). Mutant a^{m33}, which is both heterokaryon-compatible and sexually competent with *mat A* strains, is due to a single-nucleotide change that alters the protein product near the carboxy terminus (1978). Other known *mat a-1^m* mutants have simultaneously lost both mating ability and vegetative incompatibility (777, 779).

Mating Type

Symbol changed from *mt* to *mat*. See *mat*.

mb-1: male barren-1

VII. Linked to *nic-3*, *wc-1* (23%) (1582, 2156).

Perithecial development is blocked when *mb-1* is used as male parent in crosses to an *mb+* female. Many perithecia are produced. These are mostly small, brown, and without beaks or ascospores, but a few become mature and produce ascospores (2156). Perithecia are normal and fertile when *mb-1* is used as female parent, fertilized by *mb+* (2202). Events following pachytene are not completely normal in the female, however, as observed cytologically (1673). Homozygous barren (1582). Recessive in heterokaryons, complementing *mb-2* and *mb-3* (2157). One occurrence, allele 8455.

mb-2: male barren-2

IR. Between *cyh-1* (5%) and *al-1* (7%) (1582).

Perithecial development is blocked when *mb-2* is used as male parent to fertilize an *mb+* female. Many perithecia are produced, mostly small and brown, without beaks or ascospores; a few perithecia mature and produce ascospores (2156). Perithecia are normal and fully fertile when *mb-2* is used as female parent and fertilized by *mb+* (2202). Homozygous barren (1582). Recessive in heterokaryons, complementing *mb-1* and *mb-3* (2157).

mb-3: male barren-3

IR. Linked to *cyh-1* (18%), *al-1* (2%), *mb-2* (6%) (1172, 1582).

Perithecial development is blocked when *mb-3* is used as male parent to fertilize an *mb+* female. Many perithecia are produced, mostly small, brown, and without beaks or ascospores; a few perithecia mature and produce ascospores (2156). Perithecia are normal and fully fertile when *mb-3* is used as female parent and fertilized by *mb+* (2202). Development of perithecia may then be slower than normal, however (1673). Homozygous barren (1582). Recessive in heterokaryons, complementing *mb-1* and *mb-2* (2157). Six occurrences.

mcb: microcycle blastoconidiation

VR. Linked to *al-3* (3%) (1250), *cyh-2*, *sod-2* (0T/18 asci) (1447).

Cloned and sequenced: EMBL/GenBank L78009, EMBL NCMCB, GenBank NEUMCB; Orbach-Sachs clone X13C10.

Encodes the regulatory subunit of cyclic AMP-dependent protein kinase (PKA). The mutant has increased PKA activity. The gene consists of two overlapping transcriptional units. Discovered as a mutant allele extracted from a wild-collected strain (1250). Cultures grown on agar are morphologically wild type at 25–30°C. When macroconidia from agar-grown mycelial cultures are germinated in liquid shake culture at 25°C, they generate hyphae that consist of chains of spherical cellular compartments (1250) (Fig. 43). Similar morphology is observed when hyphae grown at 25°C are switched to 37°C. When conidia are germinated at 37°C, the

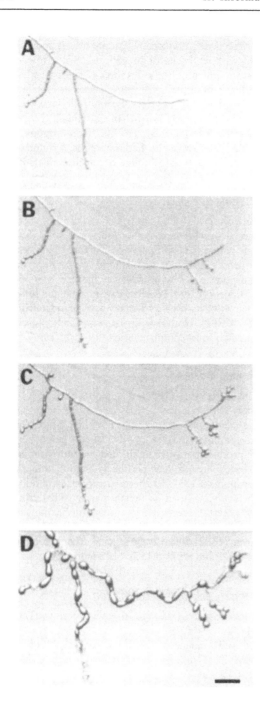

FIGURE 43 Loss of growth polarity in the *mcb* mutant at restrictive temperature. (A) Hyphae of an *mcb* strain after overnight incubation at 25°C on a dialysis membrane overlying sucrose minimal agar medium. The hyphae shown in (A) were shifted to 37°C and photographed after 1.5 (B), 3 (C), and 9 hr (D). Note that growth polarity is lost in all regions of *mcb* hyphae grown at 37°C for 9 hr, and that apolar growth continues until hyphal compartments burst (D). Bar length = 100 μm. From ref. (247) [Bruno, K. S. *et al.* (1996). *EMBO J.* **15**: 5772–5782.], with permission from Oxford University Press.

location of septa is abnormal, actin is disorganized, and conidia grow into spheres that enlarge and ultimately burst (247). Apolar growth increases the secretion of extracellular proteins (1152). Apolar growth is suppressed by *cr-1*. *mcb* may therefore provide an opportunity to select for new forward mutations at loci involved in the cAMP-dependent kinase regulatory pathway (1632). Intracellular actin levels do not increase in the *mcb* mutant under conditions that substantially increase the amount of growing surface area (2092).

mcm: microcycle microconidiation

IIL. Linked to *ro-7* (1%) (1250, 1251).

When macroconidia from agar-grown mycelial cultures are germinated in liquid shake culture at 22°C, the germlings produce uninucleate microconidia within 24 hr, reaching concentrations of 10^7/ml at 72 hr (1251). At 30°C, multinucleate arthroconidia and few or no microconidia are produced. The morphology of surface-grown cultures is normal wild type (Fig. 44).

md: mad

VR. Between *sh* (3%) and *sp* (9%) (569).

Spreading growth with a characteristic branching pattern. Modifies the banding phenotype of *cl*. Photographs (569).

mdk-1: mitotic division kinase-1

Unmapped.

Cloned by degenerate primers encoding MAPK (mitogen-activated protein kinase)-conserved domains (1267).

The predicted protein is very similar to the product of *Saccharomyces PHO85*, a *cdc2*-like kinase (1267). Characterization of genes regulating phosphorus uptake and metabolism in *Neurospora* predicts that the product of the uncloned *pgov* gene will be functionally and structurally like Pho85p (1322), suggesting that *mdk-1* and *pgov* may be allelic. Attempts to recover the gene from the Orbach-Sachs cosmid library or to clone it from the *Neurospora* chromosome have failed (1267), as has an attempt to retrieve it by tagged mutagenesis (1322). This suggests that the gene may be toxic in *Escherichia coli*.

mdk-2: mitotic division kinase-2

III or IV. By correlation with cosmids assigned to linkage groups (573).

Cloned: Orbach-Sachs clones G15B07, G15D07 (1267).

Cloned by PCR with degenerate primers encoding MAPK (mitogen-activated protein kinase)-conserved domains (1267). The predicted protein shows similarity to the product of *S. pombe dsk1* ("division suppressor kinase"),

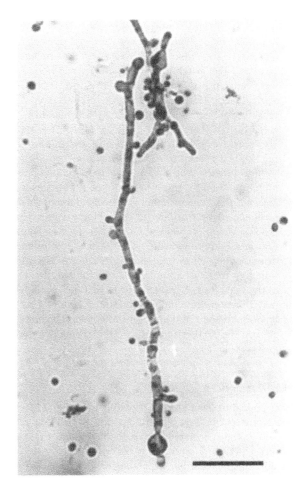

FIGURE 44 Microcycle microconidiation by an *mcm* strain ~12 hr after inoculation of air-grown macroconidia into liquid shake culture at 22°C. Microconidia are produced as lateral protuberances (arrows) on the microcycle structure produced by a germinated macroconidium. Free microconidia also are present. Bar length=25 μm. From ref. (1250), with permission from the Society for General Microbiology.

which was cloned in *S. pombe* as a multicopy suppressor of another division kinase. The mutant gene shows no obvious phenotype in *S. pombe*, however. No homologue is present in the *Saccharomyces* genome (1267).

me: methionine

Symbol changed to *met*.

mea-1: methylammonium resistant

I. Right of *acr-3* (9%) (2220).
Resistant to inhibition by methylammonium, a structural analog of ammonium. *nit-2; mea-1* double mutants show nitrogen-starved growth on ammonium (566). Confers resistance to 50 mM methylamine in nitrate medium (2220).

med: medusa

IVR. Linked to *met-5* (5%), *pan-1* (8%) (719).
A slow-growing, spreading morphological mutant forming distinctive grooves on the surface of agar (719). Photograph (719). Cell-wall peptides are reduced in amount (2249).

mei: meiotic

This name has typically been used for mutants having partial or almost complete blocks to meiosis and ascus development. Both recessive and dominant meiotic mutants are known. Crosses homozygous for recessive mutants usually produce few or no viable black ascospores; those that germinate may show much reduced meiotic recombination and greatly increased nondisjunction. Abnormal segregation results in white, inviable, and hypohaploid ascospores. Production of white ascospores provides a basis for isolation and scoring (1933). Sexual-phase recessive mutants that affect meiotic development are very common in natural populations of *Neurospora crassa* (1174). Some of the mutants designated *mei* specifically affect meiotic or premeiotic events (e.g., *mei-1* and *mei-4*), and similar mutants may have been given other names such as *asc*, *fmf-1*, or *mb*. Other meiotic-defective mutants may be more generally deficient in recombinational DNA repair [e.g., *mei-2* and *mei-3* (1838)]. These typically show hypersensitivity to genotoxic agents as well as alterations in mutation and/or mitotic recombination (Table 3). They may be isolated by sensitivity to UV or other mutagens, and they are most commonly called *uvs* or *mus*. Mutagen-sensitive mutants with meiotic effects include *uvs-3*, *-5*, and *6*; *mus-7*, *-8*, *-9* and *11*, and, with the exception of *mus-18*, all mutants from *mus-15* to *mus-25*.

mei-1: meiotic-1

IVR. Linked to *arg-2* (<1%), probably to the right (1933).
Meiosis is impaired in homozygous crosses. There is no defect in growth or DNA repair. Meiotic divisions occur and many ascospores are produced, but 70–90% are inviable and white. The viable ascospores are usually disomic for one or more linkage groups, indicating high nondisjunction at the first division (1933). Chromosome pairing is defective: axial elements of a synaptonemal complex are present, but a complete complex is rarely seen. Defects at anaphases I and II lead to four-poled second- and third-division spindles (1232). The *mei-1* mutation is present in wild-collected strain Abbott 4A (1933), which is an ancestor of many Beadle and Tatum mutants (136). Two mutants obtained in ref. (518), *asc(DL95)* and *asc(DL243)*, complemented *mei-1* and each other and showed no recombination in intercrosses, suggesting allelic complementation and/or

a gene cluster. Abnormal disjunction in these crosses resulted from abnormal pairing in meiosis I (519). *asc(DL95)* resembles *mei-1* and may well be allelic; but defects are less extreme. Mutant *DL243* differs from *mei-1* with a primary defect early, during karyogamy, and pleiotropic effects later, causing high nondisjunction during the second division and postmeiotic divisions (519).

mei-2: meiotic-2

VR. Between *inv* (18%) and *his-6* (12%) (956, 1933). Linked to *mus-11* (5%; 10%) (1840).

Uvs-6 epistasis group (958) (Table 3). Early dominant effects were eliminated by backcrossing to produce the commonly used strain *mei-2* ALS181 (1840). A second allele is *mei-2* SA60 (956). In homozygous *mei-2* crosses, meiotic divisions occur and asci appear to be normal, but they contain few mature ascospores. Chromosome pairing is greatly reduced or absent, resulting in aberrant segregations at anaphase I and often at subsequent divisions [B. C. Lu, cited in refs. (1841) and (1933)]. Meiotic crossing over similarly is reduced and nondisjunction increased, producing aneuploid ascospores, monosomic white inviable spores, and black ones disomic for one or more linkage groups (1841, 1933). Heterozygous crosses produce few white spores, and mitotic recombination is normal (1841). RIP and presumably DNA pairing are not inhibited by *mei-2* (663). Crosses homozygous for *mei-2* are used in the "sheltered RIP" procedure for determining whether the function of a cloned gene is essential for survival (839, 1335). *mei-2* is DNA-repair-defective and sensitive to MMS, histidine, and γ rays (1840). In contrast to *mei-3*, *mei-2* has normal spontaneous and UV-induced mutation (958). The mutant affects expression of a peptide that is inducible by DNA damage (911, 913). Epistasis tests with *uvs-3* suggest that *mei-2* may function in more than one type of repair (958), as is known for some DNA replication genes [e.g., CDC9 (ligase) in yeast], or possibly there may be an overlap between repair types of the Uvs-6 and Uvs-3 epistasis groups.

mei-3: meiotic-3

IL. Between *arg-1* (3%) and $T(39311)^R$. Probably right of *eth-1* (1%) and *arg-3* (1%) (1469, 1580).

Cloned and sequenced: EMBL/GenBank D29638, L02428, PIR S70629, GenBank NEUMEI3, NEUMEI3A; Orbach-Sachs clones X5D09, X10C02, X10D03, X11H02, X14G03, G1A07, G3B05, G10H05, G11B01.

Uvs-6 epistasis group (958) (Table 3). A homolog of *Escherichia coli recA*, *Saccharomyces RAD51*, and *Aspergillus uvsC* (373, 853, 2141). MEI-3 protein is located specifically in perithecia, but expression also is increased in growing mycelia after UV or MMS treatment (853). The seven known *mei-3* alleles are all recessive. The first (N289) was found because it increased the instability of duplications (1469). Some were isolated by MMS sensitivity [e.g., SA10 (958) and SC25, SC29 (520, 521)]. *mei-3* (N289) typically is sensitive to ionizing radiation, MMS, histidine, and (slightly) UV (best scored at 39°C) (1469, 1473). It is also sensitive to hydroxyurea (1840) and mitomycin-C (381). Spontaneous mutation is much increased (mutator phenotype), but further, UV-induced increases in mutation resemble those of wild type (958). Sometimes gives stop–start growth on minimal medium, made more pronounced by histidine (1467). Homozygous barren (1469). Meiosis is blocked in zygotene (1685). Duplication instability is increased (1469), possibly by mitotic deletion. *mei-3* may be deficient for mitotic recombination, as is the yeast homolog *rad51*. Homology to a gene of the yeast Rad52 group suggests that *mei-3* and *mei-2* may function in recombination and possibly in recombinational repair of double-strand breaks. The mutation affects the expression of a DNA-damage-inducible polypeptide (911, 913).

mei-4: meiotic-4

IIIR. Linked to *leu-1* (12%), probably to the left (1469).

A meiotic-specific mutant without defects in DNA repair (1469, 1840). Isolated from a double-mutant strain with *mei-3* (1473). Homozygous barren. Recessive. Expression is highly variable, depending on genetic background (1469). The more extreme genotypes block sexual development at crozier formation or karyogamy (1685). Less extreme genotypes complete a normal first meiotic division but become irregular at later divisions, producing abnormal spores [B. C. Lu and D. R. Galeazzi, cited in ref. (1685)].

mek-1: MAPK/ERK kinase

Unmapped.

Cloned: Orbach-Sachs clone G09C10 (1267).

Cloned by PCR with degenerate primers encoding conserved domains of MAPK (mitogen-activated protein kinase) and MEKK (MAP/ERK kinase kinase)-conserved domains (1267). The predicted protein shows similarity to products of *Saccharomyces MKK1* and *MKK2*, which are involved in a protein kinase C-responsive MAPK pathway. Proteins of this type phosphorylate MEK proteins in other organisms (1267).

mel-1: melon-1

VIIL. Left of *thi-3* (27%) (1604).

A hemispherical colony is formed similar to *bal* (1409). Growth is stimulated rather than depressed by sorbose

(1407). Cell-wall analysis (507). Photographs (507, 1409). Called *col(C-L2b)*. Not tested for allelism with *do*.

mel-2: melon-2

Allelic with *bal* (1585).

mel-3: melon-3

III. Linked to *leu-1* (17%) (1409).

Grows as a hemispherical colony similar to *bal*. A modifier gene also is located in III (1409).

mep: methylpurine resistant

Mutants resistant to inhibition by 6-methylpurine (6MP) are potentially useful as markers. Mutants called *mep(3)* and *mep(10)* were described by refs. (1542) and (1545) as differing phenotypically. Both have been mapped in IL. Another 6MP-resistant mutation was obtained independently in a *T(OY321)* strain (1322), mapped in IL between *leu-3* (10%) and *T(OY321)* ("tightly linked"), and shown to be recessive in partial diploids (261). Allelism of the three mutants has not been tested. Because a stock of *mep(3)* received by R. L. Metzenberg (1322) proved to be heterokaryotic for a nonresistant component, doubt is cast on the validity of the phenotypic diagnosis given next for *mep(3)*. Possibly all three *mep* mutations may be at the same IL locus.

mep(3): methylpurine resistant(3)

IL. Left of *mat* (1322).

Resistant to inhibition by 6MP. The activity of adenine phosphoribosyltransferase (PRTase) (EC 2.4.2.7) is nearly normal *in vitro*. Adenine uptake is normal. Has low hypoxanthine PRTase, as do *mep-2* and *ad-8*. Selected on 1 mM 6MP sorbose medium (1542). 6MP prevents purple pigment production by *ad-3* on low adenine, but does not prevent pigment production by *ad-3A mep(3)* double mutants, suggesting that *mep(3)* is altered in the regulation of *de novo* purine synthesis. The phenotype is consistent with lowered affinity of glutamine amidotransferase for 6MP as a feedback inhibitor (1545). Not tested for allelism with *mep-2* or with *adenine* mutants. Called *Mep^r-3* (1542), *mep-1* (107). If these results were obtained using the heterokaryotic stock received by R. L. Metzenberg (1322), the *mep(3)* and *mep(10)* mutations could be at the same locus.

mep(10): methylpurine resistant(10)

IL. Linked to *leu-3* (0/120) (1546).

Resistant to inhibition by 6MP. PRTase activity is negligible *in vitro*. Adenine uptake is normal. Purine synthesis is not inhibited by 6MP (1545). Resistance may result from the inability to convert the analog to the nucleotide

form. Has low hypoxanthine PRTase, as do *mep(3)* and *ad-8*. Selected on 1 mM 6MP sorbose medium. Not tested for allelism with *mep(3)* or with another IL-linked *mep* mutation mapped in ref. (261). Called *Mep^r-10*. Called *mep-2* in ref. (107).

met: methionine

Auxotrophs designated *met* are able to use methionine but not cysteine. Some can use the methionine precursors homocysteine and cystathionine (Fig. 45). Mutants able to use cysteine as well as methionine are designated *cys*. Reviewed in ref. (653). For regulation, see ref. (1874) and entries for individual loci. Methionine starvation of mutants in the *met* pathway results in decreased DNA methylation (1732).

met-1: methionine-1

IVR. Between *oxD* (3%), *T(B362i)^L* and *T(B362i)^R*, *mus-26* (2%), *col-4* (4%) (111, 322, 1411, 1490, 1578).

Uses methionine but not homocysteine (893) (see Fig. 45). Lacks methylene tetrahydrofolate reductase (EC 1.5.1.5), and, thus, lacks the coenzyme needed for transmethylating homocysteine (260, 1872, 1873). A report that *met-1* also lacks cystathionine synthase (1045) proved to be incorrect, due to the fact that methyltetrahydrofolate is an essential activator for cystathionine synthase (1874). Methylene THF reductase is feedback-inhibited by *S*-adenosylmethionine (260). Used in heteroallelic duplications from *T(S1229)* to assay mitotic recombination (113). Methionine starvation decreases DNA methylation (1732).

met-2: methionine-2

IVR. Linked to *ilv-3* (0/129) (1124). Between *trp-4* (6%) and *pan-1* (4%) (1410).

Uses methionine or homocysteine; accumulates cystathionine (893). Lacks cystathionase II (EC 4.4.1.8) (655) (Fig. 45). Fine-structure map (1413, 1417). Used in major studies of intralocus recombination and its polarity (1413, 1417). Complementation map (1410). Methionine starvation decreases DNA methylation (1732).

met-3: methionine-3

VR. Between *pab-1* (1%), *trp-5* (4%) and *pk* (1%) (13, 212, 569, 673, 2014).

Uses methionine, homocysteine, or cystathionine (893). Putative structural gene for cystathionine-β-synthase (EC 4.2.1.22) (1045) (Fig. 45). This enzyme is also lacking in *met-7* (1045). The enzyme is activated by methyl tetrahydrofolate and inhibited by *S*-adenosylmethionine (1045, 1874). A temperature-sensitive *met-3* allele is deficient in DNA methylation (1732).

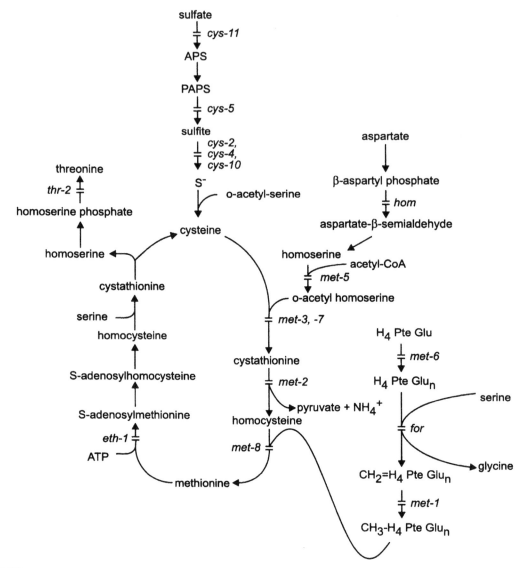

FIGURE 45 The sulfur assimilatory pathway leading from inorganic sulfate to cysteine, methionine, homoserine, and threonine. Mutants that affect specific steps are identified. For the conversion of threonine to isoleucine, see Fig. 39. It is not clear whether the polyglutamylation step controlled by *met-6* occurs only at the stage shown. Abbreviations: APS, adenosine 5'-phosphosulfate; H4PteGlu, tetrahydrofolate; PAPS, 3'-phosphoadenosine = 5'-phosphosulfate. Adapted from refs. (1284) and (1596).

met-4: methionine-4

Changed to *cys-10* (1414).

met-5: methionine-5

IVR. Between *his-4* (4%) and *nit-3* (15%) (212, 1582, 1585).

Cloned: pSV50 clone 18:5B.

Uses cystathionine, homocysteine, or methionine (656, 1411). Defective homoserine transacetylase (EC

2.3.1.31) (1045, 1426) (Fig. 45). Methionine starvation decreases DNA methylation (1732).

met-6: methionine-6

IR. Between *thi-1* (7%; 14%), *T(NM103)*, *T(ALS182)*, *T(OY343)* and *tre* (7%), *ad-9* (2%; 16%) (890, 1415, 1548, 1580, 1582, 1818, 2123).

Cloned and sequenced: EMBL/GenBank AF005040, EMBL NCAF5040, pSV50 clones 4:11C, 7:10D, 8:1F, 8:4H (68).

Structural gene for folyl polyglutamate synthetase (EC 6.3.2.17). Requires methionine and does not use precursors [(1411); N. Horowitz, cited in ref. (2294)] (Fig. 45). Used to study the polarity of intralocus recombination and to show that polarity with respect to flanking markers is not reversed when the met-6 region is inverted relative to the centromere (1416, 1417). Used to show that methionine starvation decreases DNA methylation (1732). Strains carrying allele 65108, called *mac*, were reported to differ from other met-6 mutants in having an accessory requirement for adenine and possibly cystine (556), whereas met-6 (35809) and (S2706) are stimulated by adenine only in a CO_2-enriched atmosphere [G. Roberts, cited in ref. (1417)]. met-6 (65108) (*mac*) and met-6 (35809) evidently lack different folyl polyglutamyl synthetase activities (EC 6.3.2.17) (410, 1728). This may be due to alterations in different domains of the same gene product. The cloned met-6$^+$ gene increases the level of folyl hexaglutamates when transformed into met-6 or *mac* mutants (67). A highly conserved Ser residue has been changed to Pro in allele 35809. Sequence polymorphisms among wild-type strains (68).

met-7: methionine-7

VIIR. Between qa-2 (<1%), ars (<1%), Cen-VII (<1%) and met-9 (10^{-4}), wc-1 (1%; 4%) (295, 304, 340, 1411, 1418).

Cloned and sequenced: Swissprot MET7_NEUCR, EMBL/GenBank M64066, PIR JQ1524, EMBL NCMET7A, GenBank NEUMET7A; Orbach-Sachs clones G7G03, G17H07.

Uses cystathionine, homocysteine, or methionine [(1411); N. Horowitz, cited in ref. (2294)]. Structural gene for cystathionine γ-synthase (EC 4.2.99.9) (427, 1045) (Fig. 45). Note that cystathionine-β-synthase (EC 4.2.1.22) is lacking in met-3 (1045). See met-3 for regulation. Apparently contiguous with met-9 by co-conversion. Flanking markers are recombined in most met-7$^+$ met-9$^+$ recombinants (1418). Functionally distinct from met-9, which has active cystathionine synthase (1045) but cannot use homocysteine. No mutants lacking both functions have been isolated. Allele NM251 is suppressible by ssu-1 (1418). Allele K79 is inseparable from reciprocal translocation T(K79) (1580).

met-8: methionine-8

IIIR. Between ace-2, ff-5 (1%, 4%) and ad-4 (4%), leu-1 (426, 1411, 1546, 1591, 2044).

Cloned and sequenced: GenBank A1391967; EST NC1D10. Uses methionine but not its precursors (1411). Lacks methyl THF homocysteine transmethylase (EC 2.1.1.13) (260, 1873) (Fig. 45). Methionine starvation decreases DNA methylation (1732).

met-9: methionine-9

VIIR. Between met-7 (10^{-4}) and wc-1 (1% or 2%) (1418, 1591).

Requires methionine; cannot use precursors (556, 1411, 1418). Co-converted with met-7 (1418). Functionally distinct from met-7. The met-9 mutant retains the met-7$^+$ function, producing cystathionine synthase (1045). Allele NM43 is heat-sensitive (1418). Starvation for methionine decreases DNA methylation (1732).

met-10: methionine-10

IR. Between his-2, nuc-1, T(AR173)^R and lys-4, his-3, ad-3A (1022, 1322, 1469, 1578).

Cloned and sequenced: PIR JC4255, EMBL/GenBank L40806, EMBL NCORF, GenBank NEUORF.

Requires methionine. Encodes a 475-residue polypeptide of unknown function (377). The only known allele is heat-sensitive, with requirement at 34°C but not at 25°C (1592). Does not grow at 39°C even with methionine (1469). Methionine starvation decreases DNA methylation (1732).

meth: methionine

Symbol changed to *met*.

Methionine Overproduction

See eth-1 (1028).

mfA-1: mating factor A-1

Allelic with ccg-4.

mfa-1: mating factor a-1

VR. Linked to leu-5 (0T/ 18 asci) (1447).

Cloned and sequenced: Orbach-Sachs clones X13E01, X22F11.

Encodes a pheromone precursor expressed in *mat a*. The sequence suggests an isoprenylated fungal pheromone similar to the a-factor of *Saccharomyces* (573). For other genes expressed differentially in the two mating types, see ccg-4, eat-1, eat-2, and scp.

mgk-1: glycogen synthase kinase-1

III. By correlation with cosmids assigned to linkage groups (573).

Cloned: Orbach-Sachs clone G03D09 (1267).

Cloned by PCR with degenerate primers encoding MAPK (mitogen-activated protein kinase)-conserved domains (1267). The predicted protein shows similarity to glycogen synthase kinases, which are conserved in fungi and animals where they have roles in development and gene regulation (1267).

Microconidiation

Microconidia, which are uninucleate, are developmentally and morphologically distinct from macroconidia. Microconidia in wild type are obscured by the presence of macroconidia. A simple method has been devised for obtaining large numbers of almost pure microconidia from wild type (1528). Microconidia can also be obtained in large numbers using liquid shake cultures inoculated with macroconidia of the *mcm* mutant (1251). Morphologically mutant strains *pe* and *dn* produce exceptionally high numbers of microconidia, together with macroconidia. Macroconidiation is eliminated in the *fl* (*fluffy*) mutant, but microconidiation is not. Double-mutant strains of constitution *pe fl* and *fl; dn* have long been used as sources of pure microconidia. For a photograph of microconidiophores, see Fig. 46 (1528). For an SEM photograph of microconidia erupting from hyphae, see ref. (1959). For a TEM photograph of microconidia from *pe; fl*, see ref. (1230). For a review of *Neurospora* microconidia, see ref. (1252).

mig: migration of trehalase

Allelic with *tre* (2210).

mik-1: MEK kinase-1

Unmapped.
Cloned: Orbach-Sachs clone G01H02 (1267).
Cloned by PCR with degenerate primers encoding MEKK-conserved domains (1267). The predicted protein shows similarity to the product of *Saccharomyces SLK1* (also known as *BCK1*), which is involved in a protein kinase C-responsive MAPK pathway. Proteins of this type phosphorylate MEK proteins in other organisms (1267).

mik-2: MEK kinase-2

Allelic with *nrc-1*.

mip-1: MIP1-like DNA polymerase

IIIR. Between *pro-1* and *ad-2*. Linked to *con-7, trp-1* (1T/18 asci) (1078, 1079).
Cloned and sequenced: EMBL/GenBank AF111068; Orbach-Sachs clone X25C10.
Specifies mitochondrial DNA polymerase γ (large catalytic subunit) (EC 2.7.7.7), resembling MIP1 polymerase of *Saccharomyces* and POL-G of *Escherichia coli* (1078, 1079).

mlh-1: mutL homolog-1

IV. Linked to *mtr* (0/18) (956, 2254).

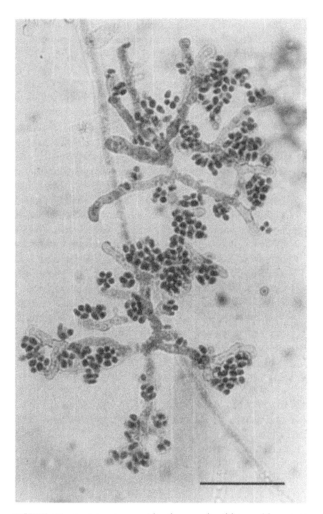

FIGURE 46   Aerial microconidiophores of wild-type *N. crassa*, formed after the removal of cellophane from the surface of an agar culture following the procedure of Pandit and Maheshwari (1528). Stained with acid fuchsin. Groups of microconidia are seen arising from very short protuberances from branches of two microconidiophores. Bar length = 50 μm. From ref. (1528).

Cloned and sequenced: Orbach-Sachs clones X14D11, X19B3, X21B5.
Homolog of the gene that encodes the *Escherichia coli* mismatch repair protein MUTL (2254). RIP-inactivated mutants show a mutator phenotype, are not sensitive to mutagens (UV, MMS, NG, cisplatin II), and are homozygous-fertile (2254).

mo: morphological

A miscellaneous group of mutants differing visibly from the wild type in vegetative morphology. The symbol *morph* has also been used. Other categories of morphological mutants have been designated *col, spco, smco,* or *moe* or have been assigned descriptive names such as *bal, fr, ro, sc.* For reviews covering morphological mutants and morphogenesis, see refs. (235), (384), (1281), (1353), (1848), (1851), and (2117). Reference (719) describes and gives photographs of numerous morphological mutants. Growth rates and hyphal diameters of 18 morphological mutants are given in ref. (384). For scanning electron microscope photographs of numerous morphological mutants, see refs. (1955), and (1958). A selective enrichment technique for obtaining morphological mutants has been described on the basis of their inability to use acetate and glutamate as sources of carbon and nitrogen (530). Mutations in *ropy* genes, which affect dynein or related molecular motors, can be obtained by selecting suppressors of *cot-1* (247, 1634). *crisp* mutations can be obtained by selecting suppressors of *mcb* (247). Mutations in *osmotic* and *spco* genes can be obtained by selecting for resistance to dicarboximide fungicides (788–790). The Fungal Genetics Stock Center lists ~30 morphological mutants that have been mapped to a linkage group, but have not been tested for allelism with already mapped morphologicals (700). These are not listed here.

mo-1: morphological-1

I. Linked to *mat* (9%) (719).
Growth from ascospores is slow (719).

mo-2: morphological-2

VII. Linked to *for* (16%), *nt* (29%) (719, 1582).
The mycelium is slow-growing and poorly pigmented. Conidia are morphologically abnormal. Recovery from ascospores is poor (1582). Scanning EM photograph (1955).

mo-3: morphological-3

Allelic with *sk*.

mo-4: morphological-4

IIIR. Right of *leu-1* (8%). Linked to *pro-1* (10%) (719).
Conidiation occurs throughout a slant. Complements *col-14*, *col-16*, *spg* (719).

mo-5: morphological-5

I. Linked to *mat* (20%) (719).
Few conidia. May make exudate on slant (719).

mo(KH160): morphological(KH160)

Changed to *shg*.

mo(P1163): morphological(P1163)

Changed to *dr*.

mo(P2402t): morphological(P2402t)

Changed to *un-20*.

mod-5: modifier of permeability-5

VI. Linked to *Cen-VI* (3% second-division segregation), *trp-2* (1973). Probably an allele of *cpc-1* in VIL, but not tested for allelism.
Improves the growth on complex media of *trp-1, -2, -3,* and *-4, aro-1, tyr-1,* and *-3, pt, met-7,* and *pyr-1*; increases sensitivity to 4-methyltryptophan and *p*-fluorophenylalanine. Recessive in heterokaryons. Attributed to a permeability change that facilitates the entry of metabolites (109, 1973). Scorable on slants of minimal medium + 4-methyl-DL-tryptophan (0.9 mg/ml, autoclaved in medium; tests read at 7 days, 34°C) (1582). *mod-5* differs from *cpc-1* (*mts*) in enabling excellent growth of *pyr-1* (H263) on complex media (323).

mod(os-5): modifier of os-5

Unmapped.
Restores normal morphology and osmotic sensitivity to *os-5* while retaining resistance to dicarboximide fungicides (786).

mod(pr): modifier of partial resistance to killing

IIIR. Between *leu-1* and *his-7*. Outside the recombination block that is associated with *Sk-2* (2121).
Enhances the resistance to killing by *Spore killer-2* that is conferred by *pr(Sk-2)* (2121).

mod(sc): modifier of scumbo

IV. Linked to *pan-1* (17%) (930).
Restricts the growth of *sc* but not of *cr-1, fr, bis, sp,* or wild type (930).

moe-1: morphological, environment sensitive-1

Probably an *sk* allele (1582).

moe-2: morphological, environment sensitive-2

VI. Linked to *trp-2* (14%), probably to the left (719).
Grows with concentric zones on minimal medium and as a restricted colony on glycerol complete medium (34°C). For photographs of allele R2532, see Figs. 23 and 24 of ref. (719). *scot* was probably present in the strain of origin.

moe-3: morphological, environment sensitive-3

IV. Left of *pan-1* (17%; 25%), *bd* (20%) (1821).
Blocks conidial germination at high temperature. Mycelial growth is colonial at high temperature if on dialysis

tubing on an agar surface, but is fairly normal if submerged. Shows a strong circadian conidiation rhythm at low temperature (1821). The effect on conidial germination (but not on vegetative growth) is counteracted by high conidial concentration or by CO_2 (360, 1821). Histidine is stimulatory, but there is disagreement as to whether it affects germination or vegetative growth (359, 1821). Partially curable by siderophores (ferricrocin). Conidia rapidly lose siderophores on contact with aqueous medium, even at permissive temperatures, suggesting an altered plasma membrane attachment site (360). Called JS134-9.

mom: mitochondrial outer membrane

Changed to *tom: translocase outer mitochondrial membrane.*

mom38: mitochondrial outer membrane 38

Changed to *tom40.*

morph: morphological

Symbol changed to *mo.*

mpp: mitochondrial processing peptidase

Unmapped.
Cloned and sequenced: Swissprot MPP1_NEUCR, EMBL/GenBank J05484, PIR A36442, EMBL NCMPPX, GenBank NEUMPPX.
Encodes the 57-kDa subunit of mitochondrial processing peptidase (EC 3.4.24.64) (1823). MPP cooperates with PEP (processing enhancing protein, encoded by *pep*) (860) in the proteolytic cleavage of matrix-targeting sequences from nuclear-encoded mitochondrial precursor proteins. (The 52-kDa subunit is encoded by *pep.*)

mrp(3): mitochondrial ribosomal protein (3)

Unmapped.
Cloned and sequenced: Swissprot ODP2_NEUCR, EMBL/GenBank J04432, PIR A30775, GenBank NEURPNUC.
Encodes a mitochondrial protein found both on ribosomes and in membrane fractions. Includes a domain similar to one found in dihydrolipoamide acetyltransferases (1097, 1750). Called *mrp-3.* Genes encoding at least two other proteins, MRP13 and MRP16, are said to be under study, and an account of a gene called *mrp-15* is said to be in preparation (2193).

msh-1: mutS homolog-1

IL. Linked to *cyt-21* (0/18) (956, 2254).
Cloned and partially sequenced.

Homolog of genes that encode *Escherichia coli* mismatch repair protein MUTS and *Saccharomyces* MSH1, MSH3, and MSH6 (956, 2254). Called *msh-X.*

msh-2: mutS homolog-2

VIIR. Linked to *for, frq* (0T/18 asci). Right of *Cen-VII* (1T/18 asci) (1447), (0/18) (956).
Cloned and sequenced: EMBL/GenBank AF030634; Orbach-Sachs cosmids X95H, X179E (956), X20E3, G1H8; YACs 8:E:6, 17:C:11 (162).
Homolog of genes that encode mismatch repair proteins *Escherichia coli* MUTS and *Saccharomyces* MSH2 (935, 2254). RIP-inactivated mutants show a mutator phenotype, are not sensitive to mutagens (UV, MMS, NG, cisplatin II), and are homozygous-fertile (2254).

msk-1: STE20-like MAP kinase-1

I. By correlation with cosmids assigned to linkage groups (573).
Cloned: Orbach-Sachs clone G03H05 (1267).
Cloned by PCR with degenerate primers encoding MEKK-conserved domains (1267). The predicted protein shows similarity to the products of *Saccharomyces* STE20, SPS1, and CLA4 and *S. pombe* shk1 and to vertebrate PAK proteins. In other systems, this class of kinases phosphorylates/activates the MEKK proteins to initiate the MAPK/kinase cascade (1267).

mt: mating type

This symbol, long used for *mating type*, was changed to *mat* in 1997 (745).

mtr: methyltryptophan resistant

IVR. Between *pdx-1* (2%) and *col-4* (1%) (196, 1990).
Cloned and sequenced: Swissprot MTR_NEUCR, EMBL/GenBank L34605, S81767, PIR A54551, EMBL NCMTR, GenBank NEUMTR; pSV50 clones 5:10A, 5:3B, 5:4H.
Structural gene for the transport of neutral aliphatic and aromatic amino acids via amino acid transport system I as defined by ref. (513), (1177), (1521), (1990), and (2233) (Fig. 47). Resistant to 4-methyltryptophan (MT) and *p*-fluorophenylalanine (FPA). Resistance is recessive in heterokaryons and in duplications from *T(S1229).* Mutation causes alteration in surface glycoproteins (2016). Used extensively for studying transport [(512); reviewed in (2231) and (2233)], mutation (541, 1981, 1982, 2192), intergenic deletion mapping (1688, 2192), and recombination (1692). Forward mutations may be obtained by selection for resistance or for defects in uptake ability; revertants are selected by the ability of an auxotroph to grow on appropriately supplemented

OUTSIDE

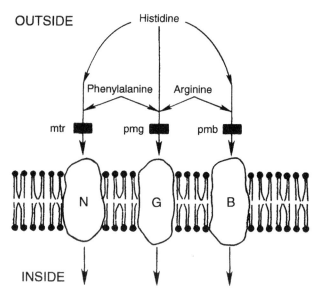

INSIDE

FIGURE 47 A schematic diagram of the three major systems for transport of amino acids across the *Neurospora* plasma membrane: neutral (N), general (G), and basic (B). Histidine is transported by all three systems. Phenylalanine is transported by the N and G systems, whereas arginine is transported by the B and G systems. Mutations that reduce or eliminate these transport activities are represented by heavy black lines intersecting the arrows and are labeled with the standard *Neurospora* designations *mtr*, *pmg*, and *pmb*. Whether the systems consist of more than one gene product is not known. N = system I (1521), G = system II, and B = system III. From ref. (2017), with permission from the National Research Council of Canada.

medium (1981). Used in one component of a heterokaryon test system for measuring the rate of recessive lethal mutation throughout the genome (1979, 1980) (Fig. 22). Suppressors have been used for selecting other resistance mutants (227, 228, 1067, 1991). FPA resistance is suppressed by *am* (1335). A light-regulatable *al-3::mtr* fusion gene, which is FPA-sensitive in the light and FPA-resistant in the dark, was used to select light-regulation mutants (286). Ascospores carrying *mtr* are slow to darken and mature; up to 50% of young ascospores from heterozygous crosses are white (318, 1582). With probable allele MN18, ascospore viability is improved by adding peptone to the crossing medium when the male parent is added (318). For scoring, medium containing 15 mg/ml FPA is recommended as an alternative to MT (541). Unlike MT, FPA is heat-stable and can be added before autoclaving . Mutants called *pmn* (=*Pm-N*, *pm n*), selected by resistance to FPA, have been shown allelic with *mtr* [(1776); also see ref. (513)]. Strains originally designated *neu^a*, *neu^r*, *neu^t* (1243) may be *mtr* alleles. Called *mt* (1177).

mts: methyltryptophan sensitive
Allelic with *cpc-1* (327, 1081).

multicent
This name has been given to strains that have centromere regions suitably marked for assigning genes or chromosome rearrangements to linkage groups. Strains, markers, and methods are described in ref. (1571). Strains *multicent-1, -3, -4,* and *-5* have been especially useful for mapping chromosome rearrangements, especially translocations. Scoring of marker phenotypes is somewhat more laborious with *multicent* strains than with *alcoy; csp-2*, which is therefore preferred for locating point mutants. For ease of crossing and stock-keeping, phenotypically wild-type heterokaryons have been constructed of *multicent-3, -4,* and *-5* in combination with the inactive-mating-type *helper* strain *a^{m1} ad-3B cyh-1*.

multicent-1
Linkage tester strains (of both mating types) with *mat, bal, acr-2, pdx-1, at, ylo-l,* and *wc-1* marking the centromere-proximal regions of linkage groups I–VII (1556, 1571).

multicent-2
A strain with *mat-a un-2, arg-5, thi-4, pyr-1, lys-1 inl,* and *nic-3 ars-1* marking six linkage groups in OR-like genetic background. This was crossed with the highly polymorphic wild-collected strain Mauriceville-1c *A* to obtain a kit of progeny for use in RFLP mapping (1334, 1339, 1340) (Appendix 3).

multicent-3
Strains of constitution *In(IL;IR)OY323, arg-5, acr-2, pdx-1, at, ylo-1,* and *wc-1* (1571). These increase efficiency over *multicent-1* by introducing the inversion as a crossover suppressor in I and by substituting *arg-5* for *bal* in II. Because the inversion is heterozygous in test crosses, markers in most of linkage group I will show close linkage to mating type.

multicent-4
mat, arg-5, acr-2, psi, at, ylo-1, wc-1 (1571). The temperature-conditional marker *psi* has been substituted for *pdx-1* in IV.

multicent-5
In(IL;IR)OY323, arg-5, acr-2, psi, at, ylo-1, wc-1 (1571). Like *multicent-4* but with the inversion present to suppress crossing over of I markers with *mat*.

mus: mutagen sensitive
This symbol was adopted in 1980 as a general designation for DNA-repair-defective mutants (1839). Locus numbers begin with *mus-7* (1010, 1017) to avoid confusion

with the previously named genes *upr-1* and *uvs-1* through -6. All of these mutants are sensitive to genotoxic agents other than UV (946, 1838, Table 3). Discrimination of *mus*, *uvs*, and *mei-2* or *mei-3* from wild type in the progeny of crosses is often readily accomplished by simple spot or stab inoculation of sorbose plates containing MMS or another genotoxic

agent (see also *uvs*). These DNA-repair genes plus *phr*, which is active in the photoreactivation of UV damage, represent several different repair types. Grouping by epistasis in double-mutant strains corresponds well with phenotypic and functional characteristics (946, 1838). Seven epistasis groups have been identified; Table 3 (946).

TABLE 3

Neurospora DNA Repair Genes: Characterization and Epistasis Relationships[a]

Epistasis group	Repair function	Gene	Characteristics of mutant[c]	Saccharomyces cerevisiae homologue
Mus-38[b]	Excision-1	*mus-38*	Sensitive to UV, 4NQO. Not sensitive to MMS, His. Partially defective in photorepair. Normal sporulation. Normal spontaneous mutation. High mutation induction by UV.	RAD1
Mus-18	Excision-2	*mus-18*	Sensitive only to UV. Not sensitive to any other mutagens. Normal sporulation. Normal spontaneous mutation. High mutation induction by UV.	None
Uvs-2	Postreplication	*uvs-2*	Highly sensitive to UV, 4NQO, MMC, MMS Sensitive to γ-rays, MNNG, EMS, NMU, ICR-170, DEO, TEB, FA, NA, aflatoxin, B1. Not sensitive to TBHP, Act-D, HC, 2AP, His, HU, Normal sporulation. Normal spontaneous mutation. High mutation induction by UV, γ rays, MNNG, MMC, 4NQO, FA, TEB, MMS, EMS, NMU, DEO, Aflatoxin B1.	RAD18
		mus-8	Highly sensitive to UV, MMS, MMC. Sensitive to X rays. Not sensitive to His, HU. Defective in meiotic mitosis. Low spontaneous and induced mutability	RAD6
Uvs-6	Recombination	*uvs-6*	High sensitivity to γ rays, UV, MMS, TBHP, His, HU. Partially dominant. Stop–start growth. Normal spontaneous and induced mutation. Blocked at early meiosis and sterile in homozygous crosses.	RAD50
		mus-23	Sensitive to MMS, UV, 4NQO, His, HU, MNNG, TBHP. Sterile in homozygous crosses.	MRE11

Epistasis group	Repair function	Gene	Characteristics of mutant[c]	Saccharomyces cerevisiae homologue
		mei-2	Sensitive to MMS, MNNG, γ rays, His, HU, TBHP. Not sensitive to UV, MMC. Normal spontaneous and induced mutation. Defective in meiotic recombination.	?
		mei-3	Sensitive to MMS, MMC, UV, 4NQO, His. High spontaneous mutation. Barren in homozygous crosses.	RAD51
		mus-25	Sensitive to MMS, His. Not sensitive to UV. Barren in homozygous crosses.	RAD54
		mus-7	Sensitive to MMS, MNNG, His. Not sensitive to UV, γ rays. Barren in homozygous crosses.	?
		mus-10	Sensitive to UV, MMS. Not sensitive to X rays. Fertile in homozygous crosses. Normal spontaneous and induced mutability.	?
Uvs-3	Unknown	uvs-3	Sensitive to UV, γ rays, MMS, MMC, 4NQO, MNNG, His, HU. Partially defective in photorepair. High spontaneous mutation. Low induced mutation. Sterile in homozygous crosses.	?
		mus-9	Sensitive to UV, MMS, X rays, His. High spontaneous mutation. Low UV-induced mutation. Sterile in homozygous crosses.	?
		mus-11	Highly sensitive to MMS, His, HU. Sensitive to UV, X rays. High spontaneous mutation. Low UV-induced mutation. Sterile in homozygous crosses.	?
Upr-1	Unknown	upr-1	Sensitive to UV, γ rays, 4NQO. Not sensitive to MMS. Partially defective in photorepair. Normal sporulation. Normal spontaneous mutation. Biphasic induced mutation.	?

Continues

TABLE 3 *Continued*

Epistasis group	Repair function	Gene	Characteristics of mutant[c]	Saccharomyces cerevisiae homologue
		mus-26	Sensitive to UV, γ rays, 4NQO. Not sensitive to MMS. Partially defective in photorepair Normal sporulation. Normal spontaneous mutation.	?
Phr	Photoreactivation	*phr*	CPD-photolyase defective	*PHR*

[a] This summary table is copied from ref. (946) [Inoue, H. (1999). DNA repair and specific-locus mutagenesis in *Neurospora crassa*. *Mutat. Res.* 437: 121–133.] Copyright 1999, with permission from Elsevier Science. It adds to Table 1 of ref. (1838), the data cited in ref. (946), unpublished data from the laboratory of H. Inoue, and the data of ref. (498). *Abbreviations:* 2AP, 2-aminopurine; MMC, mitomycin C; 4NQO, 4-nitroquinoline 1-oxide; TBHP, *t*-butyl hydroperoxide; HC, N^4-hydroxycytidine; His, histidine; HU, hydroxyurea; FA, formaldehyde; NA, nitrous acid; MNNG, *N*-methyl-*N'*-nitro-*N*-nitrosoguanidine; MNU, *N*-methyl-*N*-nitrosourea; DEO, 1,2,7,8-diepoxyoctane; TEB, 2,3,5,6-tetraethyleneimino-1,4-benzoquinone.

[b] Since this table was prepared, a new mutant, *mus-40*, has been added as a second member of the Mus-38 epistasis group (854).

[c] "Sterile" means that perithecia do not develop after fertilization. "Barren" means that perithecia are formed and develop to a certain extent, but few or no ascospores are produced.

mus-7: mutagen sensitive-7

IIR. Between *arg-5* (8%; 12%) and *nuc-2* (11%) (948, 1017).

Epistasis group Uvs-6 (Table 3). Extremely sensitive to histidine, highly sensitive to MMS, moderately sensitive to nitrosoguanidine and bleomycin, but insensitive to UV, mitomycin C, or ionizing radiation (1010, 1016). Both spontaneous and UV- or X-ray-induced mutation are normal (1010). *mus-7* resembles *mus-16* in showing high MMS-induced mutation and in other common features (954). Homozygous barren (1010), but intercrossed alleles can produce some progeny (0 *mus*⁺/300) (948). Increased mitotic chromosome instability is correlated with sensitivity to hydroxyurea (1835). Function is uncertain, considering the absence of radiation sensitivity. Three similar alleles, (FK116) (1015), (SA1) (948), and (FK107); the latter is used generally.

mus-8: mutagen sensitive-8

IVR. Linked to *mtr* (1%). Right of *pdx-1* (6%) (1010, 1017).

Cloned and sequenced: Swissprot UBC2_NEUCR, EMBL/GenBank D78372, PIR S71430, GenBank NEUMUS8.

Epistasis group Uvs-2 (1017; Table 3). A homolog of *S. pombe rhp6*, *Saccharomyces* RAD6 (1952), coding for a ubiquitin-conjugating enzyme. When transferred into *Saccharomyces*, the *Neurospora* gene shows good complementation with RAD6 for defects in damage-induced mutagenesis, but only partial complementation for sporulation defects. Highly sensitive to UV, mitomycin C, and MMS, moderately sensitive to X rays and

nitrosoguanidine, and not sensitive to inhibition by hydroxyurea or histidine (1010, 1836). Good conidial viability, but growth is reduced on minimal medium supplemented with sorbose, possibly reducing the recovery of mutants selected on that medium. Spontaneous and UV-induced mutations are moderately decreased, but X-ray mutagenesis is normal (1010). Homozygous barren (1017). Perithecia and beaks from *mus-8* × *mus-8* are well-developed, but multiple postmeiotic mitoses occur without chromosome replication, resulting in asci that contain many small nuclei, each with only a few chromosomes similar to *polymitotic* in maize (1671). The products of *mus-8* and *uvs-2* may interact, as do products of the corresponding yeast genes RAD6 and RAD18 (the homolog of *uvs-2*) (2099). Allele FK108 is usually analyzed.

mus-9: mutagen sensitive-9

IR. Between *cyh-1* (18%) and *al-2* (6%) (1010).

Epistasis group Uvs-3 (Table 3). Highly sensitive to UV, MMS, MNNG, and histidine, moderately sensitive to X rays, mitomycin C (1010), and 5-azacytidine (662), and also sensitive to bleomycin and hydrogen peroxide, especially during growth (1016). Spontaneous mutation is high (mutator phenotype); there is little or no increase in mutability by UV or X rays (1010). Homozygous sterile (1017). Mitotic chromosome instability is increased; this is correlated with increased sensitivity to hydroxyurea (1835). Levels of a nuclease that is secreted or that leaks from colonies on sorbose DNA agar are reduced (1014) (see *nuh*). Resembles *uvs-3* in most respects. Like *uvs-3*, *mus-9* is lethal in combination with *uvs-6* (1012). Three alleles are known (1015). Allele FK104, which has usually been used, has low conidial survival and shows the most pronounced mutagen sensitivity (1017).

mus-10: mutagen sensitive-10

VIIR. Right of *met-7* (7%) (1017).

Epistasis group Uvs-6 (1012; Table 3). Moderately sensitive to UV, MMS (1010), and 5-azacytidine (662). Slightly sensitive to histidine. Insensitive to X rays, MNNG, and mitomycin C. Spontaneous mutation is normal, as is mutation induced by UV or X rays (1010). Homozygous fertile, though less fertile than wild type (1017). Chromosome stability is normal (1835). Allele FK110 was reported to show lower radiation survival than allele FK105. The difference was reduced by back-crossing.

mus-11: mutagen sensitive-11

VR. Between *ad-7* (6%; 11%), *pab-2* (7%) and *inv* (4%), *un-9* (10%), *mei-2* (5%; 10%), *his-6* (6%). Linked to refs. (1016), (1017), and (1828).

Cloned and partially sequenced: Orbach-Sachs cosmid X19E8 (961).

Epistasis group Uvs-3 (1012) (Table 3). Two alleles, FK111 and FK117, are both highly sensitive to MMS, extremely sensitive to histidine (1015), significantly sensitive to X rays, γ rays, and bleomycin, and slightly sensitive to mitomycin C and hydrogen peroxide (1010, 1016). Moderately sensitive to MNNG (1835) and 5-azacytidine (662). Similar to *uvs-3* except for greater sensitivity to MMS and histidine. UV survival, which is moderately reduced, is also probably similar (1010, 1015). The mutant generally used (FK111) is more difficult to work with, being more variable (possibly a higher mutator effect). Conidial viability is similar for the two alleles. Spontaneous mutation is high, increasing the genetic variation of strains in the course of experiments (1012). (In epistasis tests, greater MMS sensitivity of one double-mutant strain compared to single mutants can be attributed to spontaneous mutation.) There is little or no increase in mutability by UV or X rays (1010). Homozygous crosses are barren. Sterility is dominant in crosses to linkage tester strains, but some progeny have been obtained from crosses of *mus-11* to prototrophic linkage tester heterokaryons. Mitotic chromosome instability is increased, correlated with sensitivity to hydroxyurea (1835).

mus-12: mutagen sensitive-12

VR. Left of *inl* (10%) (520).

Growing mycelia are highly sensitive to MMS but not to histidine (520). Conidia treated in liquid are sensitive to X rays, but not to MMS or UV (521). Spontaneous mutation is increased >10-fold. One mutant, SC15.

mus-13: mutagen sensitive-13

IR. Right of *al-2* (18%) (520).

Growing mycelia are slightly sensitive when grown on MMS medium. Growth is poor on minimal medium. Conidia are not sensitive to MMS, radiation, or histidine. Fertility is normal (521). One mutant, SC28.

mus-14: mutagen sensitive-14

VI. Linked to *lys-5* (520).

Growing mycelia are moderately sensitive to MMS, but conidia are not. Insensitive to UV, X rays, nitrosoguanidine, and histidine (521, 1836). Poor growth. Homozygous fertile (520). Spontaneous mutation is reduced (521). One mutant, SC3.

mus-15: mutagen sensitive-15

IL. Between *arg-1* (4%) and *arg-3* (<2%) (948, 1836).

Sensitive to MMS and MNNG, slightly sensitive to UV, and not sensitive to histidine. Homozygous barren (948). One mutant, SA7.

mus-16: mutagen sensitive-16

VL. Between *caf-1* (4%) and *lys-1* (9%) (954).

Isolated as a mutant sensitive to nitrogen mustard (86). Defective in DNA–DNA and DNA–protein cross-link repair. Highly sensitive to MMS and MNNG, slightly sensitive to histidine and HU, and not sensitive to UV, γ rays, or mitomycin C (confirming that mitomycin C does not form cross-links in *Neurospora* DNA; 954). Spontaneous mutation is normal; MMS-induced mutation is increased 10-fold at low doses. Homozygous sterile. Mitotic chromosome instability is increased (954). One mutant, JMB15.1

mus-17: mutagen sensitive-17

IVR. Between *pan-1* (<2%), and *cys-4* (25%) (945).

Low sensitivity to MMS, not sensitive to UV or X rays. Homozygous barren [summarized in ref. (1836)]. Two alleles, SA17 and SA21 (945).

mus-18: mutagen sensitive-18

VL. Linked to *rDNA* (0T/17 asci), *T(AR190)* (≤3%), *con-2* (2T/17 asci), *caf-1* (20%) (959, 1447).

Cloned and sequenced: EMBL/GenBank D11392, GenBank NEUUVE, PIR S55262.

Epistasis group Mus-18. (Table 3). A new type of repair gene, coding for UV endonuclease. Specific for pyrimidine dimers of both types: cyclobutane–pyrimidine dimers (CPDs) and TC(6-4) photoproducts. Cloned by transformation of highly UV-sensitive *Escherichia coli* cells (Δ*phr*; Δ*uvrA* and Δ*recA*) with *Neurospora* cDNA in a bacterial expression vector, using selection for increased UV resistance (2253). Homolog found in fission yeast (2039). The cloned *Neurospora mus-18+* gene complements *mus-18* SA8B (2253). Expressed *Neurospora* cDNA increases UV resistance of yeast *RAD1* and *RAD2* mutants and human XPA cells (2253). Expressed cDNA in *E. coli* lacking apurinic/apyrimidinic nucleases increases resistance to MMS and *t*-butyl hydroperoxide (1023). A RIP-produced disruption of *mus-18+* is phenotypically *mus-18* (2253). Mutant *mus-18* alleles show very low but exclusive UV sensitivity (959, 2253) and considerable deficiency for the removal of both types of UV pyrimidine dimers. Spontaneous mutation is normal, but UV-induced mutation is greatly increased. Fertility is normal. Conidial viability is high. Synergistic interaction for reduced UV survival in double-mutant strains with all UV-sensitive mutations tested. The double mutant of *mus-18* with *mus-38* (the yeast *RAD1* homolog) is extremely UV-sensitive and does not show any UV-dimer excision (855, 960).

mus-19: mutagen sensitive-19

IR. Between *aro-8* (8%) and *un-18* (4%) (945).
Sensitive to MMS, MNNG, γ rays, and histidine. Normal UV survival. Homozygous sterile. Two alleles, SA9 and SA19 (945, 1836). [In ref. (1836), SA6 was listed incorrectly as a *mus-19* allele, where in fact SA9 was intended (945)].

mus-20: mutagen sensitive-20

IIIR. Between *trp-1* (8%) and *phe-2* (18%) (948).
Highly sensitive to MMS and MNNG, slightly sensitive to histidine, and not sensitive to UV or γ rays (948). Highly sensitive to 5-azacytidine (662). Homozygous sterile, but a few mature perithecia may be produced (948). One allele, SA2.

mus-21: mutagen sensitive-21

IIIR. Linked to *trp-1* (20%) and *T(1) ylo-1* in *alcoy* (17%) (1016). Right of *ad-2*. Not allelic with *mus-20* (945).
Moderate-to-high MMS sensitivity, low-to-moderate sensitivity to histidine and bleomycin, and not sensitive to UV or mitomycin C (1015, 1016). Good conidial viability. Homozygous barren. Allele SC10 is sensitive to X rays, slightly sensitive to UV, and clearly sensitive to histidine at 37°C but not at 25°C. Six alleles: FK120, FK121, and FK127 with similar phenotypes (1015); FK131 and FK132 differing slightly (950); and SC10, which shows high spontaneous mutation and is female-sterile (521).

mus-22: mutagen sensitive-22

IR. Left of *cyh-1* (2%) (945).

Sensitive to MMS, but not to UV, mitomycin C, or histidine. Few perithecia are produced in homozygous crosses (1836). One mutant, SA22.

mus-23: mutagen sensitive-23

IIR. Between *fl* (<1%; 2%) and *trp-3* (9%) (85, 2190).
Cloned and sequenced: GenBank AB002530; pMOcosX clone X24:11A, Orbach-Sachs clone X24A11 (2190).
Epistasis group Uvs-6 (Table 3), other members of which have roles in recombinational repair. Homologous to *Saccharomyces MRE11* and *S. pombe rad32*, which interact with *RAD50*-like genes in double-strand-break repair and nonhomologous recombination (946, 2190). The original mutant allele, SA23, was isolated as histidine-sensitive by filtration enrichment in histidine growth medium. It is highly sensitive to a variety of mutagens, including UV, MMS, MNNG, *t*-butyl hydroperoxide, 4-NQO, histidine, and hydroxyurea (2190). X-ray sensitivity tests have not been reported. Mutants of both yeast homologs are X-ray- and MMS-sensitive. However, only *S. pombe rad32* resembles *mus-23* in also being sensitive to UV, suggesting that *mus-23* may have a similar function in nonhomologous recombinational repair of DNA. The *mus-23* mutant is defective in meiosis and ascospore formation in homozygous crosses. Expression is induced by UV or MMS (2190), similar to other members of the Uvs-6 group (947). The double mutant *mus-23; uvs-3* is inviable, similar to *uvs-6; uvs-3* (2190).

mus-24: mutagen sensitive-24

IIL. Left of *pyr-4* (8%) (945).
Sensitive to MMS, 4NQO, mitomycin C, and MNNG. Homozygous barren (1836). Mutant allele SA24.

mus-25: mutagen sensitive-25

VIIC. Between *met-7* (4%; 6%) (1016) and *un-10* (828). Cloned.
Epistasis group Uvs-6 (Table 3). Homologous to yeast *RAD54*, a gene involved in homologous recombination with sequence conserved from fungi to humans (1483). Two alleles, SA3 and FK123. Both are highly sensitive to histidine, moderately sensitive to MMS, MNNG, and γ rays, and not sensitive to UV or mitomycin C. At least FK123 is slightly sensitive to bleomycin (1016). Homozygous barren.

mus-26: mutagen sensitive-26

IVR. Between *met-1* (3%) and *col-4* (3%), *arg-2* (2%). Linked to *mus-8* (5%, probably to the left) (949).
Assigned to epistasis group Upr-1 (949, 957) (Table 3). However, a subgroup of Uvs-2 is not ruled out. Highly sensitive to UV and 4-NQO, but nearly normal for γ

rays, MMS, MNNG, and mitomycin C, suggesting a defect in some type of UV excision repair. UV survival curves plateau at moderate UV doses, a characteristic of *upr-1* (949). Decreased photoreactivation repair after UV damage is evident mainly at UV doses giving <10% survival. Dimer excision is normal, however (949, 957). Spontaneous mutation is close to control levels. Unexpectedly, UV-induced mutation in the standard forward-mutation tests was lower than in wild type (949) rather than being increased, as is typical for highly UV-sensitive mutants such as *uvs-2* (494, 497). UV reversion in congenic strains similarly showed reduced induction at high UV dose for *mus-26* and *upr-1*, but increased rates for *uvs-2*. Homozygous fertile. One mutant *mus-26* allele is known, SA3B. This is unrelated to SA3, which is a *mus-25* allele.

mus-27: mutagen sensitive-27

IIR. Right of *nuc-2* (16%) (1016).
Highly sensitive to γ rays (but less so than *uvs-6*), MMS, bleomycin (more so than *uvs-6*), and histidine, sensitive to UV only at very high doses (10% survival of *mus+*), and insensitive to mitomycin C. Homozygous fertile (1015, 1016). One mutant, FK124.

mus-28: mutagen sensitive-28

VL. Left of *lys-1* (10%) (1016).
Low sensitivity to UV, γ rays, MMS, histidine, and bleomycin. Normal viability and fertility (1015, 1016). One mutant, FK118.

mus-29: mutagen sensitive-29

VIL. Between *chol-2* (14%) and *lys-5* (15%) (1016).
Moderately sensitive to MMS and histidine, low but consistent sensitivity to γ rays and bleomycin, and not sensitive to UV or mitomycin C. Reduced conidial viability. Homozygous sterile (1015, 1016). One mutant, FK119.

mus-30: mutagen sensitive-30

IVR. Between *mtr* (8%) and *trp-4* (4%) (1016).
Sensitive to MMS, but not to UV or histidine. Homozygous fertile. One mutant, FK115.

mus-31: mutagen sensitive-31

IR. Between *Cen-I* and *ad-3B* (7%) (950).
Sensitive to MMS, but not to UV. Low fertility in homozygous crosses (950). One mutant, SA11.

mus-32: mutagen sensitive-32

IR. Between *mus-31* (3%) and *ad-3B* (4%) (950).
Sensitive to UV and MMS (950). One mutant, SA32.

mus-33: mutagen sensitive-33

VIIR. Between *met-7* (10%) and *un-10* (8%) (950).
Sensitive to UV and MMS (950). One mutant, SA33.

mus-34: mutagen sensitive-34

VR. Between *leu-5*, *ure-2* and *am*, *his-1* (3%) (950).
Sensitive to UV and MMS (950). One mutant, SA18.

mus-35: mutagen sensitive-35

VIIL. Right of *nic-3* (<9%) (950, 956).
Sensitive to MMS, but not to UV (950). One mutant, SA50.

mus-36: mutagen sensitive-36

IVR. Between *pan-1* and *uvs-2* (950).
Sensitive to MMS, but not to UV (950). One mutant, SA51.

mus-37: mutagen sensitive-37

V. Between *lys-1* and *cyh-2* (950).
Sensitive to MMS, but not to UV (950). One mutant, SA53.

mus-38: mutagen sensitive-38

IL. Left of *un-5* (7%), *leu-3* (10%) (960). Linked to *nit-2* (0/18) (855, 1447)
Cloned and sequenced: EMBL/GenBank AB009461.
Epistasis group Mus-38 (Table 3). Obtained using degenerate PCR of conserved regions from *Saccharomyces RAD1* and *RAD1* homologs (855). A mutant allele obtained by RIP was called rad1[Nc] and was used for mapping (855, 1447). A second allele, SA56, was obtained by UV induction in the highly UV-mutable, but only slightly UV-sensitive, strain *mus-18*, which is UV-endonuclease-defective. This was identified by increased UV sensitivity using filtration enrichment following UV treatment, photoreactivation repair, and periods of dark repair (liquid holding) (960). The two mutants (rad1[Nc] and SA56) do not complement in heterokaryons and are similar in sensitivity to UV, 4-NQO, mitomycin C, and cisplatin (855). Both show inefficient survival recovery with photoreactivation after UV. Spontaneous mutation is at the wild-type level. UV-induced mutation is increased 6-fold, as is typical for nucleotide excision repair mutants. *mus-38* is in the same epistasis group with *mus-40*, the *Neurospora* homolog of yeast *RAD2*, and presumably they function similarly in nucleotide excision repair. *mus-38* interacts synergistically with all previously described UV-sensitive mutants. Double mutants with *mus-18* (which is UV-endonuclease-deficient and UV-excision-defective) are supersensitive to UV and completely defective in the release from UV-irradiated DNA of thymine dimers of both types: cyclobutane–pyrimidine dimers (CPDs) and TC(6-4) photoproducts (855) (see also *mus-40*). These findings explain the puzzling inconsistency that whereas

mus-18 encodes for a very active UV endonuclease, *mus-18* mutants are barely sensitive to UV. Because two alternative excision mechanisms exist, mutants of one or the other excision repair type could be compensated for by efficient repair provided by the other [as is also found in bacteria for *Micrococcus luteus*, and in fungi for *Schizosaccharomyces pombe* (2280)].

mus-39: mutagen sensitive-39

VI. Right of *ylo-1* (3%; 8%) (950). Linked to *mus-29* (10%).

Sensitive to MMS and histidine (950, 1015). One mutant, FK133.

mus-40: mutagen sensitive-40

IL. Linked to *pzl-1* (0/18). Between *cyt-21* (1/18) and *fr*, *nit-2* (3/16) (945, 1447).

Cloned (854).

Epistasis group Mus-38. With *mus-38*, *mus-40* establishes a nucleotide excision repair epistasis group in *Neurospora*; this is the main UV-repair type in many organisms, including *Escherichia coli* and *Saccharomyces cerevisiae*. *Neurospora mus-38* is a homolog of *Saccharomyces RAD1*. Obtained using degenerate PCR of nucleotide excision repair gene *RAD2* and homologs (854). A mutant obtained by RIP is moderately sensitive to UV but is not sensitive to MMS or X rays. UV excision is defective. Double mutants with all other UV-sensitive mutations show additive or synergistic interactions; those with *mus-18* are supersensitive to UV and are completely defective in release from UV-irradiated DNA of thymine dimers of both types: cyclobutane–pyrimidine dimers (CPDs) and TC(6-4) photoproducts (854).

mus(SA5): mutagen sensitive SA5

Unmapped.

Extremely sensitive to UV and MMS. Sterile in heterozygous crosses; therefore, not mapped by genetic methods (948).

mus(SA6): mutagen sensitive SA6

IIIR. Left of *ad-2* (5%), *trp-1* (8%) (948).

Sensitive to MMS, MNNG, γ rays, and histidine. Normal UV survival. Reduced conidial viability. Homozygous barren (948). Incorrectly listed in ref. (1836) as a *mus-19* allele (945).

mus(FK125): mutagen sensitive FK125

IV or V. Linked to *T(R2355)*, *cot-1* (32%). Not linked to *mus*, *uvs*, or many other markers on IV and V (1015).

Homozygous fertile. Conidial viability is very low (2–20%). Moderately sensitive to MMS and histidine, but not to UV (1015).

mus(FK128): mutagen sensitive FK128

Unlinked to *csp-2* or to *alcoy* markers, unlike all mapped *mus* and *uvs* genes (1015).

Sensitive to MMS and histidine, but not to UV, or mitomycin C. Homozygous barren. Good conidial viability. Expression of mutagen sensitivities varies in different ascospore isolates. The original strain and the more sensitive isolates possibly contain two interacting mutations (1015).

mus(SC17): mutagen sensitive SC17

V. Left of *inl* (27%) (520).

Sensitive to MMS, but not to histidine. Sensitivity is shown by mycelium but not by conidia, and only after preincubation at 15°C. Growth is cold-sensitive on minimal medium (520).

mus(SC28): mutagen sensitive SC28

Changed to *mus-13*.

mut-1: mutator-1

IVR. Linked to *trp-4* (<3%) (541).

Spontaneous mutation rates of *mtr* and *trp-2* are increased by 10- to 80-fold (541). The mutations are predominantly −1 frameshifts.

nac: adenylate cyclase

Allelic with *cr-1*.

nada: NAD(P)ase

IV. Left of *ad-6* (18%) (1457).

Structural gene for nicotinamide adenine dinucleotide (phosphate) glycohydrolase. Normal morphology. Recessive in heterokaryons. Allele 62ts is temperature-sensitive, with altered substrate affinity (1457). Identified by the plaque test, using *Haemophilus influenzae*. Used in a study of glutamic acid decarboxylase during conidial germination (382).

nap: neutral and acidic amino acid permeability

VR. Linked to *inl* (15%) (982). Right of *ure-2* (32%) (2230).

Cloned and sequenced: PIR S47892, EMBL/GenBank AF001032.

Encodes an amino acid permease. Selected as resistant to ethionine + *p*-fluorophenylalanine (982). Causes reduced amino acid uptake by neutral, basic, and general systems. Also causes reduced uptake of uridine and glucose. The defect is not in amino acid-binding glycoproteins (1695). See ref. (2230) for aspartate uptake and resistance to inhibitors. Scored by spotting a conidial suspension on minimal medium containing 1.5% sucrose, agar, 0.3 mM ethionine, and 0.02 mM FPA.

nd: natural death

IR. Between cys-9 (<2%; 3%) and T(4540)^R, thi-1 (<2%; 4%) (162).

Progressive deterioration of mycelial growth is followed by abrupt, irreversible cessation (1903). Manifest on race tubes or after sequential transfers of macroconidia to slants. Rapid degradation of the mitochondrial chromosome results from hyperactivation of recombination between direct repeats (174, 1869). Hypersensitive to sorbose. Conidia die rapidly on slants at 4°C (1397). Recessive. Stocks are maintained as balanced heterokaryons. An aged strain can be rejuvenated through heterokaryosis or by crossing to nd^+ (1903). Mutant nd strains free of modifiers initially grow at the wild-type rate (1397). Used to examine hypotheses of senescence based on faulty protein synthesis (1183) and lipid auto-oxidation with free radical reactions (1391). Not to be confused with mitochondrial genes symbolized ND or ndh, which encode subunits of NADH ubiquinone oxidoreductase (see Appendix 4). For another Mendelian mutant that results in senescence, see sen-1.

ndc-1: nuclear division cycle-1

VR. Left of arg-4 (2%), inl (6%) (1895).

A heat-sensitive conditional mutant that fails to grow at 34°C. Recessive. The division cycle is blocked just before the initiation of DNA synthesis, when spindle pole bodies are duplicated but not separated. The effect is not nucleus-limited in heterokaryons. Used to show that asynchronous germination of macroconidia results from the arrest of nuclei that are at different stages of the division cycle when conidia became dormant and that few nuclei are in S-phase at any one time during conidiation (1895).

ndh: NADH dehydrogenase

Used as symbol and name for the seven mitochondrial genes that encode subunits of the 700-kDa NADH:ubiquinone complex of the inner mitochondrial membrane, known as respiratory complex I. See ref. (937) and Appendix 4. (These mitochondrial genes have commonly been symbolized ND.) The nuclear genes that encode the remaining subunits of complex I are called nuo: NAD-H:ubiquinone oxidoreductase. The symbol ndh is not used for nuclear genes.

ndk-1: nucleotide diphosphate kinase

VR. Between al-3 (10%; 22%) and his-6 (22%; 24%) (1485, 1486).

Cloned and sequenced: GenBank D88148.

Encodes a 15-kDa protein homologous to nucleoside diphosphate kinase (EC 2.7.4.6) of other organisms. Shows NDK activity and protein kinase activity. Not regulated at the transcriptional level by WC-1 or WC-2 (1489). Mapping is based on a mutant allele ndk-1^{P72H}, previously called psp or ps15-1, which was discovered as a segregant from wc-1 strain FGSC 3528 (1485).

ndp64: NAD(P)H dehydrogenase 64

IVR. Linked to Fsr-4 (0T/18 asci), Tel-IVR (1T/18 asci) (1318).

Cloned and sequenced: EMBL/GenBank AJ236906.

Encodes a 64-kDa mitochondrial NADH dehydrogenase (EC 1.6.5.3) associated with the inner mitochondrial membrane, perhaps in the outer face. The product of ndp64 is not a subunit of complex I (1318). Called p64.

neu: neutral amino acid transport

A synonym of mtr: methyltryptophan resistant, which has precedence.

nic: nicotinic acid

Because of permeability, nic mutants are preferably supplemented with nicotinamide rather than nicotinic acid at most medium pH values (192). To obtain good recovery of some nic mutants from crosses, crossing media should be supplemented with nicotinamide at levels 10-fold higher than those required for growth, even when the protoperithecial parent is nic^+ (1548, 1970). nt is best treated as a nic mutant for purposes of growth and scoring. For the biosynthetic pathway, see Fig. 48. For regulation, see refs. (234), (706), (1178), and (1813).

nic-1: nicotinic acid-1

IR. Between In(OY323)^R, lys-3, ace-3 (<1%) and T(UK2-26)^R, os-1 (10%; 29%) (9, 115, 278, 1122, 1548, 1578, 1592, 1971).

Cloned (25).

Uses nicotinic acid or nicotinamide, but not precursors (192, 195). Accumulates quinolinic acid (195) (Fig. 48). Affects nicotinic acid synthetase (nicotinate phosphoribotransferase, EC 2.4.2.11). Used to study intralocus recombination (1971). Called the q locus.

nic-2: nicotinic acid-2

IR. Between ad-3B (4%) and tyr-2, T(Y112M4i)^R, ace-7 (4%; 7%) (492, 1122, 1578).

Grows on nicotinic acid, nicotinamide, or high concentrations of quinolinic acid (192, 2265). Cannot use

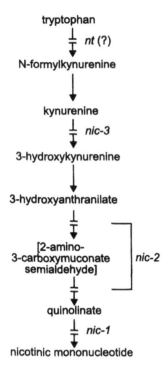

tryptophan

⊥ nt (?)

N-formylkynurenine

kynurenine

⊥ nic-3

3-hydroxykynurenine

3-hydroxyanthranilate

[2-amino-
3-carboxymuconate   nic-2
semialdehyde]

quinolinate

⊥ nic-1

nicotinic mononucleotide

FIGURE 48 The pathway from tryptophan to nicotinic mononucleotide, showing sites of gene action (191, 195, 702, 2265). The
enzymatic reactions between 3-hydroxyanthranilate and nicotinic
mononucleotide have not been demonstrated directly in *Neurospora*. From ref. (1596), with permission from the American
Society for Microbiology.

kynurenine, hydroxykynurenine, or hydroxyanthranilic
acid (191, 2265). Accumulates 3-hydroxyanthranilic
acid (191) (Fig. 48). Affects 3-hydroxyanthranilate
3,4-dioxygenase (EC 1.13.11.6). Aging cultures accumulate red-brown pigment in the medium. Used to study
intralocus recombination (1972). Translocations
T(4540) and *T(S1325)* are inseparable from *nic-2*
(1578, 1972, 1975).

nic-3: nicotinic acid-3

VIIL. Between *do* (3%), *spco-4* (1%) and *lacc* (14%), *thi-3*
(9%; 27%), *csp-2* (16%; 22%) (1017, 1582, 1585,
1592).

Uses nicotinic acid, nicotinamide, 3-hydroxyanthranilic
acid, 3-hydroxykynurenine, or high concentrations of
quinolinic acid (191, 2265). Accumulates α-N-acetyl-
kynurenine; blocked in conversion of kynurenine to
3-hydroxykynurenine by kynurenine-3-monooxygenase
(EC 1.14.13.9) (2265) (Fig. 48). Pyridine nucleotide
levels (234).

nik-1: nonidentical kinase-1

An *os-1* allele (1844).

nik-2: nonidentical kinase-2

IIR. Linked to *arg-12* (0/17) (1447).
Cloned and partially sequenced: EMBL/GenBank U50263,
U50264.
Encodes domains homologous to an SLN1-like histidine
kinase (31). The null allele has no observable phenotype
(30).

nim-1: never in mitosis-1

Unmapped.
Cloned and sequenced: Swissprot NIM1_NEUCR, EMBL/
GenBank L42573, EMBL NCNIM1A, GenBank
NEUNIM1A.
The protein product is homologous to NIMA of *Aspergillus
nidulans* (EC 2.7.1.-), which encodes a cell-cycle,
G2-specific protein kinase. Complements *Aspergillus
nimA* functionally (1653).

nit: nitrate utilization

Mutants called *nit* cannot use nitrate as the nitrogen source,
but require ammonia or other sources of reduced
nitrogen. Conveniently scored on synthetic crossing
medium (2208), in which nitrate is the sole nitrogen
source. Also scorable on slants by pH change when
grown with ammonium nitrate as the nitrogen source
and bromcresol purple (20 µg/ml) as the indicator
(1551). In most crosses, *nit* can be used as the fertilizing
parent; in crosses where a *nit* mutant is required as
female parent, crossing medium can be altered by
substituting ammonium nitrate for potassium nitrate
(320). Nitrite is toxic at low pH; test media containing
nitrite should be neutralized, and the nitrite should be
preferably be filter-sterilized (1947). For a summary of
nutritional requirements based on various authors, see
ref. (2101). *nit-1*, *nit-7*, *nit-8*, and *nit-9* involve a
molybdenum-containing cofactor common to nitrate
reductase and xanthine dehydrogenase (1155, 2101,
2102) (Figs. 8, 49, and 50). For a review of nitrate
assimilation, see ref. (722). For regulation, see reviews in
refs. (1282), (1283), (1285), and (1286). See also *gln-1,
nmr.*

nit-1: nitrate nonutilizer-1

IR. Between *ad-9* (3%; 15%), *Tp(T54M94)^M* and *cyh-1*
(6%) (890, 929, 1592).
Cannot use nitrate or hypoxanthine as the nitrogen source,
but uses nitrite, ammonia, or amino acids (1950). Does
not prevent the formation of nitrate reductase apopro-
tein (1948), but lacks molybdopterin (1095), the
molybdenum-containing cofactor common to nitrate
reductase and xanthine dehydrogenase (1155, 1441)
(Figs. 8 and 49). The nitrate reductase in *nit-1* extracts

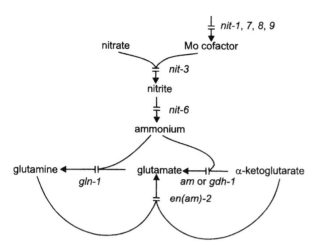

FIGURE 49 The nitrate reduction pathway showing sites of gene action. The *am* gene specifies the NADP-specific glutamate dehydrogenase, and *am* mutants do not alter the NAD-specific enzyme, which is specified by *gdh-1* (353, 566, 639, 1027, 1155, 1781, 1948, 2102). Modified from ref. (1596), with permission from the American Society for Microbiology.

does not catalyze the complete electron transport sequence from NADPH to NO₃⁻, but it does catalyze the initial part of this sequence if a suitable electron acceptor such as cytochrome *c* is provided (1948). See

ref. (385) for a model of the interaction of *nit-1* and *nit-3* gene products. For regulation, see Fig. 50 and refs. (442), (1948), and (1950).

nit-2: nitrate nonutilizer-2

IL. Between *un-5* (2%), *T(39311)ᴸ* and *lah-1*, *In(OY323)ᴸ*, *leu-3* (12%, 18%) (115, 1578, 1592, 2041).

Cloned and sequenced: Swissprot NIT2_NEUCR, EMBL/GenBank M33956, PIR A34755, EMBL NCNIT2, GenBank NEUNIT2; pSV50 clone 6:9H, Orbach-Sachs clone G1C05.

Encodes a sequence-specific DNA-binding protein that acts globally to activate many structural genes that are involved in nitrogen metabolism (Fig. 50). The mutant cannot use nitrate, nitrite, purines, or most amino acids as a nitrogen source, but will grow on ammonia, glutamine, or glutamate. Shown to bind GATA sequences (807, 2251). The DNA-binding domain has been characterized (687, 688, 690) as having the DNA sequences to which NIT-2 binds (374–376, 626). The *nit-2* mutant is missing (or has severely reduced levels of) nitrate reductase, nitrite reductase, uricase, allantoinase, allantoicase, L-amino acid oxidase, general amino acid permease, extracellular protease, an intracellular neutral PMSF-sensitive protease, and an extracellular RNase (443, 613, 831, 1198, 1712, 1951). NIT-2 does not

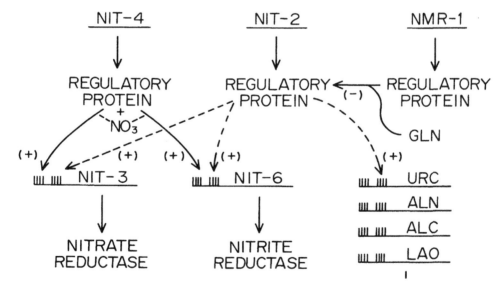

FIGURE 50 The nitrogen regulatory circuit of *N. crassa*. Expression of genes that encode nitrogen metabolic enzymes only occurs upon nitrogen catabolite derepression and simultaneous induction by a pathway-specific metabolite. The repressing metabolite appears to be glutamine, although the nature of the factor with which it interacts is unknown. NIT-2 is a sequence-specific DNA-binding protein with a single zinc finger that acts globally to activate the expression of many structural genes that encode enzymes of nitrogen metabolism. NMR is a negative-acting regulatory protein that is required to establish nitrogen repression, and it appears to act by binding directly to NIT-2 to inhibit its function. NIT-4 is a pathway-specific positive-acting regulatory factor that has a zinc cluster DNA-binding motif. NIT-4 interacts with NIT-2 to turn on the expression of *nit-3* and *nit-6*. Other specific positive-acting factors are required to express genes encoding enzymes for other nitrogen pathways. URC signifies uricase, for which the structural gene has not yet been identified in *Neurospora*. *uc-5* and *ud-1* also are subject to general nitrogen metabolite regulation. Original figure from G. A. Marzluf.

appear to be necessary for the production of xanthine dehydrogenase (770) or acid phosphatase (806). Levels of glutamate dehydrogenases also are affected (442), as are uracil and uridine uptake (266) and levels of nucleotide pools (1523). Prevents leaky growth of *am* on minimal medium (320). Allele K31 (called *pink*) originated in *Neurospora sitophila* and was introgressed into *Neurospora crassa* (638); the protein product of K31 may show altered mobility (807). Heterozygosity for the closely linked *synaptic sequence* reduces recombination in *nit-2* (313), which is subject to regulation by *rec-1* (324). Function of a nonsense (amber) mutation is restored by *ssu-1* (1610). Both *nit-2* and *nit-4* must be functional for the nitrate utilization genes *nit-3* (encoding nitrate reductase) and *nit-6* (encoding nitrite reductase) to be activated under inducing conditions (1285), and the direct interaction of NIT-2 and NIT-4 is necessary for the strong activation of gene expression (625). Nuclease-hypersensitive sites in NIT-3 depend on NIT-2 and NIT-4 (229). A cosmid that rescues the defective phenotypes of either *nmr* or *nit-2* has been identified (1949). Transformation of *Aspergillus nidulans* with the *nit-2* gene complements some of the defective phenotypes of *areA* mutants (450); *nit-2* also complements mutant phenotypes in the transformation of *nnu* mutants in *Gibberella zeae* (534). The *nit-2* mutant is complemented by its homologue *area-GF*, from *Gibberella fujikoroi* (2114). There is a possible role in photoreception (1482). Called *amr: ammonium regulation* (1712).

nit-3: nitrate nonutilizer-3

IVR. Between *met-5* (15%), *bd* (19%) and *pyr-2* (2%; 9%) (1185, 1546, 1950).

Cloned and sequenced: Swissprot NIA_NEUCR, EMBL/ Genbank X61303, PIR S16292, S34796, S37298, GenBank NCNIT3.

Structural gene for NADPH nitrate reductase (EC 1.6.6.3) (46). Cannot use nitrate as the nitrogen source, but does use nitrite, ammonia, hypoxanthine, or amino acids (46, 1948) (Fig. 49). Allele 14789 apparently produces an altered enzyme that cannot catalyze the whole electron transport sequence from NADPH to NO_3^-, but can catalyze the terminal portion of this sequence if a suitable electron donor (reduced viologen dye) is provided (1948). Molecular analyses of mutants (1494). Site-specific mutagenesis of the flavin-binding domain (750) and the heme-binding domain (1495). Nonsense (amber) mutations restorable by *ssu-1* (1610). Nuclease-hypersensitive sites in the promoter that depend on NIT-2 and NIT-4 (229). NIT-2- and NIT-4-binding sites in the promoter (375, 2053). For a model of the interaction of *nit-1* and *nit-3* gene products, see ref. (385). For regulation, see Fig. 50 and refs. (442), (691), (1948), and (1950).

nit-4: nitrate nonutilizer-4

IVR. Between *pyr-1* (1%; 6%) and *pan-1* (6%; 27%). Probably right of *col-4* (2%) (185, 1546, 1950).

Cloned and sequenced: Swissprot NIT4_NEUCR, EMBL/ GenBank M80368, PIR A41696, S20033, EMBL NCNIT4A, GenBank NEUNIT4X.

Structural gene for nitrogen assimilation transcription factor, regulator for induction by nitrate of nitrate reductase and nitrite reductase (2101, 2288–2290), turns on expression of *nit-3* and *nit-6* (Fig. 50). Cannot use nitrate or nitrite as the nitrogen source, but does use ammonia, and amino acids (185). Nuclease hypersensitive sites in NIT-3 that depend on NIT-2 and NIT-4 (229). NIT-4-binding sites in *nit-3* promoter characterized (375, 685). NIT-4 transcriptional activation domain defined (624). Interaction of NIT-2 and NIT-4 is necessary for strong activation of gene expression (625). Complements the *Aspergillus nidulans nirA1* mutation (858). The original *nit-4* allele was discovered in a wild isolate of *Neurospora intermedia* from Borneo and was introgressed into *Neurospora crassa* (185). Allele *nr15*, called *nit-5* (1950), is phenotypically identical to other *nit-4* alleles and fails to complement or recombine (0 prototrophs/2080 progeny) (2101).

nit-5: nitrate nonutilizer-5

Allelic with *nit-4*.

nit-6: nitrate nonutilizer-6

VIL. Between *chol-2* (6%), *T(OY350)* and *ser-6* (11%), *ad-8* (11%, 17%) (1172, 1582, 2277).

Cloned and sequenced: Swissprot NIR_NEUCR, EMBL/ GenBank L07391, PIR A49848, EMBL NCNIT6X, GenBank NEUNIT6X, EST SC3H7; pMOcosX clones X4:11A and X6:9F; Orbach-Sachs clone X4A11, X6F9.

Unable to use nitrate or nitrite as the nitrogen source (353). Structural gene for NAD(P)H-nitrite reductase (EC 1.6.6.4) (353, 612) (Figs. 49 and 50), which is subject to positive nitrogen metabolite repression (354). Induced by nitrite (385). Affected by the *nit-2* and *nmr* regulatory genes (1649, 2091). Potential NIT-4-binding sites in promoter (375, 685). Used to study the repression of nitrate reductase (44) and the nonenzymatic reduction of nitrate (353). Directed mutagenesis used to examine the structure–function relationship (395).

nit-7: nitrate nonutilizer-7

IIIR. Linked to *un-17* (0/63) (1546).

Cannot use nitrate or hypoxanthine as the nitrogen source. Resembles *nit-1*, *nit-8*, and *nit-9* in affecting the molybdenum-containing cofactor common to nitrate reductase and xanthine dehydrogenase (863, 2101, 2102) (Figs. 8 and 49).

nit-8: nitrate nonutilizer-8

IR. Linked to *cr-1* (0/61). Between *nic-2* (10%) and *thi-1* (3%) (1546).

Cannot use nitrate or hypoxanthine as the nitrogen source. Lacks the molybdenum cofactor for nitrate reductase and xanthine dehydrogenase (863, 2101, 2102) (Figs. 8 and 49).

nit-9: nitrate nonutilizer-9

IVR. Right of *nit-4* (9%). Linked to *nit-3* (35%; 38%) (2101).

Cloned (563).

Cannot use nitrate or hypoxanthine as the nitrogen source. Lacks the molybdenum cofactor for nitrate reductase and xanthine dehydrogenase (Figs. 8 and 49). A complex locus with three complementation groups, comparable to *cnxABC* of *Aspergillus nidulans* (863, 2101, 2102). *nit-9* alleles of groups A and B have been cloned (563).

nmr: nitrogen metabolite regulation

VR. Between *am* (3%; 7%) and *gln-1* (4%; 10%) (2100).

Cloned and sequenced: Swissprot NMR_NEUCR, EMBL/ GenBank S64286, PIR S11910.

Structural gene for a negative-acting nitrogen regulator (2282) (Figs. 49 and 50). Synthesis of nitrate reductase is derepressed on ammonium, glutamate, or glutamine. Hypostatic to *nit-2* and *nit-4*. Prototrophic. Isolated and scored by sensitivity to chlorate in the presence of glutamine (568, 2100, 2282). Mutant MS5 (1649) is an allele; a complementing cosmid also complements *nit-2* defects (1949). Allele *ms5* is different from *gln^r* (269). Levels of nitrate reductase, nitrite reductase, histidase, and acetamidase are elevated in the presence of glutamine and the respective enzyme inducer. The NMR protein binds to two distinct regions of the positive-acting NIT-2 protein, and these interactions are required to establish nitrogen catabolite repression (1527, 2250). Called *nmr-1*, MS5.

NO: Nucleolus Organizer

See *NOR*.

nop-1: new opsin-1

VIIL. Linked to *ars-1* (2T/18 asci), *nic-3* (3T/18 asci) (180).

Cloned and sequenced: EMBL/GenBank AF135863.

Encodes the first opsin identified in fungi or eukaryotic microorganisms, with sequence homology to opsins in animals and the archaea. Expression is stimulated by light and is highest under conditions favoring conidiation. NOP-1 is capable of binding and photocycling retinal (180, 181).

NOR: Nucleolus Organizer Region

VL. At *Tip-VL* between the terminal *sat* (118, 1598) and *T(AR30)*, *dgr-1*, *caf-1* (>8%), *T(AR33)*, *lys-1* (30%) (118, 1598–1601, 1617). Breakpoints of *In(UK2-y)*, *T(NM169d*, *T(ALS169)*, *T(ALS182)*, *T(AR190)*, and *T(OY321)* are in the *NOR* (1578).

Contains *rDNA*, components of which have been cloned and sequenced. See *rDNA*. *5.8S sequence*: EMBL/ GenBank X02447, M10692, NCRRN58S, EMBL NCRGB, GenBank NEURGB; *17S sequence*: EMBL/ GenBank X04971, GenBank NCRRNAS; *18S sequence (partial)*: EMBL/GenBank M11033, EMBL NCRRE, GenBank NEURRE; *25S sequence*: GenBank X01373; *ITS sequence* EMBL/GenBank M13906, EMBL NCRGITR, Genbank NEURGITR; *28S sequence (partial)*: EMBL/GenBank M38154, EMBL NCRRNA1; Orbach-Sachs clone X15E04.

The nucleolus is seen at pachytene to be at one end of the longest chromosome (1307). A dotlike terminal satellite is appended in some strains (see *sat*). Genes specifying 5.8S, 17S, and 25S ribosomal RNA (but not 5S) are located in the NOR in a series of ~150 tandem repeats (rDNA), each ~9.3 kb long (423, 672). The repeats are separated by nontranscribed spacers (423, 672). Changes in rDNA repeat number occur premeiotically as a result of crossing over between sister chromatids (262–264). The rDNA repeats are presumably located in the attenuated threads seen extending through the nucleolus when acriflavine-stained pachytene chromosomes are examined by fluorescence microscopy (1601, 1672). Nine of eleven chromosome rearrangements that appear genetically to have a terminal breakpoint in VL are in fact physically nonterminal, with a segment of rDNA repeats from the NOR exchanged reciprocally for the terminal segment of another chromosome arm (1590, 1601). *T(OY321)* divides the NOR into two portions, each of which retains the ability to form a nucleolus (1599). When crossed with normal sequence, *T(AR33)* produces duplication progeny having two copies of the NOR (1598). These undergo demagnification in such a way that different nontranscribed spacer sequences of both parental types are retained (1733, 1734). Genes specifying 5S ribosomal RNA are not included in the rDNA repeat unit but are located elsewhere in the genome as dispersed single genes flanked by heterogeneous sequences (672, 1893; see Fsr). For research applications of the NOR, see refs. (1578) and (1601).

npd: nitropropane dioxygenase

Unmapped.

Cloned and sequenced: Swissprot 2NPD_NEUCR, EMBL/ GenBank U22530.

Structural gene for 2-nitropropane dioxygenase (EC 1.13.11.32) (2063).

nrc-1: nonrepressor of conidiation-1

Unmapped.

Cloned and sequenced: EMBL/GenBank AF034090; Orbach-Sachs clones X15A05, G8E05, G11G02, G15C05, G16C02.

Encodes a protein kinase homologous to *Saccharomyces cerevisiae* STE11, which is a MEK kinase (1093). Cannot repress conidiation. Female sterile. Ascospores containing the *nrc-1* mutation have an autonomously expressed "flattened" morphology and fail to germinate. Obtained by insertional mutagenesis and identified by altered expression of a *ccg-1::tyrosinase* reporter gene (1093). The gene was cloned independently by PCR with degenerate primers encoding MEKK-conserved domains, was assigned to Orbach-Sachs clones G05:C06, G16:B08, G16:G02, and was provisionally called *mik-2* (1267). The homologs in *Saccharomyces* and *S. pombe* are involved in mating-type determination. Proteins of this type phosphorylate MEK proteins in other organisms (1267).

nrc-2: nonrepressor of conidiation-2

IIR. Linked to *Fsr-55*. Near *cya-4* (676).

Cloned and sequenced: EMBL/GenBank AF034260; Orbach-Sachs clone X7E9.

Encodes a protein kinase homologous to *Saccharomyces cerevisiae* KIN82 and YNO47W, which are serine–threonine protein kinases. The mutant cannot repress conidiation. Conidia that are produced have a separation defect. Female sterile. Identified by detecting the altered expression of a *ccg-1::tyrosinase* reporter gene (1093).

nt: nicotinic acid or tryptophan

VIIR. Between *arg-10* (2%; 12%) and *sk* (7%; 18%) (1548).

Uses nicotinic acid. May respond also to tryptophan, phenylalanine, tyrosine, quinic acid, nicotinic acid precursors, and/or tryptophan precursors, depending on genetic background (847, 1474). Best supplemented with nicotinamide and scored as a *nic* mutant. Probably deficient in tryptophan pyrrolase (tryptophan 2,3-dioxygenase) (EC 1.13.11.11) (Fig. 48), but direct evidence is lacking because tryptophan oxygenase cannot be assayed (702). Kynurenine formamidase levels are normal (702). Pyridine nucleotide levels (234).

ntf2: nuclear transport factor 2

Unmapped.

Cloned and sequenced: Swissprot NTF2_NEUCR, EMBL/GenBank Y13237, NCNUCTF2.

Homolog of the *Saccharomyces* gene encoding nuclear transport factor 2 (NTF2) (1342).

nuc: nuclease

Mutants that lack nucleases and phosphatases have been obtained in several ways: by selecting mutants unable to grow on one or more types of nucleic acid (double-stranded DNA, denatured DNA or RNA) as the sole source of phosphorus (659, 850, 967), by selecting mutants with loss or reduction in the size of a halo around the colony from which nucleic acid in the medium has been digested (1014), or by selecting mutants that lack relevant enzyme activity detected with a chromogenic substrate (747, 848, 2096). There is also some phenotype overlap with mutagen sensitivity (see *uvs-3* and *uvs-6*). *nuc-2*, *nuc-3*, *nuc-4*, *nuc-5*, *nuc-6*, *nuc-7*, *nuh-5*, and *nuh-9* are all located on linkage group IIR; it is uncertain whether each of these represents a different locus.

nuc-1: nuclease-1

IR. Between *his-2* (1%), $T(AR173)^R$ and *met-10*, *lys-4* (1%) (967, 1022, 1331).

Cloned and sequenced: Swissprot NUC1_NEUCR, EMBL/GenBank M37700, PIR A36378, EMBL NCNUC1, GenBank NEUNUC1.

Structural gene for a transcriptional activator of phosphorus acquisition genes (1022) (Fig. 51). The NUC-1 positive regulatory protein controls the expression of several unlinked target genes involved in phosphorus

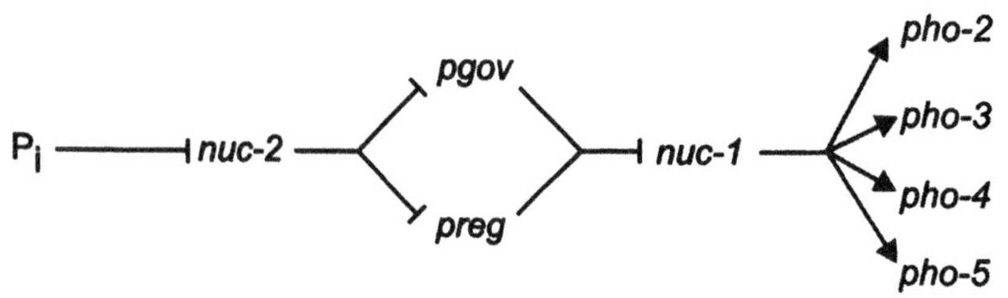

FIGURE 51 Regulation of genes concerned with phosphorus acquisition in *N. crassa*. The *nuc-2*⁺ gene encodes an ankyrin-repeat protein, *pgov*⁺ probably encodes a cyclin-dependent kinase, *preg*⁺ encodes a cyclin, and *nuc-1*⁺ encodes a helix–turn–helix transcriptional activator. Redrawn from ref. (1326), with permission. Copyright American Society of Plant Physiologists.

acquisition and metabolism. It has a helix–loop–helix motif near its C-terminus, which binds to a CACGTG target in the upstream sequences of target genes, e.g., *pho-2* and *pho-4*. The loop between the helices contains an atypical zipper domain. Helix I is required for target DNA binding, whereas the zipper and helix II are required for dimerization (1541). The mutant is defective in the production of repressible alkaline and acid phosphatases (1333, 2096), and several nucleases are absent or reduced (848). *nuc-1* mutants (other than *nuc-1^c*) are unable to use RNA or DNA as a phosphorus source (850, 967). Scored on low-phosphate medium by a staining reaction with α-naphthyl phosphate plus Diazo Blue B (747, 2096), by failure to grow on minimal medium altered to substitute 0.1 g/l RNA or DNA for an inorganic phosphate source (967, 1014), or by failure to grow on low phosphate at pH above 7 (1322). *nuc-1* is epistatic to *nuc-2* (called *pcon^c*), *preg^c*, and *pgov^c* (1325, 1326, 1331, 1333). In duplications, *nuc-1^c* is dominant to *nuc-1^+*, which is dominant to *nuc-1*. *nuc-1^c* is scored on high-phosphate medium by staining reaction with α-naphthyl phosphate plus Diazo Blue B (747, 2096) or by suppression of the phenotype of *nuc-2* on low phosphate at high pH (1331). Used to study phosphate transport (1229). For regulation model, see refs. (1325), (1326), (1331), and (1541).

nuc-2: nuclease-2

IIR. Between *aro-3*, *T(NM177)^L* and *preg* (1% or 2%), *pe* (4%). Allelic with *pcon* (0/854) (967, 1157, 1333).

Cloned and sequenced: EMBL/GenBank U51118, EMBL NC51118; EST SM3A10; Orbach-Sachs clone X2E11.

Structural gene for an ankyrin repeat protein (1540), a component of the phosphate-regulated signal transduction pathway. Unable to use RNA or DNA as the phosphorus source (967). Defective in the production of repressible alkaline and acid phosphatases (1333, 2096). Several nucleases are absent or reduced (848). Interaction with other phosphate regulatory genes (1325). Recessive to *nuc^+* in partial diploids and heterokaryons (1333). Not defective in *nuh* function (1014). Scored on low-phosphate medium by a staining reaction with α-naphthyl phosphate plus Diazo Blue B (747, 2096), by failure to grow on minimal medium altered to substitute 0.1 g/l RNA or DNA for the inorganic phosphate source (967, 1014), or by failure to grow on low phosphate at pH above 7 (1322). Used to study phosphate transport (1229). Regulator gene of repressible alkaline phosphatase (1333) and other steps in phosphorus uptake and metabolism (1322, 1325). Used to study phosphate transport (1229). For regulation model, see refs. (1325) and (1331) (Fig. 51).

nuc-3: nuclease-3

IIR. Right of *arg-12* (4%) (659).

Unable to use double- or single-stranded DNA (dsDNA, ssDNA) or RNA as the sole phosphorus source. Reduced extracellular nuclease and alkaline phosphatase levels. Nuclease activity on dsDNA is 2–12% that of wild type. Sensitive to UV, MNNG, and MMS. One of five nuclease mutants (*nuc-3* to *nuc-7*) described in refs. (659) and (1355).

nuc-4: nuclease-4

IIR. Right of *arg-12* (12%) (659).

Unable to use dsDNA as the sole phosphorus source. Reduced extracellular nuclease and alkaline phosphatase levels. Sensitive to MMS, but has antimutator activity against UV and MNNG. One of five nuclease mutants (*nuc-3* to *nuc-7*) described in refs. (659) and (1355).

nuc-5: nuclease-5

IIR. Right of *arg-12* (16%) (659).

Unable to use dsDNA as the sole phosphorus source. Reduced extracellular nuclease and alkaline phosphatase levels. Sensitive to MMS, but has antimutator activity against UV and MNNG. One of five nuclease mutants (*nuc-3* to *nuc-7*) described in refs. (659) and (1355).

nuc-6: nuclease-6

IIR. Right of *arg-12* (23%) (659).

Unable to use dsDNA, ssDNA, or RNA as the sole phosphorus source. Reduced extracellular nuclease and alkaline phosphatase levels. Sensitive to UV, MNNG, and MMS. One of five nuclease mutants (*nuc-3* to *nuc-7*) described in refs. (659) and (1355).

nuc-7: nuclease-7

IIR. Linked to *trp-3* (31%). Apparently unlinked to *arg-12* (659).

Unable to use dsDNA, ssDNA, or RNA as the sole phosphorus source. Reduced extracellular nuclease and alkaline phosphatase levels. No increased sensitivity to UV, MNNG, or MMS; in fact, has antimutator effect. One of five nuclease mutants (*nuc-3* to *nuc-7*) described in refs. (659) and (1355).

nuh: nuclease halo

Mutants show reduced nuclease halos on sorbose DNA agar plates (DNase test agar) flooded with HCl (1014). Three nuclease activities have been identified in filtrates of *Neurospora* grown in Vogel's minimal medium N (2162) + DNA + sorbose ("sorbose + DNA" medium) (667). Two of these are secreted alkaline DNases (A and

B), whose activities are also derepressed 200-fold when cells are grown in phosphate-free medium N + DNA + sucrose (1018). DNase A is a single-strand (ss) and double-strand (ds) endonuclease, specific for DNA (ss- and ds-DNase). DNase B is a periplasmic ss-specific exonuclease, active on DNA and RNA, that can also be isolated from conidia (669). The third nuclease activity corresponds to DNase C, which is an intracellular endo–exonuclease found in nuclear, mitochondrial, and vacuolar compartments and which is not secreted but is released into culture medium when cells are grown on sorbose +DNA medium (667). This enzyme has ss-specific endonuclease activity with DNA and RNA and exonuclease activity with ds-DNA, optimal at neutral pH. The *Neurospora* endo–exonuclease is immuno-chemically related to the product of *Escherichia coli* recC (668). The endo–exonuclease is implicated in DNA repair and recombination (669). Some of the isolated *nuh* mutants showed reduced levels of one or more nucleases, as judged by reduced or missing peaks of ss-DNase activity when chromatographed (667) or by reduced endo–exonuclease activity in extracts (1014). The effects of *nuh* mutations may be indirect (e.g., altered enzymes for processing or activation of precursors may reduce levels of nuclease activities; 666). Reduced halos may also result from mutations causing reduced growth, reduced secretion, or altered membrane properties (1018). One of the *nuh* mutants, originally called *nuh-4* (FK016), is allelic with the repair mutant *uvs-3* (1012). Two other repair mutants also make nuclease halos: *uvs-6* (ALS35), and *mus-9* [FK104, called *uvs(FK104)*] (1014, 1017).

nuh-1: nuclease halo-1

IIIR. Between *leu-1* (4%) and *nuh-2* (1%), *trp-1* (11%) (1014).

Deficient in nuclease released on DNase test agar with sorbose, resulting in a reduced halo around each colony. Reduced levels of alkaline, Mg^{2+}-dependent, ss-DNase activities in extracts of mycelia. Complementation is good between *nuh-1* (FK001) and the closely linked *nuh-2* (FK027). (Possibly a gene cluster?) Not sensitive to UV or MNNG (1014).

nuh-2: nuclease halo-2

IIIR. Between *leu-1* (4%), *nuh-1* (1%) and *trp-1* (11%) (1014).

Deficient in the nuclease released on DNase test agar with sorbose, resulting in a reduced halo around each colony. Normal ss-DNase activity in extracts of mycelia (1014). Not sensitive to UV or MNNG (1014).

nuh-3: nuclease halo-3

VR. Between *cyh-2* (4%) and *al-3* (17%) (1014).

Deficient in nuclease released on DNase test agar with sorbose, resulting in a reduced halo around each colony. Low levels of DNase B (1014), but normal levels of DNases A and C (667). Normal growth. Fertile in homozygous crosses. Not sensitive to UV, MNNG, or MMS (1018). Two alleles, FK003 and FK004. Closely linked (2–5%) to a complementing mutant, *nuh(FK023)* (1014).

nuh-4: nuclease halo-4

Mutant FK016 is allelic with *uvs-3*.

nuh-5: nuclease halo-5

IIR. Linked to *trp-3* (30%), close to *T(4637) al-1* (1014). Deficient in nuclease released on DNase test agar, resulting in a reduced halo around each colony. Nuclease activities in extracts are 55% of wild type (1014). Mutant FK005 is closely linked to a partially complementing mutant, *nuh(FK029)* (1008). (Possibly a complex gene or a gene cluster?)

nuh-6: nuclease halo-6

IR. Between *Cen-I* (5%) and *nic-2* (4%) (1014). Deficient in nuclease released on DNase test agar, resulting in a reduced halo around each colony. Levels of tested nucleases in extracts are normal. Not sensitive to mutagens. Growth is normal (1014).

nuh-7: nuclease halo-7

Unlinked to any mapped *nuh* gene or to *alcoy* markers (1018).

Decreases the nuclease released on DNase test agar, resulting in a reduced halo around each colony. Very low ss-DNase and reduced ds-DNase activity in filtrates, even under derepressing conditions. Early growth is normal (up to 16 hr), but growth is somewhat reduced later (3–5 days), with an increasingly high release of protein into the filtrate. Small amounts of ss- and ds-DNase are secreted early, but not later. Lack of activity is not due to the formation of an inhibitor. Overall intracellular ss-DNase activities (presumably DNase C) are 50–80% of wild type (1018). *nuh-7* (FK017) probably is not a nuclease structural mutation, but may be altered in secretion or membrane structure.

nuh-8: nuclease halo-8

IR. Right of *nic-1* (1014, 1018). Deficient in nuclease released on DNase test agar, resulting in a reduced halo around each colony. Growth and UV survival are normal, as is the enzyme profile of secreted DNases A and B in medium N+sorbose. There is virtually no derepression in phosphate-free medium N + DNA + sucrose (less than 2-fold, compared to the 200-fold normally observed). One allele, (FK018) (1018). Possibly in a cluster with *nuh(FKO40)* (1008).

nuh-9: nuclease halo-9

IIR. Linked to *arg-5*. Left of *mus-7*, *nuc-2* (1018).

Deficient in nuclease released on DNase test agar, resulting in a reduced halo around each colony. Increased growth and release of ss-DNase in sorbose + DNA medium, but growth and levels of induced ss- and ds-DNase are variably reduced in phosphate-free medium N + DNA + sucrose. The mutant is therefore difficult to use for detailed analysis (1018). Moderately sensitive to MMS. Barren in homozygous crosses.

nuh-10: nuclease halo-10

VIR. Linked to *trp-2* (1018).

Deficient in nuclease released on DNase test agar, resulting in a reduced halo around each colony. Mutant FK028 shows reduced growth and a reduced level of DNase A secretion in phosphate-free medium N + DNA + sucrose. Growth is normal, but ss-DNase activity is reduced overall in sorbose + DNA medium (1018). MMS sensitivity is not increased. Fertility is normal.

nuh(19): nuclease halo(19)

Unlinked to any mapped *nuh* gene or to *alcoy* markers (1018).

Deficient in nuclease released on DNase test agar, resulting in a reduced halo around each colony (1018).

nuh(22): nuclease halo(22)

Unlinked to any mapped *nuh* gene or to *alcoy* markers (1018).

Deficient in nuclease released on DNase test agar, resulting in a reduced halo around each colony. Nuclease profiles are normal. Poor fertility when homozygous. Sensitivity to MMS is moderately increased (1018).

nuh(41): nuclease halo(41)

Unmapped.

Deficient in nuclease released on DNase test agar, resulting in a reduced halo around each colony. Nuclease profiles are normal on sorbose + DNA medium and close to normal under derepressing conditions. Sterile in homozygous crosses. MMS sensitivity is increased slightly (1018).

nuh(42): nuclease halo(42)

Unmapped.

Deficient in nuclease released on DNase test agar, resulting in a reduced halo around each colony. Single-nuclease levels are reduced in filtrates from cells grown in either sorbose + DNA medium or phosphate-free medium N + DNA + sucrose, whereas growth is reduced only slightly in either medium (1018).

nuo: NADH:ubiquinone oxidoreductase

The mitochondrial NADH:ubiquinone complex (EC 1.6.5.3), known as respiratory complex I, has a total size of approximately 700 kDa. Complex I of *Neurospora* has been imaged at medium resolution (682, 808, 886). It consists of two arms: one membrane-embedded and one peripheral, protruding into the mitochondrial matrix. Within complex I are at least 28 nucleus-encoded and 7 mitochondrion-encoded subunits (2203). Two subcomplexes of subunits from the two sources appear to be assembled separately prior to final assembly of the native complex in the mitochondrial membrane. The nomenclature for the nuclear-encoded genes of the complex (*nuo* genes) incorporates the size of each polypeptide product in the gene name; when multiple genes specify products of similar size, they are differentiated further with letters (e.g., *nuo21.3a*, *nuo21.3b*, and *nuo21.3c* each encode a different 21.3-kDa subunit). Mitochondrion-encoded subunits include *ndh1* (41.5-kDa subunit), *ndh2*, *ndh3*, *ndh4*, *ndh4l*, *ndh5* (80-kDa subunit), and *ndh6* (see Appendix 4 for *ndh* genes). Blocking the synthesis of the mitochondrial subunits with chloramphenicol results in the synthesis of a smaller, functionally different form of complex I (683, 2183). Disruptions of each of the nuclear genes can cause mild to serious assembly defects that eliminate complex I function. Loss of complex I can slow asexual growth and reduce conidiation, but the vegetative phenotypes appear surprisingly mild. In contrast, complex I appears essential for sexual development (554). Reviews (783, 2149, 2203). See also *acp-1*.

nuo9.3: NADH:ubiquinone oxidoreductase 9.3

Unmapped.

Cloned and sequenced: Swissprot NI9M_NEUCR, EMBL/GenBank S49807, PIR A44210.

Encodes the NADH:ubiquinone oxidoreductase 9.3-kDa subunit (EC 1.6.5.3) (866).

nuo9.6: NADH:ubiquinone oxidoreductase 9.6

A synonym of *acp-1*.

nuo10.5: NADH:ubiquinone oxidoreductase 10.5

Unmapped.

Cloned and sequenced: Swissprot NI8M_NEUCR, EMBL/GenBank X69929, PIR S30186, GenBank NCMIT.

Encodes the NADH:ubiquinone oxidoreductase 10.5-kDa subunit (EC 1.6.5.3) (552).

nuo12.3: NADH:ubiquinone oxidoreductase 12.3

IR. Between *lys-4* (1T/18 asci) and *cys-9* (7T/18 asci) (1447).

Cloned and sequenced: Swissprot NUMM_NEUCR, EMBL/GenBank X68965, NCNUO, PIR S29557, S32568.

Encodes the NADH:ubiquinone oxidoreductase 12.3-kDa subunit (EC 1.6.5.3) (2150). Disruption causes assembly defects (555).

nuo14: NADH:ubiquinone oxidoreductase 14

Unmapped.

Cloned and sequenced: EMBL/GenBank Z18945, NCUBIOXRI, PIR S47150.

Encodes the NADH:ubiquinone oxidoreductase 14-kDa subunit (EC 1.6.5.3) (1445).

nuo14.8: NADH:ubiquinone oxidoreductase 14.8

Unmapped.

Cloned and sequenced: Swissprot NB4M_NEUCR, EMBL/GenBank X76344, GenBank NCUO.

Encodes the NADH:ubiquinone oxidoreductase 14.8-kDa subunit (EC 1.6.5.3) (75).

nuo17.8: NADH:ubiquinone oxidoreductase 17.8

Unmapped.

Cloned and sequenced: Swissprot NURM_NEUCR, EMBL/GenBank X71414, NCNUO178, PIR S35057; EST SM1F6.

Encodes the NADH:ubiquinone oxidoreductase 17.8-kDa subunit (EC 1.6.5.3) (75).

nuo18.3: NADH:ubiquinone oxidoreductase 18.3

Unmapped.

Cloned and sequenced: EMBL/GenBank X56226, GenBank NCNHU18.

Encodes the NADH:ubiquinone oxidoreductase 18.3-kDa subunit (EC 1.6.5.3) (2200).

nuo19.3: NADH:ubiquinone oxidoreductase 19.3

VIL. Between *gdh-1* (9T/18 asci) and *asd-1* (5T/18 asci), *Bml* (10T/18 asci) (632, 1447).

Cloned and sequenced: EMBL AJ001520, NCIRONSUL.

Encodes the NADH:ubiquinone oxidoreductase 19.3-kDa subunit (EC 1.6.5.3). Homolog of the PSST subunit of bovine complex I (1953).

nuo20.8: NADH:ubiquinone oxidoreductase 20.8

IL. Between *mat* (0 or 1T/17 asci) and *Cen-I* (3T/17 asci) (1447).

Cloned and sequenced: Swissprot NUPM_NEUCR, EMBL/GenBank M55323, PIR A36621, GenBank NEUCOMPI.

Encodes the NADH:ubiquinone oxidoreductase 20.8-kDa subunit (EC 1.6.5.3). Disruption causes assembly defects (438).

nuo20.9: NADH:ubiquinone oxidoreductase 20.9

Unmapped.

Cloned and sequenced: Swissprot NUXM_NEUCR, EMBL/GenBank X60829, PIR S27171, GenBank NCNADHC1.

Encodes the NADH:ubiquinone oxidoreductase 20.9-kDa subunit (EC 1.6.5.3) (77). Disruption causes an assembly defect (1843).

nuo21: NADH:ubiquinone oxidoreductase 21

IVR. Linked to *Tel-IVR* (1T/18 asci), *Fsr-4* (1 or 2 T/18 asci) (1447).

Cloned and sequenced: Swissprot NUYM_NEUCR, EMBL/GenBank X78082, NCNUO21, PIR S17192.

Encodes the NADH:ubiquinone oxidoreductase 21-kDa subunit (EC 1.6.5.3) (76). Erroneously called *nuo18* due to a sequencing error that gave a predicted truncated product (2200).

nuo21.3a: NADH:ubiquinone oxidoreductase 21.3a

VR. Left of *inl* (18%), *inv* (22%) (41).

Cloned and sequenced: Swissprot NUIM_NEUCR, EMBL/GenBank M32244, GenBank NEUCOM1S.

Encodes the NADH:ubiquinone oxidoreductase 21.3-kDa subunit 21.3a (EC 1.6.5.3). The polypeptide product is found in the peripheral arm of complex I. Referred to as 21.3a in reviews (2149, 2203) and as 21.3b in some primary literature (41, 79). Disruption causes an assembly defect (41).

nuo21.3b: NADH:ubiquinone oxidoreductase 21.3b

Unmapped.

Cloned and sequenced: Swissprot NUJM_NEUCR, EMBL/GenBank X56612, GenBank NCNURI, PIR S14277.

Encodes the NADH:ubiquinone oxidoreductase 21.3-kDa subunit 21.3b (EC 1.6.5.3) (1444). Polypeptide product is found in the membrane arm of complex I. Referred to as 21.3b in reviews (2149, 2203) and as 21.3a in some primary literature (79). Disruption causes an assembly defect (1443).

nuo21.3c: NADH:ubiquinone oxidoreductase 21.3c

VIR. Linked to *Tel-VIR* (0 or 1T/18 asci) (632, 1447, 2149).

Cloned and sequenced: EMBL/GenBank X95547, GenBank NCMC1.

Encodes ferredoxin-like NADH:ubiquinone oxidoreductase 21.3-kDa subunit 21.3c (EC 1.6.5.3) (553). Disruption causes an assembly defect (2149).

nuo24: NADH:ubiquinone oxidoreductase 24

VR. Linked to *cyh-2* (0T/18 asci) (632, 1447, 2149).

Cloned and sequenced: Swissprot NUHM_NEUCR, EMBL/GenBank X78083, GenBank NCNOU24.

Encodes the NADH:ubiquinone oxidoreductase 24-kDa subunit (EC 1.6.5.3) (76). Disruption causes an assembly defect (2149).

nuo29.9: NADH:ubiquinone oxidoreductase 29.9

IVR. Between *trp-4* (8 or 9T, 1NPD/18 asci) and *Tel-IVR* (8T, 1NPD/18 asci) (1447).

Cloned and sequenced: Swissprot NUFM_NEUCR, EMBL/GenBank X56237, GenBank NCNUO32, PIR S17191.

Encodes the NADH:ubiquinone oxidoreductase 29.9-kDa subunit (EC 1.6.5.3) (2140). Disruption causes an assembly defect (555).

nuo30.4: NADH:ubiquinone oxidoreductase 30.4

VIL. Linked to *Bml*, *Cen-VI*, *pan-2* (0T/18 asci) (554, 1447).

Cloned and sequenced: Swissprot NUGM_NEUCR, EMBL A35935, PIR A35935.

Encodes the NADH:ubiquinone oxidoreductase 30.4-kDa subunit (EC 1.6.5.3) (554). Disruption causes an assembly defect (554).7

nuo40: NADH:ubiquinone oxidoreductase 40

Unmapped?

Cloned and sequenced: Swissprot NUEM_NEUCR, EMBL/GenBank X56238, GenBank NCNUO40, PIR S13025.

Encodes the NADH:ubiquinone oxidoreductase 40-kDa subunit (EC 1.6.5.3). Disruption causes an assembly defect (1843); the complex lacking this subunit does not contain tightly bound NADPH (1842).

nuo41.5: NADH:ubiquinone oxidoreductase 41.5

Not a nuclear gene. See mitochondrial gene *ndh* (Appendix 4).

nuo49: NADH:ubiquinone oxidoreductase 49

Unmapped.

Cloned and sequenced: Swissprot NUCM_NEUCR, EMBL/GenBank X54508, GenBank NCNUO49, PIR S13801.

Encodes the NADH:ubiquinone oxidoreductase 49-kDa subunit (EC 1.6.5.3). Disruption causes an assembly defect (1843).

nuo51: NADH:ubiquinone oxidoreductase 51

Unmapped.

Cloned and sequenced: Swissprot NUBM_NEUCR, EMBL/GenBank X56227, PIR S17663, GenBank NCNDU51.

Encodes the NADH:ubiquinone oxidoreductase 51-kDa subunit (EC 1.6.5.3), which is the NADH-binding subunit. Disruption causes an assembly defect (616).

nuo78: NADH:ubiquinone oxidoreductase 78

IIR. Linked to *preg* (0T/15 asci) (1447).

Cloned and sequenced: Swissprot NUAM_NEUCR, EMBL/GenBank X57602, L36813, PIR S17664, S59926, GenBank NCNDU78, NEUNDS.

Encodes the NADH:ubiquinone oxidoreductase 78-kDa subunit (EC 1.6.5.3). Disruption by sheltered RIP causes an assembly defect (840).

nuo80: NADH:ubiquinone reductase 80

Not a nuclear gene. See mitochondrial gene *ndh5* (Appendix 4).

Nystatin Resistance

See *erg*.

oak: oak

VR. Between *un-9* (7%) and *T(DBL9)*, *his-6* (6%) (1593).

Conidia form in dense balls above trunklike adherent hyphal aggregates. These may sometimes stand alone on an agar surface rather than lying prone, with a shape suggesting a tiny tree.

oar1: 3-oxyacyl-[acyl carrier protein] reductase-1

Unmapped.

Cloned and sequenced: EMBL/GenBank AF042860.

Nuclear gene encoding mitochondrial protein 3-oxyacyl [acyl carrier protein]reductase-1 (EC 1.1.1.35) (251).

oli: oligomycin resistant

VIIR. Between *for*, *frq*, and *un-10* by chromosomal walk (1094). Linked to *frq-1* (<2%) (537).

Cloned and sequenced: Swissprot ATP9_NEUCR, EMBL V00864, PIR S07173, LWNCA, S43893, GenBank V00864, NCATP1, EST NC1A8 (2152).

Encodes the ATP synthase DCCD-binding subunit of the membrane proton channel F_0 (subunit c according to current nomenclature, previously designated subunit 9) (1863). The amino acid sequence (81 residues) of the purified polypeptide shows extensive homology with the corresponding proteolipid in yeast where, in contrast to *Neurospora*, it is the product of a mitochondrial gene (1866, 2171). An alternative form of this subunit (considered nonfunctional) is encoded by a *Neurospora* mitochondrial gene (937; see Appendix 4); the transcript of this gene accumulates rapidly during conidial germination (183). Specific single-amino-acid substitutions have been identified for three mutants (1865). The characterization of the cDNA encoding the precursor polypeptide yielded the first reported sequence of a cleaved N-terminal mitochondrial signal peptide (2152). *oli* mutants are resistant to oligomycin and defective in energy transduction (594). Mutants are selected effectively using the double mutant *azs; has* (called ANT-1), which is deficient in both SHAM-sensitive and azide-sensitive alternate oxidase pathways. Scored on 5 mg/ml oligomycin in liquid medium at 3 days, 30°C (596). A slightly altered period of circadian rhythm cosegregates and reverts with *oli* (536, 537). Also called *atp-9, prl-1*.

opi-1: overproduction of inositol-1

IVR. Linked to *cot-1* (9%), *T(IV;V)R2355* in *alcoy*. Unlinked to *inl* (1807).

Isolated as slow-growing suppressors of 89601, an allele that produces defective myoinositol-1-phosphate synthase (EC 5.5.1.4). The *opi-1* mutant enhances synthesis of the defective enzyme by increasing the level of its residual activity and enabling the *opi-1; inl* double-mutant strain to grow slowly. In *opi-1; inl*$^+$, the enzyme is overexpressed and inositol is excreted (1807). For an allele-specific partial suppressor of another allele, see *su(inl)*.

os: osmotic sensitive

Unable to grow on media with elevated osmotic pressure. Scorable on solid or liquid media plus 4% NaCl (1.4 M). Most mutant alleles result in a characteristic abnormal morphology, having sticky, close-cropped aerial hyphae that tend to rupture and bleed. Morphology is influenced by humidity. When conidia are produced, they are sticky and do not shake loose. Intense pigment of aggregated hyphae has led to some *os* mutants initially being called "flame" or "overproduction of carotenoids." *os* strains are useful for obtaining protoplasts (1881) and are reported to be efficient as recipients for DNA-mediated transformation in media of high osmolarity (2245). In addition to the numerous loci designated *os*, *cut* is a typical *osmotic* mutant. Osmotic sensitivity is correlated with resistance to dicarboximide fungicides (788, 790).

os-1, os-2, os-4, and *os-5* are resistant, whereas *cut* is not (788). *smco-2* (an *sc* allele), *smco-8*, and *smco-9* are osmotic-sensitive and resistant to dicarboximide fungicides (789). Mutants *sor-4 T9, os-8*, and *os-9* are osmotic-sensitive but do not show the typical *os* morphology. An *os-1* mutant forms protoplasts that grow and divide in special liquid media (1881), but *os-2, os-4*, and *os-5* do not (1844).

os-1: osmotic-1

IR. Between *nic-1* (10%; 29%), *T(UK2-26)*^{R} and *prd-4* (2%), *arg-13* (1%; 4%) (56, 1548, 1585, 1592).

Cloned and sequenced: EMBL/GenBank U53189, U50263.

Structural gene for a putative histidine kinase osmosensor (by homology to bacteria and yeast) (1844). Sensitive to high osmotic pressure. Resistant to dicarboximide fungicides (788). Readily scored by morphology on nonmoist slants or by failure to grow on media with 4% NaCl. Most *os-1* alleles form no or few conidia. Alleles NM233t and NM204t are heat-sensitive (25 vs 34°C). Cell-wall pores are 4 times larger in *os-1* than in wild type. The *os-1* mutant also has a higher exclusion threshold and a 30-fold higher galactosamine:glucosamine ratio (2105, 2106). Forms protoplasts in media of high osmolarity (609, 824). Allele B135 is an essential genotypic component of the wall-less strain slime (607). Heat-sensitive allele NM233t forms protoplasts following the incubation of macroconidia in liquid medium with polyoxin B; it then resembles the slime strain. The protoplasts divide stably at 37°C in the absence of the enzyme with a 7.5-hr redoubling time, and they show good regeneration upon return to permissive temperature (1879, 1881); (1,3)-β-glucan and chitin polymers are secreted rather than being incorporated into a cell wall (1844). Complementation can occur between alleles (1352). Allele Y256M209 was called *flm-1*. An *os-1* allele initially called *nik-1* was identified using PCR primers designed to amplify the histidine kinase consensus regions (31). The *nik-1* null mutant obtained by RIP was phenotypically *os*-like. Allelism of *nik-1* with *os-1* was subsequently inferred from the identity of the two DNA sequences (1844).

os-2: osmotic-2

IVR. Right of *cot-1* (4%) (1592, 1828).

Cloned and sequenced (2301).

Readily scored by osmotic sensitivity or morphology. Resistant to dicarboximide fungicides (788). Transformants with *os-2*$^+$ become sensitive to phenylpyrrole fungicides (2301). *T(V44o)* is inseparable from *os-2* (1578).

os-3: osmotic-3

IR. Reported to be right of *nic-2* (4%) (1304).

Because of stock loss and ambiguity, validity as a separate locus cannot be confirmed (1305, 1559).

os-4: osmotic-4

IR. Between *sn* (3%) and *T(AR173)^L*, *T(AR190)*; hence left of *un-2*, *his-2*. [Data are for allele Y256M233 (1559, 1578, 2072).]

Readily scored by osmotic sensitivity or morphology. Resistant to dicarboximide fungicides (788). Allele Y256M223, originally called *flm-2*, is preferred as a marker to NM201o, on which the locus designation was based initially (1559).

os-5: osmotic-5

IR. Linked to *al-2* (<1%). Between *T(UK1-35)*, *T(STL76)* and *Tp(T54M94)^R*, *arg-6* (1%), *al-1* (1%) (1559, 1578, 1580).

Scorable by osmotic sensitivity or morphology. Resistant to dicarboximide fungicides (788). A modifier gene *mod(os-5)* restores normal morphology and osmotic sensitivity while retaining high fungicide resistance (786).

os-6: osmotic-6

Linked to *os-1* (4%), *os-7* (6%) (1352).

An osmotic-sensitive mutant obtained among Inl$^+$ transformants (1352). Recombination and complementation evidence showed it to be nonallelic with *os-1*. Orientation with respect to the centromere was not determined (1305). Because tests for allelism with *os-3*, *os-4*, and *os-5* were made using tester strains that were later shown to be of dubious validity, it remains uncertain whether *os-6* actually represents a new locus.

os-7: osmotic-7

Linked to *os-1* (11%), *os-6* (6%) (1352).

An osmotic-sensitive mutant obtained among Inl$^+$ transformants (1352). Recombination and complementation evidence showed it to be nonallelic with *os-1*. Orientation with respect to the centromere was not determined (1352). Because tests for allelism with *os-3*, *os-4*, and *os-5* were made using tester strains that were later shown to be of dubious validity, it remains uncertain whether *os-7* actually represents a new locus.

os-8: osmotic-8

IIIR. Between *ad-2* and *trp-1* (283, 1381).

Fails to grow on 4% NaCl, but morphology and conidiation appear normal. Conidia do not shake loose when a slant is tapped, but *os-8* does not have the defects responsible for failed conidial separation in *csp* or *eas*.

Both osmotic sensitivity and conidial separation are recessive (281–283, 1381). Called SS-931.

os-9: osmotic-9

VIR. Between *ad-1*, *del* and *trp-2* (154, 254).

Fails to grow on medium with 4% sodium chloride, but growth is wild type on elevated levels of potassium chloride or glucose (1097a). Wild type morphology. Called SS-788 (252–254).

os-10: osmotic-10

IV. Linked to *cut* (5%; 21%), *col-4* (8%; 10%) (2000, 2001).

Fails to grow on medium with 4% sodium chloride. Morphologically wild type. Called SC-1018 (2001).

os-11: osmotic-11

III. Linked to *os-8* (19%) (154).

Fails to grow on medium with 4% sodium chloride, but growth is wild type on elevated levels of potassium chloride or glucose (1097a). Morphology resembles that of wild type. *os-8*, *-9*, *-10*, and *-11* all originated from the same mutant hunt, using filtration enrichment in medium with 3% NaCl following UV irradiation of conidia. Called SS-18 (154).

osb: oxysterol-binding protein

Unmapped.

Cloned and sequenced: EMBL/GenBank Y12693, GenBank NCOSBP.

Encodes oxysterol binding protein (1342).

osr-1: oxidative phosphorylation stress response-1

IV. Linked to *trp-4* (20%) (162).

Affects the stress response, which mediates the suppressiveness of those mitochondrial mutations that impair electron transport and result in senescence or respiratory deficiency. Suppressiveness is blocked because the stress response cannot be turned on in the *osr-1* mutant (163).

osr-2: oxidative phosphorylation stress response-2

II. Linked to *arg-5* (≤12%) (162).

Affects the stress response, which mediates the suppressiveness of those mitochondrial mutations that impair electron transport and result in senescence or respiratory deficiency. Suppressiveness is blocked because the stress response is turned on constitutively in the *osr-2* mutant (163).

ota: ornithine transaminase

IIIR. Between ad-4 (15%) and tyr-1 (14%) (465). Linked to pro-4 (4%) [D. West, cited in ref. (108)].

Specifies ornithine transaminase (EC 2.6.1.13) (465) (Fig. 10). Conidiates somewhat less than wild type (232). Selected by its ability to use exogenous ornithine as a precursor for arginine in the double mutant arg-5 arg-12^s. Catabolism of ornithine (to glutamic semialdehyde) is blocked, resulting in ornithine concentrations high enough to compensate for the low activity of the ornithine carbamyl transferase in arg-12^s. The ota single mutant is prototrophic, but prevents the efficient use of ornithine or arginine as the sole nitrogen source (465). Used to study flux through the arginine biosynthetic pathway (752), the utilization of endogenous vs exogenous ornithine (453), and the compartmentalization of ornithine in the cell (209, 467). The triple mutant arg-12; ota; aga is used to look at deoxyhypusine and hypusine modification of a 21-kDa polypeptide (2261) that is the eIF5A precursor (see eif5a). Sideramine production is completely blocked in the absence of ornithine in the triple mutant arg-5; ota; aga, which is used to study iron transport (2227, 2229) and iron storage (1296, 1297).

ovc: overaccumulator of carotenoids

IV. Contradictory evidence places ovc (S20-16, FGSC 4503) either in IVR between col-4 (10%) and met-5 (14%) (836) or as an allele of cut in IVL (94). The met-5 strain (FGSC 141) used in ref. (836) contained an ovc-linked, deleterious, undefined mutation that may have led to apparent linkage in IVL that is spurious (1546). For this reason, and because of phenotypic similarity, location of ovc at the cut locus is favored.

Produces more photoinduced carotenoid pigments than wild type when incubated at temperatures above 6°C, but about the same at 6°C (836). Probably allelic with cut. Strain S20-16 (FGSC 4503) is indistinguishable from cut in increased pigmentation, osmotic sensitivity, and altered morphology of aerial hyphae (94). ovc makes some pigment in the dark (834), suggesting possible allelism with one of the unmapped ccb mutants, which overaccumulate carotenoid pigments in the light (1208).

oxD: D-amino acid oxidase

IVR. Between T(S1229)^L, pdx-1 (0T/55 asci), T(S1229)^L and T(B362)^L, met-1 (3%). Not separated from cys-15 (111, 1490, 1580).

Lacks D-amino acid oxidase (EC 1.4.3.3) and is unable to use D-methionine as a sole sulfur source or to satisfy the growth requirement of met-1. Shows increased sensitivity to the toxic effects of D-phenylalanine and D-tyrosine (1490). Resistant to inhibition by D-ethionine (901). The strain of origin of allele oxD^1 simultaneously became a cysteine auxotroph, cys-15, thought to be due to a closely linked coincident lesion (1490).

Oxidase

For alternate terminal oxidases, see aod, azs, has, cni-1.

pa: pale

IR. Between cr-1 (10%) and dir (37%) (1199, 1200).

Conidia are sparse, clumped, and pale. Photograph (1200). Stock lost. (Possibly wc-2?)

pab: para-aminobenzoic acid

p-Aminobenzoate is produced from chorismate in the aromatic amino acid biosynthetic pathway (Fig. 13). The holoenzyme, p-aminobenzoate synthase (EC 4.1.3.-), has two components, ADC synthase (I) and glutamine amidotransferase (II). Typically these are encoded by two different genes. Which of the two pab genes in Neurospora specifies which activity is uncertain.

pab-1: para-aminobenzoic acid-1

VR. Between inl (1%; 10%) and trp-5, met-3 (1% or 2%) (673, 2014).

Specifies a component of p-aminobenzoate synthase (EC 4.1.3.-) (Fig. 13). Requires p-aminobenzoic acid (2058). Apparently cannot use folate (mono- or triglutamate) (2293).

pab-2: para-aminobenzoic acid-2

VR. Linked to ro-4 (0/407). Between ad-7 (8%), T(EB4)^R and mus-11 (7%), inv (3%), asn (1%; 15%) (321, 1578, 1592, 1603, 1794, 1828, 2014).

Specifies a component of p-aminobenzoate synthase (EC 4.1.3.-) (Fig. 13). Requires p-aminobenzoic acid (2304). Mutant 71301, called pab-3 (2304), was shown to be allelic (550).

pab-3: para-aminobenzoic acid-3

Allelic with pab-2.

Pad-1: Paddle-1

IIL. Between het-c and pyr-4 (1%); restriction fragment G22:H5 with het-c, X7:F1 with pyr-4 (1109).

Cloned and sequenced: EMBL/GenBank AF130355.

Sequence characteristics resemble those of heterogeneous nuclear RNA-binding proteins and mRNA-splicing factors. Pad-1 function is essential for sexual development and vegetative growth. Partially functional mutants obtained by RIP show ascus-dominant defects in ascus and ascospore formation. Some asci are swollen and paddle-shaped, with no apical pore and no ascospores. Nuclear divisions and spindle orientation are abnormal in other asci that are not paddle-shaped. Two, four, or eight ascospores may be cut out in these asci, but the ascospores are inviable. Vegetative growth is severely defective. The growth defect is recessive in heterokaryons (1109).

pan-1: pantothenic acid-1

IVR. Between *ad-6* (1% or 2%) and *cot-1* (2% or 3%) (1255, 1369, 1582). *cel-1* (1%), *col-1* (0T/47 asci), *int* (0/50), *pho-3* (<1%), *ro-1* (0/394), and *thi-5* (1%) also are closely linked.

Cloned: pSV50 clone 31:9C; Orbach-Sachs clones X3B04, X11A07, X18G05, G19F10.

Requires intact pantothenic acid for growth under standard conditions. Able to synthesize both precursors, β-alanine and pantoyllactone (2059). The ability to synthesize pantothenic acid from β-alanine + pantoyllactone is demonstrable *in vitro* but not *in vivo* unless cultures are aerated (2173, 2175, 2176). Unlike *pan-2*, *pan-1* has no effect on ascospore ripening in heterozygous crosses. Called group A. For alleles see ref. (296).

pan-2: pantothenic acid-2

VIR. Between *rib-1* (<1%; 3%) and *del* (6%), *trp-2* (11%) (299, 300, 302, 1582, 1603).

Cloned and partially sequenced: Orbach-Sachs clone X6D10 (572, 1738a).

Encodes ketopantoate hydroxymethyltransferase (EC2.1.2.11) (1738a). Unable to convert ketovaline to ketopantoic acid (296, 299, 300). Used in major studies of intralocus recombination and complementation (299–302). *pan-2* ascospores remain white or pale if the crossing medium is not supplemented, even when the protoperithecial parent is *pan-2*⁺. Asci in which gene conversion has occurred at *pan-2* thus can be recognized and isolated (2086, 2087); photographs (2086). For good recovery of *pan-2* progeny, crossing media should be supplemented with pantothenic acid (10 μg/ml), even when the protoperithecial parent is *pan*⁺. Called group B.

pat: patch

IL. Linked to *mat*, probably to the right (1988).

Growth and conidiation occur in patches and in a cyclic pattern under certain conditions (1988). Initially found in a *pro-1* 21863 strain, but the *pro-1* mutation is not necessary for the expression of *patch*. The original *patch* isolates were all sorbose-resistant (1988). A sorbose-resistant derivative that does not express the patch

phenotype is called *sor-4* (1592). It is not clear whether *pat* and *sor-4* are separate genes, or whether *patch* is not scorable in the absence of modifiers present in the parent stock [see p. 267 of ref. (1592)]. The original *patch* strain was used for the first demonstration of a circadian rhythm in fungi (223, 1628).

pcna: proliferating cell nuclear antigen

VL. Linked to rDNA (0T/18 asci) (1091).

Cloned and sequenced: Orbach-Sachs cosmid X2-H8 (1091).

Encodes the *Neurospora* homolog of proliferating cell nuclear antigen (PCNA), first identified in mammals and known to be a ring-shaped homotrimer necessary for processive DNA synthesis in eukaryotes (1091).

pcon: phosphatase control

Allelic with *nuc-2*.

pdc-1: pyruvate decarboxylase-1

Allelic with *cfp*.

pde-1: GTP-regulated cyclic phosphodiesterase-1

Unmapped. Perhaps in linkage group II (844).

Affects GTP-regulated cyclic phosphodiesterase (845).

pdx: pyridoxine

Mutants at two closely linked loci require pyridoxine, pyridoxal, or pyridoxamine for growth (1657, 1660, 1661). The first nutritional mutant ever found was a pyridoxine auxotroph (135). Pyridoxine mutants provided the first proved example of gene conversion (1663). Early *pdx* mutants were called *pdx-1* except for isolate 44204, which was called *pdx-2* on the basis of one crossover (see 1663). Mutant 44204 was subsequently reclassified as *pdx-1* on the basis of noncomplementation with certain of the alleles then designated *pdx-1*. The existence of two genes has now been reestablished, placing all pH-sensitive *pdx* mutants into *pdx-2* and non-pH-sensitive mutants into *pdx-1* (570a). Isolates originally called *pdxp* and considered alleles of *pdx-1* are pH-sensitive, growing without supplement on medium containing ammonium ions above pH 6 (2006). These now appear to be *pdx-2* alleles (570a). The two *pdx* loci are adjacent, and divergently transcribed. There is complementation between mutants (1660, 1661) and intralocus recombination (1663). Many alleles are leaky. Scoring is sharpened by the addition of 100 mg desoxypyridoxine per liter (1660). Conidia are subject to death by unbalanced growth on minimal medium (2010). A yellow pigment is excreted under certain conditions by the double mutant *pdx; En(pdx)*; see *En(pdx)*.

pdx-1: pyridoxine-1

IVR. Between *pyr-1* (1%; 10%) and *T(S1229)^L*, *pt* (2%) (87, 111, 1369, 1580, 570a).

Cloned and sequenced: Orbach-Sachs cosmid G6G8 (1441a, 1446, 570a).

Pyridoxine-requiring isolates originally classified as type a (1660, 1661, 570a). Not pH-sensitive. Shows strong sequence identity to the highly conserved gene *snz* in *saccharomyces* (224) and to *SOR1* (singlet oxygen resistance) in *Cercospora nicotianae*. SOR1 function is essential for pyridoxine synthesis both in *Cercospora* and in *Aspergillus flavus* (599). Linkage data place *pdx-1* and *snz* in the same short region in *Neurospora* (1447). Five sequenced *pdx-1* mutants are altered in the *Neurospora* gene previously identified as *snz* (1441a). Thus the *Neurospora snz* gene is allelic with *pdx-1*.

pdx-2: pyridoxine-2

IVR. Between *pyr-1* (1%; 10%) and *T(S1229)^L*, *pt* (2%). Near *pdx-1*, probably to the left (87, 111, 1369, 1580, 570a).

Cloned and sequenced: Orbach Sachs cosmid G6G8 (1441a, 1446, 570a).

Pyridoxine-requiring isolates originally classified as types b and g (1660, 1661, 570a). pH-sensitive. Homologous with *sno1* in *Saccharomyces* (599, 570a).

pe: peach

IIR. Between *nuc-2* (4%) and *arg-12* (1%; 5%) (1157, 1592).

Distinctive morphology (102, 1203). Peach-colored conidia. Short hyphae are formed more uniformly than in wild type, as a lawn close to the agar surface. Added arginine increases macroconidiation and tends to obscure scoring of *pe* at 25°C but not at 39°C. The single mutant *pe* produces both macro- and microconidia. The double mutant *pe fl* produces abundant gray microconidia and no macroconidia (102, 1389) (see *fl*). See *col-1*, *col-4*, and refs. (781) and (782) for interactions with other mutants. Called *pe^m* or *m* in some contexts.

pen-1: perithecial neck-1

VIIL. Linked to *csp-2* (4%) (1593).

When the *pen-1* mutant is used as female parent, perithecia lack necks (beaks) and ascospores are not shot, although a few ascospores may be formed. When *pen-1* is used to fertilize a *pen^+* strain, perithecial beaks are normal and crosses are fully fertile (518, 1593).

pep: processing enhancing protein

Unmapped.

Cloned and sequenced: Swissprot MPP2_NEUCR, EMBL/GenBank M20928, PIR A29881, S03968, GenBank NEUPEP.

Encodes the 52-kDa subunit of mitochondrial-processing peptidase (EC 3.4.24.34). PEP cooperates with MPP (mitochondrial-processing peptidase, encoded by *mpp*) (1823) in the proteolytic cleavage of matrix-targeting sequences from nuclear-encoded mitochondrial precursor proteins (860).

pep4: PEP4 homolog

VIIR. Linked to *for*, *frq* (0T/18 asci) (1447). On a common cosmid with *frq* (2144).

Cloned and sequenced: Swissprot CARP_NEUCR, EMBL/GenBank U36471.

Structural gene for vacuolar proteinase A, homologous with *Saccharomyces cerevisiae* PEP4 (E.C. 3.4.23.-) (2145). Contrary to the name, the substrates of *Neurospora crassa* PEP4 are proteins rather than peptides (204).

per-1: perithecial-1

VR. Between *at* (8%; 14%), *asp* (16%; 26%) and *ilv-1* (4%) (919, 1582), *ts* (25%) (1002).

Perithecial walls fail to blacken when *per-1* is the female parent (Fig. 37), regardless of the genotype of the fertilizing parent (918, 919, 1002). Alleles are of two types (918): In type I (e.g., alleles PBJ1, ABT8, AR174), *per-1* ascospores are unpigmented and perithecial walls are completely devoid of black pigment. In type II (e.g., alleles 29-278, 29-281, UG1837), *per-1* ascospores are the normal black and the initially unpigmented perithecial walls become brownish as they age. Black pigment develops in a ring around the ostiole of type II perithecia, but is pale or lacking in type I (918). Unlike the perithecial wall, the ascospore trait shows no maternal effect. White *per-1* ascospores (type I) germinate without heat shock and are usually killed by hypochlorite or by the 30-min, 60°C treatment normally used to activate ascospores (918, 1002). Expression is completely autonomous in ascospores [photographs in ref. (999)] and at least partially so in the perithecial wall (998, 999, 1002). Mosaic perithecia from heterokaryons have been used for a clonal analysis of perithecial development (998, 1002). Used to show that perithecial walls are capable of making orange carotenoids (1569). Used in an unsuccessful search for variegated-type position effects (1001). Beaks of perithecia homozygous for allele PBJl (type I) are abnormal, and ascospores are not shot properly (1673). Type I alleles were initially called *sw: snow* white (1002).

Perithecial Development Mutants

See *fs*, *ff*, *fmf*, *mb*, *mei*, *pen*, *per*, *sdv*.

Permease

See Transport.

pex2: peroxidase assembly factor 2

IL. Near *mat*. Between *cyt-21* and *mei-3* (921).

Cloned and sequenced: pSV50 cosmid 19:3E (921).

Homologous to human *pex2* and to *Podospora car-1* (161), which was used for cloning the *Neurospora* gene. Encodes a putative peroxisome assembly factor with homology to PAF1, which is implicated in the lethal Zellweger Syndrome in humans. In *Neurospora*, disruption of *pex2* by RIP probably impairs peroxisome function. The mutant is resistant to 3-AT. Growth is poor in the presence of oleic acid. Sexual development is delayed when *pex2* is used as female parent (921).

pf: puff

IVR. Right of *pyr-2* (2%). Linked to *rug* (3%) (1585, 1928). Spreading colonial growth (1585).

pfa-1: polyunsaturated fatty acids-1

VII. Linked to *csp-2* (35%) (756).

Involved in fatty acid desaturation, with reduced levels of linoleate and linolenate (753, 756) (Fig. 17).

pfa-2: polyunsaturated fatty acids-2

IV or V. Linked to *T(R2355)*, *cot-1* (32%) in cross to *alcoy* (756).

Involved in fatty acid desaturation, with reduced levels of linoleate and linolenate (753, 756) (Fig. 17).

pfa-3: polyunsaturated fatty acids-3

IV or V. Linked to *T(R235)*, *cot-1* (29%) in cross to *alcoy* (756).

Involved in fatty acid biosynthesis, with an indirect effect on the desaturation of linolenate. Accumulates free fatty acids (756) (Fig. 17). [Free fatty acids are also accumulated by mitochondrial mutant [*mi-1*] (737).]

pfa-4: polyunsaturated fatty acids-4

VII. Linked to *csp-2* (33%) (756).

Involved in fatty acid biosynthesis, with an indirect effect on the desaturation of linolenate (753, 756).

pfa-5: polyunsaturated fatty acids-5

I or II. Linked to *T(4637) al-1* (7%) in cross to *alcoy* (756).

Involved in fatty acid biosynthesis, with an indirect effect on the desaturation of linolenate (753, 756).

pgk: phosphoglycerate kinase

Unmapped.

Cloned and sequenced: Swissprot PGK_NEUCR, EMBL/GenBank X56512, GenBank NCPGKR, PIR A56616; EST NC3D6.

Structural gene for phosphoglycerate kinase (EC 2.7.2.3) (78).

pgov: phosphorus governance

IIIR. Linked to *tyr-1* (1%; 4%), probably to the right (1322).

Regulatory gene in phosphorus uptake and metabolism (1325, 1326, 1331) (Fig. 51). Probably specifies a cyclin-dependent kinase. *pgov* corresponds to a gene cloned by P. Margolis and called *mdk-1*. *pgov^c* is similar to *preg^c* phenotypically (1322, 1325). Isolated in *preg^+*/*preg^+* partial diploids. Mutants were obtained first as a suppressor of *nuc-2* and then by insertional mutagenesis. *pgov* allele c-5 is largely or completely recessive in duplications from *T(D305)* (1322). Scored on high-phosphate medium by staining reaction with α-naphthyl phosphate plus Diazo Blue B (747, 2096).

phe-1: phenylalanine-1

IL. Between *suc* (<1%), *In(H4250)^L* and *ad-5* (914, 1592).

Originally reported to grow on phenylalanine, other aromatic amino acids, leucine, or ethyl acetoacetate, with phenylalanine being the most effective; several other acids gave smaller responses (104). Utilization of phenylalanine and other compounds varies for different isolates and on different carbon sources; glycerol or ribose are preferable to sucrose (992–994, 1464). Allele NM160 does not use phenylalanine, but grows well on tyrosine or leucine (510, 1592), at least with the strains and carbon source used. *phe-1* is inhibited by basic amino acids on low phenylalanine or on leucine (104, 992). Growth on β-labeled leucine or β-labeled phenylalanine showed that neither compound is converted to the other (101). The mutant has a 19-hr circadian period, even on high levels of phenylalanine. The high levels of phenylalanine required for growth allow the synthesis of sterols by an alternate catabolic pathway, as evidenced by patterns of incorporation of radioactive phenylalanine and acetate into sterols [S. Brody, unpublished, cited in ref. (1130)]. Called *phen-1*. Allele NM160 called *tyr(NM160)* (601).

phe-2: phenylalanine-2

IIIR. Linked to *vel* (1%). Between *T(D305)^L* and *tyr-1* (2%; 4%) (87, 601, 1581).

Lacks prephenic dehydratase (EC 4.2.1.51) (87, 601) (Fig. 13). The requirement is very leaky. Grows extensively and is treacherous to score by growth on minimal vs supplement, but can be scored reliably by blue fluorescence under long-wave UV after growth on minimal medium without phenylalanine (1592). The appearance of phenylalanine in the culture medium (245) is due to spontaneous conversion of accumulated pretyrosine (988). Called *phen-2*. Allele Y16329 called *phen-3* (601).

phe-3: phenylalanine-3

Allelic with *phe-2*.

phen: phenylalanine

Symbol changed to *phe*.

pho-1: phosphorus-1

II? 20% wild-type recombinants with *nuc-2*. Independent of *nuc-1*.

Activity of the repressible alkaline phosphatase is low (Fig. 51). Complements *nuc-1* and *nuc-2* in heterokaryons. Stains pale red on low-phosphate medium with α-naphthyl phosphate + Diazo Blue B (2096).

pho-2: phosphorus-2

VR. Between *his-1* (3%) and *inl* (4%) (747, 1322).

Cloned and sequenced: EMBL/GenBank L27993, GenBank NEUALPH; pSV50 clone 23:1A.

Structural gene for P^i-repressible alkaline phosphatase (EC 3.1.3.1) (747, 1158, 1455). Scored on low-phosphate medium by a staining reaction with α-naphthyl phosphate plus Diazo Blue B (747, 2096). Encodes multiple forms (10 or more) of the enzyme from the same coding sequence (2081) (Fig. 51).

pho-3: phosphorus-3

IVR. Linked to *pan-1* (<1%) (1455).

Structural gene for repressible acid phosphatase, with phosphodiesterase activity (1455) (Fig. 51). Codominant in heterozygous *pho-3/pho-3*$^+$ duplications. Scored on low-phosphate medium by a staining reaction with bisnitrophenyl phosphate (1455).

pho-4: phosphate-4

VIIL. Linked to *nic-3* (2–6%; 1 or 2T/18 asci) (205, 1447).

Cloned and sequenced: Swissprot PHO4_NEUCR, EMBL/GenBank M31364, PIR JQ0116, GenBank NEUPHO4A; pSV50 clone 24:8F, Orbach-Sachs clones X20D10, G3D12.

Structural gene for phosphate-repressible phosphate permease (1259) (Fig. 51). Resistant to vanadate (0.1–1.0 mM) in Vogel's minimal medium with 0.15 mM phosphate and 20 mM HEPES buffer, adjusted to pH 7.5

(204). Constitutive mutation *Ab(RLM01)* is inseparable from *pho-4* (1578). Called *van*.

pho-5: phosphorus-5

IVR. Between *pyr-1* (5%; 2T/18 asci) and *trp-4* (4T/18 asci), *cot-1* (15%) (1447, 2147).

Cloned and sequenced: EMBL/GenBank L36127, GenBank NEURHAPPA; Orbach-Sachs clone G16E10.

Structural gene for repressible high-affinity phosphate permease (Fig. 51). Obtained as a suppressor in *nuc-1; pho-4*. Constitutive mutations *T(RLM02)*, *T(RLM04)*, *T(RLM06)*, *T(RLM08)*, *T(RLM09)*, and *T(RLM11)* are inseparable from *pho-5* (1578).

phr: photoreactivation deficient

IR. Right of *os-1* (15%). Linked to *con-8* (0/18), *nic-1* (4/18) (1908).

Cloned and sequenced: Swissprot PHR_NEUCR, EMBL/GenBank X58713, GenBank NCPHR, PIR S18667.

Epistasis group Phr (Table 3). Structural gene for deoxyribodipyrimidine photolyase (DNA photolyase) (EC 4.1.99.3) (600, 2252). The RIP-inactivated *phr* mutant is defective in photolyase activity for cyclobutane–pyrimidine dimers and TC(6-4) photoproducts caused by UV irradiation. Dark repair is not affected. *Neurospora* photolyase is not a blue light receptor (1908). Called *uve-1*.

pi: pile

IIL. Linked to *ro-7* (0/75). Between *T(AR30)* (28%) and *cys-3* (4%) (1173, 1578, 1592).

Slow, spreading mycelial growth. No conidia are formed (1592). Scanning EM photograph (1955): few major constriction chains. *pi* B101 grows better on minimal medium than on complex complete medium. *col-10* R2438 is a putative allele (1852). Because of growth rate, stability, and ease of handling, *pi* B101 is preferable to *col-10* R2438 as a marker. *ro-7*, in the same region, is preferable to both (1582).

pk: peak

VR. Between *met-3* (1%) and *T(EB4)*^{L}, *cot-2* (8%), *cl* (2%) (569, 1582, 1603, 2014).

Growth on an agar surface is initially colonial and flat. A mass of aerial hyphae is then sent up, which conidiates profusely (1548). Somewhat similar in morphology to *sn*, *cum*, *sp*, and *cot-4* at 25°C, but distinguishable. Hyphae branch dichotomously (1405, 1548). Increased activity of L-glutamine:D-fructose-6-phosphate amidotransferase was observed in crude extracts of one *pk* strain, but not in nine others; increased activity for this enzyme also was found in *cl* and in four other nonallelic morphological mutants (1758). Hexoseaminoglycan consists of a single component on medium without sorbose, in contrast to two components in wild type (1960). Antigenic surface mucopolyoside (532). Cell-wall analysis and photograph, allele B6 (507) and allele C-1810-1 (287). Cell-wall enzymes (633). Effect of carbon source (531). One observation suggested a functional interaction with *cl* (1965), but substantial crossing-over frequencies and recovery of the double mutant *pk cl* indicated that the loci are distinct (569). The gene was named for the vegetative mutant phenotype, which is recessive. Several alleles were called *bis* (1592). The sexual phase is also affected. Asci are thin-walled, bulbous, and nonlinear in homozygous *pk × pk* crosses (1406, 1409, 1550). Spindle orientation is abnormal in the swollen asci, and no apical pore is formed (1625, 1679). Most mutant alleles are recessive for the ascus effect, but some are dominant with variable penetrance. Sorbose-resistant mutants at various loci act as dominance modifiers of the ascus effect of dominant alleles (1757). Sexual-phase-recessive allele C-1610 and dominant allele 17-088 are both associated with reciprocal translocations (1578).

pl: plug

VR. Linked to *gran* (0/75). Between *asn* (1%; 9%) and *his-6* (16%) (1585, 1592, 2014).

Dense hyphae fill a 10-mm-diameter tube above the agar slant (1548). Because complex complete medium stimulates conidiation; scoring of morphology is clearer on minimal medium. The morphology is distinct from that of the closely linked mutant *gran* (1582).

plc-1: phospholipase C-1

Unmapped.

Cloned and partially sequenced: EMBL/GenBank U65686.

Encodes a PLC-like protein (phosphoinositide-specific phospholipase, EC 3.1.-.-). Isogenes have been identified from short PCR fragments (EMBL/Genbank U65687 and U65688). Mutants of *plc-1* obtained by RIP are highly sensitive to chemical agents that block polymerization of β-tubulin. Also sensitive to H_2O_2, which suggests that *plc-1* is related to a transduction pathway stimulated by oxidative stress (1037).

pma-1: plasma membrane ATPase-1

IIR. Between *cya-4* (2T/18 asci) and *preg* (1T/15 asci) (1447). Initially assigned incorrectly to linkage group I (813) as a result of the weak cross-hybridization shown with *eat-2* (1925).

Cloned and sequenced: Swissprot PMA1_NEUCR, EMBL/GenBank J02602, M14085, PIR A26497, PXNCP, GenBank NEUATPPM, NEUATPASE; EST NM8D11.

Structural gene for plasma membrane ATPase (proton pump) (EC 3.6.1.35) (3, 813, 1427, 1922). Extensive electrophysiological studies defining the role of the pump (762, 763, 1921, 1923, 1924, 2189). Extensive biochemical studies: residues important for catalysis (1033, 1384, 2027); ligand binding at the active site (179); role in potassium ion symport (186, 188); N-ethylmaleimide sensitivity (352, 444, 1534). Conformational changes arising from ligand binding (759, 1257). Reconstitution in vesicles (1607) and stoichiometry of proton translocation (1608). Reconstituted ATPase is functional as a single-subunit monomer (758, 1804); functional size determination by radiation inactivation (207). Comparison with mitochondrial and vacuolar proton pumps (211). Secondary-structure analysis (868, 2158) and structural determination by electron crystallography (73, 437, 1106, 2004). Studies of biosynthesis, integration into the plasma membrane and enzyme topography (1, 4–6, 869, 1195–1197, 1258, 1697, 1698, 1803, 1805, 2159). Expression and mutagenesis studies in *Saccharomyces cerevisiae* (1248, 1249). Strong resemblance to a homolog in the pathogen *Histoplasma capsulatum* (1808). Mutants have been obtained that are inhibited by concanamycin A at high pH (213).

pmb: permease basic amino acid

IVR. Between *uvs-2* (8%), $T(S4342)^R$ and *Tip-IVR* (513, 1578, 1776).

Cloned: Orbach-Sachs clone X7E5 (826).

Defective in basic L-amino acid transport system III as defined in ref. (1522); uptake of L-arginine, L-lysine, and L-histidine is reduced (1522, 2090, 2233) (Fig. 47). Used extensively for transport studies; see ref. (2231). Altered surface glycoprotein (2016). Selected as resistant to L-canavanine (1736, 2233). Scored on 1 mg/ml thiolysine or by resistance to α-difluoromethylornithine (464). Allelic with *bat* (1776), which was selected in *arg-12^s*; *pyr-3* (CPS^-ACT^+) by its ability to grow on minimal medium plus arginine when the parental double mutant was not able to grow because of arginine uptake and feedback onto the arginine-specific CPSase (2088). Probably allelic with *bm-1* (linked to *pyr-2*, 24%), which was selected by canavanine resistance (1782). Possibly allelic with *bas^a*, which was selected by the inability of *his-3* to grow on histidine plus methionine (1243). A mutant called *arg^R*, which maps right of *pyr-2* (14%) and is probably allelic with *pmb*, makes *lys-1* resistant to inhibition by L-arginine in the double mutant *lys-1; arg^R* (1082, 1083). Called Cr-10, Pm-B, *pm h*, UM-535, *can-37*.

pmg: permease general amino acid

IIL. Linked to *pyr-4* (0/207). Right of *ro-3* (4%) (29%) (1574).

Cloned and sequenced: Orbach-Sachs clone G3D3 (204).

Specifies the general amino acid permease, homologous to GAP1 of *Saccharomyces* (204). The mutant is greatly reduced in general amino acid transport [system II as defined in ref. (1521)] (Fig. 47). Uptake of arginine and phenylalanine is reduced. Selected by resistance to *p*-fluorophenylalanine in a neutral (system I), basic (system III) double mutant on medium lacking ammonium ions, where system II would be derepressed in wild type (513, 1693, 1694). A nonmetabolizable substrate has been found that is specific for this transport system (1488).

pmn: permease neutral amino acid

Allelic with *mtr*. (Also called *Pm-N*.)

Pogo (Transposable Element)

VR. Adjoins *Tel-VR*. Imperfect copies of the *Pogo* sequence are also present at nonterminal positions elsewhere in the genome and in different locations in different genetic backgrounds (1809, 1811).
Cloned and sequenced: EMBL/GenBank M37064, GenBank NEUTELVRA.
A retrotransposon-type element discovered as a 1.6-kb sequence adjoining the VR telomere. Contains direct terminal repeats and an open reading frame that is similar in sequence to reverse transcriptases (1809, 1811).

poi-1: plenty of it-1

VR. Linked to *cyh-2* (0T/18 asci) (1452).
Cloned and partially sequenced: EMBL/GenBank AA897897, AA898529; EST NP2A8.
The most abundant cDNA in starved mycelial and perithecial tissues. A novel sequence (1052).

poi-2: plenty of it-2

VIIR. Linked to *cat-2* (0 or 1T/18 asci), *cox-8* (1T/18 asci) (1452).
Cloned and partially sequenced: EMBL/GenBank AA897974, AA898152; EST NM2H2.
The second most abundant cDNA isolated from starved mycelial and perithecial tissues. A novel sequence (1052).

polD: DNA polymerase delta

VR. Between *ure-2* (<70 kb) and *am* (10 kb) (222).
Cloned and sequenced: Orbach-Sachs clone G9A10.
Encodes a sequence homologous to *Drosophila pol* δ and to DNA polymerase δ from other species (222).

por: porin

Unmapped.

Cloned and sequenced: Swissprot PORI_NEUCR, EMBL/GenBank X05824, GenBank NCPORIN, PIR S07195, MNMCP; EST NC1D11.
Porin, also called VDAC (voltage-dependent, anion-selective channel), is the major protein of the mitochondrial outer membrane (1071). Most metabolites enter the mitochondrion through the channel formed by this polypeptide (1260, 1261, 1644). The porin polypeptide is unusual because it does not require an electrochemical potential across the inner mitochondrial membrane for import (675). Also, unlike signals of most polypeptides that are targeted to mitochondria, the import signal of porin appears to be at the C-terminus, not the N-terminus (416).

pp: protoperithecia

Unmapped.
Does not form protoperithecia. Impairs ascospore germination. Enhances growth of *gpi* on glucose or sucrose (1401). Obtained as a double mutant with *gpi*. No information is available on linkage or allelism with already mapped loci that affect ascospore viability or the formation of protoperithecia (e.g., *ff, fs, gul-3, -4, -5, le-1, -2*).

pph-1: protein phosphatase-1

IV. Linked to *mtr* (0/18), arg-14 (1/18) (1447, 2271).
Cloned and sequenced: Swissprot P2A1_NEUCR, EMBL/GenBank X83593, NCPPH1, PIR S60471.
Structural gene for serine–threonine protein phosphatase PP2A catalytic subunit (EC 3.1.3.16) (2271). Essential function; the null mutant obtained by RIP can be harbored in a heterokaryon (2271). Level of mRNA is highest during conidial germination (2271). Characterization of activity (2297). Reduced expression is associated with reduced hyphal growth (2270).

pph-2: protein phosphatase-2

Allelic with *cna-1*.

pph-3: protein phosphatase-3

Unmapped.
Cloned and sequenced: Genbank AF049853 (2298).
Encodes serine–threonine protein phosphatase type 1 (PP1) (2298).

ppt-1: phosphoprotein phosphatase-1

VR. Linked to *inl* (0/18) (1447).
Cloned and sequenced: EMBL/GenBank U89985.
Encodes a protein phosphatase of the PPT/PP5 family. Transcript is abundant in dormant conidia; transcript level declines during germination (2272). For a review of the protein phosphatases identified in filamentous fungi, see ref. (535).

ppz-1: protein phosphatase Z

Changed to *pzl-1*.

pr(Sk-2): partial resistance to Sk-2 killing

IIIR. Between *ser-1* and *ad-4*. Within the recombination block that is associated with *Sk-2* (2121).

Confers partial resistance to the killing of ascospores by *Sk-2* (2121).

prd: period

Mutations at numerous loci other than *frq* have been shown to affect the period length of the circadian clock. Those named *period* were identified on this basis. Period length and/or temperature compensation are also affected by mutations at loci initially recognized on the basis of quite different characteristics, e.g., *cel, chol-1, fas, phe-1*. Reviewed in refs. (560), (1130), and (1133).

prd-1: period-1

III. Linked to *acr-2* (5%), *Cen-III* (0T/35 asci), *pro-1* (20%) (619, 710).

Altered period of circadian conidiation rhythm. One allele is known, with a 25.8-hr period at 25°C without *csp* (619, 711). Recessive. Grows at 60% of the wild-type rate. Period lengths of double mutants with *frq-1, -2, -3* (619). Temperature compensation (713). Altered membrane fatty acid composition (413); effects of fatty acid addition to the *cel* mutant are not observed in the *cel*; *prd-1* double mutant (1132). Membrane cytochromes appear to be similar to wild type (198). Name changed from *frq-5* (711). For reviews of circadian mutants, see the entry for *frq*.

prd-2: period-2

VR. Between *lys-2* and *am* (713).

Cloned (1184).

Altered period of circadian rhythm (25.5 hr at 25°C without *csp*, for allele IV-2). Slower than normal growth. Recessive (621). Temperature compensation described (713). Synergistic interactions with *frq* long-period alleles [discussed in ref. (1382)]. Called IV-2.

prd-3: period-3

I. Near *Cen-I* (713).

The period of circadian rhythm is altered (25.1 hr at 25°C without *csp*, for allele IV-4) (621). Slower than normal growth. Recessive. Temperature compensation described (713). Called IV-4.

prd-4: period-4

IR. Between *os-1* (2%) and *arg-13* (3%) (1212).

Cloned: pSV50 clone 9:2H.

The period of circadian rhythm is altered (18.0 hr at 25°C without *csp*, for allele V-7) (621). Normal growth rate. Dominant, based on heterokaryons with different ratios of wild-type and mutant nuclei (56). Temperature compensation described (713). Called V-7.

prd-5: period-5

IIR. Between *arg-12* (4%; 10%) and *un-20* (8%). Linked to *alc-1* (8%) (1186).

Altered period of circadian conidiation rhythm. One allele is known, with a 19.5-hr period at 25°C. Whereas temperature compensation is normal, expression of the mutant phenotype is temperature-dependent. The mutation is pleiotropic, affecting not only the circadian period but also morphology, sexual development, linear growth rate, tolerance to elevated temperature, and acriflavin resistance. Possibly allelic with *ff-1* (1186).

prd-6: period-6

VR. Linked to *ad-7*. Between *inl* (9%; 30%) and *pab-2* (2%), *inv* (≤1%), *asn* (7%) (1382).

The circadian period length is shortened and is temperature-sensitive. Epistatic to *prd-2* (1382).

preg: phosphatase regulation

IIR. Between $T(NM177)^L$, *nuc-2 (pcon)* (1% or 2%), and *arg-12*, $T(NM177)^R$ (1157, 1333). Linked to *cit-1* (0T/18 asci), *nuo78* (0T/17 asci) (1447).

Cloned and sequenced: Swissprot PREG_NEUCR, EMBL/GenBank L07314, PIR S52974, EMBL NCPREGPRO, GenBank NEUPREGPRO; pSV50 clones 2:8E, 2:9F.

Encodes a cyclin. Regulator of repressible alkaline phosphatase and other steps in phosphorus uptake and metabolism. Hypostatic to *nuc-1*, epistatic to *nuc-2*. The $preg^C$ constitutive mutant is recessive to its wild-type allele (1322, 1325, 1333). Scored on high phosphate medium by a staining reaction with α-naphthyl phosphate plus Diazo Blue B (747, 2096). Used to study phosphate transport (1229). For regulation model, see refs. (1325), (1326), and (1331) and Fig. 51.

Prf: Perforated

VR. Linked to *al-3* (3%), probably to the right (1678).

Many of the asci in crosses heterozygous for *Prf* produce asci with 8–12 small apical pores and variable numbers of large, multinucleate ascospores. A majority of asci cut out a single giant ascospore (Fig. 52). A few asci develop normally and produce small ascospores that can be used for mapping. *Prf* is recessive-lethal in the vegetative phase and must be maintained in a heterokaryon. The first-formed croziers appear normal and differentiate asci, but croziers that are formed later revert to mitosis

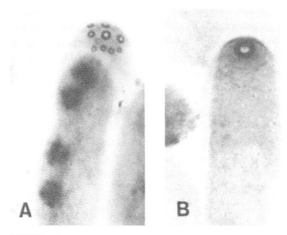

FIGURE 52 (A) The tip of an ascus from a cross heterozygous for the mutant gene *Prf: Perforated*, showing multiple apical pores. (B) A normal ascus, showing the single apical pore that is characteristic of wild type. Reprinted from ref. (1678), with permission from *Mycologia*, Vol. 79, p. 702. Copyright 1987 The New York Botanical Garden.

and become multinucleate, with synchronous nuclear divisions similar to those in *Ban* (1678).

prl-1: mitochondrial proteolipid ATP synthase

Allelic with *oli-1*.

pro: proline

Three *pro* loci are defined by mutant genes. *pro-3* and *pro-4* are putative isogenes for the first step in proline biosynthesis, the conversion of glutamate to glutamate semialdehyde. The next step, catalyzed by γ-glutamyl phosphate reductase (GPR) (EC 1.2.1.41) (glutamate-5-semialdehyde dehydrogenase) (glutamyl-γ-semialdehyde), is unrepresented by a mutant or a sequence. The final step (EC 1.5.1.2) is defined by *pro-1*. As part of the greater arginine biosynthetic pathway (Fig. 10), *pro-3* and *pro-4* originally were called *arg-8* and *arg-9*.

pro-1: proline-1

IIIR. Between *sc* (7%), *ser-1* (1%; 114 kb; 3T/38 asci) and *ace-2* (2%; 36 kb) (449, 940, 1122, 1587).
Cloned and sequenced: EMBL/GenBank U30317, PIR S57863; pSV50 clones 13:10A, 18:2C, 21:6D (449).
Structural gene for pyrroline-5-carboxylate reductase (EC 1.5.1.2) (449, 2291). Uses proline but not ornithine, citrulline, or arginine (1967) (Fig. 10). Transcription of *pro-1* appears to be independent of *cpc-1*-mediated cross-pathway control (449).

pro-3: proline-3

VR. Linked to *inl*, *pab-1* (0/74). Between *his-1* (4%) and *pk* (2%; 6%) (1585).

Uses proline, ornithine, citrulline, or arginine (1964). Blocked in the reduction of glutamic acid to glutamic semialdehyde (2165), which is catalyzed by glutamate 5-kinase (EC 2.7.2.11) (γ-glutamyl kinase) (Fig. 10). Arginine and citrulline are used via the arginine catabolic pathway (arginase and ornithine transaminase), ornithine via ornithine transaminase (310, 453, 2164, 2165). Tends to accumulate second mutations, including *arg-2*, *his-1* (116, 1932). Ability to grow on arginine is modified by *ipa* (1932) and *ota* (453). Suppressed by *arg-6* mutations that affect N-acetylglutamate kinase (383, 2204). Called *arg-8*.

pro-4: proline-4

IIIR. Linked to *thi-2* (0/78), *ota* (4%) (108, 1603).
The proline pathway is blocked in the reduction of glutamic acid to glutamic semialdehyde (2165), glutamate 5-kinase (EC 2.7.2.11) (γ-glutamyl kinase) (Figs. 10 and 11). Uses proline, ornithine, citrulline, or arginine (1967). Citrulline and arginine are used via the arginine catabolic pathway (arginase and ornithine transaminase) and ornithine via ornithine transaminase (310, 2164, 2165). Leaky. Called *arg-9*.

prol: proline

Symbol changed to *pro*.

Protein Kinases

Protein phosphorylation is required for fundamental functions such as the cell cycle, transcription, and mating. For a review of protein kinases in filamentous fungi, see ref. (535). For genes that specify protein kinases in *Neurospora*, see *cna*, *cot-1*, *mad*, *mak*, *mcb*, *mdk*, *mek*, *mgk*, *mik*, *msk*, *nik*, *nrc*, *pph-1*, *pzl-1*, and *rgb-1*. Numerous ESTs have been identified that appear to identify protein kinases.

Protein Phosphatases

The protein phosphatases identified in filamentous fungi are reviewed in ref. (535). Genes specifying protein phosphatases in *Neurospora* include *cna-1*, *cyt-4*, *pph-1*, *ppt-1*, *pzl-1*, and *rgb-1*.

prt: protease

Allelic with *pts*.

ps15-1: phosphorylation of PS15 protein

Symbol changed to *psp*.

psi-1: protein synthesis initiation-1

IVR. Between *Cen-IV* and *T(ALS159)*, *pyr-1* (4%) (1220, 1546, 1578).

Conidial germination and hyphal growth are inhibited at 35°C but are normal (or nearly so) at 20°C. Protein synthesis is reduced after a shift to restrictive temperature. Recessive in heterokaryons (1220). Scored as an irreparable heat-sensitive *un* mutant (see *un*). Germinated ascospores die after 3 days at 34°C. Used to examine the effects of inhibiting protein synthesis on heat shock (2047).

psp: phosphorylation of small protein

Allelic with *ndk-1* (1489a).

pt: phenylalanine + tyrosine

IVR. Between *pdx-1* (2%), *T(S1229)*^{L} and *col-4* (2%) (87, 111, 1580).

Evidently the structural gene for chorismate mutase (EC 5.4.99.5) (87, 601) (Fig. 13). Allele NS1 has thermolabile chorismate mutase (317). Requires phenylalanine plus tyrosine (396). Inhibited by complex complete medium. NS1 strains are temperature-sensitive, growing on minimal medium at 25°C, where they are readily scorable by blue fluorescence under long-wave UV and by browning of medium in aging cultures (2013). The original strain, S4342, contained the linked but separable insertional translocation *T(S4342)*, the presence of which should not change conclusions regarding gene order in ref. (87).

pts-1: protease-1

Unmapped. Segregates as a single gene not closely linked to *alcoy* markers (40 isolates) (831).

Structural gene for carbon-, nitrogen-, and sulfur-controlled extracellular alkaline protease (EC 3.4.21.-). The allele found in strain Groveland-1c *a* (FGSC 1945) encodes a fast electrophoretic variant that is synthesized under conditions of limiting C, N, and S. Regulation reviewed refs. (1281) and (1325). Called *prt* (831).

Punt (Transposable Element)

IVR. A RIP-modified copy (*Punt*^{RIP1}) is present in the *Fsr-63* pseudogene of Oak Ridge strains, between *Cen-IV* and *pyr-1*. Unmapped copies are also present at other locations. One of these, *dPunt*, is a defective copy that has apparently not been modified by RIP (1266).

Cloned and sequenced: EMBL/GenBank AF181821 (*Punt*^{RIP1}); AF181822 (*dPunt*).

An inactive element bounded by imperfect terminal inverted repeats and a 3-bp target-site duplication. Most closely allied to the *Fot1* family of elements thought to transpose through a DNA intermediate (439). Discovered as a 1.9-kb insertion in *Fsr-63*. Similar sequences are present in other *Neurospora crassa* strains and in *Neurospora sitophila* (1266).

Purine

See *ad*, *gua*.

put-1: putrescine-1

Symbol changed to *spe-1*.

puu-1: putrescine uptake-1

IVR. Between *arg-2* (5%) and *cot-1* (8%) (472).

Cloned and sequenced: Orbach-Sachs clone G17G02.

Calcium modulation of polyamine transport is lost (472). The mutant is temperature-sensitive (37°C) and osmotic remedial (1 M sorbitol) (451). Used to study polyamine toxicity (469, 478).

pyr: pyrimidine

All pyrimidine auxotrophs in *Neurospora* are nonspecific, responding to any pyrimidine nucleoside, nucleotide, or base. The symbol *pyr* therefore is used for genes concerned with the biosynthetic pathway. Nucleosides or nucleotides are more effective than corresponding bases as growth supplements for *pyr-1* (1223) and apparently for other *pyr* mutants. However, after a lag uracil is used nearly as effectively as uridine (1363). No cytidine- or thymidine-specific requirement exists, because *Neurospora* lacks thymidine kinase (792) and because any exogenous pyrimidine supplement is cycled back through uridine monophosphate, which provides all normal end products of pyrimidine biosynthesis (2217). For this reason, DNA cannot be specifically labeled by supplying [^{3}H]thymidine under normal circumstances [but see *tk* entry and ref. (1772)]. Mutants have been obtained (*uc-2, -3, -4,* and *-5, ud-1*) that block the pathway back through UMP and so prevent general labeling from a single precursor (2217). Cytosine in DNA can be labeled specifically by the method of ref. (2246). For a general review of pyrimidine metabolism, see ref. (1487). For systematic gene–enzyme work, see refs. (288) and (291). For the pyrimidine biosynthetic pathway, see Fig. 53. For loci concerned with pyrimidine salvage or pyrimidine uptake, see Fig. 65. Complex interactions between *lys* and *pyr* mutants have been described (910).

Regulation: Pyrimidine biosynthetic enzymes differ in their mode of regulation. The CPS(pyr)–ACT complex is feedback-inhibited by UTP and derepressed by end-product depletion, but is insensitive to repression in the fluoropyrimidine-resistant mutant *fdu-2* (267, 289). Dihydro-orotase is relatively unresponsive to end-product limitation, and dihydro-orotate dehydrogenase is induced by a precursor that is probably, by analogy with yeast, dihydro-orotate, the substrate of the enzyme. Regulation of the last two enzymes has not been studied systematically. Pyrimidine regulation of the uptake and salvage pathways of pyrimidine is discussed under

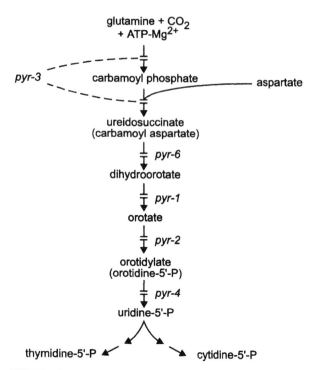

glutamine + CO_2
+ ATP-Mg^{2+}

pyr-3 - - - - carbamoyl phosphate - - - - aspartate

ureidosuccinate
(carbamoyl aspartate)

pyr-6

dihydroorotate

pyr-1

orotate

pyr-2

orotidylate
(orotidine-5'-P)

pyr-4

uridine-5'-P

thymidine-5'-P   cytidine-5'-P

FIGURE 53 The pyrimidine biosynthetic pathway, showing sites of gene action (288, 875, 1655, 2216). Carbamoyl phosphate (CAP) for arginine synthesis is made as a separate pool by a different enzyme system (see *arg-2, arg-3* in Fig. 10). Interchange between the two pools occurs only in certain mutant combinations. From ref. (1596), with permission from the American Society for Microbiology.

individual loci; see *uc-5, ud-1*. Many aspects of pyrimidine metabolism are under the control of general nitrogen metabolite regulation (266) (Fig. 50).

pyr-1: pyrimidine-1

IVR. Between *psi* (4%), *T(ALS159)* and *pdx-1* (<1%; 10%) (1369, 1580, 1582).

Requires uracil or another pyrimidine. Lacks dihydro-orotate dehydrogenase activity (EC 1.3.99.11) (288, 291) (Fig. 53). Ascospores from *pyr-1* × *pyr-1* are black if the crossing medium is supplemented with 1 mg/ml uracil, but white and inviable with 0.1 mg/ml (1254).

pyr-2: pyrimidine-2

IVR. Between *nit-3* (2%; 9%) and *rug* (3%), *T(NM152)^R* (1255, 1582, 1585, 1950).

Requires uracil or other pyrimidine. Lacks orotidine 5'-monophosphate pyrophosphorylase activity (EC 2.4.2.10) (288, 291) (Fig. 53). Needs 0.5 mg/ml uracil in medium for optimal growth. Allele 38502 is leaky.

pyr-3: pyrimidine-3

IVR. Between *T(S1229)^R arg-14, cyt-19* (5%), *T(NM152)^L* and *his-5* (1%) (457, 1377, 1580).

Requires uracil or other pyrimidine (1363). Growth is inhibited by purine nucleosides and nucleotides (1621). Structural gene for pyrimidine-specific carbamyl phosphate synthase (CPS, EC 6.3.5.5) and aspartate carbamyl transferase (ACT, also abbreviated ATC, EC 2.1.3.2) (875, 1664) (Fig. 53). Mutants may lack either or both activities, e.g., alleles KS43 (CPS$^+$ACT$^-$), KS20 (CPS$^-$ACT$^+$), and KS11 (CPS$^-$ACT$^-$) (2216). Unlike yeast, no feedback-insensitive CPS$^+$ACT$^+$ mutants have been discovered in *Neurospora* (1659). Some mutants have kinetically altered ACTase (875, 1720). Used extensively for studies of channeling and of the relation of gene structure to the two enzyme activities (455). Normally, carbamyl phosphate (CAP) produced by *pyr-3$^+$* is used solely for pyrimidine synthesis, and CAP produced by *arg-2$^+$* and *arg-3$^+$* is used for arginine synthesis, the enzymes being in different organelles; however, a deficiency of the next enzyme in either pathway permits the overflow of CAP into the other pathway [reviewed in ref. (455)]. Hence, CPS$^-$ACT$^+$ alleles are suppressed by *arg-12^s* (476), and CPS$^+$ACT$^-$ alleles can be selected as suppressors of *arg-2* and *arg-3* (1311, 1734). Some of the CPS$^+$ACT$^-$ alleles, called *pyr^{su-arg}*, suppress the arginine requirement but retain enough ACTase activity that they have no detectable pyrimidine requirement (1717, 1722). *arg-13, arg-4, -5, -6,* and *am* partly suppress CPS$^-$ATC$^+$ alleles (1313). The fertility of interallelic crosses is variable and often very poor (1311). Complementation is good between CPS$^-$ACT$^+$ and CPS$^+$ACT$^-$ mutants (476) and between some pairs of CPS$^+$ACT$^-$ mutants; otherwise, complementation is poor (1665, 2241). Complementation maps (1311, 1665, 1717, 2241). Fine-structure map (1666, 2030). Mutational analysis (1667). The direction of translation, based on enzyme types of polar mutants, is from CPS to ACT (1664). Allele 37815(t) is heat-sensitive (34 vs 25°C) (136). Allele 1298 is CO$_2$-remediable (363, 364). Strain KS12, a *pyr-1 pyr-3* double mutant, originally was called *pyr-5* (648). The different classes of *pyr-3* alleles have been called M (CPS-P-less), N (ACT-less), and MN (lacking both activities) (Figs. 11 and 53).

pyr-4: pyrimidine-4

IIL. Between *het-c* (1%), *Pad-1* (1%) and *ro-3* (1% or 2%) (1109, 1414, 1592).

Cloned and sequenced: Swissprot DCOP_NEUCR, EMBL/GenBank M13448, X05993, GenBank NCPYR4G, NEUPYR4, PIR A24398, DCNCOP.

Structural gene for orotidine 5'-monophosphate decarboxylase (EC 4.1.1.23) (288, 291, 1655) (Fig. 53). Requires uracil or other pyrimidine. Fertile crosses homozygous for *pyr-4* can be made using very high levels of uridine (15–20 mg/ml) (1424).

pyr-5: pyrimidine-5

Used for a *pyr-1 pyr-3* double-mutant strain (648).

pyr-6: pyrimidine-6

VR. Between *asn* (6%) and *un-9* (2%) (321, 1603).

Requires uracil or other pyrimidine. Lacks dihydro-orotase activity (EC 3.5.2.3) (288, 291) (Fig. 53). When given a small amount of uridine, a strain carrying the only allele (DFC37) pauses and then grows well beyond the level normally supported by the supplement; at no time is dihydro-orotase detectable (288).

pyrG: pyrimidine G

From *Aspergillus nidulans*.

Introduced into *Neurospora* in plasmid p*pyrG* and used as a selectable marker for transformation (2131). Both *pyrG* of *Aspergillus* and *pyr-4* of *Neurospora* encode orotidine-5′-phosphate carboxylase. Each gene functions in the heterologous species. The *Aspergillus* gene is not subject to RIP when single-copy transformants of *Neurospora* are used in crosses.

Pyruvate Decarboxylase

See *cfp*.

pzl-1: phosphatase-Z-like-1

IL. Linked to *mus-40* (0/18), *cyt-21* (1/18), *nit-2* (3/16) (1447).

Cloned and sequenced: EMBL/GenBank AF071751, AF071752.

Encodes a novel type-Z-like Ser/Thr protein phosphatase expressed in all vegetative stages of development (2035). Called *ppz-1*. For a review of protein phosphatases in filamentous fungi, see ref. (535).

q: quinolinic acid

Symbol changed to *nic-1*.

qa Cluster

VIIL. Between *ars-1*, *tRNA^{LEU}* and *Cen-VII*, *met-7* (0.2%; 1%) (304, 340, 1725).

Cloned and sequenced: EMBL/GenBank X14603, GenBank NCQA. Included in VII contigs GenBank AC006498.

The cluster consists of seven genes concerned with the uptake and catabolism of quinate and with regulation of

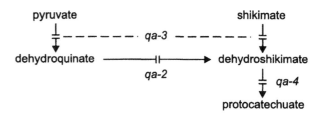

FIGURE 54 The quinate (aromatic amino acid) catabolic pathway, showing sites of gene action (304, 306, 728, 795, 983). *qa-1* is a regulatory gene affecting all three structural genes. *qa-2^+* activity (catabolic dehydroquinase) can be replaced by the product of *aro-9^+*, the equivalent gene in the aromatic biosynthetic pathway. See ref. (795) for the catabolic steps subsequent to protocatechuic acid. From ref. (1596), with permission from the American Society for Microbiology.

genes in this pathway (730, 734) (Figs. 54 and 55). The *qa-1F* and *qa-1S* genes are regulatory. The order on the conventional genetic map is: *tRNA^{LEU} qa-1F qa-1S qa-Y qa-3 qa-4 qa-2 qa-X Cen-VII met-7*.

qa-1F: quinate-1F

VIIL. Leftmost gene in the *qa* cluster, between *tRNA^{LEU}* and *qa-1S* (304, 723, 1725).

Cloned and sequenced: Swissprot QA1F_NEUCR, EMBL/GenBank X14603, GenBank NCQA, PIR F31277, S04256. Included in VII contigs GenBank AC006498.

Unable to use quinate or shikimate as the sole carbon source. Regulatory (345). Activator of the quinate catabolic pathway (723). Fast complementing. Effects of expression on chromatin structure (132). Characterization of binding sites in the *qa* cluster (131). Interaction with QA-1S; evolutionary origin (859) (Fig. 55).

qa-1S: quinate-1S

VIIL. Between *qa-1F* and *qa-Y* in the *qa* gene cluster (304, 723, 1725).

Cloned and sequenced: Swissprot QA1S_NEUCR, EMBL/GenBank X14603, GenBank NCQA, NCQA1SRA, PIR E31277, S04255. Included in VII contigs GenBank AC006498.

Unable to use quinate or shikimate as the sole carbon source. Regulatory (345). Repressor of the quinate pathway (723). Slow complementing. Effects of expression on chromatin structure (132). Effect of deletion on *qa* expression (298). Interaction with QA-1F; evolutionary origin (859) (Fig. 55).

qa-2: quinate-2

VIIL. Between *qa-4* and *qa-X* in the *qa* gene cluster (304, 723, 1725).

Cloned and sequenced: Swissprot 3DHQ_NEUCR, EMBL/GenBank X14603, GenBank NCQA, NCQAX2, PIR

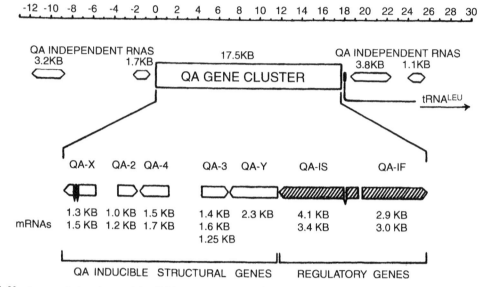

FIGURE 55 A transcriptional map of the 42-kb region containing the *qa* gene cluster and adjacent regions. The 17.3-kb region of the *qa* gene cluster is defined by the seven quinic acid-inducible mRNA transcripts shown. Other RNAs, including tRNA^LEU^, show no induction and therefore are not part of the cluster. The direction of transcription for each *qa* gene is indicated. The sizes of discrete mRNA transcripts on Northern blots detected by DNA–RNA blot hybridization are given immediately below each gene. The positions of short introns in *qa-X* and *qa-1S* are indicated. The order is reversed in this drawing relative to that of loci on the conventional genetic map, where tRNA^LEU^ adjoins *qa-1F* on the left and *qa-1S* adjoins *Cen-VII* on the right. Copied from ref. (730), with permission from the American Society for Microbiology.

A31277, S04251. Included in VII contigs, GenBank AC006498.

Structural gene for catabolic dehydroquinase (EC 4.2.1.10) (38, 306, 723, 983, 1725). The double mutant *aro-9; qa-2* is unable to use quinate or shikimate as the sole carbon source (1726). [*aro-9* lacks biosynthetic dehydroquinase. The single mutant *aro-9; qa-2*⁺ grows on minimal medium without supplement (Figs. 54 and 55).] *qa-2* is scored conveniently as an *aro* auxotroph when *aro-9* is present. The *qa-2*⁺ gene cloned in pBR322 is expressed constitutively from its own promoter in *Escherichia coli* (39, 729, 857). Used as a selectable marker for *Neurospora crassa* transformation (341) and as a reporter (435). The *qa-2* promoter has been used to regulate the expression of recombinant genes (57, 2074). Allele M246 is stable (308). Homozygous fertile with black ascospores. Position-dependent effects on gene expression (2148). Effects of rearrangements in the promoter region on activation (724). *N. crassa* and *Neurosopora africana* differ in the intergenic region between *qa-X* and *qa-2*; the differences have been examined for their effects on *qa-2* expression in *N. crassa* (65).

qa-3: quinate-3

VIIL. Between *qa-Y* and *qa-4* in the *qa* gene cluster (304, 343, 723).

Cloned and sequenced: Swissprot DHQA_NEUCR, EMBL/ GenBank X14603, GenBank NCQAX3, PIR S04253, C31277. Included in VII contigs GenBank AC006498.

Unable to use quinate or shikimate as the sole carbon source (344). Structural gene for quinate (shikimate) dehydrogenase (EC 1.1.1.24) (344). Revertants have altered enzymes. Transcription is in the direction *qa-4* to *qa-1* (307) (Figs. 54 and 55).

qa-4: quinate-4

VIIL. Between *qa-3* and *qa-2* in the *qa* gene cluster (304, 343, 723).

Cloned and sequenced: Swissprot 3SHD_NEUCR, EMBL/ GenBank M10139, X14603, GenBank NCQA4, PIR A22421, D31277, S04252. Included in VII contigs GenBank AC006498.

Unable to use quinate or shikimate as the sole carbon source (344). Structural gene for dehydroshikimate dehydrase (EC 4.2.1.-) (344, 723, 1761) (Figs. 54 and 55).

qa-X: quinate-X

VIIL. Between *qa-2* and *met-7* in the *qa* gene cluster (723).

Cloned and sequenced: Swissprot QAX_NEUCR, EMBL/ GenBank X14603, PIR B31277, S04250, GenBank NCQA. Included in VII contigs GenBank AC006498.

Part of the quinate complex. Function unknown. Figure 55.

qa-Y: quinate-Y

VIIL. Between qa-1S and qa-3 in the qa gene cluster (723).

Cloned and sequenced: Swissprot QAY_NEUCR, EMBL/ GenBank X14603, PIR G31277, S04254, GenBank NCQA. Included in VII contigs GenBank AC006498.

Encodes quinate permease (Fig. 55). Growth characteristics of deletion mutant (298).

qcr7: ubiquinol–cytochrome c reductase 7

Unmapped.

Cloned and sequenced: Swissprot UCRQ_NEUCR, EMBL/ GenBank U20790; EST SC1H8.

Encodes subunit VII of ubiquinol–cytochrome c oxidoreductase (EC 1.10.2.2), an 11-kDa ubiquinone-binding protein homologous to Saccharomyces QCR8 and mammalian subunit VII. The eight N-terminal residues of a precursor protein are cleaved to produce a mature product; these residues appear to be important for function but not for directing the polypeptide to the mitochondrion (1217).

qcr8: ubiquinol–cytochrome c reductase 8

Unmapped.

Cloned and partially sequenced: EMBL/GenBank AI399120; EST W10H7 (1452).

Specifies ubiquinol–cytochrome c oxidoreductase complex III subunit VIII (EC 1.10.2.2), an 11-kDa ubiquinone-binding protein.

qde: quelling defective

Defective in the transgene-silencing mechanism known as "quelling" (392). Quelling reduces vegetative expression when multiple copies of a gene are introduced into the genome by transformation (390, 392, 394, 1737). Quelling occurs posttranscriptionally and appears to

affect mRNA stability. For a critique relating quelling to silencing in other organisms, see ref. (130).

qde-1: quelling defective-1

Unmapped.

Cloned and partially sequenced; EMBL/GenBank AJ133528.

Specifies a product resembling RNA-dependent RNA polymerase in the tomato (391). Suppresses quelling of al-1 duplications (392).

qde-2: quelling defective-2

Unmapped.

Cloned and sequenced (310a).

Specifics a product homologous to RDE-1 of Caenorhabditis, which is essential for gene silencing by double-stranded RNA in that organism (310a). Suppresses guelling of al-1 duplications (392).

qde-3: quelling defective-3

Unmapped.

Cloned and sequenced: GenBank AF205407 (393a).

Specifies a RecQ DNA helicase (393a). Suppresses quelling of al-1 duplications (392).

qde-4: quelling defective-4

Unmapped.

Suppresses quelling of al-1 duplications (393).

R: Round spore

IR. Between aro-8 (4%) and $T(MD2)^L$, het-5, un-18 (11%) (1578, 2129).

All eight ascospores of heterozygous R/R^+ asci are round rather than ellipsoid (Fig. 56). R is thus nonautonomous

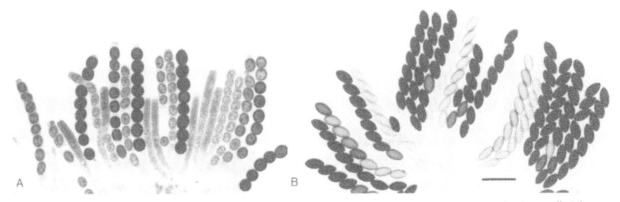

FIGURE 56 (A) Asci from a cross of wild type × Round spore. The Round spore mutation is ascus-dominant: all eight ascospores (R as well as R^+) are round. (B) Asci from a cross between two wild-type strains, for comparison. The asci with unpigmented or lightly pigmented spores are immature. Bar length = 50 µm. Photographs by N. B. Raju. From refs. (1676) and (1680), with permission from Urban & Fischer Verlag and from the British Mycological Society, respectively.

(or nearly so) in ascospores and dominant in the ascus (1367). Ascospores are round even in nonlinear asci (1546, 1966). A single germination pore is formed in spores that are completely round, two pores in spores that are slightly ovoid (1483a). Vegetative morphology is abnormal, somewhat resembling that of *peach*. Initial growth on a slant is concentrated around the inoculation point. The vegetative morphology is recessive in heterozygous duplications, such as those from *T(NM103)* (2123). Female sterile with no perithecia, but $R \times R$ crosses are productive if *R* is heterokaryotic in the female parent (1962). Used to study duplication instability (2123) and to show that ascospore development is autonomous in individual asci of different genotype within the same perithecium (999). Used to show that defective asci in crosses between inbred strains do not result from self-fertilization of *mat A × mat A* or *mat a × mat a* (1687). Recurrences have been obtained (1962, 1966). Although the *R* locus is not included in duplications from *T(MD2) × Normal*, round ascospores that appear to carry new *R* mutations are found among the progeny of crosses in which one parent is the breakdown product of such a duplication (978). For other genetically determined round ascospores, see ref. (117). Called *Rsp* (1966), but the original symbol *R* has priority. [The symbol *rsp* has been used for cytoplasmically determined respiration-deficient mutants (1741)].

r(Sk-2): resistant to Sk-2

IIIL. Between *cum* (1% or 2%) and *T(NM183)* (2%), *acr-7* (7%), *sc* (17%) (280, 2121, 2125, 2127).

Prevents killing by *Sk-2* of ascospores that are not themselves *Sk-2*. Does not confer resistance to killing by *Sk-3*. Allelism of *r(Sk-2)* with *Sk-2* is not excluded because recombination in this region is blocked when *Sk-2* is heterozygous. Recombination is not blocked by *r(Sk-2)-1*. Allele P527 was found in *Neurospora crassa* from Louisiana. Only one other known *N. crassa* wild type is resistant (2121). For reference strains and tester strains for identifying the various killers and genes conferring resistance to killing, see refs. (2127) and (2128).

r(Sk-3): resistant to Sk-3

IIIL. Between *cum* (4%) and *acr-7* (17%) (280, 2121, 2125).

Prevents killing by *Sk-3* of ascospores that are not themselves *Sk-3*. Does not confer resistance to killing by *Sk-2*. Allelism of *r(Sk-3)* with *Sk-3* is not excluded because recombination in this region is blocked when *Sk-3* is heterozygous. Recombination is not blocked by *r(Sk-3)*. Found in *Neurospora intermedia* and introgressed into *Neurospora crassa* (2121). For reference strains and tester strains for identifying the various killers and genes conferring resistance to killing, see refs. (2127) and (2128).

Radiation Sensitivity

See *Mei-2, mei-3, mus, nuh-4, phr, upr-1, uvs*. The symbol *rad* has not been used in *Neurospora*.

ran: ran-like

Unmapped.

Cloned and sequenced: EMBL/GenBank AJ000287, GenBank NCRANRNA.

Shows homology with a GTP-binding protein of the *ran* family (1341).

rap-1: ribosome-associated protein-1

IIR. Near *Cen-II*. In same cDNA with *vma-6* (~1200 bp) (1317).

Cloned and sequenced: Swissprot RSP4_NEUCR, EMBL/GenBank U36470.

Encodes a putative 40S ribosome-associated protein of unknown function (1317).

ras: ras-like

Genes belonging to the *ras* supergene family (named for an oncogene of the rat sarcoma virus) encode highly conserved G proteins that are important for signal transduction pathways regulating growth and differentiation. The *Neurospora ras*-like genes called *ras, krev,* and *ypt* have been identified as *ras*-like by DNA sequence homologies When phenotypes are examined using null mutants obtained by RIP, *ras*-like genes may prove to be allelic with already known mutants (e.g., the gene originally designated *ras-2* proved to be allelic with *smco-7*).

ras-1: ras-like-1

IVR Between *met-5* (1/18) and *nit-3* (2/28) (968).

Cloned and sequenced: Swissprot RAS1_NEUCR, EMBL/GenBank U33746, X53533, GenBank NCRAS, PIR S12892; EST SM1B4 .

Mutant strains and strains overexpressing *ras-1* are female-sterile, with defective aerial hyphae and few conidia (971, 2021). The RIP-inactivated mutant shows very slow growth (945). Called NC-ras.

ras-2: ras-like-2

Allelic with *smco-7*.

ras-3: ras-like-3

VIIR. Immediately adjacent to *thi-6*. Present in a cosmid adjacent to that carrying *cfp*. Linked near *un-10* on the RFLP map, but not borne on the same plasmid (1267).

Cloned and sequenced: Orbach-Sachs clones X18A06, G15F10, G22A09.

Encodes novel protein loops. Transcribed convergently with *thi-6* (1267).

rca-1: regulator of conidiation in Aspergillus-1

VR. Linked to *cyh-2* (0T/16 asci) (1447, 1902).

Cloned and sequenced: EMBL/GenBank AF006202.

The Myb-like functional homolog of *flb1* in *Aspergillus nidulans*. The *Neurospora* gene complements the sporulation defect of the *Aspergillus* mutant, but a deletion mutant has no visible effect on growth or conidiation in *Neurospora* other than to make early hyphal growth counter-clockwise or straight, in contrast to the typically clockwise spiral growth of *Neurospora* wild type (see Spiral Growth) (1902).

rco-1: regulator of conidiation-1

III. Linked to *acr-2* (7%; 11%) (2256).

Cloned and sequenced: Swissprot RCO1_NEUCR, EMBL/GenBank U57061; Orbach-Sachs clones X1B01, X14D07, X14F01.

Encodes a product homologous with *Saccharomyces cerevisiae* TUP1 (2256). Selected as enabling higher basal expression of a *con-10::hph* fusion gene during vegetative growth. Mycelial growth rates are reduced and conidiation is abnormal. Female sterile. Four alleles have four different conidiation morphologies. Presumptive RIP alleles are defective in the separation of conidia. Ascospores containing the null allele show counter-clockwise spiral growth when germinated, in contrast to wild type, which shows clockwise spiral growth (see Spiral Growth) (2256). Affects the expression of *con-10* and *ccg-1* (1154).

rco-3: regulator of conidiation-3

Allelic with *dgr-3* and hence, with *sor-4*. See *sor-4*.

rDNA: ribosomal DNA

VL. Located in the NOR, between *Tip-VL* and *mus-18* (0T/ 17 asci; 20 kb) (1447).

Cloned and sequenced: *5.8S sequence* EMBL/GenBank X02447, M10692, NCRRN58S, EMBL NCRGB, GenBank NEURGB; *17S sequence* EMBL/GenBank X04971, GenBank NCRRNAS; *18S sequence (partial)* EMBL/GenBank M11033, EMBL NCRRE, GenBank NEURRE; *25S sequence* (partial) GenBank X01373; *ITS sequence* EMBL/GenBank M13906, EMBL NCRGITR, Genbank NEURGITR; *28S-like sequence (partial)* EMBL/GenBank M38154, EMBL NCRRNA1; Orbach-Sachs clone X15E04.

Genes specifying 5.8S, 17S, and 25S ribosomal RNA (but not 5S) are located in the NOR in a series of ~150 tandem repeats, each ~9.3 kb long. The repeats are separated by nontranscribed spacers (423, 672). *rDNA* is a site of chromosome breakage in the breakdown of

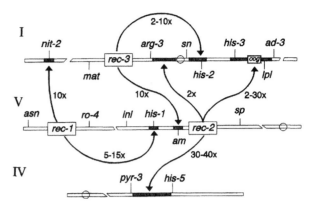

FIGURE 57 Linkage map showing the location and targets of action of *rec-1*, *rec-2*, and *rec-3*. Original figure from D. E. A. Catcheside.

duplication strains derived from translocations that involve the NOR (261). See *NOR*.

rec: recombination

trans-acting genes that regulate meiotic recombination in specific regions and constitute a class of recombination control genes first described in *Neurospora* (339, 990) (Fig. 57). Initially detected by changed intragenic recombination (up to 25×), but crossing over between loci may also be affected (up to 40×). High recombination is recessive. Products of dominant alleles (called *rec*⁺) are thought to repress the initiation of recombination at specific recombinator loci, of which *cog* is a well-studied example. A given target locus or region appears to be affected specifically by alleles at only one *rec* locus. Polarity of intragenic recombination in a target locus may be changed by the presence of the controlling *rec*⁺ allele. Control of recombination is independent from the control of gene expression (311). Three *rec* loci have been identified and characterized. Ten are estimated to be present in the *Neurospora crassa* genome. Differences at *rec* loci are present in commonly used laboratory wild types (338). For reviews, see refs. (316), (335), and (645).

rec-1: recombination-1

VR. Between *ro-4* (7%) and *asn* (5%) (325, 332).

Presence of allele *rec-1*⁺ reduces recombination within the loci *his-1* (VR) (990, 2083) and *nit-2* (IL) (320, 324) (Fig. 57), but does not affect recombination at other *his* loci (314). A recessive allele from another lineage was called *rec-z* until it was identified as *rec-1* (324).

rec-2: recombination-2

VR. Between *sp* (1%; 6%) and *am* (2%) (338, 1931).

Presence of dominant allele *rec-2*⁺ reduces recombination within the *his-3* locus (IR) and reduces crossing over in the intervals *pyr-3* to *his-5* (IVR), *his-3* to *ad-3* (IR), and *arg-3* to *sn* (IL) (336, 338, 1930) (Fig. 57). Interacts with *cog* in affecting recombination in *his-3* and crossing over between *his-3* and *ad-3* (45, 336). Used in conjunction with translocation *T(TM429) his-3* to demonstrate that *cog*⁺ acts only in *cis* in affecting recombination between sites in *his-3* (336). (See *cog*.) Recessive *rec-2* alleles from other lineages were called *rec-4*, *rec-5*, or *rec-w* until identity was demonstrated (333).

rec-3: recombination-3

IL. Between *acr-3* (1% or 2%) and *arg-3* (2%; 6%) (334, 337).

Presence of allele *rec-3*⁺ reduces recombination within the loci *his-2* (IR) and *am* (VR) but not within *gul-1*, which shows <0.3% recombination with *am* (337, 1940, 1941). Crossing over is also reduced in the interval between *sn* and *his-2* (338) (Fig. 57). Three alleles are known: *rec-3*, *rec-3*ᴸ, and *rec-3*⁺ (334). A recessive allele from another lineage was initially called *rec-x* (333).

rec-4, rec-5, rec-w: recombination-4, -5, -w

All are *rec-2* alleles.

rec-x: recombination-x

A *rec-3* allele.

rec-z: recombination-z

A *rec-1* allele.

Restless (Transposable Element)

An *Ac*-like DNA-transposable element introduced into *Neurospora* from the asexual fungus *Tolypocladium inflatum* (1042, 1043).

rg-1: ragged-1

IR. Between *T(AR173)*ᴿ (hence of *his-2*) and *lys-4* (1%; 7%) (492, 1548, 1578, 1583).

Spreading dense colonial growth with poor conidiation (1548). Increased hyphal branching, bumpy mycelial surface. Altered phosphoglucomutase (EC 5.4.2.2) (isozyme I). Accumulates glucose 1-phosphate (243). Cell-wall composition (287). Normal levels of NADPH (233) and linolenic acid (241). Photograph (235). The double mutant *rg-1 cr-1*, which grows on agar as small, discrete conidiating colonies suitable for velvet replication (1256), is not homozygous-fertile unlike *sn cr*, which it resembles phenotypically. For examples of applications, see refs. (1830) and (1979). Allele R2357 called *er: erupt*. Allele S4357 called *col-7*.

rg-2: ragged-2

IR. Linked to *mat* (15%), *rg-1* (0/650) and left of *lys-4* (12%) after introgression into *Neurospora crassa* (1546).

Altered phosphoglucomutase (EC 5.4.2.2) (isozyme II) (1357). Morphology is similar to *rg-1* (1359). Found in a *Neurspora sitophila* cross that was heterozygous for an *rg-1* allele from *N. crassa*. Interpreted from its behavior in *N. sitophila* to be unlinked to *rg-1* or to *su(rg-2)* (1359).

rgb-1: B-regulatory subunit-1

IL. Linked to *leu-4*, *mat*, *arg-3* (0/18) (1447).

Cloned and sequenced: GenBank AF077355.

Encodes the B-regulatory subunit of the type 2A phospho-protein phosphatase complex (PP2A) (2273). Intron I, which is 5′ to the presumed translocation initiation codon, contains an upstream ORF encoding 34 amino acids. Intron VI undergoes alternative splicing. RIP-induced inactivation of *rgb-1* produced progeny that grow slowly, have abnormal hyphal morphology, are female-sterile, and produce abundant arthroconidia. Microscopic and genetic analyses indicate that *rgb-1* is a regulator of the budding subroutine of macroconidiation. For a review of protein phosphatases in filamentous fungi, see ref. (535).

rgr: resistant to glutamine repression

Unmapped.

Arginine catabolism is resistant to glutamine repression (749).

rhd: rehydrin

VII? Shows sequence homology to a region in VII contigs (1041).

Cloned and partially sequenced: EMBL/GenBank AI398918, AI416419; EST W09E10, W06E2.

Putative structural gene for rehydrin (thiol-specific antioxidant, peroxiredoxin, dormancy-associated protein, non-selenium glutathione peroxidase, EC 1.11.1.7). The preceding ESTs define the 5′ and 3′ ends of a transcript with homology to a region in linkage group VII contigs (EMBL/GenBank AC006498; 1041). The sequence shows high homology to the rehydrin gene, which is highly conserved from cyanobacteria to higher plants and animals. Consistent with the probable involvement of an *rhd* gene in dormancy, the sequence is found in a subtracted conidial cDNA library (1448).

rhy-1: rhythm-1

Unmapped.

Shows circadian rhythmic conidiation with period length 23 hr at 26°C. Temperature-compensated. Arrhythmic at 30°C. Ascospore germination is very poor. Obtained by insertional mutagenesis (352a).

rib: riboflavin

Only two *rib* loci are known in *Neurospora*, compared to six that have been assigned to biosynthetic steps in *Saccharomyces* (522). Supplemented medium should be shielded from light to avoid the destruction of riboflavin.

rib-1: riboflavin-1

VIR. Between *ad-1* (3%; 6%), *Cen-VI* (1%), *glp-4* (4%), *T(AR209)* and *pan-2* (3%) (915, 1986, 2160).

Requires riboflavin (1361). Used to examine the role of flavin as a photoreceptor for carotenogenesis and for the phase-shifting and suppression of circadian conidiation (1513). Allele 51602 is heat-sensitive (34 vs 25°C), but allele C106 is not (717).

rib-2: riboflavin-2

IVR. Between *T(4342)^L* and *chol-1*. Probably left of *pyr-3* (1/24 asci) (718, 1578, 1580).

Requires riboflavin (718). Used to demonstrate the role of flavin as a photoreceptor for the phase-shifting of circadian conidiation and carotenogenesis (1513).

Ribosomal RNA

Genes specifying 5.8S, 17S, and 25S ribosomal RNA (but not 5S) are located in the nucleolus organizer region. See *rDNA*, *NOR*, and *Fsr*.

rip-1: ribosome production-1

IIR. Between *trp-3* (6%; 9%), *un-15* (1%) and *Tip-IIR* (1582).

Cloned (1818).

Scored as an irreparable heat-sensitive mutant (34 vs 25°C). Attains 2% of the wild-type linear growth rate at 35°C and 80% at 25°C (1755). Conditional defect in the accumulation of 25S rRNA and, hence, 60S cytosolic ribosomal subunits (1219, 1221, 1222, 1755). Defective ribosome biosynthesis is attributed to a defect in rRNA processing (1222). At restrictive temperature, conidia survive and apolar growth results in highly multinucleate spherical cells, with no hyphae on glycerol complete agar medium (1546, 1751). Messenger RNA synthesis is normal at 37°C. Good fertility, growth, and viability at 25°C make *rip-1* preferable to *un-15* as a marker for the right end of II. The original strain carrying both *rip-1* and *inl* was called 4M(t), and the *rip-1* allele was designated 1(t).

ro: ropy

Morphologically similar mutants, so named because the hyphae form ropelike aggregates that grow up the tube wall from agar slants, producing conidia in dense clumps at the top (1548). Hyphae are excessively branched and curled (248, 719). Microscopic analysis reveals defects in conidial germination, placement of septa, and mitochondrial morphology (1348). *ropy* mutants function as fertilizing parents, but are female-infertile unless complemented in a heterokaryon with *ro^+*. None of the *ropy* genes has been found to be essential for viability [see, for example, ref. (2093)]. Nuclear distribution in hyphae is abnormal as a result of defects in a microtubule-associated motor (248, 1634, 1729). Mutations at various *ropy* loci can be selected as partial suppressors of *cot-1* [248, 1634; reviewed in ref. (2258)]. Some combinations of allelic mutations show complementation between alleles, whereas some mutations at different *ropy* loci fail to complement one another (248). Unlinked noncomplementation most often involves *ro-1* or *ro-3*.

ro-1: ropy-l

IVR. Linked to *pan-1* (0/394), hence, between *ad-6* and *cot-1* (1255).

Cloned and sequenced: Swissprot DYHC_NEUCR, EMBL/GenBank L31504, PIR B54802, GenBank NEURO1DHC; Orbach-Sachs clone X1B06.

Morphology is typical *ropy*. Encodes the heavy chain of cytoplasmic dynein, a minus-end-directed motor molecule. Some combinations of alleles show complementation and may fail to complement some alleles at other *ropy* loci (248). Used to examine the kinesin–dynein (*kin*; *ro-1*) double mutant, which is defective in oppositely directed microtubule motors, and to compare it with the single mutants (1871). Photographs (507, 1626). The amount of cell-wall peptides is reduced (2249). Growth is limited on glycerol medium (419).

ro-2: ropy-2

IIIR. Between *trp-1* (2%, l4%) and *T(D305)^L*, *phe-2* (5%) (18, 1578, 1585). Adjoins *crps-7* (2155).

Cloned and sequenced: EMBL/GenBank U23425; Orbach-Sachs clone X20H05.

Encodes a novel 80-kDa protein that has two cysteine-rich motifs that resemble zinc-binding LIM or RING domains thought to mediate protein–protein interactions. Mutations obtained by RIP inactivation are defective in nuclear migration and are blocked in conidiation (2155).

ro-3: ropy-3

IIL. Between *pyr-4* (1% or 2%) and *T(NM149)*, *thr-2* (6%; 25%) (1580, 1585).

Cloned and sequenced: Swissprot DYNA_NEUCR, EMBL/GenBank L48661, GenBank NEURO3DYN; Orbach-Sachs clone X11C06.

Encodes the largest subunit of the dynactin complex (P150^{Glued}), which is not essential for viability but is required for normal nuclear distribution (2093). Some combinations of alleles show complementation and may fail to complement some alleles at other *ropy* loci (248). Photograph (248). Growth is limited on glycerol medium (419). The Spitzenkörper is undetectable or difficult to detect by phase contrast (1727). Called *cfl* on the map in ref. (1585).

ro-4: ropy-4

VR. Linked to *pab-2* (0/407) and carried in the same plasmid (1729). Between *ad-7* (4%), *T(EB4)^R*, and *inv* (5%) (321, 1578, 1582, 1592).

Cloned and sequenced: Swissprot ACTZ_NEUCR, EMBL/GenBank L31505, U14008, PIR A54802, PS00132, GenBank NEURO4CEN; Orbach-Sachs clone X9B02.

Structural gene for actin-related protein I (centractin). Mutations obtained by RIP or gene disruption are viable (1729). Photograph (1634). Growth is limited on glycerol medium (419). The amount of cell-wall peptides is reduced (2249). Mutants R2428 and R2520, originally called *ro-5* and *ro-8* (719, 1383), are allelic with *ro-4* (allele B38) (1592).

ro-5: ropy-5

Used for a *ro-4* allele (1592).

ro-6: ropy-6

IR. Between *T(4540)^L* *nic-2* (0/95) and *T(4540)^R*, *thi-1* (719, 1578, 1582).

Hyphae form ropelike aggregates (719).

ro-7: ropy-7

IIL. Linked to *pi* (0/75), left of *cys-3* (11%) (719, 1582).

Cloned and sequenced: Orbach-Sachs clone G18A08.

Encodes a novel 70-kDa actin-related protein (1347). The large subunits of cytoplasmic dynein and dynactin accumulate at spindle pole bodies in the mutant (1347). Growth is inhibited on glycerol medium (419). Female sterile, contrary to a misprint in ref. (1584).

ro-8: ropy-8

A *ro-4* allele.

ro-9: ropy-9

Misnamed? The existing stock is probably an allele of *dapple* (1567). See *da*.

IIL. Linked to *thr-3* (0/63), *arg-5* (8%). Right of *T(NM149)* (719, 1567).

The validity of the available strain, R2526, is suspect. It is not morphologically *ro* and is indistinguishable from *da*. Conidia of R2526 are normal sized, unlike *tng*. The mutant may have been misidentified and misnamed (unlikely), the original may have been a double mutant from which the *ropy* component was lost, or a stock may have been mislabeled (1567).

ro-10: ropy-10

IL. Between *Tip-IL* and *cyt-21* (1%), *fr* (18%), *In(SLm-1)^L* (114, 976, 1578, 1582).

Cloned and sequenced: EMBL/GenBank AF015561; Orbach-Sachs clones X7A06, G17G08.

Encodes a novel 24-kDa protein that may be necessary for dynactin stability (1347). Ascospores from rearrangement *Slm-1* × *Normal* include a class that is deficient for a short terminal segment containing *ro-10*. The deficiency progeny germinate and die unless they are rescued in a heterokaryon. The rescued deficiency can be transmitted through a cross (114).

ro-11: ropy-11

IIIR. Linked to *dow* (0/70). Between *T(AR17)^L* and *T(AR17)^R* (1347, 1349).

Cloned and sequenced: EMBL/GenBank AF015560; Orbach-Sachs clone G25B06.

Encodes a novel, nonessential 75-kDa protein involved in nuclear distribution (1347, 1349). Obtained by RIP in a cross of *Dp(AR17)* × *Normal* (1349).

rol: ropy-like

This name has been given to mutants that resemble *ropy* in growth habit on slants, but that differ from *ropy* in having hyphae that are not seen microscopically to be curled (719).

rol-1: ropy-like-1

IV. Linked to *pdx-1* (0/88) (719).

Forms mycelial aggregates similar to those of *ro* mutants, but with hyphae not curled microscopically.

rol-2: ropy-like-2

VII. Linked to *met-7* (<1%), *wc-1* (3%) (1582).

Forms mycelial aggregates similar to those of *ro* mutants, but with hyphae not curled microscopically.

rol-3: ropy-like-3

VR. Between *ilv-1* (2%) and *cot-4* (5%) (1383).

Ropylike hyphal morphology (1383). Photograph: Fig. 16 in ref. (719).

ros: rosy

Allelic with *al-3*.

Rsp: Round spore

The symbol *R* has precedence.

rug: rug

IVR. Between *pyr-2* (3%) and *T(NM152)^R*, *cys-4* (10%) (1255, 1414, 1578, 1585).

Spreading colonial morphology, with conidiating tufts. Grows better on sucrose minimal medium than on glycerol complete (1548). Formerly called *mat*. The name was changed to *rug* so that *mat* could be used as the symbol for *mating type*, in conformity with usage in other fungi (745).

s: suppressor of pyr-3

Changed to *arg-12^s*.

sar-1: surfactant resistant-1

I. Near *mat* (36).

Resistant to the surface-active agents dequalinium chloride, cetyltrimethylammonium bromide, and benzalkonium chloride. Resistant growth follows an adaptive lag phase. Wild type Em A (FGSC 627) and related *mat A* strains carry a *mat*-linked *sar* gene that may be *sar-1* (36).

sar-2: surfactant resistant-2

Unmapped. Independent of *sar-1* (36).

Resistant to surface-active agents. Phenotypically similar to *sar-1* (36).

sar-3: surfactant resistant-3

I. Near *mat* (36).

Phenotypically distinct from *sar-1*. Evidence of nonallelism is not given (36).

sat: satellite

VL. At *Tip-VL*, distal to the *NOR* (1307, 1578). Linked to *lys-1* (35%) (118).

Microscopically visible terminal satellite (trabant). Best seen at pachytene in orcein- or acriflavine-stained preparations as a tiny dot on the surface of the nucleolus at one end of the second longest chromosome (1307). Photographs (118, 1598, 1599). *Neurospora crassa* laboratory strains differ in the presence (*sat^+*) or absence (*sat^-*) of a satellite. No satellite has been found in *Neurospora intermedia* or other *Neurospora* species. The satellite was used as a marker in assigning the NOR to the left arm of linkage group V (118).

sbr: small brown

Unmapped.

Cloned and sequenced (2153).

The sequence suggests a function in transcriptional regulation. Forms compact, slow-growing colonies that darken with age. No conidia or conidiophores are produced. Sausage-shaped cell compartments produce large spherical buds. Ascospores do not germinate. The mutant allele was obtained by insertional inactivation using an *hph* construct, and the *hph* tag was used to clone *sbr* from a genomic DNA library (2153).

sc: scumbo

IIIR. Between *acr-2* (3%; 6%), *Cen-III* (4T/230 asci) and *ser-1* (4%) (103, 1546). Linked to *thi-4* (1/280), *spg* (0++/179) (931, 1587).

Irregular flat, spreading growth with knobby protrusions and abnormal conidiation, but no free conidia (Fig. 58). Mycelium usually appears yellowish rather than orange. Female fertile. Asci are nonlinear in crosses homozygous for *sc* (1623, 1624, 1969). The ascus abnormality is recessive. Cell-wall analysis (287). Cell-wall peptides are reduced in amount (2249). Resistant to vinclozolin and other dicarboximide fungicides (789). Allele R2503 was initially called *col-14*, and allele R2386 was called *smco-2*. Morphology altered by *mod(sc)*.

scon-1: sulfur control-1

VR? Probably linked to *his-6*, but inviable white spores are prevalent in *scon^c* crosses and the presence of a translocation (and the therefore pseudo-linkage) cannot be ruled out. Difficult to map because of infertility (1322).

The wild-type allele specifies a negative regulator in the sulfur catabolic regulatory circuit (Fig. 24). Allele *scon-1^c* makes production of these enzymes constitutive. Prototrophic. Non-repressible for sulfate permease and therefore resistant to chromate under conditions of repression (e.g., high methionine) when it would be expected to be sensitive. The effect in a heterokaryon is restricted to its own nucleus (259). Hypostatic to regulatory mutant *cys-3* (539). Ascospores and conidia germinate poorly (259, 1822). For reviews of the sulfur system, see refs. (1283) and (1287).

scon-2: sulfur control-2

IIIR. Between *trp-1* and *Fsr45* (1334). Linked to *tyr-1* (18%), *leu-1* (26%) (538).

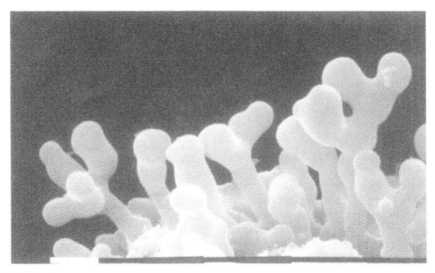

FIGURE 58   The mutant *sc: scumbo*. Bar length = 1 μm. Scanning EM photograph from M. L. Springer.

Cloned and sequenced: EMBL/GenBank U17251.

Encodes a β-transducin; contains an evolutionarily conserved F-box domain that is necessary for the regulation of *cys-3* (1110, 1111, 1516). The wild-type allele specifies a negative regulator in the sulfur catabolic regulatory circuit (Fig. 24). Prototrophic. Resembles *scon-1ᶜ* in phenotype, but action is not nucleus limited. RNA transcription that requires CYS-3 occurs during sulfur-derepression, but expression of CYS-3 under sulfur repressing conditions is not sufficient for *scon-2* expression, though it is sufficient for *ars* expression (1287, 1517). For reviews of the sulfur system, see refs. (1283) and (1287).

scot: spreading colonial temperature sensitive

VR. Between *al-3* (7%) and *his-6* (11%) (1583).

Not distinguishable from *scot⁺* at 25°C under most conditions, but above 34°C growth on solid medium is spreading colonial, with delayed, reduced conidiation, whereas growth in liquid medium is pelleted. Readily scorable on glycerol complete medium at 39°C. Photograph (980). Present in Beadle–Tatum and Rockefeller–Lindegren wild types and in numerous derivatives from them (1583). The same gene probably was discovered independently by Fincham (637) and by Emerson (608), studied by Pao (1531), and called *t: thermophobic*. Strains containing *t* showed start–stop growth on growth tubes, with growth distance depending on the carbon source (sucrose vs lactose or galactose) and concentration. The effect is greater on poorly utilized carbon sources and on the poorest, galactose, can be seen even at 25°C (637). *scot* might be responsible for the effect of temperature on *moe* mutants that originated in the RL wild type (719) and (in part) for the erratic growth of heterokaryons involving Wilson–Garnjobst *het*-tester strains (980).

scp: serine carboxypeptidase

IR. Linked near *Fsr-18* (1036).

Cloned and sequenced: Orbach-Sachs clone G3G11.

Encodes serine carboxypeptidase. Homologous to *kex-1* of *Saccharomyces*. Inactivation of *scp* results in reduced fertility in *mat A* strains, but *scp mat a* strains behave normally (1036). For other genes expressed differentially in the two mating types, see *ccg-4, eat-1, eat-2, mfa-1*.

scr: scruffy

IIR. Linked to *arg-12* (2%), probably to the left (1582).

Semicolonial growth with few aerial hyphae, dichotomization of hyphal tips, and reduced conidiation. Heat-sensitive, growing poorly with no conidia at 39°C, more like wild type at 25°C. Asci homozygous for *scr* are abnormal, with altered arrangement of ascospores. Some asci are linear. Occasional asci have fewer than eight ascospores, and some of these are large. The ascus abnormality is recessive (1623).

sdh-1: succinate dehydrogenase-1

I. Linked to *mat* (0/13) (590).

Succinate dehydrogenase (SDH) activity is 18% that of wild type. Succinate oxidase activity is low. Selected by failure to reduce nitrotetrazolium blue in the presence of succinate and phenazine methosulfate in an overlay after inositol death enrichment on acetate (590). Deficient in the HiPIP iron–sulfur center of the SDH complex (589).

sdv: sexual development

Subtractive hybridization was used to identify genes expressed specifically under conditions that induce sexual development. With two exceptions, *sdv-10* and *sdv-15*, the *sdv* genes listed here showed no detectable phenotype after being disrupted by RIP. (*sdv-10* and *-15* were renamed *asd-1* and *asd-3*.) *sdv-1*, *-2*, and *-3* are clustered on cosmid 4.3A, *sdv-4* and *-5* are on cosmid 5.6F, *sdv-6* is on cosmid 8.9A, *sdv-7*, *-8*, and *-9* are on cosmid 12.2E, *sdv-10* and *-15* are adjacent on cosmid 12.7B, *sdv-11* and *-12* are on cosmid 20.11A, and *sdv-13* and *-14* are tightly linked on cosmid 28.12G. Mapping was by RFLP (1450).

sdv-1: sexual development-1

IVL. Between *Tel-IVL* (12T, 0NPD/18 asci) and *Cen-IV*, *aod-1* (0, 1, or 2T/18 asci) (1447).
Cloned: on cosmid 4.3A with *sdv-2* and *sdv-3*.
Expressed specifically under conditions favoring sexual development (1450).

sdv-2: sexual development-2

IVL. Between *Tel-IVL* (12T, 0NPD/18 asci) and *Cen-IV*, *aod-1* (0, 1, or 2T/18 asci) (1447).
Cloned: on cosmid 4.3A with *sdv-1* and *sdv-3*.
Expressed specifically under conditions favoring sexual development (1450).

sdv-3: sexual development-3

IVL. Between *Tel-IVL* (12T, 0NPD/18 asci) and *Cen-IV*, *aod-1* (0, 1, or 2T/18 asci) (1447).
Cloned: on cosmid 4.3A with *sdv-1* and *sdv-2*.
Expressed specifically under conditions favoring sexual development (1450).

sdv-4: sexual development-4

IR. Between *al-2* (1T/18 asci) and *arg-13* (5T, 2NPD/18 asci). Linked to *vma-11* (0 or 1T/18 asci) (1447).
Cloned: on cosmid 5.8F with *sdv-5*.
Expressed specifically under conditions favoring sexual development (1450).

sdv-5: sexual development-5

IR. Between *al-2* (1T/18 asci) and *arg-13* (5T, 2NPD/18 asci). Linked to *vma-11* (0 or 1T/18 asci) (1447).
Cloned: on cosmid 5.8F with *sdv-4*.
Expressed specifically under conditions favoring sexual development (1450).

sdv-6: sexual development-6

VR. Linked to *crp-4* (0T/18 asci). Right of *inl* (9T/18 asci), *tom70* (4T/18 asci), *Fsr-20* (1T18 asci) (1447).

Cloned: on cosmid 8.9A.
Expressed specifically under conditions favoring sexual development (1450).

sdv-7: sexual development-7

IVR. Right of *trp-4* (5T/18 asci), *crp-3* (1T/18 asci) (1447).
Cloned: on cosmid 12.2E with *sdv-8*, and *sdv-9*.
Expressed specifically under conditions favoring sexual development (1450).

sdv-8: sexual development-8

IVR. Right of *trp-4* (5T/18 asci), *crp-3* (1T/18 asci) (1447).
Cloned: on cosmid 12.2E with *sdv-7* and *sdv-9*.
Expressed specifically under conditions favoring sexual development (1450).

sdv-9: sexual development-9

IVR. Right of *trp-4* (5T/18 asci), *crp-3* (1T/18 asci) (1447).
Cloned: on cosmid 12.2E with *sdv-7* and *sdv-8*.
Encodes glycogen phosphorylase (EC 2.4.1.1) (1446). Expressed specifically under conditions favoring sexual development (1450).

sdv-10: sexual development-10

Changed to *asd-1*.

sdv-11: sexual development-11

IVR. Between *pyr-1* (4T/18 asci) and *trp-4* (3T/18 asci) (1447).
Cloned: on cosmid 20.11A with *sdv-12*.
Expressed specifically under conditions favoring sexual development (1450).

sdv-12: sexual development-12

IVR. Between *pyr-1* (4T/18 asci), and *trp-4* (3T/18 asci) (1447).
Cloned: on cosmid 20.11A with *sdv-11*.
Expressed specifically under conditions favoring sexual development (1450).

sdv-13: sexual development-13

VII. Linked to *ars-1*, *Cen-VII* (0T/18 asci) (1447). On the same plasmid with *sdv-14* (1450).
Expressed specifically under conditions favoring sexual development (1450).

sdv-14: sexual development-14

VII. Linked to *ars-1*, *Cen-VII* (0T/18 asci) (1447). On the same plasmid with *sdv-13* (1450).

Expressed specifically under conditions favoring sexual development (1450).

sdv-15: sexual development-15

Changed to asd-3.

sen: senescent

VR. Between his-1 (4%) and al-3 (13%) (1441b).

Resembles the mutant *natural death* in progressive deterioration of mycelial growth followed by death. Recessive. The mutant gene was extracted from a heterokaryotic wild-collected strain of *Neurospora intermedia* by plating microconidia that were obtained as described in ref. (1528). Introgressed into *Neurospora crassa* by nine backcrosses. Stocks are maintained in a heterokaryon with the am^1 ad-3B cyh-1 helper strain, from which *sen* is extracted by crossing to wild type (1441b).

ser-1: serine-1

IIIR. Between sc (4%) and pro-1 (1%; 114 kb; 9%), ace-2 (5%) (448, 940, 1546).

Cloned: pSV50 clones 14:1C, 22:4D, 25:3G (448).

Slightly deficient in serine hydroxymethyltransferase; raised levels of 10-formyltetrahydrofolate synthetase. Lacks the ability to incorporate C_1 units from glycine. Extracts lack detectable methylfolates. Negligibly deficient in $5,10\text{-}CH_2\text{-}H_4PteGlu$ reductase (411). Uses serine or glycine (940). Inhibited on organic complete medium. Ascospores from ser-1 × ser-1 do not blacken or do so only after a long delay (1592). Morphologically normal; the abnormal morphology reported at elevated temperatures in ref. (1592) was due to *scot* (1583).

ser-2: serine-2

VR. Between met-3 (4%) (824) and $T(EB4)^L$, cot-2 (5%; 8%) (321, 322).

Uses serine (556), but not glycine (1546). Does not grow on hydrolyzed casein (556). Inhibited by thienylserine (99).

ser-3: serine-3

IL. Between cys-5 (1%) and $In(NM176)^L$, un-3 (<1%) (1578, 1592, 2129).

Uses serine and (less well) formate. Also grows fairly well on a combination of adenine, methionine, tryptophan, and lysine. Does not grow on casein hydrolysate (556). No or little response to glycine (1546). Deficient in phosphoserine phosphatase activity (EC 3.1.3.3) (139). Inhibited by leucine. Used to examine one-carbon metabolism (989). Scorability good, but vigor and leakiness vary markedly in different isolates. Homo-

zygous crosses give mostly white ascospores (1546, 1592).

ser-4: serine-4

IVR. Right of arg-2 (<1%) (1300, 1301).

Uses serine or glycine. Incompletely blocked. Not deficient for any of the enzymes involved in serine synthesis from 3-phosphoglyceric acid or glyceric acid. The intracellular pool is deficient in serine, glycine and alanine, and it accumulates threonine and homoserine (1301). Produces abundant L-amino acid oxidase, but no tyrosinase, while growing slowly on minimal medium (1301, 2139). Allele DW110 called P110.

ser-5: serine-5

IIIR. Linked to trp-1 (1%), ser-1 (12%) (1302).

Uses serine or glycine. Incompletely blocked (1302).

ser-6: serine-6

VIL. Between nit-6 (10%) and het-8 (8%), ad-8 (16%) (1034, 1172, 1582).

Responds to serine. Slight response to glycine. Extremely leaky. Scorable by slow, sparse conidiation on minimal slants or auxanographically (1546). Can be crossed without supplement on minimal crossing medium. Allele DK42 was obtained as a putative leucine regulatory mutant (1034), but a regulatory role is dubious (793). Called DK42 (1034). Probably allelic with T(OY325) ser, which has a breakpoint left of lys-5 (11%) in VIL and shows 0/28 recombination with DK42, which it resembles in leakiness. OY325 grows more profusely than DK42 on minimal slants (1582).

sf: slow-fine

I. Linked to mat (3%), cy (3% in regular perithecia) (1366, 1582).

Originally detected microscopically. Growth from ascospores or from conidia is slow at first. Fully grown cultures appear morphologically normal. Asci are irregular in some perithecia (1366). Scorable by slow growth on slants 3 or 4 days after ascospore germination. Hyphae initially grow clockwise on an agar surface (1582).

sfo: sulfonamide dependent

VII. Linked to Cen-VII (<1%) (604). Between thi-3 (6%) (1713) and hlp-1 (1%; 9%) (879).

Requires sulfonamide at 35°C and is stimulated by sulfonamide at lower temperatures (605). Overproduces p-aminobenzoic acid; hence, growth is inhibited by exogenous PABA (2293). In a *sfo; pab-1* double mutant on sulfonamide-supplemented minimal medium, PABA is stimulatory at very low concentrations but inhibitory at higher (605). In a *sfo; met-1* double mutant on sulfonamide-supplemented minimal medium, methionine is stimulatory at very low concentrations but inhibitory at higher (2294). Best tested on solid minimal growth tubes at 35°C. Suppressor mutations are frequent (604).

sg: spontaneous germination

Unmapped.

Ascospores germinate spontaneously, without heat shock. A component of the multiple-mutant combination responsible for the cell-wall-deficient "slime" phenotype. Usually associated with very poor vegetative growth. Possibly more complex than a single gene (607). (Ascospores carrying the *fluffy* mutation often germinate spontaneously.)

sh: shallow

VR. Between *ilv* (4%; 7%) and *md* (3%) (569).

Hyphae spread in a fanlike array and do not penetrate deeply into agar (1585). Photographs (569, 719).

shg: shaggy

IIIR. Linked to *trp-1* (7%), *acr-6* (0/368). Recombines with *vel, col-13* (932, 1582).

Growth is slow. Conidia are formed on the irregular aerial hyphae that are most abundant high in an agar slant. Mutants *acr-4* and *acr-6* originated in *shg*. The *acr-4* gene confers resistance to acriflavine only in combination with *shg* (932). Called *mo(KH160)*.

sit: siderophore transport

The uptake of exogenous ferricrocin and coprogen is defective. The five known *sit* mutations appear from intercrosses to be nonallelic (362). Double mutants prove that *sit-2*, *sit-3*, and *sit-5* are at different loci. Differences are present in morphology, pigment, speed of conidial germination, female fertility, and binding of siderophores. Map locations are unknown. The mutants were selected using *aga; arg-5; ota*, which is blocked in the known pathways to ornithine (463) and thus depleted of siderophores (2229) (Fig. 10). (Because exogenous arginine strongly represses ornithine synthesis, use of the *aga* single mutant in the presence of arginine was proposed for future experiments.) The five mutants, which grow well on minimal medium, take up [^{14}C]phenylalanine but are unable to take up [^{3}H]ferricrocin. Methods and characteristics of the mutants were summarized when work was suspended in 1983 (362).

The conditional germination-defective mutant *moe-1* (JS134-9) is partially curable by siderophores and may have an altered plasma membrane attachment site (360). For background on the *Neurospora* siderophore uptake system, see ref. (2228). For the molecular biology of fungal siderophores, see ref. (1162).

sit-1: siderophore transport-1

Unmapped.

Transport of siderophores is severely reduced. Binding is defective. Germination is slow. A double mutant with *sit-2* is available, and this also shows slow germination (359, 362).

sit-2: siderophore transport-2

Unmapped.

Transport of siderophores is severely reduced. Binding is defective. Conidia tend to stay connected in twos and threes. The color is deep orange. Double mutants are available with *sit-1* (FGSC 4226, 4227), *sit-3* (FGSC 4219, 4920, 4921, 4922), and *sit-5* (FGSC 4223, 4224, 4225). These all have the morphological characteristics of the *sit-2* parent (359, 362).

sit-3: siderophore transport-3

Unmapped.

Transport of siderophores is severely reduced. Binding is normal. Double mutants are available with *sit-2* and *sit-5* (FGSC 5230, 5231) (359, 362).

sit-4: siderophore transport-4

Unmapped.

Transport of siderophores is severely reduced. Binding is normal. A double mutant with *sit-5* is available (FGSC 4228, 4229) (359, 362).

sit-5: siderophore transport-5

Unmapped.

Transport of siderophores is severely reduced. Binding is normal. Double mutants are available with *sit-2*, *sit-3*, and *sit-4* (359, 362).

sk: skin

VIIR. Between *nt* (7%; 17%) and *Tip-VIIR* (1548).

Leathery, rapid surface growth, nonconidiating for most alleles (1548). "Mucilaginous substrate hyphae" (2117). Ascospores of *sk* constitution are slow to mature, but good allele ratios are obtained from crosses held for 3 weeks at 25°C. Female sterile. Allele P1718 is exceptional in making aerial hyphae and conidia, best on complete medium. Allele R2466 called *mo-3*; alleles Y6821, R2408, and R2529 called *moe-1* (719, 1582). Not to be confused with *Sk: Spore killer*.

Sk: Spore killer

When a chromosomal Spore killer element is heterozygous in a cross, it results in the death of ascospores that do not carry the element. In crosses heterozygous for a killer element, each ascus contains four viable black ascospores and four inviable unpigmented ascospores (Fig. 59). The surviving spores in 4:4 asci are all *Sk*. Ascospores are not killed in crosses homozygous for the same *Sk* element (2125). Three different killer elements are known, all from natural populations. The best studied are *Sk-2* and *Sk-3*, which were found in *Neurospora intermedia* and crossed into *Neurospora crassa* for mapping and analysis. Both are gene complexes ("haplotypes") that, when heterozygous, block meiotic recombination over an interval ~30 map units long on linkage group III. *Sk-2* and *Sk-3* are distinguished by the specificity of killing and by the specificity of their response to genes that confer resistance to killing. In crosses where killing occurs, meiosis and two postmeiotic mitoses appear to be normal, as judged by light microscopy. Sensitive

ascospores first appear abnormal after one nuclear division in the ascospore. No killing occurs in crosses homozygous for a killer nucleus of the same type, but *Sk-2* ascospores are subject to killing by *Sk-3* and vice versa. Likewise, genes conferring resistance to killing by one type do not confer resistance to the other. Less than 1% of ascospores are viable in crosses of *Sk-2* × *Sk-3* (2121, 2125). A sensitive nucleus that otherwise would be killed survives if it is included in the same ascospore with a killer nucleus (1674). If *Sk-2* and *Sk-3* nuclei are included in the same ascospore, the ascospore is not killed and both nuclei survive (1673). Because the four ascospores of *Neurospora tetrasperma* normally are heterokaryotic, sensitive nuclei are sheltered from killing in ascospores from crosses heterozygous for *Spore killer*. Crosses with *Sk-2* and *Sk-3* in *N. tetrasperma* can therefore be used to examine differences in structure and gene order within the recombination block (1686). Because the known *Sk* elements segregate at the first division of meiosis, heterozygous crosses can be used to map centromeres using physically unordered asci ejected from the perithecia (1602). For reviews, see refs. (1681), (1682), and (2126). For genes conferring resistance to killing, see *r(Sk-2)*, *r(Sk-3)*, and ref. (2124). For reference strains and tester strains for identifying the various killers and genes conferring resistance to killing,

FIGURE 59 Maturing asci from a cross heterozygous for *Spore killer-2*. Most asci contain four normal large spores and four small aborted spores. The few asci that do not show the 4:4 pattern are still immature. All the maturing ascospores are *Sk-2*. *Sk-2* segregates at the first meiotic division in all asci, indicating that there is no crossing over between the *Spore killer* complex and the centromere. Photograph by N. B. Raju. From ref. (1674), with permission from the Genetics Society of America.

see refs. (2127) and (2128). *Spore killer (Sk)* is not to be confused with *skin (sk)*, a recessive gene in VIIR.

Sk-1: Spore killer-1

Unmapped. Less than 1% second-division segregation (2125).

Resembles *Sk-2* in behavior (1674, 2125). Known only in *Neurospora sitophila*. Attempts to transfer *Sk-1* into *Neurospora crassa* or *Neurospora intermedia* have failed. The frequency of *Sk-1* in collected strains of *N. sitophila* is high relative to that of *Sk-2* and *Sk-3* in natural populations of *N. intermedia* (1674, 2125)

Sk-2: Spore killer-2

IIIL,R. A haplotype. Between *cum* (2%) and *his-7* (21%) (280). When *Sk-2* is heterozygous, crossing over is suppressed in what is normally a 30-unit interval extending from *r(Sk-2)* through *leu-1* and spanning *Cen-III* (2125). Markers *acr-7*, *acr-2*, and *leu-1* were inserted into the *Sk-2* recombination block selectively at frequencies of 10^{-5} or 10^{-6}. When these were used in crosses homozygous for *Sk-2*, crossing over in the long blocked segment was normal (280).

The four non-*Sk* ascospores in each ascus abort when *Sk-2* is heterozygous (Fig. 59). Asci with second-division segregation patterns are rare or absent. Strains carrying *Sk-2* have been found at only four sites: Borneo (2), Java, and Papua-New Guinea. Most wild *Neurospora intermedia* strains are sensitive to killing by *Sk-2*, but resistant strains are common in some geographic areas (2121). See *r(Sk-2)*. Two *Sk-2*-resistant strains have been found in *Neurospora crassa*, even though killer strains have never been identified in that species.

Sk-3: Spore killer-3

IIIL,R. A haplotype. Between *cum* (2%) and *his-7* (12%) (280). When the *Sk-3* complex is heterozygous, crossing over is suppressed in the same interval as with *Sk-2*. *Sk-3* did not then recombine with *leu-1* (none in 5×10^6). Recombination was about 10^{-5} for *acr-7* and *acr-2* (280).

Sk-3 resembles *Sk-2* in its behavior. *Sk-3* ascospores are sensitive to killing by *Sk-2*, however, and vice versa. *Sk-3* has been found only once, in *Neurospora intermedia* from Papua-New Guinea. Strains resistant to killing by *Sk-3* are present in *N. intermedia* populations in the Eastern Hemisphere (2121).

Sk(ad-3A)

Probably an *ad-3A* allele.

slime

A multiple-mutant phenotype, lacking a cell wall and growing as protoplasts or plasmodium subject to osmotic lysis. The original strain contained at least two mutations (*fz, fuzzy; sg, slow germination*) in addition to the already-present markers *arg-1*, *cr-1*, *al-1*, and *os-1*. (All markers are recessive.) Three of the markers, *fz, sg,* and *os-1*, are required for a *slime*-like phenotype (607). A variant with altered morphology, FGSC 4761, grows more rapidly (1801). The *slime* strain is altered in the production and secretion of several enzymes related to sugar catabolism (1622). Cell-wall-like material may be produced in small quantity in stocks newly resolved from a phenotypically wild-type heterokaryon. *slime* protoplasts can be induced to fuse (1453). The ability to form heterokaryons or to function as fertilizing parent is lost after continuous growth (1883). Stocks can be maintained in phenotypically normal heterokaryons of the same mating type (607) or of mixed mating type and recovered by filtration (1456). Alternatively, stocks may be kept at $-70°C$ *in situ* on agar medium (1877), in growth medium (1867), or in dimethyl sulfoxide (429). Recombinant *slime*-like progeny with inserted markers can be recovered from crosses using filtration enrichment (1456). The *slime* strain has been used for isolating and storing plasma membrane in large quantity (1287), analyzing fatty acids in the plasma membrane (681), purifying and characterizing plasma membrane H^+-ATPase (1802), extracting enzymes gently (703), isolating chitosomes (124), obtaining intact chromosomal DNAs for separation by pulsed-field gel electrophoresis (816, 1129), studying vacuoles (1275, 1800), showing that poly(A) polymerase and nuclease activities are largely located in the nucleus (1868), and creating an efficient cell-free system for protein synthesis (7, 2034). Nuclei of opposite mating type appear to be compatible in a (*tol + slime tol⁺*) heterokaryon (1456). For general methodology, see refs. (1456) and (2243). Cell-wall-free *Neurospora* resembling *slime* can be obtained following growth of an *os-1* strain in the presence of polyoxin D (1879, 1881).

slo-1: slow-1

IR. Between *mat* (14%) and *thi-1* (2%; 5%) (1548).

Growth from conidia or ascospores is slow (1548). Conidiation lags significantly behind wild type. Morphology is normal. Not tested for cytochrome deficiency or for the ability to reduce tetrazolium.

slo-2: slow-2

VII. Left of *met-7* (2%) (1592).

Growth from conidia or ascospores is slow. Conidiation lags days behind wild type (1592, 2013). Not tested for cytochrome deficiency or for the ability to reduce tetrazolium.

smco: semicolonial

This name was introduced by Garnjobst and Tatum (719) for mutants that begin growth on agar as a small colony and sooner or later produce a flare of wild-type-appearing hyphae, with or without conidia.

smco-1: semicolonial-1

I. Linked to *mat* (1%), *rg-1* (0/72) (719).
Growth initially is colonial (719).

smco-2: semicolonial-2

Allelic with *sc* (S7).

smco-3: semicolonial-3

I. Linked to *mat* (10%), *al-2* (29%). Recombines with *col-7*, *smco-1* (719).
Growth initially is colonial (719).

smco-4: semicolonial-4

IVR. Linked to *pan-1* (8%) (719).
Growth initially is colonial (719).

smco-5: semicolonial-5

I. Linked to *mat* (2%).
Semicolonial flat growth persists until 4–7 days after ascospore germination, when wild-type mycelium develops. Shows phenotypically wild-type growth when transferred, but when it is crossed by *smco-5*⁺, *smco-5* ascospores are produced that repeat the cycle. Recombines with *rg-1*, *smco-1*. Not fertile with *smco-3*; may be allelic (719).

smco-6: semicolonial-6

VR. Linked to *asn* (6%), near *pyr-6*. Right of *met-3* (14%) (321, 1383).
Growth initially is colonial (719).

smco-7: semicolonial-7

VR. Right of *ilv-1* (2%). Linked to *rol-3* (0/154), *sp* (12%, probably to the left), *chs-1* (0/18) (1021, 1383).
Cloned and sequenced: Swissprot RAS2_NEUCR, EMBL/GenBank D16137, GenBank NEUNCR2P; Orbach-Sachs clones X9E09, X11F07, X16G03, X18B08.
Growth initially is colonial. Conidiation occurs in a crescent at the top of an agar slant. Cell-wall synthesis is apparently defective. Morphology is distinct from *rol-3*, which complements *smco-7* (719).
The gene called *ras-2* is allelic with *smco-7* allele R2497, which it resembles, with which it gives 0/84 wild-type

recombinants, and which it can transform to wild type (1021, 1267). Sequence shows the gene to be a member of the *ras* superfamily. The single-base deletion found in allele R2497 is expected to truncate the protein product (1021). The null mutant obtained by RIP grows linearly on agar medium at one-tenth the wild-type rate, has fewer aerial hyphae and conidia, and periodically forms rings of short aerial hyphae with sparse conidia.

smco-8: semicolonial-8

IVR. Linked to *pan-1* (1%; 7%), *smco-4* (30%) (719).
Growth sometimes flares out at the top of an agar slant (719). Unable to grow on galactose or grows as restricted colonies (1689). Resistant to dicarboximide fungicides, especially vinclozolin (789). The amount of cell-wall peptides is reduced (2249). Complements *col-1*, *col-8* (719).

smco-9: semicolonial-9

IVR. Linked to *pan-1* (5%), *smco-4* (2%), *smco-8* (13%) (719).
Morphology is partially normalized by isomaltose or starch. Altered inhibitor of branching enzyme α-1,4-glucan-6-glycosyltransferase (2). Reduced amount of cell-wall peptides (2249). Crosses homozygous for *smco-9* produce nonlinear asci (similar to *pk*) (1969). The strain originally obtained was complex, producing two types of colonials among progeny. The component that behaved as a recessive ascus mutant was designated *smco-9* (1965). Resistant to dicarboximide fungicides, especially vinclozolin (789).

sn: snowflake

IR. Between *Cen-I* (<1%) and *os-4* (<3%), *T(AR173)*ᴸ (1605, 2072).
Spreading colonial growth. Linear growth is less than one-tenth wild type (34). Mutant progeny can be recognized microscopically immediately after ascospore germination by hyphal patterns that suggested the name *snowflake* (1365). Alleles differ in osmotic tolerance, dominance, and differentiation of conidia (2072) (Fig. 60). Large bundles of 8- to 10-nm-diameter microfilaments consisting of pyruvate decarboxylase are accumulated in vegetative cells (34, 40, 1739) (see *cfp*), but not during ascus development when PDC is associated with the cortical cytoskeleton (2085). The *sn* mutant is said not to exhibit cytoplasmic streaming (33). Meiosis and ascospore formation are normal in crosses homozygous for *sn* (1673). Female fertility is good. In visible appearance, *sn* resembles *sp* and *cum* and also *cot-4* at 25°C (1582). Used to study the development of crystalline inclusions (32). Used as a marker in a study of *rec* genes (338).

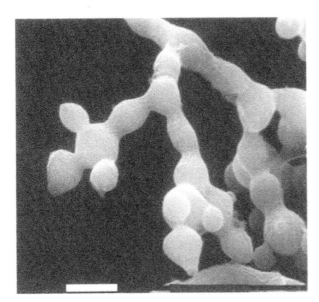

FIGURE 60 The mutant *sn: snowflake* (allele C136). Bar length = 10 μm. Scanning EM photograph from M. L. Springer.

The double mutant *sn cr-1* resembles *rg cr-1* in forming small, discrete, conidiating colonies suitable for replication. *sn cr-1* has the advantage over *rg cr-1* of being fertile in homozygous crosses (1553). *sn cr-1* has been used to select drug-resistant mutants (346). Heterokaryons between marked *sn cr-1* strains are an integral part of a system developed for determining the rate of recessive lethal mutation throughout the genome (1980) (Fig. 22).

snz: snooze

Allelic with *pdx-1*.

so: soft

IR. Between *arg-13* (2%; 12%) and *aro-8* (7%; 11%) (822, 1592).

Lightly pigmented. Distinctive morphology, with a lawn of fuzzy short aerial hyphae and conidia that are formed more uniformly than in wild type and closer to the agar surface, similar to *peach* (1548). Recurrent mutations are frequent in strains of Oak Ridge background (1011, 1014). Best scored early on short, obtuse slants. The morphological abnormality is more pronounced on sorbose–sucrose medium. Female sterile. Classified as an *age* mutant: the conidial life span is short under defined conditions, and the map location is near a site called *age* 1.3 (1394).

Hyphal fusions are not detectable microscopically. Formation of forced heterokaryons is delayed relative to wild type when *ham* is present in one or both strains, with a latent period of >24 hr for *ham* + *ham⁺* and >80 hr for *ham* + *ham*. Once heterokaryons are formed, growth rate is reduced from 4 mm/hr for wild type to 3–3.5 mm for (*ham* + *ham⁺*) and 2–2.5 mm for (*ham* + *ham*) (2224). The described *het* genes do not affect the fusion of hyphae.

A mutant called *ham: hyphal anastomosis* (2224) is allelic with *so* (809a). The phenotype of *ham* was described as follows. Hyphal fusions are not detectable microscopically. Formation of forced heterokaryons is delayed relative to wild type when *ham* is present in one or both strains, with a latent period of >24 hr for *ham* × *ham⁺* and >80 hr for *ham* + *ham*. Once heterokaryons are formed, the growth rate is reduced from 4 mm/hr for wild type to 3–3.5 mm for (*ham* + *ham⁺*) and 2–2.5 mm for (*ham* + *ham*) (2189).

sod-1: superoxide dismutase-1

IL. Between *fr* (5T/16 asci) and *Fsr-12*, *vma-4* (2 or 4T/18 asci) (1447).

Cloned and sequenced: Swissprot SODC_NEUCR, EMBL/GenBank M58687, PIR A23909, A36591, EMBL NCSOD1, GenBank NEUSOD1.

Structural gene for copper/zinc superoxide dismutase (EC 1.15.1.1) (365, 1171). A null mutant shows increased sensitivity to paraquat and oxygen as well as an increased rate of spontaneous mutation (365).

sod-2: superoxide dismutase-2

VR. Linked to *cyh-2* (0T/18 asci) (1447).

Cloned and sequenced: EMBL/GenBank AF118809.

Structural gene for mitochondrial manganese superoxide dismutase. Possibly also active in the cytoplasm (571). [The *un-1* mutant has been reported deficient in SOD-2 (1392).]

sor: sorbose resistant

Sorbose is used to induce colonial growth in platings of *Neurospora* (231, 462). [Resolution is improved if sorbose is used in conjunction with *cot-1* (332).] Numerous *sor* mutants have been obtained that show spreading growth on concentrations of sorbose that restrict the wild type (1075, 1076, 1757). With the exception of *sor-4*, scoring has usually been on minimal medium plus 0.025% filter-sterilized sorbose. Certain mutants selected for other traits may also be resistant to sorbose. For example, *sor-4^T9*, which was selected for its ability to hydrolyze starch and which is defective in extracellular amylase, is simultaneously sorbose-resistant and osmotic-sensitive (1400). Sorbose-resistant mutants at four loci act as dominance modifiers of the ascus effect of dominant *pk* alleles (1757). Most *sor* mutants have probably not been examined for possible pleiotropic amylase or osmotic phenotypes. General references (35, 1656).

sor-1: sorbose resistant-1

VIL. Left of *ylo-1* (3%) (1076).
Defective in sorbose uptake (1075) and thereby resistant to growth restriction by sorbose (1076). Recessive (1074, 1077). Symbol changed from *sor-A* (1668).

sor-2: sorbose resistant-2

VII. Linked to *nt* (31%) (1076).
Defective in sorbose uptake (1075) and thereby resistant to growth restriction by sorbose (1076). Recessive (1074, 1077). Symbol changed from *sor-B* (1668).

sor-3: sorbose resistant-3

IIIR. Linked to *ad-4* (7%) (1076).
Resistant to growth restriction by sorbose (1076). Recessive with respect to colony size. Conidiation is precocious (35). Partially recessive for percent conidial germination on sorbose test medium (1074, 1077). In heterokaryons with *sor-1* or *sor-2*, the phenotype is intermediate between the resistant single-mutant phenotype and the sensitive wild type (1074). Symbol changed from *sor-C* (1668). Called *sor^r-17* (1076).

sor-4: sorbose resistant-4

IL. Between *mat* (6%), *suc* (1%), *In(H4250)^L* and *arg-1* (1%, 4%), *Cen-I* (5%) (1400, 1592, 1988).
Cloned and sequenced (presumed allele *rco-3* has been cloned and sequenced): EMBL/GenBank U54768; Orbach-Sachs clone X10E5.
Resistant to growth restriction by sorbose. Osmotic sensitivity is comparable to that of *os-1*. Dissimilar in morphology to *os-4* (1582). Growth is slow. (Putative alleles called *dgr-3* and *rco-3* are exceptions with growth that is not slow. *rco-3* may show a lag of several hours, after which it grows at the wild-type rate on most media.) Glucoamylase and extracellular amylase are low, as is extracellular acid phosphatase activity (1400). However, *dgr-3* is reported to have high non-repressible glucoamylase (35). The glucoamylase, osmotic, and sorbose-resistance properties cosegregated in 101 isolates (1400). Modifies the dominance of *pk* alleles (1757). The growth of *gpi* on glucose or sucrose is enhanced in the double mutant *gpi-1; sor-4*. Used to obtain mutants defective in glucose phosphate isomerase (1401). Scoring of *sor-4* is clear on minimal slants with 2% sucrose + 3% sorbose after 1 and 2 days at 34°C (1592). Alleles and probable alleles have been recognized under many different guises. Called *sor(DS)* (1592, 1668), *Pk-mod-D* (1757), T9, *sor-T9* (35, 1400, 1401),

gla: glucoamylase (106), *amy* (615). *sor^r-15* is a possible allele based on map position (1076). It is not clear whether *sor-4* is a locus separate from *pat*. Allele T9 is resistant to the colonializing action of sorbose at 25°C, but not at 35°C (1402). Linked to the regulatory gene *exo-1*, which has not been tested for allelism but is stated to be on the opposite side of *mat*. Detected in the *pat; pro-1* strain that was first used to demonstrate a circadian rhythm in fungi (1988). Mutants have been obtained by plating mutagenized conidia in medium containing starch and observing a cleared zone around colonies. Mutations called *dgr*, which were selected for resistance to inhibition by deoxyglucose, are also resistant to sorbose (35). Mutations designated *dgr-3* appear to be allelic with *sor-4* (0++ recombinants in 400 intercross progeny) (35). A *sor-4* allele formerly designated *dgr-3* exhibits resistance to glucose inhibition of sexual initials (protoperithecia and melanin). A mutation called *rco-3* is allelic with *dgr-3* and, hence, presumably with *sor-4* based on its failure to complement in (*rco-3* + *sor-4*) heterokaryons (573). Summarized next is information obtained using the putative allele called *rco-3*.

The mutant previously designated *rco-3* is a putative allele of *sor-4* and is linked to *gla-2* (<10 kb) and *mat* (576). It was selected as enabling higher basal expression of a *con-10::hph* fusion gene during vegetative growth (1241). It conidiates in liquid culture without nutrient limitation and is partially resistant to L-sorbose. The expression of conidiation-specific genes is elevated in liquid culture (1241). Sequence of the cloned gene reveals similarities to sugar transporter genes, and RCO-3 is required for normal levels of glucose-transport activity (575, 1242). Expression of *qa-2* is not glucose-repressible in the mutant.

sor-5: sorbose resistant-5

V. Linked to *his-1* (1076).
Resistant to growth restriction by sorbose (1076). Mutants called *sor^r-14* and *sor^r-19* are allelic.

sor-6: sorbose resistant-6

Probably III or VI by linkage to *ylo-1* in *alcoy* (1582).
Resistant to growth restriction by sorbose (1076). Recessive (1075). Called *sor^r-6*.

sor(T9): sorbose resistant T9

A *sor-4* allele.

sp: spray

VR. Between *leu-5* (3%; 9%) and *ure-3* (8%), *am* (1%; 8%; 450 kb) (219, 220, 256, 1651, 2014).

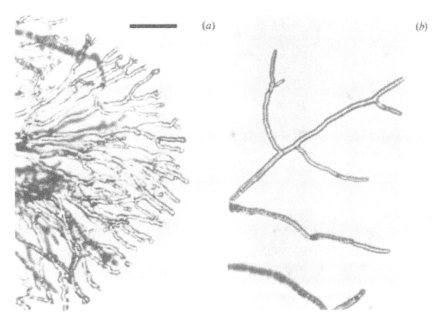

FIGURE 61 Typical morphology of the mutant *sp: spray* on minimal medium (a) and on minimal medium plus 500 mM Ca²⁺ (b). The calcium supplement effectively restores wild-type morphology. Bar length-50 μm. From ref. (533), with permission from the Society for General Microbiology.

Cloned: Orbach-Sachs clone X11G10 (317).

Initially grows as a colony flat on the agar surface. Aerial mycelium then fans upward (Fig. 61) (1548). For ultrastructure and intraconidial conidia, see references in ref. (2117). Cell-wall analysis (507). Reduced amount of cell-wall peptides (2249). Morphologically similar to *cot-4* (at 25°C), *sn,* and *cum* (1582). Excellent female fertility. Used as a lawn for mating-type tests on plates (1929). Mutant morphology is corrected by calcium (533) (Fig. 61).

spco: spreading colonial

This symbol and name were used in ref. (719) for mutants that begin growth on agar as a colony and do not remain restricted, but spread to cover the agar surface. Mutants of the series *spco-3* to *spco-15*, described in ref. (719), were sometimes named as new *spco* genes without having been tested for allelism with already named morphological mutants having similar map locations. Growth rates and other characteristics of 11 *spco* mutants were described in refs. (2107), and (2108). Used to study septa and septal plugging (2109, 2110).

spco-1: spreading colonial-1

Allelic with *col-4.*

spco-2: spreading colonial-2

Allelic with *wa.*

spco-3: spreading colonial-3

Allelic with *spco-7.*

spco-4: spreading colonial-4

VIII. Linked to *do* (<1%), *nic-3* (1%, probably to the left) (1592).

Growth on agar is initially aconidial, with fine hyphae (719). Capable of conidiating on the surface of complete medium (1546). Hyphae extend faster within agar medium than on the surface, resulting in a dense hemispherical colony, most of which is embedded (2107). Septal plugging is rare in peripheral hyphae (2109).

spco-5: spreading colonial-5

VII. Linked to *col-17* (6%), *nt* (20%) (719).

Asci are nonlinear in crosses homozygous for *spco-5* (1965). Cell-wall peptides are reduced in amount (2249). Complements *col-2* and *spco-4* (719).

spco-6: spreading colonial-6

VII. Right of *thi-3* (4%) (1546). Linked to *spco-5* (8%), *nt* (20%) (719).

Complements *col-17, col-2, spco-4,* and *spco-5* (719).

spco-7: spreading colonial-7

VI. Linked to *ad-1* (0/65). Between *ylo-1* (4%) and *T(AR209), trp-2* (16%; 21%) (1578, 1582).

Complements *moe-2*. Allele R2365 is preferred because of its excellent growth, conidiation, and fertility (1582). Allele R2365, which was originally called *spco-3* and incorrectly assigned to V (719), makes small perithecia devoid of ascospores when crossed with *spco-7* R2457 (1582).

spco-8: spreading colonial-8
IV. Linked to *pan-1* (23%) (719).
Spreading colonial morphology.

spco-9: spreading colonial-9
VR. Linked to *asn* (6%). Right of *met-3* (18%) (1383).
Multiple side branches are formed on hyphae below septa that have become occluded as a result of flooding (2110). Complements *ro-5*, *cot-2*, and *smco-6*. Morphologically distinct from *col-9*, but allelism tests were inconclusive (719, 1383).

spco-10: spreading colonial-10
VR. Between *ilv-1* (24%) and *inl* (5%) (21 asci) (719, 1383).
Spreading colonial morphology.

spco-11: spreading colonial-11
IL. Between *leu-3* (7%) and *In(NM176)^L*, *mat* (17%). Recombines with *mo-5* (18%) (719, 1578).
Conidia are occasionally produced at the top of a slant (719).

spco-12: spreading colonial-12
I. Right of *mat* (13%; 35%). Recombines with *mo-5* (5%), *spco-11* (43%) (719).
Forms colonies with a downy center and a lacy growing border (719). Hyphal growth is highly branched (2108). Nearly all septae are plugged in peripheral hyphae (2109). Cell-wall peptides are reduced in amount (2249).

spco-13: spreading colonial-13
VI. Linked to *trp-2* (16%), *Cen-VI* (1/10 asci) (719).
May be allelic with either *spco-7* or *moe-2*. Intercrosses and heterokaryons were unsuccessful with both (719).

spco-14: spreading colonial-14
II. Linked to *arg-5* (7%) (719).
Spreading colonial morphology. A few scattered conidia are produced. Complements *da* and *bal* (719).

spco-15: spreading colonial-15
III. Linked to *spg* (10%), *pro-1* (18%), *Cen-III* (0/10 asci). Recombines with *col-14*, *col-16*. Complements *mo-4*, *col-14*, and *col-16* (719).
The original *spco-15* mutation has probably been lost. A morphological mutant labeled *spco-15* (R1537, FGSC No. 2389) is not linked to III markers (1582).

spe: spermidine
The polyamines spermidine and spermine are synthesized from ornithine in the greater arginine biosynthetic pathway (Fig. 11) via ornithine decarboxylase (EC 4.1.1.17) to putrescine, spermidine synthase (EC 2.5.1.16) to spermidine, and spermine synthase (EC 2.5.1.22) to spermine (Fig. 62). The *spe-1* gene encodes the first enzyme. *spe-2* encodes S-adenosylmethionine decarboxylase (SAM decarboxylase) (EC 4.1.1.50). This catalyzes the formation of decarboxylated SAM, which reacts with putrescine to form spermidine and methylthioadenosine, a step catalyzed by spermidine synthase, which is encoded by *spe-3*. In the last step, another molecule of decarboxylated SAM reacts with spermidine to form spermine and methylthioadenosine catalyzed by spermine synthase, the gene for which is as yet uncharacterized.

spe-1: spermidine-1
VR. Between *am* (1%) and *his-1* (1%) (1626).
Cloned and sequenced: Swissprot DCOR_NEUCR, EMBL/GenBank L16920, M68969, M68970, PIR A42065, EMBL NCODC, NCODCA, NCSPE1A, GenBank NEUODC, NEUODCA, NEUSPE1A; EST NC1F9; Orbach-Sachs clones X3H05, X4E02, G3H04, G4G01, G6H12, G11H12, G17G10, G22F01, G23B02, G24C11.
Structural gene for ornithine decarboxylase (ODC) (EC 4.1.1.17) (611, 2218) (Figs. 11 and 62). Uses putrescine, spermidine, or spermine. Does not suppress *pro-4* (1310). Excretes yellow pigment into synthetic cross medium (1546). Meiosis is normal in homozygous crosses (on 5 mM spermidine); these produce mostly white ascospores and a few viable black ascospores (451, 1673). The stability of ODC is regulated by polyamine availability (96, 540). Molecular studies on regulation (1627). Effects on induction by polyamines (805). In contrast to other systems, *Neurospora crassa* ODC does not appear to be regulated by osmoticum (470). Called *put-1*.

spe-2: spermidine-2
VR. Left of *am* (6%), *spe-1* (1626).
Cloned and sequenced: EMBL/GenBank AF151380; Orbach-Sachs clone X2B09 (925).

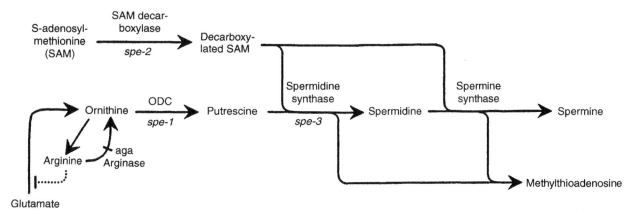

FIGURE 62 The polyamine pathway. The dotted line indicates feedback inhibition of ornithine synthesis by arginine. From ref. (464), with permission from the Genetics Society of America.

Specifies S-adenosylmethionine decarboxylase (EC 4.1.1.50) (Fig. 62). Blocked in the conversion of 5-adenosylmethionine (SAM) to decarboxylated SAM (1626). Grows on spermidine, but not on putrescine.

spe-3: spermidine-3

VR. Right of *inl* (4%) (1626). Linked to *cya-2* (1035).

Cloned and sequenced: EMBL/GenBank AB001598; Orbach-Sachs clone G22C12 (1035).

Specifies spermidine synthase (EC 2.5.1.16). Blocked in the conversion of putrescine to spermidine (1626) (Figs. 11 and 62). Cloned independently by suppression of the temperature sensitivity of *cpz-2* mycelial growth. Encodes a sequence homologous to spermidine synthase of *Saccharomyces*, human, and mouse (1035).

spg: sponge

III. Between *acr-2* (1%; 11%) and *ser-1* (8%). Linked to *sc* (<1%), *thi-4* (0/103) (1582, 1592).

Forms conidiating spreading colonies. Morphologically distinct from *sc*. Hyphae fuse to form bundles (719). Scanning EM photograph; few major constriction chains are seen (1955). Viability is good. Growth on Vogel's minimal medium (2163) is less spreading than on crossing medium (1582, 2208). Cell-wall peptides are reduced in amount (2249).

Spiral Growth

Hyphae of the laboratory wild type grow as a loose clockwise spiral in young colonies on the surface of agar (141). The pattern is made more clear if cellophane or another membrane is used to prevent growth into the agar. Strains have been described that show increased (141) or decreased (1902) tightness of coiling or anticlockwise coiling (2256); see *coil-1, coil-2, rca-1, rco-1*.

sre: siderophore regulation

Unmapped.

Cloned and sequenced: EMBL/GenBank AF087130.

Encodes a GATA-binding protein that controls iron transport by controlling the synthesis of siderophores (2302). SRE and its homologs form a subfamily of GATA factors with two zinc fingers separated by a long stretch of amino acids (2302).

ss: synaptic sequence

IL. Linked to *nit-2* (<0.2%) (313).

When alleles ss^E, ss^S, and ss^C are heterozygous (from Emerson, St. Lawrence, and Costa Rica wild types), recombination in *nit-2* is reduced by 2- to 20-fold, but crossing over in the flanking intervals *un-5* to *nit-2* or *nit-2* to *leu-3* is not affected. *ss* heterozygosity acts multiplicatively with $rec-1^+$ to reduce recombination in *nit-2* by 100-fold (313).

ssp-1: site-specific parvulin-1

Unmapped.

Cloned and sequenced: EMBL/GenBank AJ006023, EMBL NCR6023.

Encodes a homolog of site-specific parvulin (EC 5.2.1.8) (1088), which is active in isomerizing small peptides and in protein folding. Unique among known eukaryotic parvulins in containing a polyglutamine stretch between the N-terminal WW domain and the C-terminal peptidyl-prolyl *cis–trans*-isomerase (PPIase) domain.

ssu: supersuppressor

Nonsense suppressors (either proved or putative). In a few cases, putative missense mutants also were suppressed (1461). Allele-specific but not locus-specific. Often infertile or only slightly fertile (*ssu-3*) when used as female parent, forming no perithecia or empty perithecia (1859). Mutants *ssu-1*, *ssu-2*, *ssu-3*, *ssu-4*, and *ssu-5*, but not *ssu-6*, are cold-sensitive in combination with *am*, showing less than half the wild-type growth rate on minimal medium at 10°C (1752). Action of the suppressors is nucleus-restricted in heterokaryons (773). Interaction of certain supersuppressors with certain *ad-3B* alleles produces erratic stop–start growth (772). For tabular summaries of the action of individual *ssu* mutations on differential test alleles, see refs. (303), (1860), and (1862).

ssu-1: supersuppressor-1

VIIR. Between *met-7* (14%) and *su(trp-3^{td201})-1* (10%), *nt* (23%) (1859).

Allele WRN33 was selected in ref. (1858) as a suppressor of the amber mutant *am^{17}* (1861); tyrosine is inserted in a site where wild type has glutamate (1861). Used to identify suppressible amber mutants in *aro(p)*, *trp-1*, *trp-2*, *trp-3* (294, 348, 1858), *ad-3B* (1461), *arg-6* (474), *nit-2* (1610), and *nit-3* (1610). Restores full-length NIT3 to *nit-3* amber mutants (1494). Although perhaps the most efficient of known *ssu* mutants, *ssu-1* restores only about 20% of the wild-type amount of normal glutamate dehydrogenase in *am^{17}* (1861).

ssu-2: supersuppressor-2

I. Linked to *mat* (22%). Probably between *Cen-I* (7%) and *al-2* (26%). Recombines with *ssu-3* (1859).

Amber suppressor. Allele WRU35, selected in ref. (1859) as a coincident suppressor of the amber mutants *trp-3^{td140}* and *am^{17}* (258, 1861). Suppresses the same alleles as *ssu-1*. Also suppresses certain mutants in *trp-1*, *trp-2*, and *ad-3B* (1860).

ssu-3: supersuppressor-3

I. Linked to *mat* (22%). Probably between *Cen-I* (10%) and *al-2* (33%) (1859).

Amber suppressor. Allele WRU118, selected in ref. (1859) as a coincident suppressor of the amber mutants *trp-3^{td140}* and *am^{17}* (258, 1861). Fails to suppress most amber mutants that are suppressed by *ssu-1* and *ssu-2* (1860). Probably specifies the insertion of a different amino acid than is inserted by *ssu-1* or *ssu-4* [T. W. Seale and A. Kinniburgh, cited in ref. (1860)].

ssu-4: supersuppressor-4

VIIL. Between *nic-3* (28%) and *met-7* (20%) (1859).

Amber suppressor. Allele WRU18, selected in ref. (1859) as a suppressor of the amber mutants *trp-3* (td140) and *am*

(17, 258, 1861). Also suppresses certain mutants in *trp-1* and *ad-3B* (1461, 1860).

ssu-5: supersuppressor-5

III or IV (303).

Amber suppressor. Allele Y319-45 was selected as suppressor of a nonsense mutant in *aro-1(p)* (303). Also suppresses the amber mutant *trp-3^{td140}* and certain *ad-3B* alleles, but not *am^{17}* (303, 1461, 1862).

ssu-6: supersuppressor-6

VR. Linked to *his-1* (4%) (303).

Amber suppressor. Allele Y319-26 was selected as suppressor of an amber mutant in *aro-1(p)* (303). Also suppresses certain alleles in *trp-3* and *his-3*, but not *am^{17}* (303, 1461, 1862).

ssu-7: supersuppressor-7

VIL. Between *ad-8* (8%) and *ylo-1* (14%) (1859).

Amber suppressor. Allele WRU7 was selected in ref. (1859) as a coincident suppressor of the amber mutants *trp-3^{ttd140}* and *am^{17}* (258, 1861). Also suppresses certain alleles at *trp-1*, *trp-2*, and *ad-3B*. *ssu-7* shows the widest spectrum of the known supersuppressors (1860). The Y319-37 allele of *ssu-8*, which is in IR, was erroneously called *ssu-7* in some FGSC lists and in ref. (772).

ssu-8: supersuppressor-8

IR. Linked to *al-2* (2%; 8%) (303).

Amber suppressor. Allele Y319-37, selected as a suppressor of an *aro(p)* amber mutant (303). Also suppresses certain alleles in *trp-3*, *his-3* (303), and *ad-3B* (1461). Allele Y319-37 (FGSC No. 1749), also called 54-su37, was thought possibly to be an allele of *ssu-2* and was listed as *ssu-2* (?) before being designated *ssu-8* (303). It was erroneously called *ssu-7* in various FGSC lists and in ref. (772) and was erroneously stated to be in linkage group VI (the location of the real *ssu-7*).

ssu-9: supersuppressor-9

Unmapped. Locus distinct from other *ssu* genes (1860).

Amber suppressor. Allele WRU98 was selected (1860) as a coincident suppressor of the amber mutants *trp-3^{td140}* and *am^{17}* (258, 1861). Also suppresses certain alleles at *trp-1* and *trp-2* (1860).

ssu-10: supersuppressor-10

Unmapped. Locus distinct from other *ssu* genes (1860).

Amber suppressor. Allele RWU121 was selected (1860) as a coincident suppressor of the amber mutants *trp-3^(td140)* and *am^17* (258, 1861). Also suppresses certain alleles at *trp-1*, *trp-2*, and *ad-3B* (1860).

ssu(WRU79): supersuppressor (WRU79)

Unmapped. Locus distinct from other *ssu* genes (1860).

Amber suppressor. Selected (1860) as a coincident suppressor of the amber mutants *trp-3^(td140)* and *am^17* (258, 1861). Also suppresses certain alleles at *trp-l*, *trp-2*, and *ad-3B*. Spectrum resembles *ssu-1* (1860).

st: sticky

IR. Between *ad-3B* (5%) and *thi-1* (14%) (1548).

Mycelia adhere to an inoculating needle. Subtly different from wild-type in gross morphology. Exudate is sometimes present (1548). Grows poorly from conidia inoculated to minimal agar medium with ballooning of hyphal tips, which is not alleviated by mannose (1546).

Sterile Mutants

The unqualified term "sterile" has been used in a variety of ways for different situations, such as where no protoperithecia are formed, where perithecial development is blocked before ascospore formation, or where perithecia develop fully but the ascospores that are produced are inviable. To avoid ambiguity, the type and extent of sterility should be specified (1577). The term "barren" has been proposed for crosses that form perithecia but produce few or no ascospores (1685). See *ff*, *fmf*, *fs*, *mb*, *pp*.

su: suppressor

The symbol of the gene that is suppressed follows *su* in parentheses. The mutant suppressor allele is symbolized *su(. .)* and the wild-type nonsuppressor, *su(. .)^+*. *ropy* mutations can be obtained as suppressors of *cot-1* (247, 1634). *crisp* mutations can be obtained as suppressors of *mcb* (247). For suppressors of vegetative incompatibility, see *tol*, *su(het)*. *arg-12^S* acts as a suppressor of *pyr-3*.

su(arg-1)-1: suppressor-1 of arg-1

Unmapped. Not linked to *arg-1*.

Restores 23–36% of wild-type L-citrulline:L-aspartate ligase activity to *arg-1* allele 46004. Called *arg-1^R26*, *arg-1^R3*, *s-26*, *s-3* (122).

su(bal): suppressor of bal

I. Linked to *mat* (13%) (1853).

The linear growth rate of *bal* is doubled and the K_m of G6PD is decreased by the suppressor. Photographs are given of *bal* and *bal; su(bal)* (1853). Called *su-B*.

su(col-2): suppressor of col-2

IL. Tightly linked to *mat* (0/837) (1853).

Increases the linear growth rate of *col-2* 10-fold and influences the electrofocusing pattern of G6PD both in *col-2* and in *col-2^+*. The morphology of *su(col-2); col-2^+* is wild type. Photographs are given of *col-2* and *col-2; su(col-2)* (1853). Called *su-C*.

su(cot-1): suppressor of cot-1

A series of mutants selected as suppressors of the colonial growth of *cot-1* were named *gulliver* (*gul*). *ropy* mutations can be selected as suppressors of *cot-1*, and existing *ropy* mutants act as *cot-1* suppressors.

su(cr-1): suppressor of cr-1

See *hah*. Suppressors of the colonial growth of *cr-1* have also been reported in ref. (1020), with alterations in cAMP level, trehalase activity, conidiation, and pigmentation .

su(het): suppressor of heterokaryon (vegetative) incompatibility

The first described *het* suppressor, *tol* (1466), acts only on the vegetative incompatibility function of the mating-type idiomorphs (1175). Mutations since have been reported that are interpreted to suppress vegetative incompatibility at other *het* loci (51). Some of these mutations are reported to affect only single loci, others to prevent incompatibility reactions even when strains differ at several *het* loci. One mutation, thought at first to be a suppressor of *het-d*, was reinterpreted as a new *het* locus called *het-12* (1492). Characterization and mapping of an extragenic *het*-suppressor mutation, *su(het-d)*, has been reported (1345, 1346). Interpretation of the results reported in ref. (51) has been complicated by the subsequent discovery that the strains used may differ at another locus or other loci affecting heterokaryon formation (980, 2223).

su(het-d): suppressor of het-d

V. Near *lys-1* (1345, 1346).

Appears to suppress *het-d*-mediated vegetative incompatibility. The difference from wild type is less marked in young cultures (1345, 1346).

su(ile-1): suppressor of ile-1

VR. Right of *met-3* (4%) (1055).
Suppresses the requirement of *ile-1* (1055).

su(inl): suppressor of inositol

Unmapped. Not linked to *inl* (735).
An allele-specific partial suppressor that enables *inl* allele
37401 to grow suboptimally on minimal medium (735).
Does not suppress seven other *inl* alleles, including
89601, that is partially suppressed by *opi-1*.

su(mcb): suppressor of mcb

See *cr-1*. *cr-1* mutations can be selected as suppressors of
the apolar growth of *mcb* (247).

su(met-2): suppressor of met-2

Unmapped. Not tested for allelism with *su(met-7)-1* or
su(met-7)-2.
Isolated as a suppressor of *met-2^{H98}* (2012). Suppresses
leaky mutants blocked between cysteine and homocysteine. It is probably the suppressor used in ref. (2211) to
show that the suppressed strain differs from wild type in
being able to incorporate sulfur from *S*-methylcysteine
into cysteine regardless of the sulfate concentration. The
suppressor does not have significantly increased acetylhomoserine sulfhydrylase (1045).

su(met-7)-1: suppressor-1 of met-7

IR. Linked to *al-2* (1%) (727).
Selected as reversions of *met-7* (4894) in ref. (727). All
spontaneous reversions were mutations at this locus.
Also suppresses *met-2* (H98). Not tested for suppression
of other alleles or loci. Suppressed strains attain wild-
type growth on minimal medium after initial retarda-
tion, which is alleviated by methionine (727). This is
apparently the suppressor used in ref. (649) to show that
the suppressor restores cystathionase I and II activities to
4894 and H98, respectively. Called *S-1*, FGSC 39.

su(met-7)-2: suppressor-2 of met-7

Unmapped.
Suppresses *met-7* (4894). Not tested for suppression of
other mutants. Selected as a revertant of *met-7* (4894) in
ref. (727). Recovered less frequently than *su(met-7)-1*
and only after irradiation. Called *S-2*.

su([mi-1])-1: suppressor-1 of [mi-1]

IIIR. Linked to *ad-4* (14%) (1084, 1085).
Restores normal growth rate and alleviates deficient
cyanide-sensitive respiration of *[mi-1]* and other group

I mitochondrial mutants. No effect on group II or III
mitochondrial mutants (168, 400, 778, 1085). Mitochondrial ribosomal subunit ratios and cytochrome
spectrum (400). Called *sup-1*.

su([mi-1])-3: suppressor-3 of [mi-1]

II. Linked to *fl* (31%). Not allelic with *su([mi-1])-4* or
su([mi-1])-10; 20% and 25% unsuppressed progeny
from intercrosses (1084, 1085).
Restores normal growth rate and alleviates the deficient
cyanide-sensitive respiration of *[mi-1]* (*[poky]*) and
another group I cytoplasmic mutant. Does not affect
group II mitochondrial mutant *[mi-3]* (168, 400, 1085).
Mitochondrial ribosomal subunit ratios and cytochrome
spectrum (400, 1085).

su([mi-1])-4: suppressor-4 of [mi-1]

II. Linked to *fl* (22%), *arg-5* (40%). Not allelic with *su([mi-
1])-3* or *su([mi-1])-10*, 20% and 25% unsuppressed
progeny from intercrosses (1084, 1085).
Restores normal growth rate and alleviates the deficient
cyanide-sensitive respiration of *[mi-1]* (*[poky]*) and
another group I cytoplasmic mutant. Does not affect
group II cytoplasmic mutant *[mi-3]* (168, 400, 1085).
Mitochondrial ribosomal subunit ratios and cytochrome
spectrum (400). Called *sup-4* (1085).

su([mi-1])-5: suppressor-5 of [mi-1]

VIIL. Left of *nic-3* (23%) (1084, 1085). Possibly allelic with
cyt-7 (170).
Restores normal growth rate and alleviates the deficient
cyanide-sensitive respiration of *[mi-1]* (*[poky]*) and
another group I mitochondrial mutant. Does not affect
group II cytoplasmic mutant *[mi-3]* (168, 400, 1085).
Mitochondrial ribosome and cytochrome spectra (400).
Affects mitochondrial large-subunit assembly. With
wild-type mitochondria, causes cold sensitivity (399).
Called *sup-5* (1085), *su^l-5* (399).

su([mi-1])-10: suppressor-10 of [mi-1]

II. Linked to *arg-5* (29%), *fl* (24%) (1084, 1085). Not
allelic with *su([mi-1])-3* or *su([mi-1])-4*; 25% unsuppressed progeny from intercrosses (1085).
Restores normal growth rate and alleviates deficient
cyanide-sensitive respiration of *[mi-1]* (*[poky]*) and all
other group I cytoplasmic mutants. Does not affect
group II cytoplasmic mutant *[mi-3]* (168, 400, 1085).
Cytochrome spectrum (400). Called *sup-10* (1085).

su([mi-1])-14: suppressor-14 of [mi-1]

IV. Linked to *arg-2* (14%) (1084, 1085).

Restores normal growth rate and alleviates the deficient cyanide-sensitive respiration of [*mi-1*] ([*poky*]) and all other group I cytoplasmic mutants. Has no effect on group II cytoplasmic mutant [*mi-3*] (168, 400, 1085). Cytochrome spectrum (400). Called *sup-14* (1085).

su([mi-1])-f: suppressor-f of [mi-1]

VR. Left of *inl* (10%) (1915).

Restores normal growth rate. Alleviates the deficient cyanide-sensitive respiration of [*mi-1*] ([*poky*]) (1370) and all other group I mitochondrial mutants, but not of group II or III mutants (171, 739) (Appendix 4). *su([mi-1])-f* differs from other known [*mi-1*] suppressors in not restoring SHAM insensitivity (1085). Cytochrome spectrum (168). Used for studying oscillations in membrane potential (763). Called *f*.

su([mi-3])-1: suppressor-1 of [mi-3]

IR. Between *nit-1* (17%), *un-7* (≤1%) and *al-2* (1%; 4%) (739).

Restores normal growth rate and cytochrome spectrum to [*mi-3*], a group II mitochondrial mutant (172, 739). Does not suppress the mutant phenotype of group I or III mitochondrial mutants (171).

su(mtr)-1: suppressor-1 of mtr

IR. Right of *his-2* (2%) (1991).

Selected in *trp-1; mtr* by increased uptake of tryptophan (1991) or as a *his*⁺ revertant of *his-2; mtr* on excess arginine (228). Still resistant to 4-methyltryptophan, but now sensitive to *p*-fluorophenylalanine (1991). The effect of *su(mtr)-1* on *mtr* is locus-specific. *su(mtr)-1* appears to have a changed regulation for amino acid transport system II (1520). May be allelic with *lys*ᴿ (1082).

su(mtr²⁶): suppressor of mtr²⁶

VIR. Linked to *pan-2*, *trp-2*. Right of *Cen-VI* (227).

Allele-specific partial suppressor of the poor growth of *trp-1; mtr²⁶* on low levels of tryptophan. [*mtr²⁶* is a putative frameshift (227).]

su(pan-2^{Y153M66}): suppressor of pan-2^{Y153M66}

Unmapped.

Allele-specific. Not effective on three other *pan-2* alleles or on four supersuppressible alleles at *trp-3* and *am-1* (357).

su([poky]): suppressor of [poky]

See *su([mi-1])*.

su(pro-3): suppressor of pro-3

Allelic with *arg-6*.

su(suc)-1: suppressor-1 of suc

Unmapped. Unlinked to *suc* (1119).

Isolated as a revertant of *suc* KG163, which is associated with a putative inversion in IL. Grows on acetate-free minimal medium, but the growth is enhanced by acetate supplementation. Suppresses other *suc* mutants but not mutations at six *ace* loci (63). Growth is yeastlike in acetate-free, vigorously agitated, liquid culture at 37°C. Called R2, 163R2 (63, 1119).

su(suc)-2: suppressor-2 of suc

I. Linked to *suc* (3%) (1119).

Isolated as a revertant of *suc(KG163)*, which is associated with a putative inversion in IL. Grows on acetate-free minimal medium, but the growth is enhanced by acetate supplementation. Forms spheroplasts and short chains of yeastlike cells when grown at 34°C in acetate-free, vigorously agitated, liquid culture. Called 163R4 (63, 1119).

su(trp-3): suppressor of trp-3

Early work with *trp-3* suppressors is reviewed in ref. (2264). *trp-3* mutants were originally called *td*, and alleles were designated td2, td3, etc. Most suppressors of *trp-3* were assigned numbers corresponding to those of the allele in which they originally were discovered. With td2, the td number also was retained as a suppressor locus number. Thus, the first suppressor discovered in td2 is symbolized *su(trp-3^{td2})-2*. The locus number 2 does not imply that there was a previously discovered suppressor of the td2 mutation. A second suppressor gene discovered in td2 was numbered 2a and is now symbolized *su(trp-3^{td2})-2a*. In contrast, when different suppressors of the td201 mutation were found, they were treated conventionally and given consecutive locus numbers beginning with 1.

su(trp-3^{td2})-2 : suppressor-2 of trp-3^{td2}

III. Linked to *leu-1* (22%) (1126).

Allele-specific. Suppresses allele td2 but does not suppress td6, td71, or any other allele from td1 through td34 (1125, 2266). This and the other listed suppressors of td2 were not tested on later *trp-3* isolates that fall in the same complementation group as td2. Although suppressed mutants grow on minimal medium, growth is stimulated by the addition of tryptophan. Suppressed *trp-3; su* colonies are morphologically distinguishable from wild type. Tryptophan synthetase is formed at levels below wild type. Originated in *trp-3^{td2}* (originally numbered S1952) (2262, 2266). Called *su-2* (729), *su₂* (2266).

su(trp-3^{td2})-2a: suppressor-2a of trp-3^{td2}

I. Linked to *al-2* (15%) (1126).

Isolated in ref. (2263) as one of four suppressors of *trp-3^{td2}*, called *su₂ₐ*, *su₂ᵦ*, and *su₂c*, *su₂d*. Of the four, only *su₂ₐ* was mapped. All are nonallelic with *su₂* and with each other, with the possible exception of *su₂ᵦ* and *su₂c* (2263). None was tested for allelism with *su(trp-3^{td6})*, which also suppresses *trp-3^{td2}*. The *a* that follows the locus number does not refer to mating type. Called *su-2ₐ* (1126).

su(trp-3^{td3}): suppressor of trp-3^{td3}

Unmapped.

Allele-specific. Suppresses *trp-3* alleles td3, td24, and td71, but not any other mutants from td1 through td34 (1125, 2266). Originated in *trp-3^{td3}*. Probably allelic with *su(trp-3)^{td24}*. Called *su₃* (2266).

su(trp-3^{td6}) : suppressor of trp-3^{td6}

Unmapped. At a different locus from *su(trp-3^{td2})-2* and *su(trp-3^{td3})*.

Allele-specific. Suppresses *trp-3* alleles td6 and td2 but does not suppress any other allele from td1 through td34 (1125, 2266). Although suppressed mutants grow on minimal medium, growth is stimulated by the addition of tryptophan and is morphologically distinguishable from wild type. Tryptophan synthetase is formed at levels below wild type. Originated in *trp-3^{td6}*. Called *su₆* (2266).

su(trp-3^{td201})-1: suppressor-1 of trp-3^{td201}

VIIR. Between *met-7* (18%), *ssu-1* (10%; 13%) and *arg-10* (7%) (1859, 2284).

Allele-specific. Suppresses missense allele td201, but not other alleles: tdl, td6, td7, td16, td37R, td71, td138R, td141 (2284), td2, td3, or td24 (1974). A suppressed mutant has a low level of tryptophan synthetase activity, allowing slow growth on minimal medium. Enzyme activity is due to a protein physically like the wild-type enzyme (2284, 2285). Suppressor alone, without td201, grows slightly slower than wild type (1974). The suppressor is effective when in another nucleus from the suppressible td201 in a forced heterokaryon between noncomplementing alleles (*trp-3^{td16}*; *su+trp-3^{td201}*; *su⁺*) (1974). Called *Su-1* or *Su-1_{td201}* (1859), *su-YS* (1974), *su₁(201)* (2283).

su(trp-3^{td201})-2: suppressor-2 of trp-3^{td201}

VII. Linked to *su(trp-3^{td201})-1*, probably to the left (1974).

Allele-specific. Suppresses missense allele td201, but does not suppress alleles td1, td2, td3, td16, td24, or td71 (1974). A suppressed mutant has a low level of tryptophan synthetase activity, allowing slow growth on minimal medium. Enzyme activity is due to a protein physically distinguishable from wild-type tryptophan synthetase, unlike suppressors 1 and 3 (1658). Suppressor alone, without td201, grows slightly more slowly than wild type (1974). The suppressor is effective when in another nucleus from td201 in a forced heterokaryon between noncomplementing alleles (*trp-3^{16}*; *su + trp-3^{td201}*; *su⁺*) (1974). Called *su-R* (1974), *su-2* (1658).

su(trp-3^{td201})-3: suppressor-3 of trp-3^{td201}

Unmapped. Not linked to *su(trp-3^{td201})-1* or other VII markers.

Allele-specific. Suppresses missense allele td201, but not alleles td1, td6, td71, or td141. The one *su(trp-3^{td201})-3* allele tested gave less powerful suppression of td201 than did the known alleles of *su(trp-3^{td201})-1*. The suppressed mutant has enough tryptophan synthetase activity to allow slow growth on minimal medium. The restored enzyme activity is due to a protein physically like wild type. Called *su₃*, *su₃(201)* (2283).

su(trp-5): suppressor of trp-5

VIL. Closely linked to (or allelic with) *aro-6* (1055).

Possibly a feedback-negative mutant of *aro-6*, which specifies DAHP synthase (Tyr) (1055).

su(ure-1^9): suppressor of ure-1^9

Unmapped. Not linked to *ure-1* or *ure-2*.

Suppressor of *ure-1* allele 9. Does not suppress *ure-2^{47}*. Only *ure-1^9* strains from a certain lineage are suppressed, suggesting that suppression requires a cosuppressor closely linked to *ure-1^9* (255).

suc: succinate

IL. Between *acr-3* (2%) and *In(H4250)^L*, *phe-1* (1%) (1122, 1592)

Uses acetate, succinate, or any of numerous related compounds (1187). Most, but not all, strains grow better on acetate than on succinate (1119). Lacks pyruvate carboxylase activity (EC 4.1.1.31) (140, 1122, 2009). Numerous alleles are CO_2-remediable (230). Leaky, but less so on higher ammonium (2008). A special medium has been devised for selective plating [K. Munkres, unpublished; see p. 248 in ref. (1592)]. Most ascospores are poorly pigmented in homozygous *suc* × *suc* crosses. Allele KG163 shows greatly reduced recombination in the region between *leu-4* and *suc*, suggesting inversion heterozygosity (1119). *suc* would be named *ace-6* except for priority of nomenclature (1122).

Sulfur Metabolism

Reviewed in refs. (1283) and (1287). See Figs. 24 and 45.

sup: suppressor

Symbol changed to *su*.

sw: snow white

Changed to *per-1*.

T, T

The single letter italicized has been used as a gene symbol (*T*: tyrosinase, *t*: thermophobic). Translocations are symbolized by an italicized *T* followed by identifying information in parentheses. Symbols for some transposable elements also begin with *T* (e.g., *Tad*, *Tcen*, *Tgl*); unlike the symbols for translocations, those for transposons contain no parentheses. A nonitalicized T has been used as a prefix for allele numbers of mutants isolated at the Universities of Tokyo and Texas.

T: Tyrosinase

IR. Between *ad-3A* (18%), *nic-2* (12%) and *al-2*, *al-1* (12%) (351, 897).

Cloned and sequenced: Swissprot TYRO_NEUCR, EMBL/GenBank M32843, M33271, PIR A00511, A34460, YRNC, EMBL NCTYRA, NCTYRB, GenBank NEUTYRA, NEUTYRB; pSV50 clones 30:2D, 32:9A.

Tyrosinase structural gene (EC 1.14.18.1) (Fig. 13). Prototrophic. Multiple alleles from different original strains were distinguished electrophoretically and by thermolability (896, 897, 899–901, 1748). The primary structure (407 amino acids) was determined by amino acid sequencing the product of allele T^L (1163). The polypeptide contains a cysteine and a histidine linked by a thioether bridge (1165, 1167, 1169). Precursor contains a carboxy-terminal extension (1115). Two forms of enzyme were demonstrated in a heterokaryon (897). Scored by color reaction with DL-DOPA as substrate (899, 901). For regulation, see refs. (623), (898), and (899). Tyrosinase synthesis is induced by a variety of compounds, including cycloheximide and amino acid analogs. Sulfhydryl compounds can also affect tyrosinase expression (1646). The addition of copper before cycloheximide induction increases the yield of enzyme (936). Extensive biochemical and biophysical analyses of the enzyme (149, 150, 152, 1166, 1170, 1613, 1749, 2103, 2244); enzyme crystallized (692). Used as a reporter gene (1090, 1114). Tyrosinase regulatory genes are symbolized *ty*.

t: thermophobic

Probably a *scot* allele. See *scot*.

T(. . .): Translocation

A rearrangement in which segments of two or more chromosomes are interchanged or a segment has been moved from one chromosome to another. Symbolized in full by *T* followed by the involved linkage group or groups in parentheses and an identification number, all in italics. For example, *T(IR;IVR)NM172*, *T(IL → IIR)39311*. In the present text and in maps, the symbol is abbreviated with the linkage groups omitted and the identification number in parentheses. For example, *T(NM172)*, *T(39311)*. Major types of translocations are *reciprocal*, in which two terminal segments are interchanged; *insertional*, in which a segment is removed from one chromosome and inserted in another and *quasiterminal*, in which a distal segment is translocated to the tip region of another chromosome, distal to any essential gene. Insertional and quasiterminal rearrangements are useful for determining dominance and for mapping by duplication coverage (1566). See ref. (1578) for information on individual translocations.

T(OY329): Translocation (VIR → IIIR)OY329

Probably a quasiterminal translocation. [Misdiagnosed as insertional in ref. (1578)]. A segment of VIR that includes *trp-2*, *ws-1*, and *col-18* (but not *del*) apparently is translocated to *Tip-IIIR*.

ta: tufted aerial

IL. Between *un-l6* (2%) and *acr-3* (<3%) (1582, 1592).

Growth is colonial and rapid-spreading. Morphology varies with growth conditions (1592). Conidiation is best at 34°C. There may be a maternal effect, with *ta*⁺ resembling *ta* in initial cultures from ascospores (1476).

Tad: Transposable element Adiopodoume

Cloned and sequenced: GenBank/EMBL L26562, L26563, GenBank NEUGLPRCA, NEUGLPRCB.

A LINE-like retrotransposon present as an active element in a strain from the Ivory Coast and discovered when the *am* gene was inactivated by insertion of *Tad* (1065). Active *Tad* elements can be introduced into laboratory strains by crosses, heterokaryons, or transformation. In heterokaryons, an active *Tad* is transferred readily to nuclei that previously lacked it (1932); if *Tad* in the donor nucleus contained an intron, the intron is spliced out prior to transfer (1061). When a strain that previously lacked an active *Tad* acquires it, the element remains active during vegetative growth and propagation. It is inactivated by RIP when it goes through a cross, however. Laboratory strains and strains representing seven heterothallic and homothallic *Neurospora* species contain inactive *Tad*-like sequences, and these appear to have undergone RIP (1064).

tba-1: tubulin alpha-1

Unmapped.

Cloned and sequenced: Swissprot TBA1_NEUCR, EMBL/GenBank X79403, GenBank NCATUBA, PIR S45050.

Encodes the α-a chain of tubulin. Developmentally regulated; RNase protection indicates that the mRNA is not present during the first 30 min of conidial germination (1375).

tba-2: tubulin alpha-2

Unmapped.

Cloned and sequenced: Swissprot TBA2_NEUCR, EMBL/GenBank X79404, PIR S45051, GenBank NCATUBB; EST NP6G3.

Encodes the α-b chain of tubulin. Transcript is present throughout the vegetative cycle. Shows only 65% amino acid sequence homology with the product of *tba-1* (1375).

tbg: tubulin gamma

Unmapped.

Cloned and sequenced: Swissprot TBG_NEUCR, EMBL/GenBank X97753, GenBank NCGTUB.

Encodes γ-tubulin (864).

Tcen: Centromeric transposable element

VII. In *Cen-VII* and other centromere regions (275).

Sequenced (275).

A full-length *Copia*-like transposable element found in *Cen-VII*, where it is clustered with degenerate fragments of *Tgl1*, *Tg12*, and *Tad* (275).

td: tryptophan desmolase

Changed to *trp-3*, which encodes tryptophan synthetase (the renamed enzyme).

tef-1: translation elongation factor-1

IL. Linked to *leu-4*, *mat*, *arg-3* (0/18) (1447).

Cloned and sequenced: Swissprot EF1A_NEUCR, EMBL/GenBank D45837, GenBank NEUEF1A; EST NM6E1;

Orbach-Sachs clones X9B04, X9E03, X25C03, X25G03.

Structural gene for translation elongation factor-1α. The mutant isolated by RIP has a slow-growth phenotype (944). Cycloheximide inducible. Expression at the level of mRNA appears constitutive in the *lah-1* single mutant, but not in a *lah-1;cpc-1* double mutant (833).

Tel: Telomere

Neurospora telomeres consist of tandem repeats of TTAGGG (identical to the telomere sequence in humans). The prototype in VR was identified by walking from the distal marker *his-6* (1809). The oligonucleotide (TTAGGG)4 was then used as a hybridization probe to show that the repeated sequence is found only at chromosome ends and to identify restriction fragments containing telomeres and subtelomeric regions from other chromosome tips (1810).

Tel-IR: Telomere-IR

IR. Linked to *arg-13* (6T, 2NPD/18 asci) (1447).

Cloned (1447).

Telomere-derived restriction fragment M7 (1810).

Tel-IIL: Telomere-IIL

IIL. Linked to *hsp70* (2T/18 asci), *Fsr-52*, *vph-1* (14T + 1 double crossover/18 asci) (1447, 1810).

Telomere-derived restriction fragments 014, M5 (1810).

Tel-IIR: Telomere-IIR

IIR. Right of *leu-6* (1T, 1NPD/16 asci) (1447, 1810).

Telomere-derived restriction fragment 05 (1810).

Tel-IIIR: Telomere-IIIR

IIIR. Linked to *cat-3* (5T/18 asci), *Fsr-45* (3T, 1NPD/18 asci). *Fsr-45* in turn is linked to *ro-2* (3/18), *trp-1* (6/18) (1447, 1810).

Telomere-derived restriction fragments 07, M13 (1810).

Tel-IVL: Telomere-IVL

IVL. Linked to *T(AR33)* (1/20) (1334, 1811). Left of *sdv-1*, *-2*, *-3* (12T, 0NPD/18 asci) (1447).

Telomere-derived restriction fragment 010 (1810).

Tel-IVR: Telomere-IVR

IVR. Linked to *nuo21* (1T/18 asci) (13T, 2NPD/18 asci with *trp-4*) (1447).

Telomere-derived restriction fragment M3 (1810).

Tel-VR: Telomere-VR

VR. Linked to *his-6* (30 kb) (1447, 1809–1811).

Cloned and sequenced: EMBL/GenBank M37064, Gen-Bank NEUTELVRA.

Telomere-derived restriction fragment M14 (1810). Cloned by walking from *his-6* (1809). The first *Neurospora* telomere to be cloned.

Tel-VIL: Telomere-VIL

VIL. Linked to *cax* (3T/18 asci) (1447). Assignment of *Tel-VIL* (and *cax*) to VIL depends on the probable linkage of *Tel-VIL* to *tf2d* (7T, 1NPD/16 asci) (1447). The *tf2d* sequence is included in the duplicated segment generated in crosses of *T(VIL → IR)T39M777 × Normal* (1322).

Telomere-derived restriction fragments 011, M10 (1322, 1810).

Tel-VIR: Telomere-VIR

VIR. Linked to the terminal VIR breakpoint of *T(UK-T12)* (0T/9 asci; 0/13) (64) and to *nuo21.3* (0T/18 asci), *con-11a* (9T/18 asci) (1447).

Telomere-derived restriction fragment M11 (1810).

Tel-VIIL: Telomere-VIIL

VIIL. Linked to *nic-3*, *pho-4* (10T, 1NPD/18 asci) (1447).

Telomere-derived restriction fragments 09, M9 (1810).

tet: tetrazolium

IL. Linked to *acr-3* (2%), *ad-3B* (1%). Right of *mat* (7% or 8%) (739).

Detected as a difference between *A* and *a* laboratory strains in the ability to reduce tetrazolium dye. Wild-type 74-OR23-lA colonies fail to reduce dye and thus remain white, whereas 74-OR8-la colonies reduce dye to red (739). Called *Tet-R*, *Tet-W* ("red," "white"). See ref. (739) for tests on other wild types. The same gene is perhaps responsible for a *mat*-linked difference in resistance to 2,3,5-triphenyltetrazolium chloride (TTC), although linkages appear different. [TTC resistance maps left of *mei-3*, probably between *mat* and *arg-1* (707).] TTC-resistant strains include 74-OR23-lA, Em A FGSC No. 69l, Em A 5256, Lindegren lA, Lindegren 25a, and all RL wild types tested. TTC-sensitive strains include 74-OR8-la, Em a FGSC No. 692, and Em a 5297 (2180, 2181). Strain ORSa, derived by backcrossing to 74-OR23-1A, is resistant (1422). *cya*, *cyb*, and *cyt* mutants all fail to reduce TTC, and this is used as a test in the initial identification of such mutants. *cya-l* is located near *mat* (170); its relationship to *tet* is not known.

tf2d: transcription factor IID

VIL. Left of *T(T39M777)* (included in duplications from *T × Normal*) (1322). Probably linked to *cax* (6T, 1NPD/18 asci), *Tel-VIL* (7T, 1NPD/18 asci) (1447).

Specifies TATA box binding transcription factor TfIID (2134).

Tgl1: Gypsy-like-1 (Transposable Element)

Cloned and sequenced (275).

A degenerate *Gypsy*-like retrotransposon. Found in *Cen-VII* as a component of nested clusters containing this and other AT-rich immobile transposable elements with earmarks of having been subjected to RIP. Interrupted by different defective fragments of *Tad* (275).

Tgl2: Gypsy-like-2 (Transposable Element)

Cloned and sequenced (275).

A degenerate retrotransposon identified as *Gypsy*-like by sequence similarity to *Ty3-2* of *Saccharomyces*. Found in *Cen-VII* as a component of nested clusters containing this and other AT-rich immobile transposable elements with earmarks of having been subjected to RIP (275).

thi: thiamine

Thiamine biosynthesis in fungi involves the condensation of pyruvate and glyceraldehyde-3P to form the thiazole ring and then coupling of this thiazole with (hydroxymethyl)pyrimidine diphosphate by a single-gene-encoded thiamine biosynthetic bifunctional enzyme containing activities for thiamine phosphate pyrophosphorylase (EC 2.5.1.3) (TMP-PPase) and (hydroxyethyl)thiazole kinase (EC 2.7.1.50) [4-methyl-5-(β-hydroxyethyl)thiazole kinase] (THZ kinase), the last also being responsible for a salvage synthetic pathway (2062).

thi-1: thiamine-1

IR. Between *cys-9* (13%), *T(4540)^R* and *T(NM103)*, *met-6* (7%; 14%) (1414, 1578, 1592, 2123).

Uses thiamine or its precursors, pyrimidine + thiazole (2060). Adaptation to growth on minimal medium occurs after a lag. Growth tests should therefore be scored early. Adaptation is not carried over via ascospores, conidia, or small mycelial fragments. Adaptive growth is paralleled by the attainment of wild-type thiamine pyrophosphate and carboxylase levels. The function apparently concerns utilization of intact thiamine rather than its biosynthesis (584, 585). *T(17084)* is inseparable from *thi-1* (1578).

thi-2: thiamine-2

IIIR. Between his-7 (1% or 2%) and ad-2 (1%; 3%) (426, 1587).

Requires thiamine. Cannot use pyrimidine + thiazole (2060); therefore, blocked beyond the early condensation step. Does not undergo growth adaptation on minimal medium (584).

thi-3: thiamine-3

VIIL. Between nic-3 (9%; 18%), lacc (7%) and T(T54M50), ace-8 (1%) (1120, 1578, 1580, 1585, 1592, 2296). Linked to csp-2 (<1%) (1582).

Uses thiamine or thiazole (2060). Does not undergo growth adaptation on minimal medium. Growth on minimal medium is leaky at first but becomes tight with the exhaustion of endogenous thiazole (584), so that scoring is best done late.

thi-4: thiamine-4

IIIR. Between Cen-III (1T/18 asci) and ser-1, pro-1 (3%; 1T/18 asci), ace-2 (4%; 9%). Linked to acr-2, spg, sc (<1%) (449, 1447, 1585).

Cloned and sequenced: EMBL/GenBank D45894; Orbach-Sachs clone X14F01

Requires thiamine (583). Believed to be involved in the condensation of pyrimidine and thiazole precursors (28). Analog of Fusarium stress-inducible thiazole biosynthetic enzyme. Very leaky (583). A probable allele called thi-lo greatly increases the thiamine requirement of thi-1 and decreases the ability to synthesize thiamine from pyrimidine and thiazole. thi-lo has no detectable nutritional requirement in the absence of thi-1. No recombination, thi-lo × thi-4 (0/55 scorable progeny) (583).

thi-5: thiamine-5

IVR. Linked to pan-1 (1%) (1585).

Uses thiamine (907). Probably can also use 2-methyl-4-amino-5-(aminomethyl)pyrimidine (1557).

thi-6: thiamine-6

VIIR. Immediately adjacent to ras-3 (300 bp), near un-10 (1267).

Cloned and sequenced (1267).

Homologous to genes, in yeast and elsewhere, that specify the thiazole precursor of thiamine. The mutant phenotype has not been determined (1267).

thi-lo: thiamine-low

A thi-4 allele.

thr-1: threonine-1

Changed to ile-1. Because this locus specifies threonine dehydratase, the original name is inappropriate (1056, 1059) (Fig. 45). Called thre-1 (463).

thr-2: threonine-2

IIL. Between ro-3 (6%; 25%), T(NM149) and T(AR179), bal, arg-5 (3%; 18%). Adjoins thr-3 (<0.1%) (1580, 1582, 1585).

Lacks threonine synthetase (EC 4.2.99.2) (654). Requires threonine. Cannot use other amino acids (2064) (Fig. 45). Strongly inhibited by methionine (606, 2064). Known mutants are not heat-sensitive, unlike all known thr-3 alleles.

thr-3: threonine-3

IIL. Linked to thr-2 (<0.1%) (1585).

Requires threonine. Also responds slightly to α-aminobutyric acid or isoleucine. Known alleles are heat-sensitive (25 vs 34°C), tight at 34°C. Not inhibited by methionine (606, 2064).

ti: tiny

IL. Between arg-3 (1%) and T(39311)^R (1582).

Heat-sensitive. Spreading colonial morphology at 25°C or below, more restricted at 30°C, and no growth at 34°C (1548, 1582). Can be scored microscopically after ascospore germination. Cell-wall peptides are reduced in amount (2249).

tim: translocase of inner mitochondrial membrane

Translocase complex of the inner mitochondrial membrane. Reviewed in ref. (1611) (Fig. 64). Genes that specify components include tim8, tim17, and hsp70 (1194). See also tom.

tim8: translocase of inner mitochondrial membrane 8

Unmapped.

Cloned and sequenced: EMBL/GenBank AF142423.

Encodes a small zinc finger protein involved in mitochondrial carrier import (128)

tim17: translocase of inner mitochondrial membrane 17

VI. Mapped by RFLP (2046).

Cloned and sequenced (2046).

Specifies the 17-kDa protein subunit of TIM (2046). See Fig. 64.

Tip: linkage group tip

Eight of the 14 tips have been mapped genetically by determining the breakpoints of quasiterminal chromosome rearrangements. A breakpoint is known to be genetically terminal when a class of meiotic segregants is recovered that would necessarily be deficient for any essential gene distal to the breakpoint. Although quasiterminal rearrangements behave genetically as terminal, they are thought to be reciprocal translocations in which only the telomere is translocated. See refs. (1566) and (1578). VL was the first tip to be mapped, using the heteromorphic nucleolus satellite as a terminal cytological marker. Of the six tips that are not involved in quasiterminal rearrangements, all but IIIL and VIIR have been mapped by recombination using RFLPs in cloned telomere regions. See *Tel*.

Tip-IL: Left tip of linkage group I

IL. Marked by quasiterminal translocation *T(5936)*, which is linked to *ro-10* (0/38), *fr* (11%) (1580, 1582).

Tip-IR: Right tip of linkage group I

IR. Marked by quasiterminal rearrangements *In(NM176)*, *In(H4250)*, and *T(T39M777)*, which are closely linked to *un-18*, *het-5*, and *R* (279, 978, 1578, 2129).
Tel(IR) is linked to *arg-13* (1447).

Tip-IIL: Left tip of linkage group II

IIL. Left of *T(AR30)*, which is left of *pi* (28%) (1173).
Tel-IIL is linked to *hsp70* and other RFLP markers in IIL (1447, 1810).

Tip-IIR: Right tip of linkage group II

IIR. Right of *un-15*, *rip-1*.
Tel-IIR is linked to *leu-6* and other RFLP markers in IIR (1447, 1810).

Tip-IIIL: Left tip of linkage group III

IIIL. Left of *r(Sk-2)-1*, *cum*, *acr-7*.

Tip-IIIR: Right tip of linkage group III

IIIR. Marked by quasiterminal translocations *T(DBL9)*, *T(OY320)* (1578), which are right of *dow* (3/30; 2/18), *erg-3*.
Tel-IIIR is linked to IIIR RFLP markers (1447, 1810).

Tip-IVL: Left tip of linkage group IV

IVL. Marked by quasiterminal translocation *T(AR33)*, which is linked to *cys-10* (0/221), left of *acon-3* (<1%), *uvs-3* (1447, 1578, 1811).

Cloned (1810).
Tel-IVL is linked to IVL RFLP markers (1447, 1810).

Tip-IVR: Right tip of linkage group IV

IVR. Marked by quasiterminal translocations *T(AR209)*, *T(T54M50)*, *T(OY337)*, and *T(ALS179)*, which are right of *pmb*, *uvs-2* (2–6%), *cys-4* (1578).
Tel-IVR is linked to IVR RFLP markers (1447, 1810).

Tip-VL: Left tip of linkage group V

VL. Marked by *sat*, *NOR* and by quasiterminal rearrangements *T(NM169d)*, *T(ALS176)*, *T(ALS182)*, *T(OY321)*, *T(AR190)*, and *In(UK-2y)^L*, which are left of *dgr-1*, *T(AR30)*, *caf-1* (>11%), and *lys-1* (20%; 35%) (118, 1578, 1601).

Tip-VR: Right tip of linkage group V

VR. Marked by quasiterminal translocation *T(NM149)* (1578, 1809).
Cloned and sequenced: GenBank M37064.
Tel-VR is linked to *his-6*. RFLP analysis (1810). A copy of the retrotransposon *Pogo* adjoins *Tel-VR* in strains of OR background (1447, 1809, 1811).

Tip-VIL: Left tip of linkage group VI

VIL. Left of *chol-2*.
Tel-VIL is linked to RFLP markers in VIL (1447, 1810).

Tip-VIR: Right tip of linkage group VI

VIR. Marked by quasiterminal translocations *T(UK-T12)*, *T(NM103)*, and *T(ALS159)*, which are right of *ws-1*, *col-18*, *un-23*, and *trp-2* (13%) (1578, 1582).
Tel-VIR is linked to RFLP markers in VIR (64, 1447, 1809).

Tip-VIIL: Left tip of linkage group VII

VIIL. Left of *cya-8* and *T(ALS179)*.
Tel-VIIL is linked to RFLP markers in VIIL (1447, 1810).

Tip-VIIR: Right tip of linkage group VII

VIIR. Right of *sk*.

tk: thymidine kinase

Introduced from herpes simplex virus.
Fungi are deficient in the synthesis of thymidine kinase. The absence of endogenous thymidine kinase activity in wild-type *Neurospora* makes *tk* a readily detectable reporter gene. Expression of the viral thymidine kinase gene makes it possible to label DNA efficiently with radio-

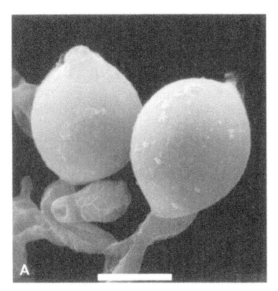

FIGURE 63 Giant conidia of the mutant *tng: tangerine*. A large conidium in the photograph to the right has partially engulfed a normal size conidium in the same conidial chain. Bar length = 10 μm. Scanning EM photographs from M. L. Springer. The photograph on the left is from ref. (1958), used here with permission from the Cold Spring Harbor Laboratory Press.

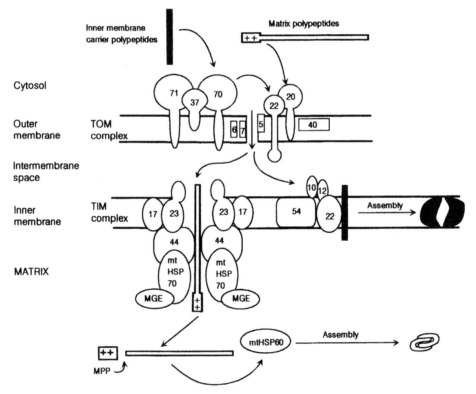

FIGURE 64 Import pathways of mitochondrial polypeptides synthesized in the cytosol. The TOM complex consists of surface-exposed components that function as preprotein receptors and of membrane-embedded components that are involved in passing the preproteins across the mitochondrial outer membrane. The TIM complexes deal with polypeptides having different final destinations. The peripheral protein TIM44 provides a docking site for HSP70 at the inner membrane. The products of the nuclear genes *hsp60* and *hsp70* are called mtHSP60 and mtHSP70 in the figure. MPP: mitochondrial processing peptidase. From ref. (461), [Davis, R.H. (2000). "Neurospora: Contributions of a Model Organism."], where it was redrawn from a figure provided by Frank Nargang. Reproduced with permission from Oxford University Press.

active thymidine and with fluorodeoxyuridine (1772). Strains expressing TK become sensitive to nucleoside drugs that are activated by TK (1772). Strains heterokaryotic for *tk* can therefore be resolved by using fluorodeoxyuridine to select conidia that contain no *tk*-bearing nuclei (1338).

Tn5: Transposon 5

Introduced from *Escherichia coli*.
Used for insertional mutagenesis (1269); see also ref. (823).

tng: tangerine

IIL. Between *pyr-4* (16%), *T(NM149)* and *arg-5* (6%; 14%). Probably left of *thr-2* (2%) (1582).
Irregular spreading growth. Hyphae are not curled (unlike *ro*) and are not as densely branched as those of *col-4*. Female infertile. Giant conidia are formed (1582) (Fig. 63), which may have many nuclei (1673). Some proconidia continue to grow beyond the normal 5- to 10-μm diameter of wild type to 100 μm or more, engulfing neighboring conidia and hyphae (1958). Microcapillary electrodes can be inserted in the spherical giant conidia and used to determine membrane potentials and resistances (187).

tnr: tetrahydroxynaphthalene reductase

VIIL. Between *nic-3* (3T/18 asci) and *ars-1* (2T/18 asci), *Cen-VII* (3T/18 asci) (1452).
Cloned and partially sequenced: GenBank AA902051, AA902055 EST NM6H10 (549).
Encodes putative tetrahydroxynaphthalene (T4HN) reductase (EC 1.1.1.252) (549).

tns: tenuous

VIIR. Between *T(5936)* and *arg-10* (1589).
Growth is thin and severely retarded. Recovered as a putative RIP allele in progeny from *Dp(VIIR→IL) 5936 × Normal* (1589).

tol: tolerant

IVR. Between *his-5* and *trp-4* (1%) (1466, 1911).
Cloned and sequenced: EMBL/GenBank AF085183; Orbach-Sachs clone X25D7.
A key regulatory gene that controls mating-type-mediated vegetative incompatibility. The *tol* gene does not affect sexual compatibility. *mat A + mat a* heterokaryons are vegetatively incompatible in standard wild-type Neurospora crassa. If the normal N. crassa allele, designated here *tol*^{C}, is replaced by a recessive allele, *tol*, heterokaryons of constitution *tol mat A + tol mat a* are fully compatible and stable, provided that other *het* loci are homokaryotic. Growth of *mat A/mat a* partial diploids is

inhibited and highly abnormal when the dominant normal allele *tol*^{C} is present, but *mat A/mat a* duplications grow normally when *tol*^{C} is replaced by a recessive *tol* allele (1466). Vegetative incompatibility mediated by *het* genes other than mating type is not affected by the presence or absence of *tol*^{C} (1175, 1466, 1560). Mutation or deletion of *tol*^{C} restores normal growth rate and morphology to slow-growing, unstable, mixed-mating-type (*tol*^{C} *mat a + tol mat A*) heterokaryons (517). A recessive *tol* allele is present in some isolates from nature, and recessive alleles have arisen spontaneously at least twice in laboratory stocks (1466, 1582, 2277). Seven mutants selected as spontaneous escapes from inhibited growth of *mat A/mat a* partial diploids were all *tol* alleles (2146). Double mutant *tol trp-4* stocks are convenient because the closely linked *trp-4* tags the *tol* allele, which otherwise requires progeny tests for scoring. *tol* has been used as an alternative to *mat a*^{m33} for maintaining stable *mat A + mat a* heterokaryons and allowing a disadvantaged component to be used as parent in a cross (1456). Homozygous *tol* may partially restore fertility to crosses involving *fmf-1*, which otherwise produce no fertile perithecia (1003). Mutations that inactivate *tol*^{C} were interpreted as recessive suppressors of the mating-type-mediated vegetative incompatibility. In Neurospora tetrasperma, which normally grows as a vegetatively compatible *mat A + mat a* heterokaryon, substitution of the N. crassa *tol* allele (*tol*^{C}) for the N. tetrasperma allele (*tol*^{T}) makes the N. tetrasperma mating types vegetatively incompatible, whereas the reverse substitution of *tol*^{T} for *tol*^{C} in N. crassa makes the N. crassa mating types vegetatively compatible (977). The normal allele in N. tetrasperma, *tol*^{T}, resembles the N. crassa recessive mutant *tol*. In both species, the vegetative incompatibility mediated by *mat A* and *mat a* thus depends on *tol* function, and suppression is due to the absence of *tol* function.

tom: translocase of outer mitochondrial membrane

Translocase of the outer mitochondrial membrane (1612). Reviewed in ref. (1459). Involved in the translocation of nuclear-coded preproteins to the mitochondrial compartment. Subunits of the complex include TOM70, TOM40, TOM22, TOM20, TOM7, and TOM6 (Fig. 64). The 13.8-nm TOM complex particle has been purified and its structure and subunit stoichiometry analyzed (1112). Subunits were formerly were called MOM (1612).

tom6: translocase of outer mitochondrial membrane 6

Unmapped.
Cloned (1112).

Encodes the 6-kDa component of the translocase. Considered to be part of the translocation pore (1112, 1459); interacts with TOM40 (1700). Formerly known as *mom8* (29, 1945) (Fig. 64).

tom7: translocase of outer mitochondrial membrane 7

Unmapped.

Cloned (1112).

Encodes the 7-kDa component of the translocase. Considered to be part of the translocation pore (1112, 1459). Formerly known as *mom7* (1049, 1373, 1945) (Fig. 64).

tom20: translocase of outer mitochondrial membrane 20

IVR. Between *pyr-1* (2T/18 asci) and *trp-4* (4T/18 asci). Linked to *pho-5* (0T/18 asci) (1447).

Cloned and sequenced: Swissprot OM20_NEUCR, EMBL/ GenBank M80528, EMBL NCMITOM, GenBank NEUMITOM.

Specifies product originally identified as a 19-kDa outer mitochondrial component (Fig. 64). Antibodies directed against TOM20 block the import of many mitochondrially destined, nuclear-encoded proteins (1943), but it is not essential for ADP/ATP carrier protein or cytochrome *c* import (840). Required for TOM22 import (1039). Sheltered RIP showed that *tom20* inactivation results in very slow growth and multiple mitochondrial defects (839, 840). Homokaryons lacking *tom20* have extremely slow stop–start growth (839). Formerly known as *mom19*.

tom22: translocase of outer mitochondrial membrane 22

VIL. Linked to *gdh-1* (0T/16 asci), *nuo19.3* (9T/18 asci) (1447).

Cloned and sequenced: Swissprot OM22_NEUCR, EMBL/ GenBank X71021, GenBank NCMOM22, PIR A40669.

Specifies an essential 22-kDa component of the mitochondrial protein import complex (1439). TOM22 is anchored to the outer membrane through a single membrane-spanning domain (1049) (Fig. 64). Functions with TOM20 as the preprotein receptor (415, 1049, 1303, 1701). A positively charged internal region of TOM22 just prior to the membrane-spanning domain has a role in targeting the TOM22 polypeptide to the outer mitochondrial membrane (1735). Abundant negative charges in the cytosolic domain may not be important for making TOM22 competent for function in translocation (1440). TOM22 influences TOM40 oligomerization (1700). It is required for the import of most mitochondrially destined proteins, including the ADP/ATP carrier. Called *mom22*.

tom40: translocase of outer mitochondrial membrane 40

VR. Linked to *cyh-2* (0T/15 asci) (1447).

Cloned and sequenced: Swissprot OM40_NEUCR, EMBL/ GenBank X56883, GenBank NCMOM38, PIR S13418.

Encodes TOM40, a major component of the translocation pore (1040, 1050), which also consists of TOM6 and TOM7 (1049, 1945) (Fig. 64). TOM40 is closely associated with polypeptides as they are translocated across the mitochondrial outer membrane (1701, 1945). Sheltered RIP shows *tom40* to be essential (1436), as is *tom22* (1439). Called *mom38*.

tom70: translocase of outer mitochondrial membrane 70

VR. Right of *inl* (6 or 7T/17 asci) (1447).

Cloned and sequenced: Swissprot OM70_NEUCR, EMBL/ GenBank X53735, AF110494, PIR A36682, GenBank NCMOM72.

Identified as encoding an outer mitochondrial membrane protein required for the import of the ATP/ADP carrier (1944, 1998) (Fig. 64). Only a fraction of total mitochondrial TOM70 associates with the TOM complex (1112). Mutants obtained by RIP are viable with a slow-growth phenotype exacerbated at elevated temperatures, and they conidiate poorly; mitochondria lacking TOM70 import the ADP/ATP carrier poorly (761). Called *mom72*.

Tp: Transposition

The name *Transposition* is used for intrachromosomal rearrangements in which an interstitial segment that contains two or more contiguous genes is removed from its original position and inserted elsewhere in the same chromosome. In crosses of *Transposition × Normal*, crossing over may produce duplications and deficiencies for defined chromosome segments and may result in dicentric bridges and acentric fragments, depending on the orientation and position of the transposed segment relative to the centromere (1605). The symbol for transpositions resembles that for insertional translocations, but with *Tp* in place of *T*.

tr^u: transport of uracil

Allelic with *uc-5*.

Transport

Amino acid transport across the plasma membrane is controlled by genes in three systems: basic (*pmb*, =*bat*); neutral (*mtr*, =*pmn*), and general (*pmg*; see also *su(mtr)*-1) (Fig. 47). Genes *arg^R*, *lys^R*, *hlp-1*, and *hlp-2* were described as being involved in the transport of amino acids. Mutants that affect (or may affect) the transport of other metabolites or ions include *car, cys-13, cys-14, fpr, glt, ipm-1, ipm-2, mea-1, sit, sor, trk, tys, uc-5,* and *ud-1.* Mutants that appear to affect more than one transport system include *cpc, fpr-6, hgu-4, nap,* and *un-3;* see ref. (2230). General regulatory loci such as *nit-2, cys-3, nuc-1, nuc-2, preg,* and *pgov* control classes of related enzymes, including the relevant permeases (Figs. 24, 49, 50, and 51). Other genes concerned with uptake are *acu-16* (acetate), *pho-4, pho-5* (phosphate), and *puu* (putrescine). Older literature on the transport of various ions and compounds is reviewed in refs. (769) and (1799), amino acid transport in ref. (2231), and peptide transport in ref. (2232). Nuclear genes called *tom* and *tim* specify subunits of translocases associated with the outer and inner membranes of mitochondria, respectively (Fig. 64). The *por* gene mediates the transport of metabolites into the mitochondrion. Ornithine transport across the mitochondrial membranes is mediated by *arg-13* (Fig. 11). See *pma* for the transport of protons across the plasma membrane and *vma* for the transport of protons across the membranes of mitochondria and other intracellular components (Fig. 66).

Transposable elements

The retrotransposon *Tad,* which was discovered in a single wild strain, is the only known active transposable element in *Neurospora.* Inactive relics of numerous transposable elements are present in the genome, however (see *Dab-1, Guest, Pogo, Punt, Tad, Tcen, Tg11, Tg12*). Centromere regions are rich in copies of defective transposable elements (see *Cen-VII*). For reviews of transposons in filamentous fungi, see refs. (439) and (1044). The paucity of active elements in *Neurospora,* in contrast to their demonstrated abundance in ameiotic filamentous fungi (439), is attributed to the RIP process, which inactivates duplicated sequences in *Neurospora* by inducing G:C to A:T mutations prior to meiosis. Consistent with this hypothesis, many of the inactive elements bear earmarks of having undergone RIP.

tre-1: trehalase-1

IR. Between *met-6* (7%) and *ad-9* (10%) (2025, 2210, 2286).

Unable to use trehalose as the carbon source. Structural gene for acid trehalase (EC 3.2.1.28) (2210). The *Neurospora* enzyme is classified as an acid trehalase of a subclass that is insensitive to Ca^{2+} and Mn^{2+} and to inhibition by ATP [reviewed in ref. (1004)]. Mutations affect thermostability and the production of an inhibitor of the wild-type enzyme (2025, 2210). Also, the invertase level is reduced by 50% and amylase is

increased. The trehalase level is reduced to 10% of the wild type in (*tre+tre+*) heterokaryons and to 1% in *tre/ tre+* duplications. Used in studying trehalose metabolism (190, 524). For a possibly related regulatory mutation, see ref. (1323). Laboratory stocks of *Neurospora crassa* and wild isolates of *Neurospora intermedia* are polymorphic for trehalase electrophoretic mobility; one such variant was called *mig* (2025, 2210, 2286).

tre-2: trehalase-2

Unmapped.

Cloned and sequenced: EMBL/GenBank AF044218.

Encodes neutral trehalase (α,α-trehalose glucohydrolase) (EC 3.2.1.28) (524). Obtained by PCR.

trk: transport of potassium

IIIR. Linked to *leu-1* (0/92) (1926).

Cloned and sequenced: GenBank AJ009758.

Encodes a major potassium transporter, homologous to the TRK transporters of *Saccharomyces cerevisiae* (842). Requires high K^+ concentration for growth. Na^+ cannot substitute. Cation transport system maximum velocity is normal; K_m is 3 times normal. Recessive. Obtained by inositolless death enrichment on low potassium (1926).

tRNA^{LEU}: leucine tRNA

VIIL. Between *ars-1* and the *qa* cluster, adjoining *qa-1F* (938).

Cloned and sequenced: EMBL/GenBank X00736, GenBank NCTRNLEU. Included in VII contig GenBank AC006498.

Encodes leucine tRNA, with the AAG anticodon. The gene contains an intervening sequence (938).

tRNA^{PHE}: phenylalanine tRNA

Unmapped.

Cloned and sequenced: EMBL/GenBank X02710, J01251, GenBank NCTRNPHE, NEUTGFF3.

Encodes phenylalanine tRNA. Contains a 16-bp intron adjacent to the anticodon (1884).

tRNA Synthetase

Cloned genes that specify tRNA synthetases include *cyt-18, cyt-20, leu-5, leu-6,* and *un-3.* The activity of Trp tRNA synthetase is greatly reduced in the *trp-5* mutant.

trp: tryptophan

For the aromatic biosynthetic pathway, see Fig. 13. Tryptophan is required in much higher concentrations than its precursors, anthranilic acid and indole. A concentration of 0.01 mg/ml precursors is sufficient, but up to 0.1 or 0.2 mg/ml tryptophan is required by some mutants. High concentrations of anthranilic acid are toxic. Most *trp* mutants grow better on 0.2 mg/ml tryptophan plus 0.2 mg/ml phenylalanine than on tryptophan alone. Also called *tryp, try*.

Regulation: Tryptophan feedback inhibits anthranilate synthase, anthranilate-PP-ribose P-phosphoribosyl transferase, and one of three isozymes of DAHP synthase (the first step in aromatic biosynthesis). Tryptophan stimulates chorismate mutase, directing chorismate to prephenate rather than to anthranilate synthesis. All four genes of the tryptophan pathway are derepressed by starvation for tryptophan. High indoleglycerol phosphate levels also cause derepression. Derepression may involve inhibition of trp tRNA synthetase (87, 318, 483, 821, 1442, 1961). Genes for tryptophan biosynthesis are derepressed coordinately with those for histidine, arginine, and lysine biosynthesis; this is called "cross-pathway regulation" (292). Reviewed in refs. (1281) and (1768); see *cpc*.

trp-1: tryptophan-1

IIIR. Between *ad-2* (1%; 7%) and *ro-2* (2%; 12%) (18, 426, 1585). Linked to *fpr-3* (<1%) (1057).

Cloned and sequenced: Swissprot TRPG_NEUCR, EMBL/GenBank J01252, PIR A01130, NNNC2, EMBL NCP1, GenBank NEUTRP1.

Structural gene for three domains and activities: anthranilate synthetase component II (EC 4.1.3.27) (the glutamine amino transferase activity; *trp-2* specifies component I of this activity), phosphoribosylanthranilate (PRA) isomerase (EC 5.3.1.24), and indoleglycerolphosphate (InGP) synthetase (EC 4.1.1.48) (347, 484) (Fig. 13). Reviewed as an example of gene fusion (425). Uses tryptophan or indole (2061). Some alleles can also use anthranilate, but others cannot (11). Alleles differ in lacking one or more of the three activities. For example, *trp-1* allele (15) lacks all three activities; *trp-1* allele (20) lacks only PRA isomerase, *trp-1C* (1) lacks only anthranilate synthetase, *trp-1* allele (25) lacks both PRA isomerase and InGP synthetase, etc. (523). To avoid confusion, note that in ref. (523) and related papers, the same "allele number" may be used for a *trp-2* mutant, a *trp-1* mutant (anthranilate-non-utilizing), and a *trp-1C* mutant (anthranilate-utilizing); mutants of the last class are listed by FGSC as *trp-1* with the allele number prefixed by C. Alleles differ in their ability to form aggregates (347, 523). The *Neurospora trp-1* gene is only partially expressed in *Escherichia coli*. Fine-structure maps (17, 523). Complementation maps (17, 330). A nonsense (amber) allele was used to demonstrate the restoration of normal enzyme aggregate by super-suppressors (348). Alleles that accumulate anthranilate

are scorable by blue fluorescence under long-wave UV after 2–5 days growth on minimal medium + indole (10 µg/ml) at 34°C (1587, 1592). Aging cultures may produce brown pigment; the blue fluorescence disappears as pigment forms.

trp-2: tryptophan-2

VIR. Between *del* (0; 13%), *T(OY329)* and *un-23* (5%; 27%), *T(OY320)*, *ws-1* (38%) (1578, 1603, 1604, 1992).

Cloned (257): pSV50 clone 2:5D.

Structural gene for anthranilate synthase component I (EC 4.1.3.27) (1038) (Fig. 13). Uses kynurenine, anthranilic acid, indole, or tryptophan (191). Kynurenine is utilized by conversion to anthranilate (846). TRP-2 catalyzes anthranilate synthesis with ammonia, but not with glutamine as the amino donor (62). Anthranilate synthetase (glutamine-linked) is specified in collaboration with *trp-1* in the trifunctional TRP-1$^+$ TRP-2$^+$ enzyme aggregate (347, 523) (Fig. 13; see *trp-1* entry). A nonsense (amber) allele was used to isolate super-suppressors (1859) and to study the enzyme complex restored by supersuppressors (348).

trp-3: tryptophan-3

IIR. Between *fl* (2%; 6%), *mus-23* (6%, 9%) and *un-5* (10%), *rip-1* (9%) (85, 1582, 1592, 1828, 2190).

Cloned and sequenced: Swissprot TRP_NEUCR, EMBL/GenBank J04594, PIR A28162, A32959, EMBL NCTRP3A, GenBank NEUTRP3A; pSV50 clone 5:4G.

Structural gene for tryptophan synthetase (EC 4.2.1.20) (2264) (called tryptophan desmolase in early literature). Uses tryptophan (1362); some alleles also use indole (11). Tryptophan synthetase catalyzes three reactions: indoleglycerol phosphate (InGP) → tryptophan, indole → tryptophan, and InGP → indole (Fig. 13). In *Neurospora*, all three reactions are catalyzed by a single trifunctional protein, which is specified by a single gene (1289, 2264). Mutants lack InG → tryptophan activity, but differ with respect to the other activities. For example, *trp-3^{td141}* is blocked in InGP utilization but can use indole; *trp-3^{td100}* can synthesize indole but cannot convert it to tryptophan, and *trp-3^{td140}* is a polar mutant lacking all three activities. [See refs. (2029) and (1127) for references and characteristics of other mutants.] Used extensively for studies of gene structure in relation to enzyme activity (482, 1127, 2264). The active enzyme is a homo-oligomer (1289); the monomer has two domains (257, 1288, 1290). Biochemical studies of complementation between alleles *in vivo* (1127, 1128) and *in vitro* (2028). Complementation maps (11, 12, 16, 1127). Fine-structure maps (12, 1024, 1127, 2029). Reviewed as an example of gene fusion (425). *trp-3* mutant C83 provided the first proved example in

Neurospora of gene-controlled loss of enzyme activity (1362); *trp-3* mutant S1952 provided the first example of allele-specific suppression restoring a functional wild-type-like enzyme (2262). Allele td140 is an amber mutant suppressible by *ssu-1* (1859, 1860). Certain classes of *trp-3* mutants are osmotic-remediable (1128). Called *td, tryp-3*.

trp-4: tryptophan-4

IVR. Between *his-5* (3%; 7%), *tol* (1%) and *leu-2* (1% or 2%) (103, 1255, 1466, 1911, 1928).

Cloned: pSV50 clone 32:2G; Orbach-Sachs clones X4E10, X10A04, X17C11, G6C09, G6E12.

Uses tryptophan or indole (1477). Deficient in anthranilate-PP-ribose-P-phosphoribosyl transferase (EC 2.4.2.18) (2199) (Fig. 13). Scorable by blue fluorescence (anthranilate) in medium under long-wave UV after 2–5 days growth on minimal medium + indole (10 μg/ml), at 34°C. Initial stocks of the first *trp-4* mutant were inhibited by suboptimal concentrations of tryptophan (1477), but derivatives have been obtained that are free of this problem (1973).

trp-5: tryptophan-5

VR. Between *inl* (4%), *pab-1* and *met-3* (4%) (13, 21, 1582).

Uses anthranilate, indole, or tryptophan (21). Tryptophanyl-tRNA synthetase activity (EC 6.1.1.2) is 5% that of wild type (1442). Anthranilate synthetase and tryptophan synthetase are derepressed (1442). Not temperature-sensitive.

trx: thioredoxin

III. Present in the same DNA fragment with *acr-2* but transcribed in the opposite direction (27).

Cloned and sequenced: Swissprot THIO_NEUCR, EMBL/GenBank D45892, EMBL NCTRX, GenBank NEUTRX.

Encodes a protein having extensive homology to thioredoxins from various organisms (27).

try: tryptophan

Changed to *trp*.

tryp: tryptophan

Changed to *trp*.

ts: tan spore

VR. Linked to *inl* (4%) (1428).

Ascospores are slow to mature, remaining light brown when wild-type ascospores have blackened. Expressed autonomously, allowing visual scoring in heterozygous asci. Only a small minority of *ts* ascospores, those that have darkened with age, are capable of germination. Photograph of asci (1428). Used to study multiple (1431) and selective (597) fertilization, preferential segregation (1429), and factors affecting crossing-over frequency (1138, 1430).

tu: tuft

IIR. Between *pe* (8%) and *fl* (19%) (1203).

Conidia are formed mostly in clusters at the shallow end of agar slants (1203). One of the first *Neurospora* genes to be mapped (stock lost).

tub^R: tubercidin resistant

Unmapped. Unlinked to *pyr-1* (1244).

Selected for resistance to tubercidin (7-deaza-adenosine), a competitive inhibitor of adenosine (1244). Adenosine kinase activity is reduced to ~10% that of wild type. Transport of adenosine into the cell is normal. No mutant stock survives (1244).

Tubulin

See *Bml, tba*, and *tbg*.

tub-2: tubulin-2

Has sometimes been used for *Bml*, which has precedence.

ty: tyrosinase

The symbol *ty* is used for genes that regulate tyrosinase. The structural gene for tyrosinase is symbolized *T*. Mutants at all of these loci are prototrophic, growing on unsupplemented minimal medium.

ty-1: tyrosinase-1

IIIR. Linked to *tyr-1* (0%; 6%), *dow* (21%) (1582, 2179).

Tyrosinase is repressed. Recessive (899, 901). Uninducible by inhibitors of protein synthesis in the sexual phase, but inducible in vegetative culture (898, 901). Prototrophic. Velvetlike vegetative morphology. Infertile as female (no or few perithecia), but fertile as male. Infertility is recessive in a heterokaryon (899). Low ascospore viability (880). Scored by color reaction with DL-DOPA as substrate (899, 901) or by morphology and female sterility.

ty-2: tyrosinase-2

IR. Right of *al-2* (880).

Tyrosinase is repressed. Prototrophic, recessive (899, 901). Inducible in vegetative culture by protein-synthesis inhibitors such as cycloheximide, but uninducible in the sexual phase (898, 901). Described as having short aerial hyphae (880), although earlier said to be morphologically normal (899). Female infertile, but fertile as male. Infertility is recessive in a heterokaryon (899). Scored by color reaction with DL-DOPA as substrate (899, 901), or by morphology, and female sterility.

ty-3: tyrosinase-3

IIIR. Between *Cen-III* and *ad-4* [W. L. Chan, Ph.D. Thesis, University of Malaya, Kuala Lumpur, 1977, cited in ref. (881)]. Not allelic with *T* or *ff-3* [368; N. H. Horowitz and H. MacLeod, cited in ref. (623)].

Tyrosinase is repressed. Uninducible in vegetative culture by inhibitors of protein synthesis such as cycloheximide (623, 881). Originally reported to be female-sterile and morphologically abnormal. These properties were later shown to be due to a second, nonallelic mutation, *ff-3*. The *ty-3* single mutant is female-fertile and morphologically normal (368). Scored by color reaction with DL-DOPA as substrate (899, 901). The original strain, called T22, contained *ty-4* in addition to *ty-3* and *ff-3*.

ty-4: tyrosinase-4

Unmapped.

Tyrosinase repressed. Uninducible by total starvation in sodium phosphate buffer, but can be induced by cycloheximide. Found in strain T22, which also contains *ty-3* and *ff-3* [W. L. Chan, Ph.D. Thesis, University of Malaya, Kuala Lumpur, 1977, cited in ref. (881)]; also present in some Emerson wild types (881).

tyr-1: tyrosine-l

IIIR. Between *vel* (3%; 5%), *phe-2* (2%; 4%) and *un-17* (4%) (601, 1322, 1582, 1592).

Requires tyrosine (2057). Lacks prephenate dehydrogenase activity (EC 1.3.1.13) (87, 601) (Fig. 13). Shows phenotypic adaptation after a lag, attaining wild-type growth rate on minimal medium. Adaptation is not carried through conidia (2057). Inhibition of prephenate dehydrogenase by phenylalanine examined (312). Allele UTl45 was called *tyr-3* (601).

tyr-2: tyrosine-2

IR. Between *T(4540)^L nic-2* and *T(Y112M4i)^R*, *ace-7*, *cr-1* (2%) (1578, 1585).

Requires tyrosine (1585). Prephenate dehydrogenase activity is decreased (601). *pe*-like morphology. Female sterile. *tyr⁻* ascospores do not become fully pigmented. Growth is suboptimal even on fully supplemented medium. Adapts to growth on minimal medium after

several days. Scoring by growth on slants is treacherous, but the mutant can be recognized by its darkening of tyrosine-supplemented minimal medium.

tyr-3: tyrosine-3

Used for an allele of *tyr-1*.

tyr-s: tyrosine sensitive

Symbol changed to *tys*.

tys: tyrosine sensitive

I. Right of *mat* (6%) (1582, 2234).

Growth is inhibited 80% by 0.07 mM L-tyrosine. Growth on minimal medium also is inhibited by glycyl-leucyl-tyrosine and by various tyrosine analogues. Uptake of L-[¹⁴C]-*p*-tyrosine is increased slightly, but this is not proposed as the cause of the inhibition. The primary defect is unknown (2234). Used to obtain the oligopeptide transport mutant *glt*, which is resistant to glycyl-leucyl-tyrosine but not to tyrosine (2236). Called *tyr-s* (2234, 2236).

ubi: ubiquitin

Unmapped.

Cloned and sequenced: Swissprot UBIQ_NEUCR, EMBL/GenBank U01220, X13140, NCUBQ, PIR S05323, UQNC; EST NP3F9.

Structural gene for polyubiquitin (2036).

ubi::crp-6: ubiquitin/cytoplasmic ribosomal protein-6

Unmapped.

Cloned and sequenced: EMBL/GenBank U01221, X15338, PIR UQNCR, GenBank NCUBICRP6.

Encodes ubiquitin fused to ribosomal protein S27a. The regulatory region appears to be shared with *crp-5*, which encodes ribosomal protein S26. Transcription is divergent, and the two genes are expressed coordinately (2056). Called *ubi-3* (2037).

uc: uracil salvage (or uracil uptake)

Used for mutants affecting the thymidine salvage pathway (*uc-1, -2, -3, -4*) or the transport of pyrimidine bases (*uc-5*) (2217) (Fig. 65). The symbol *pyr* is used for genes in the biosynthetic pathway.

uc-1: uracil salvage-1

II. Linked to *pyr-4* (31%) (2215, 2217).

Altered thymidine salvage pathway (Fig. 65). Able to use thymidine, thymine, 5-(hydroxymethyl)uracil, or

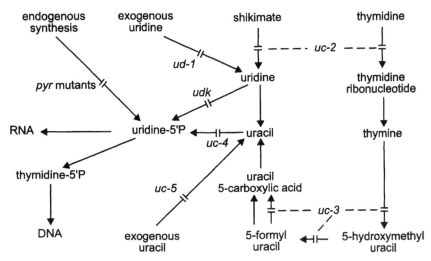

FIGURE 65   The uracil salvage pathway, by which many exogenously supplied pyrimidines are cycled via uracil to uridine 5'-phosphate and thus into the latter part of the pyrimidine biosynthetic pathway. The figure shows the sites of action of the salvage-pathway *uc* genes: the *ud-1* and *uc-5* genes, which control the uptake of exogenous uridine and uracil, and *udk*, which controls uridine kinase (265, 791, 1211, 1897, 2217). For endogenous pyrimidine synthesis, see Fig. 53. From ref. (1596), with permission from the American Society for Microbiology.

5-formyluracil as the sole pyrimidine source in germinating conidia of *pyr-4*, which is blocked in pyrimidine biosynthesis. The double mutant *uc-1⁺ pyr-4* can use these compounds only if a primer of uridine or cytidine is supplied (2217). *uc-1* causes elevated activities of enzymes that oxidatively demethylate thymine, including iso-orotate decarboxylase, and thus it is a putative regulatory gene in thymidine salvage (791, 1927).

uc-2: uracil salvage-2

I. Linked to *mat* (0/12 asci) (2215, 2217).
Thymidine salvage pathway defect (Fig. 65). Unable to use thymidine or deoxyuridine as the sole pyrimidine source for *pyr-4* in the presence of *uc-1*, but can use thymine, 5-(hydroxymethyl)uracil, 5-formyluracil, uracil, or uridine (2217). Reduced activity of the 2'-hydroxylase reactions thymidine → thymine ribonucleoside and deoxyuridine → uridine (1897).

uc-3: uracil salvage-3

Unmapped.
Thymidine salvage pathway defect (Fig. 65). Unable to use thymidine, thymine, or (hydroxymethyl)cytosine to support the growth of *pyr-4* in the presence of *uc-1*, but can use deoxyuridine, 5-formyluracil, uracil, or uridine. Excretes thymine into the medium when grown in the presence of thymidine (2217). Lacks thymine 7-hydroxylase (EC 1.14.11.6), which catalyzes the reactions thymin → 5-(hydroxymethyl)uracil → 5-formyluracil → uracil-5-carboxylic acid. Apparently there is an

alternate enzyme for 5-formyluracil → uracil-5-carboxylic acid; this would explain the growth on 5-formyluracil (1211, 1897).

uc-4: uracil salvage-4

VR. Between *inl* (12%) and *his-6* (11%) (265). [Not in I, as was reported (2215, 2217) on the basis of 1T/8 asci with *mat*.]
Defective in the thymidine salvage pathway (Fig. 65). Deficient in phosphoribosyl transferase (EC 2.4.2.9) (267). Unable to use thymidine, thymine, 5-(hydroxymethyl)uracil, 5-formyluracil, or uracil to support *pyr-1*, even in the presence of a uridine primer (2217). The *uc-4* single mutant converts 50% of supplied uridine to uracil and excretes it into the medium (2217). Resistant to 5-fluorouracil (265).

uc-5: uracil uptake-5

IVR. Right of *cot-1* (32%) (265).
Apparently defective in the transport of pyrimidine bases (Fig. 65). Unable to use any free pyrimidine base to support the growth of the pyrimidine auxotroph *pyr-1*, although it can use both ribose and deoxyribose nucleosides (2217). Resistant to 5-fluorouracil (265). The *pyr-1 uc-5* double mutant has been used to study uptake inhibition of structural analogs (441). Selected by the inability to use uracil to supplement *pyr* auxotrophs or by resistance to 5-fluorouracil on ammonia-free minimal medium (265). Uracil uptake is decreased by NH₃ (1245) or other good nitrogen sources or by a *nit-2* mutation (266). Called *tr^u*.

ud-1: uridine uptake-1

VIIR. Between *met-7* (27%) and *arg-10* (10%) (265).

Unable to use pyrimidine nucleosides to support the growth of a pyrimidine auxotroph (*pyr-1*) although it can use any free pyrimidine base. Probably defective in pyrimidine nucleoside transport (2217) (Fig. 65). Apparently also defective in purine nucleoside transport (440). Resistant to 5-fluorodeoxyuridine and 5-fluorouridine (265). Resistance is recessive in heterokaryons. Shows interallelic complementation (267). Scored by spotting conidial suspension on medium containing 4×10^{-5} M fluorodeoxyuridine, filter-sterilized (885). Mutant CIF-dUrd 7, selected by resistance and called *fdu-1* (885), is allelic (265). Uridine uptake is under general nitrogen metabolite regulation (266).

udk: uridine kinase

VR. Left of *uc-4* (29%) (265).

Deficient in uridine kinase (EC 2.7.1.48) (267) (Fig. 65). The *udk uc-4* double mutant is resistant to 5-fluoro-deoxyuridine and 5-fluorouridine (265), but *udk* alone is not resistant to any analogue.

ufa-1: unsaturated fatty acids-1

IVR. Linked to *met-5* (7%) (1593).

Lacks stearate desaturase activity. The requirement is satisfied by 16- or 18-carbon fatty acids having a double bond at either the Δ^9 or Δ^{11} position or 16 carbons with a *trans*-double bond at Δ^9 or multiple *cis*-double bonds interrupted by methylene bridges (1849, 1850). Reverts readily, and reversions overgrow the culture on sub-optimal supplement (232). Tween 80 (0.1%) is satisfactory for maintenance. Stock viability is better on Tween 80 than on fatty acids suspended in the detergent Tergitol NP-40 (1850). For the biosynthetic pathway, see ref. (1850) and Fig. 17.

ufa-2: unsaturated fatty acids-2

IV or V. Linked to *T(R2355), inl* in cross to *alcoy* (1850).

Lacks stearate desaturase (EC 1.4.99.5) (Fig. 17). Also affects auxiliary electron transport reactions. The requirement is satisfied by 16- or 18-carbon fatty acids having a *cis*-double bond at either the Δ^9 or Δ^{11} position or 16 carbons with a *trans*-double bond at Δ^9 or multiple *cis*-double bonds interrupted by methylene bridges (1849, 1850). Linkage is similar to that of *ufa-1*. No intercross data. The designation as a second *ufa* locus is based solely on complementation with *ufa-1* (1850). Highly revertible; stocks are lost readily on suboptimal medium (232).

ufa-3: unsaturated fatty acids-3

I or II. Linked to *T(4637) al-1* in cross to *alcoy* (757).

Lacks stearate desaturase activity. Requires an unsaturated fatty acid for growth. Highly revertible (753, 757).

ufa-4: unsaturated fatty acids-4

I or II. Linked to *T(4637) al-1* in cross to *alcoy* (757).

Affects fatty acid biosynthesis and, presumably, electron transport. Highly revertible. Growth is very slow even with fatty acid supplementation (757).

un: unknown

Used for conditional mutants for which the function is unknown. The name was introduced by early *Neurospora* workers for temperature-sensitive conditional mutants that are irreparable by supplementation at the restrictive temperature (34°C or higher). Originally referred to as "unknown requirement" on the unfounded supposition that the mutants would prove to be reparable auxotrophs. Some heat-sensitive genes have been characterized sufficiently before naming to be assigned more specific names, such as *ndc, rip, psi, eth-1,* and *fs-2*. Certain heat-sensitive mutants with altered morphology at restrictive temperature have been named *cot* or *scot*. Little is known of the molecular or cellular basis of most *un* mutants. At least some are deficient in amino acid transport (1029, 2089). The *un* mutants have been useful as genetic markers. Scoring may require growth comparisons at two temperatures, preferably using small conidial inocula. Many *un* mutants do not achieve normal growth rates even at the permissive temperature (usually 25°C). Temperature-sensitive auxotrophs with conventional requirements have sometimes have been classed initially as *un* because complex complete medium is either inadequate or inhibitory. For example, among mutants called *un* in a study in ref. (951), nos. 3, 13, and 14 proved to be temperature-sensitive *thr* auxotrophs, nos. 6, 20, and 30 were *his*, and nos. 19 and 35 were *asn* (962). Locus numbers were assigned to the first eight *un* mutants bin ref. (1586). The letter T in allele numbers of numerous *un* mutants designates Tokyo, not translocation.

un-1: unknown-1

IR. Between *cr-1* (5%; 9%) and *nd, T(4540)^R* . Linked to *cys-9* (0/72), *bs* (9%) (1578, 1582, 1592, 1603).

Function unknown. Heat-sensitive. No growth at 34°C (909). Linear growth is suboptimal at permissive temperature (1392). Reported deficient in mitochondrial superoxide dismutase SOD2 (1392). Called *un(44409)*.

un-2: unknown-2

IR. Between *sn*, *os-4*, *T(AR173)^L* and *T(AR190)*, *his-2* (1%) (1578).

Function unknown. Heat-sensitive (907), growing at 25°C but not at 39°C. Scorable but leaky at 34°C. Called *un(46006)*.

un-3: unknown-3

Mutant 55701, first called *un(55701)* and then *un-3*, was shown to have a complex phenotype with both a defect in growth, which is temperature-sensitive, and a defect in transport, which is not. The heat-sensitive growth defect is allelic with an independently described mutation, *cyt-20*. The transport defect confers resistance to ethionine and *p*-fluorophenylalanine. The lesions responsible for temperature sensitivity and analogue resistance may be separable by recombination (1104). When ascospores from *un-3^+ × un-3* 55701 were plated on minimal medium plus ethionine at 35°C, ethionine-resistant colonies were obtained at a frequency of 0.1% (309, 1104), suggesting but not proving that recombination occurs and that separate genes may be involved. The order of the two putative mutations relative to the *mat* locus was not determined. Although the symbol *un-3* might be considered to have priority over *cyt-20* in designating the temperature-sensitive component, we propose to retain *cyt-20* for the independently arising mutation that has no transport defect and to continue using *un-3* for the original 55701 mutant, which has long been treated as a single gene and which will undoubtedly continue to be useful as a marker. If the ethionine-resistance trait proves to be due to mutation in a separate gene, an appropriate name for the second component would then be *eth-2*, and the 55701 genotype might be given explicitly as *cyt-20 eth-2*. The following description is based on strains of the original 55701 mutant, which is defective in both functions.

IL. Between *ser-3* (1%), *In(NM176)^L* and *mat* (≤0.1%) (917, 1470, 1578). No recombination (0/21,200) or complementation with *cyt-20* allele 1 (1322).

Cloned and sequenced: Swissprot SYV_NEUCR, EMBL/ GenBank M64703, GenBank NEUXXX.

Includes the structural gene for valyl-tRNA synthetase mitochondrial precursor. Growth is heat-sensitive (909). Unlike *cyt-20* (1631), growth of *un-3* (55701) ceases sharply between 28.5 and 30°C (1322), and there is no gross deficiency in cytochromes *b* and *aa₃* or in protein synthesis, suggesting that valyl-tRNA synthetase may have another function in addition to protein synthesis (1104). The mutant is multiply transport-deficient at permissive temperatures: resistant to ethionine and *p*-fluorophenylalanine at 25°C (1028, 1029), with increased fragility of protoplasts (1029) and a reduced rate of uptake of citrulline (2089) and aspartate (1029, 2230).

Used to tag the *mat* genes (916) and as a flanker in an attempt to resolve *mat* into components by recombination (1470). Used in cloning *mat A* (746, 2167). Mutations that are probably *un-3* alleles are selected as citrulline-resistant mutants of *pyr-3 arg-12^s*; most mutants selected in this way show complementation between alleles (2089). Growth at 25°C is aided by 0.3 mg/ml sodium acetate. May be scored by slow growth at 25°C on minimal medium without acetate (916). See *cyt-10*.

un-4: unknown-4

VIL. Between *lys-5* (2%; in a common cosmid) and *T(T39M777)*, *un-13* (3%) (1578, 1815, 1986).

Cloned by complementation (1815).

Function unknown. Heat-sensitive, 34 vs 25°C (906). Called *un(66204)*.

un-5: unknown-5

IL. Between *fr* (6%), *T(OY330)^L* and *T(39311)^L*, *nit-2* (2%) (1555, 1592).

Function unknown. Heat-sensitive, 34 vs 25°C. Inhibited by histidine and tryptophan. Osmophilic at 26–28°C (1117, 1118). Called *un(b39)*.

un-6: unknown-6

IIIR. Between *sc* (21%) and *leu-1* (1546, 1582, 1592).

Function unknown. Heat-sensitive, 34 vs 25°C (909). Called *un(83106)*.

un-7: unknown-7

IR. Between *cyh-1*, *T(STL76)* and *al-1* (3%) (1578, 1586, 1603).

Cloned: Orbach-Sachs clones G14C06, G18H05.

Function unknown. Heat-sensitive, 34C vs 25°C (947). Dies slowly at restrictive temperature (952). Allele T35M50 is mutant 31, called TS31 in refs. (951) and (952).

un-8: unknown-8

IVR. Linked to *pyr-1* (0/47). Between *T(ALS159)* and *col-4* (5%); hence, right of *psi-1* (1578, 1582, 1586).

Function unknown. Heat-sensitive (951). No growth at 34°C. Morphology is abnormal at 25°C, unlike *psi*, which is closely linked. Allele T27M9 is mutant 1 in ref. (951).

un-9: unknown-9

VR. Between *pyr-6* (3%) and *oak* (7%), *his-6* (5%; 9%) (1593, 1603).

Function unknown. Growth is heat-sensitive, 34 vs 25°C (951). Allele T54M96 is mutant 42 in ref. (951).

un-10: unknown-10

VIIR. Between *frq* (9%) and *arg-10*, *nt* (22%) (1582). Right of *for*, *frq*, *oli* (by chromosomal walk) (1094). The order *wc-1 un-10 oli frq for*, shown previously, is apparently incorrect.
Cloned (1815).
Function unknown. Heat-sensitive, 34 vs 25°C (951). Allele T42M45 is mutant 11 in ref. (951).

un-11: unknown-11

VR. Left of *his-1* (<1%) (1322).
Function unknown. Growth is heat-sensitive, 34 vs 25°C (951). Allele T42M30 is mutant 10 in ref. (951).

un-12: unknown-12

IVR. Linked to *pdx-1* (5%), *col-4* (0/73) (1603).
Function unknown. Heat-sensitive, 34 vs 25°C (951). Allele T51M118 is mutant 17 in ref. (951).

un-13: unknown-13

VIL. Between *lys-5* (4%), *un-4* (3%), *ylo-1* (2%; 38 kb), *T(IBj5)^L* and *cpc-1* (1% or 2%; 105 kb), *T(IBj5)^R* (1578, 1603, 2182).
Cloned by complementation and partially sequenced (both genomic and cDNA) (1815, 2182): Orbach-Sachs clone G55F.
The sequence suggests that *un-13* encodes a tRNA synthetase (1815). Heat-sensitive, 34 vs 25°C (951). Allele T42M24 is mutant 9 in ref. (951).

un-14: unknown-14

IIIR. Between *acr-2*, *thi-4* (8%; 20%) and *met-8*, *leu-1* (5%) (1546, 1582, 1603).
Function unknown. Heat-sensitive, 34 vs 25°C (951). Allele T54M55 is mutant 36 in ref. (951).

un-15: unknown-15

IIR. Between *trp-3* (10%) and *rip-1* (1%) (1546, 1582, 1586).
Function unknown. Heat-sensitive, 34 vs 25°C (962). Grows poorly at permissive temperature. *rip-1* is therefore preferred to *un-15* as a marker for the right end of II (1584). Semidominant at 39°C. Transfers grow after 4 days at 34°C (1546). Allele T54M50 (962).

un-16: unknown-16

IL. Between *mat* (<1%) and *ta* (1%), *acr-3* (<3%) (1582, 1603).

Function unknown. Heat-sensitive, 34°C vs 25°C (951). Allele T42M69 is mutant 16 in ref. (951).

un-17: unknown-17

IIIR. Between *tyr-1* (4%) and *T(UK8-18)^R*. Linked to *nit-7* (0/63) (1546, 1578, 1603).
Heat-sensitive, 34 vs 25°C (951). Ascospores containing *un-17* are white or slow to mature (1546). Shows rapid, exponential death at 35°C, which is averted by cycloheximide or conditions allowing no protein synthesis. Cold-sensitive and osmotic-remediable at 11°C (1390). Phospholipid synthesis is altered (952). When grown above 34°C, the mutant accumulates the same phospholipid intermediates as does chol-2 (753) (Fig. 18). Reported to be deficient in extracellular superoxide dismutase SOD4 (EC 1.15.1.1) (1392). Allele T51M171 is mutant 25 or TS25 in refs. (951) and (952).

un-18: unknown-18

IR. Between *R* (11%), *T(MD2)^L*, *T(NM169d)^L* and *T(MD2)^R*, *Tip-IR* (1578, 1580).
Cloned and sequenced: GenBank AB006052.
Encodes the second largest subunit of RNA polymerase I (EC 2.7.7.6). Heat-sensitive (951). No growth at 34°C. Growth at 25°C is substantial but not wild-type and is better on complete than on minimal medium. The period length of circadian rhythm is increased by ~2 hr at temperatures between 22 and 32°C (1498). Allele T54M94 is mutant 41 in ref. (951).

un-19: unknown-19

VR. Linked to *al-3* (9%), *un-11* (14%) (1546).
Function unknown. Growth is heat-sensitive, 34 vs 25°C (1546).

un-20: unknown-20

IIR. Between *aro-1* (5%; 9%), *ff-1* (4%) and *ace-1* (15%) (210, 1582, 1592, 2044).
Function unknown. Heat-sensitive, 39 vs 25°C. Best scored at 39°C on minimal medium. Leaky. Some flat, aconidiate, unpigmented growth occurs even at restrictive temperature (1582). Hyphae bleed at 39°C. Called *mo(P2402t)* (1592, 2044).

un-21: unknown-21

IIIR. Between *acr-2* and *met-8*, *un-6* (1546, 1582).
Function unknown. Heat-sensitive, 34 vs 25°C (951). Allele T53M26 is mutant 29 in ref. (951).

un-22: unknown-22

VII. Linked to *met-7* (1%), *un-10* (>20%) (1582, 1992).

Function unknown. Heat-sensitive, growing at 20–28°C, but not at 37°C. Called *un(6lC)*, *un(62C)* (1992).

un-23: unknown-23

VIR. Between *trp-2* (5%; 27%) and *col-18* (3/28), *T(OY320)*, *T(UK14-1)*, *ws-1* (1546, 1578, 1992).

Function unknown. Heat-sensitive, growing at 25°C but not at 28°C. Called *un(64D)* (1992).

un-24: unknown-24

IIL. Between $T(AR18)^L$, *hsp70* and *het-6* (0/222; 14 kb), $T(AR18)^R$; hence, between *cys-3* and *T(P1869)*, *cot-5* (1421, 1935).

Cloned and sequenced: GenBank AF171697; Orbach-Sachs cosmids G8G1 and X14C1 contain both *het-6* and *un-24*.

Encodes the large subunit of ribonucleotide reductase (1936). A variable domain confers vegetative incompatibility when $un-24^{OR}$ is transformed into spheroplasts of a heterotypic strain, $un-24^{PA}$ (1937). Growth is heat-sensitive, 34 vs 25°C. The growth defect at restrictive temperature is partially overcome on media of high osmotic pressure (1936). Obtained by RIP in progeny from *Dp(AR18)* × *Normal sequence* (976, 1589, 1936). *un-24* and *het-6* are in linkage disequilibrium, with no recombination coupling phase in strains from natural populations (1351).

un-25: unknown-25

VIL. Right of *cpl-1* (4%) (1815). Linked to *ylo-1* (0/86) (1582).

Cloned and sequenced (1815).

The sequence suggests that *un-25* encodes ribosomal protein L13. Not complemented by DNA from either $un-4^+$ or $un-13^+$ (1815). Heat-sensitive, 34 vs 25°C (951). Allele T51M154 is mutant 22 in ref. (951).

un(STL6)

Allelic with *fls*.

upr-1: ultraviolet photoreactivation-1

IL. Between *mat* (2%) and *arg-1* (7%) (2132).

Epistasis group Upr-1 (Table 3). The original allele shows moderate sensitivity to UV (494, 2133), very low sensitivity to ionizing radiation (1846), moderate or low sensitivity to MNNG, 4-NQO, ICR-170, (953) and nitrous acid (2132), and no sensitivity to MMS (1009) or histidine (1469). (See *mus* and *uvs* entries for analyses of other genes conferring mutagen sensitivity.) A second *upr-1* allele has been obtained among supersensitive double mutants by mutagenizing the highly UV-mutable, slightly UV-sensitive *mus-18* (960; see mus-38 for method). Homozygous fertile. Defective photoreactivation of UV damage *in vivo*, but the photoreactivation enzyme is functional *in vitro* (2132, 2133). Survival after UV irradiation is increased by photoreactivation only at UV levels giving <10% survival, as was also found for *mus-26* (957), which is in the same epistasis group and which is phenotypically similar (949, 957). Earlier results, at least in tests with UV (947), had suggested that *upr-1* and *uvs-2* belonged to the same epistasis group. Although such a grouping was plausible considering other similarities (it figured in the literature for some time without being rechecked; 1012, 1836), results now favor placing *upr-1* with *mus-26* in a separate group (949, 957, 1838). A subgroup is possible.

ure: urease

Mutants at four loci, numbered *ure-1* to *-4*, lack all detectable urease activity (Figs. 11 and 65). Possibly all four are structural genes; urease from other organisms comprises numerous distinct subunits of a hetero-oligomeric protein (156). *ure* mutant D2, which is tightly linked to *ure-1*, fails to complement mutants at any of the four loci, suggesting a regulatory role (862). Mutants of another type possess partial activity, but are readily scorable by poor growth on urea as the N-source. These probably represent additional loci in V (A7, S3), IV (E3, E7), or elsewhere (C5, K3, R2) (862, 914). They are not given separate entries here. Mutants have been isolated by methods based on the inability to generate ammonia from urea, using pH indicators (156, 862, 1087). The following methods for scoring isolates are probably generally applicable, but all have not been tested on mutants at all loci. Method 1: Little growth on filter-sterilized urea as the sole N source (862) (not good if amino acids must be added). Method 2: 5- to 20-min scoring test touching bits of filter paper, dipped in urea–bromthymol blue buffer, to conidia or aerial growth (1086). Method 3: Color change when grown on slants of synthetic crossing medium (2208) containing phenol red and urea at 3 mg/ml as the sole N source, autoclaved, and scored after 4–5 days at 34°C (1055, 1582).

ure-1: urease-1

VR. Between *ure-2* (3%), *am* (1%; 150 kb), *ace-5* (<1%) and *his-1* (<1%) (217, 1086, 1122).

No urease activity (EC 3.5.1.5) (862, 1087) (Figs 11 and 65). Some revertants show altered heat stability of urease, suggesting a structural role (156). Used to study the metabolic fate of arginine by measuring urea accumulation (1156). Used for arginine tracer experiments

(424) and flux (657). Used to determine the relative contributions of arginine and purines in urea formation (454). For the role of urease in purine catabolism, see *xdh-1*. For scoring methods, see *ure*. Possible allele D2 fails to complement *ure-1, -2, -3, -4* (862). Called *ure(9)* (1086).

ure-2: urease-2

VR. Between *sp* (8%) and *am* (2%), *ure-1* (3%) (220, 1086).

No urease activity (EC 3.5.1.5) (862, 1087) (Figs. 11 and 65). The enzyme from interallelic complementation and from some revertants shows altered heat stability, suggesting a structural role (156). Shows hyperinducibility of purine catabolic enzymes uricase, allantoicase, and allantoinase (1712). Reversion and complementation data (156). For scoring methods, see *ure*. Called *ure(47)* (1086).

ure-3: urease-3

IIR. Between *arg-12* (7%; 12%) and *un-20* (8%), *ace-1* (15%) (1582). Because allele B1 is closely linked to translocation *T(IR;II)B1* in the original strain, *ure-3* was at first assigned incorrectly to IR (156, 862). Point-mutant allele F29 showed linkage in IIR, not IR (1582).

No urease activity (EC 3.5.1.5) (Fig. 8). Some revertants show altered urease thermostability, suggesting a structural role (156). For scoring methods, see *ure*.

ure-4: urease-4

IR. Left of *ad-3B* (3%). Probably right of *his-3* (1%) (156).

No urease activity (EC 3.5.1.5) (862) (Fig. 8.) Some revertants show altered heat stability of urease, suggesting a structural role (156) For scoring methods, see *ure*.

ure(D2): urease(D2)

VR. At or near *ure-1* (0/45) (862).

No urease activity. Fails to complement representatives of all four *ure* loci; possible regulatory gene (862).

usg-1: upstream gene-1

VR. Located 3.5-kb upstream of the distal regulatory sequence of *am* (670).

A gene of unknown function, transcribed in the same direction as *am* (670).

uve-1: UV-endonuclease-1

Changed to *phr*.

uvs: ultraviolet sensitive

The *uvs* (and *upr*) mutants of *Neurospora* are defective in DNA repair and have highly pleiotropic phenotypes (946, 1838) (Table 3). Genes designated *uvs-1* through *uvs-6* were identified prior to 1980; after that time, *mus* (*mutagen sensitive*) was adopted as the symbol to designate mutants defective in DNA repair, beginning with *mus-7* (1839). Thus, whereas a given *mus* mutant may not be sensitive to UV, any UV-sensitive mutant isolated since 1980 has been given a *mus* designation. The sensitivity of *uvs* mutants is typically recessive. Most of these mutants are also sensitive to various genotoxic chemicals that can be incorporated into agar media for tests during cell division. Scoring of qualitative differences in UV sensitivity is accomplished most readily by spot tests in which the growth of cells on exposed plates and control plates is compared (1830, 1993). For the more accurate assessment, survival curves are used, especially when single and double mutants are compared (1836). Sensitivities to different genotoxic agents may be associated in patterns and can be diagnostic for certain DNA repair types or epistatic groups (see the introduction to *mus* genes). In general, the increases in UV sensitivity observed for UV sensitive mutants in filamentous fungi such as *Neurospora* have been relatively low. This has made genetic analysis labor-intensive. Thus, *uvs-2*, which has the highest UV sensitivity for any single *Neurospora crassa* mutant, is only 20-fold more sensitive to UV than is the wild type (1833, 1838). In contrast, 100- to 200-fold increases in sensitivity are reported for mutants of *Escherichia coli* and *Saccharomyces cerevisiae*. Some of this difference is now attributed to the alternative repair pathways (e.g., *mus-18* and *mus-38*) available in filamentous fungi (also in fission yeast and probably in higher eukaryotes). Properties of the *uvs* and related mutants are summarized in refs. (946), (948), (953), (1010), (1015), (1835), (1836), (1838), and (1839). For properties of double mutants, see refs. (855), (947), (957)–(960), (1012), and (1017).

uvs-1: ultraviolet sensitive-1

Unmapped (355).

Repeated efforts have failed to confirm that *uvs-1* is a single Mendelian mutant. The presumed mutant strain is therefore no longer used in investigations (1836, 1838).

uvs-2: ultraviolet sensitive-2

IVR. Between *cys-4* (5%) (1993) and *pmb* (8%) [S. Ogilvie-Villa, cited in refs. (513) and (1776)], *T(S4342)^R*, *T(AR209)*, *T(T54M50)*, *T(ALS179)* (2–6%) (1578).

Cloned and sequenced: Swissprot UVS2_NEUCR, EMBL/GenBank D11458, PIR S34825, GenBank NEUUVS2.

Epistasis group Uvs-2 (Table 3). *uvs-2* has the highest UV sensitivity of all *Neurospora uvs* mutants [see *uvs* and refs. (494) and (1993)]. The allele usually used was discovered in stocks of mixed ancestry and may be present in other laboratory stocks (1984, 1993). Recessive in heterokaryons (1993). *uvs-2* is sensitive to ionizing radiation (1833, 1846), MNNG (953, 1833), 4NQO, nitrous acid (1984), ICR-170 (953), and mitomycin C (1012). Highly sensitive to MMS (1009), as was found for several other *uvs* mutants. A second allele (SA4) with similarly high UV and MMS sensitivity was isolated among mutants identified by MMS sensitivity (948). Shows slight or no sensitivity to histidine (1473). Normal spontaneous mutation and high UV-induced mutation (88, 494, 497, 502). Homozygous fertile. Meiosis and meiotic crossing over are unaffected. Used to show that the DNA repair system is induced by a small dose of UV (1983) and to demonstrate post replication repair (273).

The *uvs-2*$^+$ gene was isolated by transformation of *uvs-2* with a genomic library and selection for resistance on MMS medium. Encodes a structural gene homologous to *Saccharomyces RAD18* (2099) and *Aspergillus uvsH* (2281). However, the amino acid identity with RAD18 is only 25%, and there is no functional complementation of the yeast *RAD18* mutant with *uvs-2* cDNA cloned in a yeast expression vector (1952). No major changes in the two-dimensional profile of polypeptides were detected when the mutant was compared to wild type under uninduced conditions; but differences in some inducible peptides were evident (911–913). Originally, the finding that most of the phenotypic properties of *uvs-2* resembled those of yeast nucleotide excision repair mutants suggested that *uvs-2* was involved in the excision of UV dimers. Supporting this hypothesis, no excised dimers were detectable by chromatographic tests of *uvs-2* mutants (2247). Two observations were contradictory, however: the observed X-ray sensitivity of *uvs-2* and the finding in *uvs-2* strains (and in all tested UV-sensitive *Neurospora* strains) that UV-dimer repair occurs at the wild-type level when unrepaired dimers are identified by UV-dimer specific endonuclease or antibody (89, 90, 1239). These observations are consistent with the discovery that the excision of dimers in *Neurospora* occurs not only by the same mechanism identified for nucleotide excision repair genes in yeast and human cells but also by a UV-dimer-specific endonucleolytic mechanism that is roughly similar but distinct from that found in *Micrococcus luteus* (see *mus-18*). The homology of *uvs-2* to *Saccharomyces RAD18*, which is active in postreplication repair (2099), and the discovery in *Neurospora* of two bona fide nucleotide excision repair genes, *mus-38* and *mus-40,* which are homologous to yeast *RAD1* and *RAD2*, respectively (855), finally provided the expected parallels with the yeast repair system. Thus, whereas *uvs-2* is known not to

be involved in a known process of excision, the actual function carried out by genes of the Uvs-2 epistasis group (Table 3), or by their yeast homologs *RAD18* and *RAD6*, remains unknown.

uvs-3: ultraviolet sensitive-3

IVL. Linked to *cys-10* (3%; 7%), probably to the left (1014, 1830).

Cloned and partially sequenced (961).

Epistasis group Uvs-3 (Table 3) (1012). Sensitive to UV (494, 1010, 1830, 1831) and highly sensitive to MMS (1009, 1012). Moderately sensitive to 5-azacytidine (662). The commonly used allele ALS11 has low and variable conidial viability (502, 1830). A second *uvs-3* allele, FK016 (1012), has approximately 2-fold better conidial viability and forms more uniform colonies. Both alleles are sensitive to histidine, MNNG, X rays, γ rays, and mitomycin C (381, 953, 1010, 1831, 1846). However, ALS11 is significantly more sensitive than FK016 to X rays, γ rays, and mitomycin C (1012). Other agents have also been tested for their effects on survival and mutation (953). *uvs-3* displays a mutator phenotype (502, 1010), resulting in spontaneous reversions and variable sensitivities in specific isolates (1012, 1830). There is no induction of mutation by UV (494) or by γ rays and chemical mutagens (953, 1846). Therefore, the mutant is probably defective in error-prone repair (1834). Homozygous barren; sexual development is blocked before karyogamy (1685, 1830). Morphological mutants are frequent in backcrosses and ascospore viability is low; few asci contain eight black spores. Duplication instability is increased significantly, and this is correlated with increased sensitivity to hydroxyurea (1835). Sometimes gives stop–start growth on minimal medium or minimal medium + histidine (1467). Photoreactivation is impaired *in vivo*, but the photoreactivation enzyme functions *in vitro* (960, 1832). Levels of endo–exonuclease are low, probably accounting for the reduced nuclease halo phenotype (1014). The mutant appears deficient in proteolytic conversion of nuclease precursors to active intra- and extracellular DNases, but this effect could be indirect (666). Protease activity may be reduced (1723). The mutant shows a high constitutive level of repair in rescuing a heterokaryotic component that carries potentially lethal mutagen-induced damage (1983). Excised dimers were reported to be detectable only after a delay and to be released at a reduced rate (2247). However, alternate tests indicated that, under specific conditions, constitutive levels of dimer disappearance from DNA were as high in *uvs-3* as the induced level in wild type (88, 1838). One significant charge change in a polypeptide is observed in constitutive vs inducible expression of polypeptides detected in two-dimensional profiles of near-isogenic strains of both *uvs-3* alleles compared to wild type (911–913). The

double mutant *uvs-3; uvs-6* is inviable, as are the double
mutants *mus-9; uvs-6* (947, 1012) and *uvs-3; mei-3*
(958). However, when double mutants are tested for
MMS survival, a highly synergistic interaction is found
for *uvs-3* with *uvs-2* and with *mus-8* (1012). Similarly,
for the *uvs-3; upr-1* double mutant, reduction of UV
survival is at least additive; measurements of mutation
rates also indicate interaction between the two. Spontaneous mutation levels are low for *uvs-3; upr-1* (resembling *upr-1*); there is no increase after UV induction
(resembling *uvs-3*) (947). These results establish the Uvs-
3 group (including *mus-9*) separate from the Uvs-2
group (including *mus-8*) and separate from the Uvs-6
group (including *mei-3*). Additive (or stronger) interactions with *uvs-3* have also been demonstrated for *mus-
18* and *mus-38*, which interact synergistically with each
other and which represent different epistatis groups and
repair types (959, 960). Allele FK016 was isolated as a
nuclease halo mutant and originally called *nuh-4* (1014).
A strain used in ref. (667) and incorrectly called *nuh-4*
was actually *uvs(FK104)*, a *mus-9* allele. This error
explains what appeared to be contradictions concerning
uvs-3 phenotypes (1012).

uvs-4: ultraviolet sensitive-4

IIIR. Left of *ad-4* (4%) (1830).
Epistasis group Uvs-2 (1838) (Table 3). Sensitive to UV
(494, 1830, 1831) and histidine (1467, 1835); moderately sensitive to MMS; slightly sensitive to MNNG
(1010, 1831) and no increased sensitivity to ionizing
radiation (1846) or mitomycin C (381). Gives stop–start
growth (1467). Spontaneous mutation is probably
normal (502); UV-induced mutation is reduced (494).
For induction of mutation by other agents, see ref.
(1846). Homozygous fertile, although ascospore viability and early growth are severely impaired in homozygous crosses (1830). No effect on mitotic gene
conversion or recombination (1829). Chromosome
stability in duplication strains is normal (1835). Defects
in mitochondrial DNA (856). Difficult to work with.
Survival of conidia is poor, even on silica gel. Recessive
in heterokaryons. Stocks are best maintained in a
heterokaryon (1576).

uvs-5: ultraviolet sensitive-5

IIIR. Linked to *vel* (1%) (1830).
Sensitive to UV (494, 1830, 1831), MNNG, and ICR-170
(953). Not sensitive to ionizing radiation (1846), 4NQO
(953), or histidine (1467). Spontaneous mutation is
normal (502); UV-induced mutation is reduced (494).
For induction of mutation by other agents, see refs. (953)
and (1846). Recovery from heterozygous crosses is low.
Meiotic recombination is not affected. Homozygous
barren (1830), with meiosis blocked at pachytene

(1685). No increase in mitotic chromosomal instability
(1835). Abnormal mitochondrial DNA (856). Difficult
to work with. Growth is slow (1838). Consequently,
uvs-5 is no longer used in investigations. Stocks are best
maintained in a heterokaryon (1576).

uvs-6: ultraviolet sensitive-6

IR. Linked to *met-6* (<1%). Between *thi-1* (3%; 8%) and
ad-9 (4%) (1476, 1837).
Epistasis group Uvs-6 (828, 958, 1012) (Table 3). Reduced
conidial viability (1010). Switches to stop–start growth
after initial normal growth (1467). Not completely
recessive in heterokaryotic conidia (1900). Increased
sensitivity to UV (494, 1837, 1846), MNNG (953,
1010), and histidine (1466, 1467). Also sensitive to ICR-
170, 4NQO (953), MMS, and possibly mitomycin C
(1010). Spontaneous mutation is normal in the *ad-3*
haploid mutation system (502); a 2-fold increase in
spontaneous mutation found for recessive lethals in a
heterokaryon test system was not statistically significant
and may have resulted from an unusual chance
occurrence of several large clones (1013). UV-induced
mutation is normal in both mutation test systems (494,
1013). For mutation induction by other agents, see refs.
(953) and (1846). Homozygous barren, with a block at
crozier differentiation (1685). Increased duplication
instability (1473). The induction of a specific polypeptide in response to UV is altered compared to wild type,
as was also found for *mei-2* and *mei-3* (911–913).
Possibly defective in nuclease levels or in secretion,
giving reduced nuclease halos on DNase test medium
(667, 1014); DNase A is much reduced in culture
filtrates of cells grown in sorbose + DNA medium (see
nuh for nuclease assays). Protease activity may be
reduced (1723). Double mutants of *uvs-6* with the two
uvs-3 alleles ALS11 and FK016 are inviable (947, 1012).
Double mutants with representative members of other
epistasis groups showed additive or synergistic interactions, however (947, 957, 959, 960). Homology of at
least two genes of the *Neurospora* Uvs-6 epistasis group
with genes of the yeast Rad52 group suggests that the
function may be in recombinational repair (828, 853).

uvs(FK104): ultraviolet sensitive(FK104)

A *mus-9* allele (1014). Mislabeled as *nuh-4* in ref. (667).

vac-5: vacuolar protein-5

Changed to *htl*.

val: valine

VR. Right of *at* (10%). Linked to *ilv-2* (0/135) (1582).
Alleles 33026 and 33050 appear to be valine auxotrophs,
not requiring or responding to added isoleucine (907,

1582). Not defective in valyl-tRNA synthetase (610), as are the only *val* mutants known in bacteria; it may therefore define a gene specifying the unique conversion step from α-keto-isovalerate to valine, branched-chain amino acid aminotransferase (EC 2.6.1.42). The requirement is somewhat leaky, not temperature-sensitive. An *ilv* mutant strain with an incomplete isolation number, *ilv(?6201)*, was incorrectly designated *val (45201)* and was the source of linkage data for a locus erroneously shown as *val* on maps prior to 1980.

van: vanadate resistant

Changed to *pho-4*.

var-1: variant-1

Unmapped. Segregates as a single gene (169).

The rate of growth is reduced by 40% and aerial hyphae are decreased, resulting in a shaved appearance. Functional protoperithecia are absent. Lysed areas appear and spread in old cultures. Not rejuvenated in heterokaryons. Originated in an experiment on uninterrupted growth (169).

vel: velvet

IIIR. Between *ro-2* (6%), *T(D305)^L* and *tyr-1* (3%; 5%). Linked to *phe-2* (l%) (1582, 1587, 1592).

Soft, colonial growth (1548). Flat at first and then elevated, with conidia sometimes formed later, in aerial puffs.

Good female fertility. *col-13* (R2471) and *col-15* (R2531) are putative alleles. Mutations in strains R2471 and R2531, named *col-13* and *col-15* in ref. (719), are probably *vel* alleles, resembling *vel* in phenotype and mapping in the same IIIR region [see entries in ref. (1596)].

vis(3717): visible

Inseparable from translocation *T(I;III)3717* (1578). Aconidial flat morphology (1580). No corresponding point mutant is known.

vma: vacuolar membrane ATPase

This multiple-subunit complex acts as a proton pump to acidify the vacuole and related intracellular compartments (208). The vacuolar membrane ATPase complex is organized into a hydrophobic sector, V_0, that conducts protons through the membrane, and a protruding hydrophilic sector, V_1, that hydrolyzes ATP. Reviewed in ref. (1271) (Fig. 66).

vma-1: vacuolar membrane ATPase-1

VR. Between *sp* (8%) and *am* (2%; ~40 kb). Linked to *vma-3* (~100 kb), *cyh-2*, *leu-5* (0T/17) (1447, 1920).

Cloned and sequenced: Swissprot VATA_NEUCR, EMBL/GenBank J03955, EMBL NCVMA1A, GenBank NEUVMA1A; pSV50 clone 14:4A.

Specifies vacuolar ATPase (EC 3.6.1.3), 67-kDa subunit (subunit A) (215) (Fig. 66). This subunit contains the site

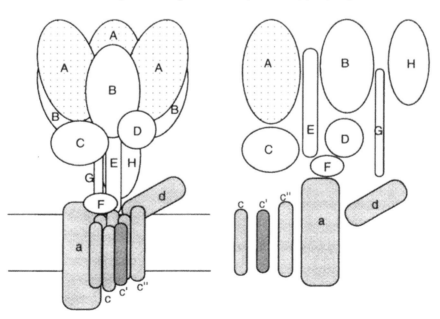

FIGURE 66 A diagram of the vacuolar membrane ATPase. Subunits in the hydrophilic portion have been given capital letters and those associated with the membrane lower-case letters. Symbols: A, Vma-1; B, Vma-2; D, Vma-8; E, Vma-4; F, Vma-7; G, Vma-10; H, Vma-13; a, Vph-1; c, Vma-3; c', Vma-11; d, Vma-6 (1271).

at which ATP is bound and hydrolyzed. Although the
vma-1$^+$ gene initially appeared to be essential (628),
mutants disrupted by RIP and lacking detectable
vacuolar ATPase activity proved to be viable even
though tip growth was inhibited, multiple branching
was induced, and they were unable to differentiate
conidia or perithecia (211a). The mutants were completely resistant to concanamycin C.

vma-2: vacuolar membrane ATPase-2

IIR. Between *preg* (6T/18 asci) and *Fsr-3* (1T/18 asci), *eas*
(2T/18 asci) (1447).

Cloned and sequenced: Swissprot VATB_NEUCR, EMBL/
GenBank J03956, EMBL NCVMA2A, GenBank
NEUVMA2A.

Encodes vacuolar membrane ATPase (EC 3.6.1.3), 57-kDa
subunit (subunit B) (Fig. 66). This subunit appears to
contain an ATP-binding site, but nucleotides bound at
this site are not hydrolyzed during catalysis (206).

vma-3: vacuolar membrane ATPase-3

VR. Between *his-1* and *al-3*. Linked to *vma-1* ($\sim$100 kb),
cyh-2, *leu-5* (0T/17 asci) (1447, 1920).

Cloned and sequenced: Swissprot VATL_NEUCR, EMBL/
GenBank L07105, EMBL NCVMA3A, GenBank
NEUVMA3A; pSV50 clone 3:4G.

Encodes vacuolar ATPase (EC 3.6.1.3), 16-kDa V$_0$
c-proteolipid subunit. This highly hydrophobic subunit
is embedded in the membrane and probably forms part
of the proton-conducting pathway through the membrane (1920) (Fig. 66).

vma-4: vacuolar membrane ATPase-4

IL. Linked to *nit-2* (0 or 1/16), *Fsr-12* (0T/18 asci). Left of
mat (4T/18 asci) (1447).

Cloned and sequenced: EMBL/GenBank U17641, EMBL
NC17641; pSV50 clone 14:5H.

Specifies vacuolar ATPase (EC 3.6.1.3), 26-kDa E-subunit.
This hydrophilic subunit may form part of the stalk that
connects the hydrophilic ATP-binding sector with the
integral membrane sector (214) (Fig. 66).

vma-6: vacuolar membrane ATPase-6

II. Near *Cen-II*. In the same cDNA fragment with *rap-1*
($\sim$1200 bp) (1317).

Cloned and sequenced: EMBL/GenBank U36470; Orbach-
Sachs clone X2B08.

Encodes vacuolar ATP synthase (EC 3.6.1.3), 41-kDa
subunit (Fig. 66). This polypeptide is tightly associated
with the membrane, but does not appear to have any
membrane-spanning hydrophobic regions (1317).

vma-7: vacuolar membrane ATPase-7

IVR. Linked near *Fsr-62* (2073).

Cloned and sequenced: GenBank AF099136; Orbach-Sachs
clones G17C10, G20D10.

Encodes vacuolar ATPase (EC 3.6.1.34), 13-kDa F subunit
(2073) (Fig. 66).

vma-8: vacuolar membrane ATPase-8

IVR. Right of *trp-4* (3T/18 asci). Linked to *Fsr-13* (0T/18
asci) (1447).

Cloned and sequenced: EMBL/GenBank AF053230;
Orbach-Sachs clone G15H03.

Encodes vacuolar ATP synthase (EC 3.6.1.34), 28-kDa
D-subunit in the hydrophilic sector of the ATPase.

vma-10: vacuolar membrane ATPase-10

IIR. Linked near *arg-12*.

Cloned and sequenced: EMBL/GenBank U84904; Orbach-
Sachs clones G4E11, G6F12.

Encodes vacuolar ATPase (EC 3.6.1.3), 13-kDa G-subunit
(Fig. 66). This subunit is in the hydrophilic sector and
may function to connect the membrane-embedded and
peripheral sectors of the enzyme (941).

vma-11: vacuolar membrane ATPase-11

IR. Between *al-2* (1T/18 asci) and *arg-13* (5T, 2NPD/18
asci). Linked to *sdv-4, -5* (0 or 1T/18 asci) (1447).

Cloned and sequenced: EMBL/GenBank AF162776;
Orbach-Sachs clones X2A03, X2H07, G6B11,
G25A04.

Vacuolar ATP synthase (EC 3.6.1.3), 16-kDa subunit. High
sequence similarity to *vma-3*. This highly hydrophobic
subunit is embedded in the membrane and probably
forms part of the proton-conducting pathway through
the membrane (1645) (Fig. 66).

vph-1: vacuolar pH-sensitive ATPase-1

IIL. Left of *Cen-II* (5T/18 asci), *Fsr-32* (4T/18 asci). Linked
to *Fsr-52* (0T/18 asci) (1447).

Cloned and sequenced: EMBL/GenBank U36396.

Encodes vacuolar membrane ATPase 98-kDa subunit. The
N-terminal half of this polypeptide appears to be
hydrophilic, and the C-terminal half has six or seven
putative membrane-spanning regions (210) (Fig. 66).
Named to conform with its yeast homologue, *VPH1*.

vvd: vivid

VIL. Between *T(IBj5)*^{L}, *ylo-1* and *T(IBj5)*^{R}, *cpc-1* (817,
1578, 1589). Not in a cosmid containing *ylo-1* and *lys-5*
(1815).

Intense color. The level and speed of synthesis of carotenoids is elevated in al^+ ylo^+ strains, in the ylo-1 mutant, and in strains with pigmenting mutant alleles of the *albino* genes. Mutant *vvd* alleles have been obtained independently in three laboratories (573, 818–820, 1589). Expression of *con-6* and *con-10* in the light is increased when the *vvd* mutation is present (573).

wa: washed

VR. Between *al-3, inl* (6%), *pk, T(EB4)^L* and *ad-7* (4%) (1546, 1578, 1585).

Surface growth is spreading and thin on minimal medium, with sparse conidiation (1585). Restricted and colonial on glycerol complete medium. Called *spco-2* (719).

wc: white collar

Carotenoids are absent from mycelia, whereas conidia become pigmented (although slowly). Named for the appearance of the mutant in agar slants, where a nonconidiating rim at the top of the slant remains white. A double mutant of *wc* with *fluffy* or another nonconidiating gene remains white and would be classed as

albino. The *wc-1* and *wc-2* genes encode regulatory proteins that are involved in blue-light-induced transcriptional control (1205).

wc-1: white collar-1

VIIR. Between *met-9* (1%; 4%) and *for* (6%) (1417, 1585, 1592).

Cloned and sequenced: Swissprot WC1_NEUCR, EMBL/ GenBank X94300, GenBank NCDNAWC1, PIR S69206.

Blue-light-response regulator, encoding a GATA-family zinc-finger protein that appears to act as a transcription factor with the *wc-2* gene product (Fig. 67) (91, 93, 1206, 1225). Mutants are blind (515, 1141, 1208). Homodimerization and heterodimerization (with WC-2) demonstrated *in vitro* (92); the complex of WC-1 and WC-2 has been observed *in vivo* (2040). Light-dependent hyperphosphorylation of WC-1 is related to protein turnover (2040). Plays an essential role in establishing the circadian oscillator (144, 431). Useful as a genetic marker (1418, 1556). Scoring is clearest at high temperatures (34°C). *T(P73B159)* is inseparable from *wc-1* (1578).

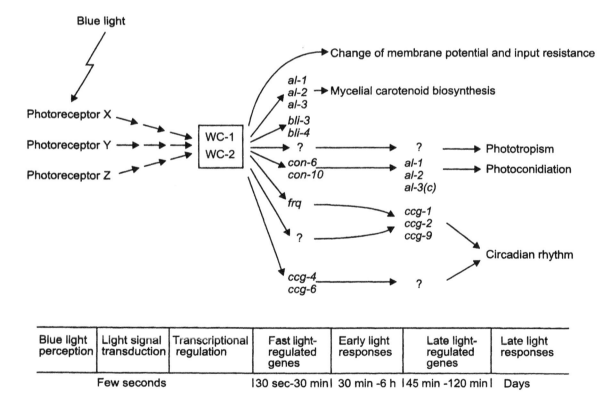

FIGURE 67 A model of photoresponses mediated by WC-1 and WC-2. Photoreceptors (unidentified) respond to light and trigger the activation of WC-1 and WC-2. This leads to the activation of both fast- and late-responding genes. In addition to inducing transcriptional responses for the genes indicated (and other genes indicated by question marks), signals mediated by WC-1 and WC-2 result in other alterations, such as a change in membrane potential. Redrawn from ref. (1206).

wc-2: white collar-2

IR. Between *thi-1* (8%), *T(ALS182)* and *ad-9* (7%; 20%) (1582).

Cloned and sequenced: Swissprot WC2_NEUCR, EMBL/ GenBankY09119, GenBank NCWC2.

Resembles *wc-1* phenotypically. A regulator of the blue light response, encoding a GATA-family zinc-finger protein that appears to act as a transcription factor, possibly interacting with WC-1 (91, 1206, 1207, 1225). Mutants are blind (515, 1141, 1208). Homodimerization and heterodimerization (with WC-1) *in vitro* (92); the complex of WC-1 and WC-2 has been observed *in vivo* (2040). Essential in establishing the circadian oscillator (144, 431).

ws-1: white spore-1

VIR. Between *trp-2* (38%), *T(UK14.1)*, *T(OY320)* and *Tip-IVR* (1578, 1618).

Ascospores fail to darken or do so slowly. Expressed autonomously. Black spots appear on some *ws-1* ascospores. In aged crosses, a few percent of *ws-1* ascospores darken and are capable of germination. Fertile and prototrophic, with normal tyrosinase activity vegetatively. Photograph of segregating asci (1618). Second-division segregation frequencies of 80 and 96% (1138, 1618) are significantly greater than two-thirds, providing evidence for chiasma interference.

ws-2: white spore-2

VI. Linked to *ylo-1* (16%), *trp-2* (2%), *Cen-VI* (9%; 24%). Recombines with *ws-1* (1102, 1138, 1616).

Ascospores are white initially, browning with age. Expressed autonomously (1592).

xdh-1: xanthine dehydrogenase-1

II. Linked to *pe* (14%), *alc-1* (24%) (1712).

Lacks xanthine dehydrogenase (EC 1.1.1.204) (XDH; xanthine oxidase). Defective in purine catabolism, of which XDH catalyzes the first step (Fig. 8). Unable to use hypoxanthine as the sole nitrogen source. Mutants *nit-1*, *-7*, *-8*, and *-9* are deficient in XDH activity because of a defect in the molybdenum-containing cofactor that is common to XDH and nitrate reductase (1441, 2101, 2102).

ylo: yellow

This name is used for mutants with yellow carotenoids. Ease of recognition of carotenoid differences depends on the light source. Hue appears to differ under various fluorescent lamps or in daylight. The difference from wild type is less marked in young cultures. In addition to mutations at loci designated *ylo*, some alleles of *al-1* produce lemon yellow carotenoids (especially ALS4 and RES-25), although in reduced quantity (2018, 2020, 2084). Some morphological mutants may appear yellowish or pale rather than wild-type orange. Saturation is modified by *vvd* and by genes such as *eas* and *os*.

ylo-1: yellow-1

VIL. Between *cys-1* (8%), *T(IBj5)^L* and *un-13* (2%; 38 kb), *cpc-1* (1%; 3%; 143 kb), *Bml* (1815, 2182).

Cloned and sequenced: Orbach-Sachs clone G17E11 (1815, 2182).

Encodes a putative aldehyde dehydrogenase (1815). Mutants are defective in the genesis of carotenoids. Both conidia and mycelia become yellow. The exact nature of the defect is not known; no one specific stage of carotenoid biosynthesis appears to be blocked. Small amounts of several carotenoids, including β-carotene, are accumulated, but neurosporaxanthin, the major carotenoid pigment in wild type, is not (748, 1785). A role has been suggested for *ylo-1* in the conversion of either lycopene to 3,4-dehydrolycopene or torulene or γ-carotene to neurosporaxanthin (748) (Fig. 7). The structural similarity between neurosporaxanthin and retinoic acid suggests that late steps of carotenoid synthesis in *Neurospora* may resemble those in retinoic acid metabolism (1815). The appearance of mutant strains is nearly wild type in young cultures, but a color difference becomes clear with age. The apparent color and the ease of scoring depend on the illumination source, being quite different under fluorescent and incandescent lamps than in daylight. Undefined modifiers affect intensity.

ylo-2: yellow-2

IL. Between *suc*, *In(H4250)^L* and *arg-1* (1%) (1592).

Conidia are yellow and rather sparse (1592). The yellow color may be a secondary effect of abnormal development. First shown to be distinct from *ylo-1* in ref. (1031). Inferior to *ylo-1* as a marker.

ylo-3: yellow-3

A slow-conidiating allele of *fl* (1594).

ylo-4: yellow-4

A pigmenting *al-1* allele, called *age-3* as well as *ylo-4* (1394).

ypt-1: ypt-like-1

IIR. Linked to *eas* (0T/17 asci) (867).

Cloned and sequenced: Swissprot YPT1_NEUCR, EMBL/ GenBank S51252, PIR S30096; EST NP2C4.

Belongs to the *ypt* subfamily of *ras*-like genes, encoding a small GTP-binding protein thought to be involved in intracellular transport or secretion. Obtained by screening cDNA libraries with *Zmypt* genes from maize. Mutant phenotype not known (867).

zip-1: *zip-like-1*

IL. Linked to *cyt-21* (0T/18 asci) (1446, 1452).

Cloned and partially sequenced. GenBank AA898030, AA901718. cDNA clone NP5E10.

Encodes a bzip protein related to the human transcription factor AP-1 (*c-jun*) and to the CCAAT/enhancer-binding protein of *Caenorhabditis* (1446, 1634a).

znr: *zinc resistant*

Three nonidentical strains show 2- to 3-fold resistance to Zn^{2+} in liquid medium. They differ, however, in inhibition of hyphal growth rate by zinc (*znr-3* > *znr-2* > *znr-1*) and in cross-resistance to cobalt and copper (1699). The mutant strains called *znr-1* and *znr-2* may be complex, having been obtained after repeated subculture on selective medium and being associated with a IV-linked chromosome rearrangement.

znr-1: *zinc-resistant-1*

IV? Associated with a chromosome aberration linked to IV. Not cross-resistant to cobalt or copper (1699).

znr-2: *zinc resistant-2*

IV? Associated with a chromosome aberration linked to IV. Cross-resistant to both cobalt and copper (1699).

znr-3: *zinc resistant-3*

I. Linked to *mat* (1699).

Selected for growth on 16 mM Zn^{2+} in agar medium. The rate of zinc uptake is reduced. Cross-resistant to cobalt (1699).

Neurospora Genetic Nomenclature[*],[†]

Introduction
1. Gene Names and Symbols
  1.1. Names
  1.2. Symbols
  1.3. Dominance and Recessiveness
  1.4. Gene Loci Recognized by DNA Sequence
  1.5. Genes in Ectopic Positions
  1.6. Gene Fusions and Transposable-Element
    Insertions
  1.7. Priority and Synonyms
  1.8. Changing Gene Names and Symbols
  1.9. Multilocus Genotypes
  1.10. Species Other Than *N. crassa*
2. Alleles
  2.1. Symbols
  2.2. Pseudogenes
  2.3. Isolation Numbers (Allele Numbers)
  2.4. Culture Collection Accession Numbers
3. Heterokaryons
4. Distinguishing Generations in a Pedigree
5. Genotypes and Phenotypes
6. Gene Products
7. Distinguishing "Locus" and "Gene"
8. Linkage Groups and Chromosomes
9. Chromosome Rearrangements
10. Wild Types
  10.1. Laboratory Wild Types
  10.2. Strains from Nature
11. Transposable Elements
12. Mitochondria
  12.1. Mutant Mitochondrial Genomes
  12.2. Individual Mitochondrial Genes
  12.3. Mitochondrial Plasmids
13. Use of Italics
References

* From *Fungal Genetics Newsletter* **46**: 35–42 (1999).
† David D. Perkins (Department of Biological Sciences, Stanford University, Stanford CA 94305-5020), with the advice of Rowland Davis, Jay Dunlap, Alan Radford, Matthew Sachs, and Tony Griffiths.

INTRODUCTION

Rules for the genetic nomenclature of *Neurospora crassa* have been evolving since 1941. This document is a guide to present usage. The basic *Neurospora* conventions antedate genetic nomenclature of bacteria and other microorganisms and follow *Drosophila* nomenclature more closely. Rules for *Neurospora* were last summarized in 1982 (16). The expanded version given here incorporates subsequent refinements and changes (e.g., 5, 7, 13). It aims to provide a system that is adaptable to new developments in molecular genetics, while retaining the gene names and working vocabulary that have become established over the years.

Most of the existing *Neurospora* nomenclature conforms to the rules set forth here, but some does not. To avoid confusion, we have usually refrained from changing established names and symbols that do not conform to current usage. Past practice and the retention of names and conventions that flout the present rules should not be taken as an excuse for bad practice in the future.

What may be considered standards for *Neurospora* nomenclature are provided by usage in the current compendium and in the latest lists and maps published by the Fungal Genetics Stock Center (FGSC). The present document has benefited from recent descriptions of *Drosophila* and maize nomenclature (6, 2). The detailed rules for *Drosophila* published by *Flybase* (6) may provide guidance for problems that are not considered here.

1. GENE NAMES AND SYMBOLS

1.1. Names

Like *Drosophila*, *Neurospora* has a relatively well-defined wild-type phenotype. In the formative years with both organisms, existence of a gene was recognized when a mutation occurred that deviated from the wild type. The gene was then named using a word that described the mutant phenotype. Gene loci recognized on the basis of naturally occurring variants (e.g., mating-type idiomorphs, vegetative incompatibility genes, isozyme markers) were named according to the phenotype affected. Descriptive gene names were given in preference to using numbers or nondescriptive names. They are informative, easier to remember, and less likely to result in confusion with other loci.

In choosing what aspect of the phenotype to use as a basis for naming a mutant gene, preference was given to the most convenient and useful manifestation. For example, all arginine auxotrophs were named "arginine" rather than being given different names based on the earliest utilizable precursor (citrulline, ornithine, etc.) or based on the enzyme that was rendered nonfunctional. A gene specifying the molybdenum cofactor that is shared by nitrate reductase and xanthine dehydrogenase was named "nitrate-8" rather than "molybdenum cofactor" or "xanthine dehydrogenase" because the mutant is scored as a nitrate nonutilizer. These considerations still hold.

Gene names should be concise and informative. Each name must be unique and must not have been used previously for a *Neurospora* gene. Gene names or symbols should not be prefixed with the word *Neurospora* or the letters *n* or *nc* to indicate that a gene is from *Neurospora*. To do so would be redundant. Sequence database identification code entries, which often begin with NC or NEU, are not gene names or symbols, nor do they establish priority.

Different loci bearing the same name and the same base symbol should be numbered sequentially beginning with 1, e.g., *arg-1, arg-2, arg-3*. If a name applies to only one locus, use of the number 1 is optional. For example, the gene that specifies invertase is symbolized *inv* rather than *inv-1*. Arbitrary strain-identification numbers should not be converted into locus numbers.

Regulatory genes have usually been given the same name and symbol as the structural genes they regulate (e.g., *nit-2, leu-3, cys-3*), but this is not always true (e.g., *pcon, pgov, scon, ty*).

When new names, symbols, locus numbers, or allele number prefixes are to be assigned, it is essential to avoid duplication by consulting the most recent FGSC stock list and the lists that accompany the current genetic maps.

1.2. Symbols

Symbols preferably are three-letter abbreviations of the gene name, but they may consist of two letters or (rarely) one or four. Symbols are written in lower case italics (e.g., *inv*), except when the name is based on a mutant allele that is dominant. The first letter is then capitalized (e.g., *Asm*). Nonallelic genes that have the same descriptive name and symbol are distinguished from one another by numbers that are separated from the base symbol by a hyphen (e.g., *al-1, al-2, al-3*). This use of hyphens in *Neurospora* and *Drosophila* differs sharply from the convention in many other organisms, where a locus number (or letter) is not

separated from the base symbol. Hyphens are used only to separate the locus number from the base symbol to which it is appended.

When a gene name contains a number that is necessary for identifying the product or phenotype, the product-identifying number is included as an integral part of the base symbol, with digits unseparated from the letters by a hyphen (e.g., *tom22; nuo78*). A hyphen then can be used if it is needed to distinguish locus numbers from numerals belonging to the gene name (e.g., *hsp70-1, hsp70-2*) (ref. 13).

Roman numerals should be avoided in gene symbols.

Suppressors are symbolized by using the letters *su*, immediately followed by the symbol of the suppressed gene in parentheses. If nonallelic suppressors of the same gene are known, locus numbers follow the parentheses [e.g., *su(met-7)-1, su(met-7)-2*]. As in *Drosophila, su$^+$* designates the wild-type gene and *su* the mutant suppressor allele. For allele-specific suppressors, the allele number is included as a superscript of the locus symbol [e.g., *su(trp-3^{td201})-2*]. Enhancers are symbolized in a similar way [e.g., *en(am)-1*].

The mating-type genes, formerly called *mt*, are now symbolized *mat* (7). The *mat* locus is occupied by either of two nonhomologous sequences, *mat A* or *mat a*. These are called idiomorphs rather than alleles (10). In most contexts, the abbreviated symbol *A* or *a* is used. Only when the context requires complete specification of the idiomorph is it necessary to use the more ponderous symbols *mat A-1, mat A-2, mat A-3* (for A) or *mat a-1* (for a). Unless the context requires that *mat* be shown in the actual order of its location on the map, the mating-type symbol follows all other symbols when multilocus genotypes are written (e.g., *leu-3 cr-1 al-2 A* or *cot-1 a*)

Chromosomal loci other than genes usually have the initial letter of the symbol capitalized (e.g., *Cen, Tel, In, T*). The initial letter is also usually capitalized in symbols for active or relic transposons (e.g., *Tad, Pogo*).

1.3. Dominance and Recessiveness

When a gene is named for a mutant phenotype that is recessive to the wild type, the name and symbol are written in lower-case letters (e.g., *al: albino*). The initial letter is capitalized when the mutant phenotype is dominant (e.g., *Ban: Banana*). The initial letter is not capitalized when a gene is named for alleles that show codominance (e.g., *het: heterokaryon incompatibility*).

Mutant phenotypes may be expressed in either the vegetative phase, the sexual phase, or both. Some mutant genes are known to be dominant in the sexual phase but recessive in vegetative tissues. The initial letter of the name and symbol is then capitalized if the gene name is based on the dominant mutant phenotype (e.g., *R, Asm*), but the initial letter is not capitalized if the name is based on the recessive mutant phenotype (e.g., *mei-3, pk^D*).

Dominance or recessiveness usually is not known at the time new vegetative-phase mutants are named. In the absence of that information, lower-case symbols are used routinely because recessive loss-of-function mutations are the most common type to be detected phenotypically. Tests for dominance in the vegetative phase may employ either heterokaryons or heterozygous partial diploids. Partial diploids are preferred because they ensure a 1:1 allele ratio, whereas the ratio of nuclear types in heterokaryons may depart widely from equality. Partial diploids are obtained as duplication progeny from crosses heterozygous for insertional or quasiterminal rearrangements.

Mutant genes that are recognized by their expression in the perithecia of heterozygous crosses immediately are known to be dominant (e.g, *R*, *Asm*). Recessive sexual-phase mutations are less likely to be detected because they must be present in both parents of a cross in order to be expressed. Many of the known sexual-phase recessives were recognized in crosses homozygous for mutant genes affecting mutagen sensitivity and DNA repair (e.g., *uvs*, *mus*). These had already been detected and named as recessive vegetative-phase mutants. Other recessive sexual-phase mutants have come from backcrosses in experiments specifically designed to detect them (9). Still others were discovered accidentally in crosses between inbred parents (e.g., *mei-1*, *mei-3*).

1.4. Gene Loci Recognized by DNA Sequence

We need no longer depend on mutant differences. cDNA libraries and sequencing now make it possible to recognize genes for which no variant product or phenotype has been detected. These "anonymous" genes can be placed on the genetic map by using them as probes in RFLP mapping. In the absence of a known mutant phenotype, gene names may be based on the time or site of expression (e.g; *con*). The null mutant of such a gene may (e.g., *asd-1*) or may not (e.g., *con-11*) reveal a conveniently recognizable mutant phenotype on which to base a descriptive name. If the null mutant is lethal (as with *tom19* and *tom22*, for example), if it is phenotypically wild type, or if the mutant phenotype remains undetermined, it is appropriate and informative to base the name on sequence homology with a gene or gene family, the function of which is known in another organism (e.g., *ras*: *ras-like*, *pzl*: *phosphatase-z-like*). This should be done, however, only if the sequence makes a strong prediction of function. A *Neurospora* gene should not be named for the overt phenotype of its homolog in another organism if that phenotype is developmentally complex and far removed from the primary gene product. Manifestation of the genes may have diverged in the two organisms, resulting in quite different phenotypes. For example, mutations in homologous genes appear to be responsible for cerebrohepatorenal anomalies in humans and for the failure of premeiotic nuclear fusion in the croziers of *Podospora* (3) and *Neurospora* (18).

If neither phenotype nor homology is known, a gene may be given a generic symbol indicating anonymity. The symbol *anon* is used in *Drosophila*, with some distinguishing suffix, and this is recommended for *Neurospora*. An alternative that has been proposed is *eat* (*encodes anonymous transcript*) (17). The meaning of *eat* is not obvious from the symbol, however. Generic names and symbols of this type, which represent a category of mutants rather than a specific mutant, have a long history of use in *Neurospora*. Best known is the use of *un* for temperature-sensitive genes of unknown function. Other generic categories are *ccg* for clock-controlled genes, *con* for genes expressed during conidiation, and *sdv* for genes expressed under conditions favoring sexual development

When a mutant phenotype or a definitive sequence homology is discovered for an anonymous mutant, the option exists of changing the name to something more definitive. For example, if the null allele of a gene initially called *anon(NP6C9)* were found to result in restricted colonial growth, the name could be changed to *col-x*.

Different *anon* genes are best distinguished using isolation numbers, as in the example, because if the genes were numbered serially, a clearing house would be needed to avoid using the same number repetitively.

1.5. Genes in Ectopic Positions

A *Neurospora* gene that has been integrated ectopically is designated by appending *(EC)* to the gene symbol, e.g., *am(EC)*. The genotype of a strain with a gene deleted from its normal position and a wild-type copy of the same gene inserted elsewhere in the genome thus would be symbolized *am; am⁺(EC)*.

1.6. Gene Fusions and Transposable-Element Insertions

A double colon is employed to indicate the genes or elements that have been joined. For example, *mtr::Asm-1⁺(EC)* or *am::Tad*. Symbols cannot be expected to convey full information about complex constructs or genotypes. This is best done in the text or by using a figure.

1.7. Priority and Synonyms

Where differences exist in published names for the same gene, the symbol and name adopted are those that were used when the gene was first reported to be at a previously unknown chromosomal locus — usually when the gene was first mapped. For example, when sequence comparisons and allelism tests revealed that mutations called *ccg-2* and *bli-4* are allelic with the already established gene *eas*, the two newly coined names became inactive synonyms of *eas*. Inactive synonyms should never be used later for another gene.

Where the same symbol has inadvertently been used for two genes with different names, the symbol that was published first is retained and a different symbol is assigned to symbolize the name of the other locus. For example, if the long-established symbol *nd* (*natural death*) were accidentally used for a gene that specified NADH ubiquinone reductase, it would be necessary to find a new symbol for the latter.

Priority is established by publication in a refereed journal or book or in an article in *Fungal Genetics Newsletter*. Use of a gene name or symbol in conference abstracts, *Dissertation Abstracts International*, or in unpublished theses does not establish priority, although names and symbols reported there may be adopted in the absence of any conflict. Sequence databank codes are not gene symbols and do not confer priority.

1.8. Changing Gene Names and Symbols

The name and symbol of a mutant gene should be changed only for compelling reasons, as when the original name is found to be incorrect or misleading. Reasons for making the change should be clearly stated, as was done in changing *met-4* to *cys-10* or *ol* to *cel*. For examples, see ref. 13.

Gene loci that bear generic names, such as *anon*, *ccg*, *con*, *eat*, *sdv*, or *un*, present a special case. When a definitive phenotype, function, or homology is discovered for such a gene, the question arises whether to propose substituting a more informative name for the generalized original. An investigator may propose to abandon the generic name, as was done when *un(STL6)* was changed to *fls*, *sdv-10* to *asd-1*, and *con-8* to *phr*, for example. On the other hand, the decision may be to retain the generic name, as was done when gene products were identified for *un-18* and *un-24*. The decision whether to rename an *un* mutant may be influenced by the consideration that scoring and recognition are best accomplished on the basis of temperature sensitivity. Decisions whether to change a name will depend not only on the anticipated usefulness of a more specific name but also on how firmly the original name is established and how widely it has been used.

1.9. Multilocus Genotypes

When more than one locus in the same linkage group is to be shown, symbols are written in the linear order of loci on the conventional linkage map and are separated from one another by single spaces, e.g., *cr-1 al-2 nic-1*. Commas are not used. When a genotype includes markers from different linkage groups, the groups are separated by semicolons and spaces, e.g., *cr-1 al-2; am inl; nic-3*, for markers in linkage groups I, V, and VII. In designating multilocus genotypes, wild type is implied for a specific gene if no symbol is given for the locus. For example, *cr-1 al-2 A × nic-1 a* implies *cr-1 al-2 nic-1⁺ A × cr-1⁺ al-2⁺ nic-1 a*. If there is no ambiguity when genotypes are written out, a simple unraised + sign may be used to indicate the relevant wild-type allele, for example, *cr-1 al-2 + A × + + nic-1 a*. Acronyms may be used to represent complex genotypes, e.g., *alcoy, multicent*.

1.10. Species Other Than N. crassa

Names and symbols for genes in other *Neurospora* species should be identical to those of their N. *crassa* homologs when the homology is securely known. A nonconforming name should be changed unless this seems inadvisable because of long-established usage in the other species.

2. ALLELES

2.1. Symbols

Where there is a standard wild-type allele in a defined laboratory strain, the locus symbol without a superscript represents the mutant allele. The same symbol with a + superscript designates the wild-type allele (Bml^+). In designating genotypes, the symbol + is reserved for the wild-type allele. Multiple alleles, alleles differing in their resistance to a toxic agent, or allelic genes having no definitive wild type are distinguished by appropriate superscripts (e.g., frq^1, frq^2, frq^3; $cyh-1^R$, $cyh-1^S$; $het-6^{OR}$, $het-6^{PA}$; a^{m1}, a^{m33}). Intragenic deletions are treated as alleles at the gene locus. Deletions of an entire single gene or a large portion of it are designated by prefixing the symbol with a capital Greek delta, e.g., Δam.

When superscripting is impossible, as in ASCII, the superscripted text is enclosed in square brackets. Thus, frq^7 would be written *frq[7]*.

2.2. Pseudogenes

If DNA sequence indicates that a locus is occupied by a defective member of a gene family, the locus is named as a member of that gene family. If it is a pseudogene, that fact may be shown by appending the letters *ps* to the base symbol as a superscript (e.g., $Fsr63^{ps}$, a 5S RNA pseudogene). An active gene may be found later that is allelic with such a pseudogene (see, for example, *Fsr33*).

2.3. Isolation Numbers (Allele Numbers)

Allelic mutations bear identical locus symbols and locus numbers. Each new mutation originating independently at a gene locus is assigned a unique isolation number (often called allele number), even though it is phenotypically indistinguishable from the mutants known previously. Isolation numbers are commonly prefixed by letters indicating the laboratory of origin. A list of letters already used as prefixes is published in the FGSC catalog following the section of single-mutant stocks. Isolation numbers are not part of the gene symbol and are not displayed except when necessary to distinguish between alleles. The number may then be shown in parentheses after the full locus symbol, e.g., *pyr-3*(KS43). When a new mutant gene has not yet been assigned a locus number, pending tests for allelism with similar genes at previously established loci, the mutant may be designated temporarily by an appropriate letter symbol followed immediately by the allele number in parentheses, e.g., *ilv(STL6)*.

2.4. Culture Collection Accession Numbers

Culture collections assign an arbitrary number to each stock as it is acquired. This number is usually prefaced by initials of the organization, e.g., FGSC, ATCC, CMI, or CBS. The same strain may have more than one accession number if it is included in two or more collections. Care should be taken to distinguish accession numbers from isolation (allele) numbers. The culture collection accession number may be provided to identify the exact source of a strain that was used. It should not be used in place of the isolation number to identify what allele was used.

When depositing strains in a culture collection, investigators should make sure that each strain is given a unique identification number and that allele numbers are provided for all mutant genes.

3. HETEROKARYONS

Genotype symbols for the component nuclei of a heterokaryon are separated by a plus sign. Parentheses are used to enclose symbols defining the entire heterokaryon, for example, (*col-2 A + ad-3B cyh-1 A*) (13).

4. DISTINGUISHING GENERATIONS IN A PEDIGREE

Haploid parent and progeny strains in a series of crosses are called p_1, f_1, etc., using lower-case letters. Backcross generations also are designated using lower-case letters, b_1, b_2, etc. (The progeny from $f_1 \times p_1$ are designated b_1, progeny from $b_1 \times p_1$ are b_2, etc.) This deviation from the upper-case symbols used with diploid organisms (P_1, F_1, etc.) was introduced in 1924 for haploid gametophytes of the bryophyte *Sphaerocarpus* (1). Lower-case letters were first used for *Neurospora* by Dodge in 1928 (4).

When subscripting is impossible, as in ASCII, subscripted text is enclosed in doubled square brackets, e.g., b_2 would be written b[[2]].

5. GENOTYPES AND PHENOTYPES

Genotype symbols are italicized, and phenotype symbols are not. Genotype designations include not only genes but

also centromeres (e.g., *Cen-VII*), *telomeres* (*Tel-V*), the nucleolus organizer (*NO*) or nucleolus organizer region (*NOR*), and chromosome rearrangements [e.g., *T(III;-V)AR177*), *In(IL;IR)OY323*), *Dp(VL→IVL)AR33*; see ref. (14)]. The locus of a gene is designated by using the base symbol without a superscript. Acronyms and abbreviations for complex genotypes are italicized (e.g., *alcoy*, *multicent*).

A phenotype symbol is obtained by converting each relevant base gene symbol to nonitalic, capitalizing the initial letter, and adding "+", "−", or another allele designation as a superscript. For example, a strain of genotype *al-2 arg-6* is phenotypically Al⁻ Arg⁻, a strain of genotype *al-2⁺ arg-6⁺* is phenotypically Al⁺ Arg⁺; a partial-diploid strain of genotype *al-2 arg-6/al-2⁺ arg-6⁺* is phenotypically Al⁺ Arg⁺, and a *cpl-1ˢ* strain is phenotypically Cpl^S. Double mutants for the same function may be designated phenotypically using the shared base symbol. Thus, the phenotype of an *arg-2; arg-3* strain is written Arg⁻.

6. GENE PRODUCTS

The protein products of genes are represented by the same characters used to designate the genes encoding them, written in nonitalic capital letters (5). For example, the protein product of the *preg* gene is PREG and that of *inv* is INV. If the name of a gene product is written in full, capitalization is unnecessary.

7. DISTINGUISHING "LOCUS" AND "GENE"

"The words 'locus' and 'gene' should not be treated as synonymous. A locus can be defined as a chromosomal site of variable size at or within which is located a gene, a restriction site, a knob, a breakpoint, an insertion, or other distinguishable feature" (ref. 2). A gene is a DNA sequence that is regularly or conditionally transcribed at some time during the life cycle.

8. LINKAGE GROUPS AND CHROMOSOMES

Linkage groups, which are defined genetically, are designated by Roman numerals I–VII. These are not italicized except when they are included in the symbol for a chromosome rearrangement. Chromosomes, which are defined microscopically or physically, are designated by Arabic numerals. For correspondences between linkage groups and chromosomes, see refs. (12) and (15). Most genes are readily assigned to a linkage group, but obtaining direct information about their physical chromosome location is difficult. Linkage group numbers are therefore used rather than chromosome numbers in the symbols for chromosome rearrangements and for identifying electrophoretically separated chromosomal DNAs (12).

9. CHROMOSOME REARRANGEMENTS

Intragenic rearrangements or single-gene deletions are represented as alleles at the relevant locus. For rearrangements involving chromosome segments that contain two or more genes ("segmental rearrangements") the base symbols are *T* (translocation), *In* (inversion), *Tp* (transposition within the same chromosome), *Dp* (duplication), and *Df* (deficiency, synonymous with deletion) (13). The base symbol is followed in parentheses by Roman numerals indicating the relevant linkage group or groups. (*L* or *R* may be added to indicate the linkage-group arm.) The final element in the symbol is an identification number. The entire symbol is italicized, with no intervening spaces. In symbols for reciprocal translocations, linkage group numbers are separated by a semicolon [e.g., *T(IIIR;VR)P1226)*]. With insertional or quasiterminal rearrangements, the linkage groups are separated by an arrow indicating which is the donor and which is the recipient of the transferred segment [e.g., *T(IL→IIR)39311*]. Progeny from insertional or quasi-terminal rearrangements may contain two copies of the transposed segment. These are symbolized, as, for example, *Dp(IL→IIR)39311*.

When a rearrangement has two breaks in the same linkage group, superscripts may be used to distinguish left and right breakpoints [e.g., *In(OY323)^L*, *In(OY323)^R*; *T(39311)^L*, *T(39311)^R*]. If a rearrangement breakpoint is inseparable from the mutant phenotype of an associated gene, the gene symbol follows the rearrangement symbol and is separated from it by a space with no comma [e.g., *T(IR;IIR)4637 al-1*]. For further details, see ref. (14).

10. WILD TYPES

10.1. Laboratory Wild Types

Most experimental work has employed *N. crassa* markers and strains that were derived from relatively few wild-type progenitors. Names of laboratory wild-type strains are commonly abbreviated and are not italicized. Best known are Oak Ridge (OR), Standard (ST; commonly called St. Lawrence), Emerson (Em), and Rockefeller–Lindegren (RL), each of which exists in opposite mating types. Each of these has been used extensively in the past, in one or another laboratory and each has contributed mutant alleles that are still in use as markers. Cryptic genetic polymorphisms exist between all these strains and some-

times between *A* and *a* strains that bear the same name. The highly inbred, heterokaryon-compatible Oak Ridge strains OR23-1A and ORS-6a have been used most widely and have now been adopted as standards by most laboratories. For a pedigree showing the derivation of OR and other wild-type strains, see ref. (11).

10.2. Strains from Nature

A newly acquired strain is given an identification number by the laboratory first putting it in stock. It may then be deposited in culture collections where it is given different accession numbers. For example, a *N. crassa* strain from Panama was numbered CZ30.7 by the original collector, QM 4839 by the US Army Quartermaster Corps, and FGSC 1132 by the Fungal Genetics Stock Center. To avoid confusion, the original identification number CZ30.7 should be specified in designating what strain was used.

11. TRANSPOSABLE ELEMENTS

Names and symbols are italicized and the first letter is capitalized, e.g., *Pogo*, *Punt*, *Tad*, *Tourist*. This conforms with usage for chromosomally integrated mobile or relic transposable elements in maize and in other fungi.

12. MITOCHONDRIA

12.1. Mutant Mitochondrial Genomes

The names and symbols of mutant mitochondrial genomes (and strains carrying them) are enclosed in nonitalicized square brackets to distinguish them from nuclear genes. The symbols, but not the brackets, are italicized. For example, [*mi-2*], [*poky*], [*stp*].

12.2. Individual Mitochondrial Genes

The symbols proposed for fungi by Hudspeth (8) are recommended. Gene symbols are written unhyphenated, using lower-case italics and Arabic numerals, e.g., *cox1*, *atp6*. The corresponding gene products are symbolized using nonitalicized upper-case letters.

12.3. Mitochondrial plasmids

Plasmid names begin with a lower-case letter "p" followed by the name in capital letters, e.g., pKALILO. When additional members of a family are found, they are assigned numbers sequentially, e.g., pKALILO-2. Symbols incorporate the first three letters of the name, e.g., pKAL-2. Names and symbols of mitochondrial plasmids are not italicized.

13. USE OF ITALICS

The following are italicized: Gene names and symbols, including superscripts; symbols for individual rearrangements; symbols for centromeres, telomeres, and the nucleolus organizer; acronyms that represent genotypes; names and symbols of mitochondrial genes and mutant mitochondria (the latter are enclosed in nonitalicized brackets); and symbols for transposable elements.

The following are not italicized: Names and symbols of phenotypes and gene products; linkage group numbers (except when the Roman numeral is an integral part of the symbol for a rearrangement); names and abbreviations for wild-type strains; names of chromosome rearrangement types ("translocation," "inversion," etc.) when written in the text; and names and symbols of mitochondrial plasmids.

REFERENCES

1. Allen, C. E. (1924). Gametophytic inheritance in *Sphaerocarpos*. I. Intraclonal variation, and the inheritance of the tufted character. *Genetics* **9**: 530–587.
2. Beavis, W., and Maize Nomenclature Committee (1995). A standard for maize genetic nomenclature. *Maize Genet. Coop. Newsl.* **69**: 182–184.
3. Berteaux-Lecellier, V., M. Picard, C. Thompson-Coffe, D. Zickler, A. Panvier-Adoutte, and J.-M. Simonet (1995). A nonmammalian homolog of the *PAF1* gene (Zellweger syndrome) discovered as a gene involved in caryogamy in the fungus *Podospora anserina*. *Cell* **81**: 1043–1051.
4. Dodge, B. O. (1928). Production of fertile hybrids in the ascomycete *Neurospora*. *J. Agric. Res.* **36**: 1–14.
5. Dunlap, J. C., M. Sachs, and J. Loros (1996). A recommendation for naming proteins in Neurospora. *Fungal Genet. Newsl.* **43**: 72. (available at http://www.fgsc.net).
6. FlyBase (1997). Genetic nomenclature for *Drosophila melanogaster*. *Drosophila Information Service* **79**: 13–36 (available at http://flybase.bio.indiana.edu).
7. Glass, N. L., and C. Staben (1997). *Neurospora* mating type symbol *mt* revised to *mat*. *Fungal Genet. Newsl.* **44**: 64 (available at http://www.fgsc.net).
8. Hudspeth, M. E. S. (1992). The fungal mitochondrial genome—A broader perspective. *In* "Handbook of Applied Mycology. Volume 4: Fungal Biotechnology" (D. K. Arora, R. P. Elander, and K. G. Mukerji, eds.). pp. 213–241. Marcel Dekker, New York.
9. Leslie, J. F., and N. B. Raju (1985). Recessive mutations from natural populations of *Neurospora crassa* that are expressed in the sexual diplophase. *Genetics* **111**: 795–777.
10. Metzenberg, R. L., and N. L. Glass (1990). Mating type and mating strategies in *Neurospora*. *BioEssays* **12**: 53–59.
11. Newmeyer, D., D. D. Perkins, and E. G. Barry (1987). An annotated pedigree of *Neurospora crassa* laboratory wild-types, showing the probable origin of the nucleolus satellite and showing that certain stocks are not authentic. *Fungal Genet. Newsl.* **34**: 46–51.
12. Orbach, M. J., D. Volrath, R. W. Davis, and C. Yanofsky (1988). An electrophoretic karyotype of *Neurospora crassa*. *Mol. Cell. Biol.* **8**: 1469–1473.
13. Perkins, D. D (1996). Recommendation regarding *Neurospora* genetic nomenclature. *Fungal Genet. Newsl.* **43**: 73–75 (available at http://www.fgsc.net).
14. Perkins, D. D (1997). Chromosome rearrangements in *Neurospora* and other filamentous fungi. *Adv. Genet.* **36**: 239–398.
15. Perkins, D. D., and E. G. Barry (1977). The cytogenetics of *Neurospora*. *Adv. Genet.* **19**: 133–285.
16. Perkins, D. D., A. Radford, D. Newmeyer, and M. Björkman. 1982. Chromosomal loci of *Neurospora crassa*. *Microbiol. Rev.* **46**: 426–570 (available at http://www.fgsc.net).
17. Randall, T. A., and R. L. Metzenberg (1998). The mating type locus of *Neurospora crassa*: Identification of an adjacent gene and characterization of transcripts surrounding the idiomorphs. *Mol. Gen. Genet.* **259**: 615–621.
18. Howe, K., and M. A. Nelson, Personal communication.

Genetic Maps and Mapped Loci[*]

Loci are displayed on the drawn maps only if their order is established with reasonable certainly, either from meiotic crossing over in multiple-point crosses, duplication coverage, or physical mapping. Loci are listed below the maps but are not displayed if their order is uncertain relative to the loci that are shown, except that closely linked loci may sometimes be shown on the map in parentheses even if their relative order is unknown. Evidence for linkage is given in the lists and in the entry for each locus in the text. Crossover values and interval lengths are approximations. The numbers on which they are based are often small, and the various stocks used for mapping differ in genes that affect meiotic recombination. Linkage group I is estimated to be at least 200 map units long, and the total length for all seven groups probably exceeds 1000.

Rearrangement breakpoints are shown only if their map positions have been determined precisely. Insertional and quasiterminal rearrangements are usually the best mapped. Many of these duplication-generating rearrangements have been instrumental in determining the order of flanking gene loci by means of duplication-coverage tests. When a rearrangement involves more than one break in the same chromosome, the breakpoints are distinguished by super-scripts *L* (left), *R* (right), or *M* (middle). Symbols for rearrangements with quasiterminal breakpoints are shown below the relevant linkage-group tip. Rearrangement symbols are abbreviated. For example, *T(IL→IIR)39311* is shortened to *T(39311)*. Full symbols can be found in the FGSC Stock List or in Perkins, D. D., *Adv. Genet.* **36**: 239–398 (1997), where descriptions and linkage relations of 355 chromosome rearrangements are given. (This reference is available at *www.fgsc.net.)*

Accompanying the map of each linkage group is a list of all the loci known with reasonable certainty to be located in that group. The listing is alphabetical by symbol. Neither the maps nor the lists include anonymous RFLP and RAPD markers. Recombination percentages and numerical fractions are from random ascospore progeny. When data are from asci, the numbers of tetratype (T) and nonparental ditype (NPD) asci are given, together with the total number. In statements using the format "locus *c* is between *a*, *b* and *d*, *e*," the word "and" separates markers to the left of *c* from markers to the right. If synonyms exist (text Table 2), only the preferred symbol is given. See entries in the main text for gene names, additional information, and references. Over 1000 loci have been assigned to linkage group.

---

[*] From *Fungal Genetics Newsletter* **47** (2000).

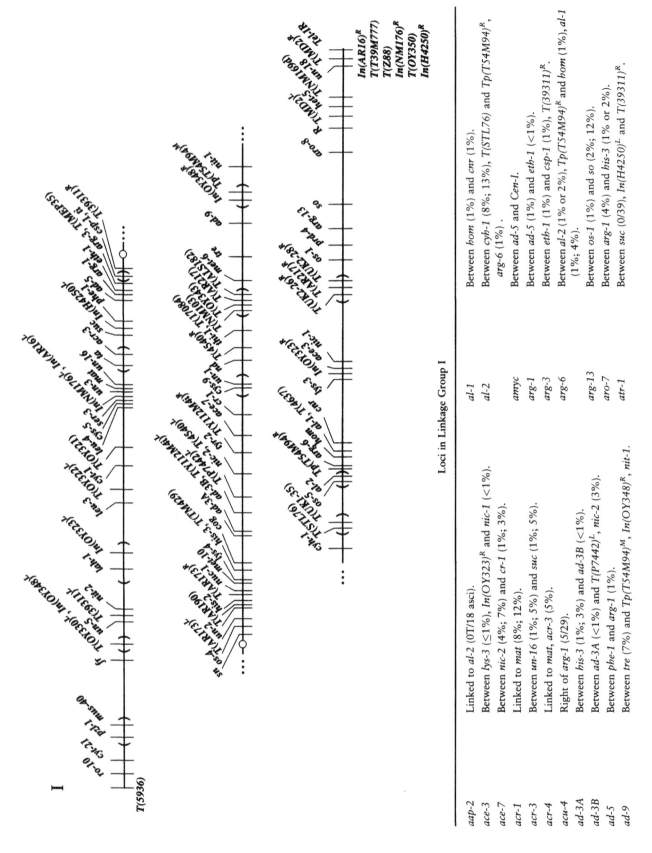

Loci in Linkage Group I

Locus	Description
aap-2	Between suc (0/39), In(H4250)^L and T(39311)^R.
ace-3	Between lys-3 (≤1%), In(OY323)^R and nic-1 (<1%).
ace-7	Between nic-2 (4%; 7%) and cr-1 (1%; 3%).
acr-1	Linked to mat (8%; 12%).
acr-3	Between un-16 (1%; 5%) and suc (1%; 5%).
acr-4	Linked to mat, acr-3 (5%).
aca-4	Right of arg-1 (5/29).
ad-3A	Between his-3 (1%; 3%) and ad-3B (<1%).
ad-3B	Between ad-3A (<1%) and T(P7442)^L, nic-2 (3%).
ad-5	Between phe-1 and arg-1 (1%).
ad-9	Between tre (7%) and Tp(T54M94)^M, In(OY348)^R, nit-1.
al-1	Linked to al-2 (0T/18 asci).
al-2	Between bom (1%) and cnr (1%).
amyc	Between cyh-1 (8%; 13%), T(STL76) and Tp(T54M94)^R, arg-6 (1%).
arg-1	Between ad-5 and Cen-I.
arg-3	Between ad-5 (1%) and eth-1 (<1%).
arg-6	Between eth-1 (1%) and csp-1 (1%), T(39311)^R.
arg-13	Between al-2 (1% or 2%), Tp(T54M94)^R and bom (1%), al-1 (1%; 4%).
aro-7	Between os-1 (1%) and so (2%; 12%).
atr-1	Between arg-1 (4%) and his-3 (1% or 2%).

aza-1	Left of mat (23%).	
aza-2	Linked to mat (2%), aza-1 (39%).	
B^m	Left of nit-2 (30%).	
Ban	Left of mat (14%).	
bs-1	Linked to un-1 (9%), probably to the right.	
ca-1	Linked to cys-9 (10%), un-1 (10%).	
ca-3	Linked to ca-1 (5%), cys-9, un-1 (10%).	
ccg-4	Between mat (0/18; 3T/18 asci) and Cen-1 (2T/18 asci).	
Cen-1	Between arg-3 (2%), T(39311)^R and sn (<1%), os-4.	
cfs(OY305)	Between mat and al-2.	
cfs(OY306)	Linked to al-2, probably to the right.	
cfs(OY307)	Between mat and al-1.	
chs-4	Linked to leu-4, mat, arg-3 (0/18).	
cnr	Between hom (1%), al-1 and In(OY323)^R, lys-3.	
cog	Between his-3 (1%; 3%; 3 kb) and ad-3A (2%; 7%).	
col-12	Linked to mat (17%; 22%).	
con-8	Linked to pbr (0/18), nic-1 (4/18).	
cpd-2	Right of arg-1 (6%).	
cpd-3	Right of arg-1 (11%).	
cr-1	Between ace-7 (1%; 3%) and cys-9 (3%), un-1 (5%).	
cr-2	Between T(NM103) and al-2 (18%).	
cr-3	Between T(4540)^R and cr-2.	
crp-7	Linked near con-8.	
csb	Between thi-1 (12%; 20%) and ad-9 (5%).	
csp-1	Between arg-3 (1%) and T(39311)^R .	
csr-1	Between his-2 (20%) and cyb-1 (1%; 3%).	
cy	Linked to ad-5 (1/54), probably to the left.	
cya-1	Linked to mat (6%).	
cyh-1	Between nit-1 (6%) and T(STL76), al-2 (8%; 13%).	
cys-5	Between leu-4 and ser-3 (<1%).	
cys-9	Between cr-1 (3%) and nd, T(4540)^R, thi-1 (13%).	
cys-11	Linked to cys-5 (<1%), probably to the right.	
cys-12	Linked to al-1 or al-2 (0/76). Right of ad-9 (12%).	
cys-13	Right of nic-2, tyr-2, T(Y112M4i)^R.	
cyt-1	Between leu-3 (5%; 8%) and T(OY321), leu-4.	
cyt-4	Between T(AR173) breakpoints.	
cyt-18	Linked to al-2 (10%), nic-1 (1%; 5%).	
cyt-20	Linked to mat (<1%).	
cyt-21	Between ro-10 (1%) and nit-2 (5/18).	
dgr-2	Right of arg-1 (2%). Linked to csp-1 (1%).	
dgr-4	Linked to al-2 (1%), probably to the left.	
dir	Right of pa.	
dot	Linked to ad-9 (0/44). Right of thi-1 (2%).	
eat-1	Right of mat. Between mat and eat-2.	
eat-2	Right of mat. Between eat-1 and eat-3.	
eat-3	Right of mat. Between eat-2 and eat-4.	
eat-4	Right of mat and eat-3.	
eat-5	Left of mat. Between eat-6 and mat.	
eat-6	Left of mat. Between un-3 and eat-5.	
En(pdx)	Linked to mat (5%), probably to the left.	
erg-4	Linked to al-1 (10%).	
esr-2	Linked to mat (10%).	
eth-1	Between arg-1 (<1%) and arg-3 (<1%).	
exo-1	Linked to mat (7%), probably to the left.	
ff-3	Right of os-1 (3%).	
fls	Between nit-1 (5%; 19%) and al-1 (6%; 19%).	
fmf-1	Between mat (2%; 15%) and cr-1 (2%).	
fpr-5	Left of al-2 (25%).	
fr	Between ro-10 (18%) and T(OY330)^L, un-5 (6%).	
fs-3	Left of mat (16%).	
Fsp-2	Right of nic-2 (6%).	
Fsr-1	Linked to met-6, ad-9, al-2, arg-6 (0/18).	
Fsr-12	Linked to nit-2 (0 or 1/16), vma-4 (0T/18 asci).	
Fsr-18	Linked to nic-1 (0/18), arg-13 (2T/18 asci).	
Fsr-30	Linked to mat (0T/18 asci).	
Fsr-33	Linked to lys-4 (0 or 1T/18 asci).	
gap-1	Between mat (6%) and Cen-1 (4%).	
gla-2	Linked to mat (3/68), sor-4 (≤10 kb), his-3 (3/65).	
glp-1	Probably between ad-9 (2%) and nit-1 (11%).	
glp-5	Left of cr-1 (15%).	
grp78	Linked to mat (0T/18 asci).	
gsp	Left of mat (10%).	
gua-1	Probably between his-2 (3%) and cr-1 (3%).	
hda-2	Linked to arg-13 (3%).	
het-5	Between R, T(MD2)^L and T(NM169d, un-18, T(MD2)^R.	
his-2	Between un-2 (<1%), T(AR190) and T(AR173)^R, nuc-1 (<1%).	

Continues

Continued

his-3	Between met-10, lys-4 (1%) and cog (1%; 3%), ad-3A (1%; 3%).
hom	Between Tp(T54M94)^R, arg-6 (1%) and al-1 (<1%).
hpp	Between al-2 and nic-1.
hsp30	Linked to mei-3, Cen-1, un-2, his-2 (0T/18 asci).
htl	Right of ad-3 (1%; 10%).
In(AR16)^L	Between ser-3 and un-3.
In(AR16)^R	At Tip-IR.
In(NM176)^L	Between ser-3 and un-3.
In(NM176)^R	At Tip-IR.
In(OY323)^L	Between nit-2 and leu-3.
In(OY323)^R	Between lys-3 and ace-3, nic-1.
In(OY348)^L	Between fr and un-5.
In(OY348)^R	Between ad-9 and nit-1.
In(H4250)^L	Between suc and phe-1.
In(H4250)^R	At Tip-IR.
ipa	Between mat (20%) and arg-1 (1%).
krev-1	Linked to nit-2 (0/18).
lah-1	Between nit-2 (2%) and leu-3 (5%).
leu-3	Between lah-1, In(OY323)^L and T(OY322)^L, cyt-1 (5%; 8%), T(OY321).
leu-4	Between cyt-1, T(OY321) and cys-5 (≪1%).
lis-1	Between ad-3 (6%) and al-1 (16%).
lpl	Right of his-3, cog (adjoining).
lys^R	Between his-3 and nic-2.
lys-3	Between al-1 (9%) and In(OY323)^R, ace-3, nic-1.
lys-4	Between nuc-1 (1%), met-10 and his-3 (1%).
mad-1	By cosmid assignment.
mat	Between un-3 (≤0.1%) and un-16 (<1%). Between eat-5 and eat-1.
mb-2	Between cyb-1 (5%) and al-1 (7%).
mb-3	Linked to cyb-1 (18%), al-1 (2%), mb-2 (6%).
mea	Right of acr-3 (9%).
mei-3	Between arg-1 (3%) and T(39311)^R.
mep(3), mep(10)	Left of mat.
met-6	Between T(NM103), T(ALS182), T(OY343) and tre (7%), ad-9 (2%, 16%).
met-10	Between his-2, T(AR173)^R, nuc-1 and lys-4, his-3.
mo-1	Linked to mat (9%).
mo-5	Linked to mat (20%).

msb-1	Linked to cyt-21 (0/18).
msk-1	By cosmid assignment.
mus-9	Between cyb-1 (18%) and al-2 (6%).
mus-13	Right of al-2 (18%).
mus-15	Between arg-1 (4%) and arg-3 (<2%).
mus-19	Between aro-8 (8%) and un-18 (4%).
mus-22	Left of cyb-1 (2%).
mus-31	Between Cen-1 and ad-3B (7%).
mus-32	Between mus-31 (3%) and ad-3B (4%).
mus-38	Left of un-5 (7%), leu-3 (10%). Linked to nit-2 (0/18).
mus-40	Linked to pzl-1 (0/18). Between cyt-21 (1/18) and fr, nit-2 (3/16).
nd	Between cys-9 (<2%; 3%) and T(4540)^R, thi-1 (<2%; 4%).
nic-1	Between lys-3 ace-3 (<1%) and T(UK2-26)^R, os-1 (10%; 30%).
nic-2	Between ad-3B (4%) and tyr-2, T(Y112M4i)^R, ace-7 (4%; 7%).
nit-1	Between ad-9 (3%; 15%), Tp(T54M94)^M and cyb-1 (6%).
nit-2	Between un-5 (2%), T(39311)^L and lah-1, In(OY323)^L, leu-3 (12%, 18%).
nit-8	Linked to cr-1 (0/61). Between nic-2 (10%) and thi-1 (3%).
nuc-1	Between his-2 (1%), T(AR173)^R and met-10, lys-4 (1%).
nub-6	Between Cen-1 (5%) and nic-2 (4%).
nub-8	Right of nic-1.
nuo12.3	Between lys-4 (1T/18 asci) and cys-9 (7T/18 asci).
nuo20.8	Between mat (0 or 1T/17 asci) and Cen-1 (3T/17 asci).
os-1	Between nic-1 (10%; 29%), T(UK2-26)^R and prd-4 (2%), arg-13 (1%; 4%).
os-4	Between sn (3%) and T(AR173)^L.
os-5	Between T(STL76) and T(54M94)^R, arg-6 (1%), al-1 (1%). Linked to al-2.
os-6	Linked to os-1 (4%), os-7 (6%).
os-7	Linked to os-1 (11%), os-6 (6%).
pa	Between cr-1 (10%) and dir (37%).
pat	Linked to mat, probably to the right.
pex-2	Linked to mat. Between cyt-21 and mei-3.
phe-1	Between suc (<1%), In(H4250)^L and ad-5.
phr	Linked to con-8 (0/18). Right of os-1 (15%).
prd-3	Near Cen-1.
prd-4	Between os-1 (2%) and arg-13 (3%).

Gene	Description
pzl-1	Linked to mus-40 (0/18). Between cyt-21 (1/18) and nit-2 (3/16).
R	Between aro-8 (4%) and T(MD2)^L, un-18 (11%).
rec-3	Between acr-3 (1% or 2%) and arg-3 (2%; 6%).
rg-1	Between his-2, T(AR173)^R and lys-4 (1%; 7%).
rg-2	Linked to rg-1 (0/650). Left of lys-4 (12%).
rgb-1	Linked to leu-4, mat, arg-3 (0/18).
ro-6	Between T(4540)^L nic-2 (0/95) and T(4540)^R, thi-1.
ro-10	Between Tip-IL and cyt-21 (1%), fr (18%).
sar-1, sar-3	Near mat.
scp	Linked to Fsr-18.
sdb-1	Linked to mat (0/13).
sdv-4, sdv-5	Between al-2 (1T/18 asci) and arg-13 (5T, 2NPD/18 asci).
ser-3	Between cys-5 (<1%) and In(NM176)^L, un-3 (<1%).
sf	Linked to mat (3%), cy (3%).
slo-1	Between mat (14%) and thi-1 (2%; 5%).
smco-1	Linked to mat (1%), rg-1 (0/72).
smco-3	Linked to mat (10%), al-2 (29%).
smco-5	Linked to mat (2%).
sn	Between Cen-I (<1%) and os-4 (<3%). T(AR173)^L.
so	Between arg-13 (2%; 12%) and aro-8 (7%; 11%).
sod-1	Between fr (5T/16 asci) and Fsr-12, vma-4 (2 or 4T/18 asci).
sor-4	Linked to phe-1 (<1%). Between In(H4250)^L and arg-1 (1%; 4%).
spco-11	Between leu-3 (7%) and In(NM176)^L, mat (17%).
spco-12	Right of mat (13%; 35%).
ss	Linked to nit-2 (<0.2%).
ssu-2	Linked to mat (22%), Cen-I (7%), al-2 (26%).
ssu-3	Linked to mat (22%), Cen-I (10%), al-2 (33%).
ssu-8	Linked to al-2 (2%; 8%).
st	Between ad-3B (5%) and thi-1 (14%).
su(bal)	Linked to mat (13%).
su(col-2)	Linked to mat (0/837).
su(met-7)-1	Linked to al-2 (1%).
su([mi-3])-1	Between un-7 (≤1%) and al-2 (1%; 4%).
su(mtr)-1	Right of his-2 (2%).
su(suc)-2	Linked to suc (3%).
su(trp-3^{td2})-2a	Linked to al-2 (15%).
suc	Between acr-3 (2%) and In(H4250)^L, phe-1 (1%).
T	Between ad-3A (18%), nic-2 (12%) and al-2, al-1 (12%).

Translocation	Description
T(UK1-35)	Between cyb-1 and al-2.
T(MD2)^L	Between R and het-5.
T(MD2)^R	Between un-18 and Tip-IR.
T(UK2-26)^L	Between met-6 and ad-9.
T(UK2-26)^R	Between nic-1 and os-1.
T(T39M777)	At Tip-IR. Linked to R (0/85).
T(STL76)	Between cyb-1 and os-5.
T(Z88)	At Tip-IR. Linked to un-18 (0/50).
T(NM103)	Between thi-1 and met-6.
T(Y112M4i)^L ad-3A	Breakpoint at ad-3A.
T(Y112M4i)^R	Between tyr-2 and ace-7 .
T(NM169d)	Between het-5 and un-18 .
T(NM173)^L	Between sn, os-4 and un-2.
T(NM173)^R	Between his-2 and nuc-1.
T(ALS182)	Between thi-1 and met-6.
T(ALS190)	Between un-2 and his-2 .
T(AR217)^L	Between thi-1, cr-3 and met-6.
T(AR217)^R	Between nic-1 and os-1.
T(OY321)	Between cyt-1 and leu-4.
T(OY322)^L	Between leu-3 and cyt-1.
T(OY322)^R	Between un-16 and suc.
T(OY330)^L	Between fr and un-5.
T(OY330)^R	Between un-16 and suc.
T(OY343)	Between thi-1 and met-6.
T()Y350)	At Tip-IR. Linked to un-18 (2/44).
T(4540)^L nic-2	Breakpoint at nic-2.
T(4540)^R	At Tip-IL.
T(5936)	Between un-1, cys-9 and thi-1.
T(P7442)^L	Between ad-3B and nic-2.
T(P7442)^R	Between cys-9 and thi-1.
T(39311)^L	Between un-5 and nit-2.
T(39311)^R	Between arg-3, csp-1 and Cen-I.
ta	Between un-16 (2%) and acr-3 (<3%) .
tef-1	Linked to leu-4, mat, arg-3 (0/18).
Tel-IR	Linked to arg-13 (6T, 2NPD/18 asci).
tet	Linked to acr-3 (2%), ad-3B (1%).
thi-1	Between cys-9 (13%), T(4540)^R and T(NM103), met-6 (7%; 14%).
ti	Between arg-3 and T(39311)^R.

Continues

Continued

Name	Description
Tip-1L	Marked by T(S936). Left of ro-10, fr.
Tip-1R	Marked by In(H4250) and other rearrangements. Right of R, un-18.
Tp(T54M94)^L	Between arg-3 (2%) and Cen-I.
Tp(T54M94)^M	Between ad-9 and nit-1.
Tp(T54M94)^R	Between al-2 and arg-6.
tre-1	Between met-6 (7%) and ad-9 (10%).
ty-2	Right of al-2.
tyr-2	Between T(4540)^L nic-2 and T(Y112M4i)^R, ace-7, cr-1 (2%).
tys	Right of mat (6%).
uc-2	Linked to mat (0/12 asci), upr-1. Between mat (2%) and arg-1 (7%).
un-1	Between cr-1 (5%; 9%) and nd, T(4540)^R. Linked to cys-9 (0/72).
un-2	Between sn, os-4, T(AR173)^L and nd, T(AR190), his-2 (1%).

Name	Description
un-3	Between ser-3 (1%), In(NM176)^L and mat (≤0.1%).
un-5	Between fr (6%), T(OY330)^L and T(39311)^L, nit-2 (2%).
un-7	Between cyb-1, T(STL76) and al-1 (3%).
un-16	Between mat (<1%) and ta (1%), acr-3 (<3%).
un-18	Between R (11%), T(MD2)^L, T(NM169d) and T(MD2)^R, Tip-IR.
upr-1	Between mat (2%) and arg-1 (7%).
ure-4	Linked to his-3 (1%), probably to the right. Left of ad-3B (3%).
uvs-6	Linked to met-6 (<1%).
vma-4	Linked to Fsr-12 (0T/18 asci). Left of mat (4T/18 asci).
vma-11	Between al-2 (1T/18 asci) and arg-13 (5T, 2 NPD/18 asci).
wc-2	Between thi-1 (8%), T(ALS182) and ad-9 (7%; 20%).
ylo-2	Between suc, In(H4250)^L and arg-1 (1%).
zip-1	Linked to cyt-21 (0T/18 asci).
znr-3	Linked to mat.

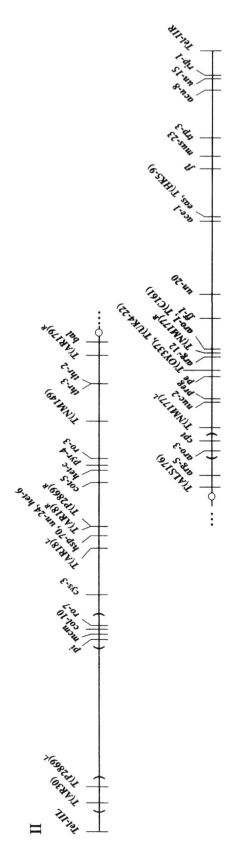

II

Loci in Linkage Group II

aag-1	Linked to *acu-5* (26%), *arg-5* (33%).
ace-1	Between *un-20* (15%) and *eas* (1%), *fl* (11%).
ace-9	Between *nuc-2* (2%) and *arg-12* (3%).
acr-5	Between *arg-5* (6%) and *pe* (9%).
acu-5	Linked to *arg-5* (6%), *aro-3* (7%).
acu-8	Between *trp-3* (8%) and *un-15* (3%).
acu-12	Right of *Cen-II*. Linked to *trp-3*.
acu-13	Linked to *trp-3* (23%).
alc-1	Linked to *pe* (10%).
aod-2	Linked to *arg-5* (7%), *thr-3* (16%).
arg-5	Between *bal* (1%; 9%), *T(ALS176)* and *aro-3*, *pe* (6%, 18%).
arg-12	Between *pe* (1%; 5%) and *T(NM177)^R*, *aro-1* (<1%).
aro-1	Between *arg-12* (<1%), *T(NM177)^R* and *ff-1* (4%; 6%).
aro-3	Between *arg-5* (1%; 3 %) and *T(NM177)^L*, *nuc-2*.
arp3	Near *arg-12*.
atp-2	Linked to *Cen-II*, *arg-5* (0T/18 asci).
bal	Between *T(AR179)* and *T(ALS176)*, *arg-5* (1%; 7 %). Probably left of *Cen-II*.
bli-4	Linked to *Fsr-34* (0T/18 asci). Between *eas* (1T/18 asci) and *leu-6* (9T, 1NDP/18 asci).
cdr-1	Left of *arg-12* (35%).
Cen-II	Between *T(AR179)^R* and *T(ALS176)*, *arg-5* (0T/18 asci).
cit-1	Linked to *preg* (0T/18 asci).
col-10	Left of *cys-3* (14%). At or near *pi*.
con-6	Left of *arg-12* (2 or 3/18).
cot-5	Between *T(P2869)*, *T(AR18)^R* and *het-c* (3%).
cpd-4	Right of *arg-12* (7%).
cpt	Between *arg-5* (3%) and *T(NM177)^L*, *pe* (6%).
cul-1	Between *arg-5* (3%; 6%) and *arg-12* (12%).
cul-2	Between *arg-12* (10%), *un-20* (4%) and *fl* (9%).
cya-4	Between *arg-5* (2/39; 1T/18 asci) and *preg* (1T + 1 double exchange/18 asci).
cyb-3	Between *Tip-IIL* and *ro-3* (9%).
cyc-1	Between *thr-3* (38%) and *trp-3* (18%).
cys-3	Between *ro-7*, *pi* (4%) and *T(AR18)^L* .
da	Linked to *thr-3* (3%); right of *T(NM149)*.
dim-4	Linked to *arg-12*, *aro-1*, *ure-3*.
eas	Between *ace-1* (1%) and *fl* (9%).
en(am)-2	Linked to *pe*.
ff-1	Between *aro-1* (5%) and *un-20* (4%).
fl	Between *ace-1*, *eas* (9%) and *mus-23* (1%; 2%), *trp-3* (3%).
fs-2	Probably linked to *cot-5* (14/48).
Fsp-1	Right of *pe* (4%).
Fsr-3	Between *vma-2* (1T/18 asci) and *eas* (1T/18 asci).
Fsr-17	Between *trp-3* (1/16) and *leu-6* (0/16; 9T, 1NPD/18 asci).
Fsr-21	Between *arg-12* (1/17) and *trp-3* (1/17; 4/41).
Fsr-32	Between *pyr-4* (10/41) and *Cen-II* (1T/18 asci), *arg-5* (2/41).
Fsr-34	Between *arg-12* (1/18), *eas* (1T/18 asci) and *trp-3* (1/17). Linked to *Fsr-21* (0/59).
Fsr-52	Between *pyr-4* (2/41) and *Cen-II* (5T/18 asci), *arg-5* (0/41).

214

Locus	
Fsr-54	Between *pyr-4* (2/41), *Fsr-52* (3/18, 84/1) and *arg-5* (10/41).
Fsr-55	Between *arg-5* (1T/18 asci; 2/39) and *preg* (4/39). Linked to *cya-4* (0T/18 asci).
glp-2	Between *T(ALS176). arg-5* (8%) and *T(NM177)^L, pe* (7%).
gpd-1	Linked to *arg-12* (0/18), *vma-2* (1T/18 asci), *eas* (3T/18 asci).
bbs	Left of *con-6* (1 or 2/18), *arg-12* (3 or 4/18).
hda-2	Linked to *arg-13* (2T/18 asci).
het-c	Between *cot-5* (3%) and *Pad-1, pyr-4* (1%), *ro-3*.
het-d	Right of *fl* (25%).
het-6	Between *cys-3, T(AR18)^L, cys-3, hsp70, un-24* and *T(AR18)^R, T(2869), cot-5*.
bH3	Between *Fsr-55* and *Fsr-3*. On same plasmid with *aro-9*.
hH4-1	Adjoins *bH3*.
hsp70	Between *T(AR18)^L* and *un-24* .
leu-6	Between *trp-3* (2/16), *Fsr-17* (5T/18 asci), and *Tel-IIR* (1T, 1NPD/16 asci).
lp	Right of *thr-3* (10%).
mcm	Linked to *ro-7* (1%).
mus-7	Between *arg-5* (8%; 12%) and *nuc-2* (11%).
mus-23	Between *fl* (<1%; 2%) and *trp-3* (9%).
mus-24	Left of *pyr-4* (8%).
mus-27	Right of *nuc-2* (16%).
nik-2	Linked to *arg-12* (0/17).
nrc-2	Linked to *cya-4, Fsr-55*.
nuc-2	Between *aro-3, T(NM177)^L* and *preg* (1% or 2%), *pe* (4%).
nuc-3	Right of *arg-12* (4%).
nub-5	Linked to *trp-3* (30%), *T(4637)*.
nub-9	Linked to *arg-5*. Left of *mus-7, nuc-2*.
nuo78	Linked to *preg* (0T/5 asci).
osr-2	Linked to *arg-5* (≤12%).
Pad-1	Between *het-c* and *pyr-4* (1%).
pe	Between *nuc-2* (4%) and *arg-12* (1%, 5%).
pi	Linked to *ro-7* (0/75). Between *T(AR30)* (28%) and *cys-3* (4%).
pma-1	Between *cya-4* (2T/18 asci) and *preg* (1T/15 asci).
pmg	Linked to *pyr-4* (<1%).
prd-5	Between *arg-12* (4%; 10%) and *un-20* (8%).
preg	Between *T(NM177)^L, nuc-2* (1% or 2 %) and *arg-12, T(NM177)^R*.
pyr-4	Between *het-c* (1%), *Pad-1* (1%) and *ro-3* (1% or 2 %).
rap-1	Adjacent to *vma-6*. Near *Cen-II*.
rip-1	Between *un-15* (1%) and *Tip-IIR*.
ro-3	Between *pyr-4* (1% or 2 %) and *T(NM149), thr-2* (6%; 25%).
ro-7	Linked to *pi* (0/75). Left of *cys-3* (11%).
scr	Linked to *arg-12* (2%), probably to the left.
spco-14	Linked to *arg-5* (7%).
su([mi-1])-3	Linked to *fl* (31%).
su([mi-1])-4	Linked to *fl* (22%), *arg-5* (40%).
su([mi-1])-10	Linked to *arg-5* (29%), *fl* (24%).
T(UK4-22)	Linked to *pe* (0/46) and *arg-12, aro-1* (0/23).
T(AR18)^L	Between *ro-7, cys-3* and *hsp70*.
T(AR18)^R	Between *het-6* and *cot-5*. Within 5.6 kb of *T(P2869)*.
T(AR30)	Between *Tip-IIL* and *pi* (28%).
T(NM149)	Between *ro-3* and *thr-2, thr-3*.
T(ALS176)	Between *Cen-II* and *arg-5*.
T(NM177)^L	Between *aro-3 cpt* and *nuc-2*.
T(NM177)^R	Between *arg-12* and *aro-1*.
T(OY337)	Between *pe* and *arg-12*.
T(P2869)^L	Between *Tip-IIL* and *ro-7*.
T(P2869)^R	Between *het-6* and *cot-5*. Within 5.6 kb of *T(AR18)^R*.
Tel-IIL	Linked to *hsp70* (2T/18 asci).
Tel-IIR	Right of *leu-6* (1T, 1NPD/18 asci).
Tip-IIL	Left of *T(AR30), pi* (≥28%).
Tip-IIR	Right of *un-15, rip-1*.
thr-2	Between *ro-3, T(NM149)* and *T(AR179), bal, arg-5* (3%; 18%).
thr-3	Linked to *thr-2* (<0.1%).
tng	Between *pyr-4* (16%), *T(NM149)* and *arg-5* (6%; 14%). Linked to *thr-2* (2%).
trp-3	Between *fl* (2%; 6%), *mus-23* (6%) and *un-15* (10%).
tu	Between *pe* (8%) and *fl* (19%).
uc-1	Linked to *pyr-4* (31%).
un-15	Between *trp-3* (10%) and *rip-1* (1%).
un-20	Between *aro-1* (5%; 9%), *ff-1* (4%) and *ace-1* (15%).
un-24	Between *T(AR18)^L, hsp70* and *bet-6, T(AR18)^R*.
ure-3	Between *arg-12* (7%; 12%) and *un-20* (8%).
vma-2	Between *preg* (6T/18 asci) and *Fsr-3* (1T/18 asci), *eas* (2T/18 asci).
vma-6	Near *Cen-II*. In same cDNA fragment with *rap-1*.
vma-10	Linked near *arg-12*.
vph-1	Left of *Cen-II* (5T/18 asci), *Fsr-32* (4T/18 asci). Linked to *Fsr-52* (0T/18 asci).
xdh-1	Linked to *pe* (14%).
ypt-1	Linked to *eas* (0T/17 asci).

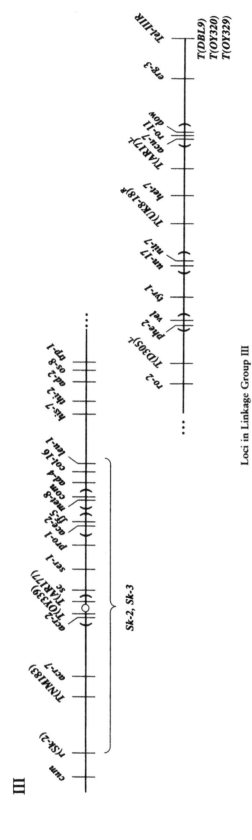

III

Sk-2, Sk-3

Loci in Linkage Group III

aab-1	Linked to trp-1.
ace-2	Between pro-1 (1%; 9%; 36kb) and com (5%), ad-4 (4%; 7%).
acon-2	Linked to vel (6%), tyr-1 (9%; 14%).
acr-2	Linked to Cen-III. Between acr-7 and T(T54M140b), thi-4 (0/286), sc (3%; 6%).
acr-6	Linked to shg (<1%).
acr-7	Between r(Sk-2)-1 (7%), T(NM183) (5%) and acr-2 (12%), Cen-III.
acu-7	Linked to dow (0/72).
ad-2	Between thi-2 (1%) and trp-1 (1%; 7%).
ad-4	Between met-8 (1%; 4%), com (0; 5%) and col-16, leu-1 (1%;3%).
aza-3	Linked to trp-1 (14%).
cat-1	Left of trp-1 (17%).
cat-3	Linked to Fsr-45 (2T/18 asci), Tel-IIIR (5T/18 asci).
Cen-III	Between acr-7 and thi-4 (1T/18 asci), sc (2T/24 asci).
col-16	Between pro-1 (10%), ad-4 and leu-1 (1%).
com	Between ace-2 (5%) and ad-4 (<1 %; 5%).
con-1	Left of Cen-III (6T/18 asci).
con-7	Between ace-2 (4T/18 asci), cyt-8 (3T/18 asci) and ad-2, trp-1 (0/17; 1T/18 asci).
cor-1	Linked to dow (1%), probably to the right.
cox-4	Linked to acr-2 (1 or 2/15).
cpk	Linked to un-17 (5%).
cr-5	Linked to ad-4 (19%).
crp-2	Linked to Cen-III (0T/18 asci).
cum	Between Tip-IIIL and r(Sk-2)-1 (4%), acr-7 (5 %; 18%).
cyt-8	Near ad-4. Between ace-2 (1 or 2T/18 asci) and ad-2 (4T/18 asci), trp-1 (3/18).
cyt-22	Linked to r(Sk-2)-1 (30%), cum (6%; 30%), acr-7 (19%). Left of cyt-8 (4/18).
dim-1	Linked to mus-20, trp-1.
Dip-1	Between ro-2 (7%) and phe-2 (1%).
dow	Between un-17 (23%), T(AR17)^L and T(AR17)^R, erg-3 (10%; 14%).
erg-3	Between dow (10%; 14%), T(AR17)^R and Tip-IIIR.
ff-5	Between pro-1 (2%) and met-8 (1%).
ff-6	Near ty-1.
fpr-3	Linked to trp-1 (<1%), thi-2 (5%).
Fsr-45	Between ro-2 (3/18) and cat-3 (2T, 1NPD/18 asci), Tel-IIIR (3T, 1NPD/18 asci).
gluc-1	Linked to dow (10%).
gna-1	Linked to ro-2 (0/17).
gnb-1	Linked to trp-1 (0/18).
het-7	Between T(D305)^L, T(UK8-18)^R and T(AR17)^L.
hH4-2	Linked to trp-1 (0/18).
his-7	Between leu-1 (8%; 20%) and thi-2 (1% or 2%).
lao	Linked to un-17 (5%).
leu-1	Between ad-4 (1%; 5%), col-6 (1%) and his-7 (8%).

216

mei-4	Linked to *leu-1* (12%), probably to the left.
mel-3	Linked to *leu-1* (17%).
met-8	Between *ace-2*, *ff-5* (1%; 4%), and *ad-4* (4%).
mgk-1	By cosmid assignment to chromosomal DNA. Tentative.
mip-1	Linked to *trp-1* (1T/18 asci).
mo-4	Right of *leu-1* (8%).
mod(pr)	Between *leu-1* and *his-7*.
mus-20	Between *trp-1* (8%) and *phe-2* (18%).
mus-21	Linked to *trp-1* (20%). Right of *ad-2*.
mus(SA6)	Left of *ad-2* (5%), *trp-1* (8%).
nit-7	Linked to *un-17* (0/63).
nub-1	Between *leu-1* (4%) and *nub-2* (<1%), *trp-1* (11%).
nub-2	Between *leu-1* (4%), *nub-1* (<1%) and *trp-1* (11%).
os-8	Between *ad-2* and *trp-1*.
os-11	Linked to *os-8* (19%).
ota	Between *ad-4* (15%) and *tyr-1* (14%). Linked to *pro-4* (4%).
pgov	Linked to *tyr-1* (1%; 4%), probably to the right.
phe-2	Linked to *vel* (1%). Between *T(D305)^L* and *tyr-1* (2%; 4%).
pr(Sk-2)	Between *Cen-III*, *ser-1* and *ad-4*, *leu-1*, *mod(pr)*.
prd-1	Linked to *acr-2* (5%), *Cen-III* (0/35 asci), *pro-1* (20%).
pro-1	Between *ser-1* (1%; 114 kb) and *ace-2* (2%; 36kb).
pro-4	Linked to *thi-2* (0/78), *ota* (4%).
r(Sk-2)	Between *cum* (2%) and *T(NM183)* (2%), *acr-7* (7%).
r(Sk-3)	Between *cum* (4%) and *acr-7* (17%).
rco-1	Linked to *acr-2* (7%; 17%).
ro-2	Between *trp-1* (2%; 14%) and *T(D305)^L*, *phe-2* (5%).
ro-11	Linked to *dow* (0/70). Between *T(AR17)^L* and *T(AR17)^R*.
sc	Between *acr-2* (3%; 6%), *Cen-III* (4T/230) and *ser-1* (4%). Linked to *thi-4*, *spg* (<1%).
scon-2	Between *trp-1* and *Fsr-45*. Linked to *tyr-1* (18%).
ser-1	Between *sc* (4%) and *pro-1* (1%; 9%; 114 kb).
ser-5	Linked to *trp-1* (1%), *ser-1* (12%).
sbg	Linked to *trp-1* (7%), *acr-6* (0/368).
Sk-2	A haplotype, *r(Sk-2)* through *leu-1*. Between *cum* (2%) and *his-7* (21%).
Sk-3	A haplotype, *r(Sk-3)* through *leu-1*. Between *cum* (2%) and *his-7* (12%).
sor-3	Linked to *ad-4* (7%).

spco-15	Linked to *spg* (10%), *Cen-III* (0T/10 asci).
spg	Linked to *sc thi-4* (<1%). Between *acr-2* (1%; 11%) and *ser-1* (8%).
su([mi-1])-1	Linked to *ad-4* (14%).
su(trp-3^{td2})-2	Linked to *leu-1* (22%).
T(UK8-18)^L	Between *his-7* and *trp-1*.
T(UK8-18)^R	Between *nit-7 un-17* and *het-7*.
T(DBL9)	At *Tip-IIIR*. Linked to *dow* (2/28).
T(AR17)^L	Between *tyr-1 het-7* and *dow*.
T(AR17)^R	Between *dow* and *erg-3*.
T(T54M140b)	Between *acr-2* and *thi-4*, *spg* (0/34).
T(AR177)	Near *Cen-III*, probably to the right.
T(NM183)	Between *r(Sk-2)-1* (2%) and *acr-7* (5%).
T(D305)^L	Between *ro-2* and *phe-2*.
T(D305)^R	Right of *dow*.
T(OY320)	At *Tip-IIIR*. Linked to *dow* (3/30).
T(OY329)	At *Tip-IIIR*. Linked to *trp-1* (4/42).
T(OY339)	Near *Cen-III*, probably to the left.
Tel-IIIR	Linked to *cat-3* (5T/18 asci).
thi-2	Between *his-7* (1% or 2%) and *ad-2* (1%; 3%).
thi-4	Between *Cen-III* (1T/18 asci) and *pro-1* (3%; 1T/18 asci). Linked to *spg*, *sc* (<1%).
Tip- IIIL	Left of *r(Sk-2)*, *cum*, *acr-7*.
Tip-IIIR	Marked by *T(OY320)*. Right of *dow*, *erg-3*.
trk	Linked to *leu-1* (0/92).
trp-1	Between *ad-2* (1%; 7%) and *ro-2* (2%; 12%).
trx	Adjoins *acr-2*.
ty-1	Linked to *tyr-1* (0%; 6%).
ty-3	Between *Cen-III* and *ad-4*.
tyr-1	Between *vel* (3%; 5%), *phe-2* (2%; 4%) and **un-17** (4%).
un-6	Between *sc* (21%) and *leu-1*.
un-14	Between *acr-2*, *thi-4* (8%; 20%) and **met-8**, *leu-1* (5%).
un-17	Between *tyr-1* (4%) and *T(UK8-18)^R*. Linked to *nit-7* (0/63).
un-21	Between *acr-2* and **met-8**, *un-6*.
uvs-4	Left of *ad-4* (4%).
uvs-5	Linked to *vel* (1%).
vel	Linked to *phe-2* (1%). Between *ro-2* (6%), *T(D305)^L* and *tyr-1* (3%; 5%).

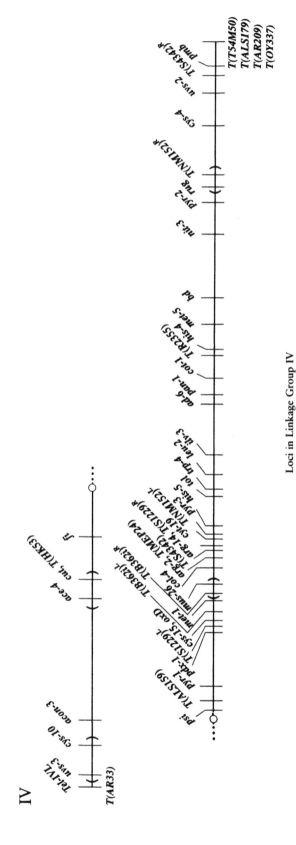

218

Loci in Linkage Group IV

Locus	Description
ace-4	Between cys-10 (19%; 27%) and fi (10%).
acon-3	Between cys-10 (1%; 6%) and cut (33%).
acu-2	Between leu-2 (11%) and pan-1 (6%).
ad-6	Between ilv-3 (9%) and pan-1 (1% or 2 %).
ads	Linked to col-4.
aod-1	Linked to Cen-IV (0T/17 asci).
arg-2	Between col-4 (1% or 2%) and T(S4342)ᴸ, pyr-3 (1%; 3%).
arg-14	Between arg-2 (1%), T(S4342)ᴸ, and cyt-19, (T(NM152)ᴸ, pyr-3 (1%).
Asd-4	Linked near net-5
bd	Between met-5 (9%) and nit-3 (19%).
bld	Right of pdx-1 (4%).
bli-3	Linked to Fsr-3 (0T/15 asci), vma-8 (0T/18 asci) Right of trp-4 (1T/15 asci).
car	Linked to cys-10 (1%).
cel-1	Linked to pan-1 (1%), cot-1 (0/17).
cel-2	Linked to cot-1 (1%), bd.
Cen-IV	Between cut, fi and psi-1, T(ALS159).
chol-1	Linked to ad-6 (1%), probably to the right.
chol-4	Linked to cot-1 (32%), chol-1 (25%).
chs-2	Linked to nit-3 (6/18), Fsr-4 (0/17), uvs-2 (0/18).
chs-5	Between arg-14 (2/18) and cot-1 (1/18).
coil	Between arg-2 (2%) and leu-2 (12%).
col-1	Linked to pan-1 (0/47 asci), cot-1 (3%).
col-4	Between met-1 (4%), mus-26 (3%) and arg-2 (1% or 2%).
col-5	Linked to cot-1 (1%), probably to the right.
col-6	Linked to Cen-IV (0T/28 asci), pan-1 (21%).
col-8	Linked to pan-1 (4%; 13%).
con-5	Linked to cot-1 (0/18).
con-9	Linked to cot-1 (0/18).
con-10	Linked to cot-1 (0/18).
con-13	Adjoins con-10.
cot-1	Between pan-1 (2%) and his-4 (1%; 6%).
cot-3	Left of pan-1 (16%; 25%). Probably right of arg-2.
cpd-1	Right of pyr-2 (19%).
cr-4	Linked to pdx-1 (6%), cot-1 (22%).
cre-1	Right of trp-4 (5 or 6 T, 1NPD/17 asci).
crib-1	Linked to met-1 (6%).
crp-3	Right of trp-4 (3T, 1NPD/18 asci).
cut	Between cys-10 (28%, 37%) and fi (4%; 10%).
cya-5	Right of pan-1 (2%).

Gene	Description	Gene	Description
cya-6	Right of pan-1 (2%).	moe-3	Left of pan-1 (17%; 25%), bd (20%).
cya-9	Left of trp-1 (9%).	mtr	Between pdx-1 (2%) and col-4 (1%).
cys-4	Between rug (10%), T(NM152)^R, and uvs-2 (5%).	mus-8	Linked to mtr (1%).
cys-10	Between Tip IVL and acon-3 (1%; 6%).	mus-17	Between pan-1 (2%) and cys-4 (25%).
cys-14	Linked to cot-1 (21%).	mus-26	Between met-1 (2%) and col-4 (3%), arg-2 (2%).
cys-15	Between pdx-1 (0T/55 asci); T(S1229)^L and arg-14, met-1 (3%).	mus-30	Between mtr (8%) and trp-4 (4%).
cyt-5	Left of trp-4 (9%). Linked to cyt-19 (<1%; 5%; 10%).	mus-36	Between pan-1 and uvs-2.
cyt-19	Between T(S4342)^L, arg-14 and T(NM152)^L, pyr-3 (5%).	mut-1	Linked to trp-4 (<3%).
dn	Linked to rug (1%). Right of pyr-2 (4%).	nada	Left of ad-6 (18%).
fdu-2	Right of cys-4 (2%).	ndp64	Linked to Fsr-4 (0T/18 asci), Tel-IVR (1T/18 asci).
fl	Between ace-4 (10%), cut (4%; 10%) and pyr-1 (10%; 15%).	nit-3	Between met-5 (15%), bd (19%) and pyr-2 (2%; 9%).
fld	Left of his-5 (2%).	nit-4	Between pyr-1 (1%; 6%) and pan-1 (6%; 27%).
Fsr-4	Linked to Tel-IVR (1T/18 asci), chs-2, uvs-2 (0/17), nuo21 (2T/15 asci).	nit-9	Right of nit-4 (9%).
Fsr-13	Between trp-4 (3/15), Fsr-51 (1/15) and cot-1.	nuo21	Linked to Tel-IVR (1T/18 asci), Fsr-4 (1 or 2T/18 asci).
Fsr-51	Between arg-14 (2/18), trp-4 (2T/18 asci) and cot-1 (1/18).	nuo29.9	Between trp-4 (8 or 9T, 1NPD/18 asci) and Tel-IVR (8T, 1NPD/18 asci).
Fsr-62	Linked to Fsr-63 (0/18), mtr (0/18), arg-14 (1/18).	opi-1	Linked to T(R2355), cot-1 (9%).
Fsr-63	Between Cen-IV (2T/15 asci) and pyr-1 (1T/18 asci), arg-14 (1/18).	os-2	Right of cot-1 (4%).
gpi-1	Linked to ad-6 (10%).	os-10	Linked to cut (5%; 21%), col-4 (8%; 10%).
grey	Linked to cot-1 (4%).	osr-1	Linked to trp-4 (20%).
gua-2	Linked to cot-1 (5%).	oxD	Between pdx-1 (0T/55 asci), T(S1229)^L and T(B62)^L, met-1 (3%).
gul-3	Linked to cot-1 (10%), pyr-2 (7%).	pan-1	Between ad-6 (1% or 2%) and cot-1 (2% or 3%).
his-4	Between cot-1 (1%; 4%) and met-5 (4%).	pdx-1	Between pyr-1 (1%; 10%) and T(S1229)^L, pt (2%).
his-5	Between pyr-3 (1%) and tol, trp-4 (3%; 7%). Linked to cyt-19 (5%; 9%).	pdx-2	Near pdx-1, probably to the left.
hss-1	Linked to cys-4 (2%).	pf	Right of pyr-2 (2%). Linked to rug (3%).
htb-1	Linked to bd (27%).	pho-3	Linked to pan-1 (<1%).
ilv-3	Linked to met-2 (0/129). Between leu-2 (4%) and ad-6 (9%).	pho-5	Between pyr-1 (5%; 2T/18 asci) and trp-4 (4T/18 asci), cot-1 (15%).
int	Linked to pan-1 (0/50).	pmb	Between uvs-2 (8%), T(S4342)^R and Tip-IVR.
le-1	Linked to pan-1 (1% or 2%). Right of cot-1 (14%).	pph-1	Linked to mtr (0/18). arg-14 (1/18).
leu-2	Between trp-4 (2%) and ilv-3 (4%).	psi-1	Between Cen-IV and T(ALS159), pyr-1 (4%).
med	Linked to met-5 (5%), pan-1 (8%).	pt	Between pdx-1 (2%), T(S1229)^L and col-4 (2%).
mei-1	Linked to arg-2 (<1%), probably to the right.	puu-1	Between arg-2 (5%) and cot-1 (8%).
met-1	Between oxD (3%), T(B362i)^L and T(B362i)^R, mus-26 (2%), col-4 (4%).	pyr-1	Between psi-1 (4%), T(ALS159) and pdx-1 (<1%; 10%).
met-2	Linked to ilv-3 (0/129). Between trp-4 (6%) and pan-1 (4%).	pyr-2	Between nit-3 (2%; 9%) and rug (3%), T(NM152)^R.
met-5	Between his-4 (4%) and nit-3 (15%).	pyr-3	Between arg-14 T(S1229)^R, cyt-19 (5%), T(NM152)^L and his-5 (1%).
mlb-1	Linked to mtr (0/18).	ras-1	Between met-5 (1/18) and nit-3 (2/18).
mod(sc)	Linked to pan-1 (17%).	rib-2	Between T(S4342)^L and chol-1.
		ro-1	Linked to pan-1 (<1%).

Continues

Continued

rol-1	Linked to *pdx-1* (0/88).
rug	Between *pyr-2* (3%) and T(NM152)^R, *cys-4* (10%).
sdv-1, -2, and *-3*	Between *Tel-IVL* (12T, 0NPD/18 asci) and *Cen-IV* (≤2T/18 asci).
sdv-7, -8, and *-9*	Right of *trp-4* (5T/18 asci), *crp-3* (1T/18 asci).
sdv-11, sdv-12	Between *pyr-1* (4T/18 asci) and *trp-4* (3T/18 asci).
ser-4	Right of *arg-2* (<1%).
smco-4	Linked to *pan-1* (8%).
smco-8	Linked to *pan-1* (1%; 7%), *smco-4* (30%).
smco-9	Linked to *pan-1* (5%), *smco-4* (2%), *smco-8* (13%).
spco-8	Linked to *pan-1* (23%).
spco-9	Linked to *asn* (6%).
su([mi-1])-14	Linked to *arg-2* (14%).
T(UK2-32)	Between *his-4* and *nit-3*.
T(AR33)	At *Tip-IVL*. Linked to *cys-10* (0/70).
T(T54M50)	At *Tip-IVR*. Linked to *uvs-2* (0/49).
T(NM152)^L	Between *arg-14, cyt-19* and *pyr-3*.
T(NM152)^R	Between *rug* and *cys-4*.
T(ALS159)	Between *psi* and *pdx-1*.
T(ALS179)	At *Tip-IVR*. Linked to *uvs-2* (1/51).
T(AR209)	At *Tip-IVR*. Linked to *uvs-2* (1/29).
T(OY337)	At *Tip-IVR*. Linked to *rug* (3/50).
T(B362i)^L	Between *met-1* and *col-4*.
T(B362i)^R	Between *cys-15* and *met-1*.
T(S1229)^L	Between *pdx-1* and *cys-15*.
T(S1229)^R	At *arg-14*.
T(R2355)	Between *cot-1* (1/63) and *his-4*.
T(S4342)^L	Between *arg-2* and *arg-14*.
T(S4342)^R	Between *uvs-2* and *pmb, Tip-IVR*.
Tel-IVL	Linked to T(AR33) (1/20), *sdv-1, -2, -3* (12T, 0NPD/18 asci).
Tel-IVR	Linked to *nuo21* (1T/18 asci).
thi-5	Linked to *pan-1* (1%).
Tip-IVL	Marked by T(AR33). Left of *uvs-3, acon-3, cys-10*.
Tip-IVR	Marked by T(AR209) and other translocations. Right of *cys-4, uvs-2, pmb*.
tol	Between *his-5* and *trp-4* (1%).
tom20	Linked to *pho-5* (0T/18 asci).
trp-4	Between *his-5* (3%; 7%), *tol* (1%) and *leu-2* (1% or 2%).
uc-5	Right of *cot-1* (32%).
ufa	Linked to *met-5* (7%).
un-8	Linked to *pyr-1* (0/47). Between T(ALS159) and *col-4* (5%).
un-12	Linked to *pdx-1* (5%) *col-4* (0/73).
uvs-2	Between *cys-4* (5%) and T(S4342)^R, *pmb* (8%).
uvs-3	Linked to *cys-10* (3%; 7%), probably to the left.
vma-7	Near *Fsr-62*.
vma-8	Right of *trp-4* (3T/18 asci).

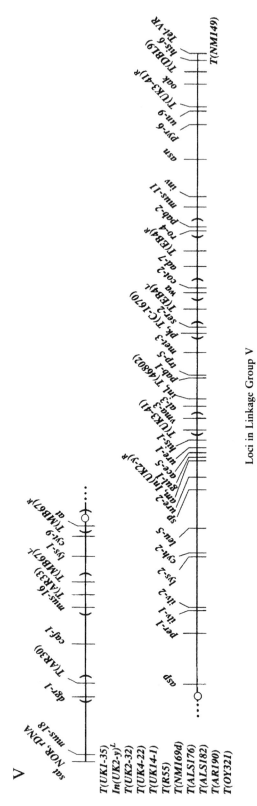

Loci in Linkage Group V

ace-5	Between gul-1 (<1%) and ure-1 (<1%).
act	Linked near inl.
acu-1	Right of asn (21%).
acu-3	Between inl (7%) and asn (20%).
ad-7	Between cot-2 (4%) and T(EB4)^R, ro-4 (4%), pab-2 (8%).
al-3	Between pbo-2 and inl (1%; 1T/18 asci).
am	Between ure-2 (2%) and gul-1 (<1%), ace-5 (<1%), ure-1 (1%; 150 kb).
amt	Between crp-4 (8T/18 asci) and Tel-VR (2T/18 asci).
arg-4	Between sp (1%; 11%) and inl (2%; 4%).
Asc(KH2A83)	Left of al-3, probably right of his-1.
Asm-1	Linked to al-3 (0T/18 asci).
asn	Between inv (4%; 9%) and pyr-6 (6%).
asp	Between at (0%; 3%) and per-1 (16%; 26%).
at	Linked to Cen-V (0/57 asci). Between cyt-9 (2%; 5%) and asp (0%; 3%).
caf-1	Between dgr-1 (8%), T(AR30) (19%) and mus-16 (4%), T(AR33) (5%; 12%).
ccg-1	Between Cen-V (4T/18 asci) and ilv-2 (4T/18 asci).
ccg-6	Linked to al-3, inl (0 or 1T/18 asci).
ccg-8	Linked to lys-1, Cen-V (0T/18 asci).
Cen-V	Between lys-1 (5%), cyt-9 (2%), T(MB67)^R and asp. Linked to at (0T/57 asci).
chol-3	Between at (18%) and al-3 (19%; 47%).
chs-1	Between con-4a, Fsr-9 (1/18) and am (4/18).
chs-3	Between am (1 or 2/18) and inl (1 or 2/18).
cit-2	Linked to inl (1T/18 asci).
cl	Right of pk (1% or 2%).
cmd	Between inl (3T/18 asci), cya-2 (2T/18 asci) and tom70 (1 or 2T/17 asci).
cna-1	Linked to cyb-2.
col-9	Between inl (16%) and asn (5%).
con-2	Linked to rDNA (2/18; 4T/18 asci) and con-4a (2/18), lys-1 (8T/18 asci).
con-4	Between rDNA (4/18), con-2 (2/18) and chs-1 (1 or 2/18), am (4/18).
cot-2	Between ser-2 (5%), T(EB4)^L and ad-7 (4%).
cot-4	Between rol-3 (5%) and inl (10%).
cpc-3	Linked to cyb-2 (0 or 1/16).
crp-4	Right of inl (9T/18 asci), Fsr-20 (1T/18 asci). Linked to sdv-6 (0T/18 asci).
cya-2	Linked to al-3 (3%), inl (0T/18 asci).
cya-7	Right of Cen-V.
cyb-1	Between al-3 (24%) and his-6 (10%).
cyb-2	Between lys-2 (<1%) and leu-5 (1% or 2%).
cyt-9	Between lys-1 (5%) and T(MB67)^R, Cen-V (4T/57 asci), at (5%; 2T/55 asci).

Gene	Description
dgr-1	Between T(UK2-33), In(UK2-y)^L and T(AR30), *caf-1* (7%).
dim-3	Between *am* and *inl*.
en(am)-1	Between *am* (8%) and *inl* (1%).
erg-1	Between *pk* (2%) and *asn* (9%).
erg-2	Left of *inl* (6%).
esr-1	Between *pyr-6* and *inl* (7%).
fdb	Contiguous with *leu-5*.
fkr-1	Between *leu-5* (4%) and *ure-2* (13%).
fkr-2	Between *inl* (10%) and *cot-2* (3%).
fpr-1	Linked to *cyb-2* (<1%).
fpr-4	Right of *inl* (11%).
Fsr-9	Linked to *Cen-V*, *lys-1* (1T/18 asci), *con-4*, *Fsr-16* (0T/18 asci).
Fsr-16	Linked to *Cen-V*, *lys-1* (1T/14 asci).
Fsr-20	Right of *ro-4* (1/18).
gln-1	Linked to *inl* (2%), probably to the right.
glp-6	Left of *inl* (30%).
gna-2	Linked to *inl* (0/18).
gran	Between *pab-2* (1%; 8%) and *his-6* (8%; 27%).
gul-1	Between *am* (<0.1%) and *ace-5* (<1%).
hah	Linked to *inl* (13%).
hgu-4	Between *cyb-2* (7%) and *ure-2* (10%).
bH1	Linked to *ccg-9*, *pho-4*.
his-1	Between *am* (3%), *ure-1* (1%) and *pho-2* (3%), *al-3*, *inl* (1%; 10%).
his-6	Between *oak* (6%) and T(NM149) (0/499), *Tel-VR* (30 kb).
hsp80	Between *inl* (3T/18 asci), *cya-2* (2T/18 asci) and *tom70* (1/17 asci).
hsp83	Linked to *cmd*, *hsp80* (0/18 asci).
lasc	Right of *Cen-V*.
ilv-1	Between *per-1* (4%) and *ilv-2* (<1%; 9%), *lys-2* (4%; 7%).
ilv-2	Between *ilv-1* (<1%; 9%) and *lys-2*.
In(UK2-y)^L	Breakpoint in NOR.
In(UK2-y)^R am	Breakpoint at *am*.
inl	Between *al-3* (1%; 1/18 asci) and *pab-2* (1%; 10%).
inv	Between *ro-4* (5%; 8%), *pab-2* (3%), *mus-11* (4%) and *asn* (4%; 9%).
leu-5	Between *cyb-2* (1%) and *sp* (3%; 9%).
lis-3	Right of *inl* (4%).
lys-1	Between *caf-1* (4%; 7%) and *cyt-9* (5%), *at* (1%; 7%; 20%).
lys-2	Between *ilv-1*, *ilv-2* (4%; 7%) and *cyb-2* (<1%), *leu-5* (9%).
mc	Linked to *cyb-2*.
mcb	Linked to *al-3* (3%; 1T/18 asci), *cyb-2*, *sod-2* (0T/18 asci).
md	Between *sb* (3%) and *sp* (9%).
mei-2	Between *inv* (18%) and *his-6* (12%).
met-3	Between *pab-1* (1%), *trp-5* (4%) and *pk* (1%).
mfa-1	Linked to *leu-5* (0T/18 asci).
mus-11	Between *ad-7* (6%; 11%), *pab-2* (7%) and *inv* (4%), *un-9* (10%), *mei-2* (5%; 10%).
mus-12	Left of *inl* (10%).
mus-16	Between *caf-1* (4%) and *lys-1* (9%).
mus-18	Linked to *rDNA* (0T/17 asci), *con-2* (2/17 asci), T(AR190) (≤3%).
mus-28	Left of *lys-1* (10%).
mus-34	Between *leu-5*, *ure-2* and *am*, *bis-1* (3%).
mus-37	Between *lys-1* and *cyb-2*.
mus(SCl7)	Left of *inl* (27%).
nap	Linked to *inl* (15%). Right of *ure-2* (32%).
ndc-1	Left of *arg-4* (2%), *inl* (6%).
ndk-1	Between *al-3* (10%; 22%) and *his-6* (22%; 24%).
nmr	Between *am* (3%; 7%) and *gln-1* (4%; 10%).
NOR	Between *Tip-VL*, *sat* and T(AR30), *dgr-1*. Involved in numerous rearrangements.
nub-3	Between *cyb-2* (4%) and *al-3* (17%).
nuo21.3a	Left of *inl* (18%), *inv* (22%) .
nuo24	Linked to *cyb-2* (0T/18 asci).
oak	Between *un-9* (7%) and T(DBL9), *his-6* (6%).
pab-1	Between *inl* (1%; 10%) and *trp-5*, *met-3* (1% or 2%).
pab-2	Linked to *ro-4* (0/407). Between *ad-7* (8%), T(EB4)^R and *inv* (3%).
pcna	Linked to *rDNA* (0T/18 asci).
per-1	Between *asp* (16%; 26%) and *ilv-1* (4%).
pho-2	Between *his-1* (3%) and *inl* (4%).
pk	Between *met-3* (1%) and T(EB4)^L, *cot-2* (8%).
pl	Linked to *gran* (0/75). Between *asn* (1%; 9%) and *his-6* (16%).
Pogo	Adjoins *Tel-VR*.
poi-1	Linked to *cyb-2* (0T/18 asci).
pold	Between *ure-2* (<70 kb) and *am* (10 kb).
ppt-1	Linked to *inl* (0/18).
prd-2	Between *lys-2* and *am*.

Continues

223

prd-6	Linked to *ad-7*. Between *inl* (9%; 30%) and *pab-2* (2%).
Prf	Linked to *al-3* (3%), probably to the right.
pro-3	Linked to *inl*, *pab-1* (0/74). Between *his-1* (4%) and *pk* (2%; 6%).
pvl(PCNA)	Linked to NOR.
pyr-6	Between *asn* (6%) and *um-9* (2%).
rca-1	Linked to *cyb-2* (0T/16 asci).
rDNA	Between *Tip-VL* and *mus-18* (0T/17 asci; 20 kb), *dgr-1*. In the NOR.
rec-1	Between *ro-4* (7%) and *asn* (5%).
rec-2	Between *sp* (1%; 6%) and *am* (2%).
ro-4	Linked to *pab-2* (0/407). Between *ad-7* (4%), *T(EB4)*^{R} and *inv* (5%).
rol-3	Between *ilv-1* (2%) and *cot-4* (5%).
sat	At *Tip-VL*, distal to NOR.
scon-1	Probably linked to *his-6*.
scot	Between *al-3* (7%) and *his-6* (11%).
sdv-6	Linked to *crp-4* (0T/18 asci). Right of *inl* (9T/18 asci).
sen	Between *his-1* (4%) and *al-3* (13%).
ser-2	Between *met-3* (4%) and *T(EB4)*^{L}, *cot-2* (5%; 8%).
sb	Between *ilv* (4%; 7%) and *md* (3%).
smco-6	Linked to *asn* (6%) near *pyr-6*.
smco-7	Right of *ilv-1* (2%). Linked to *rol-3* (0/154), *sp* (12%).
sod-2	Linked to *cyb-2* (0T/18 asci).
sor-5	Linked to *his-1*.
sp	Between *leu-5* (3%; 9%) and *ure-2* (8%), *am* (1%; 8%; 450 kb).
spco-9	Linked to *asn* (6%), right of *met-3* (18%).
spco-10	Between *ilv-1* (24) and *inl* (5%).
spe-1	Between *am* (1%) and *his-1* (1%).
spe-2	Left of *am* (6%), *spe-1*.
spe-3	Right of *inl* (4%).
ssu-6	Linked to *his-1* (4%).
su(het-d)	Near *lys-1*.
su(ile-1)	Right of *met-3* (4%).
su([mi-1])-f	Left of *inl* (10%).
T(UK1-35)	Breakpoint in NOR.
T(UK2-32)	Breakpoint in NOR.
T(UK3-41)^L	Between *his-1* and *al-3*.
T(UK3-41)^R	Between *pyr-6* and *oak*.
T(EB4)^L	Between *pk*, *ser-1* and *cot-2*, *wa*.
T(EB4)^R	Between *ad-7* and *ro-4*, *pab-2*.
T(UK4-22)	Breakpoint in NOR.
T(DBL9)	Between *oak* and *Tip-R*. Left of *his-6* (0/48).
T(UK14-1)	Breakpoint in NOR.
T(AR30)	Between *Tip-VL* and *caf-1*.
T(AR33)	Between *caf-1* and *lys-1*.
T(RS5)	Breakpoint in NOR.
T(MB67)^L	Between *caf-1* and *lys-1*.
T(MB67)^R	Between *cyt-9* and *Cen-V*.
T(NM149)	Linked to *his-6* (0/499).
T(NM169d)	Linked to NOR at *Tip-VL*.
T(ALS176)	Breakpoint in NOR.
T(ALS182)	Breakpoint in NOR.
T(AR190)	Breakpoint in NOR.
T(OY321)	Breakpoint in NOR.
T(OY325)	Between NOR and *dgr-1*.
Tel-VR	Linked to *his-6* (30 kb).
Tip-VL	Marked by *sat*, NOR, rearrangements. Left of *mus-18*, *dgr-1*.
Tip-VR	Marked by *T(NM149)*. Right of *his-6*.
tom40	Linked to *cyb-2* (0T/15 asci).
tom70	Right of *inl* (6 or 7T/17 asci).
trp-5	Between *inl*, *pab-1* and *met-3* (4%).
ts	Linked to *inl* (4%).
uc-4	Between *inl* (12%) and *his-6* (11%).
udk	Left of *uc-4* (29%).
un-9	Between *pyr-6* (3%) and *oak* (7%), *his-6* (5%; 7%).
un-11	Left of *his-1* (<1%).
un-19	Linked to *al-3* (9%), *un-11* (14%).
ure-1	Linked to *ace-5* (<1%) and *his-1* (<1%).
ure-2	Between *sp* (8%) and *am* (2%), *ure-1* (3%).
usg-1	Linked to *am* (3.5 kb).
val	Linked to *ilv-2* (0/135).
vma-1	Between *leu-5* (1/18 asci), *sp* (8%) and *am* (2%; ~40 kb), *al-3* (1/18 asci).
vma-3	Between *his-1* and *al-3*. Linked to *leu-5* (0T17 asci), *vma-1* (~1100 kb), *cyb-2*.
wa	Between *pk*, *T(EB4)*^{L} and *ad-7* (4%).

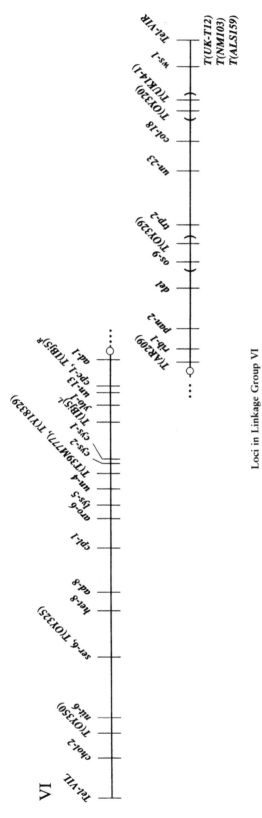

Loci in Linkage Group VI

Locus	Description
acu-6	Left of cys-1 (3%).
ad-1	Between ylo-1 (6%) and Cen-VI (1% or 2%), glp-4 (0; 2%).
ad-8	Between ser-6 (15%) and cpl-1 (7%; 11%), aro-6 12%).
age-2	Right of ws-1 (8%).
aro-6	Between ad-8 (8%; 12%), cpl-1 (5%) and lys-5 (3%).
asd-1	Between nuo19.3 (5T/18 asci) and cmt (2 or 3T/18 asci), Bml (5 or 6T/18 asci).
asd-3	Adjoins asd-1.
Bml	Between cys-2 (2%), cpc-1 and ad-1, Cen-VI.
cax	Linked to Tel-VIL (3T/18asci).
Cen-VI	Between ad-1 (1% or 2%) and T(AR209), rib-1 (1%; 4%), pan-2 (2%).
chol-2	Between Tip-VIL and nit-6 (6%; 8%).
cbr	Between chol-2 (10%) and pan-2.
cmt	Between asd-1 (2T/18 asci) and Bml, Cen-VI (1/18; 1T/18 asci).
col-18	Between un-23 (3/28) and T(OY320).
con-3	Linked to Bml (0/18).
con-11	Right of Cen-VI (6T/17 asci). Linked to trp-2 (0/18).
cpc-1	Between ylo-1 (1%; 3%; 143 kb), un-13 (105 kb) and Cen-VI.
cpl-1	Between ad-8 (7%; 11%) and aro-6 (5%), lys-5 (6%), un-25 (2%; 4%).
cya-3	Between chol-2 (10%) and cyt-2 (10%).

Locus	Description
cys-1	Between cys-2 (1%; 3%) and ylo-1 (8%).
cys-2	Between un-4 (4%), T(T39M777), T(Y18329) and cys-1 (1%; 3%).
cyt-2	Between cya-3 (10%) and lys-5 (6%).
cyt-15	Between pan-2 (4%) and, trp-2 (3%).
del	Between pan-2 (6%) and os-9, T(OY329), trp-2 (0; 13%).
edr-1	Linked to ad-1, pan-2 (<1%).
edr-2	Left of ad-1 (19%).
fpr-6	Between pan-2 and trp-2.
Fsr-50	Linked to Bml (0/18), Cen-VI, pan-2 (0T/18 asci).
gdb-1	Linked to tom22 (0T/16 asci).
glp-4	Between ad-1 (0; 2%) and rib-1 (3%; 4%).
gul-5	Linked to trp-2 (10%).
het-8	Between ser-6 (8%) and ad-8 (3%; 12%).
het-9	Between T(AR209) and T(OY329)^L, hence, between Cen-VI and trp-2.
hspp-1	Linked to Bml, Cen-VI, pan-2 (0T/18 asci).
lgd	Linked to chol-2 (7/32). Between Tip-VIL and T(OY350).
lis-2	Between chol-2 (11%) and trp-2 (26%).
lys-5	Between cyt-2 (6%), aro-6 (3%) and un-4 (2%) .
mod-5	Linked to Cen-VI (3% 2nd division), trp-2.
moe-2	Linked to trp-2 (14%), probably to the left.

226

mus-14	Linked to *lys-5*.	*T(AR209)*	Between *Cen-VI* and *rib-1*.
mus-29	Between *chol-2* (14%) and *lys-5* (15%).	*T(OY320)*	Between *un-23* and *ws-1*.
mus-39	Right of *ylo-1* (3%; 8%). Linked to *mus-29* (10%).	*T(OY329)*	Between *del* and *trp-2*.
nit-6	Between *chol-2* (6%), *T(OY350)* and *ser-6* (11%).	*T(OY343)*	At *Tip-VIR*.
nub-10	Linked to *trp-2*.	*T(OY347)*	At *Tip-VIL*.
nuo19..3	Between *gdb-1* (9T/18 asci) and *asd-1* (5T/18 asci), *Bml* (10T/18 asci).	*T(OY350)*	Between *chol-2* and *nit-6*.
		T(Y18329)	Between *un-4* and *cys-1*.
nuo21.3c	Linked to *Tel-VIR* (0 or 1T/18 asci).	*Tel-VIL*	Linked to *cax* (3T/18 asci).
nuo30.4	Linked to *Bml, Cen-VI, pan-2* (0T/18 asci).	*Tel-VIR*	Linked to *T(UK-T12)* (0/13; 0T/9 asci), *nuo21.3c* (0T/18 asci).
os-9	Between *ad-1, del* and *trp-2*.	*tf2d*	Left of *T(T39M777)*.
pan-2	Between *rib-1* (<1%; 3%) and *del* (6%), *trp-2* (11%).	*tim17*	Linked to RFLP markers.
rib-1	Between *Cen-VI* (1%), *glp-4* (4%), *T(AR209)* and *pan-2* (3%).	*Tip-VIL*	Left of *chol-2*.
ser-6	Between *nit-6* (10%) and *het-8* (8%), *ad-8* (16%).	*Tip-VIR*	Marked by *T(NM103)* and others. Right of *us-1, col-18*.
sor-1	Left of *ylo-1* (3%).	*tom22*	Linked to *gdb-1* (0T/16 asci), *nuo19.3* (9T/18 asci).
spco-7	Linked to *ad-1* (0/65). Between *ylo-1* (4%) and *T(AR209)*.	*trp-2*	Between *del* (0; 13%), *T(OY329)* and *un-23* (5%; 27%).
spco-13	Linked to *trp-2* (16%), *Cen-VI* (1/10 asci).	*un-4*	Between *lys-5* (2%; in the same plasmid) and *T(T39M777, un-13* (3%).
ssu-7	Between *ad-8* (8%) and *ylo-1* (14%).		
su(mtr²⁶)	Linked to *pan-2, trp-2*. Right of *Cen-VI*.	*un-13*	Between *ylo-1* (2%; 38 kb) and *cpc-1* (1% or 2%; 105 kb).
su(trp-5)	Near *aro-6* (perhaps an allele).	*un-23*	Between *trp-2* (5%; 27%) and *col-18* (3/28), *T(OY320)*, *T(UK14-1), ws-1*.
T(IBj5)ᴸ	Between *cys-1* and *ylo-1*.		
T(IBj5)ᴿ cpc-1	At *cpc-1*.	*un-25*	Between *cpl-1* (4%) and *ylo-1* (0/86).
T(UK-T12)	At *Tip-VIR*.	*vvd*	Between *T(IBj5)ᴸ, ylo-1* and *T(IBj5)ᴿ cpc-1*.
T(UK14-1)	Between *un-23* and *ws-1*.	*ws-1*	Between *trp-2* (38%), *un-23*, *T(OY320)*, *T(UK14-1)* and *Tip-VIR*.
T(T39M777)	Between *un-4* and *cys-1*.	*ws-2*	Linked to *ylo-1* (16%), *trp-2* (2%).
T(NM103)	At *Tip-VIR*.	*ylo-1*	Between *cys-1* (8%), *T(IBj5)ᴸ* and *un-13* (2%; 38 kb), *cpc-1* (1%; 3%; 143 kb).
T(ALS159)	At *Tip-VIR*.		

Loci in Linkage Group VII

Locus	Description
ace-8	Between *thi-3* (1%), *T(T54M50)* and *qa* cluster (2% or 3%).
acu-9	Linked to *nic-3* (29%).
acu-11	Linked to *arg-10* (21%).
adb	Linked to *do* (0/53), *spco-4* (4%). Between *cyt-7* (9%) and *nic-3* (4%; 11%).
aga	Between *uvc-1* (2%) and *arg-10* (24%; 27%).
aln-1	Linked in VII.
arg-10	Between *T(S946)*, *arg-11* (1% or 2%) and *nt* (1%; 12%).
arg-11	Between *T(S946)* and *arg-10* (1% or 2%).
ars-1	Between *thi-3* (2%; 5%) and *qa*, *Cen-VII*, *met-7* (1%).
asd-2	Linked to *nic-3* (15%), *ars-1* (36%).
bn	Linked to *met-7* (0/93), *uvc-1* (1%). Right of *T(T54M50)*, *thi-3* (2%).
cat-2	Linked to *for*, *frq* (8 to 10T, 0NPD/18 asci), *cox-8* (4 to 6T, 1 NPD/18 asci).
ccg-9	Linked to *nic-3* (2T/18 asci).
cdr-2	Between *nic-3* and *met-7*.
Cen-VII	Between *ars-1*, *qa* cluster and *met-7* (1%).
cfp	Between *for* (1 or 2T/18 asci) and *T(S936)*. Near *ras-3*, *un-10*.
col-2	Linked to *met-7* (1%), probably to the left. Right of *T(T54M50)*.
col-17	Linked to *nt* (14%), *spco-5* (6%).
cox-8	Linked to *un-10* (3/17), *lacc* (4/18).
cpc-2	Linked to *arg-10* (18%). Right of *T(S936)*.
csp-2	Linked to *thi-3* (<1%), probably to the right. Left of *T(T54M50)*, *ace-8*.
cya-8	Between *Tip-VIIL* and *T(ALS179)* (5%), *cyt-7* (7%).
cyt-6	Linked to *uvc-1* (2%).
cyt-7	Between *T(ALS179)* and *adb* (9%), *nic-3* (18%).
dim-2	Between *uvc-1* and *un-10*, *arg-10*.
do	Linked to *spco-4* (<1%). Left of *nic-3* (1%; 3%).

Locus	Description
dr	Between *for* (3%) and *T(T5936)*, *arg-10* (12%).
for	Between *uvc-1* (6%) and *frq* (2%), *oli* (5% or 6%), *dr* (3%).
frq	Between *uvc-1*, *for* (3%) and *oli* (<2%), *un-10*.
gul-4	Linked to *nic-3* (17%).
het-e	Between *cya-8*, *T(ALS179)* and *nic-3* (28%).
het-10	Between *dr*, *T(S936)* and *Tip-VIIR*.
hH1	Linked to *ccg-9*, *pho-4*.
hH2A	Linked to *Cen-VII* (1T/18 asci).
hH2B	Linked to *Cen-VII*. Adjoins *hH2A*.
hlp-1	Between *sfo* (1%; 9%) and *nt* (28%; 37%).
hlp-2	Between *hlp-1* (8%; 25%) and *nt* (29%).
ile-1	Between *ars* (1%) and *uvc-1* (3%).
kyn-1	Linked to *nic-3* (30%), *uvc-1* (20%).
lacc	Between *nic-3* (14%) and *thi-3* (7%).
le-2	Linked to *met-7* (7%).
mak-2	By cosmid assignment to chromosomal DNA. Tentative.
mb-1	Linked to *nic-3*, *uvc-1* (23%).
mel-1	Left of *thi-3* (27%).
met-7	Between *Cen-VII* (<1%) and *met-9* (10^{-4}), *uvc-1*.
met-9	Between *met-7* (10^{-4}) and *uvc-1* (1% or 2%).
mgk-1	By cosmid assignment to chromosomal DNA. Tentative.
mo-2	Linked to *for* (16%), *nt* (29%).
msh-2	Linked to *for*, *frq* (0T/18 asci). Right of *Cen-VII* (1T/18 asci).
mus-10	Right of *met-7* (7%).
mus-25	Between *met-7* (4%; 6%) and *un-10* (8%).
mus-33	Between *met-7* (10%) and *un-10* (8%).
mus-35	Right of *nic-3* (<9%).
nic-3	Between *do* (3%), *spco-4* (1%) and *lacc* (14%), *thi-3* (9%; 27%), *csp-2* (16%; 22%).
nop-1	Linked to *nic-3* (3T/18 asci), *ars-1* (2T/18 asci).

nt	Between *arg-10* (2%; 12%) and *sk* (7%; 18%).
oli	Between *for*, *frq* (<2%) and *un-10*.
pen-1	Linked to *csp-2* (4%).
pep-4	Linked to *for*, *frq* (0T/18 asci).
pfa-1	Linked to *csp-2* (35%).
pfa-4	Linked to *csp-2* (33%).
pho-4	Linked to *nic-3* (2%; 6%; 1 or 2T/18 asci).
poi-2	Between *cat-2* (0 or 1T/18 asci) and *cox-8* (1T/18 asci).
qa cluster	Between *ars-1*, *tRNA^LEU* and *Cen-VII*, *met-7* (≤1%).
ras-3	Adjacent to *thi1-6* (300 bp). Near *un-10*.
rol-2	Linked to *met-7* (<1%), *uvc-1* (3%).
sdv-13, -14	Linked to *ars-1*, *Cen-VII* (0/18).
sfo	Between *thi-3* (6%) and *hlp-1* (1%; 9%). Linked to *Cen-VII* (<1%).
sk	Between *nt* (7%; 17%) and *Tip-VIIR*.
slo-2	Left of *met-7* (2%).
sor-2	Linked to *nt* (31%).
spco-4	Linked to *do* (<1%), *nic-3* (1%).
spco-5	Linked to *col-17* (6%), *nt* (20%).
spco-6	Right of *thi-3* (4%). Linked to *spco-5* (8%), *nt* (20%).
ssu-1	Between *met-7* (14%) and *su(trp-3^td201)-1* (10%), *nt* (23%).
ssu-4	Between *nic-3* (28%) and *met-7* (20%).

su([mi-1])-5	Left of *nic-3* (23%).
su(trp-3^td201)-1	Between *met-7* (18%), *ssu-1* (10%; 13%) and *arg-10* (7%).
su(trp-3^td201)-2	Linked to *su(trp-3^td201)-1*, probably to the left.
T(T54M50)	Between *thi-3*, *csp-2* and *ace-8*.
T(Z88)	Between *dr* and *arg-10*.
T(ALS179)	Between *cya-8* and *cyt-7*.
T(5936)	Between *dr* and *arg-10*.
Tel-VIIL	Linked to *nic-3*, *pho-4* (10T, 1NPD/18 asci).
thi-3	Linked to *csp-2* (<1%). Between *nic-3* (9%; 28%), *lacc* (7%) and *T(T54M50)*, *ace-8* (1%).
thi-6	Adjacent to *ras-3* (300 bp). Near *un-10*.
Tip-VIIL	Left of *cya-8*.
Tip-VIIR	Right of *sk*.
tnr	Between *nic-3* (3T/18 asci) and *ars-1* (2T/18 asci), *Cen-VII* (3T/18 asci).
tns	Between *T(5936)* and *arg-10*.
tRNA^LEU	Between *ars-1* and *qa-1* cluster.
ud-1	Between *met-7* (27%) and *arg-10* (10%).
un-10	Between *for*, *frq*, *oli* and *arg-10*.
un-22	Linked to *met-7* (1%).
uvc-1	Between *met-9* (1%; 4%) and *for* (6%).

Appendix 3

Data for RFLP Mapping

Tables are reproduced from Nelson, M. A., and D. D. Perkins (2000). Restriction polymorphism maps of *Neurospora crassa*: 2000 update. *Fungal Genet. Newsl.* **47**. These incorporate data received until January 2000. This and future versions with new data will be posted on the FGSC web site, *http://www.fgsc.net.*

Principle: Restriction fragment length polymorphisms (RFLPs) can be used to determine the approximate map location of any cloned piece of DNA. The genomes of strains that differ strongly in their genetic background may differ at many thousands of nucleotide base pairs and, thus, at thousands of potential restriction sites. Two such strains nevertheless may be capable of mating vigorously and productively. Such a cross is heterozygous, and therefore marked, at each of the restriction sites, and progeny will inherit the restriction trait from one or the other parent. Once progeny have been isolated and DNA from them has been digested with an appropriate restriction enzyme, electrophoresed on gels, and blotted, the blots can be probed repeatedly and scored for the segregation of probed DNA sequences that are anonymous or that represent known genes. By comparing the segregation among progeny of a hitherto unmapped piece of *Neurospora* DNA with the segregation of previously mapped gene sequences or anonymous DNA sequences, recombinant progeny can be identified. Linkage can then be detected and gene order can be established on the basis of recombination frequency, just as in a cross with conventional markers.

Application to Neurospora: To establish an RFLP mapping system for *N. crassa*, R. L. Metzenberg and associates crossed a laboratory strain of nominally Oak Ridge genetic background that carried several conventional mutant markers to Mauriceville-1c, a phenotypically wild-type strain from nature [Metzenberg *et al.* (1984). *Neurospora Newsl.* **31**: 35–39; Metzenberg *et al.* (1985). *Proc. Natl. Acad. Sci. USA* **82**: 2067–2071; Metzenberg and Grotelueschen (1995). *Fungal Genet. Newsl.* **42**: 82–90]. Many cryptic DNA sequence polymorphisms differentiate the two strains. For the first cross, 38 progeny from 18 ordered asci were analyzed. Because nonsister spores from the same half of the ascus were selected, first-division

segregation can be distinguished from second-division. For the second cross, 18 random ascospore progeny were analyzed. Permanent stocks of the progeny from these two crosses are available from the Fungal Genetics Stock Center. These have been widely distributed for use in mapping.

The first cross is preferred for RFLP mapping for several reasons: Resolution is better because more loci have already been scored; distance from the centromere can be estimated, regardless of what linkage group is involved; and double crossovers within intervals can be recognized, as can putative gene conversions or scoring errors.

For 17 asci from the first cross, two nonsister spores from the same half of the ascus were selected to be scored for markers. (Genotypes of spores in the unscored half-ascus can, of course, be inferred.) For one of the asci (E), all four ascospore pairs were included, with the object of providing an internal check of each probed DNA sequence for heterozygosity and Mendelian segregation. Progeny are denoted as A1, A4, B6, B7, etc., where the letter identifies the ascus of origin and the number specifies position in the ascus. In the second cross, 18 progeny were tested that came from randomly isolated ascospores, together with the parent strains of the cross. The 20 isolates are designated by numbers. Isolate 1 is the Oak Ridge-derived parent, and isolate 6 is the Mauriceville parent. "M" and "O" indicate that a segregant is like the Mauriceville parent or the Oak Ridge-derived parent, respectively. A dash indicates that the isolate was not scored or that scoring was equivocal for technical reasons. For the two parental strains of the second cross, "(O)" and "(M)" are parenthesized; these scorings are true by definition.

The tables incorporate not only the markers used originally but also data obtained subsequently for many other loci and contributed from many sources. Loci are designated in a variety of ways. To specify the sequences used for probing (left column), gene symbols are used whenever they have been assigned. Designations such as 22:10G (consisting of number, colon, number, letter) refer to a particular cosmid in the Vollmer–Yanofsky clone library. Numbers beginning with AP or R specify RAPD markers [Williams *et al.* (1991). In "More Gene Manipula-tions in Fungi" (J. W. Bennett and L. Lasure, eds.), pp. 431–439. Academic Press]. When the RFLP pattern of a known RAPD marker is identical to that of a cloned gene, the RAPD is not shown. These redundant RAPDs are identified in earlier versions of the RFLP maps (Nelson *et al.* (1998). *Fungal Genet. Newsl.* **45**: 44–54). The *ssr* entries are for simple-sequence repeats, previously called Ncr (I. de la Serna and B. Tyler, personal communication). Five-digit numbers beginning with zeroes denote DNA segments from random regions of the *Neurospora* genome.

Loci are listed in their apparent order on the map. Assignment to linkage group, and often to a particular region, is unequivocal for most of the markers. Some assignments remain provisional, however, and some loci may be positioned incorrectly within linkage groups. Numbers are small and sampling error is high in both crosses. Caution should be used in interpreting the quantitative results and in translating recombination frequencies into map distances.

Attempts have been made to streamline this version of the RFLP mapping data, making them easier to use. Entries that previously were listed separately have been pooled into a single consensus line if their RFLP patterns are identical (or probably identical), except for one or two lanes that were not scored. As a result, each line is separated from its neighbors by one or more crossovers. In previous versions of the RFLP map data, entries were listed on separate lines if they contained discordant scorings that are best attributed to gene conversions or scoring errors. A suspected conver-sion or error can be recognized either as an apparent 4-strand double exchange (NPD) in an interval containing few or no single exchanges or as an apparent double crossover consisting of coincident single crossovers in flanking regions when few or no single crossovers were present in either region. In the present version, apparent conversions or errors have been identified and corrected or deleted from the entries containing them. Finally, the positions of gaps have been indicated within which recombination is high. Doing so highlights the need to fill the gaps and presents a more accurate picture of distance relationships. A few distally located blocks of markers appear to segregate independently of the linkage group to which they are attached (these are indicated as gaps; see, for example, IVL). Linkage-group assignments of these appar-ently unlinked loci are based on their linkage to rearrange-ment breakpoints or to known loci, in crosses other than those shown here. Results of another OR × M cross in which 41 segregants were scored for 12 markers in linkage group II were reported by Metzenberg *et al.* [*Proc. Natl. Acad. Sci. USA* **82**: 2067–2071 (1985)].

An on-line RFLP mapping program was developed for use with the first cross to facilitate the placement of new loci. This program is available through the University of New Mexico Neurospora Genome Project Home Page (*http://www.unm.edu/~ngp/*).

First Cross
LG1

	A1	A4	B6	B7	C1	C4	D5	D7	E1	E3	E5	E7	F1	F3	G1	G4	H5	H7	I6	I8	J1	J4	K1	K4	L1	L4	M5	M8	N2	N3	O2	O4	P1	P4	Q2	Q4	R1	R4
AP10.1, AP10.4	M	O	M	O	M	O	M	O	M	O	M	O	M	M	M	O	O	M	M	M	M	O	O	O	O	O	O	O	O	M	M	M	O	O	M	O	O	M
ft	M	O	M	O	M	O	M	O	M	O	M	O	M	M	M	O	O	M	M	M	M	O	O	M	O	O	O	O	–	–	O	M	O	M	M	–	O	M
cyt-21, zip-1, 2:7H	M	M	M	O	M	M	M	O	M	M	M	O	O	M	M	O	O	M	M	M	M	O	O	M	O	O	O	O	M	M	O	M	O	M	M	O	O	M
(GAP)																																						
AP11h.4	M	O	M	M	O	O	M	M	M	M	O	O	O	M	M	O	O	M	M	M	M	O	O	O	O	O	O	O	O	M	M	M	O	M	O	O	M	M
sod-1	–	O	M	M	M	M	M	M	M	M	O	O	M	M	M	O	O	M	O	M	M	O	O	O	O	O	O	O	O	M	O	M	O	M	M	M	M	M
Fsr-12, uma-4	M	O	M	M	M	M	M	M	M	M	O	O	M	M	M	O	O	M	M	M	M	O	O	O	O	O	O	O	O	M	O	M	O	M	O	O	M	M
clone 106	M	O	M	M	M	O	M	M	M	M	O	O	M	M	M	M	M	M	O	M	M	O	O	O	O	O	O	O	O	M	O	M	O	M	O	O	M	M
AP4a.4	M	O	M	M	M	O	M	M	M	M	O	O	M	M	M	M	O	M	O	M	M	O	O	O	M	O	O	O	O	O	O	M	O	M	O	O	M	M
matA/a, Fsr-30, grp78, pex2, G1:7F, X10:11B, 8:1HL, 17:2B, 22:10G	M	O	M	–	–	–	M	M	M	M	O	O	M	M	M	M	M	M	M	M	M	–	O	O	O	O	O	O	–	–	O	M	O	M	O	O	–	M
R58.2, AP36a.2	M	O	M	M	O	O	M	M	M	M	O	O	M	M	M	M	O	M	O	M	M	O	O	O	O	O	O	O	O	O	O	M	O	O	O	O	M	M
nuo20.8, NP2G2	M	M	M	M	O	O	M	M	M	M	O	O	M	M	M	M	M	M	O	M	M	O	O	O	O	O	O	O	O	O	O	M	O	O	O	O	M	M
X10:E5	M	M	M	M	O	O	M	M	M	M	O	O	M	M	M	M	M	M	M	M	M	O	O	O	O	O	O	O	O	O	O	M	O	O	O	O	M	M
ccg-4	M	M	M	M	O	O	M	M	M	M	O	O	M	M	M	M	M	M	M	M	M	O	O	O	M	O	O	O	O	O	O	M	O	O	O	O	M	M
00036	M	M	M	M	O	O	M	M	M	M	O	O	M	M	M	M	M	M	M	M	M	O	O	O	M	O	O	O	O	O	O	M	O	O	O	O	M	M
AP31a.1, R15.1	M	M	M	M	O	O	M	M	M	M	O	O	M	M	M	M	M	M	M	M	M	O	O	O	M	O	O	O	O	O	O	M	O	O	O	O	M	M
Cen-1, his-2, bsp30, met-3, un-2, 3:8BL	M	M	M	M	O	O	M	M	M	M	O	O	M	M	M	M	M	M	M	M	M	O	O	O	M	O	O	O	O	O	O	M	O	O	O	O	M	M
Fsr-33, lys-4	M	M	M	M	O	O	M	M	M	M	O	O	M	M	M	M	M	M	M	M	M	O	O	O	M	O	O	O	O	O	O	M	O	O	O	O	M	M
nuo 12.3	M	M	M	M	O	O	M	M	M	M	O	O	M	M	M	M	M	M	–	M	M	O	O	O	M	O	O	O	O	O	O	M	O	O	O	O	M	M
AP5a.4, AP12i.1	M	M	M	M	O	M	M	M	M	M	O	O	M	M	M	M	M	M	O	M	M	O	O	O	M	–	O	O	O	O	M	M	O	O	O	O	–	M
12.8B	M	M	M	M	O	M	M	M	M	M	O	O	M	M	M	M	M	M	O	M	O	O	O	O	M	O	O	O	O	O	M	M	O	O	O	O	M	M
cys-9	M	M	M	M	M	M	M	M	M	M	O	O	M	M	M	M	M	M	–	M	M	O	O	O	M	O	O	O	O	O	M	M	O	O	M	O	M	M
ht1	M	M	M	M	M	–	M	M	M	M	O	O	M	M	M	M	M	M	O	M	O	O	O	O	M	O	O	O	O	O	M	M	O	O	O	O	M	M
24.8B	M	M	M	M	M	O	M	M	M	M	O	O	M	M	M	M	M	M	O	M	M	O	O	M	M	O	O	O	O	O	M	M	O	M	O	O	O	M
AP16.1	M	M	M	M	M	O	M	M	M	M	O	O	M	M	M	M	M	M	O	M	M	O	O	M	M	O	O	O	O	O	M	M	O	M	M	O	O	M
aap-2, al-2, Fsr-1, 12:11EL	M	M	M	M	M	O	M	M	M	M	O	O	M	M	M	O	M	M	M	M	M	O	O	M	M	O	O	O	O	M	M	–	O	O	O	O	O	M
sdv-4, sdv-5, uma-11, 5:6F	M	O	M	O	M	O	O	M	M	M	O	O	M	O	O	O	O	M	M	M	O	O	O	M	O	O	O	O	O	M	O	O	O	M	O	O	O	M
(GAP)																																						
Fsr-18, scp	M	O	M	O	M	O	O	O	M	M	O	O	M	O	O	M	O	M	O	M	M	O	O	O	O	O	M	O	O	M	O	M	O	M	O	O	O	M
21:H5	M	O	M	O	M	O	O	O	O	M	M	O	M	O	O	M	M	M	O	M	O	O	O	O	O	O	M	O	O	M	O	M	O	M	O	O	O	M
arg-13	M	O	O	O	O	O	O	O	O	M	M	O	O	O	O	M	M	M	O	M	O	O	O	O	O	O	M	O	O	M	O	M	O	M	O	O	O	M
(GAP)																																						
Tel-IR, un-18	M	M	M	O	O	O	O	M	O	M	M	O	O	M	M	M	O	M	O	M	O	O	O	O	M	M	M	O	O	M	O	M	O	O	M	O	M	M

LGII

	R4	R1	Q4	Q2	P4	P1	O4	O2	N3	N2	M8	M5	L4	L1	K4	K1	J4	J1	I8	I6	H7	H5	G4	G1	F3	F1	E7	E5	E3	E1	D7	D5	C4	C1	B7	B6	A4	A1
Tel-IIL	M	O	M	O	O	M	M	M	M	O	M	O	M	O	M	O	M	O	O	M	O	M	O	M	M	O	–	M	O	M	O	O	O	M	M	O	M	M
hsp70	M	O	M	O	O	M	M	O	M	O	M	O	M	O	M	O	M	M	O	M	O	M	O	M	M	O	O	M	O	M	O	O	O	M	M	O	M	M
(GAP)	O	O	M	M	O	O	M	M	M	O	M	M	M	O	M	O	M	O	O	M	O	M	O	M	O	O	O	O	M	M	O	O	–	–	O	O	–	–
00008	O	O	M	M	O	O	M	O	O	O	M	M	–	O	M	O	M	O	O	M	O	M	O	M	O	O	O	O	M	M	O	O	M	M	O	O	M	O
pSK2-1A	O	O	M	M	O	O	M	O	O	O	M	M	O	O	M	O	M	O	O	M	O	M	O	M	O	O	O	O	M	M	O	O	M	M	O	O	M	O
3-9A	O	O	M	M	O	O	M	O	O	O	M	M	O	O	M	O	M	O	M	M	M	M	O	M	O	O	O	O	M	M	O	O	M	M	O	O	M	O
Fsr-52, upb-1	O	O	M	M	O	O	M	O	M	O	M	M	O	O	M	O	M	M	M	M	M	M	O	M	O	O	O	O	M	M	O	O	M	M	O	O	M	O
Fsr-32	O	O	M	M	O	O	M	O	M	O	M	M	O	O	M	O	M	M	M	M	M	M	O	M	O	O	O	O	M	M	O	O	M	M	O	O	M	O
Cen-II, arg-5, atp-2, 25:1D, G8:11H	O	O	M	M	O	O	M	M	M	O	M	M	M	M	M	M	M	M	M	M	M	M	M	M	O	O	O	O	M	M	O	O	M	M	O	O	M	M
cya-4, Fsr-55	O	O	M	M	O	O	M	M	M	O	M	M	M	M	M	M	M	M	M	M	M	M	M	M	O	O	O	O	M	M	O	O	M	M	O	O	M	M
AP3.2, AP8u.4	O	O	M	M	O	O	M	M	M	O	M	M	M	M	M	M	M	M	M	M	M	M	M	M	O	O	O	O	M	M	O	O	M	M	O	O	M	M
pma-1, 12:11C	O	O	M	M	O	O	M	O	M	O	M	M	M	M	M	M	M	M	M	M	M	M	M	M	O	O	O	O	M	M	O	O	M	M	O	O	M	M
cit-1, nuo78, preg	O	O	M	M	O	O	O	O	M	O	M	M	M	M	M	M	M	M	M	M	M	M	M	M	O	O	O	O	M	M	O	O	–	–	O	O	M	M
ssr-5	O	O	M	M	O	O	O	M	M	O	M	M	M	M	M	M	M	M	M	M	M	M	M	M	O	O	O	O	M	M	O	O	M	M	O	O	M	M
ssr-2	O	O	M	M	O	O	O	M	M	O	M	M	M	M	M	O	M	O	M	M	M	M	M	M	O	O	O	O	M	M	O	O	M	M	O	O	M	M
bH3, bH4-1	O	O	M	M	O	O	O	M	M	O	M	M	M	M	M	–	M	M	M	M	M	M	M	M	O	O	O	O	–	M	O	O	M	M	O	O	M	M
X18:9A	O	M	M	M	O	–	O	O	M	O	M	M	M	M	M	M	M	M	M	M	M	M	M	M	O	O	O	O	–	M	O	O	M	M	O	O	M	O
11:F3, 21:D3	M	O	M	M	O	–	O	O	M	O	M	M	M	M	M	M	M	M	M	M	M	M	M	M	O	O	O	O	M	M	O	O	M	M	O	O	M	O
gpd-1	M	O	M	M	O	O	O	O	M	O	M	M	M	M	M	M	M	M	M	M	M	M	M	M	O	O	O	O	M	M	O	O	M	M	O	O	M	O
uma-2, DB0001	O	O	M	M	O	O	O	O	M	O	O	M	M	M	M	M	M	M	M	M	M	M	M	M	O	O	M	O	M	M	O	O	M	M	O	O	M	O
Fsr-3, ypt-1	O	O	M	M	O	O	O	O	–	O	O	M	M	M	M	M	M	M	M	M	M	M	M	M	O	O	M	O	M	M	O	O	M	M	O	O	M	O
X24:A11	O	O	M	M	O	O	O	M	M	O	O	M	M	M	M	M	M	M	M	M	M	M	M	M	O	O	M	O	O	M	O	O	M	M	O	O	M	O
eas=ccg-2=bli-7	O	O	M	M	O	O	O	O	M	O	O	M	M	M	M	O	M	M	M	M	M	M	M	M	O	O	M	O	O	M	M	M	M	M	O	O	O	O
bli-4, Fsr-34	O	O	M	M	O	O	O	O	M	O	O	M	M	M	M	O	M	M	M	M	M	M	M	M	O	O	M	O	O	M	M	M	M	M	O	O	O	O
Fsr-17, 8-4GL	M	O	M	M	M	O	O	O	M	O	O	M	M	O	M	O	M	M	M	M	M	M	O	O	O	O	O	O	O	M	M	M	O	M	M	O	–	O
AP32c.2, R32.2	M	O	–	M	M	O	O	M	–	–	O	M	O	O	M	O	M	M	M	M	O	M	O	O	M	O	O	M	O	M	M	M	O	M	M	M	M	O
leu-6	M	O	M	O	M	O	O	O	–	–	O	M	O	O	M	O	M	M	O	O	O	M	O	O	M	O	O	M	O	M	M	M	O	M	M	M	M	O
Tel-IIR	M	O	M	M	M	O	O	O	M	M	O	M	M	O	O	O	M	O	O	O	O	M	O	O	M	O	O	M	O	M	M	M	O	M	O	M	M	O

234

LG III

	R4	R1	Q4	Q2	P4	P1	O4	O2	N3	N2	M8	M5	L4	L1	K4	K1	J4	J1	I8	I6	H7	H5	G4	G1	F3	F1	E7	E5	E3	E1	D7	D5	C4	C1	B7	B6	A4	A1
con-1	M	O	O	O	O	O	M	M	M	O	O	O	M	O	M	O	O	M	O	O	M	M	M	–	M	M	M	M	O	O	M	M	M	M	O	–	M	O
AP31a.4, R15.4	M	M	O	O	O	O	M	M	O	O	O	O	M	M	O	O	M	M	–	O	M	M	M	M	M	M	M	M	O	O	M	M	M	M	O	O	O	O
Cen-III, crp-2	M	M	O	O	O	O	M	M	O	M	O	O	M	M	O	O	M	M	O	O	M	M	M	M	M	M	M	M	O	O	M	M	M	M	O	O	O	O
thi-4, pSK7-81C	M	M	O	O	O	O	M	M	M	M	O	O	M	M	O	O	M	M	O	O	M	M	M	M	M	M	M	M	O	O	M	M	M	M	O	O	O	O
pro-1, ser-1	M	M	O	O	O	O	M	M	M	M	O	O	M	M	O	O	M	M	M	O	M	M	M	M	O	M	M	M	O	O	M	M	M	M	O	O	O	O
ace-2	M	M	O	O	O	O	M	M	M	M	O	O	M	M	O	O	M	M	M	O	M	M	M	M	O	M	M	M	O	O	M	M	O	M	O	O	O	O
LZE4, LZE5, RRK	M	M	O	O	O	O	M	M	M	M	O	O	M	M	O	O	O	M	O	O	M	M	M	O	M	M	M	M	O	O	M	M	M	M	O	O	O	O
cox-4, cyt-8	M	M	O	O	O	O	M	M	M	M	O	O	M	M	O	O	M	M	M	O	O	M	M	M	O	M	M	M	O	O	M	M	M	O	O	O	O	O
X14:1F	M	M	O	O	O	O	M	M	M	M	O	O	M	M	O	O	M	M	O	O	O	M	M	M	O	M	M	M	O	O	O	M	M	O	O	O	O	O
aab-1, con-7, mip-1, G22:1H	M	M	O	O	O	M	M	M	M	M	O	O	M	M	O	O	M	M	M	O	O	M	M	M	O	M	M	M	O	O	O	M	M	O	O	O	O	O
ad-2, trp-1, DB002	M	M	O	O	O	M	M	M	O	M	O	O	M	M	O	O	M	M	M	O	O	M	M	M	O	M	M	M	O	–	O	M	M	O	O	O	O	O
32:2G	M	M	O	O	O	M	M	M	M	M	O	O	M	M	O	O	M	M	M	O	O	M	M	M	O	M	M	M	O	–	O	M	M	O	O	O	O	O
AP32a.2	M	M	O	O	–	O	M	M	O	M	M	M	M	M	O	O	M	M	M	–	O	M	–	M	O	M	M	M	O	–	O	M	M	O	O	M	O	O
LZC1	M	M	O	O	O	M	O	O	M	–	M	M	O	O	O	O	M	M	M	O	O	M	M	M	O	M	M	M	O	O	O	M	O	O	O	O	O	O
(GAP)																																						
00032	O	M	M	O	O	M	M	O	O	O	M	O	O	O	O	O	O	M	O	M	O	O	O	M	O	M	M	O	M	–	M	O	M	O	O	O	M	M
Fsr-45	O	M	M	O	O	M	M	O	O	O	M	O	O	O	O	O	O	M	O	M	O	O	M	M	O	M	M	O	M	O	M	O	O	O	O	M	O	M
cat-3	O	M	M	O	O	M	M	O	M	O	M	O	O	O	M	M	O	M	O	M	O	O	M	M	M	O	M	O	M	O	M	O	O	O	O	M	O	M
ssr-6	O	O	M	O	O	–	M	O	M	–	M	O	M	O	M	M	–	M	–	M	–	O	M	O	M	O	–	O	–	O	M	O	O	–	O	M	O	M
AP8f.1	O	M	O	O	O	M	M	M	O	O	M	O	O	O	M	O	O	M	O	M	O	O	M	M	M	O	M	O	O	O	M	O	M	O	O	M	O	M
AP12f.4	O	M	M	O	O	M	M	M	O	O	M	O	O	O	M	O	O	M	O	M	O	O	M	O	M	O	M	O	O	O	M	O	M	O	O	M	O	M
Tel-IIIR	O	M	M	O	O	M	M	M	O	O	M	O	O	O	M	O	O	M	O	M	O	O	M	O	M	–	–	O	M	O	M	O	M	O	O	M	O	M

LG V

	R4	R1	Q4	Q2	P4	P1	O4	O2	N3	N2	M8	M5	L4	L1	K4	K1	J4	J1	I8	I6	H7	H5	G4	G1	F3	F1	E7	E5	E3	E1	D7	D5	C4	C1	B7	B6	A4	A1
AP8w.4	M	M	M	M	M	O	M	M	O	O	O	M	M	M	M	M	O	M	M	O	O	M	O	M	M	M	O	M	O	M	O	M	O	M	O	M	M	O
rDNA, X15:4E	M	O	M	M	M	O	M	M	O	O	O	M	O	M	M	M	O	M	M	O	O	M	O	M	M	M	O	M	O	M	O	M	O	O	O	M	M	O
AP5c.3, AP13.3	O	O	M	M	M	O	M	M	O	O	O	M	O	M	M	M	O	M	M	O	O	M	O	O	M	M	O	M	O	M	O	M	O	O	O	O	M	O
con-2	O	O	M	M	M	O	M	M	O	O	O	M	O	O	M	M	O	M	M	O	O	M	O	O	M	M	O	M	O	O	O	O	O	O	O	O	M	O
AP4a.6	O	O	M	M	M	M	M	M	O	O	O	M	O	O	M	M	M	M	M	O	M	M	O	O	M	M	M	M	O	O	O	O	O	O	O	O	M	O
Cen-V, ccg-8, Fsr-9, Fsr-16, lys-1, pSK21d	O	M	M	M	M	M	M	M	O	O	M	M	O	O	O	M	M	M	O	O	M	M	O	O	O	M	M	M	O	O	O	O	O	O	O	O	M	M
AP36c.2, R64.2	O	M	M	M	M	M	M	M	O	O	M	M	O	O	O	M	M	M	O	O	M	M	O	O	O	M	M	M	O	O	O	O	O	M	O	O	M	M
ccg-1 = grg-1, AP12g.1	O	M	M	O	M	M	M	O	M	O	M	M	O	O	O	M	M	M	O	O	M	M	O	O	O	M	M	M	O	O	O	O	O	O	O	M	M	M
itv-2	O	M	M	O	–	–	–	O	–	O	–	–	O	–	O	M	–	O	O	O	M	–	–	–	–	–	–	–	–	–	O	O	–	O	O	–	O	M
X11:D2	M	M	M	M	M	M	M	O	M	O	M	M	O	O	O	M	M	O	O	O	M	M	O	O	O	M	M	M	O	M	M	O	O	O	O	O	M	M
AP8w.7	M	M	M	O	M	M	M	O	M	O	M	M	O	O	O	M	M	O	O	O	M	M	O	O	O	M	M	M	O	O	M	O	O	M	O	O	M	M
cpc-3, cyh-2, mcb, nuo24, poi-1, rca-1, sod-2, tom40, vma-1, vma-3, 14:10G, G16:11H, X23:C8	M	M	M	O	M	M	M	M	M	O	M	M	M	M	O	M	M	O	O	O	M	M	O	O	O	M	M	M	O	O	M	O	O	M	O	O	M	M
leu-5, mfa	M	M	M	O	M	M	M	O	M	O	M	M	O	O	O	M	M	O	O	O	M	M	O	O	O	M	M	M	O	O	M	O	O	M	O	O	M	M
al-3, Asm-1, ccg-6, 18:10A, 23:1A	M	M	M	O	M	M	M	O	M	O	M	M	M	O	O	M	M	O	M	O	M	M	M	O	O	M	M	M	O	O	M	O	O	M	O	O	M	M
inl	M	M	M	O	M	M	M	O	M	M	M	M	M	O	O	M	M	O	M	O	M	M	M	O	O	M	M	M	O	O	M	O	O	M	O	O	M	M
cit-2	M	M	M	O	M	M	M	O	M	M	M	M	M	O	M	M	M	O	M	O	M	M	M	O	O	M	M	M	O	O	M	O	O	M	O	O	M	M
cox-6, cya-2	M	M	M	O	M	M	M	O	M	M	M	M	M	O	M	M	M	O	M	O	M	M	M	O	O	M	M	M	O	O	M	O	O	M	O	O	M	M
cmd, bsp80	M	M	M	O	M	M	M	O	M	M	M	M	M	O	M	M	M	O	M	O	M	M	M	M	M	M	M	M	O	O	M	O	O	M	O	O	M	M
tom70	–	M	M	M	M	M	M	O	M	M	M	M	M	O	M	O	M	O	O	O	M	M	M	–	M	M	M	M	O	O	M	–	O	M	O	–	–	M
AP34b.3, R42.3	O	M	M	O	O	M	M	O	M	M	M	M	M	O	M	M	M	O	O	O	O	M	M	M	M	M	M	M	O	O	M	O	O	M	M	O	O	O
Fsr-20	O	O	M	O	O	M	M	O	M	M	M	M	M	O	M	M	M	O	O	O	O	M	M	M	M	M	M	M	O	O	M	O	O	M	M	O	O	O
crp-4, sdv-6, 8:9A	O	M	M	M	O	M	M	O	M	M	M	M	M	O	M	M	M	O	O	O	O	M	M	–	M	M	M	M	O	O	M	O	O	M	M	O	O	O
clone 129	O	O	M	M	O	M	M	O	O	M	M	M	M	O	M	M	M	O	O	O	O	M	M	M	M	M	M	M	O	O	M	O	O	M	M	O	O	O
AP35a.2, R47.2	O	M	M	M	O	M	M	O	O	M	M	M	M	O	M	M	M	M	O	O	O	M	M	M	M	M	M	M	M	O	M	O	O	M	M	O	O	O
AP8w.1	O	–	M	M	O	M	M	O	O	M	M	M	M	O	M	M	M	M	O	O	O	M	M	–	M	M	M	M	M	O	M	O	O	M	M	O	M	O
amt	O	M	M	M	O	M	M	O	O	M	M	M	M	O	M	M	M	M	O	O	O	M	M	M	M	M	M	M	M	O	M	O	O	M	M	O	M	O
AP5a.3	O	–	M	M	O	M	O	O	O	M	O	O	M	O	M	M	M	M	O	O	O	M	M	M	M	M	M	O	M	O	M	O	O	M	M	O	M	O
Tel-VR	O	M	M	M	O	M	O	M	O	M	M	O	M	O	M	M	M	M	O	O	O	M	M	M	M	M	M	O	M	O	M	O	O	M	M	O	M	O

LG VI

	A 1	A 4	B 6	B 7	C 1	C 4	D 5	D 7	E 1	E 3	E 5	E 7	F 1	F 3	G 1	G 4	H 5	H 7	I 6	I 8	J 1	J 4	K 1	K 4	L 1	L 4	M 5	M 8	N 2	N 3	O 2	O 4	P 1	P 4	Q 2	Q 4	R 1	R 4
Tel-VIL	M	O	O	M	M	O	O	M	M	O	O	M	O	M	M	M	O	M	O	M	O	M	O	M	O	M	O	O	O	M	O	O	O	M	M	O	M	O
cax	O	O	O	M	M	O	O	M	M	O	O	M	O	M	O	O	O	M	O	M	O	M	M	M	O	—	O	O	O	M	M	O	O	M	M	O	O	O
tf2d	O	M	M	—	—	—	O	—	O	—	O	M	M	M	M	O	O	O	O	O	O	M	O	O	M	O	O	O	M	M	M	M	—	O	O	O	O	M
AP8d.5	M	M	—	M	M	O	O	M	—	—	O	—	O	M	M	M	M	M	O	O	O	M	O	O	M	M	O	O	O	O	M	M	—	M	O	O	M	O
R249.1	M	O	O	O	O	O	—	O	M	—	O	M	M	O	O	O	M	O	O	O	M	O	—	O	M	—	—	O	O	O	O	M	M	M	O	O	M	M
gdb-1, tom22, X25:7C	O	O	O	M	M	O	M	M	M	—	M	M	M	M	M	M	M	O	O	O	O	O	O	O	M	O	O	O	O	O	M	M	M	O	O	O	M	M
AP3b.1	M	O	M	M	M	M	M	M	M	O	M	M	M	O	M	M	M	O	O	O	M	M	O	O	M	O	O	O	O	O	M	M	M	O	O	O	M	M
nuo19.3	M	O	O	O	M	M	M	M	M	O	M	M	M	O	M	M	M	O	O	O	O	M	O	O	M	O	O	O	O	O	M	M	M	O	O	O	M	M
AP8g.3	O	—	M	M	M	M	M	M	M	O	M	M	M	O	M	M	M	O	O	O	O	O	O	O	M	O	O	O	O	O	M	M	M	O	O	O	M	M
tim17	O	O	M	M	M	M	M	M	M	O	M	M	M	O	O	M	M	O	O	O	O	M	O	O	M	O	O	O	O	—	M	M	M	M	O	O	M	M
asd-1, asd-3, 12:7B	O	O	M	M	M	M	O	M	M	O	M	M	M	O	O	M	M	O	O	O	M	M	O	O	M	O	O	O	O	M	M	M	M	M	O	O	M	M
cmt, G9:12G	O	M	M	M	M	M	O	M	M	O	M	M	M	M	M	M	M	O	O	O	M	M	O	O	M	M	O	O	M	O	M	M	M	M	O	O	M	M
AP31a.7, R15.7	O	M	M	M	M	M	M	M	M	O	M	M	M	O	O	O	M	M	O	O	M	M	O	O	M	M	O	O	O	O	M	M	M	M	O	O	M	M
Cen-VI, Bml, Fsr-50, hsttp-1, nuo30.4, pan-2, G2:3A, G7:5F, G8:1C, G8:6D, G12:10C, X5:9H, X7:8E, X13:5A, X13:5H, X16:12D, X24:8E, pHL200	M	M	M	M	M	M	M	M	M	O	M	M	M	—	O	—	M	M	O	O	M	M	O	O	M	M	O	O	M	O	M	M	M	M	O	O	M	M
con-11	O	O	—	—	O	M	M	M	M	O	M	M	M	O	O	O	O	O	O	M	O	O	O	O	O	O	O	M	M	M	O	O	M	M	O	O	M	M
AP9b.3	O	O	M	M	M	M	M	O	O	O	M	O	O	O	M	O	O	O	O	M	O	O	O	O	O	O	O	M	O	O	M	M	M	O	M	O	M	M
Tel-VIR, nuo21.3c, DA122	M	M	O	M	M	M	O	M	O	M	M	M	O	M	O	O	O	O	O	M	O	O	O	M	O	M	O	M	O	M	M	M	M	O	M	O	M	M

LG VII

	A 1	A 4	B 6	B 7	C 1	C 4	D 5	D 7	E 1	E 3	E 5	E 7	F 1	F 3	G 1	G 4	H 5	H 7	I 6	I 8	J 1	J 4	K 1	K 4	L 1	L 4	M 5	M 8	N 2	N 3	O 2	O 4	P 1	P 4	Q 2	Q 4	R 1	R 4
Tel-VIIL	M	M	M	M	M	O	M	O	O	M	O	M	O	O	O	M	M	O	M	O	O	M	O	M	M	M	M	O	M	O	O	O	O	O	M	O	O	O
5:5A	M	M	M	M	M	O	M	M	O	M	O	M	O	O	O	M	M	O	M	O	O	M	O	M	M	M	M	O	M	O	O	O	O	O	M	O	O	O
(GAP)	M	—	O	O	M	M	M	M	O	O	M	—	O	O	O	M	O	M	M	O	O	O	O	M	M	M	M	M	O	O	M	O	O	O	M	M	O	O
00003	M	O	O	O	M	M	M	M	O	O	M	M	O	O	O	M	O	O	M	O	O	O	O	M	M	M	M	M	O	O	M	M	O	O	M	M	O	O
ccg-9	M	M	O	O	M	O	M	M	O	O	M	M	O	O	O	M	O	O	M	O	O	O	O	M	M	M	M	M	O	O	M	M	O	O	M	M	O	O
pho-4	M	M	O	O	M	M	M	M	O	O	M	M	O	O	O	M	O	O	M	O	O	O	O	M	M	M	M	M	O	O	M	M	O	O	M	M	O	O
nic-3, 3:4CL	O	O	O	O	M	M	M	M	O	O	M	M	O	O	O	O	O	O	O	O	O	O	O	O	M	M	M	M	O	O	M	M	O	O	M	M	O	O
AP11c.3	O	O	O	O	M	O	M	O	O	O	M	M	O	O	O	O	O	O	M	O	O	O	O	O	M	M	M	M	O	O	M	M	O	O	M	M	O	O
nop-1, tnr	O	O	O	O	M	O	M	M	O	O	M	M	O	O	O	O	O	O	M	O	O	O	O	O	M	M	M	M	O	O	M	M	O	O	M	M	O	O
AP34a.1, R40.1	O	O	O	O	M	M	M	O	O	O	—	M	O	O	O	O	O	O	O	O	O	O	O	O	M	—	M	M	O	O	M	M	O	O	M	M	O	M
AP39a.2, R198.2	O	O	O	O	M	O	M	M	O	O	M	M	O	O	O	O	O	O	O	O	O	O	O	O	M	M	M	M	O	O	M	M	O	O	M	M	O	M
Cen-VII, ars-1, sdv-13, sdv-14, 28:12G, X23:9G	O	O	O	O	O	M	M	M	O	O	M	M	O	O	O	M	O	O	O	O	—	O	O	O	M	M	O	O	O	O	M	M	O	O	M	M	O	M
for, frq, msb-2, pep-4	O	O	O	O	M	M	M	O	O	O	M	M	O	O	O	M	O	O	O	O	O	O	O	O	M	M	O	O	O	O	M	M	O	O	M	M	O	M
cfp	O	O	O	O	M	M	M	M	O	O	M	M	O	O	O	M	O	M	M	O	O	O	M	M	M	M	O	O	M	M	M	M	O	O	M	M	M	M
cat-2	O	O	O	O	M	M	M	O	O	O	M	M	M	O	M	M	O	O	M	O	O	M	M	M	M	M	O	O	M	M	M	—	O	O	O	O	M	M
poi-2	O	M	O	O	M	M	M	O	O	M	O	M	M	O	M	M	O	O	M	O	M	M	M	M	O	—	O	O	M	M	—	—	O	O	O	O	M	M
cox-8, ssr-1, pRB22	O	O	O	O	M	M	M	M	O	M	O	M	O	M	M	M	O	O	M	O	M	M	M	M	O	O	O	O	O	O	M	M	O	O	O	O	M	M
NP4A9	M	O	O	O	M	M	M	O	O	M	O	M	M	M	M	M	O	M	M	O	M	M	M	M	O	O	O	O	O	O	—	—	O	O	O	O	O	O
AP5i.2	M	M	O	M	O	M	M	M	O	M	O	M	M	M	M	M	O	M	M	O	M	M	M	O	M	M	O	O	O	O	M	M	O	—	M	O	O	O

Second Cross

LG I

	1	2	3	4	5	6	7	8	9	10	11	12	13	14	15	16	17	18	19	20
cyt-21, msh-1	(O)	O	M	O	O	(M)	M	M	O	M	O	O	O	O	O	O	O	O	M	M
mus-40, pzl-1	(O)	O	M	O	M	(M)	M	M	O	M	O	O	O	O	O	O	O	O	M	M
Fsr-12	(O)	O	M	M	M	(M)	M	O	O	M	M	O	O	O	O	O	M	O	-	-
krev-1, mus-38, nit-2	(O)	O	M	M	M	(M)	M	O	O	M	M	O	O	O	O	O	O	O	M	O
matA/a, arg-3, chs-4, leu-4, rgb-1, tef-1, G3:F4	(O)	M	M	M	M	(M)	O	O	O	M	M	M	O	O	O	O	O	O	O	O
Fsr-33	(O)	M	M	M	M	(M)	O	O	O	M	M	M	O	O	O	O	O	O	M	O
al-2, arg-6, cr-1, Fsr-1, met-6, un-7, 26:4H	(O)	M	M	M	M	(M)	O	O	O	O	M	M	M	O	O	O	O	O	O	O
31:2B	(O)	M	M	M	M	(M)	O	O	O	O	M	O	M	O	O	O	O	O	O	O
20:5G	(O)	M	M	M	M	(M)	O	O	O	O	M	M	M	-	O	-	O	-	-	O
Fsr-18, nic-1	(O)	M	M	M	M	(M)	O	O	O	O	M	M	O	O	O	O	O	O	O	O
con-8, phr	(O)	O	M	M	M	(M)	M	O	O	M	O	M	O	O	O	O	O	O	O	O

LG II

	1	2	3	4	5	6	7	8	9	10	11	12	13	14	15	16	17	18	19	20
cys-3	(O)	O	M	M	O	(M)	O	M	M	O	M	O	O	O	O	M	O	O	M	O
00008	(O)	O	M	M	O	(M)	O	M	O	O	O	O	O	M	O	M	O	M	O	O
Fsr-32, Fsr-52	(O)	O	O	M	O	(M)	O	M	O	O	O	O	O	O	M	O	O	O	O	O
Fsr-54	(O)	O	O	M	M	(M)	O	M	O	O	O	O	M	O	M	O	O	O	M	O
hbs	(O)	O	O	M	M	(M)	O	M	O	O	O	O	M	O	M	O	M	M	M	O
con-6	(O)	O	O	M	O	(M)	O	M	O	O	O	O	M	O	M	O	O	O	M	M
Fsr-55	(O)	O	O	M	M	(M)	M	M	O	O	O	O	M	O	M	O	O	M	M	M
alc-1, arg-12, gpd-1, nik-2	(O)	O	O	M	M	(M)	M	M	O	M	O	M	O	M	O	O	M	O	M	M
Fsr-21, Fsr-34	(O)	O	O	M	M	(M)	M	M	O	O	M	M	M	O	O	M	M	O	M	M
trp-3	(O)	O	O	M	M	(M)	M	O	O	O	M	M	M	O	O	M	O	O	M	M
G5:H2	(O)	O	O	M	M	(M)	M	O	O	O	M	M	M	M	O	O	O	O	M	M
Fsr-17, leu-6	(O)	O	O	M	M	(M)	M	O	O	O	M	M	M	M	O	O	M	O	M	O

LG III

	1	2	3	4	5	6	7	8	9	10	11	12	13	14	15	16	17	18	19	20
cyt-22	(O)	M	O	M	M	(M)	O	O	M	M	O	M	M	M	O	M	O	M	M	M
con-1	(O)	O	O	M	M	(M)	O	O	M	M	O	O	M	M	O	M	O	O	O	M
cyt-8	(O)	O	O	M	M	(M)	O	O	M	M	O	O	M	M	O	M	M	O	O	M
DA25	(O)	M	O	M	M	(M)	O	O	M	M	O	O	M	M	O	M	O	O	O	M
con-7, gnb-1, trp-1, LZE4	(O)	M	O	M	M	(M)	O	O	M	M	O	O	M	M	O	M	O	M	O	M
gna-1	(O)	M	O	M	O	(M)	O	O	M	M	O	O	M	M	O	O	O	M	O	M
ro-2	(O)	O	O	M	O	(M)	O	M	M	M	O	O	M	M	O	M	O	O	O	M
Fsr-45	(O)	O	O	M	O	(M)	O	M	M	O	O	O	M	O	O	M	M	O	O	M

LG IV

	1	2	3	4	5	6	7	8	9	10	11	12	13	14	15	16	17	18	19	20
Tel-IVL	(O)	O	M	M	M	(M)	O	M	M	M	O	O	O	O	O	O	O	O	O	M
SPIAE	(O)	O	M	M	M	(M)	O	M	M	M	O	O	O	O	O	O	M	O	O	O
Fsr-62, Fsr-63, mlh-1, mtr, mus-8, pph-1, 4:E12, 5:3B, 5:4H, 22:G3, 23:B5, G16:C5, X12:B4	(O)	M	M	M	O	(M)	M	M	M	M	M	O	O	O	O	O	O	O	O	O
cyt-5, X5:H4	(O)	M	M	M	O	(M)	M	M	M	O	M	O	O	O	O	O	O	O	O	O
arg-14, YAC 6H3	(O)	M	M	M	O	(M)	M	M	M	M	O	M	O	O	O	O	O	O	O	O
X15:A10	(O)	-	-	O	O	(M)	M	M	M	M	M	M	M	O	O	O	O	O	O	O
G17:B4, G17:G6	(O)	M	O	M	O	(M)	M	M	M	M	O	O	O	O	O	O	O	O	O	O
Fsr-51	(O)	M	M	M	O	(M)	M	M	M	M	O	O	O	O	O	O	O	O	O	O
chs-5	(O)	M	M	M	O	(M)	M	M	M	M	O	M	O	O	O	O	O	O	O	O
con-5, con-9, con-10, cot-1, 7:A1, G25:D4	(O)	M	M	M	M	(M)	M	M	M	M	O	O	O	O	O	O	O	O	O	O
met-5	(O)	M	M	M	M	(M)	M	M	O	M	O	O	O	O	O	O	O	O	O	O
ras-1	(O)	M	M	M	M	(M)	M	M	O	M	O	O	O	O	O	O	O	O	M	O
nit-3	(O)	M	M	M	M	(M)	M	M	O	M	O	O	O	M	O	O	O	M	M	O
chs-2, Fsr-4	(O)	O	M	M	M	(M)	M	M	O	O	O	O	M	M	M	O	M	M	M	M
gna-3, uvs-2	(O)	O	M	M	M	(M)	M	M	O	O	M	O	M	M	M	O	M	M	M	M

Appendix 4

The *Neurospora crassa* Mitochondrial Genome*

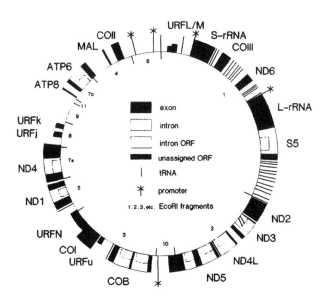

APPENDIX 4: FIGURE 1. Physical map of the mitochoridrial DNA of wild-type strain 74-OR23-1A. The corresponding genes are identified in Appendix 4, Table 1. For a map of restriction endonuclease cleavage sites of the same strain, see Taylor and Smolich [*Curr. Genet.* 9: 597–603 (1995)]. From Griffiths, Collins, and Nargang (1995), with permission from Springer-Verlag.

APPENDIX 4: TABLE 1

Mitochondrial Genes of *Neurospora crassa*

Gene[a]	Equivalent[b]	Encoded product
Respiratory chain		
cox1	coI	Cytochrome *c* oxidase subunit 1
cox2	coII	Cytochrome *c* oxidase subunit 2
cox3	coIII	Cytochrome *c* oxidase subunit 3
cob	cob	Apocytochrome *b*
ndh1	ND1	NADH dehydrogenase subunit 1
ndh2	ND2	NADH dehydrogenase subunit 2
ndh3	ND3	NADH dehydrogenase subunit 3
ndh4	ND4	NADH dehydrogenase subunit 4
ndh4l	ND4L	NADH dehydrogenase subunit 4L
ndh5	ND5	NADH dehydrogenase subunit 5
ndh6	ND6	NADH dehydrogenase subunit 6
ATP synthesis		
atp6	ATP6	ATP synthase subunit 6
atp8	ATP8	ATP synthase subunit 8
atp9[c]	MAL	(ATP synthase subunit 9)[c]
Translational apparatus		
rnl	L-rRNA	Large ribosomal subunit RNA
rns	S-rRNA	Small ribosomal subunit RNA
S-5	S5	Small ribosomal subunit protein
tsl[d]	tRNA	tRNA
Unidentified reading frame		
URF	URF j	
	URF k	
	URF L/M	
	URF N	
	URF u	
Open reading frame		
ORF[e]	ORF	

[a] Symbols used for fungal mitochondrial genes by Hudspeth (1992) and recommended for *Neurospora*.

[b] Symbols used for *Neurospora* mitochondrial genes in Fig. 1 and in Griffiths *et al.* (1995).

[c] The *Neurospora* atp9 gene is translationally inactive. The translationally active gene for ATP synthase subunit 9 is the nuclear gene *oli*, which is closer in sequence to the translationally active mitochondrial gene *ATP9* of *Saccharomyces*.

[d] "tRNA synthesis locus" (Hudspeth, 1992).

[e] Several intron-encoded ORFs have been identified.

*Based on a review by Griffiths, A. J. F., R. A. Collins, and F. E. Nargang (1995). Mitochondrial genetics of *Neurospora*. *In* "The Mycota II, Genetics and Biotechnology" (U. Kück, ed.), pp. 93–105. Springer-Verlag, Berlin, Heidelberg. See also Hudspeth, M. E. S. (1992). The fungal mitochondrial genome—A broader perspective. *In* "Handbook of Applied Mycology. Volume 4: Fungal Biotechnology" (D. K. Arora, R. P. Elander, and K. G. Mukerji, eds.), pp. 213–241. Marcel Dekker, New York.

241

APPENDIX 4: TABLE 2

Mutant Mitochondrial Genomes of *Neurospora crassa*[a]

Group	Mutant	Growth	Cytochromes	Suppression by *su([mi-1])-f*	*su([mi-1])-1*
I	[poky] (=[mi-1])	Lag	Low aa_3 and b, high c	+	−
	[SG-1,-3]	Lag	Low aa_3 and b, high c	+	−
	[stp-B1]	Lag	Low aa_3 and b, high c	+	−
	[exn-1,2,3,4]	Lag	Low aa_3 and b, high c	+	−
	[C93]	Temperature-sensitive	Low aa_3 and b, high c	?	?
II	[mi-3]	Less lag	Low aa_3, high c, normal b	−	+
	[exn-5]	Loss lag	Low aa_3, high c, normal b	−	+
III	[stp]	Stop–start	Low aa_3, and b, high c	−	−
	[stp-A,-A18,-B2,-C]	stop–start	Low aa_3, and b, high c	−	−
	[abn-1,-2]	Stop–start	Low aa_3, and b, high c	−	−
	[E35]	Stop–start	?	?	?
	[ER-3]	Stop–start	Low b and c, no aa_3	?	?
	[stp-107]	Stop–start	?	?	?

[a] The mutations include nucleotide-pair substitutions, deletions, duplications, insertions, and complex rearrangements. The strains with mutant mitochondria are cyanide-insensitive and salicyl hydroxamide-sensitive. Thus, they must rely on an alternate oxidase system. Except for Group II, they are all defective in mitochondrial protein synthesis. Inclusion of [C93] in Group I is tentative. Mutant names: *SG*, slow growth; *stp*, stopper; *exn*, extranuclear; *mi*, maternally inherited; *abn*, abnormal growth. Adapted from a table of Griffiths *et al.* (1995) and based on references given in that review.

Appendix 5

Expressed Sequences from Different Stages of the *Neurospora* Life Cycle: Putative Identification of cDNAs

Shown here are *Neurospora* Expressed Sequence Tags that encode proteins previously characterized in *Neurospora* as well as those with significant sequence identity to proteins in other organisms. RNA was isolated at different stages of the life cycle or under different conditions: germinating conidia (C), mycelia (M), perithecia (P), and unfertilized mycelial mats floating on liquid Westergaard–Mitchell crossing medium (W). For details of methodology, see Nelson *et al.* (1997). Expressed sequences from conidial, mycelial, and sexual stages of *Neurospora crassa. (Fungal Genet. Biol.* **21**: 348–363). The list reproduced here is from Dolan P., D. O. Natvig, and M. A. Nelson (2000). *Fungal Genet. Newsl.* **47**.

The EGAD cellular role classification scheme [White and Kerlavage (1996). *Methods Enzymol.* **266**: 24–40] has been used whenever possible. Those identified *Neurospora* ORFs lacking homologs in the EGAD classification were classified as appropriate under secondary metabolism, etc. Only those sequences with BLASTX *P/E* values of 10^{-5} or lower (highly or moderately significant) are reported in this table, except as noted. Fourteen sequences falling within this range were determined to reflect spurious matches (e.g., proline-rich regions) and not actual homology; those sequences were not included in this table. As additional sequence information becomes available, expanded versions of this table can be accessed at the New Mexico Genome Project Web site: http://www.unm.edu/~ngp/.

Putative Identification of *Neurospora* cDNAs

Clone ID[a]	MatchAcc[b]	Identification[c]	P/E value[d]	Tissues[e]
		I. Cell Division		
	Apoptosis			
NP4B8	NP_035181.1	Programmed cell death 6 protein (Mm)[i]	4.0×10^{-32}	1P
	Cell Cycle			
SC6F9	gil762850	Cell cycle regulator p21 protein (Sp)	2.0×10^{-6}	1C
NM3G1	pirllS43279	Cell division control protein 3 (Ca)	1.1×10^{-80}	1M
NP4C10	P36618	Cell division control protein 16 (Sp)	7.0×10^{-18}	1P
SP6G3	gil4097884	Cell division control protein 42 (Gc)	2.0×10^{-44}	1P
NC2F1	pirllS34027	Cell division control protein 47 (Sc)	1.3×10^{-50}	1C
NM5H9	P41733	Cell division control protein 91 (Sc)	3.2×10^{-19}	1M
SC2G4	pirllS49206	G1 cyclin CLN1 (Ca)	5.0×10^{-8}	1C
SC2D8	P05453	G1, to S-phase transition protein 1 homolog (Hs)	2.4×10^{-9}	1C
NP4C2	NP_010018	Involved in cell cycle, cdc 50p (Sc)	1.0×10^{-48}	1P
NM1D8	NP_011747	Mitochondrial protein, prohibitin (Sc)	2.0×10^{-18}	1M
SP6F11	P32334	MSB2 protein (Sc)	2.0×10^{-6}	1P
NC2A11	gil2944404	Serine/threonine protein phosphatase type 1 (Nc)	4.0×10^{-49}	2C, 1W
	Chromosome Structure			
NC2A6	P46672	GU4 nucleic-binding protein 1 (Sc)	9.8×10^{-12}	1C
SC3B9	P08844	Histone H2A (En)	1.6×10^{-68}	1C
NP4A11	P07041	Histone H3 (Nc)	3.3×10^{-87}	2C, 1P, 2W
NC1B2	P04914	Histone H4 (Nc)	1.9×10^{-51}	1C, 1M, 2W
SM2H6	Q09330	MLO3 protein (Sp)	2.0×10^{-7}	1M
SC5D8	P11633	Non-histone chromosomal protein 6B (Sc)	2.0×10^{-29}	1C
NP5B7	gil172034	Nucleosome assembly protein (Sc)	7.7×10^{-48}	1P
	DNA Synthesis–Replication			
W08D11	NP_013968	DNA-binding protein, mtDNA-stabilizing protein (Sc)	4.0×10^{-40}	1C, 1W
W09H12	CAA03898.1	MCM7-like protein (Sp)	9.7×10^{-15}	1W
SC3F6	Q03392	Proliferating cell nuclear antigen, PCNA (Sp)	1.1×10^{-50}	1C
		II. Cell Signaling–Cell Communication		
	Channels Transport Proteins			
SM1G3	P53386	Aquaporin-like protein YPR192w (Sc)	5.3×10^{-7}	1M
NM4H11	P33970	Halorhodopsin (light-induced chloride pump) (Hh)	8.3×10^{-9}	1M
NP2B6	gil1063415	K+ channel protein (At)	2.1×10^{-67}	1P
W01B10	gil2654088	Potassium transporter (At)	1.4×10^{-12}	1W
NM8G7	CAA73031.1	Putative organic cation transporter (Dm)	9.8×10^{-10}	1M
	Effectors/Modulators			
W17G8	Q01631	Adenylate cyclase (Nc)	3.9×10^{-88}	1P, 1W
SC1B12	P26364	Adenylate kinase 2 (Sc)	3.7×10^{-15}	1C
NP3C6	AAD42978.1	Adenyl cyclase-associated protein (Ca)	4.0×10^{-25}	1P
NM5H5	CAA17814.2	β-Catenin family member (Sp)	2.0×10^{-12}	1M
W09C12	Q02052	Calmodulin (Nc)	8.7×10^{-86}	1P, 2W
	Hormones–Growth Factors			
W13B2	gil2224892	Gibberelin 7-oxidase (Cum)	4.0×10^{-7}	1W
	Intracellular Transducers			
SM1B4	P22126	24-kDa ras-like protein (Nc)	1.1×10^{-40}	1M
NC4D7	pirlS57839	CPC2 protein (Nc)	3.7×10^{-27}	1C

Clone ID[a]	MatchAcc[b]	Identification[c]	P/E value[d]	Tissues[e]
NC5F1	P39958	GDP-dissociation inhibitor (Sc)	3.9×10^{-44}	1C
NP3A7	Q05425	Guanine nucleotide-binding protein α2-subunit (Nc)	1.8×10^{-108}	1P
W10F8	Q01369	Guanine nucleotide-binding protein, β-subunit-like protein (Nc)	3.7×10^{-98}	1W
SC7C4	Q10107	Multicopy suppressor of ras1 (Sp)	8.0×10^{-18}	1C
SM3A10	gil1399532	nuc-2; nuclease-2 (Nc)	9.9×10^{-68}	1M
NP3E5	CAB05920	palA (En)	2.0×10^{-52}	1P
SM4D6	CAA22174.1	Putative signal transduction protein (Sp)	5.0×10^{-17}	1M
SP6F6	Q09914	Rho1 protein (Sp)	5.0×10^{-54}	1P
SC2E3	P40319	SUR4 protein (Sc)	1.0×10^{-22}	1C
NC2C11	NP_013741.1	Tap42p, physically associates with PP2A and SIT4 protein phosphatase catalytic subunits (Sc)	8.7×10^{-19}	2C
SM2H4	P35169	TOR1, phosphatidylinositol 3-kinase (Sc)	2.2×10^{-45}	1M
Metabolism				
NM6A6	P17296	Metapyrocatechase 2 (catechol 2,3-dioxygenase II) (Ae)	6.7×10^{-12}	1M
NM4F4	P37297	Phosphatidylinositol 4-kinase (Sc)	5.8×10^{-72}	1M, 2P
Protein Modification				
SC1C7	gil806859	14-3-3 protein (Trh)	2.9×10^{-93}	3C, 1M, 2W
SM1H12	O14408	Calcium/calmodulin-dependent protein kinase (Ma)	3.0×10^{-45}	2M
NC2B3	gil516040	cAMP-dependent protein kinase catalytic subunit (Mg)	6.3×10^{-46}	1C
NC1F3	Q08466	Casein kinase II, α-catalytic subunit (At)	2.8×10^{-47}	1C
NM1F4	P46594	Halotolerance protein HAL2 (Os)	2.7×10^{-11}	1M
NM5B1	CAB11500.1	MAP kinase kinase kinase (Sp)	1.0×10^{-12}	1M
SM2H12	gil2832241	Nonphototropic hypocotyl 1 (At)	2.7×10^{-10}	1M
W08A7	AAF09475	Osmotic sensitivity MAP kinase (Pg)	1.0×10^{-52}	2W
SM1D8	gil1322070	Protein kinase, cAMP-dependent regulatory subunit (Nc)	1.5×10^{-22}	1M
SC2G1	CAA72731.1	Protein kinase C homolog (Nc)	1.1×10^{-69}	2C
SP6F4	gil2654106	Protein kinase NRC-2 (Nc)	4.0×10^{-50}	1P
NM4G1	CAA22609.1	Protein kinase skp1p (Sp)	1.0×10^{-80}	1M
W13A11	AAD15987.1	Protein phosphatase 2A regulatory B-subunit (Nc)	5.0×10^{-5}	1W
SC3A2	gil458284	Serine/threonine protein kinase (Trr)	3.7×10^{-30}	1C
NC5F10	P05323	Serine/threonine protein phosphatase PP2A-α catalytic subunit (Hs)	2.8×10^{-87}	1C
SP1F10	gil1706961	Shk1 kinase-binding protein (Sp)	4.8×10^{-18}	1P
SP6D9	NP_014366	Tyrosine phosphatase Siw 14p (Sc)	2.0×10^{-12}	1P
Receptors				
W08E3	P30536	Peripheral-type benzodiazepine receptor, mitochondrial (Hs)	8.8×10^{-10}	1W
SP6G1	gil1655907	Protein–tyrosine phosphatase CRYP-2 (Galg)	4.0×10^{-12}	1P

III. Cell Structure–Cytoskeleton

Cell Wall				
SC5D3	Q12114	Chitin biosynthesis protein CHS5 (Sc)	1.0×10^{-23}	1C
NC4G7	P29070	Chitin-UDP acetylglucosaminyl transferase 3 (Nc)	4.6×10^{-50}	1C, 1W
SC7C2	gil2613108	Class V chitin synthase (Um)	2.0×10^{-28}	1C
SM3F10	AAB47060	Exochitinase (Trh)	1.0×10^{-53}	2M
SM1F2	gil3608406	GEL1 protein (Aspfu)	2.0×10^{-76}	2M
NP5B11	gil1261823	Glycine-rich protein (Nc)	4.0×10^{-31}	1P
W01A6	P34226	skt5 protein (Sc)	3.8×10^{-11}	1W
NC5H6	NP_014035.1	Soluble cell-wall proteins Scw4p, Scw10p, Scw11p (Sc)	7.0×10^{-13}	3C, 1W

Continues

Continued

Clone ID[a]	MatchAcc[b]	Identification[c]	P/E value[d]	Tissues[e]
		Cytoskeletal		
W17E2	gil178045	Actin, γ (Hs)	8.0×10^{-13}	1W
NM4E10	CAB66436.1	Actin-related protein (Sp)	6.0×10^{-18}	1M
NP4C4	P18091	α-Actinin, sarcomeric (Dm)	2.4×10^{-47}	1P
NP5C11	P40234	Casein kinase I, homolog Cki2 (Sp)	2.0×10^{-64}	1P
NM4B4	Q03048	Cofilin (Sc)	1.2×10^{-24}	1M, 1W
W13D12	O13923	Coronin-like protein (Sp)	2.1×10^{-45}	1W
W06D1	P14315	F-Actin capping protein, β-subunit isoforms (Galg)	1.6×10^{-19}	1W
SC2E6	P32599	Fimbrin (ABP67) (Sc)	1.3×10^{-66}	1C
NP5E9	AAF18567	Myosin-related protein homolog MlpA (En)	8.0×10^{-18}	1P
NC1G12	P78774	Probable arp2/3 complex 41-kDa subunit (Sp)	1.0×10^{-42}	1C
NC3C2	P39825	Profilin 1 (Sp)	5.5×10^{-33}	1C
NC3A1	CAB39803.1	Putative actin polymerization complex protein (Sp)	4.0×10^{-36}	1C, 2M
SM2C5	P32368	Recessive suppressor of secretory defect, SAC1 (Sc)	4.7×10^{-23}	1M
NP5B3	gil6119698	Tropomyosin, α- and β-chains (Sp)	7.9×10^{-30}	1C, 2P
SC1G3	P38669	Tubulin, α, B-chain (Nc)	1.4×10^{-64}	1C, 1P
NC3D10	P05220	Tubulin, β (Nc)	7.4×10^{-94}	3C, 1P
SP3B12	pirllS51342	Verprolin (Sc)	1.8×10^{-6}	1P
		Motility		
SM4C3	NP_001118.1	Adaptin, β-subunit (Hs)	9.0×10^{-29}	1M
		IV. Cell–Organism Defense		
		Carrier Proteins–Membrane Transport		
SC5E7	BAA33011.1	Flavohemoglobin (Fo)	4.0×10^{-32}	2C
		Detoxification		
SM1E4	BAA08308.1	acr-2; acriflavine resistant (Nc)	6.7×10^{-82}	1M
NM9C10	P38918	Aflatoxin B1 aldehyde reductase (AFB1-AR) (Rn)	3.7×10^{-21}	1M
SP7C3	Q02068	Aliphatic nitrilase (Rhrh)	2.0×10^{-30}	1M, 2P
NP3D9	gil603050	CAP20 (plays role in infection of host) (Gc)	7.6×10^{-39}	2M, 2P
NN2E10	P78574	Catalase A (Aspfu)	9.0×10^{-34}	1M
NM6H12	CAA74698	Catalase/peroxidase (Strr)	4.0×10^{-47}	1M
NM3D4	CAA26793.1	Copper metallothionein (Nc)	6.7×10^{-45f}	1M
SM1G12	P39179	HM-1 killer toxin resistance protein (Sc)	4.6×10^{-33}	1M
SM3H12	pirllS70702	Maackiain detoxification protein 1 (Nh)	2.0×10^{-10}	1M
SM2D2	P35724	Manganese resistance protein, MNR2 (Sc)	4.9×10^{-29}	1M
NC2D6	P38356	Metal homeostasis protein Bsd2p (Sc)	1.5×10^{-9}	1C
NC4D11	gil487426	Pisatin demethylase (Nh)	1.6×10^{-23}	1C, 2M
SM2B10	Q39172	Probable NADP-dependent oxidoreductase P1 (At)	3.1×10^{-11}	1M
SM4C9	NP_014140	Putative copper-binding–homeostasis protein Abx 1 (Sc)	1.0×10^{-6}	1M
SC3A6	CAA21951	Rehydrin-like protein (Ca)	8.0×10^{-36}	2C, 1P, 3W
SP7A9	gil1117921	Rod1p, involved in drug resistance (Sc)	8.0×10^{-26}	1P
NM7A11	CAB66461.1	Similarity to *Saccharomyces cerevisiae* kti12 protein (Sp)	5.0×10^{-5}	1M
		DNA Repair		
NM6F2	P40235	Casein kinase I homolog (HHP1) (Sp)	1.4×10^{-95}	2M
NM1H10	P38632	DNA repair protein MMS21 (Sc)	4.0×10^{-6}	1M
NM4F3	gil703466	Exonuclease I (Sp)	2.6×10^{-52}	1M

Clone ID[a]	MatchAcc[b]	Identification[c]	P/E value[d]	Tissues[e]
	Stress Response			
W01C2	CAA35682.1	Cyclophilin A (Nc)	4.7×10^{-104}	4C, 2P, 4W
NC2C5	Q92249	Erp38 (Nc)	1.2×10^{-15}	1C
NC4A7	P20080	FK506-binding protein (Nc)	3.9×10^{-76}	2C, 2P, 2W
SM2F5	P40581	Glutathione peroxidase (Sc)	1.0×10^{-26}	2M
NM2D3	CAA70214.1	grp78 homolog (endoplasmic reticulum Hsp70) (Nc)	2.0×10^{-36}	1M
NC2E9	P38523	GRPE protein homolog precursor (Sc)	2.4×10^{-40}	1C
W06B9	NP_010978	Heat-regulated protein Hig 1p (Sc)	7.0×10^{-15}	1W
SC6H9	CAA67431	Heat shock protein 70 (En)	3.0×10^{-72}	1C, 1P
W01G6	AAF34607	Heat strees protein 80-1 (Nc)	5.0×10^{-61}	1W
NM4A1	P12807	Peroxisomal copper amine oxidase (Hp)	2.3×10^{-71}	1C, 1M
SC5D10	AAD42074.1	Peroxisomal membrane protein (Pnci)	2.0×10^{-34}	2C, 1W
NC4F4	Q03178	pir1 protein precursor (Sc)	8.6×10^{-9}	1C
NP6H2	CAB50926.1	RICI protein (Phi)	4.0×10^{-12}	1P
W13G6	P80645	Sulfate starvation-induced protein (Ec)	1.0×10^{-5}	1W
NM5H1	P39076	T-complex protein 1, β-subunit (Sc)	1.2×10^{-82}	2M
SC4A3	P39077	T-complex protein 1, γ-subunit (Sc)	7.9×10^{-26}	1C
NP3F8	P23618	Thiazole biosynthetic enzyme (stress-inducible protein sti35) (Fo)	8.6×10^{-86}	5C, 2M, 1P, 14W
		V. Metabolism		
	Amino Acid			
NC1H9	P25605	Acetolactate synthase, small subunit homolog (Sc)	1.2×10^{-50}	5C, 1M, 2P
NC1G8	pir‖A53429	Acetylglutamate kinase (arg-6) (Nc)	5.2×10^{-6}	1C
NC4E1	gi‖1066330	Adenosine-5′-phosphosulfate kinase (Pnch)	2.7×10^{-65}	2C
W17B9	P50514	Argininosuccinate lyase (Sp)	4.7×10^{-32}	2W
NC3G2	P22768	Argininosuccinate synthase (Sc)	2.6×10^{-16}	1C
W07A10	P55325	Aspergillopepsin A precursor (Aspn)	2.6×10^{-13}	1W
W01H2	Q99145	ATP phosphoribosyltransferase (his 1) (Yl)	3.2×10^{-23}	1W
NP2D4	AAD10616.1	β-Isopropylmalate dehydrogenase (Nc)	6.4×10^{-47}	1P
SC5B10	P22572	Carbamoyl phosphate synthase (arg-2) (Nc)	2.0×10^{-58}	3C
NC3A12	P03965	Carbamoyl phosphate synthase, arginine-specific, large chain (Sc)	7.2×10^{-28}	1C
SM1F11	P08456	CDP-diacylglycerol–serine O-phosphatidyltransferase (Sc)	1.7×10^{-48}	1C, 1M
SP1E12	AAF11089.1	Cephalosporin acylase (Dr)	2.0×10^{-13}	1P
SC5B3	Q12640	Chorismate synthase (Nc)	4.0×10^{-62}	1C
NC5F6	P46794	Cystathionine β-synthase (Dd)	5.6×10^{-6}	1C
NC3A10	P31373	Cystathionine γ-lyase (Sc)	2.4×10^{-59}	1C, 1M
SC7G11	P78568	δ-1-pyrroline-5-carboxylate dehydrogenase (Ab)	5.0×10^{-12}	2C
NM6A1	P09624	Dihydrolipoamide dehydrogenase (Sc)	4.8×10^{-57}	1M, 1W
SP1B12	BAA18999.1	Farnesylcysteine carboxylmethyltransferase (Sp)	2.1×10^{-31}	1P
NP2H3	CAA70219.1	Fructosyl amino acid oxidase (Pnj)	3.6×10^{-15}	1P
NM2B7	P00369	Glutamate dehydrogenase (Nc)	2.8×10^{-52}	1M, 1W
W09G5	NP_009807	Glutamine amidotransferase (Sc)	1.0×10^{-53}	1M, 1W
SP1C9	NP_010110.1	Glutamate synthase (NADH-dependent) (Sc)	9.0×10^{-27}	1P
NC5G11	gi‖1322275	Glutamine synthetase (Gc)	3.1×10^{-73}	4C, 1W
SM4H11	AAD35304.1	Glycine cleavage system H protein (Tm)	4.0×10^{-21}	1M
W08H2	gi‖2853023	Histidine-3 protein (Nc)	3.7×10^{-91}	1W

Continues

Continued

Clone ID[a]	MatchAcc[b]	Identification[c]	P/E value[d]	Tissues[e]
SC6D3	CAA11503	Homocitrate synthase (Pnch)	5.0×10^{-19}	2C, 2M
SC2C6	P31116	Homoserine dehydrogenase (Sc)	1.2×10^{-18}	1C
NC3A9	pir‖S57097	Indoleamine–pyrrole 2,3-dioxygenase homolog (Sc)	2.6×10^{-13}	1C
SC1H9	P38674	Ketol-acid reductoisomerase precursor (Nc)	1.2×10^{-65}	3C
NM8H12	pir‖S40296	L-Arginine:glycine amidinotransferase (Ssd)	1.3×10^{-17}	2M, 1P
NC1D10	P05694	Methionine synthase (Sc)	9.4×10^{-81}	1C
SM4F2	Q92413	Ornithine aminotransferase (En)	3.0×10^{-37}	1M
NC1F9	P27121	Ornithine decrboxylase 1 (Nc)	1.4×10^{-85}	1C, 1P
SM4D5	P07547	Pentafunctional arom polypeptide (En)	4.0×10^{-28}	1M
NM8D7	AAF04875.1	Putative alanine aminotransferase (At)	7.0×10^{-39}	1M, 1P
SC5E11	Q10270	Putative phosphoadenosine phosphosulfate reductase (Sp)	6.0×10^{-5}	1C
W10A5	P38999	Saccharopine dehydrogenase (Sc)	9.7×10^{-65}	2W
W13G12	P48466	*S*-Adenosylmethionine synthetase (Nc)	6.3×10^{-40}	1W
W10G12	P34898	Serine hydroxymethyltransferase, cytosolic (Nc)	4.4×10^{-70}	2C, 1W
NM1H5	pir‖A53651	Sulfate adenylyltransferase (Pnch)	2.7×10^{-99}	1M, 1P
NP3D4	gi‖601846	T-cell reactive protein (Ci)	4.5×10^{-31}	2P
W07F11	P13228	Tryptophan synthase (Nc)	4.4×10^{-51}	1W
W17A5	P07259	URA2 protein (Sc)	5.6×10^{-38}	1W
Cofactors				
SP4F2	gi‖1465774	Cofactor C (Hs)	2.8×10^{-7}	1P
NC1B12	CAA87397	GTP cyclohydrolase I (Sc)	6.3×10^{-64}	1C
NM2H1	P10867	L-Gulonolactone oxidase (Rn)	9.7×10^{-27}	1M
NC3F9	gi‖2598964	Molybdopterin cofactor biosynthetic protein (En)	2.0×10^{-12}	1C
W17D11	P38681	Nitrite reductase (NAD(P)H) (Nc)	1.5×10^{-56}	3C, 2W
W01E6	P42882	nmt1 protein homolog thiamine biosynthesis enzyme (Aspp)	2.6×10^{-112}	11C, 6M, 1P, 72W
SM3D2	CAB16409	Pyridoxal reductase (Sp)	1.0×10^{-26}	1M
SC7F11	AAD49809.1	Pyroa-pyridoxine biosynthesis protein (En)	2.0×10^{-46}	1C
Energy/TCA Cycle				
SC3F7	P19414	Aconitate hydratase (Sc)	5.8×10^{-66}	5C, 1M, 2P
W13E4	P11943	Acyl carrier protein, mitochondrial precursor (Nc)	5.9×10^{-58}	1C, 1W
SC6F12	CAA12224	ATP citrate lyase (Som)	3.0×10^{-29}	1C, 1M, 1P
SC1H4	P37211	ATP synthase α-chian, mitochondrial precursor (Nc)	3.9×10^{-88}	3C, 2M
SC3G5	P23704	ATP synthase β-chain, mitochondrial (Nc)	3.4×10^{-103}	3C, 1M, 1P
W13F5	P56525	ATP synthase δ-chian, mitochondrial precursor (Nc)	1.8×10^{-61}	1W
SP1E7	P49377	ATP synthase γ-subunit, mitochondrial (Sc)	2.0×10^{-34}	3C, 2P, 5W
SC5G11	P05626	ATP synthase, subunit 4, mitochondrial precursor (Sc)	5.9×10^{-8}	1C
W17C6	P00842	ATP synthase, subunit 9, mitochondrial (Nc)	4.0×10^{-62}	2C, 5W
SC5D6	gi‖3172115	β-Ketoadipate enol–lactone hydrolase (Acb)	3.0×10^{-5}	1C
SM1H6	CAB02709	C01G10.7 (similar to citrate lyase β-chain) (Ce)	3.1×10^{-15}	1M
NM6F5	P34085	Citrate synthase, mitochondrial precursor (Nc)	1.1×10^{-99}	1M
NP6C12	P43635	Citrate synthase 3 (Sc)	5.7×10^{-27}	1P
NP2A2	CAA20783	Cytochrome *c* oxidase, subunit VIa (Sp)	1.0×10^{-17}	2P
SC7G9	P00427	Cytochrome *c* oxidase polypeptide VI precursor (Sc)	4.0×10^{-23}	2C, 1M
NM1B3	gi‖2443751	Fumarase (At)	2.0×10^{-60}	2M, 1P
SM2H1	P27680	Hexaprenyldihydroxybenzoate methyltransferase (Sc)	2.1×10^{-5}	1M
NP3E9	P28299	Isocitrate lyase (Nc)	1.5×10^{-49}	1M, 1P

Clone ID[a]	MatchAcc[b]	Identification[c]	P/E value[d]	Tissues[e]
W08E4	P17505	Malate dehydrogenase, mitochondrial precursor (Sc)	1.2×10^{-38}	1P, 2W
NM8D3	CAB41986.1	64-kDa mitochondrial NADH dehydrogenase (Nc)	7.0×10^{-54}	2M
W06E1	P11913	Mitochondrial processing peptidase β-subunit precursor (Nc)	7.6×10^{-48}	1W
NP3A11	P36060	NADH-cytochrome B5 reductase precursor (Sc)	1.8×10^{-14}	1M, 1P
SC3A5	Q03015	NADH-ubiquinone oxidoreductase 12-kDa subunit precursor (Nc)	1.3×10^{-54}	2C
SM1F6	P42116	NADH-ubiquinone oxidoreductase 17.8-kDa subunit precursor (Nc)	2.2×10^{-83}	1M
SP6B10	P25711	NADH-ubiquinone oxidoreductase 21-kDa subunit precursor (Nc)	7.0×10^{-47}	1P
SC6G8	P24917	NADH-ubiquinone oxidoreductase 51-kDa subunit precursor (Nc)	2.0×10^{-35}	1C
SP4F8	O13931	Putative ATP synthase J chain, mitochondrial (Sp)	3.0×10^{-6}	1P
NC3D2	P33287	Pyruvate decarboxylase (cfp gene product) (Nc)	2.0×10^{-107}	10C, 1M, 2W
SC5F1	P37298	Succinate dehydrogenase membrane anchor subunit precursor (Sc)	1.0×10^{-5}	1C
NC1F7	Q00711	Succinate dehydrogenase (ubiquinone), flavoprotein subunit of complex II (Sc)	8.4×10^{-57}	1C
SC6G6	O42772	Succinate dehydrogenase (ubiquinone), iron–sulfur protein (Mg)	4.0×10^{-48}	1C, 1M
SC5H11	O13750	Succinyl-CoA ligase, α-chain precursor (Sp)	6.0×10^{-18}	1C, 1M
W13H10	P53312	Succinyl-CoA synthetase, β-chain precursor (Sc)	8.0×10^{-25}	2W
W01C7	pir‖S56285	Sulfite reductase (Sc)	9.2×10^{-36}	1M, 2W
NM4C4	P00128	Ubiquinol–cytochrome C reductase complex, 14-kDa protein (complex III, subunit VII) (Sc)	1.3×10^{-32}	1M
NC2A1	P00127	Ubiquinol–cytochrome C reductase complex, 17-kDa protein (mitochondrial hinge protein) (Sc)	8.7×10^{-13}	1C, 1M
W10H7	P48503	Ubiquinol–cytochrome C reductase complex subunit VIII (Nc)	2.7×10^{-65}	1C, 1W
SC7F10	P07056	Ubiquinol–cytochrome C reductase Rieske iron–sulfur protein (Nc)	4.0×10^{-50}	1C, 1M
	Lipid			
NC4G6	P15937	Acetyl-CoA hydrolase (acu-8) (Nc)	2.5×10^{-11}	1C
SC1A4	Q04677	Acetyl-coenzyme A acetyltransferase (Ct)	3.9×10^{-52}	3C
SP6B1	CAA75926.1	Acetyl-coenzyme A carboxylase (En)	8.0×10^{-78}	1C, 1P
W09H5	BAA13434	Acetylesterase (Aspa)	8.0×10^{-67}	1W
NM4B3	NP_005460.1	Peroxisomal acyl-CoA thioesterase (Hs)	5.0×10^{-15}	1M
SC7E12	gi‖3859560	Acyl-protein thioesterase (Hs)	8.0×10^{-8}	1C
NM7A4	CAA96522.1	AMP-binding protein (Bn)	2.0×10^{-27}	1M
NP3D7	gi‖2970667	β-Ketoacyl reductase (Psae)	3.5×10^{-11}	1P
NC1F6	gi‖1161339	C-4 sterol methyl oxidase (Sc)	4.2×10^{-38}	2C
SP6F8	gi‖1478048	Cytochrome 450 mono-oxygenase (Dm)	4.0×10^{-6}	1P
SP6G6	BAA10929	Cytochrome P450-like (Nt)	1.0×10^{-7}	1P
SC1E5	BAA11409	Cytochrome P450 nor2 (Cl)	1.4×10^{-49}	1C
NM5B12	P25087	Δ (24) sterol C-methytransferase (Sc)	3.3×10^{-38}	1M
NM1B4	P19262	Dihydrolipoamide succinyltransferase, mitochondrial (Sc)	8.3×10^{-97}	1M
NP3F2	NP_010580	Dihydrosphingosine phosphate lyase (Sc)	2.0×10^{-13}	1P
SM3E6	CAB10453	Enoyl-CoA hydratase (At)	2.0×10^{-20}	1M, 1P
NP4C6	P15368	Fatty acid synthase, subunit α (Pnp)	2.2×10^{-24}	1P
SM1B1	P30839	Fatty aldehyde dehydrogenase, microsomal, class 3 (Rn)	1.4×10^{-13}	1M

Continues

Continued

Clone ID[a]	MatchAcc[b]	Identification[c]	P/E value[d]	Tissues[e]
NM1C7	gil348167	Glycerol kinase (Hs)	8.4×10^{-44}	1M
NC5B9	pirllA32937	Glycerol-3-phosphate dehydrogenase (Dm)	4.9×10^{-10}	1C, 1P
NM3A1	P18900	Hexaprenyl pyrophosphate synthetase (Sc)	1.1×10^{-49}	1M
NP3B8	P28811	3-Hydroxyisobutyrate dehydrogenase (Psae)	8.1×10^{-8}	1P
SP7C9	Q10132	Isopentyl diphosphate Δ isomerase (Sp)	3.0×10^{-41}	1P
NM5E11	Q05493	3-Ketoacyl-CoA thiolase, peroxisomal (Yl)	1.0×10^{-26}	1M
SP7D3	AAD49559.1	Linoleate diol synthase precursor (Gg)	7.0×10^{-27}	1P
SP1D3	P30624	Long-chain fatty acid-CoA ligase (Sc)	5.9×10^{-65}	3P
SP3B11	Q02253	Methylmalonate semialdehyde dehydrogenase precursor (Rn)	9.3×10^{-13}	1P
NM4H8	pirllS54786	Multifunctional β-oxidation protein (Nc)	2.7×10^{-133}	1M, 1P
SC5D5	gil3152731	Myoinositol 1-phosphate synthase (Hv)	1.0×10^{-5}	1C
NC4H10	CAB10120.1	Putative aldose reductase (Sp)	1.0×10^{-15}	1C
SP4C7	CAB11656.1	Putative oxysterol-binding protein (Sp)	9.0×10^{-30}	1P, 1W
SC3C11	CAA91416.1	Similar to γ-butyrobetaine, 2-oxoglutarate dioxygenase (Ce)	7.0×10^{-9}	1C
NP3H4	CAB52620.1	Similar to phosphatidic acid phosphatase (Sp)	4.0×10^{-13}	1P
SM4H12	pirllS52745	Stearoyl-coenzyme A desaturase (Ac)	9.0×10^{-50}	2C, 1M
SP3B10	CAA66277.1	Sterol carrier protein x (Dm)	6.0×10^{-21}	1P
W10D7	P24640	Triacylglycerol lipase (Mor)	1.6×10^{-10}	1W
NP4D5	pirllS57337	Trichodiene oxygenase 4 (Fs)	4.1×10^{-34}	1P
	Nucleotide			
W07E5	P49435	Adenine phosphoribosyltransferase (Sc)	4.0×10^{-40}	1M, 1W
W17C5	CAA75628	Adenosine kinase (Pp)	9.2×10^{-34}	2W
SC2C3	P27604	Adenosylhomocysteinase (Ce)	9.5×10^{-49}	1C, 1M, 1W
SM1G2	P32518	Deoxyuridine 5′-triphosphate nucleotidohydrolase (Le)	1.4×10^{-46}	1M
SC1F7	gil522302	Endonuclease (Mg)	7.4×10^{-21}	1C
SM2E5	P38913	FAD synthetase (FMA adenylyltransferase) (Sc)	8.6×10^{-25}	1M
W07E2	CAA72985.1	GTPase (Sp)	3.0×10^{-14}	1W
SP6C3	P19117	Inorganic pyrophosphatase (Sp)	6.0×10^{-67}	1C, 1P, 2W
SP4E4	P08466	Mitochondrial nuclease (Sc)	2.8×10^{-9}	1P
NM2D1	Q05927	5′-Nucleotide precursor (ECTO-nucleotidase) (Bt)	2.0×10^{-15}	1P
SC6B6	Q99148	Phosphoribosylamine–glycine ligase (GARS) (Yl)	2.0×10^{-21}	1C, 1P
SC2D12	P54113	Phosphoribosylaminoimidazolecarboxamide formyltransferase (Sc)	2.4×10^{-40}	3C, 2W
SC5A4	Q01930	Phosphoribosylaminoimidazole carboxylase (Pm)	2.6×10^{-23}	1C
NC2G7	P27602	Phosphoribosylaminoimidazolesuccinocarboxamide synthase (Cm)	2.9×10^{-35}	1C
SC2A9	P38972	Phosphoribosylformylglycinamidine synthase (Sc)	9.8×10^{-5}	1C
W17G1	P04161	Phosphoribosylglycinamide formyltransferase (Sp)	2.0×10^{-18}	2W
W07E10	P50095	Probable inosine-5′-monophosphate dehydrogenase (Sc)	7.0×10^{-35}	1W
NP5A12	NP_010376	Putative member of nontransport group of ATP-binding cassette (ABC) superfamily, Rli1 (Sc)	9.0×10^{-65}	1P
NC5G9	P23921	Ribonucleoside diphosphate reductase, M1 chain (Hs)	9.4×10^{-45}	1C
NC3E7	AAD40852.1	Sirtuin, type 4 (Hs)	9.0×10^{-9}	1C
NP2C3	NP_003738	Tankyrase, TRF-1-interacting ankyring-related ADP-ribose polymerase (Hs)	7.0×10^{-11}	1P
SC1G12	Q00511	Uricase (Aspfl)	1.8×10^{-41}	1C
SM3A6	P27515	Uridine kinase (Sc)	1.0×10^{-47}	1M

Clone ID[a]	MatchAcc[b]	Identification[c]	P/E value[d]	Tissues[e]
Protein Modification				
SP7C6	P42158	Casein kinase 1, δ-isoform (At)	7.0×10^{-21}	1P
Secondary Metabolism				
NP4F6	P32021	1-Aminocyclopropane-1-carboxylate oxidase (Pss)	2.2×10^{-10}	1P
SP4A5	BAA12723	Dihydroflavonol 4-reductase (Rh)	5.8×10^{-13}	1P
NP2G2	P16543	Granaticin polyketide synthase (Strv)	3.8×10^{-6}	1P
NP4A9	P55441	Hypothetical mono-oxygenase Y4FC (Rhz)	7.0×10^{-32}	1P
NP3C2	AAA33590	Laccase (Nc)	2.2×10^{-79}	3P
SP4G5	pir‖S60224	Melanin biosynthetic polyketide synthase PKSI (Ctl)	2.1×10^{-83}	7M, 25P
NC3C12	AAD37457.1	NonF, non-actin biosynthesis (Strg)	3.0×10^{-12}	1C
NP5A6	P23262	Salicylate hydroxylase (Pspt)	3.1×10^{-7}	1P
NP2G9	pir‖S41412	Tetrahydroxynaphthalene reductase (Mg)	1.8×10^{-88}	2M, 7P
Sugar–Glycolysis				
NC4C1	P41747	Alcohol dehydrogenase 1 (Aspfl)	1.4×10^{-66}	7C, 2M, 2P, 1W
SC3H8	P38426	α, α-Trehalose phosphate synthase (Sc)	1.7×10^{-28}	1C
NM9B11	P21567	α-Amylase precursor (Sf)	1.0×10^{-23}	1C, 1M
NC1E1	P32775	1,4-α-Glucan branching enzyme (Sc)	9.7×10^{-25}	3C
NM1E6	Q12558	α-Glucosidase precursor (maltase) (Aspo)	6.0×10^{-73}	1M, 1C
SM1G4	pir‖JC4836	α-Glucuronidase (Trr)	1.7×10^{-27}	1M
SC2A7	Q00310	α-1,2-Mannosyltransferase (Ca)	4.8×10^{-14}	1C
SM3H7	BAA29031	Avicelase III (Aspac)	3.0×10^{-48}	1M
SP6F3	CAA05375.1	β-1,3-Exoglucanase (Trh)	3.0×10^{-24}	1C, 1P, 1W
SC5H10	gi‖1491929	1,3-β-D-Glucan synthase catalytic subunit (En)	1.0×10^{-50}	1C
NP3B6	gi‖493580	β-D-Glucoside glycohydrolase (Trr)	1.7×10^{-92}	5M, 1P
SP4D6	Q00023	Cellulose growth-specific protein precursor (Ab)	4.0×10^{-15}	1P
W01G4	gi‖1154950	Choline dehydrogenase (Rr)	5.3×10^{-13}	1W
NM2B6	CAB16581	Dihydroxyacetone kinase (Sp)	5.0×10^{-12}	2M
NP4G12	P31382	Dolichyl phosphate mannose protein mannosyltransferase (Sc)	3.8×10^{-58}	1P
SC1H2	Q12560	Enolase (Aspo)	8.0×10^{-73}	10C, 1P, 2W
NM1D6	Q07103	Formate dehydrogenase, NAD-dependent (Nc)	1.0×10^{-43}	1M
W08E12	P53444	Fructose-1,6-bisphosphate aldolase (Nc)	4.3×10^{-73}	4C, 1M, 3W
NP3H2	P32604	Fructose 2,6-bisphosphatase (Sc)	6.0×10^{-90}	2M, 1P
NP4D10	P08431	Galactose-1-phosphate uridylyltransferase (Sc)	1.3×10^{-8}	1P
NM6H8	P49426	Glucan 1,3-β-glucosidase (Cc)	6.5×10^{-37}	1M, 3P
NM6B8	P14804	Glucan 1,4-α-glucosidase (Nc)	3.0×10^{-100}	3M
W10H9	Q92407	Glucokinase (Aspn)	1.5×10^{-51}	1W
W07H8	P53704	Glucosamine–fructose-6-phosphate aminotransferase (Ca)	7.7×10^{-8}	1W
W07G12	pir‖S54720	Glucose-6-phosphate dehydrogenase (Aspn)	4.0×10^{-74}	1C, 2W
NC2H9	gi‖1532189	Gluceraldehyde-3-phosphate dehydrogenase (Nc)	1.3×10^{-120}	25C, 2M, 1P, 47W
SM1F7	pir‖S61144	Glycogen phosphorylase (Sc)	2.3×10^{-75}	2C, 5M, 4P
SC5B4	CAA08922.1	Hexokinase (Aspn)	6.0×10^{-38}	2C, 1P
WO1C8	P41734	Isoamyl acetate-hydrolyzing esterase (Sc)	5.0×10^{-8}	4W
WO1C1	gi‖2266941	Isocitrate dehydrogenase (NAD$^+$ specific), mitochondrial subunit 1 precursor (Ac)	1.2×10^{-20}	1C, 1W
NM1D7	gi‖606352	Maltodextrin phosphorylase (Ec)	7.6×10^{-16}	1M
SC5A11	Q02418	Mannitol-1-phosphate 5-dehydrogenase (Sm)	3.0×10^{-25}	6C

Continues

Continued

Clone ID[a]	MatchAcc[b]	Identification[c]	P/E value[d]	Tissues[e]
SP1A12	BAA76558.1	Mok12, an α-glucan synthase (Sp)	1.0×10^{-27}	1P
WO7A4	AAD43564.1	Pectate lyase (Cg)	6.0×10^{-24}	1W
SM3H9	O43112	Phosphoenolpyruvate carboxykinase (ATP) (Kl)	3.0×10^{-68}	1M
SC6G11	P38720	6-Phosphogluconate dehydrogenase (Sc)	1.0×10^{-48}	1C, 1P
NC3D6	P38667	Phosphoglycerate kinase (Nc)	2.2×10^{-116}	7C, 2W
NC2D12	gil1673879	Phosphoglycerate mutase (Mp)	2.3×10^{-27}	6C
W10B3	P31353	Phosphomannomutase (Ca)	4.1×10^{-52}	1W
NM2H6	P29951	Phosphomannose isomerase (En)	1.9×10^{-35}	1M
SC1C5	CAA19114	Putative betaine aldehyde dehydrogenase precursor (Sp)	2.0×10^{-5}	1C
SC5D9	CAA18655.1	Putative mannose-1-phosphate guanyltransferase (Sp)	4.0×10^{-41}	1C
SC5C2	P16387	Pyruvate dehydrogenase, E1 comp., α-subunit (Sc)	2.2×10^{-35}	1C, 1M, 2W
SC5E2	P32473	Pyruvate dehydrogenase, E1 comp., β-subunit (Sc)	2.0×10^{-24}	1C
NM6D2	gil1016358	Pyruvate formate lyase-activating protein (Tl)	1.3×10^{-12}	1M
NC4C8	P31865	Pyruvate kinase (Trr)	2.1×10^{-51}	3C, 1M, 1P
W13F7	P46969	Ribulose phosphate 3-epimerase (Sc)	2.6×10^{-28}	1W
SM1H9	P87218	Sorbitol utilization protein, SOU1 and SOU2 (Ca)	9.5×10^{-37}	1M, 1W
NC3G5	P45055	Transaldolase (Hi)	2.6×10^{-65}	1C, 1M, 2W
SC5C12	CAA21811.1	Transketolase (Sp)	4.0×10^{-69}	3C, 1M, 2W
NC2D9	P04828	Triosephosphate isomerase (En)	9.6×10^{-68}	8C, 1M, 1W
NC5C3	pirllA54926	UDP glucose 6-dehydrogenase, 52-kDa subunit (Bt)	1.2×10^{-19}	1C
Transport				
W09A5	P02723	ADP/ATP carrier protein (Nc)	1.9×10^{-104}	2C, 3M, 5P, 15W
W10D1	P40260	Ammonium transporter MEP1 (Sc)	2.1×10^{-17}	2M, 1W
SP1A4	CAB06078	AmMst-1, monosaccharide transporter (Am)	8.0×10^{-33}	2P
NM1H8	AAC09237.1	Annexin XIV (Nc)	1.0×10^{-128}	3M
NM8C11	CAA65259.1	Canalicular multidrug resistance protein, cMrp (Hs)	4.3×10^{-26}	1M, 1P
NP4A3	CAB44434.1	Carnitine/acyl carnitine carrier (En)	3.0×10^{-55}	1P
SM1B7	P53048	General α-glucoside permease (Sc)	6.2×10^{-21}	1M
SC7G7	P32836	GTP-binding nuclear protein GSP2/CNRs (Sc)	2.0×10^{-28}	1C
NM5F6	gil1139591	H^+/Ca^{2+} exchange protein, vacuolar (Sc)	1.5×10^{-26}	1M
SM4A3	AAF26275.1	Hexose transporter (Aspp)	1.0×10^{-36}	2M, 1P
SP3C1	O74713	High-affinity glucose transporter (Ca)	3.0×10^{-6}	1C, 1P
NM3A10	Q09887	Hypothetical amino acid permease (Sp)	2.2×10^{-7}	1M
NC3G12	Q02821	Importin α-subunit (Sc)	2.5×10^{-13}	1C
NM4E7	AAD53168.1	Iron-transporter Fth 1p (Sc)	8.0×10^{-29}	2M
NM7D6	AAA97590.1	Lpz11p, hypothetical protein similar to mitochondrial carrier protein family (Sc)	2.9×10^{-16}	1M
NP5D11	P35848	Mitochondrial import receptor subunit (Nc)	1.0×10^{-84}	1P
NP3E4	NP_012579	Mitochondrial matrix protein involved in protein import; subunit of SceI endonuclease (Sc)	2.0×10^{-63}	2C, 1M, 1P
W06D9	P23641	Mitochondrial phosphate carrier protein (Sc)	1.9×10^{-8}	1W
NM5B3	AAD44697.1	MUM2 (Hs)	9.0×10^{-18}	1M
NM7B1	CAA90827	Nuclear transport protein Nip1 (Sc)	4.8×10^{-21}	1M
NC1D11	P07144	Outer mitochondrial membrane protein porin (Nc)	1.6×10^{-114}	3C, 1M, 1W
SP2A11	P46030	Peptide transporter PTR2 (Ca)	1.6×10^{-18}	1P
SP6E7	P07038	Plasma membrane ATPase (proton pump) (Nc)	1.0×10^{-81}	2M, 2P

Clone ID[a]	MatchAcc[b]	Identification[c]	P/E value[d]	Tissues[e]
NC3E5	P40024	Probable ATP-dependent transporter, ABC transporter protein (Sc)	1.7×10^{-53}	1C
NM6G6	O74431	Probable cation-transporting ATPase (Sp)	1.0×10^{-33}	1M
W17H1	CAB65616.1	Probable membrane transporter (Sp)	5.0×10^{-14}	1W
SP6A4	gil2197050	Putative 20-kDa subunit of the V-ATPase (Nc)	2.0×10^{-79}	3M, 4P
SP6B8	gil3885836	Putative cercosporin transporter (Ck)	1.0×10^{-12}	1P
NP6E6	CAA21891.1	Putative Golgi membrane protein-sorting protein (Sp)	7.0×10^{-25}	1P
SP4F1	CAA21303.1	Putative Golgi uridine diphosphate *N*-acetylglucosamine transporter (Sp)	2.0×10^{-10}	1P
SC5E1	P38988	Putative mitochondrial carrier protein YHM1/SHM1 (Sc)	3.0×10^{-25}	1C, 1P
NM4A10	P38702	Putative mitochondrial carrier protein YHR002W (Sc)	1.6×10^{-28}	1M
NM4G4	gil805291	Putative tartrate transporter (Av)	5.6×10^{-30}	1M
SC5A7	O13879	Putative transporter C1B3.15C (Sp)	1.0×10^{-5}	1C
W17B2	CAA19115.1	RANBP7/importin-β/Cse 1p superfamily protein (Sp)	1.3×10^{-43}	1W
SP6C8	BAA13080.1	RAN/spil-binding protein (Sp)	1.0×10^{-41}	1P, 1W
NM1C1	NP_015353	Similar to human polyposis locus protein (YPD) (Sc)	4.0×10^{-24}	1C, 2M
NM7A7	gil1066487	Similar to mitochondrial ADP/ATP carrier protein (Sc)	6.0×10^{-9}	1M
NM6B3	CAB52718.1	Similar to yeast component of COPII coat of ER–Golgi vesicles, SEC24 (Sp)	5.0×10^{-33}	1M
W01B12	P39111	Vacuolar ATP synthase 14-kDa subunit (Sc)	1.7×10^{-25}	1W
SM4C6	P31413	Vacuolar ATP synthase 16-kDa proteolipid subunit (Nc)	7.0×10^{-40}	1M, 1W
NM7G11	P11592	Vacuolar ATP synthase catalytic subunit A (Nc)	2.5×10^{-61}	2C, 1M
NM7B4	P11593	Vacuolar ATP synthase subunit B (Nc)	9.1×10^{-79}	1M, 1P
SC6H4	AAD45120.1	V-type ATPase subunit c' (Nc)	6.0×10^{-40}	1C
NC5G10	gil1814380	V-type ATPase subunit G (Nc)	3.8×10^{-43}	1C
NM1E10	NP_013231	Zrt2p, low-affinity zinc transport protein (Sc)	3.0×10^{-19}	1M

VI. Protein Synthesis

Posttranslational Modification–Targeting

NM4D4	P36581	Calnexin homolog precursor (Sp)	8.6×10^{-53}	1C, 1M, 1W
SC1F12	P35605	Coatomer β-subunit (Bt)	1.5×10^{-43}	1C
SP6B11	P87140	Coatomer γ-subunit (Sp)	9.0×10^{-49}	1C, 1M, 1P
W07H12	BAA34384.1	Cyclophilin (Trm)	5.0×10^{-56}	1C, 1W
NM6G8	P32469	Diphthine synthase, DPH5 (Sc)	6.0×10^{-47}	1M
NC3C7	JC2291	Disulfide isomerase (Hui)	5.5×10^{-59}	2C, 1W
W13F2	P28748	GTP-binding nuclear protein SPI1 (Sp)	4.6×10^{-7}	3W
NM4C2	P36586	GTP-binding protein, YPT1-related protein 5, ypt5 (Sp)	3.6×10^{-61}	1C, 2M
NP2C4	P33723	GTP-binding peotein, YPT1-related protein (Nc)	2.8×10^{-95}	1P
SM2A1	P36863	GTP-binding protein YPTV4 (Vc)	5.0×10^{-36}	1M
SC2D6	CAA17784.1	Hypothetical ubiquitin system (Sp)	4.0×10^{-7}	1C
NP6C4	pirllA23543	Methylumbellifery acetate deacetylase (Hs)	6.7×10^{-61}	2C, 1M, 2P
NC1F1	P33755	Np14 protein (Sp)	2.9×10^{-39}	1C
NP5F8	Q99144	Peroxisomal targeting signal receptor (peroxisomal protein pay32) (Yl)	4.3×10^{-10}	1P
NP4C9	NP_015173	Similar to phosphotyrosyl phosphatase activator Rrd2p (Sc)	2.0×10^{-26}	1P
SC3D1	gil1723924	Probable ubiquitin–protein ligase HUL5 (Sc)	3.0×10^{-26}	1C
SM4B11	P53024	Protein transport protein SEC13 (Pip)	5.0×10^{-40}	2M
W09C6	Q10243	Putative 35.8-kDa vacuolar sorting protein C4G9.13C (Sp)	1.4×10^{-21}	1W

Continues

Continued

Clone ID[a]	MatchAcc[b]	Identification[c]	P/E value[d]	Tissues[e]
NP4F1	gil1619843	rab2-like (Ce)	2.9×10^{-10}	1P
NM4H12	P46638	ras-related protein Rab11b (Mm)	3.0×10^{-96}	1M, 2W
SC5C11	Q40195	ras-related protein Rab11E (Lj)	5.0×10^{-11}	1C
SP4E3	AAD29715.1	Ring-box peotein 1 (Hs)	2.0×10^{-45}	1P
NC4H5	P39940	rsp5 protein (Sc)	2.0×10^{-21}	1C
W17F1	P45816	SEC14 cytosolic factor (Yl)	7.3×10^{-56}	1W
NM4H10	P32916	Signal recognition particle receptor (Sc)	8.6×10^{-46}	1M
W10B7	gil2507637	SNARE protein Ykt6 (Hs)	4.3×10^{-15}	2W
SP6F9	CAA39056.1	Ubiquitin-activating enzyme E1 (Sc)	3.0×10^{-49}	1P, 1W
NC2H3	CAB52608.1	Ubiquitin carboxyl-terminal hydrolase-like protein (Sp)	2.0×10^{-32}	2C
NM1B5	P46595	Ubiquitin-conjugating enzyme (Sp)	1.2×10^{-96}	2C, 6M, 1P
NP3F9	pirllUQNC	Ubiquitin precursor (Nc)	1.2×10^{-99}	2C, 3M, 1P, 5W
NC1C11	pirllUQNCR	Ubiquitin–ribosomal protein S27a (Nc)	9.8×10^{-71}	1C
NM8G8	gil1244555	UDP-Glc:glycoprotein, glucosyltransferase (Sp)	7.8×10^{-57}	1M
NM6H3	gil790621	Ufd1p (ubiquitin fusion degradation) (Sc)	7.3×10^{-30}	1M
NM1G1	gill1477468	Vacuolar protein-sorting homologue r-vps33a (Rn)	3.3×10^{-14}	1M
Protein Turnover				
SM2A12	P37898	Alanine/arginine aminopeptidase (Sc)	1.4×10^{-45}	1M
SM2A9	BAA00258.1	Alkaline protease (Aspo)	1.3×10^{-40}	1M, 1P
SM3B3	P36774	ATP-dependent protease LA2 (Mx)	2.0×10^{-23}	1M
NM9C12	AAC96121	Carboxypeptidase Y precursor (Pa)	7.0×10^{-37}	3M, 2P, 1W
SC3D10	gil2408232	Lysosomal pepstatin-insensitive protease (Hs)	5.3×10^{-5}	1C
NC1B9	P23724	Potential proteasome subunit C5 (Sc)	1.1×10^{-46}	1C
SC2B12	P21243	Proteasome component C7-α (Sc)	2.6×10^{-23}	1C
W13F9	P40303	Proteasome component PRE6 (Sc)	7.4×10^{-8}	1W
SM2E10	Q09841	Proteasome component PUP1 precursor (Sp)	7.5×10^{-68}	1M
SM4E12	P32379	Proteasome component PUP2 (Sc)	2.0×10^{-25}	1M
W06F9	P53616	Proteasome component SUN4 (Sc)	5.2×10^{-23}	1C, 2W
NP6D4	P23639	Proteasome component Y7 (macropain subunit Y7) (Sc)	9.1×10^{-20}	1P
NM7B2	P38886	26S proteasome regulatory component SUN1 (Sc)	9.3×10^{-40}	1M, 1P
NM6F3	P43122	Putative porotease QR17 (Sc)	1.1×10^{-7}	1M
SM3D1	Q09682	Putative proteasome component C9/Y13 (Sp)	5.0×10^{-49}	1M
SP1B11	gil1469396	Secreted aspartic proteinase precursor (Gc)	5.9×10^{-6}	1P
NM5D6	P33295	Subtilisin-like serine protease PEPC precursor (Aspn)	1.0×10^{-78}	1C, 3M, 6P
NM8F11	NP_015433	Subunit of regulatory particle of proteasome Rpn7p (Sc)	3.0×10^{-7}	1M
NM4F2	NP_011981.1	Vacuolar aminopeptidase (Sc)	3.5×10^{-34}	1C, 1M
Ribosomal Proteins[g]				
NP2B1	Q01291	40S ribosomal proteins (assorted) (Nc)	1.7×10^{-131}	71C, 11M, 6P, 58W
NC5F7	P14126	60S ribosomal proteins (assorted) (Sc)	2.0×10^{-98}	73C, 13M, 1P, 53W
tRNA Synthesis–Metabolism				
SC2B2	P04802	Aspartyl-tRNA synthetase (Sc)	7.3×10^{-20}	1C
NC1F12	gil171768	Isoleucyl-tRNA synthetase (Sc)	9.7×10^{-95}	1C
SM1D7	P10857	Leucyl-tRNA synthetase, cytoplasmic (Nc)	3.1×10^{-70}	1M
NM7D3	CAA19575.1	tRNA splicing endonuclease subunit (Sp)	1.0×10^{-9}	1M
NC2B1	Q12109	Tryptophanyl-tRNA synthetase (Sc)	9.7×10^{-30}	1C

Clone ID[a]	MatchAcc[b]	Identification[c]	P/E value[d]	Tissues[e]
Translation Factors				
NM7H7	Q64252	Eukaryotic translation initiation factor EIF-3, P48 (Hs)	3.0×10^{-9}	1M
NP5G8	P46943	GTP-binding protein Guf1 (Sc)	3.0×10^{-45}	1P
SP6G12	Q10425	Probable eukaryotic translation initiation factor EIF-3, P90 subunit (Sp)	3.0×10^{-5}	1P
SM4C5	Q09689	Probable eukaryotic translation factor EIF-5 (Sp)	5.0×10^{-8}	1M
NM6E1	gil961482	Translation elongation factor 1, α (Nc)	1.9×10^{-118}	9C, 5M, 4P, 9W
NC1A10	P34826	Translation elongation factor 1, β (Oc)	2.3×10^{-40}	2C, 1M
NM6D9	pirllS29345	Translation elongation factor 1, γ (Sc)	9.7×10^{-36}	3C, 1M
NM3E7	P32324	Translation elongation factor 2 (Sc)	1.2×10^{-92}	8C, 3M, 3P
NP5E5	P47943	Translation initiation factor 4A (Sp)	1.6×10^{-105}	2C, 2P, 3W
NM6C9	P23588	Translation initiation factor 4B (Hs)	7.7×10^{-6}	1M, 1W
SP7B8	gil3253159	Translation initiation factor EIF-2C (Oc)	3.0×10^{-9}	1P
SC6E9	P79083	Translation initiation factor EIF-3, P39 subunit (Sp)	2.0×10^{-26}	2C
NC5E11	gil2351380	Translation initiation factor EIF-3, P40 subunit (Hs)	3.6×10^{-28}	1C
VII. RNA Synthesis				
RNA Polymerases				
NM5H4	P27999	DNA-directed RNA polymerase II, 14.2 kDa (Sc)	4.8×10^{-12}	1M
SC6D4	P37382	DNA-directed RNA polymerase II, 33 kDa (Sp)	3.0×10^{-41}	1C
RNA Processing				
SP7E5	NP_014287	ATP-dependent RNA helicase of DEAD box family (Sc)	3.0×10^{-44}	1P
W13A6	P15646	Fibrillarin (nucleolar protein 1) (Sc)	4.7×10^{-64}	1W
W17F11	Q06975	GAR1 protein (Sp)	9.3×10^{-33}	1W
NM9H12	BAA25324.1	Moc2 RNA helicase (Sp)	4.0×10^{-74}	1M, 1P
NM6B12	gil495128	Nuclear poly(C)-binding protein (mCBP) (Mm)	6.1×10^{-12}	1M
NC3F12	gil459650	Poly (A)-binding protein (Tc)	3.0×10^{-8}	1C, 1M
W08D5	Q07478	Probable ATP-dependent RNA helicase P47 homolog (Sc)	1.9×10^{-9}	1W
SC1G4	P32843	RNA12 protein (Sc)	4.2×10^{-8}	1C
NM3H6	gil172438	RNA-binding protein (Sc)	6.6×10^{-10}	1M
NC2A10	pirllS31443	RNA-binding protein, glycine-rich (At)	9.1×10^{-18}	2C, 2P
NM3D5	Q00539	RNA-binding protein involved in mitochondrial RNA splicing, NAM8 (Sc)	5.4×10^{-41}	1M
SP4F4	CAA21234.1	RNA-binding protein, putative pre-mRNA splicing factor (Sp)	2×10^{-21}	2M, 1P
W08D9	P20449	RNA helicase Dbp5 (Sc)	2.0×10^{-10}	1W
Transcription Factors				
NM5E12	P07250	Arginine metabolism regulation protein III (Sc)	1.2×10^{-8}	1M
SP4A1	gil1517923	Ascospore maturation 1 protein (Nc)	2.0×10^{-70}	1P, 2W
NM5F11	P36627	Cellular nucleic acid-binding protein (byr3) (Sp)	1.3×10^{-33}	1M
W01D5	P11115	Cross-pathway control protein 1 (Nc)	6.1×10^{-71}	6C, 9M, 3P, 3W
SP1B4	Q04832	DNA-binding protein Hexbp (Lm)	2.0×10^{-13}	1P
W13E1	NP_009930	FMN-binding protein (Sc)	4.0×10^{-11}	1W
NM4D1	CAA19036.1	Fungal ZN(2)–Cys(6) binuclear cluster zinc finger transcription factor (Sp)	4.0×10^{-6}	1M
NP3F1	gil1176420	Hmp1 (Um)	7.2×10^{-7}	1P
W17H5	Q99160	Homeobox protein HOY1 (Yl)	1.2×10^{-12}	1W

Continues

Continued

Clone ID[a]	MatchAcc[b]	Identification[c]	P/E value[d]	Tissues[e]
W08G4	P43588	MPR1 protein (Sc)	7.8×10^{-13}	1W
NP3E6	P33181	Probable sucrose utilization protein SUC1 (Ca)	9.8×10^{-11}	1P
NM5D7	Q09818	Putative general negative regulator of transcription (Sp)	1.6×10^{-29}	1M
NC1B4	CAA22288	Putative mitosis and maintenance of ploidy protein (Sp)	2.0×10^{-7}	1C
NM1F10	CAB11180.1	Putative snf2 family helicase (Sp)	4.0×10^{-62}	1M
NC5D8	CAB10003.1	Putative transcriptional activator (Sp)	4.0×10^{-5}	1C
NM7E9	gil2367591	Putative transcriptional regulator (Mg)	1.1×10^{-34}	1M, 1W
SC5A10	P78706	rco-1 gene product (Nc)	7.3×10^{-52}	1C, 1P
NC1H4	gil1947129	Similar to CCAAT/enhancer-binding protein (Ce)	3.8×10^{-7}	1C, 3M, 10P
W10A9	gil2826519	STE12 α (Fn)	8.7×10^{-21}	1W
NP6F9	gil1147800	Sug2p, putative transcriptional coactivator (Sc)	9.9×10^{-94}	1P
NM7H5	P47192	Synaptobrevin-related protein (formerly called homeotic protein HAT24) (At)	1.4×10^{-14}	1M
SP4D5	NP_011967	Tra1p (ATM/Mec1/TOR1+2-related) (Sc)	1.0×10^{-8}	1P
W08B1	CAB11717.1	Transcription factor BTF3 homolog (Sp)	7.4×10^{-34}	1M, 1W
W07H6	Q01371	White collar 1 protein (Nc)	4.6×10^{-7}	1W
NP5F3	CAA67549.1	Zinc finger protein (Ai)	3.8×10^{-67}	1C, 2M, 1P
SP1E6	gil498734	Zinc finger protein HZF8, Krueppel-related (Hs)	7.1×10^{-11}	1P

Unclassified[h]

Clock-Controlled Genes

Clone ID[a]	MatchAcc[b]	Identification[c]	P/E value[d]	Tissues[e]
SM4C4	Q01358	BLI-3 protein (Nc)	1.0×10^{-44}	1M
NP5G7	P22151	ccg-1, glucose-repressible gene protein (Nc)	2.2×10^{-44}	11M, 6P, 7W
NC2C2	Q04571	ccg-2 (hydrophobin precursor; rodlet protein; blue light induced protein 7) (Nc)	2.3×10^{-57}	17C
NM9C1	gil1184781	ccg-4, putative polypeptide 1 or 2 (Nc)	4.3×10^{-59}	1C, 18M, 5P
NM9D7	gil1184784	ccg-6, putative polypeptide (Nc)	3.6×10^{-23}	5C, 2M, 2P, 13W

Related to Putative ORF in *Saccharomyces cerevisiae*

Clone ID[a]	MatchAcc[b]	Identification[c]	P/E value[d]	Tissues[e]
SM4B6	NP_009348	ORF YAL053w, hypothetical 87.5-kDa protein	1.0×10^{-18}	1M
NM5F5	P38248	ORF YBR0727, hypothetical 48.3-kDa protein homologous to Sps2p	6.0×10^{-9}	1M
NP5E3	P38286	ORF YBR159W	2.8×10^{-19}	1P
W13F11	P38297	ORF YBR179c, hypothetical ATP-binding protein	5.3×10^{-13}	1C, 1W
SM4D11	P25618	ORF YCR017c, hypothetical 107.9-kDa protein	4.0×10^{-35}	1M
SM3F3	CAA98605	ORF YDL045w-a	5.0×10^{-8}	1M
NM9A6	pirllS51251	ORF YDR100w, probable membrane protein YD8557.09	1.1×10^{-12}	1M
NM6A12	gil1078218	ORF YDR105c, hypothetical protein YD9727.01c	2.9×10^{-29}	1M
SC2E5	NP_010573	ORF YDR287w, hypothetical protein with similarity to inositol monophosphatases	1.6×10^{-17}	1C
NP2B7	NP_010617	ORF YDR330w	1.0×10^{-38}	1P
SM3G10	gil1230675	ORF YDR348c, hypothetical protein	9.0×10^{-12}	1M
SM1D3	pirllS69637	ORF YDR470c, hypothetical protein with similarity to chromosome segregation protein Cse1p	1.1×10^{-70}	1M
NM8G11	pirllS56248	ORF YFL006w	8.7×10^{-27}	1M, 1P
SC2E9	P53173	ORF YGL054c, hypothetical 15.9-kDa protein	1.2×10^{-15}	1C

Clone ID[a]	MatchAcc[b]	Identification[c]	P/E value[d]	Tissues[e]
SP1E11	P53134	ORF YGL114w, hypothetical 80.0-kDa protein	1.4×10^{-35}	1P
NM6B4	CAA97021.1	ORF YGR033c	5.6×10^{-13}	1M
NM7F2	P32793	ORF YHR016c, SH3 domain-containing protein	4.5×10^{-13}	1M
NM5G5	P38860	ORF YHR168w, hypothetical GTP-binding protein	3.8×10^{-26}	1M
NP4E4	P47111	ORF YJR044c, probable membrane protein	3.7×10^{-32}	1P
SC1G9	P47179	ORF YJR151c, hypothetical 118.4-kDa protein	1.2×10^{-18}	1C
NM2E7	pir‖S37791	ORF YKL160w	4.4×10^{-14}	1M
NP3F11	CAA97471	ORF YLL023c	1.2×10^{-13}	2P
SM3G11	gi‖609375	ORF YLR228c, probable membrane protein	4.0×10^{-8}	1M
SP6C2	pir‖S59397	ORF YLR251w, probable membrane protein	2.0×10^{-18}	1P
NP3A2	Q06063	ORF YLR450w, hypothetical 41.7-kDa protein	8.1×10^{-7}	1P
SC7E5	NP_013755	ORF YMR041c, hypothetical 38.2-kDa protein	2.0×10^{-7}	1C
NM3E5	pir‖S55125	ORF YMR178w, putative protein YM8010.08	1.3×10^{-39}	1M
SC2F7	Q04336	ORF YMR196w, hypothetical 126.6-kDa protein	3.7×10^{-77}	2C
SM4A12	Q03655	ORF YMR215w, hypothetical 56.8-kDa protein	4.0×10^{-41}	3M, 1P
W17A10	Q05016	ORF YMR226c, hypothetical oxidoreductase	7.0×10^{-23}	1M, 2W
NC2B4	P40157	ORF YNL212w, hypothetical 88.8-kDa protein	1.8×10^{-47}	1C
W07G1	P40345	ORF YNR008w, hypothetical 75.4-kDa protein	2.9×10^{-26}	1W
NP3F12	CAA99053.1	ORF YOLO48c	5.1×10^{-11}	1P
W01C11	pir‖S66771	ORF YOLO78w, hypothetical protein	5.0×10^{-5}	1W
NP2E2	gi‖1420338	ORF YOR131c	7.9×10^{-9}	1P
SP1C8	pir‖S67089	ORF YOR197w	3.8×10^{-26}	1M, 1P
W17G6	pir‖S57544	ORF YPR011c, probable membrane protein	1.7×10^{-20}	1W
SP6D5	pir‖S54084	ORF YPR063c, probable membrane protein	1.0×10^{-5}	1P
NM6F11	NP_015480.1	ORF YPR154w, SH3 domain-containing protein	1.3×10^{-7}	3M

Related to Other Putative ORFs

Clone ID[a]	MatchAcc[b]	Identification[c]	P/E value[d]	Tissues[e]
W08A4	BAA29511	440 amino acid long hypothetical protein (Pyh)	8.0×10^{-14}	1W
W06B9	Z99167.1	Chromosome I cosmid C3G6 (Sp)	8.0×10^{-12f}	4W
SC5F11	AAD35882.1	Conserved hypothetical protein (AE001747) (Tm)	9.0×10^{-11}	1C, 2M
SM2D11	CAA22272.1	Conserved hypothetical protein (AL034381) (Sp)	6.0×10^{-14}	1M
SC2E4	CAB53730.1	Conserved hypothetical protein (AL110295) (Sp)	5.0×10^{-9}	1C
SC1A9	CAA21253.1	Conserved hypothetical protein, Phd finger (Sp)	2.0×10^{-17}	1C
NM5D9	CAB16281.1	Conserved hypothetical protein (Z99167) (Sp)	2.0×10^{-27}	1M
NM1E1	gi‖868225	F35D11.3 gene product (Ce)	1.4×10^{-21}	1M
SC6C6	gi‖4226060	H04M03.4 protein (Ce)	3.0×10^{-7}	1C
SM3D3	Q22700	Hypothetical 6.3-kDa protein T23F2.3 (Ce)	8.0×10^{-14}	1M
W07A7	Q10167	Hypothetical 8.2-kDa protein C26A3.14C (Sp)	9.9×10^{-12}	1W
SM2E11	Q09896	Hypothetical 13.5-kDa protein C24B11.09 (Sp)	3.4×10^{-41}	1M
W07C6	P54607	Hypothetical 24.7-kDa protein, CSPB-GLPP intergenic (Bs)	3.1×10^{-7}	1W
SP4E5	O13725	Hypothetical 26.5-kDa protein C15A10.05c (Sp)	1.0×10^{-23}	1P
NP3B11	P53806	Hypothetical 26.6-kDa protein F54E7.7 (Ce)	2.0×10^{-7}	1P
W07E3	Q10010	Hypothetical 26.6-kDa protein T19C3.4 (Ce)	1.2×10^{-5}	1W
W13A9	Q10446	Hypothetical 27.0-kDa protein C12B10.13 (Sp)	4.1×10^{-20}	1W
NM2D12	Q09686	Hypothetical 28.0-kDa protein C13C85.04 (Sp)	5.5×10^{-36}	1M
SM2E3	P46218	Hypothetical 31.5-kDa protein (Sa)	6.0×10^{-6}	1M
W17E12	pir‖JC4256	Hypothetical 32.0-kDa protein (Nc)	1.4×10^{-67}	1W
SC5E3	Q10212	Hypothetical 34.8-kDa protein C4H3.04C (Sp)	3.0×10^{-9}	1C

Continues

Continued

Clone ID[a]	MatchAcc[b]	Identification[c]	P/E value[d]	Tissues[e]
SC2H7	Q10562	Hypothetical 40.6-kDa protein CY31.34 precursor (Mt)	1.7×10^{-5}	1C
SM4D8	Q09895	Hypothetical 43.7-kDa protein C24B11.08c (Sp)	3.0×10^{-7}	1M
NC1E7	Q09906	Hypothetical 49.3-kDa protein (Sp)	2.7×10^{-29}	1C
SC3G4	Q10478	Hypothetical 51.8-kDa protein C17C9.06 (Sp)	5.7×10^{-14}	1C
SM1G9	Q09744	Hypothetical 63.9-kDa protein C12C2.03C (Sp)	3.4×10^{-36}	1M
SP4A7	Q10211	Hypothetical 74.5-kDa protein (Sp)	6.2×10^{-28}	2P
NM5G4	Q09778	Hypothetical 103.4-kDa protein in C22F3.13 (Sp)	3.9×10^{-5}	1M
SP4C1	Q10064	Hypothetical 420.8-kDa protein C1F5.11C (Sp)	1.2×10^{-12}	1P
NP4D9	AAF01525.1	Hypothetical protein (AC0009991) (At)	3.0×10^{-24}	1P
W01B2	CAA17792.1	Hypothetical protein (AL022070) (Sp)	3.3×10^{-13}	1W
W17A11	CAA18310.1	Hypothetical protein (AL022245) (Sp)	1.1×10^{-37}	2W
SM4A10	CAA20062	Hypothetical protein (AL031154) (Sp)	3.0×10^{-29}	1M
SM3D8	CAB40177.1	Hypothetical protein (AL049559) (Sp)	2.0×10^{-19}	1M, 1P
NP4G3	CAB60706.1	Hypothetical protein (AL132798) (Sp)	1.0×10^{-34}	1P
SC7A11	CAB61580.1	Hypothetical protein (AL133206) (Hs)	5.0×10^{-13}	1P
SC1F8	Q10342	Hypothetical protein (C19G10.16) (Sp)	4.0×10^{-9}	1C
NC1H3	BAA18808	Hypothetical protein (D90917) (Syn)	3.0×10^{-33}	1C
SC7H7	CAB11476.1	Hypothetical protein (Z98762) (Sp)	1.0×10^{-9}	1C
NM8D1	CAB16230.1	Hypothetical protein (Z99192) (Sp)	4.0×10^{-12}	1M
NM6D5	BAA34509.1	KIAA0789 protein (Hs)	4.0×10^{-12}	1M
W06E6	CAA20238	Membrane protein with histidine-rich charge clusters (Hs)	7.0×10^{-13}	1W
NM1A1	gil1145409	ORF of unknown function (Sp)	1.8×10^{-7}	1M
SM2C6	CAB02772.1	Predicted using Genefinder (Z81039) (Ce)	1.0×10^{-7}	1M
SP6C1	CAB05297.1	Predicted using Genefinder (Z82285) (Ce)	8.0×10^{-6}	1P
NM3F5	O53426	Proline-rich antigen homolog (Mt)	5.0×10^{-6}	1M
SM4F10	BAA12197	Similar to pir:S52731 (Sp)	3.0×10^{-26}	1M
SC7G8	CAB02797.1	Similar to yeast hypothetical protein YEY6 like (Ce)	5.0×10^{-22}	1C
SM2A6	gil3676056	Unknown ORF (En)	5.0×10^{-18}	1M
SM2A3	AAD32806.1	Unknown protein (AC007660) (At)	1.0×10^{-10}	1M
NM6A11	AAF01586.1	Unknown protein (AC009895) (At)	4.0×10^{-13}	1M
SP6D4	BAA19234.1	YNL157 homolog (Sp)	4.3×10^{-7}	1P

Other Genes, Unclassified

SC5A3	gil2944191	Acetyl xylan esterase II precursor (Pnpur)	6.0×10^{-9}	1C
NP3C8	gil604427	AcoB protein (En)	2.2×10^{-14}	1P
NM2D11	CAA21790	Carbonic anhydrase (Sp)	1.0×10^{-24}	1M
NP3C7	AAF13817.1	CARP (Oc)	8.0×10^{-6}	1P
NM3D8	P50197	2,5-Dichloro-1,5-cyclohexadiene-1,4-diol dehydrogenae (Spp)	1.0×10^{-21}	1M
NM6F12	gil606960	Lectin (Ab)	4.0×10^{-13}	7M, 1P
SC5B5	P53998	LET1 protein (Km)	4.0×10^{-17}	1C
SC1H6	P42058	Minor allergen Alt A VII (Aa)	7.0×10^{-64}	2C, 1P
NC1F4	gil2253310	myb-1 (Nc)	1.0×10^{-28}	1C, 1M
SM2C11	gil1353701	N33 protein (Hs)	2.5×10^{-7}	1M
NM9B2	P43076	pH-responsive protein 1 precursor (Ca)	7.4×10^{-42}	2M
NM5C6	CAB45367.1	Putative hydrolase (Strc)	2.0×10^{-12}	1M, 1W
W10D9	CAA04959	rAsp f4 (allergen) (Aspfu)	2.7×10^{-11}	1W
SM2B5	P40900	Sexual differentiation process protein isp4 (Sp)	6.9×10^{-29}	1M

Clone ID[a]	MatchAcc[b]	Identification[c]	P/E value[d]	Tissues[e]
SC2A3	CAA10960.1	Small glutamine-rich tetratricopeptide protein (SGT) (Rn)	9.9×10^{-6}	1C
W01H10	P35691	Translationally controlled tumor protein homolog (Sc)	1.1×10^{-44}	1W
SP1E5	CAA73975	Vipl protein (p53-related protein) (Sp)	1.1×10^{-29}	1P

[a]A single representative clone ID is given in those cases in which multiple (duplicate) cDNAs have been identified.

[b]MatchAcc generally indicates the best match (identified by its accession number) to a sequence in the NCBI nonredundant protein databse; however, in those cases in which the best match was to an unidentified open reading frame, a less optimal match to an identified sequence is shown.

[c]Identification refers to the reported match in the NCBI protein database. The organism of the best match is indicated in parentheses (see list of abbreviations below).

[d]The BLASTX P/E value is that obtained with the respective *Neurospora* cDNA clone and the sequence identified in the MatchAcc and Identification columns.

[e]The tissues from which the respective cDNAs were isolated are identified, where C indicates conidial, M is mycelial, P is perithecial, and W is Westergaard (unfertilized sexual tissue). The number preceding these abbreviations indicates the number of duplicate cDNAs isolated from each tissue.

[f]The BLASTN P value is reported. The corresponding BLASTX P value was greater than 10^{-5} and so was not considered significant (see text).

[g]The following 40S ribosomal proteins were identified: MRP2, P40 homolog B, RP10, RP41, S2-18, S20-22, S24-28, S30-31, S33, and the putative ribosome-associated protein similar to ribosomal protein SA. The identified 60S ribosomal proteins included: L1-5, L7, L9-19, L22-23, L25-30, L32, L35, L37, L39-40, and the acidic ribosomal proteins P0, P1 and P2.

[h]The unclassified genes do not include genes identified only by putative homology to ESTs.

[i]Abbreviations of organisms: Aa, *Alternaria alternata*; Ab, *Agaricus bisporus*; Ac, *Ajellomyces capsulatus*; Acb, *Acinetobacter* sp.; Ae, *Alcaligenes eutrophus*; Ai, *Ascobolus immersus*; Am, *Amanita muscaria*; Aspa, *Aspergillus awamorii*; Aspac, *Aspergillus aculeatus*; Aspfl, *Aspergillus flavus*; Aspfu, *Aspergillus fumigatus*; Aspn, *Aspergillus niger*; Aspo, *Aspergillus oryzae*; Aspp, *Aspergillus parasiticus*; At, *Arabidopsis thaliana*; Av, *Agrobacterium vitis*; Bn, *Brassica napus*; Bs, *Bacillus subtilis*; Bt, *Bos taurus*; Ca, *Candida albicans*; Cc, *Cochliobolus carbonum*; Ce, *Caenorhabditis elegans*; Cg, *Colletotrichum gloeosporioides (valvae)*; Ci, *Coccidioides immitis*; Ck, *Cercospora kikuchii*; Cl, *Cylindrocarpon lichenicola*; Cm, *Candida maltosa*; Ct, *Candida tropicalis*; Ctl, *Colletotrichum lagenarium*; Cum, *Cucurbita maxima*; Dd, *Dictyostelium discoideum*; Dm, *Drosophila melanogaster*; Dr, *Deinococcus radiodurans*; Ec, *Escherichia coli*; En, *Emericella nidulans*; Fn, *Filobasidiella neoformasn*; Fo, *Fusarium oxysporum*; Fs, *Fusarium sporotrichioides*; Galg, *Gallus gallus*; Gc, *Glomerella cingulata*; Gg, *Gaeumannomyces graminis (graminis)*; Hh, *Halobacterium halobium*; Hi, *Haemophilus influenzae*; Hp, *Hansenula polymorpha*; Hs, *Homo sapiens*; Hui, *Humicola insolens*; Hv, *Hordeum vulgare*; Km, *Kluyveromyces marxianus* var. *lactis*; Le, *Lycopersicon esculentum*; Lj, *Lotus japonicus*; Lm, *Leishmania major*; Ma, *Metarhizium anisopliae*; Mg, *Magnaporthe grisea*; Mm, *Mus musculus*; Mor, *Moraxella* sp.; Mp, *Mycoplasma pneumoniae*; Mt, *Mycobacterium tuberculosis*; Mx, *Myxococcus xanthus*; Nc, *Neurospora crassa*; Nh, *Nectria haematococca*; Nt, *Nicotiana tabacum*; Oc, *Oryctolagus cuniculus*; Os, *Oryza sativa*; Pa, *Pichia angusta*; Pg, *Pyricularia grisea*; Phi, *Phytophthora infestans*; Pip, *Pichia pastoris*; Pm, *Pichia methanolica*; Pnch, *Penicillium chrysogenum*; Pnci, *Penicillium citrinum*; Pnj, *Penicillium janthinellum*; Pnp, *Penicillium patulum*; Pnpur, *Penicillium purporogenum*; Pp, *Physcomitrella patens*; Psae, *Pseudomonas aeruginosa*; Pspc, *Pseudomonas paucimobilis*; Pspt, *Pseudomonas putida*; Pss, *Pseudomonas syringae*; Pyh, *Pyrococcus horikoshii*; Rh, *Rosa hybrida*; Rhrh, *Rhodococcus rhodochrous*; Rhz, *Rhizobium* sp.; Rn, *Rattus norvegicus*; Rr, *Rattus rattus*; Sa, *Sulfolobus acidocaldarius*; Sc, *Saccharomyces cerevisiae*; Sf, *Saccharomycopsis fibuligera*; Sm, *Streptococcus mutans*; Som, *Sordaria macrospora*; Sp, *Schizosaccharomyces pombe*; Spp, *Sphingomonas paucimobilis*; Ssd, *Sus scrofa domestica*; Strc, *Streptomyces coelicolor*; Strg, *Streptomyces griseus*; Strr, *Streptomyces reticuli*; Strv, *Streptomyces violaceoruber*; Syn, *Synechocystis* sp.; Tc, *Trypanosoma cruzi*; Tl, *Thermococcus litoralis*; Tm, *Thermatoga maritima*; Trh, *Trichoderma harzianum*; Trm, *Trichophyton mentagrophytes*; Trr, *Trichoderma reesei*; Um, *Ustilago maydis*; Vc, *Volvox carteri*; Yl, *Yarrowia lipolytica*.

References

1. Aaronson, L. R., K. M. Hager, J. W. Davenport, S. M. Mandala, A. Chang, D. W. Speicher, and C. W. Slayman (1988). Biosynthesis of the plasma membrane H^+-ATPase of *Neurospora crassa. J. Biol. Chem.* **263**: 14552–14558.
2. Abramsky, T., and E. L. Tatum (1976). Differential inhibition of branching enzyme in a morphological mutant and in wild-type *Neurospora*. Influence of carbon source in the growth medium. *Biochim. Biophys. Acta* **421**: 106–114.
3. Addison, R. (1986). Primary structure of the *Neurospora* plasma membrane H^+-ATPase deduced from the gene sequence. Homology to Na^+/K^+-, Ca^{2+}-, and K^+-ATPase. *J. Biol. Chem.* **261**: 14896–14901.
4. Addison, R. (1990). Studies on the sedimentation behavior of the *Neurospora crassa* plasma membrane H^+-ATPase synthesized *in vitro* and integrated into homologous microsomal membranes. *Biochim. Biophys. Acta* **1030**: 127–133.
5. Addison, R. (1991). GTP is required for the integration of a fragment of the *Neurospora crassa* H^+-ATPase into homologous microsomal vesicles. *Biochim. Biophys. Acta* **1065**: 130–134.
6. Addison, R. (1993). The initial association of a truncated form of the *Neurospora* plasma membrane H^+-ATPase and of the precursor of yeast invertase with microsomes are distinct processes. *Biochim. Biophys. Acta* **1152**: 119–127.
7. Addison, R. (1998). A cell-free translation–translocation system reconstituted with subcellular fractions from the wall-less variant *fz;sg;os-1V* of *Neurospora crassa. Fungal Genet. Biol.* **24**: 345–353.
8. Aguirre, J., and W. Hansberg (1986). Oxidation of *Neurospora crassa* glutamine synthetase. *J. Bacteriol.* **166**: 1040–1045.
9. Ahmad, M. (1964). A study of the *lys-3* locus in *Neurospora crassa. Neurospora Newsl.* **6**: 5.
10. Ahmad, M., M. U. Ahmad, and A. Zaman (1966). Mapping of locus *lysine-2* in *Neurospora crassa. Proc. Pak. Acad. Sci.* **3**: 1–12.
11. Ahmad, M., and D. G. Catcheside (1960). Physiological diversity among tryptophan mutants in *Neurospora crassa. Heredity* **15**: 55–64.
12. Ahmad, M., M. K. U. Choudhury, and S. M. Islam (1969). Complementation and recombination between indole-utilizing tryptophan-3 mutants of *Neurospora crassa. Heredity* **24**: 656–660.
13. Ahmad, M., and S. Haque (1980). A more precise mapping of *tryp-5* in *Neurospora crassa. Neurospora Newsl.* **27**: 34.
14. Ahmad, M., S. Haque, M. U. Ahmad, A. Zaman, M. Mohiuddin, Y. A. Saeed, and A. Khairul (1980). Studies on the organization of genes controlling lysine biosynthesis in *Neurospora crassa*. IV. Segregation patterns, maturity and viability of ascospores and conidiation of some *lys-5* mutants. *Pak. J. Bot.* **12**: 57–67.
15. Ahmad, M., S. Haque, A. Mozmadar, A. Baset, M. Fayaz, A. Badrul, A. Rahman, and B. C. Saha (1979). Studies on the organization of genes controlling lysine biosynthesis in *Neurospora crassa*. II. Organization of locus *lys-1. Pak. J. Bot.* **11**: 21–32.
16. Ahmad, M., and N. Islam (1969). Interallelic complementation at the tryptophan-3 locus in *Neurospora crassa. Heredity* **24**: 651–655.
17. Ahmad, M., M. Khalil, N. A. Khan, and A. Mozmadar (1964). Structural and functional complexity at the tryptophan-1 locus in *Neurospora crassa. Genetics* **49**: 925–933.
18. Ahmad, M., and S. H. Mirdha (1968). Linkage data for four linkage group III markers in *Neurospora crassa. Neurospora Newsl.* **13**: 22.
19. Ahmad, M., A. Mozmadar, A. Baset, M. Fayaz, A. Badrul, M. A. Rahman, and B. Saha (1977). Studies on the organization of genes controlling lysine biosynthesis in *Neurospora crassa*. I. Isolation and characterization of lysine mutants belonging to four loci. *Pak. J. Bot.* **9**: 99–106.
20. Ahmad, M., A. Mozmadar, A. Baset, M. Fayaz, M. A. Rahman, and B. C. Saha (1979). Studies on the organization of genes controlling lysine biosynthesis in *Neurospora crassa*. III. Studies on the organization of loci *lys-3* and *lys-4. Pak. J. Bot.* **11**: 179–184.
21. Ahmad, M., A. Mozmadar, and S. Hendler (1968). A new locus in the tryptophan pathway of *Neurospora crassa. Genet. Res.* **12**: 103–107.
22. Ahmed, A. (1968). Organization of the histidine-3 region of *Neurospora. Mol. Gen. Genet.* **103**: 185–193.
23. Ahmed, A., M. E. Case, and N. H. Giles (1964). The nature of complementation among mutants in the histidine-3 region of *Neurospora crassa. Brookhaven Symp. Biol.* **17**: 53–65.

24. Aisemberg, G. O., E. Grotewold, G. E. Taccioli, and N. D. Judewicz (1989). A major transcript in the response of *Neurospora crassa* to protein synthesis inhibition by cycloheximide. *Exp. Mycol.* **13**: 121–128.

25. Akins, R. A., and A. M. Lambowitz (1985). General method for cloning *Neurospora crassa* nuclear genes by complementation of mutants. *Mol. Cell. Biol.* **5**: 2272–2278.

26. Akins, R. A., and A. M. Lambowitz (1987). A protein required for splicing group I introns in *Neurospora* mitochondria is mitochondrial tyrosyl-tRNA synthetase or a derivative thereof. *Cell* **50**: 331–345.

27. Akiyama, M., and H. Nakashima (1996). Molecular cloning of the *acr-2* gene which controls acriflavine sensitivity in *Neurospora crassa. Biochim. Biophys. Acta* **1307**: 187–192.

28. Akiyama, M., and H. Nakashima (1996). Molecular cloning of *thi-4*, a gene necessary for the biosynthesis of thiamine in *Neurospora crassa. Curr. Genet.* **30**: 62–67.

29. Alconada, A., F. Gärtner, A. Hönlinger, M. Kübrich, and N. Pfanner (1995). Mitochondrial receptor complex from *Neurospora crassa* and *Saccharomyces cerevisiae. Methods Enzymol.* **260**: 263–286.

30. Alex, L. A., Personal communication.

31. Alex, L. A., K. A. Borkovich, and M. I. Simon (1996). Hyphal development in *Neurospora crassa*: Involvement of a two-component histidine kinase. *Proc. Natl. Acad. Sci. USA* **93**: 3416–3421.

32. Allen, E. D. (1976). Development of crystalline inclusions ("ergosterol crystals") in *Neurospora crassa. Protoplasma* **90**: 297–306.

33. Allen, E. D., R. Aiuto, and A. S. Sussman (1980). Effects of cytochalasins on *Neurospora crassa*. I. Growth and ultrastructure. *Protoplasma* **102**: 63–75.

34. Allen, E. D., R. J. Lowry, and A. S. Sussman (1974). Accumulation of microfilaments in a colonial mutant of *Neurospora crassa. J. Ultrastruct. Res.* **48**: 455–464.

35. Allen, K. E., M. T. McNally, H. S. Lowendorf, C. W. Slayman, and S. J. Free (1989). Deoxyglucose-resistant mutants of *Neurospora crassa*: Isolation, mapping, and biochemical characterization. *J. Bacteriol.* **171**: 53–58.

36. Al-Saqur, A., and B. R. Smith (1980). Resistance to surface active drugs of wild-type strains and newly isolated mutants of *Neurospora crassa. Neurospora Newsl.* **27**: 16.

37. Altmiller, D. H., and R. P. Wagner (1970). Deficiency of dihydroxy acid dehydratase in the mitochondria of the *iv-1* mutants of *Neurospora crassa. Biochem. Genet.* **4**: 243–251.

38. Alton, N. K., F. Buxton, V. Patel, N. H. Giles, and D. Vapnek (1982). 5'-Untranslated sequences of two structural genes in the *qa* gene cluster of *Neurospora crassa. Proc. Natl. Acad. Sci. USA* **79**: 1955–1959.

39. Alton, N. K., J. A. Hautala, N. H. Giles, S. R. Kushner, and D. Vapnek (1978). Transcription and translation in *E. coli* of hybrid plasmids containing the catabolic dehydroquinase gene from *Neurospora crassa.Gene* **4**: 241–259.

40. Alvarez, M. E., A. L. Rosa, E. D. Temporini, A. Wolstenholme, G. Panzetta, L. Patrito, and H. J. Maccioni (1993). The 59-kDa polypeptide constituent of 8–10-nm cytoplasmic filaments in *Neurospora crassa* is a pyruvate decarboxylase. *Gene* **130**: 253–258.

41. Alves, P. C., and A. Videira (1994). Disruption of the gene coding for the 21.3-kDa subunit of the peripheral arm of complex I from *Neurospora crassa. J. Biol. Chem.* **269**: 7777–7784.

42. Ames, B. N. (1957). The biosynthesis of histidine: D-Erythroimidazoleglycerol phosphate dehydrase. *J. Biol. Chem.* **228**: 131–143.

43. Ames, B. N. (1957). The biosynthesis of histidine: L-Histidinol phosphate phosphatase. *J. Biol. Chem.* **226**: 583–593.

44. Amy, N. K., and R. H. Garrett (1979). Immunoelectrophoretic determination of nitrate reductase in *Neurospora crassa. Anal. Biochem.* **95**: 97–107.

45. Angel, T., B. Austin, and D. G. Catcheside (1970). Regulation of recombination at the *his-3* locus in *Neurospora crassa. Aust. J. Biol. Sci.* **23**: 1229–1240.

46. Antoine, A. D. (1974). Purification and properties of the nitrate reductase isolated from *Neurospora crassa* mutant *nit-3*. Kinetics, molecular weight determination and cytochrome involvement. *Biochemistry* **13**: 2289–2294.

47. Aramayo, R., and R. L. Metzenberg (1996). Gene replacements at the *his-3* locus of *Neurospora crassa. Fungal Genet. Newsl.* **43**: 9–13.

48. Aramayo, R., and R. L. Metzenberg (1996). Meiotic transvection in fungi. *Cell* **86**: 103–113.

49. Aramayo, R., Y. Peleg, R. Addison, and R. Metzenberg (1996). *Asm-1*+, a *Neurospora crassa* gene related to transcriptional regulators of fungal development. *Genetics* **144**: 991–1003.

50. Arends, H., and W. Sebald (1984). Nucleotide sequence of the cloned mRNA and gene of the ADP/ATP carrier from *Neurospora crassa. EMBO J.* **3**: 377–382.

51. Arganoza, M. T., J. Ohrnberger, J. Min, and R. A. Akins (1994). Suppressor mutants of *Neurospora crassa* that tolerate allelic differences at single or at multiple heterokaryon incompatibility loci. *Genetics* **137**: 731–742.

52. Armaleo, D., M. Fischer, and S. R. Gross (1985). Effect of α-isopropylmalate on the synthesis of RNA and protein in *Neurospora. Mol. Gen. Genet.* **200**: 346–349.

53. Arnaise, S., D. Zickler, and N. L. Glass (1993). Heterologous expression of mating-type genes in filamentous fungi. *Proc. Natl. Acad. Sci. USA* **90**: 6616–6620.

54. Arnold, J. (2000). Physical map of *Neurospora crassa. Fungal Genet. Newsl.* **47S**: 16 (Abstr.).

55. Aronson, B. D., K. A. Johnson, and J. C. Dunlap (1994). Circadian clock locus frequency: Protein encoded by a single open reading frame defines period length and temperature compensation. *Proc. Natl. Acad. Sci. USA* **91**: 7683–7687.

56. Aronson, B. D., K. A. Johnson, Q. Liu, and J. C. Dunlap (1992). Molecular analysis of the *Neurospora* clock: Cloning and characterization of the *frequency* and *period-4* genes. *Chronobiol. Int.* **9**: 231–239.

57. Aronson, B. D., K. A. Johnson, J. J. Loros, and J. C. Dunlap (1994). Negative feedback defining a circadian clock: Autoregulation of the clock gene *frequency. Science* **263**: 1578–1584.

58. Arpaia, G., A. Carattoli, and G. Macino (1995). Light and development regulate the expression of the *albino-3* gene in *Neurospora crassa. Dev. Biol.* **170**: 626–635.

59. Arpaia, G., F. Cerri, S. Baima, and G. Macino (1999). Involvement of protein kinase C in the response of

Neurospora crassa to blue light. *Mol. Gen. Genet.* **262**: 314–322.

60. Arpaia, G., J. J. Loros, J. C. Dunlap, G. Morelli, and G. Macino (1993). The interplay of light and the circadian clock. Independent dual regulation of clock-controlled gene *ccg-2(eas)*. *Plant Physiol.* **102**: 1299–1305.

61. Arpaia, G., J. J. Loros, J. C. Dunlap, G. Morelli, and G. Macino (1995). Light induction of the clock-controlled gene *ccg-1* is not transduced through the circadian clock in *Neurospora crassa*. *Mol. Gen. Genet.* **247**: 157–163.

62. Arrayo-Begovich, A., and J. A. DeMoss (1973). The isolation of the components of the anthranilate synthetase complex from *Neurospora crassa*. *J. Biol. Chem.* **248**: 1262–1267.

63. Asahi, T., and H. Kuwana (1992). Dimorphism and cell-wall components in *Neurospora crassa*. *Mol. Genet (Life Sci. Adv.)* **11**: 203–207.

64. Asch, D. K., G. Frederick, J. A. Kinsey, and D. D. Perkins (1992). Analysis of junction sequences resulting from integration at nonhomologous loci in *Neurospora crassa*. *Genetics* **130**: 737–748.

65. Asch, D. K., M. Orejas, R. F. Geever, and M. E. Case (1991). Comparative studies of the quinic acid (*qa*) cluster in several *Neurospora* species with special emphasis on the *qa-x-qa-2* intergenic region. *Mol. Gen. Genet.* **230**: 337–344.

66. Ashby, B., J. C. Wooton, and J. R. S. Finchman (1974). Slow conformational changes of a *Neurospora* glutamate dehydrogenase studied by protein fluorescence. *Biochem. J.* **14**: 317–329.

67. Atkinson, I. J., F. E. Nargang, and E. A. Cossins (1995). Folylpolyglutamate synthesis in *Neurospora crassa*: Transformation of polyglutamate-deficient mutants. *Phytochemistry* **38**: 603–608.

68. Atkinson, I. J., F. E. Nargang, and E. A. Cossins (1998). Folylpolyglutamate synthesis in *Neurospora crassa*: Primary structure of the folylpolyglutamate synthetase gene and elucidation of the *met-6* mutation. *Phytochemistry* **49**: 2221–2232.

69. Attar, R. M., E. Grotewold, G. E. Taccioli, G. O. Aisemberg, H. N. Torres, and N. D. Judewicz (1989). A cycloheximide-inducible gene of *Neurospora crassa* belongs to the cytochrome P-450 superfamily. *Nucleic Acids Res.* **17**: 7535–7536.

70. Atwood, K. C., and F. Mukai (1954). Survival and mutation in *Neurospora* exposed at nuclear detonations. *Am. Nat.* **88**: 295–314.

71. Atwood, K. C., and F. Mukai (1955). Nuclear distribution in conidia of *Neurospora* heterokaryons. *Genetics* **40**: 438–443.

72. Atwood, K. C., F. Mukai, and T. H. Pittenger (1958). "Punch tube" and "squirting" methods for *Neurospora*. *Microb. Genet. Bull.* **16**: 34–35.

73. Auer, M., G. A. Scarborough, and W. Kühlbrandt (1998). Three-dimensional map of the plasma membrane H⁺-ATPase in the open conformation. *Nature* **392**: 840–843.

74. Austin, B., R. M. Hall, and B. M. Tyler (1990). Optimized vectors and selection for transformation of *Neurospora crassa* and *Aspergillus nidulans* to bleomycin and phleomycin resistance. *Gene* **93**: 157–162.

75. Azevedo, J. E., J. Abrolat-Scharff, C. Eckerskorn, and S. Werner (1993). Cloning, in vitro mitochondrial import and membrane assembly of the 17.8-kDa subunit of complex I from *Neurospora crassa*. *Biochem. J.* **293**: 501–506.

76. Azevedo, J. E., M. Duarte, J. A. Belo, S. Werner, and A. Videira (1994). Complementary DNA sequences of the 24-kDa and 21-kDa subunits of complex I from *Neurospora*. *Biochim. Biophys. Acta* **1188**: 159–161.

77. Azevedo, J. E., U. Nehls, C. Eckerskorn, H. Heinrich, H. Rothe, H. Weiss, and S. Werner (1992). Primary structure and mitochondrial import *in vitro* of the 20.9-kDa subunit of complex I from *Neurospora crassa*. *Biochem. J.* **288**: 29–34.

78. Azevedo, J. E., M. Tropschug, and S. Werner (1992). Primary structure and *in vitro* expression of the *N. crassa* phosphoglycerate kinase. *DNA Seq.* **2**: 265–267.

79. Azevedo, J. E., and A. Videira (1994). Characterization of a membrane fragment of respiratory chain complex I from *Neurospora crassa*. Insights on the topology of the ubiquinone-binding site. *Int. J. Biochem.* **26**: 505–510.

80. Baasiri, R. A., X. Lu, P. S. Rowley, G. E. Turner, and K. A. Borkovich (1997). Overlapping functions for two G-protein subunits in *Neurospora crassa*. *Genetics* **147**: 137–145.

81. Bachman, B. J., and W. N. Strickland (1965). "*Neurospora* Bibliography and Index." Yale University Press, New Haven, CT.

82. Backus, M. P. (1939). The mechanics of conidial fertilization in *Neurospora sitophila*. *Bull. Torrey Bot. Club* **66**: 63–76.

83. Badgett, T. C., and C. S. Staben, Personal communication.

84. Bailey, L. A., and D. J. Ebbole (1997). Characterization of the *fluffy* gene of *Neurospora crassa*. *19th Fungal Genet. Conf. Abstr.* 14.

85. Bailey, L. A., and D. J. Ebbole (1998). The *fluffy* gene of *Neurospora crassa* encodes a Gal4p-type C6 zinc cluster protein required for conidial development. *Genetics* **148**: 1813–1820.

86. Baker, J. M., J. H. Parish, and J. P. E. Curtis (1984). DNA–DNA and DNA–protein cross-linking and repair in *Neurospora crassa* following exposure to nitrogen mustard. *Mutat. Res.* **132**: 171–179.

87. Baker, T. I. (1968). Phenylalanine–tyrosine biosynthesis in *Neurospora crassa*. *Genetics* **58**: 351–359.

88. Baker, T. I. (1983). Inducible nucleotide excision repair in *Neurospora*. *Mol. Gen. Genet.* **190**: 295–299.

89. Baker, T. I., C. E. Cords, C. A. Howard, and R. J. Radloff (1990). The nucleotide excision repair epistasis group in *Neurospora crassa*. *Curr. Genet.* **18**: 207–209.

90. Baker, T. I., R. J. Radloff, C. E. Cords, S. R. Engel, and D. L. Mitchell (1991). The induction and repair of (6–4) photoproducts in *Neurospora crassa*. *Mutat. Res.* **255**: 211–228.

91. Ballario, P., and G. Macino (1997). White collar proteins: PASsing the light signal in *Neurospora crassa*. *Trends Microbiol.* **5**: 458–462.

91a. Ballario, P., G. Morelli, E. Sporeno, and G. Macino (1989). Cosmids from the Vollmer–Yanofsky library identified with a chromosome VII probe. *Fungal Genet. Newsl.* **36**: 38–39.

92. Ballario, P., C. Talora, D. Galli, H. Linden, and G. Macino (1998). Roles in dimerization and blue light photoresponse of the PAS and LOV domains of *Neurospora crassa* white collar proteins. *Mol. Microbiol.* **29**: 719–729.

93. Ballario, P., P. Vittorioso, A. Magrelli, C. Talora, A. Cabibbo, and G. Macino (1996). White collar-1, a central regulator of blue light responses in *Neurospora*, is a zinc finger protein. *EMBO J.* **15**: 1650–1657.

94. Banks, C. W., S. N. Bennett, and W. A. Krissinger (1997). Allelism of the mutants *ovc* and *cut* of *Neurospora crassa*. *Fungal Genet. Newsl.* **44**: 10.

95. Barbato, C., M. Calissano, A. Pickford, N. Romano, G. Sandmann, and G. Macino (1996). Mild RIP—An alternative method for *in vovo* mutagenesis of the *albino-3* gene in *Neurospora crassa*. *Mol. Gen. Genet.* **252**: 353–361.

96. Barnett, G. R., M. Seyfzadeh, and R. H. Davis (1988). Putrescine and spermidine control degradation and synthesis of ornithine decarboxylase in *Neurospora crassa*. *J. Biol. Chem.* **263**: 10005–10008.

97. Barnett, W. E., and F. J. de Serres (1963). Fixed genetic instability in *Neurospora crassa*. *Genetics* **48**: 717–723.

98. Barra, J. L., M. R. Mautino, and A. L. Rosa (1996). A dominant negative effect of *eth-1*, a mutant allele of the *Neurospora crassa* S-adenosylmethionine synthetase-encoding gene conferring resistance to the methionine toxic analogue ethionine. *Genetics* **144**: 1455–1462.

99. Barratt, R. W., Personal communication.

100. Barratt, R. W. (1963). Effect of environmental conditions on the NADP-specific glutamic acid dehydrogenase in *Neurospora crassa*. *J. Gen. Microbiol.* **33**: 33–42.

101. Barratt, R. W., R. C. Fuller, and S. W. Tanenbaum (1956). Amino acid interrelationhips in certain leucine- and aromatic-requiring strains of *Neurospora crassa*. *J. Bacteriol.* **71**: 108–114.

102. Barratt, R. W., and L. Garnjobst (1949). Genetics of a colonial microconidiating mutant strain of *Neurospora crassa*. *Genetics* **34**: 351–369.

103. Barratt, R. W., D. Newmeyer, D. D. Perkins, and L. Garnjobst (1954). Map construction in *Neurospora crassa*. *Adv. Genet.* **6**: 1–93.

104. Barratt, R. W., and W. N. Ogata (1954). A strain of *Neurospora* with an alternative requirement for leucine or aromatic amino acids. *Am. J. Bot.* **41**: 763–771.

105. Barratt, R. W., and W. N. Ogata (1978). *Neurospora* stock list, ninth revision. *Neurospora Newsl.* **25**: 29–96.

106. Barratt, R. W., and W. N. Ogata (1979). First supplement to stock list, ninth revision. *Neurospora Newsl.* **26**: 29–37.

107. Barratt, R. W., and W. N. Ogata (1980). *Neurospora* stock list, tenth revision. *Neurospora Newsl.* **27**: 39–121.

108. Barratt, R. W., and A. Radford (1970). Genetic markers, linkage groups, and enzymes in *Neurospora crassa*—Supplement 1. *Neurospora Newsl.* **16**: 19–22.

109. Barratt, R. W., and P. St. Lawrence (1969). Antimetabolite inhibition of *mod-5*. *Neurospora Newsl.* **15**: 15.

110. Barry, E. G., Personal communication.

111. Barry, E. G. (1960). A complex chromosome rearrangement in *Neurospora crassa*. Ph.D. Thesis, Stanford University. Diss. Abstr. **21**: 3233–3234.

112. Barry, E. G. (1960). Genetic analysis of an insertional translocation in *Neurospora crassa*. *Genetics* **45s**: 974.

113. Barry, E. G. (1978). Interallelic somatic recombination in *Neurospora crassa*. *Proc. 14th Int. Congr. Genet.* Moscow, p. 224, Abstr. part 1.

114. Barry, E. G. (1992). A combination inversion and translocation in *Neurospora crassa* with inviable deficiency progeny that can be rescued in heterokaryons. *Genetics* **132**: 403–412.

115. Barry, E. G., and J. F. Leslie (1982). An interstitial pericentric inversion in *Neurospora*. *Can J. Genet. Cytol.* **24**: 693–703.

116. Barry, E. G., and M. J. Marsho (1968). Mutants affecting arginine utilization by *arg-8* mutants of *Neurospora crassa*. *Proc. 12th Int. Congr. Genet.*, Tokyo, 17.

117. Barry, E. G., D. Newmeyer, D. D. Perkins, and B. C. Turner (1972). Genetically determined round ascospores in *Neurospora crassa*. *Neurospora Newsl.* **19**: 17.

118. Barry, E. G., and D. D. Perkins (1969). Position of linkage group V markers in chromosome 2 of *Neurospora crassa*. *J. Hered.* **60**: 120–125.

119. Barthelmess, I. B., Personal communication.

120. Barthelmess, I. B. (1982). Mutants affecting amino acid cross-pathway control in *Neurospora crassa*. *Genet. Res.* **39**: 169–185.

121. Barthelmess, I. B. (1986). Regulation of amino acid synthetic enzymes in *Neurospora crassa* in the presence of high concentrations of amino acids. *Mol. Gen. Genet.* **203**: 533–537.

122. Barthelmess, I. B., C. F. Curtis, and H. Kacser (1974). Control of the flux to arginine in *Neurospora crassa*: Derepression of the last three enzymes of the arginine pathway. *J. Mol. Biol.* **87**: 303–316.

123. Barthelmess, I. B., and M. Tropschug (1993). FK506-binding protein of *Neurospora crassa* (NcFKBP) mediates sensitivity to the immunosuppressant FK506; resistant mutants identify two loci. *Curr. Genet.* **23**: 54–58.

124. Bartnicki-Garcia, S., C. E. Bracker, E. Lippman, and J. Ruiz-Herrera (1984). Chitosomes from the wall-less "slime" mutant of *Neurospora crassa*. *Arch. Microbiol.* **139**: 105–112.

125. Bates, W. K. (1967). Inheritance of lactose growth characteristics in *Neurospora*. *Genetics* **56**: 543–544 (Abstr.).

126. Bates, W. K., S. C. Hedman, and D. O. Woodward (1967). Comparative inductive responses of two β-galactosidases of *Neurospora*. *J. Bacteriol.* **93**: 1631–1637.

127. Bates, W. K., and D. O. Woodward (1964). Neurospora β-galactosidase: Evidence for a second enzyme. *Science* **146**: 777–778.

128. Bauer, M. F., A. Adam, M. Brunner, and S. Hofmann (1999). Direct database submission.

129. Bauerle, R., and H. R. Garner (1963). The metabolism of canavanine in a threonineless mutant of *Neurospora crassa*. *Genetics* **48**: 882–883.

130. Baulcombe, D. C. (1999). RNA makes RNA makes no protein. *Curr. Biol.* **9**: R599–R601.

131. Baum, J. A., R. Geever, and N. H. Giles (1987). Expression of *qa-1F* activator protein: Identification of upstream binding sites in the *qa* gene cluster and localization of the DNA-binding domain. *Mol. Cell. Biol.* **7**: 1256–1266.

132. Baum, J. A., and N. H. Giles (1986). DNase I hypersensitive sites within the inducible *qa* gene cluster of *Neurospora crassa*. *Proc. Natl. Acad. Sci. USA* **83**: 6533–6537.

133. Baylis, J. R., and A. G. DeBusk (1967). Estimation of the frequency of multinucleate conidia in microconidiating strains. *Neurospora Newsl.* **11**: 9.

134. Beadle, G. W., and V. L. Coonradt (1944). Heterocaryosis in *Neurospora crassa. Genetics* **29**: 291–308.

135. Beadle, G. W., and E. L. Tatum (1941). Genetic control of biochemical reactions in *Neurospora. Proc. Natl. Acad. Sci. USA* **27**: 499–506.

136. Beadle, G. W., and E. L. Tatum (1945). *Neurospora.* II. Methods of producing and detecting mutations concerned with nutritional requirements. *Am. J. Bot.* **32**: 678–686.

137. Beatty, N. P., M. L. Smith, and N. L. Glass (1994). Molecular characterization of mating-type loci in selected homothallic species of *Neurospora, Gelasinospora* and *Anixiella. Mycol. Res.* **98**: 1309–1316.

138. Beauchamp, P. M., E. W. Horn, and S. R. Gross (1977). Proposed involvement of an internal promoter in regulation and synthesis of mitochondrial and cytoplasmic leucyl-tRNA synthetases of *Neurospora. Proc. Natl. Acad. Sci. USA* **74**: 1172–1176.

139. Beck, F. W. J. (1981). *Neurospora crassa* phosphoserine phosphatase: The effect of pH on the hydrolysis of phosphoserine by enzymes present in wild-type and *ser(JBM5)* extracts and partial purification and characterization of wild-type phosphoserine phosphatase. M.S. Thesis, California State University, Northridge, CA.

140. Beever, R. E. (1973). Pyruvate carboxylase and *N. crassa suc* mutants. *Neurospora Newsl.* **20**: 15–16.

141. Beever, R. E. (1980). A gene influencing spiral growth of *Neurospora crassa* hyphae. *Exp. Mycol.* **4**: 338–342.

142. Beever, R. E., and G. P. Dempsey (1978). Function of rodlets on the surface of fungal spores. *Nature* **272**: 608–610.

143. Beever, R. E., and J. R. Fincham (1973). Acetate-non-utilizing mutants of *Neurospora crassa: acu-6*, the structural gene for PEP carboxykinase and inter-allelic complementation at the *acu-6* locus. *Mol. Gen. Genet.* **126**: 217–226.

144. Bell-Pedersen, D. (1998). Keeping pace with *Neurospora* circadian rhythms. *Microbiology* **144**: 1699–1711.

145. Bell-Pedersen, D., J. C. Dunlap, and J. J. Loros (1992). The *Neurospora* circadian clock-controlled gene, *ccg-2*, is allelic to *eas* and encodes a fungal hydrophobin required for formation of the conidial rodlet layer. *Genes Dev.* **6**: 2382–2394.

146. Bell-Pedersen, D., J. C. Dunlap, and J. J. Loros (1996). Distinct *cis*-acting elements mediate clock, light, and developmental regulation of the *Neurospora crassa eas (ccg-2)* gene. *Mol. Cell. Biol.* **16**: 513–521.

147. Bell-Pedersen, D., M. L. Shinohara, J. J. Loros, and J. C. Dunlap (1996). Circadian clock-controlled genes isolated from *Neurospora crassa* are late night- to early morning-specific. *Proc. Natl. Acad. Sci. USA* **93**: 13096–13101.

148. Bell-Pederson, D., N. Garceau, and J. J. Loros (1996). Circadian rhythms in fungi. *J. Genet.* **75**: 387–401.

149. Beltramini, M., and K. Lerch (1982). Fluorescence properties of *Neurospora* tyrosinase. *Biochem. J.* **205**: 173–180.

150. Beltramini, M., and K. Lerch (1983). The reconstitution reaction of *Neurospora* apotyrosinase. *Biochem. Biophys. Res. Commun.* **110**: 313–319.

151. Beltramini, M., K. Lerch, and M. Vas'ak (1984). Metal substitution of *Neurospora* copper metallothionein. *Biochemistry* **23**: 3422–3427.

152. Beltramini, M., B. Salvato, M. Santamaria, and K. Lerch (1990). The reaction of CN^- with the binuclear copper site of *Neurospora* tyrosinase: Its relevance for a comparison between tyrosinase and hemocyanin active sites. *Biochim. Biophys. Acta* **1040**: 365–372.

153. Benarous, R., C. M. Chow, and U. L. RajBhandary (1988). Cytoplasmic leucyl-tRNA synthetase of *Neurospora crassa* is not specified by the *leu-5* locus. *Genetics* **119**: 805–814.

154. Bennett, S. N., Personal communication.

155. Bennett, S. N. (1976). Genetic studies of protoperithecial development and polyol utilization in *Neurospora.* Ph.D. Thesis, University of Georgia, Athens, GA.

156. Benson, E. W., and H. B. Howe, Jr. (1978). Reversion and interallelic complementation at four urease loci in *Neurospora crassa. Mol. Gen. Genet.* **165**: 277–282.

157. Bergquist, A., E. A. Eakin, R. T. Eakin, and R. P. Wagner (1974). Growth, respiratory, and cytochrome characteristics of certain of the isoleucine–valine mutants of *Neurospora crassa. Biochem. Genet.* **12**: 39–49.

158. Berlin, V., and C. Yanofsky (1985). Isolation and characterization of genes differentially expressed during conidiation of *Neurospora crassa. Mol. Cell. Biol.* **5**: 849–855.

159. Bernardini, D., and G. Turian (1978). Recherches sur la différenciation conidienne de *Neurospora crassa.* VII. Régulation alcooligène et capacité de conidiation (souche sauvage et mutant *"fluffy"*). *Ann. Microbiol. (Paris)* **129B**: 551–559.

160. Bernstein, H. (1961). Imidazole compounds accumulated by purine mutants of *Neurospora crassa. J. Gen. Microbiol.* **25**: 41–46.

161. Berteaux-Lecellier, V., M. Picard, C. Thompson-Coffe, D. Zickler, A. Panvier-Adoutte, and J. M. Simonet (1995). A nonmammalian homolog of the *PAF1* gene (Zellweger syndrome) discovered as a gene involved in caryogamy in the fungus *Podospora anserina. Cell* **81**: 1043–1051.

162. Bertrand, H., Personal communication.

163. Bertrand, H. (1995). Senescence is coupled to induction of an OXPHOS stress response by mitochondrial DNA mutations in *Neurospora. Can. J. Bot.* **73**: S198–S204.

164. Bertrand, H., C. A. Argan, and N. A. Szakacs (1983). Genetic control of the biogenesis of cyanide-insensitive respiration in *Neurospora crassa. In* "Mitochondria" (R. J. Schweyen, K. Wolf, and F. Kaudewitz, eds.), pp. 495–507. Walter de Gruyter Co., Berlin.

165. Bertrand, H., C. A. Argan, J. Vanderleyden, and N. A. Szakacs (1979). Genetic control of inducible cyanide-insensitive respiratory activity in *Neurospora crassa. Genetics* **91s**: 9.

166. Bertrand, H., P. Bridge, R. Collins, G. Garriga, and A. Lambowitz (1982). RNA splicing in *Neurospora* mitochondria. Characterization of new nuclear mutants with defects in splicing the mitochondrial large rRNA. *Cell* **29**: 517–526.

167. Bertrand, H., and R. A. Collins (1978). A regulatory system controlling the production of cytochrome aa_3 in *Neurospora crassa. Mol. Gen. Genet.* **166**: 1–13.

168. Bertrand, H., and J. Kohout (1977). Nuclear suppressors of the (*poky*) cytoplasmic mutant in *Neurospora crassa*. II. Mitochondrial cytochrome systems. *Can. J. Genet. Cytol.* **19**: 81–91.

169. Bertrand, H., K. J. McDougall, and T. H. Pittenger (1968). Somatic cell variation during uninterrupted growth of *Neurospora crassa* in continuous growth tubes. *J. Gen. Microbiol.* **50**: 337–350.

170. Bertrand, H., F. E. Nargang, R. A. Collins, and C. A. Zagozeski (1977). Nuclear cytochrome-deficient mutants of *Neurospora crassa*: Isolation, characterization, and genetic mapping. *Mol. Gen. Genet.* **153**: 247–257.

171. Bertrand, H., and T. H. Pittenger (1972). Isolation and classification of extranuclear mutants of *Neurospora crassa*. *Genetics* **71**: 521–533.

172. Bertrand, H., N. A. Szakacs, F. E. Nargang, C. A. Zagozeski, R. A. Collins, and J. C. Harrigan (1976). The function of mitochondrial genes in *Neurospora crassa*. *Can. J. Genet. Cytol.* **18**: 397–409.

173. Bertrand, H., and S. Werner (1979). Cytochrome *c* oxidase subunits in nuclear and extranuclear cytochrome-*aa₃*-deficient mutants of *Neurospora crassa*. *Eur. J. Biochem.* **98**: 9–18.

174. Bertrand, H., Q. Wu, and B. L. Seidel-Rogol (1993). Hyperactive recombination in the mitochondrial DNA of the *natural death* nuclear mutant of *Neurospora crassa*. *Mol. Cell. Biol.* **13**: 6778–6788.

175. Beth-Din, A., and O. Yarden (2000). The *Neurospora crassa ch-3* gene encodes an essential class I chitin synthase. *Mycologia* **92**: 65–73.

176. Bhattacharya, L., and J. F. Feldman (1971). A rapid screening procedure for female-sterile mutants. *Neurospora Newsl.* **18**: 11.

177. Bibbins, M., N. J. Cummings, and I. F. Connerton (1998). *DAB1*: A degenerate retrotransposon-like element from *Neurospora crassa*. *Mol. Gen. Genet.* **258**: 431–436.

178. Bibbins-Martinez, M. (1997). Molecular genetic studies on the control of acetate metabolism in *Neurospora crassa*. Ph.D. Thesis, University of Reading, UK.

179. Bidwai, A. P., N. A. Morjana, and G. A. Scarborough (1989). Studies on the active site of the *Neurospora crassa* plasma membrane H⁺-ATPase with periodate-oxidized nucleotides. *J. Biol. Chem.* **264**: 11790–11795.

180. Bieszke, J. A., E. L. Braun, L. E. Bean, S. Kang, D. O. Natvig, and K. A. Borkovich (1999). The *nop-1* gene of *Neurospora crassa* encodes a seven-transmembrane helix retinal-binding protein homologous to archaeal rhodopsins. *Proc. Natl. Acad. Sci. USA* **96**: 8034–8039.

181. Bieszke, J. A., E. N. Spudich, K. L. Scott, K. A. Borkovich, and J. L. Spudich (1999). A eukaryotic protein, NOP-1, binds retinal to form an archaeal rhodopsin-like photochemically reactive pigment. *Biochemistry* **38**: 14138–14145.

182. Bistis, G. N. (1983). Chemotropic interactions between trichogynes and conidia of opposite mating type in *Neurospora crassa*. *Mycologia* **73**: 959–975.

183. Bittner-Eddy, P., A. F. Monroy, and R. Brambl (1994). Expression of mitochondrial genes in the germinating conidia of *Neurospora crassa*. *J. Mol. Biol.* **235**: 881–897.

184. Black, K., and C. Yamashiro, Personal communication.

185. Blakely, R. M., and A. M. Srb (1962). Studies of the genetics and physiology of a nitrate-nonutilizing strain of *Neurospora*. *Neurospora Newsl.* **2**: 5–6.

186. Blatt, M. R., A. Rodriguez-Navarro, and C. L. Slayman (1987). Potassium–proton symport in *Neurospora*: Kinetic control by pH and membrane potential. *J. Membr. Biol.* **98**: 169–189.

187. Blatt, M. R., and C. L. Slayman (1983). KCl leakage from microelectrodes and its impact on the membrane parameters of a nonexcitable cell. *J. Membr. Biol.* **72**: 223–234.

188. Blatt, M. R., and C. L. Slayman (1987). Role of "active" potassium transport in the regulation of cytoplasmic pH by nonanimal cells. *Proc. Natl. Acad. Sci. USA* **84**: 2737–2741.

189. Bobrowicz, P., R. Lowe, D. Bell-Pedersen, and D. J. Ebbole, Personal communication.

190. Bonini, B. M., M. J. Neves, J. A. Jorge, and H. F. Terenzi (1995). Effects of temperature shifts on the metabolism of trehalose in *Neurospora crassa* wild-type and a trehalase-deficient (*tre*) mutant. Evidence against the participation of periplasmic trehalase in the catabolism of intracellular trehalose. *Biochim. Biophys. Acta* **1245**: 339–347.

191. Bonner, D. M. (1948). The identification of a natural precursor of nicotinic acid. *Proc. Natl. Acad. Sci. USA* **34**: 5–9.

192. Bonner, D. M., and G. W. Beadle (1946). Mutant strains of *Neurospora* requiring nicotinamide or related compounds for growth. *Arch. Biochem.* **11**: 319–328.

193. Bonner, D. M., Y. Suyama, and J. A. DeMoss (1960). Genetic fine structure and enzyme formation. *Fed. Proc.* **19**: 926–930.

194. Bonner, D. M., E. L. Tatum, and G. W. Beadle (1943). The genetic control of biochemical reactions in *Neurospora*: A mutant strain requiring isoleucine and valine. *Arch. Biochem.* **3**: 71–79.

195. Bonner, D. M., and C. Yanofsky (1949). Quinolinic acid accumulation in the conversion of 3-hydroxyanthranilic acid to niacin in *Neurospora*. *Proc. Natl. Acad. Sci. USA* **35**: 576–581.

196. Boone, D. M., and D. R. Stadler (1970). Reciprocal and nonreciprocal recombination between closely linked markers. *Neurospora Newsl.* **16**: 12–13.

197. Borck, K., and H. D. Braymer (1974). The genetic analysis of resistance to benomyl in *Neurospora crassa*. *J. Gen. Microbiol.* **85**: 51–56.

198. Borgeson, C. E., and B. J. Bowman (1990). Mutations that affect circadian rhythms in *Neurospora crassa* can alter the reduction of cytochromes by blue light. *J. Biol. Rhythms* **5**: 291–301.

199. Borkovich, K. A., Personal communication.

200. Borkovich, K. A., and R. L. Weiss (1987). Purification and characterization of arginase from *Neurospora crassa*. *J. Biol. Chem.* **262**: 7081–7086.

201. Bottorff, D. A., H. Bertrand, S. Parmaksizoglu, and F. E. Nargang (1990). The *cyt-12-1* mutant of *Neurospora crassa* affects the coding sequence of the cytochrome *c* structural gene. *Genet. Soc. Canada Bull.* **21** (**Suppl.**): 65 (Abstr.).

202. Bottorff, D. A., S. Parmaksizoglu, E. G. Lemire, J. W. Coffin, H. Bertrand, and F. E. Nargang (1994). Mutations

in the structural gene for cytochrome *c* result in deficiency of both cytochrome *aa₃* and cytochrome *c* in *Neurospora crassa*. *Curr. Genet.* **26**: 329–335.

203. Bowen, A. R., J. L. Chen-Wu, M. Momany, R. Young, P. J. Szaniszlo, and P. W. Robbins (1992). Classification of fungal chitin synthases. *Proc. Natl. Acad. Sci. USA* **89**: 519–523.

204. Bowman, B. J., Personal communication.

205. Bowman, B. J., K. E. Allen, and C. W. Slayman (1983). Vanadate-resistant mutants of *Neurospora crassa* are deficient in a high-affinity phosphate transport system. *J. Bacteriol.* **153**: 292–296.

206. Bowman, B. J., R. Allen, M. A. Wechser, and E. J. Bowman (1988). Isolation of genes encoding the *Neurospora* vacuolar ATPase. Analysis of *vma-2* encoding the 57-kDa polypeptide and comparison to *vma-1*. *J. Biol. Chem.* **263**: 14002–14007.

207. Bowman, B. J., C. J. Berenski, and C. Y. Jung (1985). Size of the plasma membrane H⁺-ATPase from *Neurospora crassa* determined by radiation inactivation and comparison with the sarcoplasmic reticulum Ca²⁺-ATPase from skeletal muscle. *J. Biol. Chem.* **260**: 8726–8730.

208. Bowman, B. J., and E. J. Bowman (1996). Mitochondrial and vacuolar ATPases. *In* "The Mycota: Biochemistry and Molecular Biology" (R. Brambl and G. A. Marzluf, eds.), Vol. III, pp. 57–83. Springer-Verlag, Heidelberg.

209. Bowman, B. J., and R. H. Davis (1977). Cellular distribution of ornithine in *Neurospora*: Anabolic and catabolic steady states. *J. Bacteriol.* **130**: 274–284.

210. Bowman, E. J., Personal communication.

211. Bowman, E. J. (1983). Comparison of the vacuolar membrane ATPase of *Neurospora crassa* with the mitochondrial and plasma membrane ATPases. *J. Biol. Chem.* **258**: 15238–15244.

211a. Bowman, E. J., R. Kendle, and B. J. Bowman (2000). Disruption of *vma-1*, the gene encoding the catalytic subunit of the vacuolar H⁺-ATPase, causes severe morphological changes in *Neurospora crassa*. *J. Biol. Chem.* **275**: 167–176.

212. Bowman, E. J., and T. E. Knock (1992). Structures of the genes encoding the α- and β-subunits of the *Neurospora crassa* mitochondrial ATP synthase. *Gene* **114**: 157–163.

213. Bowman, E. J., F. J. O'Neill, and B. J. Bowman (1997). Mutations of *pma-1*, the gene encoding the plasma membrane H⁺-ATPase of *Neurospora crassa*, suppress inhibition of growth by concanamycin A, a specific inhibitor of vacuolar ATPases. *J. Biol. Chem.* **272**: 14776–14786.

214. Bowman, E. J., A. Steinhardt, and B. J. Bowman (1995). Isolation of the *vma-4* gene encoding the 26-kDa subunit of the *Neurospora crassa* vacuolar ATPase. *Biochim. Biophys. Acta* **1237**: 95–98.

215. Bowman, E. J., K. Tenney, and B. J. Bowman (1988). Isolation of genes encoding the *Neurospora* vacuolar ATPase. Analysis of *vma-1* encoding the 67-kDa subunit reveals homology to other ATPases. *J. Biol. Chem.* **263**: 13994–14001.

216. Bowring, F. J., and D. E. Catcheside (1991). The initiation site for recombination *cog* is at the 3′-end of the *his-3* gene in *Neurospora crassa*. *Mol. Gen. Genet.* **229**: 273–277.

217. Bowring, F. J., and D. E. Catcheside (1995). The orientation of gene maps by recombination of flanking markers for the *am* locus of *Neurospora crassa*. *Curr. Genet.* **29**: 27–33.

218. Bowring, F. J., and D. E. Catcheside (1996). Gene conversion alone accounts for more than 90% of recombination events at the *am* locus of *Neurospora crassa*. *Genetics* **143**: 129–136.

219. Bowring, F. J., and D. E. Catcheside (1999). Evidence for negative interference. Clustering of crossovers close to the *am* locus in *Neurospora crassa* among *am* recombinants. *Genetics* **152**: 965–969.

220. Bowring, F. J., and D. E. A. Catcheside (1994). Some observations concerning *sp* and *ure-2* in *Neurospora*. *Fungal Genet. Newsl.* **41**: 85.

221. Bowring, F. J., and D. E. A. Catcheside (1998). Analysis of conversion tracts associated with recombination events at the *am* locus of *Neurospora crassa*. *Curr. Genet.* **34**: 43–49.

222. Bowring, F. J., and D. E. A. Catcheside (1999). Recombinational landscape across a 650-kb contig on the right arm of linkage group V in *Neurospora crassa*. *Curr. Genet.* **36**: 270–274.

223. Brandt, W. H. (1953). Zonation in a prolineless strain of Neurospora. *Mycologia* **45**: 194–208.

224. Braun, E. L., E. K. Fuge, P. A. Padilla, and M. Werner-Washburne (1996). A stationary-phase gene in *Saccharomyces cerevisiae* is a member of a novel, highly conserved gene family. *J. Bacteriol.* **178**: 6865–6872.

225. Braun, E. L., S. Kang, M. A. Nelson, and D. O. Natvig (1998). Identification of the first fungal annexin: Analysis of annexin gene duplications and implications for eukaryotic evolution. *J. Mol. Evol.* **47**: 531–543.

226. Brett, M., G. K. Chambers, A. A. Holder, J. R. Fincham, and J. C. Wootton (1976). Mutational amino acid replacements in *Neurospora crassa* NADP-specific glutamate dehydrogenase. *J. Mol. Biol.* **106**: 1–22.

227. Brink, N. G. (1972). Tryptophan transport in *Neurospora crassa* by various types of *mtr* revertants. *J. Gen. Microbiol.* **73**: 153–160.

228. Brink, N. G., B. Kariya, and D. R. Stadler (1969). The detection of reverse mutations at the *mtr* locus in *Neurospora* and evidence for possible intragenic (second site) suppressor mutations. *Genetics* **63**: 281–290.

229. Brito, N., C. Gonzalez, and G. A. Marzluf (1993). Hypersensitive sites in the 5′-promoter region of *nit-3*, a highly regulated structural gene of *Neurospora crassa*. *J. Bacteriol.* **175**: 6755–6759.

230. Broadbent, J. A., and H. P. Charles (1965). Some carbon dioxide-requiring mutants of *Neurospora crassa*. *J. Gen. Microbiol.* **39**: 63–74.

231. Brockman, H. E., and F. J. de Serres (1963). "Sorbose toxicity" in *Neurospora*. *Am. J. Bot.* **50**: 709–714.

232. Brody, S., Personal communication.

233. Brody, S. (1970). Correlation between reduced nicotinamide adenine dinucleotide phosphate levels and morphological changes in *Neurospora crassa*. *J. Bacteriol.* **101**: 802–807.

234. Brody, S. (1972). Regulation of pyridine nucleotide levels and ratios in *Neurospora crassa*. *J. Biol. Chem.* **247**: 6013–6017.

235. Brody, S. (1973). Metabolism, cell walls, and morphogenesis. In "Developmental Regulation: Aspects of Cell Differentiation" (S. J. Coward, ed.), pp. 107–154. Academic Press, New York, NY.

236. Brody, S. (1992). Circadian rhythms in *Neurospora crassa*: The role of mitochondria. *Chronobiol. Int.* **9**: 222–230.

237. Brody, S., and B. Allen (1972). The effects of branched-chain fatty acid incorporation into *Neurospora crassa* membranes. *J. Supramol. Struct.* **1**: 125–134.

238. Brody, S., C. Dieckmann, and S. Mikolajczyk (1985). Circadian rhythms in *Neurospora crassa*: The effects of point mutations on the proteolipid portion of the mitochondrial ATP synthetase. *Mol. Gen. Genet.* **200**: 155–161.

239. Brody, S., and S. A. Martins (1979). Circadian rhythms in *Neurospora crassa*: Effects of unsaturated fatty acids. *J. Bacteriol.* **137**: 912–915.

240. Brody, S., and S. Mikolajczyk (1988). *Neurospora* mitochondria contain an acyl carrier protein. *Eur. J. Biochem.* **173**: 353–359.

241. Brody, S., and J. F. Nyc (1970). Altered fatty acid distribution in mutants of *Neurospora crassa. J. Bacteriol.* **104**: 780–786.

242. Brody, S., and E. L. Tatum (1966). The primary biochemical effect of a morphological mutation in *Neurospora crassa. Proc. Natl. Acad. Sci. USA* **56**: 1290–1297.

243. Brody, S., and E. L. Tatum (1967). Phosphoglucomutase mutants and morphological changes in *Neurospora crassa. Proc. Natl. Acad. Sci. USA* **58**: 923–930.

244. Brody, S., K. Willert, and L. Chuman (1988). Circadian rhythms in *Neurospora crassa*: The effects of mutations at the *ufa* and *cla-1* loci. *Genome* **30 (Suppl. 1)**: 1988.

245. Brooks, C. J., B. G. DeBusk, and A. G. DeBusk (1973). Cellular compartmentation of aromatic amino acids in *Neurospora crassa*. II. Synthesis and misplaced accumulation of phenylalanine in *phen-2* auxotrophs. *Biochem. Genet.* **10**: 105–120.

246. Broquist, H. P. (1971). Lysine biosynthesis. *Methods. Enzymol.* **17**: 112–129.

247. Bruno, K. S., R. Aramayo, P. F. Minke, R. L. Metzenberg, and M. Plamann (1996). Loss of growth polarity and mislocalization of septa in a *Neurospora* mutant altered in the regulatory subunit of cAMP-dependent protein kinase. *EMBO J.* **15**: 5772–5782.

248. Bruno, K. S., J. H. Tinsley, P. F. Minke, and M. Plamann (1996). Genetic interactions among cytoplasmic dynein, dynactin, and nuclear distribution mutants of *Neurospora crassa. Proc. Natl. Acad. Sci. USA* **93**: 4775–4780.

249. Buchanan, J. M. (1960). The enzymatic synthesis of the purine nucleotides. *Harvey Lectures* **54(1958–1959)**: 104–130.

250. Buerger, F., R. Schneider, B. Brors, and H. Weiss (1997). Direct database submission.

251. Buerger, F., R. Schneider, B. Brors, and H. Weiss (1998). Direct database submission.

252. Buntin, S. M. (1993). Genetic analysis and characterization of SS-788 and SS-1044, new osmotic-sensitive mutants of *Neurospora crassa*. M.S. Thesis, Georgia Southern University, Statesboro, GA.

253. Buntin, S. M., S. N. Bennett, and W. A. Krissinger (1995). Further genetic analysis and characterization of SS-788, a new *osmotic* mutant of *Neurospora crassa. Georgia J. Sci.* **53**: 26 (Abstr.).

254. Buntin, S. M., S. N. Bennett, and W. A. Krissinger (1995). A new locus for osmotic-sensitive mutants of *Neurospora crassa. Fungal Genet. Newsl.* **42A**: 79.

255. Buremalm, K. E., and H. G. Kølmark (1971). Genetic analysis of a suppressor of urease deficiency in *Neurospora crassa. Hereditas* **69**: 249–262.

256. Bürk, R. R. (1964). The location of *i*, enhancer of *am*. *Neurospora Newsl.* **6**: 27.

257. Burns, D. M., and C. Yanofsky (1989). Nucleotide sequence of the *Neurospora crassa trp-3* gene encoding tryptophan synthetase and comparison of the *trp-3* polypeptide with its homologs in *Saccharomyces cerevisiae* and *Escherichia coli. J. Biol. Chem.* **264**: 3840–3848.

258. Burns, P. A., J. H. Kinnaird, and J. R. S. Fincham (1984). *Neurospora crassa* suppressors act on *amber. Neurospora Newsl.* **31**: 20.

259. Burton, E. G., and R. L. Metzenberg (1972). Novel mutation causing derepression of several enzymes of sulfur metabolism in *Neurospora crassa. J. Bacteriol.* **109**: 140–151.

260. Burton, E. G., and R. L. Metzenberg (1975). Regulation of methionine biosynthesis in *Neurospora crassa. Arch. Biochem. Biophys.* **168**: 219–229.

261. Butler, D. K. (1992). Ribosomal DNA is a site of chromosome breakage in aneuploid strains of *Neurospora. Genetics* **131**: 581–592.

262. Butler, D. K., and R. L. Metzenberg (1989). Premeiotic change of nucleolus organizer size in *Neurospora. Genetics* **122**: 783–791.

263. Butler, D. K., and R. L. Metzenberg (1990). Expansion and contraction of the nucleolus organizer region of *Neurospora*: Changes originate in both proximal and distal segments. *Genetics* **126**: 325–333.

264. Butler, D. K., and R. L. Metzenberg (1993). Amplification of the nucleolus organizer region during the sexual phase of *Neurospora crassa. Chromosoma* **102**: 519–525.

265. Buxton, F. P., and A. Radford (1982). Isolation and mapping of fluoropyrimidine-resistant mutants of *Neurospora crassa. Mol. Gen. Genet.* **185**: 129–131.

266. Buxton, F. P., and A. Radford (1982). Nitrogen metabolite repression of fluoropyrimidine resistance and pyrimidine uptake in *Neurospora crassa. Mol. Gen. Genet.* **186**: 259–262.

267. Buxton, F. P., and A. Radford (1982). Partial characterization of 5-fluoropyrimidine-resistant mutants of *Neurospora crassa. Mol. Gen. Genet.* **185**: 132–135.

268. Caeser-Ton That, T. C., C. Rossier, F. Barja, G. Turian, and U.-P. Roos (1988). Induction of multiple germ tubes in *Neurospora crassa* by antitubulin agents. *Eur. J. Cell Biol.* **46**: 68–79.

269. Calderón, J., and L. M. Martinez (1993). Regulation of ammonium ion assimilation enzymes in *Neurospora crassa nit-2* and *ms-5* mutant strains. *Biochem. Genet.* **31**: 425–439.

270. Calderón, J., L. M. Martinez, and J. Mora (1990). Isolation and characterization of a *Neurospora crassa* mutant altered in the α-polypeptide of glutamine synthetase. *J. Bacteriol.* **172**: 4996–5000.

271. Calderón, J., L. Olvera, L. M. Martinez, and G. Dávila (1997). A *Neurospora crassa* mutant altered in the regulation of L-amino acid oxidase. *Microbiology* **143**: 1969–1974.

272. Calhoun, F., and J. Howe (1968). Genetic analysis of eight-spored asci produced by gene *E* in *Neurospora tetrasperma*. *Genetics* **60**: 449–459.

273. Calza, R. E., and A. L. Schroeder (1982). Postreplication repair in *Neurospora crassa*. *Mol. Gen. Genet.* **185**: 111–119.

274. Camacho, J. A., C. Obie, B. Biery, B. K. Goodman, C. A. Hu, S. Almashanu, G. Steel, R. Casey, M. Lambert, G. A. Mitchell, and D. Valle (1999). Hyperornithinaemia-hyperammonaemia–homocitrullinuria syndrome is caused by mutations in a gene encoding a mitochondrial ornithine transporter. *Nature Genet.* **22**: 151–158.

275. Cambareri, E. B., R. Aisner, and J. Carbon (1998). Structure of the chromosome VII centromere region in *Neurospora crassa*: Degenerate transposons and simple repeats. *Mol. Cell. Biol.* **18**: 5465–5477.

276. Cambareri, E. B., H. M. Foss, M. R. Rountree, E. U. Selker, and J. A. Kinsey (1996). Epigenetic control of a transposon-inactivated gene in *Neurospora* is dependent on DNA methylation. *Genetics* **143**: 137–146.

277. Cambareri, E. B., and J. A. Kinsey (1994). A simple and efficient system for targeting DNA to the *am* locus of *Neurospora crassa*. *Gene* **142**: 219–224.

278. Cameron, H., K. Hsu, and D. Perkins (1966). Crossing-over frequency following inbreeding in *Neurospora*. *Genetica* **37**: 1–6.

279. Cameron, H. R., Personal communication.

280. Campbell, J. L., and B. C. Turner (1987). Recombination block in the Spore killer region of *Neurospora*. *Genome* **29**: 129–135.

281. Canady, C. (1996). Microscopic examination of conidial variation of an osmotic-sensitive, conidial mutant, *os-8*, and other osmotic-sensitive or conidiation mutant strains of *Neurospora crassa*. *Georgia J. Sci.* **54**: 26 (Abstr.).

282. Canady, C. (1997). Analysis of the *os-8* locus of *Neurospora crassa*. *Georgia J. Sci.* **55**: 31 (Abstr.).

283. Canady, C. E. (1994). Characterization and genetic analysis of the *os-8* locus of *Neurospora crassa*. M.S. Thesis, Georgia Southern University, Statesboro, GA.

284. Capelli, N., F. Barja, D. van-Tuinen, J. Monnat, G. Turian, and R. Ortega-Perez (1997). Purification of a 47-kDa calmodulin-binding polypeptide as an actin-binding protein from *Neurospora crassa*. *FEMS Microbiol. Lett.* **147**: 215–220.

285. Caprara, M. G., G. Mohr, and A. M. Lambowitz (1996). A tyrosyl-tRNA synthetase protein induces tertiary folding of the group I intron catalytic core. *J. Mol. Biol.* **257**: 512–531.

286. Carattoli, A., F. Kato, M. Rodriguez-Franco, W. D. Stuart, and G. Macino (1995). A chimeric light-regulated amino acid transport system allows the isolation of blue light regulator (*blr*) mutants of *Neurospora crassa*. *Proc. Natl. Acad. Sci. USA* **92**: 6612–6616.

287. Cardemil, L., and G. Pincheira (1979). Characterization of the carbohydrate component of fraction I in the *Neurospora crassa* cell wall. *J. Bacteriol.* **137**: 1067–1072.

288. Caroline, D. F. (1969). Pyrimidine synthesis in *Neurospora crassa*: Gene–enzyme relationships. *J. Bacteriol.* **100**: 1378–1384.

289. Caroline, D. F., and R. H. Davis (1969). Pyrimidine synthesis in *Neurospora crassa*: Regulation of enzyme activities. *J. Bacteriol.* **100**: 1378–1384.

290. Caroline, D. F., R. W. Harding, H. Kuwana, T. Satyanarayana, and R. P. Wagner (1969). The *iv-3* mutants of *Neurospora crassa*. II. Activity of acetohydroxy acid synthetase. *Genetics* **62**: 487–494.

291. Caroline, D. J. F. (1968). Pyrimidine synthesis in *Neurospora crassa*. Ph.D. Thesis, University of Michigan, Ann Arbor, MI.

292. Carsiotis, M., and R. F. Jones (1974). Cross-pathway regulation: Tryptophan-mediated control of histidine and arginine biosynthetic enzymes in *Neurospora crassa*. *J. Bacteriol.* **119**: 889–892.

293. Caru, M., V. Cifuentes, G. Pincheira, and A. Jimenez (1989). Molecular cloning and expression in *Saccharomyces cerevisiae* and *Neurospora crassa* of the invertase gene from *Neurospora crassa*. *J. Appl. Bacteriol.* **67**: 401–410.

294. Case, M., and N. Giles (1968). Evidence for nonsense mutations in the *arom* gene cluster of *Neurospora crassa*. *Genetics* **60**: 49–58.

295. Case, M. E., Personal communication.

296. Case, M. E. (1957). A genetic study of pantothenic acid mutants in *Neurospora crassa*. Ph.D. Thesis, Yale University, New Haven, CT.

297. Case, M. E., L. Burgoyne, and N. H. Giles (1969). *In vivo* and *in vitro* complementation between DHQ synthetase mutants in the *arom* gene cluster of *Neurospora crassa*. *Genetics* **63**: 581–588.

298. Case, M. E., R. F. Geever, and D. K. Asch (1992). Use of gene replacement transformation to elucidate gene function in the *qa* gene cluster of *Neurospora crassa*. *Genetics* **130**: 729–736.

299. Case, M. E., and N. H. Giles (1958). Evidence from tetrad analysis for both normal and aberrant recombination between allelic mutants in *Neurospora crassa*. *Proc. Natl. Acad. Sci. USA* **44**: 378–390.

300. Case, M. E., and N. H. Giles (1958). Recombination mechanisms at the *pan-2* locus in *Neurospora crassa*. *Cold Spring Harbor Symp. Quant. Biol.* **23**: 119–135.

301. Case, M. E., and N. H. Giles (1960). Comparative complementation and genetic maps of the *pan-2* locus in *Neurospora crassa*. *Proc. Natl. Acad. Sci. USA* **46**: 659–676.

302. Case, M. E., and N. H. Giles (1964). Allelic recombination in *Neurospora*: Tetrad analysis of a three-point cross within the *pan-2* locus. *Genetics* **49**: 529–540.

303. Case, M. E., and N. H. Giles (1974). Revertants and secondary *arom-2* mutants induced in noncomplementing mutants in the *arom* gene cluster of *Neurospora crassa*. *Genetics* **77**: 613–626.

304. Case, M. E., and N. H. Giles (1976). Gene order in the *qa* gene cluster of *Neurospora crassa*. *Mol. Gen. Genet.* **147**: 83–89.

305. Case, M. E., N. H. Giles, and C. H. Doy (1972). Genetical and biochemical evidence for further interrelationships between the polyaromatic synthetic and the quinate-

shikimate catabolic pathways in *Neurospora crassa*. *Genetics* **71**: 337–348.

306. Case, M. E., J. A. Hautala, and N. H. Giles (1977). Characterization of *qa-2* mutants of *Neurospora crassa* by genetic, enzymatic, and immunological techniques. *J. Bacteriol.* **129**: 166–172.

307. Case, M. E., C. Pueyo, J. L. Barea, and N. H. Giles (1978). Genetical and biochemical characterization of *qa-3* mutants and revertants in the *qa* gene cluster of *Neurospora crassa*. *Genetics* **90**: 69–84.

308. Case, M. E., M. Schweizer, S. R. Kushner, and N. H. Giles (1979). Efficient transformation of *Neurospora crassa* by utilizing hybrid plasmid DNA. *Proc. Natl. Acad. Sci. USA* **76**: 5259–5263.

309. Casey, C., and A. M. Lambowitz, Personal communication.

310. Castaneda, M., J. Martuscelli, and J. Mora (1967). The catabolism of L-arginine by *Neurospora crassa*. *Biochim. Biophys. Acta* **141**: 276–286.

310a. Catalanotto, C., G. Azzalin, G. Macino, and C. Cogoni (2000). Gene silencing in worms and fungi. *Nature* **404**: 245.

311. Catcheside, D. (1968). Regulation of the *am-1* locus in *Neurospora*: Evidence of independent control of allelic recombination and gene expression. *Genetics* **59**: 443–452.

312. Catcheside, D. (1969). Prephenate dehydrogenase from *Neurospora*: Feedback activation by phenylalanine. *Biochem. Biophys. Res. Commun.* **36**: 651–656.

313. Catcheside, D. (1981). Genes in *Neurospora* that suppress recombination when they are heterozygous. *Genetics* **98**: 55–76.

314. Catcheside, D., and B. Austin (1969). The control of allelic recombination at histidine loci in *Neurospora crassa*. *Am. J. Bot.* **56**: 685–690.

315. Catcheside, D., P. Storer, and B. Klein (1985). Cloning of the *ARO* cluster gene of *Neurospora crassa* and its expression in *Escherichia coli*. *Mol. Gen. Genet.* **199**: 446–451.

316. Catcheside, D. E. (1986). A restriction and modification model for the initiation and control of recombination in *Neurospora*. *Genet. Res.* **47**: 157–165.

317. Catcheside, D. E. A., Personal communication.

318. Catcheside, D. E. A. (1966). Genetic and biochemical studies in *Neurospora crassa*. Ph.D. Thesis, University of Birmingham, Birmingham, UK.

319. Catcheside, D. E. A. (1969). A new carotenoid mutant of *Neurospora*. *Neurospora Newsl.* **15**: 3–4.

320. Catcheside, D. E. A. (1970). Control of recombination within the *nitrate-2* locus of *Neurospora crassa*: An unlinked dominant gene which reduces prototroph yields. *Aust. J. Biol. Sci.* **23**: 855–865.

321. Catcheside, D. E. A. (1973). New linkage data for group V markers in *N. crassa*. *Neurospora Newsl.* **20**: 43–44.

322. Catcheside, D. E. A. (1974). Gene order in linkage group V of *Neurospora crassa*. *Neurospora Newsl.* **21**: 24.

323. Catcheside, D. E. A. (1978). A pleiotropic mutation in *Neurospora* conferring sensitivity to analogues of amino acids, purines and pyrimidines. *Neurospora Newsl.* **25**: 17–18.

324. Catcheside, D. E. A. (1974). A second locus subject to recombination control by the *rec-1*⁺ gene in *Neurospora crassa*. *Aust. J. Biol. Sci.* **27**: 561–575.

325. Catcheside, D. E. A. (1976). Map location of *rec-1* in *N. crassa*. *Neurospora Newsl.* **23**: 23.

326. Catcheside, D. E. A. (1978). A pleiotropic mutation in *Neurospora* conferring sensitivity to analogues of amino acids, purines and pyrimidines. *Neurospora Newsl.* **25**: 17–18.

327. Catcheside, D. E. A. (1992). The nomenclature of *mts* and *cpc* mutants of *Neurospora*. *Fungal Genet. Newsl.* **38**: 71.

328. Catcheside, D. E. A., P. J. Yeadon, and A. Petersen (1998). Direct database submission.

329. Catcheside, D. G. (1960). Complementation among histidine mutants of *Neurospora crassa*. *Proc. Roy. Soc. London. B Biol. Sci.* **153**: 179–194.

330. Catcheside, D. G. (1964). Interallelic complementation. *Brookhaven Symp. Biol.* **17**: 1–14.

331. Catcheside, D. G. (1965). Multiple enzymic functions of a gene in *Neurospora crassa*. *Biochem. Biophys. Res. Commun.* **18**: 648–651.

332. Catcheside, D. G. (1966). A second gene controlling allelic recombination in *Neurospora crassa*. *Austr. J. Biol. Sci.* **19**: 1039–1046.

333. Catcheside, D. G. (1974). Fungal genetics. *Annu. Rev. Genet.* **8**: 279–300.

334. Catcheside, D. G. (1975). Occurrence in wild strains of *Neurospora crassa* of genes controlling genetic recombination. *Aust. J. Biol. Sci.* **28**: 213–225.

335. Catcheside, D. G. (1977). "The Genetics of Recombination." Edward Arnold Ltd., London.

336. Catcheside, D. G., and T. Angel (1974). A *histidine-3* mutant, in *Neurospora crassa*, due to an interchange. *Aust. J. Biol. Sci.* **27**: 219–229.

337. Catcheside, D. G., and B. Austin (1971). Common regulation of recombination at the *amination-1* and *histidine-2* loci in *Neurospora crassa*. *Aust. J. Biol. Sci.* **24**: 107–115.

338. Catcheside, D. G., and D. Corcoran (1973). Control of nonallelic recombination in *Neurospora crassa*. *Aust. J. Biol. Sci.* **26**: 1337–1353.

339. Catcheside, D. G., A. P. Jessup, and B. R. Smith (1964). Genetic controls of allelic recombination in *Neurospora*. *Nature* **202**: 1242–1243.

340. Centola, M., and J. Carbon (1994). Cloning and characterization of centromeric DNA from *Neurospora crassa*. *Mol. Cell. Biol.* **14**: 1510–1519.

341. Chakraborty, B. N., P. M. Ouimet, G. M. Sreenivasan, C. A. Curle, and M. Kapoor (1995). Sequence repeat-induced disruption of the major heat-inducible HSP70 gene of *Neurospora crassa*. *Curr. Genet.* **29**: 18–26.

342. Chakravarti, D. N., B. Chakravarti, and P. Chakrabarti (1981). Incorporation of photolabile azido fatty acid probes in *Neurospora crassa* and *Saccharomyces cerevisiae*. *Experientia* **37**: 353–354.

343. Chaleff, R. (1971). Evidence for a gene cluster controlling the inducible quinate catabolic pathway in *Neurospora crassa*. *Genetics* **68s**: 10–11.

344. Chaleff, R. S. (1974). The inducible quinate–shikimate catabolic pathway in *Neurospora crassa*: Genetic organization. *J. Gen. Microbiol.* **81**: 337–355.

345. Chaleff, R. S. (1974). The inducible quinate–shikimate catabolic pathway in *Neurospora crassa*: Induction and regulation of enzyme synthesis. *J. Gen. Microbiol.* **81**: 357–372.

346. Chalmers, J. H., Jr. (1973). A novel method of selecting drug-sensitive mutants of *Neurospora crassa. Genetics* **74s:** 43.

347. Chalmers, J. H., Jr., and J. A. DeMoss (1970). Genetic control of a multienzyme complex: Subunit structures of mutationally altered forms. *Genetics* **65:** 213–221.

348. Chalmers, J. H., Jr., and T. W. Seale (1971). Super-suppressible mutants in *Neurospora*: Mutants at the *tryp-1* and *tryp-2* loci affecting the structure of the multienzyme complex in the tryptophan pathway. *Genetics* **67:** 353–363.

349. Chalmers, J. H., Jr., and P. St. Lawrence (1979). *cpl-1*: A *Neurospora* mutant sensitive to chloramphenicol. *Neurospora Newsl.* **27:** 8–9.

350. Chambers, J. A. A. (1980). The *i* phenotype of *Neurospora crassa. Neurospora Newsl.* **27:** 17.

351. Chan, W. L. (1977). Relationship of tyrosinase and L-amino acid oxidase with development of protoperithecium in *Neurospora*. Ph.D. Thesis, University of Malaya, Kuala Lumpur, Malaysia.

352. Chang, A., and C. W. Slayman (1990). A structural change in the *Neurospora* plasma membrane [H⁺]-ATPase induced by N-ethylmaleimide. *J. Biol. Chem.* **265:** 15531–15536.

352a. Chang, B., and H. Nakashima (1998). Isolation of temperature-sensitive rhythm mutant in *Neurospora crassa. Genes Genet. Syst.* **73:** 71–73.

353. Chang, H. C.-P., G. J. Mulkins, J. C. Dyer, and G. J. Sorger (1975). Enzymatic and nonenzymatic reduction of nitrite by extracts of *Neurospora crassa. J. Bacteriol.* **123:** 755–758.

354. Chang, H. C.-P., and G. J. Sorger (1976). Effect of ammonium ions on the induction of nitrite reductase in *Neurospora crassa. J. Bacteriol.* **126:** 1002–1004.

355. Chang, L.-T., and R. W. Tuveson (1967). Ultraviolet-sensitive mutants in *Neurospora crassa. Genetics* **56:** 801–810.

356. Chang, S., and C. Staben (1994). Directed replacement of *mt A* by *mt a-1* effects a mating-type switch in *Neurospora crassa. Genetics* **138:** 75–81.

357. Chang, S. T., and D. G. Catcheside (1973). An allele-specific suppressor of *pan-2* mutants in *Neurospora crassa. Genetics* **74:** s44 (Abstr.).

358. Chang, T., R. L. Metzenberg, and B. Weisblum, Personal communication.

359. Charlang, C. W., and N. P. Williams, Personal communication.

360. Charlang, G., and N. P. Williams (1977). Germination-defective mutant of *Neurospora crassa* that responds to siderophores. *J. Bacteriol.* **132:** 1042–1044.

361. Charlang, G. W. (1979). An improved glycerol minimal medium. *Neurospora Newsl.* **26:** 20.

362. Charlang, G. W., and N. P. Williams (1983). Siderophore transport mutants (*sit*) in *Neurospora crassa. Neurospora Newsl.* **30:** 6–7.

363. Charles, H. P. (1962). Response of *Neurospora* mutants to carbon dioxide. *Nature* **195:** 359–360.

364. Charles, H. P., and J. A. Broadbent (1964). Carbon dioxide mutants in *Neurospora. Nature (London)* **201:** 1004–1006.

365. Chary, P., D. Dillon, A. L. Schroeder, and D. O. Natvig (1994). Superoxide dismutase (*sod-1*) null mutants of *Neurospora crassa*: Oxidative stress sensitivity, sponta-neous mutation rate and response to mutagens. *Genetics* **137:** 723–730.

366. Chary, P., and D. O. Natvig (1989). Evidence for three differentially regulated catalase genes in *Neurospora crassa*: Effects of oxidative stress, heat shock, and development. *J. Bacteriol.* **171:** 2646–2652.

367. Chaure, P. T., and I. F. Connerton (1995). Derepression of the glyoxylate cycle in mutants of *Neurospora crassa* accelerated for growth on acetate. *Microbiology* **141:** 1315–1320.

368. Cheah, S. C., and C. C. Ho (1976). Two genetic regulatory systems for tyrosinase synthesis in *Neurospora crassa. Abstr. Annu. Meeting Soc. Microbiol.*, 108.

369. Chen, B., A. R. Kubelik, S. Mohr, and C. A. Breitenberger (1996). Cloning and characterization of the *Neurospora crassa cyt-5* gene. A nuclear-coded mitochondrial RNA polymerase with polyglutamine repeat. *J. Biol. Chem.* **271:** 6537–6544.

370. Chen, H., J. Crabb, and J. Kinsey (1998). The *Neurospora aab-1* gene encodes a CCAAT-binding protein homologous to yeast HAP5. *Genetics* **148:** 123–130.

371. Chen, H., and J. A. Kinsey (1994). Sequential gel mobility shift scanning of 5'-upstream sequences of the *Neurospora crassa am* (GDH) gene. *Mol. Gen. Genet.* **242:** 399–403.

372. Chen, Y. F., C. C. Ho, and A. L. Demain (1976). Genetic improvement of production of amylases in *Neurospora intermedia. Genetics* **83s:** 14.

373. Cheng, R., T. I. Baker, C. E. Cords, and R. J. Radloff (1993). *mei-3*, a recombination and repair gene of *Neurospora crassa*, encodes a RecA-like protein. *Mutat. Res., DNA Repair* **294:** 223–234.

374. Chiang, T. Y., and G. A. Marzluf (1994). DNA recognition by the NIT2 nitrogen regulatory protein: Importance of the number, spacing, and orientation of GATA core elements and their flanking sequences upon NIT2 binding. *Biochemistry* **33:** 576–582.

375. Chiang, T. Y., and G. A. Marzluf (1995). Binding affinity and functional significance of NIT2 and NIT4 binding sites in the promoter of the highly regulated *nit-3* gene, which encodes nitrate reductase in *Neurospora crassa. J. Bacteriol.* **177:** 6093–6099.

376. Chiang, T. Y., R. Rai, T. G. Cooper, and G. A. Marzluf (1994). DNA binding site specificity of the *Neurospora* global nitrogen regulatory protein NIT2: Analysis with mutated binding sites. *Mol. Gen. Genet.* **245:** 512–516.

377. Chow, C. M., S. Kang, R. L. Metzenberg, and U. L. RajBhandary (1995). Sequence of the *met-10⁺* locus of *Neurospora crassa*: Homology to a sequence of unknown function in *Saccharomyces cerevisiae* chromosome 8. *Gene* **162:** 111–115.

378. Chow, C. M., R. L. Metzenberg, and U. L. RajBhandary (1989). Nuclear gene for mitochondrial leucyl-tRNA synthetase of *Neurospora crassa*: Isolation, sequence, chromosomal mapping, and evidence that the *leu-5* locus specifies structural information. *Mol. Cell. Biol.* **9:** 4631–4644.

379. Chow, C. M., and U. L. RajBhandary (1989). Regulation of the nuclear genes encoding the cytoplasmic and mitochondrial leucyl-tRNA synthetases of *Neurospora crassa. Mol. Cell. Biol.* **9:** 4645–4652.

380. Chow, C. M., and U. L. RajBhandary (1993). Developmental regulation of the gene for formate dehydrogenase in *Neurospora crassa. J. Bacteriol.* **175:** 3703–3709.

381. Chow, T. Y.-K., and M. J. Fraser (1978). Sensitivity of DNA-repair deficient mutants of *Neurospora* to histidine and to mitomycin C. *Microb. Genet. Bull.* **45:** 4–5.

382. Christensen, R. L., and J. C. Schmit (1980). Regulation and glutamic acid decarboxylase during *Neurospora crassa* conidial germination. *J. Bacteriol.* **144:** 983–990.

383. Chung, J. Y., S.-K. Chae, and R. L. Weiss (1997). Arginine feedback resistant mutation in *Neurospora*. *19th Fungal Genet. Conf.* Asilomar, CA. 16 (Abstr.).

384. Clutterbuck, A. J. (1978). Genetics of vegetative growth and asexual reproduction. *In* "The Filamentous Fungi, Vol. 3: Developmental Mycology" (J. E. Smith and D. R. Berry, eds.), pp. 240–256. John Wiley & Sons, New York, NY.

385. Coddington, A. (1976). Biochemical studies on the *nit* mutants of *Neurospora crassa*. *Mol. Gen. Genet.* **145:** 195–206.

386. Coddington, A., J. Fincham, and T. Sundaram (1966). Multiple active varieties of *Neurospora* glutamate dehydrogenase formed by hybridization between two inactive mutant proteins *in vivo* and *in vitro*. *J. Mol. Biol.* **17:** 503–512.

387. Coddington, A., and J. R. S. Fincham (1965). Proof of hybrid enzyme formation in a case of interallelic complementation in *Neurospora crassa*. *J. Mol. Biol.* **12:** 152–161.

388. Coffin, J., R. Dhillon, R. Ritzel, and F. Nargang (1997). The *Neurospora crassa cya-5* nuclear gene encodes a protein with a region of homology to the *Saccharomyces cerevisiae* PET309 protein and is required in a post-transcriptional step for the expression of the mito-chondrially encoded COXI protein. *Curr. Genet.* **32:** 273–280.

389. Coggins, J., M. Boockock, S. Chaudhuri, J. Lambert, J. Lumsden, G. Nimmo, and D. Smith (1987). The *arom* multifunctional enzyme from *Neurospora crassa*. *Methods Enzymol.* **142:** 325–341.

390. Cogoni, C., J. T. Irelan, M. Schumacher, T. J. Schmidhauser, E. U. Selker, and G. Macino (1996). Transgene silencing of the *al-1* gene in vegetative cells of *Neurospora* is mediated by a cytoplasmic effector and does not depend on DNA–DNA interactions or DNA methylation. *EMBO J.* **15:** 3153–3163.

391. Cogoni, C., and G. Macino (1999). Gene silencing in *Neurospora crassa* requires a protein homologous to RNA-dependent RNA polymerase. *Nature* **399:** 166–169.

392. Cogoni, C., and M. Macino (1997). Isolation of quelling-defective (*qde*) mutants impaired in posttranscriptional transgene-induced gene silencing in *Neurospora crassa*. *Proc. Natl. Acad. Sci. USA* **94:** 10233–10238.

393. Cogoni, C., and M. Macino (1997). Transgene induced gene silencing "*quelling*" in *Neurospora crassa*. *19th Fungal Genet. Conf.* Asilomar, CA, 26 (Abstr.).

393a. Cogoni, C., and G. Macino (1999). Posttranscriptional gene silencing in *Neurospora* by a recA DNA helicase. *Science* **286:** 2342–2344.

394. Cogoni, C., N. Romano, and G. Macino (1994). Suppression of gene expression by homologous transgenes. *Antonie Van Leeuwenhoek* **65:** 205–209.

395. Colandene, J. D., and R. H. Garrett (1996). Functional dissection and site-directed mutagenesis of the structural gene for NAD(P)H-nitrite reductase in *Neurospora crassa*. *J. Biol. Chem.* **271:** 24096–24104.

396. Colburn, R. W., and E. L. Tatum (1965). Studies of a phenylalanine tyrosine-requiring mutant of *Neurospora crassa* (strain S4342). *Biochim. Biophys. Acta* **97:** 442–448.

397. Collett, M. A., J. J. Loros, and J. C. Dunlap, Personal communication.

398. Collinge, A. J., M. H. Fletcher, and A. P. J. Trinci (1978). Physiology and cytology of septation and branching in a temperature-sensitive colonial mutant (*cot-1*) of *Neurospora crassa*. *Trans. Brit. Mycol. Soc.* **71:** 107–121.

399. Collins, R. A. (1979). Mitochondrial assembly in wild-type and mutant strains of *Neurospora crassa*. Ph.D. Thesis, University of Regina, Regina, Saskatchewan, Canada.

400. Collins, R. A., and H. Bertrand (1978). Nuclear suppressors of the [*poky*] cytoplasmic mutant in *Neurospora crassa*. III. Effects on other cytoplasmic mutants and on mitochondrial ribosome assembly in [*poky*]. *Mol. Gen. Genet.* **161:** 267–273.

401. Collins, R. A., H. Bertrand, R. J. LaPolla, and A. M. Lambowitz. (1979). Mitochondrial ribosome assembly in *Neurospora crassa*: Mutants with defects in mitochondrial ribosome assembly. *Mol. Gen. Genet.* **177:** 73–84.

402. Collins, R. A., and A. M. Lambowitz (1985). RNA splicing in *Neurospora* mitochondria. Defective splicing of mito-chondrial mRNA precursors in the nuclear mutant *cyt18-1*. *J. Mol. Biol.* **184:** 413–428.

403. Coniordos, N., and G. Turian (1973). Recherches sur la différenciation conidienne de *Neurospora crassa*. IV. Modifications chimio-structurales de la paroi chez le type sauvage et chez deux mutants aconidiens. *Ann. Microbiol. (Paris)* **124A:** 5–28.

404. Connerton, I. F., Personal communication.

405. Connerton, I. F. (1990). Premeiotic disruption of the *Neurospora crassa* malate synthase gene by native and divergent DNAs. *Mol. Gen. Genet.* **223:** 319–323.

406. Connerton, I. F., N. E. Owen, P. T. Chaure, and L. D. S. Gainey. (1992). Function and control of acetate metabolism in *Neurospora crassa*. *In* "Molecular Biology of Filamentous Fungi" (U. Stahl and P. Tudzynski, eds.), pp. 201–225. VCH, Waldheim.

407. Coppin, E., R. Debuchy, S. Arnaise, and M. Picard (1997). Mating types and sexual development in filamentous ascomycetes. *Microbiol. Mol. Biol. Rev.* **61:** 411–428.

408. Cornelius, G., and H. Nakashima (1987). Vacuoles play a decisive role in calcium homeostasis in *Neurospora crassa*. *J. Gen. Microbiol.* **133:** 2341–2347.

409. Corrochano, L. M., F. R. Lauter, D. J. Ebbole, and C. Yanofsky (1995). Light and developmental regulation of the gene *con-10* of *Neurospora crassa*. *Dev. Biol.* **167:** 190–200.

410. Cossins, E., and P. Chan (1983). Folylpolyglutamate synthetase activities of *Neurospora crassa*: Nature of products formed by soluble and particulate enzymes in the wild type and polyglutamate-deficient mutants. *Adv. Exp. Med. Biol.* **163:** 183–197.

411. Cossins, E., P. Chan, and G. Combepine (1976). One-carbon metabolism in *Neurospora crassa* wild type and in mutants partially deficient in serine hydroxymethyltrans-ferase. *Biochem. J.* **160:** 305–314.

412. Cossins, E. A., and S. H. Y. Pang (1980). Loss of cytosolic serine hydroxymethyltransferase in a formate mutant of *Neurospora crassa*. *Experientia* **36**: 289–290.

413. Cote, G. G., and S. Brody (1987). Circadian rhythms in *Neurospora crassa*: A clock mutant, *prd-1*, is altered in membrane fatty acid composition. *Biochim. Biophys. Acta* **904**: 131–139.

414. Coulter, K. R., and G. A. Marzluf (1998). Functional analysis of different regions of the positive-acting CYS3 regulatory protein of *Neurospora crassa*. *Curr. Genet.* **33**: 395–405.

415. Court, D., F. Nargang, H. Steiner, R. Hodges, W. Neupert, and R. Lill (1996). Role of the intermembrane-space domain of the preprotein receptor Tom22 in protein import into mitochondria. *Mol. Cell. Biol.* **16**: 4035–4042.

416. Court, D. A., R. Kleene, W. Neupert, and R. Lill (1996). Role of the N- and C-termini of porin in import into the outer membrane of *Neurospora* mitochondria. *FEBS Lett.* **390**: 73–77.

417. Courtright, J. B., Personal communication.

418. Courtright, J. B. (1975). Characteristics of a glycerol utilization mutant of *Neurospora crassa*. *J. Bacteriol.* **124**: 497–502.

419. Courtright, J. B. (1975). Growth of *ropy* mutants on glycerol and acetate. *Neurospora Newsl.* **22**: 5.

420. Courtright, J. B. (1976). Genetic control of glycerol-3-phosphate dehydrogenase synthesis in *Neurospora*. *In* "Genetics and Biogenesis of Chloroplasts and Mitochondria" (T. E. A. Bücher, ed.), pp. 881–884. North Holland Publishing Co., Amsterdam.

421. Courtright, J. B. (1976). Induction of enzymes of the glycerophosphate pathway in *leu-5* mutants of *Neurospora crassa*. *Biochem. Genet.* **14**: 1057–1063.

422. Cox, J. A., C. Ferraz, J. G. Demaille, R. O. Perez, D. van Tuinen, and D. Marme (1982). Calmodulin from *Neurospora crassa*. General properties and conformational changes. *J. Biol. Chem.* **257**: 10694–10700.

423. Cox, R. A., and K. Peden (1979). A study of the organization of the ribosomal ribonucleic acid gene cluster of *Neurospora crassa* by means of restriction endonuclease analysis and cloning in bacteriophage λ. *Mol. Gen. Genet.* **174**: 17–24.

424. Cramer, C. L., L. E. Vaughn, and R. H. Davis (1980). Basic amino acids and inorganic polyphosphates in *Neurospora crassa*: Independent regulation of vacuolar pools. *J. Bacteriol.* **142**: 945–952.

425. Crawford, I. P. (1980). Gene fusions in the tryptophan pathway: Tryptophan synthetase and phosphoribosyl anthranilate isomerase:indole glycerolphosphate synthetase. *In* "Multifunctional Proteins" (H. Bisswanger and Schmincke-Ott, eds.), pp. 151–173. John Wiley and Sons, New York, NY.

426. Crawford, J. E. (1967). New data on linkage group III markers in *Neurospora crassa*. *Aust. J. Biol. Sci.* **20**: 121–125.

427. Crawford, J. M., R. F. Geever, D. K. Asch, and M. E. Case (1992). Sequence and characterization of the *met-7* gene of *Neurospora crassa*. *Gene* **111**: 265–266.

428. Creaser, E., and M. Gardiner (1969). Proteins immunologically related to *Neurospora* histidinol dehydrogenase. *J. Gen. Microbiol.* **55**: 417–423.

429. Creighton, M. O., and J. R. Trevithick (1973). Method for freezing slime. *Neurospora Newsl.* **20**: 32.

430. Crocken, B. J., and J. F. Nyc (1964). Phospholipid variations in mutant strains of *Neurospora crassa*. *J. Biol. Chem.* **239**: 1727–1730.

431. Crosthwaite, S. K., J. C. Dunlap, and J. J. Loros (1997). *Neurospora wc-1* and *wc-2*: Transcription, photoresponses, and the origins of circadian rhythmicity. *Science* **276**: 763–769.

432. Crotti, L. B., H. F. Terenzi, J. A. Jorge, M. de Lourdes, and M. L. Polizeli (1998). Regulation of pectic enzymes from the *exo-1* mutant strain of *Neurospora crassa*: Effects of glucose, galactose, and galacturonic acid. *J. Basic Microbiol.* **38**: 181–188.

433. Crotti, L. B., H. F. Terenzi, J. A. Jorge, and M. L. Polizeli (1998). Characterization of galactose-induced extracellular and intracellular pectolytic activities from the *exo-1* mutant strain of *Neurospora crassa*. *J. Indust. Microbiol. Biotechnol.* **20**: 238–243.

434. Cruz, A. K., H. F. Terenzi, and J. A. Jorge (1988). Cyclic AMP-dependent, constitutive thermotolerance in the adenylate cyclase-deficient *cr-1* (crisp) mutant of *Neurospora crassa*. *Curr. Genet.* **13**: 451–454.

435. Cujec, T. P., and B. M. Tyler (1996). Nutritional and growth control of ribosomal protein mRNA and rRNA in *Neurospora crassa*. *Nucleic. Acids Res.* **24**: 943–950.

436. Cybis, J., and R. H. Davis (1975). Organization and control in the arginine biosynthetic pathway of *Neurospora*. *J. Bacteriol.* **123**: 196–202.

437. Cyrklaff, M., M. Auer, W. Kuhlbrandt, and G. A. Scarborough (1995). 2D structure of the *Neurospora crassa* plasma membrane ATPase as determined by electron cryomicroscopy. *EMBO J.* **14**: 1854–1857.

438. da Silva, M. V., P. C. Alves, M. Duarte, N. Mota, A. Lobo da Cunha, T. A. Harkness, F. E. Nargang, and A. Videira (1996). Disruption of the nuclear gene encoding the 20.8-kDa subunit of NADH: Ubiquinone reductase of *Neurospora* mitochondria. *Mol. Gen. Genet.* **252**: 177–183.

439. Daboussi, M. J. (1997). Fungal transposable elements and genome evolution. *Genetica* **100**: 253–260.

440. Dalke, A. P. (1980). The transport and utilization of purine and pyrimidine nucleosides and bases by germinating conidia of *Neurospora crassa*. Ph.D. Thesis, Texas A&M University, College Station, TX.

441. Dalke, P., and J. M. Magill (1979). Specificity of uracil uptake in *Neurospora crassa*. *J. Bacteriol.* **139**: 212–219.

442. Dantzig, A. H., F. L. Wiegmann, Jr., and A. Nason (1979). Regulation of glutamate dehydrogenases in *nit-2* and *am* mutants of *Neurospora crassa*. *J. Bacteriol.* **137**: 1333–1339.

443. Dantzig, A. H., W. K. Zurowski, T. M. Ball, and A. Nason (1978). Induction and repression of nitrate reductase in *Neurospora crassa*. *J. Bacteriol.* **133**: 671–679.

444. Davenport, J. W., and C. W. Slayman (1988). The plasma membrane H+-ATPase of *Neurospora crassa*. Properties of two reactive sulfhydryl groups. *J. Biol. Chem.* **263**: 16007–16013.

445. David, M., and U. L. RajBhandary, Personal communication.

446. Dávila, G., S. Brom, Y. Mora, R. Palacios, and J. Mora (1983). Genetic and biochemical characterization of

glutamine synthetase from *Neurospora crassa* glutamine auxotrophs and their revertants. *J. Bacteriol.* **156**: 993–1000.

447. Dávila, G., F. Sanchez, R. Palacios, and J. Mora (1978). Genetics and physiology of *Neurospora crassa* glutamine auxotrophs. *J. Bacteriol.* **134**: 693–698.

448. Davis, C. R., R. R. Kempainen, M. S. Srodes, and C. R. McClung (1994). Correlation of the physical and genetic maps of the centromeric region of the right arm of linkage group III of *Neurospora crassa*. *Genetics* **136**: 1297–1306.

449. Davis, C. R., M. A. McPeek, and C. R. McClung (1995). Molecular characterization of the *proline-1* (*pro-1*) locus of *Neurospora crassa*, which encodes Δ^1-pyrroline-5-carboxylate reductase. *Mol. Gen. Genet.* **248**: 341–350.

450. Davis, M. A., and M. J. Hynes (1987). Complementation of *areA⁻* regulatory gene mutations of *Aspergillus nidulans* by the heterologous regulatory gene *nit-2* of *Neurospora crassa*. *Proc. Natl. Acad. Sci. USA* **84**: 3753–3757.

451. Davis, R. H., Personal communication.

452. Davis, R. H. (1962). A mutant form of ornithine transcarbamylase found in a strain of *Neurospora* carrying a pyrimidine–proline suppressor gene. *Arch. Biochem. Biophys.* **97**: 185–191.

453. Davis, R. H. (1968). Utilization of exogenous and endogenous ornithine by *Neurospora crassa*. *J. Bacteriol.* **96**: 389–395.

454. Davis, R. H. (1970). Sources of urea in *Neurospora*. *Biochim. Biophys. Acta* **215**: 412–414.

455. Davis, R. H. (1972). Metabolite distribution in cells. *Science* **178**: 835–840.

456. Davis, R. H. (1975). Compartmentation and regulation of fungal metabolism: Genetic approaches. *Annu. Rev. Genet.* **9**: 39–65.

457. Davis, R. H. (1979). The genetics of arginine biosynthesis in *Neurospora crassa*. *Genetics* **93**: 557–575.

458. Davis, R. H. (1986). Compartmental and regulatory mechanisms in the arginine pathways of *Neurospora crassa* and *Saccharomyces cerevisiae*. *Microbiol. Rev.* **50**: 280–313.

459. Davis, R. H. (1995). Genetics of *Neurospora*. *In* "The Mycota: Genetics and Biotechnology" (U. Kück, ed.), Vol. II, pp. 3–18. Springer-Verlag, Berlin.

460. Davis, R. H. (1996). The fungal genetic system: A historical overview. *J. Genet.* **75**: 245–253.

461. Davis, R. H. (2000). "Neurospora: Contributions of a Model Organism." Oxford University Press, Oxford, UK.

462. Davis, R. H., and F. J. de Serres (1970). Genetic and microbiological research techniques for *Neurospora crassa*. *Methods Enzymol.* **27A**: 79–143.

463. Davis, R. H., M. B. Lawless, and L. A. Port (1970). Arginaseless *Neurospora*: Genetics, physiology, and polyamine synthesis. *J. Bacteriol.* **102**: 299–305.

464. Davis, R. H., P. Lieu, and J. L. Ristow (1994). *Neurospora* mutants affecting polyamine-dependent processes and basic amino acid transport mutants resistant to the polyamine inhibitor, α-difluoromethylornithine. *Genetics* **138**: 649–655.

465. Davis, R. H., and J. Mora (1968). Mutants of *Neurospora crassa* deficient in ornithine-δ-transaminase. *J. Bacteriol.* **96**: 383–388.

466. Davis, R. H., and T. J. Paulus (1983). Uses of arginaseless cells in the study of polyamine metabolism (*Neurospora crassa*). *Methods Enzymol.* **94**: 112–117.

467. Davis, R. H., and J. L. Ristow (1983). Control of the ornithine cycle in *Neurospora crassa* by the mitochondrial membrane. *J. Bacteriol.* **154**: 1046–1053.

468. Davis, R. H., and J. L. Ristow (1987). Arginine-specific carbamoyl phosphate metabolism in mitochondria of *Neurospora crassa*. *J. Biol. Chem.* **262**: 7109–7117.

469. Davis, R. H., and J. L. Ristow (1991). Polyamine toxicity in *Neurospora crassa*: Protective role of the vacuole. *Arch. Biochem. Biophys.* **285**: 306–311.

470. Davis, R. H., and J. L. Ristow (1995). Osmotic effects on the polyamine pathway of *Neurospora crassa*. *Exp. Mycol.* **19**: 314–319.

471. Davis, R. H., J. L. Ristow, and C. L. Ginsburgh (1981). Independent localization and regulation of carbamyl phosphate synthetase A polypeptides of *Neurospora crassa*. *Mol. Gen. Genet.* **181**: 215–221.

472. Davis, R. H., J. L. Ristow, A. D. Howard, and G. R. Barnett (1991). Calcium modulation of polyamine transport is lost in a putrescine-sensitive mutant of *Neurospora crassa*. *Arch. Biochem. Biophys.* **285**: 297–305.

473. Davis, R. H., and W. M. Thwaites (1963). Structural gene for ornithine transcarbamylase in *Neurospora*. *Genetics* **48**: 1551–1558.

474. Davis, R. H., and R. L. Weiss (1983). Identification of nonsense mutants in *Neurospora*: Application to the complex *arg-6* locus. *Mol. Gen. Genet.* **192**: 46–50.

475. Davis, R. H., and R. L. Weiss (1988). Novel mechanisms controlling arginine metabolism in *Neurospora*. *Trends Biochem. Sci.* **13**: 101–104.

476. Davis, R. H., and V. V. Woodward (1962). The relationship between gene suppression and aspartate transcarbamylase activity in *pyr-3* mutants of *Neurospora*. *Genetics* **47**: 1075–1083.

477. Davis, R. H., and J. D. Zimmerman (1965). A mutation of *Neurospora* affecting the assimilation of exogenous metabolites. *Genetics* **52s**: 439.

478. Davis, R. L. (1996). Polyamines in fungi. *In* "The Mycota: Biochemistry and Molecular Biology" (R. Brambl and G. A. Marzluf, eds.), Vol. III, pp. 347–356. Springer-Verlag, Heidelberg.

479. De Fabo, E. C., R. W. Harding, and W. Shropshire, Jr. (1976). Action spectrum between 260 and 800 nm for the photoinduction of carotenoid biosynthesis in *Neurospora crassa*. *Plant Physiol.* **57**: 440–445.

480. de la Serna, I., D. Ng, and B. M. Tyler (1999). Carbon regulation of ribosomal genes in *Neurospora crassa* occurs by a mechanism which does not require *cre-1*, the homologue of the *Aspergillus* carbon catabolite repressor, *creA*. *Fungal Genet. Biol.* **26**: 253–269.

481. de la Serna, I. L., T. P. Cujec, Y. Shi, and B. M. Tyler (2000). Non-coordinate regulation of 5S rRNA genes and the gene encoding the 5S rRNA-binding ribosomal protein homolog in *Neurospora crassa*. *Mol. Gen. Genet.* In Press.

482. de Moss, J. A. (1962). Studies on the mechanism of the tryptophan synthetase reaction. *Biochim. Biophys. Acta* **2**: 279–293.

483. de Moss, J. A. (1965). The conversion of shikimic acid to anthranilic acid by extracts of *Neurospora crassa. J. Biol. Chem.* **240**: 1231–1235.

484. de Moss, J. A., and J. Wegman (1965). An enzyme aggregate in the tryptophan pathway of *Neurospora crassa. Proc. Natl. Acad. Sci. USA* **54**: 241–247.

485. de Serres, F. (1957). A genetic analysis of an insertional translocation involving the *ad-3* region in *Neurospora crassa. Genetics* **42**: 366–367 (Abstr.).

486. de Serres, F. J., Personal communication.

487. de Serres, F. J., Jr. (1956). Studies with purple adenine mutants in *Neurospora crassa*. I. Structural and functional complexity in the *ad-3* region. *Genetics* **41**: 668–676.

488. de Serres, F. J. (1962). Heterocaryon incompatibility factor interaction in tests between *Neurospora* mutations. *Science* **138**: 1342–1343.

489. de Serres, F. J. (1963). Studies with purple adenine mutants in *Neurospora crassa*. V. Evidence for allelic complementation among *ad-3B* mutants. *Genetics* **48**: 351–360.

490. de Serres, F. J. (1964). Mutagenesis and chromosome structure. *J. Cell. Comp. Physiol.* **64 (Suppl. 1)**: 33–42.

491. de Serres, F. J. (1966). Carbon dioxide stimulation of the *ad-3* mutants of *Neurospora crassa. Mutat. Res.* **3**: 420–425.

492. de Serres, F. J. (1969). Comparison of the complementation and genetic maps of closely linked nonallelic markers on linkage group I of *Neurospora crassa. Mutat. Res.* **8**: 43–50.

493. de Serres, F. J. (1971). Studies with purple adenine mutants in *Neurospora crassa*. VI. The effects of differences in genetic background on *ad-3A* × *ad-3B* crosses. *Genetics* **68**: 383–400.

494. de Serres, F. J. (1980). Mutagenesis at the *ad-3A* and *ad-3B* loci in haploid UV-sensitive strains of *Neurospora crassa*. II. Comparison of dose–response curves for inactivation and mutation induced by UV. *Mutat. Res.* **71**: 181–191.

495. de Serres, F. J. (1989). X-ray-induced specific locus mutation in the *ad-3* region of two-component heterokaryons of *Neurospora crassa*. II. More extensive genetic tests reveal an unexpectedly high frequency of multiple-locus mutations. *Mutat. Res.* **210**: 281–290.

496. de Serres, F. J. (1991). Utilization of the specific-locus assay in the *ad-3* region of two-component heterokaryons of *Neurospora* for risk assessment of environmental chemicals. *Mutat. Res.* **250**: 251–274.

497. de Serres, F. J. (1992). Development of a specific-locus assay in the *ad-3* region of two-component heterokaryons of *Neurospora*: A review. *Environ. Mol. Mutagen.* **20**: 225–245.

498. de Serres, F. J., and H. Brockman (1986). Genetic characterization of the mutagenic activity of environmental chemicals at specific loci in two-component heterokaryons of *Neurospora crassa*. *In* "Progress in Clinical and Biological Research, Vol. 209A, Genetic Toxicology of Environmental Chemicals: Basic Principles and Mechanisms of Action" (C. Ramel, B. Lambert, and J. Magnusson eds.), pp. 209–218. Alan R. Liss, Inc, New York.

499. de Serres, F. J., and H. E. Brockman (1991). Qualitative differences in the spectra of genetic damage in 2-aminopurine-induced *ad-3* mutants between nucleotide excision repair proficient and deficient strains of *Neurospora crassa. Mutat. Res.* **251**: 41–58.

500. de Serres, F. J., and H. E. Brockman (1993). Comparison of the spectra of genetic damage in N^4-hydroxycytidine-induced *ad-3* mutations between nucleotide excision repair-proficient and -deficient heterokaryons of *Neurospora crassa. Mutat. Res.* **285**: 145–163.

501. de Serres, F. J., H. E. Brockman, W. E. Barnett, and H. G. Kølmark (1967). Allelic complementation among nitrous acid-induced *ad-3B* mutants of *Neurospora crassa. Mutat. Res.* **4**: 415–424.

502. de Serres, F. J., H. Inoue, and M. E. Schüpbach (1980). Mutagenesis at the *ad-3A* and *ad-3B* loci in haploid UV-sensitive strains of *Neurospora crassa*. I. Development of isogenic strains and spontaneous mutability. *Mutat. Res.* **71**: 53–65.

503. de Serres, F. J., and H. G. Kølmark (1958). A direct method for determination of forward mutation rates in *Neurospora crassa. Nature* **182**: 1249–1250.

504. de Serres, F. J., and H. V. Malling (1971). Measurement of recessive lethal damage over the entire genome and at two specific loci in the *ad-3* region of a two-component heterokaryon of *Neurospora crassa. In* "Chemical Mutagens: Principles and Methods for Their Detection" (A. Hollaender ed.), Vol. 2, pp. 311–342. Plenum Press, New York.

505. de Serres, F. J., and H. V. Malling (1994). Forward-mutation tests on the antitumor agent ICR-170 in *Neurospora crassa* demonstrate that it induces gene/point mutations in the *ad-3* region and an exceptionally high frequency of multiple-locus *ad-3* mutations with closely linked sites of recessive lethal damage. *Mutat. Res.* **310**: 15–36.

506. de Serres, F. J., H. V. Malling, H. E. Brockman, and T. M. Ong (1997). Quantitative and qualitative comparison of spontaneous and chemically induced specific-locus mutation in the *ad-3* region of heterokaryon 12 of *Neurospora crassa. Mutat. Res.* **37S**: 53–72.

507. De Terra, N., and E. L. Tatum (1963). A relationship between cell-wall structure and colonial growth in *Neurospora crassa. Am. J. Bot.* **50**: 669–677.

508. Debets, A. J. M., and A. J. F. Griffiths (1998). Polymorphism of *het* genes prevents resource plundering in *Neurospora crassa. Mycol. Res.* **102**: 1343–1349.

509. Debets, F., X. Yang, and A. J. Griffiths (1994). Vegetative incompatibility in *Neurospora*: Its effect on horizontal transfer of mitochondrial plasmids and senescence in natural populations. *Curr. Genet.* **26**: 113–119.

510. DeBusk, A. G., Personal communication.

511. DeBusk, R. M., Personal communication.

512. DeBusk, R. M., D. T. Brown, A. G. DeBusk, and R. D. Penderghast (1981). Alternate mechanism for amino acid entry into *Neurospora crassa*: Extracellular deamination and subsequent keto acid transport. *J. Bacteriol.* **146**: 163–169.

513. DeBusk, R. M., and A. G. DeBusk (1980). Physiological and regulatory properties of the general amino acid transport system of *Neurospora crassa. J. Bacteriol.* **143**: 188–197.

514. DeDeken, R. H. (1963). Biosynthèse de l'arginine chez la levure. I. Le sort de la N-α-acétylornithine. *Biochim. Biophys. Acta* **78**: 606–616.

515. Degli-Innocenti, F., and V. E. A. Russo (1984). Isolation of new *white collar* mutants of *Neurospora crassa* and studies on their behavior in the blue light induced formation of protoperithecia. *J. Bacteriol.* **159**: 757–761.

516. DeLange, A. M. (1981). The mutation *SK(ad-3A)* cancels the dominance of *ad-3A⁺* over *ad-3A* in the ascus of *Neurospora. Genetics* **97**: 237–246.

517. DeLange, A. M., and A. J. Griffiths (1975). Escape from mating-type incompatibility in bisexual (*A + a*) *Neurospora* heterokaryons. *Can. J. Genet. Cytol.* **17**: 441–449.

518. DeLange, A. M., and A. J. Griffiths (1980). Meiosis in *Neurospora crassa*. I. The isolation of recessive mutants defective in the production of viable ascospores. *Genetics* **96**: 367–378.

519. DeLange, A. M., and A. J. Griffiths (1980). Meiosis in *Neurospora crassa*. II. Genetic and cytological characterization of three meiotic mutants. *Genetics* **96**: 379–398.

520. DeLange, A. M., and N. C. Mishra (1981). The isolation of MMS- and histidine-sensitive mutants in *Neurospora crassa. Genetics* **97**: 247–259.

521. DeLange, A. M., and N. C. Mishra (1982). Characterization of MMS-sensitive mutants of *Neurospora crassa. Mutat. Res.* **96**: 187–199.

522. Demain, A. L. (1972). Riboflavin oversynthesis. *Annu. Rev. Microbiol.* **26**: 369–388.

523. DeMoss, J. A., R. W. Jackson, and J. H. Chalmers, Jr. (1967). Genetic control of the structure and activity of an enzyme aggregate in the tryptophan pathway of *Neurospora crassa. Genetics* **56**: 413–424.

524. d'Enfert, C., B. M. Bonini, P. D. Zapella, T. Fontaine, A. M. da Silva, and H. F. Terenzi (1999). Neutral trehalases catalyse intracellular trehalose breakdown in the filamentous fungi *Aspergillus nidulans* and *Neurospora crassa. Mol. Microbiol.* **32**: 471–483.

525. Denor, P. F. (1979). Genetic and enzymatic analysis of the *glp* system in *Neurospora crassa. Genetics* **91s**: 26.

526. Denor, P. F. (1981). Studies on the inducible glycerol dissimilatory system of *Neurospora crassa*. Ph.D. Thesis, Marquette University, Milwaukee, WI.

527. Denor, P. F., and J. B. Courtright (1978). Isolation and characterization of glycerol-3-phosphate dehydrogenase-defective mutants of *Neurospora crassa. J. Bacteriol.* **136**: 960–968.

528. Deutsch, A., A. Dress, and L. Rensing (1993). Formation of morphological differentiation patterns in the ascomycete *Neurospora crassa. Mech. Dev.* **44**: 17–31.

529. Dharmananda, S., and J. F. Feldman (1979). Spatial distribution of circadian clock phase in aging cultures of *Neurospora crassa. Plant Physiol. (Bethesda)* **63**: 1049–1054.

530. Dicker, J. W., A. G. DeBusk, and G. Turian (1970). A nutritional method for the isolation of morphological mutants of *Neurospora crassa. Experientia* **26**: 1154–1155.

531. Dicker, J. W., N. Oulevey, and G. Turian (1969). Amino acid induction of conidiation and morphological alterations in wild type and morphological mutants of *Neurospora crassa. Arch. Mikrobiol.* **65**: 241–257.

532. Dicker, J. W., R. Peduzzi, and G. Turian (1975). Recherches sur la différenciation conidienne de *Neurospora crassa*. VI. Anomalies fonctionnelles du mutant aconidien conditionnel *"amycelial." Ann. Microbiol. (Paris)* **126A**: 409–420.

533. Dicker, J. W., and G. Turian (1990). Calcium deficiencies and apical hyperbranching in wild type and the "frost" and "spray" morphological mutants of *Neurospora crassa. J. Gen. Microbiol.* **136**: 1413–1420.

534. Dickman, M. B., and J. F. Leslie (1992). The regulatory gene *nit-2* of *Neurospora crassa* complements a *nnu* mutant of *Gibberella zeae* (*Fusarium graminearum*). *Mol. Gen. Genet.* **235**: 458–462.

535. Dickman, M. B., and O. Yarden (1999). Serine/threonine protein kinases and phosphatases in filamentous fungi. *Fungal Genet. Biol.* **26**: 99–117.

536. Dieckmann, C., and S. Brody (1980). Circadian rhythms in *Neurospora crassa*: Oligomycin-resistant mutations affect periodicity. *Science* **207**: 896–898.

537. Dieckmann, C. L. (1980). Circadian rhythms in *Neurospora crassa*: A biochemical and genetic study of the involvement of mitochondrial metabolism in periodicity. Ph.D. Thesis, University of California at San Diego, San Diego, CA.

538. Dietrich, P. S., Personal communication.

539. Dietrich, P. S., and R. L. Metzenberg (1973). Metabolic suppressors of a regulatory mutant in *Neurospora. Biochem. Genet.* **8**: 73–84.

540. DiGangi, J. J., M. Seyfzadeh, and R. H. Davis (1987). Ornithine decarboxylase from *Neurospora crassa*. Purification, characterization, and regulation by inactivation. *J. Biol. Chem.* **262**: 7889–7893.

541. Dillon, D., and D. Stadler (1994). Spontaneous mutation at the *mtr* locus in *Neurospora*: The molecular spectrum in wild type and a mutator strain. *Genetics* **138**: 61–74.

542. Din, A. B., C. A. Specht, P. W. Robbins, and O. Yarden (1996). *chs-4*, a class IV chitin synthase gene from *Neurospora crassa. Mol. Gen. Genet.* **250**: 214–222.

543. Din, A. B., and O. Yarden (1994). The *Neurospora crassa chs-2* gene encodes a nonessential chitin synthase. *Microbiology* **140**: 2189–2197.

544. Dobinson, K. F., M. Henderson, R. L. Kelley, R. A. Collins, and A. M. Lambowitz (1989). Mutations in nuclear gene *cyt-4* of *Neurospora crassa* result in pleiotropic defects in processing and splicing of mitochondrial RNAs. *Genetics* **123**: 97–108.

545. Dobosy, J., and E. U. Selker, Personal communication.

546. Dodge, B. O. (1934). A lethal for ascus abortion in *Neurospora. Mycologia* **26**: 360–376.

547. Dodge, B. O. (1935). A recessive factor lethal for ascospore formation in *Neurospora. Bull. Torrey Bot. Club* **62**: 117–128.

548. Doermann, A. H. (1944). A lysineless mutant of *Neurospora* and its inhibition by arginine. *Arch. Biochem.* **5**: 373–384.

549. Dolan, P., and M. A. Nelson, Personal communication.

549a. Dolan, P. L., D. O. Natvig and M. A. Nelson (2000). *Neurospora* proteome 2000. *Fungal Genet. Newsl.* 47.

550. Drake, B. (1956). Evidence for two loci governing *p*-aminobenzoic acid synthesis in *Neurospora crassa. Genetics* **41s**: 640 (Abstr.).

551. Drygas, M. E., A. M. Lambowitz, and F. E. Nargang (1989). Cloning and analysis of the *Neurospora crassa*

gene for cytochrome *c* heme lyase. *J. Biol. Chem.* **264**: 17897–17906.

552. Duarte, M., J. A. Belo, and A. Videira (1993). Primary structure of the nuclear-encoded 10.5-kDa subunit of complex I from *Neurospora crassa*. *Biochim. Biophys. Acta* **1172**: 327–328.

553. Duarte, M., M. Finel, and A. Videira (1996). Primary structure of a ferredoxin-like iron–sulfur subunit of complex I from *Neurospora crassa*. *Biochim. Biophys. Acta* **1275**: 151–153.

554. Duarte, M., N. Mota, L. Pinto, and A. Videira (1998). Inactivation of the gene coding for the 30.4-kDa subunit of respiratory chain NADH dehydrogenase: Is the enzyme essential for *Neurospora? Mol. Gen. Genet.* **257**: 368–375.

555. Duarte, M., R. Sousa, and A. Videira (1995). Inactivation of genes encoding subunits of the peripheral and membrane arms of *Neurospora* mitochondrial complex I and effects on enzyme assembly. *Genetics* **139**: 1211–1221.

556. Dubes, G. R. (1953). Investigations of some "unknown" mutants of *Neurospora crassa*. Ph.D. Thesis, California Institute of Technology, Pasadena, CA.

557. Dubins, J. S., L. K. Overton, R. R. Cobb, and F. J. de Serres (1989). Classical and molecular genetic analyses of *his-3* mutants of *Neurospora crassa*. II. Southern blot analyses and molecular mechanisms of mutagenicity. *Mutat. Res.* **215**: 39–47.

558. Dunlap, J. (1999). Molecular bases for circadian clocks. *Cell* **96**: 271–290.

559. Dunlap, J. C. (1990). Closely watched clocks: Molecular analysis of circadian rhythms in *Neurospora* and *Drosophila*. *Trends Genet.* **6**: 159–165.

560. Dunlap, J. C. (1993). Genetic analysis of circadian clocks. *Annu. Rev. Physiol.* **55**: 683–628.

561. Dunlap, J. C. (1996). Genetic and molecular analysis of circadian rhythms. *Annu. Rev. Genet.* **30**: 579–601.

562. Dunlap, J. C. (1998). Common threads in eukaryotic circadian systems. *Curr. Opin. Genet. Dev.* **8**: 400–406.

563. Dunn-Coleman, N. S. (1984). Cloning and preliminary characterization of a molybdenum cofactor gene of *Neurospora crassa*. *Curr. Genet.* **8**: 589–595.

564. Dunn-Coleman, N. S., and R. H. Garrett (1980). The role of glutamine synthetase and glutamine metabolism in nitrogen metabolite repression, a regulatory phenomenon in the lower eukaryote *Neurospora crassa*. *Mol. Gen. Genet.* **179**: 25–32.

565. Dunn-Coleman, N. S., and R. H. Garrett (1981). Effect of the *gln-1b* mutation on nitrogen metabolite repression in *Neurospora crassa*. *J. Bacteriol.* **145**: 884–888.

566. Dunn-Coleman, N. S., E. A. Robey, A. B. Tomsett, and R. H. Garrett (1981). Glutamate synthase levels in *Neurospora crassa* mutants altered with respect to nitrogen metabolism. *Mol. Cell. Biol.* **1**: 158–164.

567. Dunn-Coleman, N. S., A. B. Tomsett, and R. H. Garrett (1979). Nitrogen metabolite repression of nitrate reductase in *Neurospora crassa*: Effect of the *gln-1a* locus. *J. Bacteriol.* **139**: 697–700.

568. Dunn-Coleman, N. S., A. B. Tomsett, and R. H. Garrett (1981). The regulation of nitrate assimilation in *Neurospora crassa*: Biochemical analysis of the *nmr-1* mutants. *Mol. Gen. Genet.* **182**: 234–239.

569. Durkee, T. L., A. S. Sussman, and R. J. Lowry (1966). Genetic localization of the *clock* mutant and a gene modifying its band-size in *Neurospora*. *Genetics* **53**: 1167–1175.

570. Durrenberger, F., and J. Kronstad (1999). The *ukc1* gene encodes a protein kinase involved in morphogenesis, pathogenicity and pigment formation in *Ustilago maydis*. *Mol. Gen. Genet.* **261**: 281–289.

570a. Dvorachek, W. H., and D. O. Natvig, Personal communication.

571. Dvorachek, W. H., and D. O. Natvig (1995). Manganese superoxide dismutase activity in *Neurospora crassa*. (p. 61.) *18th Fungal Genet. Conf.* Asilomar, CA.

572. Ebbole, D., Personal communication; see *http://www.seqnet.dl.ac.uk/research/fgsc/walkdata/walkdata2.html*.

573. Ebbole, D. J., Personal communication.

574. Ebbole, D. J. (1990). Vectors for construction of translocational fusions to β-galactosidase. *Fungal Genet. Newsl.* **37**: 15–16.

575. Ebbole, D. J. (1998). Carbon catabolite repression of gene expression and conidiation in *Neurospora crassa*. *Fungal Genet. Biol.* **25**: 15–21.

576. Ebbole, D. J., and L. Madi (1996). Identification of a gene from *Neurospora crassa* with similarity to a glucoamylase gene from *Schwanniomyces occidentalis*. *Fungal Genet. Newsl.* **43**: 23–24.

577. Ebbole, D. J., J. L. Paluh, M. Plamann, M. S. Sachs, and C. Yanofsky (1991). *cpc-1*, the general regulatory gene for genes of amino acid biosynthesis in *Neurospora crassa*, is differentially expressed during the asexual life cycle. *Mol. Cell. Biol.* **11**: 928–934.

578. Eberhart, B. (1980). A *Neurospora* mutant resistant to 2-deoxy-D-glucose. *Neurospora Newsl.* **27**: 19–20.

579. Eberhart, B. M., Personal communication.

580. Eberhart, B. M. (1962). Methods for screening for glucosidaseless mutant strains of *Neurospora*. *Microb. Genet. Bull.* **18**: 27–28.

581. Eberhart, B. M., and R. S. Beck (1970). Localization of the β-glucosidases in *Neurospora crassa*. *J. Bacteriol.* **101**: 408–417.

582. Eberhart, B. M., D. F. Cross, and L. R. Chase (1964). β-Glucosidase system of *Neurospora crassa*. I. β-glucosidase and cellulase activities of mutant and wild-type strains. *J. Bacteriol.* **87**: 761–770.

583. Eberhart, B. M., and E. L. Tatum (1959). A gene modifying the thiamine requirement of strains of *Neurospora crassa*. *J. Gen. Microbiol.* **20**: 43–53.

584. Eberhart, B. M., and E. L. Tatum (1961). Thiamine metabolism in *Neurospora crassa*. *Am. J. Bot.* **48**: 702–711.

585. Eberhart, B. M., and E. L. Tatum (1963). Thiamine metabolism in wild-type and mutant strains of *Neurospora crassa*. *Arch. Biochem. Biophys.* **101**: 378–387.

586. Eberle, J., and V. E. A. Russo (1992). *Neurospora crassa* blue-light-inducible gene *bli-7* encodes a short hydrophobic protein. *J. DNA Sequencing Mapping* **3**: 131–141.

587. Edson, C. M., and S. Brody (1976). Biochemical and genetic studies on galactosamine metabolism in *Neurospora crassa*. *J. Bacteriol.* **126**: 799–806.

588. Edwards, D. L. (1978). Cyanide-insensitive respiratory systems in *Neurospora*. In "Functions of Alternative Terminal Oxidases" (H. Degn, D. Lloyd, and G. Hill, eds.), pp. 21–29. Pergamon Press Ltd., Oxford, UK.

589. Edwards, D. L., and D. M. Belsole (1980). Deficiency of a HiPIP center in a succinate dehydrogenase mutant of *Neurospora crassa. Fed. Proc.* **39**: 2013 (Abstr.).

590. Edwards, D. L., D. M. Belsole, H. J. Guzik, and B. W. Unger (1979). Selection of succinic dehydrogenase mutants of *Neurospora crassa. J. Bacteriol.* **137**: 900–904.

591. Edwards, D. L., J. Chalmers, H. J. Guzik, and J. T. Warden (1976). Assembly of the cyanide-insensitive respiratory pathway in *Neurospora crassa. In* "Genetics and Biogenesis of Chloroplasts and Mitochondria" (T. E. A. Bücher, ed.), pp. 865–872. North Holland Publishing Co., Amsterdam.

592. Edwards, D. L., and F. Kwiecinski (1973). Altered mitochondrial respiration in a chromosomal mutant of *Neurospora crassa. J. Bacteriol.* **116**: 610–618.

593. Edwards, D. L., F. Kwiecinski, and J. Horstmann (1973). Selection of respiratory mutants of *Neurospora crassa. J. Bacteriol.* **114**: 164–168.

594. Edwards, D. L., and B. Unger (1980). Defective mitochondrial energy transduction in an oligomycin-resistant mutant of *Neurospora crassa. Biochem. Int.* **1**: 262–269.

595. Edwards, D. L., and B. W. Unger (1978). Cyanide- and hydroxamate-resistant respiration in *Neurospora crassa. J. Bacteriol.* **133**: 1130–1134.

596. Edwards, D. L., and B. W. Unger (1978). Nuclear mutations conferring oligomycin resistance in *Neurospora crassa. J. Biol. Chem.* **253**: 4254–4258.

597. Egashira, T., and K. Nakamura (1972). Genes influencing selective fertilization in *Neurospora crassa. Genetics* **70**: 511–519.

598. Eggerding, C., J. A. Randall, and M. L. Sargent (1975). An altered invertase in the *cot-2* mutant of *Neurospora crassa. J. Gen. Microbiol.* **89**: 102–112.

599. Ehrenshaft, M., P. Bilski, M. Y. Li, C. F. Chignell, and M. E. Daub (1999). A highly conserved sequence is a novel gene involved in *de novo* vitamin B_6 biosynthesis. *Proc. Natl. Acad. Sci. USA* **96**: 9374–9378.

600. Eker, A. P. M. (1994). DNA photolyase from the fungus *Neurospora crassa*. Purification, characterization and comparison with other photolyases. *Photochem. Photobiol.* **60**: 125–133.

601. el-Eryani, A. A. (1969). Genetic control of phenylalanine and tyrosine biosynthesis in *Neurospora crassa. Genetics* **62**: 711–723.

602. Elgren, T. E., and D. E. Wilcox (1989). A unique low-frequency Raman band associated with metal binding to metallothionein. *Biochem Biophys. Res. Commun.* **163**: 1093–1099.

603. Elovson, J. (1975). Purification and properties of the fatty acids synthetase complex from *Neurospora crassa*, and the nature of the *fas* mutation. *J. Bacteriol.* **124**: 524–533.

604. Emerson, M. R., and S. Emerson (1951). The sulfonamide-requiring character (*sfo*) as a marker for chromosome VII of *Neurospora crassa. Microb. Genet. Bull.* **4**: 7–9.

605. Emerson, S. (1947). Growth responses of a sulfonamide-requiring mutant strain of *Neurospora. J. Bacteriol.* **54**: 195–207.

606. Emerson, S. (1950). Competitive reactions and antagonisms in the biosynthesis of amino acids by *Neurospora. Cold Spring Harbor Symp. Quant. Biol.* **14**: 40–48.

607. Emerson, S. (1963). Slime a plasmodoid variant in *Neurospora crassa. Genetica* **34**: 162–182.

608. Emerson, S., and J. E. Cushing (1946). Altered sulfonamide antagonism in *Neurospora. Fed. Proc.* **5**: 379–389.

609. Emerson, S., and M. R. Emerson (1958). Production, reproduction and reversion of protoplast-like structures in the *osmotic* strain of *Neurospora crassa. Proc. Natl. Acad. Sci. USA* **44**: 668–671.

610. Evans, J., Personal communication.

611. Eversole, P., J. J. DiGangi, T. Menees, and R. H. Davis (1985). Structural gene for ornithine decarboxylase in *Neurospora crassa. Mol. Cell. Biol.* **5**: 1301–1306.

612. Exley, G. E., J. D. Colandene, and R. H. Garrett (1993). Molecular cloning, characterization, and nucleotide sequence of *nit-6*, the structural gene for nitrite reductase in *Neurospora crassa. J. Bacteriol.* **175**: 2379–2392.

613. Facklam, T. J., and G. A. Marzluf (1978). Nitrogen regulation of amino acid catabolism in *Neurospora crassa. Biochem. Genet.* **16**: 343–354.

614. Fan, G. Y., F. Maldonado, Y. Zhang, R. Kincaid, M. H. Ellisman, and L. N. Gastinel (1996). *In vivo* calcineurin crystals formed using the baculovirus expression system. *Microsc. Res. Tech.* **34**: 77–86.

615. Fass, D. N. (1969). Glucoamylase of *Neurospora*: A regulated exoenzyme. Ph.D. Thesis, Florida State University, Tallahassee, FL.

616. Fecke, W., V. D. Sled, T. Ohnishi, and H. Weiss (1994). Disruption of the gene encoding the NADH-binding subunit of NADH: ubiquinone oxidoreductase in *Neurospora crassa*. Formation of a partially assembled enzyme without FMN and the iron–sulphur cluster N-3. *Eur. J. Biochem.* **220**: 551–558.

617. Feldman, J. F. (1982). Genetic approaches to circadian clocks. *Annu. Rev. Plant Physiol.* **33**: 583–608.

618. Feldman, J. F. (1983). Genetics of circadian clocks. *BioScience* **33**: 426–431.

619. Feldman, J. F., and C. A. Atkinson (1978). Genetic and physiological characteristics of a slow-growing circadian clock mutant of *Neurospora crassa. Genetics* **88**: 255–265.

620. Feldman, J. F., and J. C. Dunlap (1983). *Neurospora crassa*: A unique system for studying circadian rhythms. *In* "Photochemical and Photobiological Reviews" (C. Kendrick, ed.), Vol. 7. Plenum Press, New York.

621. Feldman, J. F., G. Gardner, and R. Denison (1979). Genetic analysis of the circadian clock of *Neurospora. In* "Naito Foundation Symposium on Biological Rhythms and Their Central Mechanism (M. Suda, O. Hayaishi, and H. Nakagawa, eds.), pp. 55–66. North-Holland Publishing Co., Amsterdam.

622. Feldman, J. F., and M. N. Hoyle (1973). Isolation of circadian clock mutants of *Neurospora crassa. Genetics* **75**: 605–613.

623. Feldman, J. F., and J. P. Thayer (1974). Cyclic AMP-induced tyrosinase synthesis in *Neurospora crassa. Biochem. Biophys. Res. Commun.* **61**: 977–982.

624. Feng, B., and G. A. Marzluf (1996). The regulatory protein NIT4 that mediates nitrate induction in *Neurospora crassa* contains a complex tripartite activation domain with a novel leucine-rich, acidic motif. *Curr. Genet.* **29**: 537–548.

625. Feng, B., and G. A. Marzluf (1998). Interaction between major nitrogen regulatory protein NIT2 and pathway-specific regulatory factor NIT4 is required for their synergistic activation of gene expression in *Neurospora crassa*. *Mol. Cell. Biol.* **18**: 3983–3990.

626. Feng, B., X. Xiao, and G. A. Marzluf (1993). Recognition of specific nucleotide bases and cooperative DNA binding by the *trans*-acting nitrogen regulatory protein NIT2 of *Neurospora crassa*. *Nucleic Acids Res.* **21**: 3989–3996.

626a. Feng, B., H. Haas, and G. A. Marzluf (2000). A new GATA factor, ASD, regulates sexual development in *Neurospora crassa*. *Fungal Genet. Newsl.* **47S**: 8 (Abstr.).

627. Ferea, T., E. T. Contreras, T. Oung, E. J. Bowman, and B. J. Bowman (1994). Characterization of the *cit-1* gene from *Neurospora crassa* encoding the mitochondrial form of citrate synthase. *Mol. Gen. Genet.* **242**: 105–110.

628. Ferea, T. L., and B. J. Bowman (1996). The vacuolar ATPase of *Neurospora crassa* is indispensable: Inactivation of the *vma-1* gene by repeat-induced point mutation. *Genetics* **143**: 147–154.

629. Ferreira, A. V., Z. An, R. L. Metzenberg, and N. L. Glass (1998). Characterization of *mat A-2*, *mat A-3* and Δ *mat A* mating-type mutants of *Neurospora crassa*. *Genetics* **148**: 1069–1079.

630. Ferreira, A. V. B., and N. L. Glass (1994). The *Neurospora crassa* idiomorph *A* contains three genes which code for proteins that resemble transcription factors. *Fifth Int. Congr. Mycol.*, Vancouver, BC, Canada, 63 (Abstr.).

631. Ferreira, A. V. B., S. Saupe, and N. L. Glass (1996). Transcriptional analysis of the *mt A* idiomorph of *Neurospora crassa* identifies two genes in addition to *mt A-1*. *Mol. Gen. Genet.* **250**: 767–774.

632. Ferreirinha, F., T. Almeida, M. Duarte, and A. Videira (1998). Chromosomal mapping of genes encoding subunits of complex I from *Neurospora crassa*. *Fungal Genet. Biol.* **45**: 10.

633. Fevre, M., G. Turian, and J.-P. Larpent (1975). Bourgeonnements et croissance hyphale fongiques. Homologies structurales et functionnelles modèles *Neurospora* et *Saprolegnia*. *Physiol. Veg.* **13**: 23–38.

634. Fincham, J. R. (1976). Recombination in the *am* gene of *Neurospora crassa* — A new model for conversion polarity and an explanation for a marker effect. *Heredity* **36**: 81–89.

635. Fincham, J. R., and A. J. Baron (1977). The molecular basis of an osmotically reparable mutant of *Neurospora crassa* producing unstable glutamate dehydrogenase. *J. Mol. Biol.* **110**: 627–642.

636. Fincham, J. R., I. F. Connerton, E. Notarianni, and K. Harrington (1989). Premeiotic disruption of duplicated and triplicated copies of the *Neurospora crassa am* (glutamate dehydrogenase) gene. *Curr. Genet.* **15**: 327–334.

637. Fincham, J. R. S., Personal communication.

638. Fincham, J. R. S. (1951). A comparative genetic study of the mating-type chromosomes of two species of *Neurospora*. *J. Genet.* **50**: 221–229.

639. Fincham, J. R. S. (1962). Genetically determined multiple forms of glutamic dehydrogenase in *Neurospora crassa*. *J. Mol. Biol.* **4**: 257–274.

640. Fincham, J. R. S. (1963). Complementation at the *am* locus of *Neurospora crassa*: A reaction between different mutant forms of glutamic dehydrogenase. *J. Mol. Biol.* **6**: 361–373.

641. Fincham, J. R. S. (1967). Recombination within the *am* gene of *Neurospora crassa*. *Genet. Res.* **9**: 42–62.

642. Fincham, J. R. S. (1981). Further properties of the *i (en-am-1)* mutant. *Neurospora Newsl.* **28**: 11.

643. Fincham, J. R. S. (1990). Generation of new functional mutant alleles by premeiotic disruption of the *Neurospora crassa am* gene. *Curr. Genet.* **18**: 441–445.

644. Fincham, J. R. S., and J. B. Boylen (1957). *Neurospora crassa* mutants lacking argininosuccinase. *J. Gen. Microbiol.* **16**: 438–448.

645. Fincham, J. R. S., P. R. Day, and A. Radford (1979). "Fungal Genetics," 4th ed. Blackwell Scientific Publications, Oxford, UK.

646. Fincham, J. R. S., and J. A. Pateman (1957). Formation of an enzyme through complementary action of mutant "alleles" in separate nuclei in a heterocaryon. *Nature* **179**: 741–742.

647. Fincham, J. R. S., and J. A. Pateman (1957). A new allele at the *am* locus in *Neurospora crassa*. *Genet.* **52**: 829–834.

648. Finck, D., Y. Suyama, and R. H. Davis (1965). Metabolic role of the *pyrimidine-3* locus of *Neurospora*. *Genetics* **52**: 829–834.

649. Fischer, G. A. (1957). The cleavage and synthesis of cystathionine in wild-type and mutant strains of *Neurospora crassa*. *Biochim. Biophys. Acta* **25**: 50–55.

650. Fisher, C. R. (1967). Determination of the enzymatic functions controlled by the *ad-3A* and *ad-3B* loci in *Neurospora crassa*. *Genetics* **56**: 560 (Abstr.).

651. Flavell, R. B., and J. R. Fincham (1968). Acetate-non-utilizing mutants of *Neurospora crassa*. I. Mutant isolation, complementation studies, and linkage relationships. *J. Bacteriol.* **95**: 1056–1062.

652. Flavell, R. B., and J. R. Fincham (1968). Acetate-non-utilizing mutants of *Neurospora crassa*. II. Biochemical deficiencies and the roles of certain enzymes. *J. Bacteriol.* **95**: 1063–1068.

653. Flavin, M. (1975). Methionine biosynthesis. *In* "Metabolic Pathways" (D. Greenburg, ed.), 3rd ed., Vol. 1, Academic Press, New York, pp. 457–503.

654. Flavin, M., and C. Slaughter (1960). Purification and properties of threonine synthetase of *Neurospora*. *J. Biol. Chem.* **235**: 1103–1108.

655. Flavin, M., and C. Slaughter (1964). Cystathionine cleavage enzymes of *Neurospora*. *J. Biol. Chem.* **239**: 2212–2219.

656. Fling, M., N. H. Horowitz, and V. Reinking (1957). Linkage of some methionine mutants of *Neurospora*. *Microb. Genet. Bull* **15**: 12–13.

657. Flint, H. J., D. J. Porteous, and H. Kacser (1980). Control of the flux in the arginine pathway of *Neurospora crassa*. The flux from citrulline to arginine. *Biochem. J.* **190**: 1–15.

658. Folco, D., and A. L. Rosa, Personal communication.

659. Forsthoefel, A. M., and N. C. Mishra (1983). Biochemical genetics of *Neurospora* nuclease I: Isolation and characterization of nuclease (*nuc*) mutants. *Genet. Res.* **41**: 271–286.

660. Foss, H. M., C. J. Roberts, K. M. Claeys, and E. U. Selker (1993). Abnormal chromosome behavior in *Neurospora*

mutants defective in DNA methylation. *Science* **262**: 1737–1741.

661. Foss, H. M., C. J. Roberts, K. M. Claeys, and E. U. Selker (1995). Chromosome behavior in *Neurospora* mutants defective in DNA methylation: Correction. *Science* **267**: 316–317.

662. Foss, H. M., C. J. Roberts, and E. U. Selker (1998). Mutations in the *dim-1* gene of *Neurospora crassa* reduce the level of DNA methylation. *Mol. Gen. Genet.* **259**: 60–71.

663. Foss, H. M., and E. U. Selker (1991). Efficient DNA pairing in a *Neurospora* mutant defective in chromosome pairing. *Mol. Gen. Genet.* **231**: 49–52.

664. Fosså, A., A. Beyer, E. Pfitzner, B. Wenzel, and W. H. Kunau (1995). Molecular cloning, sequencing and sequence analysis of the *fox-2* gene of *Neurospora crassa* encoding the multifunctional β-oxidation protein. *Mol. Gen. Genet.* **247**: 95–104.

665. Franklin, N. C. (1954). Studies on gene function in lactose mutants of *Neurospora crassa*. Ph.D. Thesis, Yale University, New Haven, CT.

666. Fraser, M., T. Chow, and E. Käfer (1980). Nucleases and their control in wild-type and *nuh* mutants of *Neurospora*. *Basic Life Sci.* **15**: 63–74.

667. Fraser, M. J. (1979). Alkaline deoxyribonucleases released from *Neurospora crassa* mycelia: Two activities not released by mutants with multiple sensitivities to mutagens. *Nucleic Acids Res.* **6**: 231–246.

668. Fraser, M. J., H. Koa, and T. Y. Chow (1990). *Neurospora* endo-exonuclease is immunochemically related to the *recC* gene product of *Escherichia coli*. *J. Bacteriol.* **172**: 507–510.

669. Fraser, M. J., and R. L. Low (1993). Fungal and mitochondrial nucleases. *In* "Nucleases" (R. J. Roberts, S. M. Linn, and S. Lloyd, eds.), 2nd ed., pp. 171–207. Cold Spring Harbor Laboratory Press, Cold Spring Harbor, NY.

670. Frederick, G. D., and J. A. Kinsey (1990). Distant upstream regulatory sequences control the level of expression of the *am* (GDH) locus of *Neurospora crassa*. *Curr. Genet.* **18**: 53–58.

671. Frederick, G. D., and J. A. Kinsey (1990). Nucleotide sequence and nuclear protein binding of the two regulatory sequences upstream of the *am* (GDH) gene in *Neurospora*. *Mol. Gen. Genet.* **221**: 148–154.

672. Free, S. J., P. W. Rice, and R. L. Metzenberg (1979). Arrangement of the genes coding for ribosomal ribonucleic acids in *Neurospora crassa*. *J. Bacteriol.* **137**: 1219–1226.

673. Freese, E. (1957). Uber die Feinstruktur des Genoms im Bereich eines PAB Locus von *Neurospora crassa*. *Z. Indukt. Abstammungs-Vererbungsl.* **88**: 388–406.

674. Freitag, D. G., P. M. Ouimet, T. L. Girvitz, and M. Kapoor (1997). Heat shock protein 80 of *Neurospora crassa*, a cytosolic molecular chaperone of the eukaryotic stress 90 family, interacts directly with heat shock protein 70. *Biochemistry* **36**: 10221–10229.

675. Freitag, H., M. Janes, and W. Neupert (1982). Biosynthesis of mitochondrial porin and insertion into the outer mitochondrial membrane of *Neurospora crassa*. *Eur. J. Biochem.* **126**: 197–202.

676. Freitag, M., Personal communication.

677. Freitag, M., N. Dighde, and M. S. Sachs (1996). A UV-induced mutation that affects translational regulation of the *Neurospora arg-2* gene. *Genetics* **142**: 117–127.

678. Freitag, M., and M. S. Sachs (1995). A simple dot blot assay to measure hygromycin B phosphotransferase activity in whole-cell extracts of *Neurospora crassa*. *Fungal Genet. Newsl.* **42**: 26–28.

679. Friedman, K. J. (1977). Role of lipids in the *Neurospora crassa* membrane. I. Influence of fatty acid composition on membrane lipid phase transitions. *J. Membr. Biol.* **32**: 33–47.

680. Friedman, K. J. (1977). Role of lipids in the *Neurospora crassa* membrane. II. Membrane potential and resistance studies; the effect of altered fatty acid composition on the electrical properties of the cell membrane. *J. Membr. Biol.* **36**: 175–190.

681. Friedman, K. J., and D. Glick (1978). Fatty acid composition of *Neurospora* plasma membrane. *Neurospora Newsl.* **25**: 18.

682. Friedrich, T., A. Abelmann, B. Brors, V. Guenebaut, L. Kintscher, K. Leonard, T. Rasmussen, D. Scheide, A. Schlitt, U. Schulte, and H. Weiss (1998). Redox components and structure of the respiratory NADH:ubiquinone oxidoreductase (complex I). *Biochim. Biophys. Acta* **1365**: 215–219.

683. Friedrich, T., G. Hofhaus, W. Ise, U. Nehls, B. Schmitz, and H. Weiss (1989). A small isoform of NADH:ubiquinone oxidoreductase (complex I) without mitochondrially encoded subunits is made in chloramphenicol-treated *Neurospora crassa*. *Eur. J. Biochem.* **180**: 173–180.

684. Froehner, S. C., and K.-E. Eriksson (1974). Induction of *Neurospora crassa* laccase with protein synthesis inhibitors. *J. Bacteriol.* **120**: 450–457.

685. Fu, Y. H., B. Feng, S. Evans, and G. A. Marzluf (1995). Sequence-specific DNA binding by NIT4, the pathway-specific regulatory protein that mediates nitrate induction in *Neurospora*. *Mol. Microbiol.* **15**: 935–942.

686. Fu, Y. H., and G. A. Marzluf (1990). *cys-3*, the positive-acting sulfur regulatory gene of *Neurospora crassa*, encodes a sequence-specific DNA-binding protein. *J. Biol. Chem.* **265**: 11942–11947.

687. Fu, Y. H., and G. A. Marzluf (1990). *nit-2*, the major positive-acting nitrogen regulatory gene of *Neurospora crassa*, encodes a sequence-specific DNA-binding protein. *Proc. Natl. Acad. Sci. USA* **87**: 5331–5335.

688. Fu, Y. H., and G. A. Marzluf (1990). Site-directed mutagenesis of the "zinc finger" DNA-binding domain of the nitrogen regulatory protein NIT2 of *Neurospora*. *Mol. Microbiol.* **4**: 1847–1852.

689. Fu, Y. H., J. V. Paietta, D. G. Mannix, and G. A. Marzluf (1989). *cys-3*, the positive-acting sulfur regulatory gene of *Neurospora crassa*, encodes a protein with a putative leucine zipper DNA-binding element. *Mol. Cell. Biol.* **9**: 1120–1127.

690. Fu, Y. H., and G. Marzluf (1990). *nit-2*, the major nitrogen regulatory gene of *Neurospora crassa*, encodes a protein with a putative zinc finger DNA-binding domain. *Mol. Cell. Biol.* **10**: 1056–1065.

691. Fu, Y. H., and G. A. Marzluf (1988). Metabolic control and autogenous regulation of *nit-3*, the nitrate reductase

structural gene of *Neurospora crassa*. *J. Bacteriol.* **170:** 657–661.

692. Fuentes, A. M., M. A. J. Taylor, J. Jenkins, and I. Connerton (1993). Rapid purification and crystallization of *Neurospora crassa* tyrosinase. *Fungal Genet. Newsl.* **40:** 38–39.

693. Fujimoto, T., T. Ohnishi, and H. Nakashima (1993). Newly isolated mutants of *Neurospora crassa* that can grow optimally in a low-calcium medium. *Plant Cell Physiol.* **34:** 209–213.

694. Fujimura, M. (1994). Mode of action of diethofencarb to benzimidazole-resistant strains in *Neurospora crassa*. *J. Pest. Sci.* **19:** S219–S228.

695. Fujimura, M., T. Kamakura, H. Inoue, and I. Yamaguchi (1994). Amino acid alterations in the β-tubulin gene of *Neurospora crassa* that confer resistance to carbendazim and diethofencarb. *Curr. Genet.* **25:** 418–422.

696. Fujimura, M., K. Oeda, H. Inoue, and T. Kato (1992). A single amino acid substitution in the β-tubulin gene of *Neurospora* confers both carbendazim resistance and diethofencarb sensitivity. *Curr. Genet.* **21:** 399–404.

697. Fuller, R. C., R. W. Barratt, and E. L. Tatum (1950). The relationship between hexachlorocyclohexane and inositol in *Neurospora*. *J. Biol. Chem.* **186:** 823–827.

698. Fuller, R. C., and E. L. Tatum (1956). Inositol phospholipid in *Neurospora* and its relationship to morphology. *Am. J. Bot.* **43:** 361–365.

699. Fungal Genetics Stock Center (1996). "Catalog of Strains," 6th ed. *Fungal Genet. Newsl.* **43** (Suppl.).

700. Fungal Genetics Stock Center (1998). Catalog of strains, 7th ed. *Fungal Genet. Newsl.* **45** (Suppl.).

701. Furakawa, K., and K. Hasunuma (1984). Isolation and partial characterization of a dominant mutant defective in the production of viable ascospores in *Neurospora crassa*. *Jpn. J. Genet.* **59:** 181–194.

702. Gaertner, F. H. (1981). A response to a letter from David Perkins concerning the question of whether *nt* mutants involve tryptophan pyrrolase. *Neurospora Newsl.* **28:** 12.

703. Gaertner, F. H., and K. W. Cole (1970). Sedimentation properties of anthranilate synthetase from osmotic lysates of a wall-less variant of *Neurospora*. *Biochem. Biophys. Res. Commun.* **41:** 1192–1197.

704. Gaertner, F. H., and K. W. Cole (1973). Properties of chorismate synthase in *Neurospora crassa*. *J. Biol. Chem.* **248:** 4602–4609.

705. Gaertner, F. H., and K. W. Cole (1977). A cluster gene: Evidence for one gene, one polypeptide, five enzymes. *Biochem. Biophys. Res. Commun.* **75:** 259–264.

706. Gaertner, F. H., K. W. Cole, and G. R. Welch (1971). Evidence for distinct kynureninase and hydroxykynureninase activities in *Neurospora crassa*. *J. Bacteriol.* **108:** 902–909.

707. Galeazzi, D. R., Personal communication.

708. Galsworthy, S. B. (1974). Biochemical aspects of temperature sensitivity in *Neurospora*. Ph.D. Thesis, University of Wisconsin, Madison, WI.

709. Garceau, N., Y. Liu, J. J. Loros, and J. C. Dunlap (1997). Alternative initiation of translation and time-specific phosphorylation yields multiple forms of the essential clock protein FREQUENCY. *Cell* **89:** 469–476.

710. Gardner, G. F., Personal communication.

711. Gardner, G. F., and J. F. Feldman (1979). Gene interaction among circadian clock mutants of *Neurospora crassa*. *Abstr. Annu. Meeting. Am. Soc. Microbiol*, 104.

712. Gardner, G. F., and J. F. Feldman (1980). The *frq* locus in *Neurospora crassa*: A key element in circadian clock organization. *Genetics* **96:** 877–886.

713. Gardner, G. F., and J. F. Feldman (1981). Temperature compensation of circadian period length in clock mutants of *Neurospora crassa*. *Plant Physiol.* **68:** 1244–1248.

714. Garnjobst, L., Personal communication.

715. Garnjobst, L. (1953). Genetic control of heterocaryosis in *Neurospora crassa*. *Am. J. Bot.* **40:** 607–614.

716. Garnjobst, L. (1955). Further analysis of genetic control of heterocaryosis in *Neurospora crassa*. *Am. J. Bot.* **42:** 444–448.

717. Garnjobst, L. (1962). A temperature-independent mutation at the *rib-1t* locus in *Neurospora crassa*. *Genetics* **47:** 281–283.

718. Garnjobst, L., and E. L. Tatum (1956). A temperature-independent riboflavin-requiring mutant of *Neurospora crassa*. *Am. J. Bot.* **43:** 149–157.

719. Garnjobst, L., and E. L. Tatum (1967). A survey of new morphological mutants in *Neurospora crassa*. *Genetics* **57:** 579–604.

720. Garnjobst, L., and E. L. Tatum (1970). New crisp genes and crisp-modifiers in *Neurospora crassa*. *Genetics* **66:** 281–290.

721. Garnjobst, L., and J. F. Wilson (1956). Heterocaryosis and protoplasmic incompatibility in *Neurospora crassa*. *Proc. Natl. Acad. Sci. USA.* **42:** 613–618.

722. Garrett, R. H., and N. K. Amy (1978). Nitrate assimilation in fungi. *Adv. Microb. Physiol.* **18:** 1–65.

723. Geever, R. F., L. Huiet, J. A. Baum, B. M. Tyler, V. B. Patel, B. J. Rutledge, M. E. Case, and N. H. Giles (1989). DNA sequence, organization and regulation of the *qa* gene cluster of *Neurospora crassa*. *J. Mol. Biol.* **207:** 15–34.

724. Geever, R. F., T. Murayama, M. E. Case, and N. H. Giles (1986). Rearrangement mutations on the 5'-side of the *qa-2* gene of *Neurospora* implicate two regions of *qa-1F* activator–protein interaction. *Proc. Natl. Acad. Sci. USA* **83:** 3944–3948.

725. Geng, A. M., and A. M. Srb (1976). Phenocopies of *Neurospora* mutants induced by biotin deficiency. *Neurospora Newsl.* **23:** 25–26.

726. Germann, U. A., G. Muller, P. E. Hunziker, and K. Lerch (1988). Characterization of two allelic forms of *Neurospora crassa* laccase. Amino- and carboxyl-terminal processing of a precursor. *J. Biol. Chem.* **263:** 885–896.

727. Giles, N. H. (1951). Studies on the mechanism of reversion in biochemical mutants of *Neurospora crassa*. *Cold Spring Harbor Symp. Quant. Biol.* **16:** 283–313.

728. Giles, N. H. (1978). The organization, function and evolution of gene clusters in eukaryotes. *Am. Nat.* **112:** 641–657.

729. Giles, N. H., N. K. Alton, M. E. Case, J. A. Hautala, J. W. Jacobson, S. R. Kushner, V. B. Patel, W. R. Reinert, P. Stroman, and D. Vapnek (1978). The organization of the *qa* gene cluster in *Neurospora crassa* and its expression in *Escherichia coli*. *Stadler Genet. Symp.* **10:** 49–63.

730. Giles, N. H., M. E. Case, J. Baum, R. Geever, L. Huiet, V. Patel, and B. Tyler (1985). Gene organization and

regulation in the *qa* (quinic acid) gene cluster of *Neurospora crassa. Microbiol. Rev.* **49**: 338–358.

731. Giles, N. H., M. E. Case, C. W. Partridge, and S. I. Ahmed (1967). A gene cluster in *Neurospora crassa* coding for an aggregate of five aromatic synthetic enzymes. *Proc. Natl. Acad. Sci. USA* **58**: 1453–1460.

732. Giles, N. H., F. de Serres, and C. W. H. Partridge (1955). Comparative studies of X-ray-induced forward and reverse mutation. *Ann. NY Acad. Sci.* **59**: 536–552.

733. Giles, N. H., F. J. de Serres, Jr., and E. Barbour (1957). Studies with purple adenine mutants in *Neurospora crassa*. II. Tetrad analyses from a cross of an *ad-3A* mutant with an *ad-3B* mutant. *Genetics* **42**: 608–617.

734. Giles, N. H., R. F. Geever, D. K. Asch, J. Avalos, and M. E. Case (1991). The Wilhelmine E. Key 1989 invitational lecture: Organization and regulation of the *qa* (quinic acid) genes in *Neurospora crassa* and other fungi. *J. Hered.* **82**: 1–7.

735. Giles, N. H., and C. W. H. Partridge (1953). The effect of a suppressor on allelic inositol-less mutants in *Neurospora crassa. Proc. Natl. Acad. Sci. USA.* **39**: 479–488.

736. Giles, N. H., C. W. H. Partridge, and N. J. Nelson (1957). The genetic control of adenylsuccinase in *Neurospora crassa. Proc. Natl. Acad. Sci. USA.* **43**: 305–317.

737. Gilham, N. W. (1978). "Organelle Heredity." Raven Press, New York.

738. Gillie, O. J., Personal communication.

739. Gillie, O. J. (1970). Methods for the study of nuclear and cytoplasmic variation in respiratory activity of *Neurospora crassa*, and the discovery of three new genes. *J. Gen. Microbiol.* **61**: 379–395.

740. Gillies, C. B. (1979). The relationship between synaptinemal complexes, recombination nodules and crossing over in *Neurospora crassa* bivalents and translocation quadrivalents. *Genetics* **91**: 1–17.

741. Glass, N. L., J. Grotelueschen, and R. L. Metzenberg (1990). *Neurospora crassa A* mating-type region. *Proc. Natl. Acad. Sci. USA* **87**: 4912–4916.

742. Glass, N. L., and G. A. Kuldau (1992). Mating type and vegetative incompatibility in filamentous ascomycetes. *Annu. Rev. Phytopathol.* **30**: 201–224.

743. Glass, N. L., and L. Lee (1992). Isolation of *Neurospora crassa A* mating-type mutants by repeat-induced point (RIP) mutation. *Genetics* **132**: 125–133.

744. Glass, N. L., and M. A. Nelson (1994). Mating-type genes in mycelial ascomycetes. *In* "The Mycota" (R. Brambl and G. A. Marzluf, eds.), Vol. I, pp. 295–306. Springer-Verlag, Heidelberg.

745. Glass, N. L., and C. Staben (1997). *Neurospora* mating-type symbol *mt* revised to *mat. Fungal Genet. Newsl.* **44**: 64.

746. Glass, N. L., S. J. Vollmer, C. Staben, J. Grotelueschen, R. L. Metzenberg, and C. Yanofsky (1988). DNAs of the two *mating-type* alleles of *Neurospora crassa* are highly dissimilar. *Science* **241**: 570–573.

747. Gleason, M. K., and R. L. Metzenberg (1974). Regulation of phosphate metabolism in *Neurospora crassa*: Isolation of mutants deficient in the repressible alkaline phosphatase. *Genetics* **78**: 645–659.

748. Goldie, A. H., and R. E. Subden (1973). The neutral carotenoids of wild-type and mutant strains of *Neurospora crassa. Biochem. Genet.* **10**: 275–284.

749. González, A., M. Tenorio, G. Vaca, and J. Mora (1983). *Neurospora crassa* mutant impaired in glutamine regulation. *J. Bacteriol.* **155**: 1–7.

750. González, C., N. Brito, and G. A. Marzluf (1995). Functional analysis by site-directed mutagenesis of individual amino acid residues in the flavin domain of *Neurospora crassa* nitrate reductase. *Mol. Gen. Genet.* **249**: 456–464.

751. Good, N., R. Heilbronner, and H. K. Mitchell (1950). ε-Hydroxynorleucine as a substitute for lysine for *Neurospora. Arch. Biochem.* **28**: 464–465.

752. Goodman, I., and R. L. Weiss (1980). Control of arginine metabolism in *Neurospora*: Flux through the biosynthetic pathway. *J. Bacteriol.* **141**: 227–234.

753. Goodrich-Tanrikulu, M., Personal communication.

754. Goodrich-Tanrikulu, M., D. Jacobson, S. Schwartz, W. D. Stuart, A. Stafford, J.-T. Lin, and T. McKeon (1998). RIP of a homolog of the *Saccharomyces cerevisiae* gene for the β-subunit of fatty acid synthase yields *cel*-like mutants. *Neurospora 1998*, Asilomar, CA, 20 (Abstr.).

755. Goodrich-Tanrikulu, M., D. J. Jacobson, A. E. Stafford, J.-T. Lin, and T. A. McKeon (1999). Characterization of *Neurospora crassa* mutants isolated following repeat-induced point mutations of the β-subunit of fatty acid synthase. *Curr. Genet.* **36**: 147–152.

756. Goodrich-Tanrikulu, M., L. T. Lin, A. E. Stafford, M. I. Makapugay, T. A. McKeon, and G. Fuller (1995). Novel *Neurospora crassa* mutants with altered synthesis of polyunsaturated fatty acids. *Microbiology.* **141**: 2307–2314.

757. Goodrich-Tanrikulu, M., A. E. Stafford, J. T. Lin, M. I. Makapugay, G. Fuller, and T. A. McKeon (1994). Fatty acid biosynthesis in novel *ufa* mutants of *Neurospora crassa. Microbiology* **140**: 2683–2690.

758. Goormaghtigh, E., C. Chadwick, and G. A. Scarborough (1986). Monomers of the *Neurospora* plasma membrane H$^+$-ATPase catalyze efficient proton translocation. *J. Biol. Chem.* **261**: 7466–7471.

759. Goormaghtigh, E., L. Vigneron, G. A. Scarborough, and J. M. Ruysschaert (1994). Tertiary conformational changes of the *Neurospora crassa* plasma membrane H$^+$-ATPase monitored by hydrogen/deuterium exchange kinetics. A Fourier transformed infrared spectroscopy approach. *J. Biol. Chem.* **269**: 27409–27413.

760. Gorovits, R., O. Propheta, M. Kolot, V. Dombradi, and O. Yarden. (1999). A mutation within the catalytic domain of COT1 kinase confers changes in the presence of two COT1 isoforms and in Ser/Thr protein kinase and phosphatase activities in *Neurospora crassa. Fungal Genet. Biol.* **27**: 264–274.

761. Grad, L. I., A. T. Descheneau, W. Neupert, R. Lill, and F. E. Nargang (1999). Inactivation of the *Neurospora crassa* mitochondrial outer membrane protein TOM70 by repeat-induced point mutation (RIP) causes defects in mitochondrial protein import and morphology. *Curr. Genet.* **36**: 137–146.

762. Gradmann, D., U. P. Hansen, W. S. Long, C. L. Slayman, and J. Warncke (1978). Current–voltage relationships for the plasma membrane and its principal electrogenic pump in *Neurospora crassa*: I. Steady-state conditions. *J. Membr. Biol.* **39**: 333–367.

763. Gradmann, D., and C. L. Slayman (1975). Oscillations of an electrogenic pump in the plasma membrane of Neurospora. *J. Membr. Biol.* **23**: 181–212.

764. Gratzner, H., and D. N. Sheehan (1969). *Neurospora* mutant exhibiting hyperproduction of amylase and invertase. *J. Bacteriol.* **97**: 544–549.

765. Gratzner, H. G. (1972). Cell wall alterations associated with the hyperproduction of extracellular enzymes in *Neurospora crassa. J. Bacteriol.* **111**: 443–446.

766. Grayburn, W. S., and E. U. Selker (1989). A natural case of RIP: Degeneration of the DNA sequence in an ancestral tandem duplication. *Mol. Cell. Biol.* **9**: 4416–4421.

767. Greer, W. L. (1981). Guanosine metabolism in *Neurospora crassa*. Ph.D. Thesis, University of Western Ontario, London, Ontario, Canada.

768. Greer, W. L., and A. M. Wellman (1980). Isolation and characterization of guanine auxotrophs in *Neurospora crassa. Can. J. Microbiol.* **26**: 1412–1415.

769. Griffin, D. H. (1981). "Fungal Physiology". Wiley, New York.

770. Griffith, A. B., and R. H. Garrett (1988). Xanthine dehydrogenase expression in *Neurospora crassa* does not require a functional *nit-2* regulatory gene. *Biochem. Genet.* **26**: 37–52.

771. Griffiths, A. J. (1970). Topography of the *ad-3* region of *Neurospora crassa. Can. J. Genet. Cytol.* **12**: 420–424.

772. Griffiths, A. J. (1975). Growth patterns of adenine-3B; supersuppressor strains of *Neurospora crassa. Can. J. Genet. Cytol.* **17**: 227–240.

773. Griffiths, A. J. (1976). Evidence for nuclear restriction of supersuppressor gene products in *Neurospora* heterokaryons. *Can. J. Genet. Cytol.* **18**: 35–38.

774. Griffiths, A. J., A. M. Delange, and J. H. Jung (1974). Identification of a complex chromosome rearrangement in *Neurospora crassa. Can. J. Genet. Cytol.* **16**: 805–822.

775. Griffiths, A. J. F., Personal communication.

776. Griffiths, A. J. F. (1979). A *Neurospora* experiment for an introductory biology course. *Neurospora Newsl.* **26**: 12.

777. Griffiths, A. J. F. (1982). Null mutants of the *A* and *a* mating-type alleles of *Neurospora crassa. Can. J. Genet. Cytol.* **24**: 167–176.

778. Griffiths, A. J. F., R. A. Collins, and F. E. Nargang (1995). Mitochondrial genetics of *Neurospora*. *In* "The Mycota: Genetics and Biotechnology" (U. Kück, ed.), Vol. II, pp. 93–105. Springer-Verlag, Berlin.

779. Griffiths, A. J. F., and A. M. DeLange (1978). Mutations of the *a* mating type in *Neurospora crassa. Genetics* **88**: 239–254.

780. Griffiths, A. J. F., and A. Rieck (1981). Perithecial distribution patterns in standard and variant strains of *Neurospora crassa. Can. J. Bot.* **59**: 2610–2617.

781. Grigg, G. W. (1958). The genetic control of conidiation in a heterokaryon of *Neurospora crassa. J. Gen. Microbiol.* **19**: 15–22.

782. Grigg, G. W. (1960). Temperature-sensitive genes affecting conidiation in *Neurospora. J. Gen. Microbiol.* **22**: 667–670.

783. Grigorieff, N. (1999). Structure of the respiratory NADH:ubiquinone oxidoreductase (complex I). *Curr. Opin. Struct. Biol.* **9**: 476–483.

784. Grindle, M. (1973). Sterol mutants of *Neurospora crassa*: Their isolation, growth characteristics and resistance to polyene antibiotics. *Mol. Gen. Genet.* **120**: 283–290.

785. Grindle, M. (1974). The efficacy of various mutagens and polyene antibiotics for the induction and isolation of sterol mutants of *Neurospora crassa. Mol. Gen. Genet.* **130**: 81–90.

786. Grindle, M., and G. H. Dolderson (1986). Effects of a modifier gene on the phenotype of a dicarboximide-resistant mutant of *Neurospora crassa. Trans. Brit. Mycol. Soc.* **87**: 457–460.

787. Grindle, M., and R. Farrow (1978). Sterol content and enzyme defects of nystatin-resistant mutants of *Neurospora crassa. Mol. Gen. Genet.* **165**: 305–308.

788. Grindle, M., and W. Temple (1982). Fungicide resistance of *os* mutants of *Neurospora crassa. Neurospora Newsl.* **29**: 16–17.

789. Grindle, M., and W. Temple (1983). Fungicide resistance of *smco* mutants of *Neurospora crassa. Neurospora Newsl.* **30**: 7–8.

790. Grindle, M., and W. Temple (1985). Sporulation and osmotic sensitivity of dicarboximide-resistant mutants of *Neurospora crassa. Trans. Brit. Mycol. Soc.* **84**: 369–372.

791. Griswold, W. R., V. O. Madrid, P. M. Shaffer, D. C. Tappen, C. S. G. Pugh, and M. T. Abbott (1976). Regulation of thymidine metabolism in *Neurospora crassa. J. Bacteriol.* **125**: 1040–1047.

792. Grivell, A. R., and J. F. Jackson (1968). Thymidine kinase: Evidence for its absence from *Neurospora crassa* and some other micro-organisms, and the relevance of this to the specific labelling of deoxyribonucleic acid. *J. Gen. Microbiol.* **54**: 307–317.

793. Gross, S. R., Personal communication.

794. Gross, S. R. (1952). Heterokaryosis between opposite mating types in *Neurospora crassa. Am. J. Bot.* **39**: 574–577.

795. Gross, S. R. (1959). Enzymatic autoinduction and the hypothesis of intracellular permeability barriers in *Neurospora. Trans. NY Acad. Sci.* **22**: 44–48.

796. Gross, S. R. (1962). On the mechanism of complementation at the *leu-2* locus of *Neurospora. Proc. Natl. Acad. Sci. USA* **48**: 922–930.

797. Gross, S. R. (1962). A selection method for mutants requiring sulfur-containing compounds for growth. *Neurospora Newsl.* **1**: 4–5.

798. Gross, S. R. (1965). The regulation of synthesis of leucine biosynthetic enzymes in *Neurospora. Proc. Natl. Acad. Sci. USA* **54**: 1538–1546.

799. Gross, S. R. (1969). Genetic regulatory mechanisms in the fungi. *Annu. Rev. Genet.* **3**: 395–424.

800. Gross, S. R., and A. Fein (1960). Linkage and function in *Neurospora. Genetics* **45**: 885–904.

801. Gross, S. R., and H. S. Gross (1961). Some features of complementation at the *leucine-4* locus of *Neurospora. Genetics* **46s**: 868–869.

802. Gross, S. R., and E. W. Horn (1980). Regulation of cytoplasmic and mitochondrial leucyl-transfer ribonucleic acid synthetases in *Neurospora crassa. J. Bacteriol.* **144**: 447–450.

803. Gross, S. R., M. T. McCoy, and E. B. Gilmore (1968). Evidence for the involvement of a nuclear gene in the

production of the mitochondrial leucyl-tRNA synthetase of *Neurospora*. *Proc. Natl. Acad. Sci. USA* **61**: 253–260.

804. Gross, S. R., and R. E. Webster (1963). Some aspects of interallelic complementation involving leucine biosynthetic enzymes of *Neurospora*. *Cold Spring Harbor Symp. Quant. Biol.* **28**: 543–547.

805. Grotewold, E., G. O. Aisemberg, G. E. Taccioli, and N. D. Judewicz (1990). Genes responsive to the alteration of polyamine biosynthesis in *Neurospora crassa*. *Cell. Biol. Int. Rep.* **14**: 69–78.

806. Grove, G., and G. A. Marzluf (1980). Nitrogen regulation of acid phosphatase in *Neurospora crassa*. *J. Bacteriol.* **141**: 1470–1473.

807. Grove, G., and G. A. Marzluf (1981). Identification of the product of the major regulatory gene of the nitrogen control circuit of *Neurospora crassa* as a nuclear DNA-binding protein. *J. Biol. Chem.* **256**: 463–470.

808. Guenebaut, V., A. Schlitt, H. Weiss, K. Leonard, and T. Friedrich (1998). Consistent structure between bacterial and mitochondrial NADH:ubiquinone oxidoreductase (complex I). *J. Mol. Biol.* **276**: 105–112.

809. Haas, F., M. B. Mitchell, B. N. Ames, and H. K. Mitchell (1952). A series of histidineless mutants of *Neurospora crassa*. *Genetics* **37**: 217–226.

809a. Haedo, S. D., Personal communication.

810. Haedo, S. D., E. D. Temporini, M. E. Alvarez, H. J. Maccioni, and A. L. Rosa (1992). Molecular cloning of a gene (*cfp*) encoding the cytoplasmic filament protein P59Nc and its genetic relationship to the snowflake locus of *Neurospora crassa*. *Genetics* **131**: 575–580.

811. Hafker, T., D. Techel, G. Steier, and L. Rensing (1998). Differential expression of glucose-regulated (*grp78*) and heat-shock-inducible (*hsp70*) genes during asexual development of *Neurospora crassa*. *Microbiology* **144**: 37–43.

812. Hagemann, A., M. Freitag, and E. U. Selker, Personal communication.

813. Hager, K. M., S. M. Mandala, J. W. Davenport, D. W. Speicher, E. J. Benz, Jr., and C. W. Slayman (1986). Amino acid sequence of the plasma membrane ATPase of *Neurospora crassa*: Deduction from genomic and cDNA sequences. *Proc. Natl. Acad. Sci. USA* **83**: 7693–7697.

814. Hager, K. M., and C. Yanofsky (1990). Genes expressed during conidiation in *Neurospora crassa*: Molecular characterization of *con-13*. *Gene* **96**: 153–159.

815. Halaban, R. (1975). Glucose transport-deficient mutant of *Neurospora crassa* with an unusual rhythmic growth pattern. *J. Bacteriol.* **121**: 1056–1063.

816. Hall, J. L., Z. Ramanis, and D. J. Luck (1989). Basal body/centriolar DNA: Molecular genetic studies in *Chlamydomonas*. *Cell* **59**: 121–132.

817. Hall, M. D., and S. N. Bennett, Personal communication.

818. Hall, M. D., S. N. Bennett, and W. A. Krissinger (1993). Characterization of a newly isolated pigmentation mutant of *Neurospora crassa*. *Georgia J. Sci.* **51**: 27 (Abstr.).

819. Hall, M. D., W. A. Krissinger, and S. N. Bennett (1993). SS-692, a newly isolated pigmentation mutant of *Neurospora crassa*. *Fungal Genet. Newsl.* **40A**: 67 (Abstr.).

820. Hall, M. D., W. A. Krissinger, and S. N. Bennett (1994). Characterization and genetic analysis of SS-692, a newly isolated pigmentation mutant of *Neurospora crassa*. *Georgia J. Sci.* **52**: 31 (Abstr.).

821. Halsall, D. M., and D. E. Catcheside (1971). Structural genes for DAHP synthase isoenzymes in *Neurospora crassa*. *Genetics* **67**: 183–188.

822. Halsall, D. M., and C. H. Doy (1969). Studies concerning the biochemical genetics and physiology of activity and allosteric inhibition mutants of *Neurospora crassa* 3-deoxy-D-*arabino*-heptulosonate 7-phosphate synthase. *Biochim. Biophys. Acta* **185**: 432–446.

823. Hamer, L., and S. Gilger (1997). Bacterial transposons containing markers for fungal gene disruption. *Fungal Genet. Newsl.* **44**: 19–23.

824. Hamilton, J. G., and J. Calvet (1964). Production of protoplasts in an *osmotic* mutant of *Neurospora crassa* without added enzyme. *J. Bacteriol.* **88**: 1084–1086.

825. Hammill, A., Personal communication.

826. Han, H. Y., M. S. Lee, and K. H. Min (1995). Sib selection and restriction fragment length polymorphisms for cloning the *pmb* gene in *Neurospora crassa*. *Molecules Cells* **5**: 442–447.

827. Han, J. S. (1997). Mutagenic activity and specificity of hydrogen peroxide in the *ad-3* forward-mutation test in two-component heterokaryons of *Neurospora crassa*. *Mutat. Res.* **374**: 169–184.

828. Handa, N., P. Ballario, A. Cabibbo, G. Macino, and H. Inoue (1992). Cloning of *mus-25*, a DNA repair gene of *Neurospora crassa*. *Jpn. J. Genet.* **67**: 574 (Abstr.).

829. Hanson, B., and S. Brody (1979). Lipid and cell-wall changes in an inositol-requiring mutant of *Neurospora crassa*. *J. Bacteriol.* **138**: 461–466.

830. Hanson, B. A. (1980). Inositol-limited growth, repair, and translocation in an inositol-requiring mutant of *Neurospora crassa*. *J. Bacteriol.* **143**: 18–26.

831. Hanson, M. A., and G. A. Marzluf (1975). Control of the synthesis of a single enzyme by multiple regulatory circuits in *Neurospora crassa*. *Proc. Natl. Acad. Sci. USA* **72**: 1240–1244.

832. Harashima, T., and H. Inoue (1995). *Neurospora crassa* *cpc1* and additional regulatory component are required for induction of laccase gene expression by cycloheximide treatment. *Jpn. J. Genet.* **70**: 768 (Abstr.).

833. Harashima, T., and H. Inoue (1998). Pleiotropic deficiencies of the laccase-derepressed mutant *lah-1* are caused by constitutively increased expression of the cross-pathway control gene *cpc-1* in *Neurospora crassa*. *Mol. Gen. Genet.* **258**: 619–627.

834. Harding, R. W., Personal communication.

835. Harding, R. W. (1974). The effect of temperature on photoinduced carotenoid biosynthesis in *Neurospora crassa*. *Plant Physiol.* **54**: 142–147.

836. Harding, R. W., D. Q. Philip, B. Z. Drozdowicz, and N. P. Williams (1984). A *Neurospora crassa* mutant which over-accumulates carotenoid pigments. *Neurospora Newsl.* **31**: 23–25.

837. Harding, R. W., and W. Shropshire, Jr. (1980). Photocontrol of carotenoid biosynthesis. *Annu. Rev. Plant Physiol.* **31**: 217–238.

838. Harding, R. W., and R. V. Turner (1981). Photoregulation of the carotenoid biosynthetic pathway in albino and white collar mutants of *Neurospora crassa*. *Plant Physiol.* **68**: 745–749.

839. Harkness, T. A., R. L. Metzenberg, H. Schneider, R. Lill, W. Neupert, and F. E. Nargang (1994). Inactivation of the *Neurospora crassa* gene encoding the mitochondrial protein import receptor MOM19 by the technique of "sheltered RIP." *Genetics* **136:** 107–118.

840. Harkness, T. A., F. E. Nargang, I. van der Klei, W. Neupert, and R. Lill (1994). A crucial role of the mitochondrial protein import receptor MOM19 for the biogenesis of mitochondria. *J. Cell Biol.* **124:** 637–648.

841. Harnisch, U., H. Weiss, and W. Sebald (1985). The primary structure of the iron–sulfur subunit of ubiquinol–cytochrome *c* reductase from *Neurospora*, determined by cDNA and gene sequencing. *Eur. J. Biochem.* **149:** 95–99.

842. Haro, R., L. Sainz, F. Rubio, and A. Rodríguez-Navarro (1999). Cloning of two genes encoding potassium transporters in *Neurospora crassa* and expression of the corresponding cDNAs in *Saccharomyces cerevisiae*. *Mol. Microbiol.* **31:** 511–520.

843. Harold, C. E., and M. Fling (1952). Two mutants of *Neurospora crassa* which utilize formate or formaldehyde for growth. *J. Biol. Chem.* **194:** 339–406.

844. Hasanuma, K., Personal communication.

845. Hasanuma, K. (1988). Genetic and biochemical analysis of GTP-regulated cyclic phosphodiesterase in *Neurospora crassa* [in Japanese with English title]. *Jpn. J. Genet.* **63:** 556.

846. Haskins, F. A., and H. K. Mitchell (1949). Evidence for a tryptophan cycle in *Neurospora*. *Proc. Natl. Acad. Sci. USA* **35:** 500–506.

847. Haskins, F. A., and H. K. Mitchell (1952). An example of the influence of modifying genes in *Neurospora*. *Am. Nat.* **86:** 231–238.

848. Hasunuma, K. (1978). Control of the activity of intracellular nucleases in *Neurospora crassa*. *Mol. Gen. Genet.* **160:** 259–265.

849. Hasunuma, K. (1984). A simple method to isolate mutants in repressible cyclic phosphodiesterase in *Neurospora crassa*. *Neurospora Newsl.* **31:** 33–34.

850. Hasunuma, K., and T. Ishikawa (1972). Properties of two nuclease genes in *Neurospora crassa*. *Genetics* **70:** 371–384.

851. Hasunuma, K., and Y. Shinohara (1985). Characterization of *cpd-1* and *cpd-2* mutants which affect the activity of orthophosphate regulated cyclic phosphodiesterase in *Neurospora*. *Curr. Genet.* **10:** 197–203.

852. Hasunuma, K., and Y. Shinohara (1986). Mutations affecting cyclic phosphodiesterases and adenylate cyclase in *Neurospora*. *Curr. Genet.* **10:** 893–901.

853. Hatakeyama, S., C. Ishii, and H. Inoue (1995). Identification and expression of *Neurospora crassa mei-3* gene which encodes a protein homologous to Rad51 of *Saccharomyces cerevisiae*. *Mol. Gen. Genet.* **249:** 439–446.

854. Hatakeyama, S., Y. Ito, C. Ishii, and H. Inoue (1998). Isolation and identification of *Neurospora RAD1* and *RAD2* homologs: Evidence for two excision repair systems in *Neurospora crassa*. *Neurospora 1998*, Asilomar, CA, 21 (Abstr.).

855. Hatakeyama, S., Y. Ito, A. Shimane, C. Ishii, and H. Inoue (1998). Cloning and characterization of the yeast RAD1 homolog gene (*mus-38*) from *Neurospora crassa*: Evidence for involvement in nucleotide excision repair. *Curr. Genet.* **33:** 276–283.

856. Hausner, G., S. Stoltzner, S. K. Hubert, K. A. Nummy, and H. Bertrand (1995). Abnormal mitochondrial DNA in *uvs-4* and *uvs-5* mutants of *Neurospora crassa*. *Fungal Genet. Newsl.* **42A:** 59.

857. Hautala, J. A., M. Schweizer, N. H. Giles, and S. R. Kushner (1979). Constitutive expression in *Escherichia coli* of the *Neurospora crassa* structural gene encoding the inducible enzyme catabolic dehydroquinase. *Mol. Gen. Genet.* **172:** 93–98.

858. Hawker, K. L., P. Montague, G. A. Marzluf, and J. R. Kinghorn (1991). Heterologous expression and regulation of the *Neurospora crassa nit-4* pathway-specific regulatory gene for nitrate assimilation in *Aspergillus nidulans*. *Gene* **100:** 237–240.

859. Hawkins, A. R., H. K. Lamb, J. D. Moore, and C. F. Roberts (1993). Genesis of eukaryotic transcriptional activator and repressor proteins by splitting a multidomain anabolic enzyme. *Gene* **136:** 49–54.

860. Hawlitschek, G., H. Schneider, B. Schmidt, M. Tropschug, F.-U. Hartl, and W. Neupert (1988). Mitochondrial protein import: Identification of processing peptidase and of PEP, a processing-enhancing protein. *Cell* **53:** 795–806.

861. Hays, S., M. Freitag, and E. U. Selker., Personal communication.

862. Haysman, P., and H. B. Howe, Jr. (1971). Some genetic and physiological characteristics of urease-defective strains of *Neurospora crassa*. *Can. J. Genet. Cytol.* **13:** 256–269.

863. Heck, I. S., and H. Ninnemann (1995). Molybdenum cofactor biosynthesis in *Neurospora crassa*: Biochemical characterization of pleiotropic molybdoenzyme mutants *nit-7, nit-8, nit- 9A, B,* and *C*. *Photochem. Photobiol.* **61:** 54–60.

864. Heckman, S., M. Schliwa, and E. Kube-Granderath (1997). Primary structure of *Neurospora crassa* γ-tubulin. *Gene* **199:** 303–309.

865. Hedman, S. C., and S. Vanderschmidt (1981). Germination of microconidia from selected *Neurospora* strains. *Neurospora Newsl.* **28:** 14.

866. Heinrich, H., J. E. Azevedo, and S. Werner (1992). Characterization of the 9.5-kDa ubiquinone-binding protein of NADH:ubiquinone oxidoreductase (complex I) from *Neurospora crassa*. *Biochemistry* **31:** 11420–11424.

867. Heintz, K., K. Palme, T. Diefenthal, and V. E. Russo (1992). The *Ncypt1* gene from *Neurospora crassa* is located on chromosome 2: Molecular cloning and structural analysis. *Mol. Gen. Genet.* **235:** 413–421.

868. Hennessey, J. P., Jr., and G. A. Scarborough (1988). Secondary structure of the *Neurospora crassa* plasma membrane H⁺-ATPase as estimated by circular dichroism. *J. Biol. Chem.* **263:** 3123–3130.

869. Hennessey, J. P., Jr., and G. A. Scarborough (1990). Direct evidence for the cytoplasmic location of the NH_2- and COOH-terminal ends of the *Neurospora crassa* plasma membrane H⁺-ATPase. *J. Biol. Chem.* **265:** 532–537.

870. Henningsen, U., and M. Schliwa (1997). Reversal in the direction of movement of a molecular motor. *Nature* **389:** 93–95.

871. Henry, S. A., and A. D. Keith (1971). Saturated fatty acid requirer of *Neurospora crassa*. *J. Bacteriol.* **106**: 174–182.

872. Henstrand, J. M., N. Amrhein, and J. Schmid (1995). Cloning and characterization of a heterologously expressed bifunctional chorismate synthase/flavin reductase from *Neurospora crassa*. *J. Biol. Chem.* **270**: 20447–20452.

873. Hernández, G., Y. Mora, and J. Mora (1986). Regulation of glutamine synthesis by glycine and serine in *Neurospora crassa*. *J. Bacteriol.* **165**: 133–138.

874. Higuchi, S., J. Tamura, P. R. Giri, J. W. Polli, and R. L. Kincaid (1991). Calmodulin-dependent protein phosphatase from *Neurospora crassa*: Molecular cloning and expression of recombinant catalytic subunit. *J. Biol. Chem.* **266**: 18104–18112.

875. Hill, J. M., and V. W. Woodward (1968). Genetic control of aspartate transcarbamylase by the *pyr-3* locus of *Neurospora crassa*. *Arch. Biochem. Biophys.* **125**: 1–12.

875a. Hillyer, C., Personal communication.

876. Hinnebusch, A. G. (1997). Translational regulation of yeast *GCN4*: A window on factors that control initiator-tRNA binding to the ribosome. *J. Biol. Chem.* **272**: 21661–21664.

877. Hitz, H.-R. (1963). Die Wirkung von Hexachlorcyclohexanen bei *Neurospora crassa* inositol-less. *Arch. Mikrobiol.* **45**: 217–246.

878. Ho, C. C., Personal communication.

879. Ho, C. C. (1969). Mutants of *Neurospora crassa* permeable to histidinol. *Genetics* **62**: 725–733.

880. Ho, C. C. (1972). Mutations blocking development of the protoperithecium in *Neurospora*. *Neurospora Newsl.* **19**: 15–16.

881. Ho, C. C., and S. E. Toh (1979). Multiple genetic control of regulation of tyrosinase synthesis in *Neurospora crassa*. 2nd Malaysian Microbiology Symposium, pp. 63–65. University of Malaya, Kuala Lumpur, Malaysia.

882. Hoang-Van, K., C. Rossier, F. Barja, and G. Turian (1989). Characterization of tubulin isotypes and of β-tubulin mRNA of *Neurospora crassa* and effects of benomyl on their developmental time course. *Eur. J. Cell Biol.* **49**: 42–47.

883. Hoffman, G. R. (1972). The development of a mutagenicity testing system based upon new drug-resistant mutants of *Neurospora crassa*. Ph.D. Thesis, University of Tennessee, Knoxville, TN.

884. Hoffmann, G. R., and H. V. Malling (1974). Mutants of *Neurospora crassa* resistant to 8-azaguanine. *J. Gen. Microbiol.* **83**: 319–326.

885. Hoffmann, G. R., H. V. Malling, and T. J. Mitchell (1973). Genetics of 5-fluorodeoxyuridine-resistant mutants of *Neurospora crassa*. *Can. J. Genet. Cytol.* **15**: 831–844.

886. Hofhaus, G., H. Weiss, and K. Leonard (1991). Electron microscopic analysis of the peripheral and membrane parts of mitochondrial NADH dehydrogenase (complex I). *J. Mol. Biol.* **221**: 1027–1043.

887. Hogg, R. W., and H. P. Broquist (1968). Homocitrate formation in *Neurospora crassa*. Relation to lysine biosynthesis. *J. Biol. Chem.* **243**: 1839–1845.

888. Holder, A. A., J. C. Wootton, A. J. Baron, G. K. Chambers, and J. R. Fincham (1975). The amino acid sequence of *Neurospora* NADP-specific glutamate dehydrogenase. Peptic and chymotryptic peptides and the complete sequence. *Biochem. J.* **149**: 757–773.

889. Holloway, B. W. (1955). Genetic control of heterokaryosis in *Neurospora crassa*. *Genetics* **40**: 117–129.

890. Holm, K., E. Nilheden, and H. G. Kølmark (1976). Genetic and enzymatic analysis of a glycerol kinase deficient mutant in *Neurospora crassa*. *Mol. Gen. Genet.* **144**: 11–15.

891. Homann, V., K. Mende, C. Arntz, V. Ilardi, G. Macino, G. Morelli, G. Bose, and B. Tudzynski (1996). The isoprenoid pathway: Cloning and characterization of fungal FPPS genes. *Curr. Genet.* **30**: 232–239.

892. Horowitz, N. H. (1946). The isolation and identification of a natural precursor of choline. *J. Biol. Chem.* **162**: 413–419.

893. Horowitz, N. H. (1947). Methionine synthesis in *Neurospora*. *J. Biol. Chem.* **171**: 255–264.

893a. Horowitz, N. H. (1991). Fifty years ago: The *Neurospora* revolution. *Genetics* **127**: 631–635.

894. Horowitz, N. H. (1999). *Neurospora* and the molecular revolution. *Genetics* **151**: 3–4.

895. Horowitz, N. H., D. Bonner, and M. B. Houlahan (1945). The utilization of choline analogues by cholineless mutants of *Neurospora*. *J. Biol. Chem.* **159**: 145–151.

896. Horowitz, N. H., and M. Fling (1953). Genetic determination of tyrosinase thermostability in *Neurospora*. *Genetics* **38**: 360–374.

897. Horowitz, N. H., and M. Fling (1956). Studies of tyrosinase production by heterocaryon of *Neurospora*. *Proc. Natl. Acad. Sci. USA* **42**: 498–501.

898. Horowitz, N. H., M. Fling, H. M. Feldman, M. L. Pall, and S. C. Froehner (1970). Derepression of tyrosinase synthesis in *Neurospora* by amino acid analogs. *Dev. Biol.* **21**: 147–156.

899. Horowitz, N. H., M. Fling, H. L. Macleod, and N. Sueoka (1960). Genetic determination and enzymatic induction of tyrosinase in *Neurospora*. *J. Mol. Biol.* **2**: 96–104.

900. Horowitz, N. H., M. Fling, H. L. Macleod, and N. Sueoka (1961). A genetic study of two new structural forms of tyrosinase in *Neurospora*. *Genetics* **45**: 1015–1024.

901. Horowitz, N. H., M. Fling, H. L. Macleod, and Y. Watanabe (1961). Structural and regulative genes controlling tyrosinase synthesis in *Neurospora*. *Cold Spring Harbor Symp. Quant. Biol.* **26**: 233–238.

902. Horowitz, N. H., and H. Macleod (1960). The DNA content of *Neurospora* nuclei. *Microb. Genet. Bull.* **17**: 6–7.

903. Horowitz, N. H., and S. C. Shen (1952). *Neurospora* tyrosinase. *J. Biol. Chem.* **197**: 513–520.

904. Horowitz, N. H., and A. M. Srb (1948). Growth inhibition of *Neurospora* by canavanine, and its reversal. *J. Biol. Chem.* **174**: 371–378.

905. Horowitz, N. W., and G. W. Beadle (1943). A microbiological method for the determination of choline by use of a mutant of *Neurospora*. *J. Biol. Chem.* **150**: 325–333.

906. Houlahan, M. B. (1944). The genetics of a group of biochemical mutants of *Neurospora crassa*. M.A. Thesis, Stanford University, Stanford, CA.

907. Houlahan, M. B., G. W. Beadle, and H. G. Calhoun (1949). Linkage studies with biochemical mutants of *Neurospora crassa*. *Genetics* **34**: 493–507.

908. Houlahan, M. B., and H. K. Mitchell (1947). A suppressor in *Neurospora* and its use as evidence of allelism. *Proc. Natl. Acad. Sci. USA* 33: 223–229.

909. Houlahan, M. B., and H. K. Mitchell (1948). The accumulation of acid-labile, inorganic phosphate by mutants of *Neurospora*. *Arch. Biochem.* 19: 257–264.

910. Houlahan, M. B., and H. K. Mitchell (1948). Evidence for an interrelation in the metabolism of lysine, arginine and pyrimidines in *Neurospora*. *Proc. Natl. Acad. Sci. USA* 34: 465–470.

911. Howard, C. A., and T. I. Baker (1986). Identification of DNA repair and damage-induced proteins from *Neurospora crassa*. *Mol. Gen. Genet.* 203: 462–467.

912. Howard, C. A., and T. I. Baker (1988). Relationship of histidine sensitivity to DNA damage and stress-induced responses in mutagen-sensitive mutants of *Neurospora crassa*. *Curr. Genet.* 13: 391–399.

913. Howard, C. A., and T. I. Baker (1989). Inducible responses to DNA-damaging or stress-inducing agents in *Neurospora crassa*. *Curr. Genet.* 15: 47–55.

914. Howe, H. B., Jr., Personal communication.

915. Howe, H. B., Jr. (1956). Crossing over and nuclear passing in *Neurospora crassa*. *Genetics* 41: 610–622.

916. Howe, H. B., Jr. (1961). Determining mating type in *Neurospora* without crossing tests. *Nature* 190: 1036.

917. Howe, H. B., Jr. (1962). Precision of mating-type determination without crossing tests in *Neurospora*. *Microb. Genet. Bull.* 18: 12–13.

918. Howe, H. B., Jr. (1976). Phenotypic diversity among alleles at the *per-1* locus of *Neurospora crassa*. *Genetics* 82: 595–603.

919. Howe, H. B., Jr., and E. W. Benson (1974). A perithecial color mutant of *Neurospora crassa*. *Mol. Gen. Genet.* 131: 79–83.

920. Howe, H. B., Jr., and J. E. Pyle (1982). Female sterility and ascospore lethality in *Neurospora crassa*. *Abstr. Annu. Meeting. Am. Soc. Microbiol*, 127.

921. Howe, K. A., and M. A. Nelson, Personal communication.

922. Howell, A. N. (1972). Biochemistry and genetics of respiratory-deficient mutants of *Neurospora crassa*. Ph.D. Thesis, University of Wisconsin, Madison, WI.

923. Howell, N., C. A. Zuiches, and K. D. Munkres (1971). Mitochondrial biogenesis in *Neurospora crassa*. I. An ultrastructural and biochemical investigation of the effects of anaerobiosis and chloramphenicol inhibition. *J. Cell Biol.* 50: 721–736.

924. Howlett, B. C., J. F. Leslie, and D. D. Perkins (1993). Putative multiple alleles at the vegetative incompatibility loci *het-c* and *het-8* in *Neurospora crassa*. *Fungal Genet. Newsl* 40: 40–42.

925. Hoyt, M. A., L. J. Williams-Abbott, J. W. Pitkin, and R. H. Davis (2000). Cloning and expression of the S-adenosylmethionine decarboxylase gene of *Neurospora crassa* and processing of its product. *Mol. Gen. Genet.* 263: 664–673.

926. Hsu, K. S., Personal communication.

927. Hsu, K. S. (1962). Drug-resistant mutants. *Neurospora Newsl.* 1: 5.

928. Hsu, K. S. (1962). Genetic basis of drug-resistant mutants in *Neurospora*. *Genetics* 47s: 961.

929. Hsu, K. S. (1963). The genetic basis of actidione resistance in *Neurospora*. *J. Gen. Microbiol.* 32: 341–347.

930. Hsu, K. S. (1963). A modifier of the morphological mutant scumbo in *Neurospora crassa*. *Neurospora Newsl.* 4: 7–9.

931. Hsu, K. S. (1965). Acriflavin resistance controlled by chromosomal genes in *Neurospora*. *Neurospora Newsl.* 8: 4–6.

932. Hsu, K. S. (1973). New acriflavin-resistant mutants and a gene affecting conidiation, which are expressed only in the presence of a morphological mutant. *Neurospora Newsl.* 20: 39.

933. Huang, P. C. (1964). Recombination and complementation of albino mutants in *Neurospora*. *Genetics* 491: 453–469.

934. Hubbard, S. C., and S. Brody (1975). Glycerophospholipid variation in choline and inositol auxotrophs of *Neurospora crassa*. Internal compensation among zwitterionic and anionic species. *J. Biol. Chem.* 250: 7173–7181.

935. Huber, D. H., G. Hausner, B. Seidel-Rogol, and B. Bertrand (1997). Cloning and characterization of a *Neurospora crassa* MutS mismatch repair homolog. *19th Fungal Genet. Conf.*, Asilomar, CA, 43 (Abstr.).

936. Huber, M., and K. Lerch (1987). The influence of copper on the induction of tyrosinase and laccase in *Neurospora crassa*. *FEBS Lett.* 219: 335–338.

937. Hudspeth, M. E. S. (1992). The fungal mitochondrial genome—A broader perspective. In "Handbook of Applied Mycology. Volume 4: Fungal Biotechnology" (D. K. Arora, R. P. Elander, and K. G. Mukerji, eds.), pp. 213–241. Marcel Dekker, New York.

938. Huiet, L., B. M. Tyler, and N. H. Giles (1984). A leucine tRNA gene adjacent to the *qa* gene cluster of *Neurospora crassa*. *Nucleic Acids Res.* 12: 5757–5765.

939. Hummelt, G., and J. Mora (1980). NADH-dependent glutamate synthase and nitrogen metabolism in *Neurospora crassa*. *Biochem. Biophys. Res. Commun.* 92: 127–133.

940. Hungate, F. P. (1946). The biochemical genetics of a mutant of *Neurospora crassa* requiring serine or glycine. Ph.D. Thesis, Stanford University, Stanford, CA.

941. Hunt, I. E., and B. J. Bowman (1997). The intriguing evolution of the "b" and "G" subunits in F-type and V-type ATPases: Isolation of the *vma-10* gene from *Neurospora crassa*. *J. Bioenerg. Biomembr.* 29: 533–540.

942. Hurd, M. S. (1962). An hereditary growth-rate-retarding factor of *Neurospora crassa*. M.A. Thesis, Stanford University, Stanford, CA.

943. Hwang, C. S., and P. E. Kolattukudy (1995). Isolation and characterization of genes expressed uniquely during appressorium formation by *Colletotrichum gloeosporioides* conidia induced by the host surface wax. *Mol. Gen. Genet.* 247: 282–294.

944. Ichi-Ishi, A., and H. Inoue (1995). Cloning, nucleotide sequence, and expression of *tef-1*, the gene encoding translation elongation factor 1 α (EF-1 α) of *Neurospora crassa*. *Jpn. J. Genet.* 70: 273–287.

945. Inoue, H., Personal communication.

946. Inoue, H. (1999). DNA repair and specific-locus mutagenesis in *Neurospora crassa*. *Mutat. Res.* 437: 121–133.

947. Inoue, H., R. C. Harvey, D. F. Callen, and F. J. de Serres (1981). Mutagenesis at the *ad-3A* and *ad-3B* loci in haploid UV-sensitive strains of *Neurospora crassa*. V. Comparison of dose–response curves of single- and double-mutant strains with wild type. *Mutat. Res.* 84: 49–71.

948. Inoue, H., and C. Ishii (1984). Isolation and characterization of MMS-sensitive mutants of *Neurospora crassa*. *Mutat. Res.* **125**: 185–194.

949. Inoue, H., and C. Ishii (1985). A new ultraviolet-light-sensitive mutant of *Neurospora crassa* with unusual photoreactivation property. *Mutat. Res.* **152**: 161–168.

950. Inoue, H., and C. Ishii (1990). Designation of newly identified mutagen-sensitive mutations in *Neurospora crassa* and their linkage data. *Fungal Genet. Newsl.* **37**: 20.

951. Inoue, H., and T. Ishikawa (1970). Macromolecule synthesis and germination of conidia in temperature-sensitive mutants of *Neurospora crassa*. *Jpn. J. Genet.* **45**: 357–369.

952. Inoue, H., and T. Ishikawa (1975). Death resulting from unbalanced growth in a temperature-sensitive mutant of *Neurospora crassa*. *Arch. Microbiol.* **104**: 1–6.

953. Inoue, H., T. M. Ong, and F. J. de Serres (1981). Mutagenesis at the *ad-3A* and *ad-3B* loci in haploid UV-sensitive strains of *Neurospora crassa*. IV. Comparison of dose–response curves for MNNG, 4NQO and ICR-170 induced inactivation and mutation induction. *Mutat. Res.* **80**: 27–41.

954. Inoue, H., and A. L. Schroeder (1988). A new mutagen-sensitive mutant in *Neurospora*, *mus-16*. *Mutat. Res.* **194**: 9–16.

955. Irelan, J. T., and E. U. Selker (1997). Cytosine methylation associated with repeat-induced point mutation causes epigenetic gene silencing in *Neurospora crassa*. *Genetics* **146**: 509–523.

956. Ishii, C., Personal communication.

957. Ishii, C., and H. Inoue (1989). Epistasis, photoreactivation and mutagen sensitivity of DNA repair mutants *upr-1* and *mus-26* in *Neurospora crassa*. *Mutat. Res.* **218**: 95–103.

958. Ishii, C., and H. Inoue (1994). Mutagenesis and epistatic grouping of the *Neurospora* meiotic mutants, *mei-2* and *mei-3*, which are sensitive to mutagens. *Mutat. Res.* **315**: 249–259.

959. Ishii, C., K. Nakamura, and H. Inoue (1991). A novel phenotype of an excision repair mutant in *Neurospora crassa*: Mutagen sensitivity of the *mus-18* mutant is specific to UV. *Mol. Gen. Genet.* **228**: 33–39.

960. Ishii, C., K. Nakamura, and H. Inoue (1998). A new UV-sensitive mutant that suggests a second excision repair pathway in *Neurospora crassa*. *Mutat. Res.* **408**: 171–182.

961. Ishii, C., and A. L. Schroeder, Personal communication.

962. Ishikawa, T., Personal communication.

963. Ishikawa, T. (1962). Genetic studies of *ad-8* mutants in *Neurospora crassa*. I. Genetic fine structure of the *ad-8* locus. *Genetics* **47**: 1147–1161.

964. Ishikawa, T. (1962). Genetic studies of *ad-8* mutants in *Neurospora crassa*. II. Interallelic complementation at the *ad-8* locus. *Genetics* **47**: 1755–1771.

965. Ishikawa, T. (1965). A molecular model for an enzyme based on the genetic and complementation analyses at the *ad-8* locus in *Neurospora*. *J. Mol. Biol.* **13**: 586–591.

966. Ishikawa, T. (1977). Growth inhibition by adenine in temperature-sensitive mutants. *Neurospora Newsl.* **24**: 7.

967. Ishikawa, T., A. Toh-e, I. Uno, and K. Hasunuma (1969). Isolation and characterization of nuclease mutants in *Neurospora crassa*. *Genetics* **63**: 75–92.

968. Ito, S., T. Ichihara, Y. Matsui, A. Toh-e, and H. Inoue (1995). Structural and functional analyses of *Neurospora crassa ras* gene. *Jpn. J. Genet.* **70**: 768 (Abstr.).

969. Ito, S., Y. Matsui, A. Toh-e, T. Harashima, and H. Inoue (1997). Isolation and characterization of the *krev-1* gene, a novel member of *ras* superfamily in *Neurospora crassa*: Involvement in sexual cycle progression. *Mol. Gen. Genet.* **255**: 429–437.

970. Ito, S., Y. Matsui, A. Toh-e, and H. Inoue (1991). Isolation and characterization of a new member of the *ras* super-family of *Neurospora crassa*. *Jpn. J. Genet.* **66**: 768 (Abstr.).

971. Ito-Harashima, S., M. Sugisaki, T. Ichihara, Y. Matsui, A. Toh-e, and H. Inoue (1999). Analyses of structures and *ras* superfamily of *Neurospora crassa*. *Fungal Genet. Newsl.* **46 (Suppl.)**: 131 (Abstr.).

972. Ivey, F. D., P. N. Hodge, G. E. Turner, and K. A. Borkovich (1996). The Gα$_i$ homologue *gna-1* controls multiple differentiation pathways in *Neurospora crassa*. *Mol. Biol. Cell* **7**: 1283–1297.

973. Ivey, F. D., Q. Yang, and K. A. Borkovich (1999). Positive regulation of adenylyl cyclase activity by a Gα$_i$ homolog in *Neurospora crassa*. *Fungal Genet. Biol.* **26**: 48–61.

974. Jackson, C. R. (1992). Characterization and genetic analysis of a newly isolated *crisp*-like mutant and an osmotic-sensitive mutant of *Neurospora crassa*. M.S. Thesis, Georgia Southern University, Statesboro, GA.

975. Jackson, C. R., S. N. Bennett, and W. A. Krissinger (1993). *J-6*, a new *crisp* mutant of *Neurospora crassa*. *17th Int. Congr. Genet.*, Birmingham, 182 (Abstr).

976. Jacobson, D. J., Personal communication.

977. Jacobson, D. J. (1992). Control of mating type hetero-karyon incompatibility by the *tol* gene in *Neurospora crassa* and *N. tetrasperma*. *Genome* **35**: 347–353.

978. Jacobson, D. J. (1992). New round spore mutations in *Neurospora crassa* accompanying changes in a duplication closely linked to the *R* locus. *Fungal Genet. Newsl.* **39**: 24.

979. Jacobson, D. J., K. Beurkens, and K. L. Klomparens (1998). Microscopic and ultrastructural examination of vegetative incompatibility in partial diploids heterozygous at *het* loci in *Neurospora crassa*. *Fungal Genet. Biol.* **23**: 45–56.

980. Jacobson, D. J., J. Ohrnberger, and R. A. Akins (1995). The Wilson–Garnjobst heterokaryon incompatibility tester strains of *Neurospora crassa* contain modifiers which influence growth rate of heterokaryons and distort segregation ratios. *Fungal Genet. Newsl.* **42**: 34–40.

981. Jacobson, E. S., G. S. Chen, and R. L. Metzenberg (1977). Unstable *S*-adenosylmethionine synthetase in an ethionine-resistant strain of *Neurospora crassa*. *J. Bacteriol.* **132**: 747–748.

982. Jacobson, E. S., and R. L. Metzenberg (1968). A new gene which affects uptake of neutral and acidic amino acids in *Neurospora crassa*. *Biochim. Biophys. Acta* **156**: 140–147.

983. Jacobson, J. W., J. A. Hautala, M. E. Case, and N. H. Giles (1975). Effect of mutations in the *qa* gene cluster of *Neurospora crassa* on the enzyme catabolic dehydroquinase. *J. Bacteriol.* **124**: 491–496.

984. Jarai, G., and G. A. Marzluf (1991). Sulfate transport in *Neurospora crassa*: Regulation, turnover, and cellular

localization of the CYS-14 protein. *Biochemistry* **30**: 4768–4773.

985. Jarai, G., B. Yagmai, Y. H. Fu, and G. A. Marzluf (1990). Regulation of branched-chain amino acid biosynthesis in *Neurospora crassa*: Cloning and characterization of the *leu-1* and *ilv-3* genes. *Mol. Gen. Genet.* **224**: 383–388.

986. Jeenes, D. J., R. Pfaller, and D. B. Archer (1997). Isolation and characterisation of a novel stress-inducible PDI family gene from *Aspergillus niger*. *Gene* **193**: 151–156.

987. Jenkins, M. B., and H. R. Garner (1967). Studies of a homoserineless bradytroph of *Neurospora crassa*: Demonstration of an altered aspartate β-semialdehyde dehydrogenase. *Biochim. Biophys. Acta* **141**: 287–295.

988. Jensen, R. A., L. Zamir, M. Saint Pierre, N. Patel, and D. L. Pierson (1977). Isolation and preparation of pretyrosine, accumulated as a dead-end metabolite by *Neurospora crassa*. *J. Bacteriol.* **132**: 896–903.

989. Jeong, S. S., and V. Schirch (1996). Role of cytosolic serine hydroxymethyltransferase in one-carbon metabolism in *Neurospora crassa*. *Arch. Biochem. Biophys.* **335**: 333–341.

990. Jessup, A. P., and D. G. Catcheside (1965). Interallelic recombination at the *his-1* locus in *Neurospora crassa* and its genetic control. *Heredity* **20**: 237–256.

991. Jha, K. K., Personal communication.

992. Jha, K. K. (1965). Notes on *phen-1* mutants. *Neurospora Newsl.* **7**: 15–18.

993. Jha, K. K. (1967). Nature of a leucine-requiring strain derived from a *phen-1* stock. *Neurospora Newsl.* **11**: 4.

994. Jha, K. K. (1969). Growth characteristics of *phen-1* mutants. *Neurospora Newsl.* **14**: 3.

995. Jha, K. K. (1972). Genes conferring resistance to 8-aza-adenine in *Neurospora crassa* and the variability of resistant alleles in the *aza-1* locus with respect to excretion of purines. *Mol. Gen. Genet.* **114**: 156–167.

996. Jha, K. K. (1972). An unlinked mutation affecting control of purine metabolism in a revertant of an *ad-7* auxotroph of *Neurospora crassa* lacking the phosphoribosylpyrophosphate amidotransferase. *Mol. Gen. Genet.* **114**: 168–172.

997. Johnson, N., and N. H. Giles (1956). Genetic studies of adenineless mutants in *Neurospora crassa*. *Microb. Genet. Bull.* **13**: 15–16.

998. Johnson, T. E. (1976). Analysis of pattern formation in *Neurospora* perithecial development using genetic mosaics. *Dev. Biol.* **54**: 23–36.

999. Johnson, T. E. (1977). Mosaic analysis of autonomy of spore development in *Neurospora*. *Exp. Mycol.* **1**: 253–258.

1000. Johnson, T. E. (1978). Isolation and characterization of perithecial development mutants in *Neurospora*. *Genetics* **88**: 27–47.

1001. Johnson, T. E. (1979). The search for position effects in *Neurospora*. *Neurospora Newsl.* **26**: 14.

1002. Johnston, T. E. (1975). Perithecia development and pattern formation in *Neurospora crassa*. Ph.D. Thesis, University of Washington, Seattle, WA.

1003. Johnston, T. E. (1979). A *Neurospora* mutation that arrests perithecial development as either male or female parent. *Genetics* **92**: 1107–1120.

1004. Jorge, J. A., M. de Lourdes, T. M. Polizeli, J. M. Thevelein, and H. F. Terenzi (1997). Trehalases and trehalose hydrolysis in fungi. *FEMS Microbiol. Lett.* **154**: 165–171.

1005. Jorge, J. A., and H. F. Terenzi (1980). An enzymatic alteration secondary to adenylyl cyclase deficiency in the *cr-1* (crisp) mutant of *Neurospora crassa*: Nicotinamide adenine dinucleotide (phosphate) glycohydrolase overproduction. *Dev. Biol.* **74**: 231–238.

1006. Juretic, D. (1976). Cyanide-resistant respiration of a *Neurospora crassa* membrane mutant. *J. Bacteriol.* **126**: 542–543.

1007. Juretic, D. (1977). Lecithin requirement for the sporulation process in *Neurospora crassa*. *J. Bacteriol.* **130**: 524–525.

1008. Käfer, E., Personal communication.

1009. Käfer, E. (1978). Sensitivity to methylmethane sulfonate (MMS) in *Neurospora*. *Neurospora Newsl.* **25**: 19.

1010. Käfer, E. (1981). Mutagen sensitivities and mutator effects of MMS-sensitive mutants in *Neurospora*. *Mutat. Res.* **80**: 43–64.

1011. Käfer, E. (1982). Backcrossed mutant strains which produce consistent map distances and negligible interference. *Neurospora Newsl.* **29**: 41–44.

1012. Käfer, E. (1983). Epistatic grouping of DNA repair deficient mutants in *Neurospora*: Comparative analysis of two *uvs-3* alleles and *uvs-6*, and their *mus* double-mutant strains. *Genetics* **105**: 19–33.

1013. Käfer, E. (1984). UV-induced recessive lethals in *uvs* strains of *Neurospora* which are deficient in UV mutagenesis. *Mutat. Res.* **128**: 137–46.

1014. Käfer, E., and M. Fraser (1979). Isolation and genetic analysis of nuclease halo (*nuh*) mutants in *Neurospora*. *Mol. Gen. Genet.* **169**: 117–127.

1015. Käfer, E., and D. Luk (1988). Properties and strains of additional DNA repair-defective mutants in known and new genes of *Neurospora crassa*. *Fungal Genet. Newsl.* **35**: 11–13.

1016. Käfer, E., and D. Luk (1989). Sensitivity to bleomycin and hydrogen peroxide of DNA repair defective mutants in *Neurospora crassa*. *Mutat. Res.* **217**: 75–81.

1017. Käfer, E., and E. Perlmutter (1980). Isolation and genetic analysis of MMS-sensitive *mus* mutants of *Neurospora*. *Can. J. Genet. Cytol.* **22**: 535–552.

1018. Käfer, E., and G. R. Witchell (1984). Effects of *Neurospora* nuclease halo (*nuh*) mutants on secretion of two phosphate-repressible alkaline deoxyribonucleases. *Biochem. Genet.* **22**: 403–417.

1019. Kaldenhoff, R., and V. E. Russo (1993). Promoter analysis of the *bli-7/eas* gene. *Curr. Genet.* **24**: 394–399.

1020. Kana-uchi, A. (1998). Suppressor mutants of an adenylyl cyclase mutant. *Neurospora 1998*, Asilomar, CA, 22 (Abstr.).

1021. Kana-uchi, A., C. T. Yamashiro, S. Tanabe, and T. Murayama (1997). A *ras* homologue of *Neurospora crassa* regulates morphology. *Mol. Gen. Genet.* **254**: 427–432.

1022. Kang, S., and R. L. Metzenberg (1990). Molecular analysis of *nuc-1+*, a gene controlling phosphorus acquisition in *Neurospora crassa*. *Mol. Cell. Biol.* **10**: 5839–5848.

1023. Kanno, S., S. Iwai, M. Takao, and A. Yasui (1999). Repair of apurinic/apyrimidinic sites by UV damage endonuclease; a repair protein for UV and oxidative damage. *Nucleic Acids Res.* 27: 3096–3103.

1024. Kaplan, S., Y. Suyama, and D. M. Bonner (1964). Fine structure analysis at the *td* locus of *Neurospora crassa*. *Genetics* 49: 145–158.

1025. Kapoor, M., Personal communication.

1026. Kapoor, M., C. A. Curle, and C. Runham (1995). The hsp70 gene family of *Neurospora crassa*: Cloning, sequence analysis, expression, and genetic mapping of the major stress-inducible member. *J. Bacteriol.* 177: 212–221.

1027. Kapoor, M., Y. Vijayaraghavan, R. Kadonaga, and K. E. LaRue (1993). NAD⁺-specific glutamate dehydrogenase of *Neurospora crassa*: Cloning, complete nucleotide sequence, and gene mapping. *Biochem. Cell Biol.* 71: 205–219.

1028. Kappy, M. S., and R. L. Metzenberg (1965). Studies on the basis of ethionine resistance in *Neurospora*. *Biochim. Biophys. Acta* 107: 425–433.

1029. Kappy, M. S., and R. L. Metzenberg (1967). Multiple alterations in metabolite uptake in a mutant of *Neurospora crassa*. *J. Bacteriol.* 94: 1629–1637.

1030. Kappy, M. S., and R. L. Metzenberg (1967). Studies on the basis of ethionine resistance in *Neurospora*. Ph.D. Thesis, University of Wisconsin, Madison, WI.

1031. Kapular, A. M., Personal communication to FGSC.

1032. Karpova, O. I., N. M. Anan'eva, T. M. Ermokina, and I. A. Krasheninnikov (1986). Peculiarities of the primary structure of histone H2b of the mold fungus *Neurospora crassa*. *Biochemistry (USSR)* 51: 675–686 (English translation of the Russian original in *Biokhimiya* 51: 788–800).

1033. Kasher, J. S., K. E. Allen, K. Kasamo, and C. W. Slayman (1986). Characterization of an essential arginine residue in the plasma membrane H⁺-ATPase of *Neurospora crassa*. *J. Biol. Chem.* 261: 10808–10813.

1034. Kashmiri, S. V., and S. R. Gross (1970). Mutations affecting the regulation of production of the enzymes of leucine synthesis in *Neurospora*. *Genetics* 64: 423–440.

1035. Katagiri, S., K. Onai, and H. Nakashima (1998). Spermidine determines the sensitivity to the calmodulin antagonist, chlorpromazine, for the circadian conidiation rhythm but not for the mycelial growth in *Neurospora crassa*. *J. Biol. Rhythms* 13: 452–460.

1036. Kato, E. E. (1998). Characterization of a serine carboxypeptidase in *Neurospora crassa*, homologous to the *KEX1* gene of *S. cerevisiae*. Ph.D. Thesis, University of Hawaii, Honolulu, HI.

1037. Kawahashi, Y., M. Sugisaki, A. Toh-e, and H. Inoue (1999). Cloning and functional analysis of the gene coding a PLC homologue in *Neurospora crassa*. *Fungal Genet. Newsl.* 46 (Suppl.): 132 (Abstr.).

1038. Keesey, J. K., Jr., J. Paukert, and J. A. Demoss (1981). Subunit structure of anthranilate synthase from *Neurospora crassa*: Preparation and characterization of a protease-free form. *Arch. Biochem. Biophys.* 207: 103–109.

1039. Keil, P., and N. Pfanner (1993). Insertion of MOM22 into the mitochondrial outer membrane strictly depends on surface receptors. *FEBS Lett.* 321: 197–200 [published erratum appears in *FEBS Lett.* 326(1–3): 299].

1040. Keil, P., A. Weinzierl, M. Kiebler, K. Dietmeier, T. Söllner, and N. Pfanner (1993). Biogenesis of the mitochondrial receptor complex. Two receptors are required for binding of MOM38 to the outer membrane surface. *J. Biol. Chem.* 268: 19177–19180.

1041. Kelkar, H. S., T. J. Hagen, J. K. Wunderlich, M. J. Weise, J. R. Wagner, and J. Arnold (1999). Direct database submission.

1042. Kempken, F., and D. E. A. Catcheside, Personal communication.

1043. Kempken, F., and U. Kück (1996). *restless*, an active Ac-like transposon from the fungus *Tolypocladium inflatum*: Structure, expression, and alternative RNA splicing. *Mol. Cell. Biol.* 16: 6563–6572.

1044. Kempken, F., and U. Kück (1998). Transposons in filamentous fungi — Facts and perspectives. *Bioessays* 20: 652–659.

1045. Kerr, D. S., and M. Flavin (1970). The regulation of methionine synthesis and the nature of cystathionine γ-synthase in *Neurospora*. *J. Biol. Chem.* 245: 1842–1855.

1046. Ketter, J. S., G. Jarai, Y. H. Fu, and G. A. Marzluf (1991). Nucleotide sequence, messenger RNA stability, and DNA recognition elements of *cys-14*, the structural gene for sulfate permease II in *Neurospora crassa*. *Biochemistry* 30: 1780–1787.

1047. Kidd, G. L. (1987). Coregulated synthesis of leucine and histidine biosynthetic enzymes, and the cloning and analysis of the *leu-2* gene of *Neurospora crassa*. Ph.D. Thesis, Duke University, Durham, NC.

1048. Kidd, G. L., and S. R. Gross (1984). Specific regulatory interconnection between the leucine and histidine pathways of *Neurospora crassa*. *J. Bacteriol.* 158: 121–127.

1049. Kiebler, M., P. Keil, H. Schneider, I. J. van der Klei, N. Pfanner, and W. Neupert (1993). The mitochondrial receptor complex: A central role of MOM22 in mediating preprotein transfer from receptors to the general insertion pore. *Cell* 74: 483–492.

1050. Kiebler, M., R. Pfaller, T. Söllner, G. Griffiths, H. Horstmann, N. Pfanner, and W. Neupert (1990). Identification of a mitochondrial receptor complex required for recognition and membrane insertion of precursor proteins. *Nature* 348: 610–616.

1051. Kientsch, R., and S. Werner (1976). Cold sensitivity of mitochondrial biogenesis in a nuclear mutant of *Neurospora crassa*. In "Genetics and Biogenesis of Chloroplasts and Mitochondria" (T. E. A. Bücher, ed.), pp. 247–252. North Holland Publishing Co., Amsterdam.

1052. Kim, H., and M. A. Nelson, Personal communication.

1053. Kinnaird, J. H., P. A. Burns, and J. R. Fincham (1991). An apparent rare-codon effect on the rate of translation of a *Neurospora* gene. *J. Mol. Biol.* 221: 733–736.

1054. Kinnaird, J. H., D. F. Revell, I. F. Connerton, I. Hasleham, and J. R. S. Fincham (1992). Alternate modes of mRNA processing in a 3'-splice site mutant of *Neurospora crassa*. *Curr. Genet.* 22: 37–40.

1055. Kinsey, J. A., Personal communication.

1056. Kinsey, J. A. (1970). Isoleucineless mutants of *Neurospora* lacking threonine deaminase activity. *Genetics* 64s: 33–34.

1057. Kinsey, J. A. (1975). Three new *p*-fluorophenylalanine-resistant (*fpr*) mutants. *Neurospora Newsl.* 22: 20.

1058. Kinsey, J. A. (1977). Direct selective procedure for isolating *Neurospora* mutants defective in nicotinamide adenine dinucleotide phosphate-specific glutamate dehydrogenase. *J. Bacteriol.* **132**: 751–756.

1059. Kinsey, J. A. (1979). Isolation of *Neurospora crassa* bradytrophs. *J. Bacteriol.* **140**: 1133–1136.

1060. Kinsey, J. A. (1985). A spontaneous mutation of the *am* gene of *Neurospora* results from a small insertion. *Genetics* **110**: s290 (Abstr.).

1061. Kinsey, J. A. (1993). Transnuclear retrotransposition of the *Tad* element of *Neurospora*. *Proc. Natl. Acad. Sci. USA* **90**: 9384–9387.

1062. Kinsey, J. A., and J. R. Fincham (1979). An unstable allele of the *am* locus of *Neurospora crassa*. *Genetics* **93**: 577–586.

1063. Kinsey, J. A., J. R. Fincham, M. A. Siddig, and M. Keighren (1980). New mutational variants of *Neurospora* NADP-specific glutamate dehydrogenase. *Genetics* **95**: 305–316.

1064. Kinsey, J. A., P. W. Garrett-Engele, E. B. Cambareri, and E. U. Selker (1994). The *Neurospora* transposon *Tad* is sensitive to repeat-induced point mutation (RIP). *Genetics* **138**: 657–664.

1065. Kinsey, J. A., and J. Helber (1989). Isolation of a transposable element from *Neurospora crassa*. *Proc. Natl. Acad. Sci. USA* **86**: 1929–1933.

1066. Kinsey, J. A., and B. S. Hung (1981). Mutation at the *am* locus of *Neurospora crassa*. *Genetics* **99**: 405–414.

1067. Kinsey, J. A., and D. R. Stadler (1969). Interaction between analogue resistance and amino acid auxotrophy in *Neurospora*. *J. Bacteriol.* **97**: 1114–1117.

1068. Kirchner, J., S. Seiler, S. Fuchs, and M. Schliwa (1999). Functional anatomy of the kinesin molecule *in vivo*. *EMBO J.* **18**: 4404–4413.

1069. Kiritani, K. (1962). Linkage relationships among a group of isoleucine and valine requiring mutants of *Neurospora crassa*. *Jpn. J. Genet.* **37**: 42–56.

1070. Kiritani, K., and R. P. Wagner (1970). α-Hydroxy-β-keto acid reductoisomerase (*Neurospora crassa*). *Methods Enzymol.* **17A**: 745–750.

1071. Kleene, R., N. Pfanner, R. Pfaller, T. A. Link, W. Sebald, W. Neupert, and M. Tropschug (1987). Mitochondrial porin of *Neurospora crassa*: cDNA cloning, *in vitro* expression and import into mitochondria. *EMBO J.* **6**: 2627–2633.

1072. Klein, J. L. (1976). Studies of mitochondrial membrane assembly in a chromosomal mutant of *Neurospora crassa*. Ph.D. Thesis, State University of New York at Albany, Albany, NY.

1073. Klein, J. L., D. L. Edwards, and S. Werner (1975). Regulation of mitochondrial membrane assembly in *Neurospora crassa*: Transient expression of a respiratory mutant phenotype. *J. Biol. Chem.* **250**: 5852–5858.

1074. Klingmüller, W. (1967). Analyse der Sorbose-Resistenz von *Neurospora crassa* an Heterokaryen sorboseresistenter Mutanten; ein Beitrag zur Genetik des aktiven Transports. I. *Mol. Gen. Genet.* **100**: 117–139.

1075. Klingmüller, W. (1967). Die Aufnahme der Zucker Sorbose, Fructose, and Glucose durch Sorbose-resistente Mutanten von *Neurospora crassa*. *Z. Naturforsch. Sect. C Biosci.* **22b**: 327–335.

1076. Klingmüller, W. (1967). Kreuzungs-Analyse Sorboseresistenter Mutanten von *Neurospora crassa*. *Mol. Gen. Genet.* **100**: 109–116.

1077. Klingmüller, W., and F. Kaudewitz (1967). "No growth" complementation in forced heterokaryons from sorbose-resistant (transport-defective) *Neurospora crassa* mutants. *Z. Naturforsch. Sect. C. Biosci.* **22b**: 1024–1027.

1078. Ko, T., B. L. Seidel-Rogol, and H. Bertrand, Personal communication.

1079. Ko, T., B. L. Seidel-Rogol, and H. Bertrand (1999). Cloning and nucleotide sequence of the catalytic subunit of DNA polymerase-γ of *Neurospora crassa*. *Fungal Genet. Newsl.* **46 (Suppl.)**: 55 (Abstr.).

1080. Koch, J., and I. B. Barthelmess (1986). Are the *cpc-1* and *mts-1* mutations of *Neurospora crassa* allelic? *Neurospora Newsl.* **33**: 30–32.

1081. Koch, J., and I. B. Barthelmess (1988). *mts(MN9)*, a *cpc-1* allele involved in a translocation. *Neurospora Newsl.* **35**: 22–23.

1082. Koh, C. L. (1973). Genetic control of the utilization of basic amino acids by their auxotrophs in *Neurospora crassa*. M.S. Thesis, University of Malaya, Kuala Lumpur, Malaysia.

1083. Koh, C. L., and C. C. Ho (1972). Mutations affecting the utilization of basic amino acids in *Neurospora*. *Genetics* **71s**: 31–32.

1084. Kohout, J. (1975). Nuclear suppression of *poky*-like cytoplasmic mutations in *Neurospora crassa*. M.S. Thesis, University of Regina, Regina, Saskatchewan, Canada.

1085. Kohout, J., and H. Bertrand (1976). Nuclear suppressors of the [*poky*] cytoplasmic mutant in *Neurospora crassa*. *Can. J. Genet. Cytol.* **18**: 311–324.

1086. Kølmark, H. G. (1969). Genetic studies of *urease* mutants in *Neurospora crassa*. *Mutat. Res.* **8**: 51–63.

1087. Kølmark, H. G. (1969). Urease-defective mutants in *Neurospora crassa*. *Mol. Gen. Genet.* **104**: 219–234.

1088. Kops, O., C. Eckerskorn, S. Hottenrott, G. Fischer, H. Mi, and M. Tropschug (1998). *ssp1*, a site-specific parvulin homolog from *Neurospora crassa* active in protein folding. *J. Biol. Chem.* **273**: 31971–31976.

1089. Kore-eda, S., T. Murayama, and I. Uno (1991). Isolation and characterization of the adenylate cyclase structural gene of *Neurospora crassa*. *Jpn. J. Genet.* **66**: 317–334.

1090. Kothe, G., M. Deak, and S. J. Free (1993). Use of the *Neurospora* tyrosinase gene as a reporter gene in transformation experiments. *Fungal Genet. Newsl.* **40**: 43–45.

1091. Kothe, G., and E. U. Selker, Personal communication.

1092. Kothe, G. O., and S. J. Free (1998). Calcineurin subunit B is required for normal vegetative growth in *Neurospora crassa*. *Fungal Genet. Biol.* **23**: 248–258.

1093. Kothe, G. O., and S. J. Free (1998). The isolation and characterization of *nrc-1* and *nrc-2*, two genes encoding protein kinases that control growth and development in *Neurospora crassa*. *Genetics* **149**: 117–130.

1094. Kouzminova, E., and E. U. Selker (1999). A mistake in the published map of linkage group VII. *Fungal Genet. Newsl.* **46**: 14–15.

1095. Kramer, S., R. V. Hageman, and K. V. Rajagopalan (1984). *In vitro* reconstitution of nitrate reductase activity of the *Neurospora crassa* mutant *nit-1*: Specific incorporation of molybdopterin. *Arch. Biochem. Biophys.* **233**: 821–829.

1096. Kreader, C. A., and J. A. Heckman (1987). Isolation and characterization of a *Neurospora crassa* ribosomal gene homologous to *cyh-2* of yeast. *Nucleic Acids Res.* **15**: 9027–9041.

1097. Kreader, C. A., C. S. Langer, and J. E. Heckman (1989). A mitochondrial protein from *Neurospora crassa* detected both on ribosomes and membrane fractions: Analysis of the gene, the message, and the protein. *J. Biol. Chem.* **264**: 317–327.

1097a. Krissinger, W. A., and S. N. Bennett (2000). Mapping and osmotic sensitivity of the mutants *os-9* (allele SS-788 and allele SS-462) and SS-18. *Fungal Genet. Newsl.* **47S**: 28 (Abstr.).

1098. Kronstad, J. W., and C. Staben (1997). Mating type in filamentous fungi. *Annu. Rev. Genet.* **32**: 245–276.

1099. Krüger, D., J. Koch, and I. B. Barthelmes (1990). *cpc-2*, a new locus involved in general control of amino acid synthetic enzymes in *Neurospora crassa*. *Curr. Genet.* **18**: 211–215.

1100. Kruschwitz, H., D. McDonald, E. Cossins, and V. Schirch (1993). Purification of *Neurospora crassa* cytosolic serine hydroxymethyltransferase. *Adv. Exp. Med. Biol.* **338**: 719–722.

1101. Kruschwitz, H. L., D. McDonald, E. A. Cossins, and V. Schirch (1994). 5-Formyltetrahydropteroylpolyglutamates are the major folate derivatives in *Neurospora crassa* conidiospores. *J. Biol. Chem.* **269**: 28757–28763.

1102. Kruszewska, A., Personal communication.

1103. Kubelik, A. R. (1989). Mitochondrial ribosome assembly in wild-type and mutant strains of *Neurospora crassa*. Ph.D. Thesis, St. Louis University, St. Louis, MO.

1104. Kubelik, A. R., B. Turcq, and A. L. Lambowitz (1991). The *Neurospora crassa cyt-20* gene encodes cytosolic and mitochondrial valyl-tRNA synthetases and may have a second function in addition to protein synthesis. *Mol. Cell. Biol.* **11**: 4022–4035.

1105. Kuffner, R., A. Rohr, A. Schmiede, C. Krull, and U. Schulte (1998). Involvement of two novel chaperones in the assembly of mitochondrial NADH:ubiquinone oxidoreductase (complex I). *J. Mol. Biol.* **283**: 409–417.

1106. Kühlbrandt, W., M. Auer, and G. A. Scarborough (1998). Structure of the P-type ATPases. *Curr. Opin. Struct. Biol.* **8**: 510–516.

1107. Kuiper, M. T., R. A. Akins, M. Holtrop, H. de Vries, and A. M. Lambowitz (1988). Isolation and analysis of the *Neurospora crassa cyt-21* gene. A nuclear gene encoding a mitochondrial ribosomal protein. *J. Biol. Chem.* **263**: 2840–2847.

1108. Kuiper, M. T. R., M. Holtrop, H. Vennema, A. M. Lambowitz, and H. de Vries (1988). A 3′ splice site mutation in a nuclear gene encoding a mitochondrial ribosomal protein in *Neurospora crassa*. *J. Biol. Chem.* **263**: 2848–2852.

1109. Kuldau, G. A., N. B. Raju, and N. L. Glass (1998). Repeat-induced point mutations in *Pad-1*, a putative RNA splicing factor from *Neurospora crassa*, confer dominant lethal effects on ascus development. *Fungal Genet. Biol.* **23**: 169–180.

1110. Kumar, A., and J. V. Paietta (1995). The sulfur controller-2 negative regulatory gene of *Neurospora crassa* encodes a protein with β-transducin repeats. *Proc. Natl. Acad. Sci. USA* **92**: 3343–3347.

1111. Kumar, A., and J. V. Paietta (1998). An additional role for the F-box motif: Gene regulation within the *Neurospora crassa* sulfur control network. *Proc. Natl. Acad. Sci. USA* **95**: 2417–2422.

1112. Künkele, K. P., S. Heins, M. Dembowski, F. E. Nargang, R. Benz, M. Thieffry, J. Walz, R. Lill, S. Nussberger, and W. Neupert (1998). The preprotein translocation channel of the outer membrane of mitochondria. *Cell* **93**: 1009–1019.

1113. Kunugi, S., Y. Uehara-Kunugi, F. von der Haar, J. Schischkoff, W. Freist, U. Englisch, and F. Cramer (1986). Biochemical comparison of the *Neurospora crassa* wild type and the temperature-sensitive and leucine auxotroph mutant *leu-5*. Purification of the cytoplasmic and mitochondrial leucyl-tRNA synthetases and comparison of the enzymatic activities and the degradation patterns. *Eur. J. Biochem.* **158**: 43–49.

1114. Kupper, U., M. Linden, K. Z. Cao, and K. Lerch (1990). Expression of tyrosinase in vegetative cultures of *Neurospora crassa* transformed with a metallothionein promoter/protyrosinase fusion gene. *Curr. Genet.* **18**: 331–335.

1115. Kupper, U., D. M. Niedermann, B. C. Schilling, and K. Lerch (1990). ATP-induced protyrosinase synthesis and carboxyl-terminal processing in *Neurospora crassa*. *Pigment Cell Res.* **3**: 207–213.

115a. Kusuda, M., H. Yajima, and H. Inoue (2000). Characterization and expression of a *Neurospora crassa* ribosomal protein gene, *crp-7*. *Curr Genet.* **37**: 119–124.

1116. Kuwana, H. (1953). Studies on the morphological mutant *"cut"* in *Neurospora crassa*. *Cytologia (Tokyo)* **18**: 235–239.

1117. Kuwana, H. (1960). Studies on a temperature-sensitive irreparable mutant of *Neurospora crassa*. *Jpn. J. Genet.* **35**: 49–57.

1118. Kuwana, H. (1961). Studies on a temperature-sensitive irreparable mutant of *Neurospora crassa*. II. Osmotic nature of the mutant b39a. *Jpn. J. Genet.* **36**: 187–199.

1119. Kuwana, H., and S. Imaeda (1976). An acetate-requiring mutant strain of *Neurospora crassa* due to a possible paracentric inversion. *Jpn. J. Genet.* **51**: 337–345.

1120. Kuwana, H., and M. Kubota (1983). Pyruvate kinase deficiency in a new acetate-requiring mutant, *ace-8*, of *Neurospora crassa*. *Jpn. J. Genet.* **58**: 579–589.

1121. Kuwana, H., and M. Miyano (1980). Map location of *ace-5*. *Neurospora Newsl.* **27**: 34.

1122. Kuwana, H., and R. Okumura (1979). Genetics and some characteristics of acetate-requiring strains in *Neurospora crassa*. *Jpn. J. Genet.* **54**: 235–244.

1123. Kuwana, H., and K. Tanaka (1987). The structural gene for pyruvate kinase in *Neurospora crassa*. *Jpn. J. Genet.* **62**: 283–290.

1124. Kuwana, H., and R. P. Wagner (1969). The *iv-3* mutants of *Neurospora crassa*. I. Genetic and biochemical characteristics. *Genetics* **62**: 479–485.

1125. Lacy, A. M. (1959). A genetic analysis of the *td* locus in *Neurospora crassa*. Ph.D. Thesis, Yale University, New Haven, CT.

1126. Lacy, A. M. (1962). Linkage data on two *tryp-3* suppressors. *Neurospora Newsl.* 1: 15.
1127. Lacy, A. M. (1965). Structural and physiological relationships within the *td* locus in *Neurospora crassa. Biochem. Biophys. Res. Commun.* 18: 812–823.
1128. Lacy, A. M., S. Mellen, and K. Pomerance (1968). Genetics and biochemistry of osmotic-remedial *td* mutants of *Neurospora. Proceedings 12th Int. Congr. Genet.*, Tokyo, Japan, 21.
1129. Lai, E. (1989). Preparation of *Neurospora crassa* chromosomes from a cell-wall-less strain. *Nucleic Acids Res.* 17: 10510.
1130. Lakin-Thomas, P., G. G. Coté, and S. Brody (1990). Circadian rhythms in *Neurospora crassa*: Biochemistry and genetics. *Crit. Rev. Microbiol.* 17: 365–416.
1131. Lakin-Thomas, P. L. (1998). Choline depletion, *frq* mutations, and temperature compensation of the circadian rhythm in *Neurospora crassa. J. Biol. Rhythms* 13: 268–277.
1132. Lakin-Thomas, P. L., and S. Brody (1985). Circadian rhythms in *Neurospora crassa*: Interactions between clock mutations. *Genetics* 109: 49–66.
1133. Lakin-Thomas, P. L., S. Brody, and G. G. Cote (1997). Temperature compensation and membrane composition in *Neurospora crassa. Chronobiol. Int.* 14: 445–454.
1134. Lambowitz, A. M., Personal communication.
1135. Lambowitz, A. M., J. R. Sabourin, H. Bertrand, R. Nickels, and L. McIntosh (1989). Immunological identification of the alternative oxidase of *Neurospora crassa* mitochondria. *Mol. Cell. Biol.* 9: 1362–1364.
1136. Landman, O. E. (1950). Formation of lactase in mutants and parental strains of *Neurospora. Genetics* 35: 673–674.
1137. Landman, O. E., and D. M. Bonner (1952). *Neurospora* lactase. I: Properties of lactase preparations from a lactose-utilizing and a lactose- non-utilizing strain. *Arch. Biochem. Biophys.* 41: 254–265.
1138. Landner, L. (1971). Genetic control of recombination in *Neurospora crassa*: Correlated regulation in unlinked chromosome intervals. *Heredity* 27: 385–392.
1139. Lansbergen, J. C., and R. E. Subden (1972). A constitutive carotenoid synthesis mutant of *Neurospora crassa (car conᶜ)). Microb. Genet. Bull.* 34: 17–18.
1140. Lara, M., L. Blanco, M. Campomanes, E. Calva, R. Palacios, and J. Mora (1982). Physiology of ammonium assimilation in *Neurospora crassa. J. Bacteriol.* 150: 105–112.
1141. Lauter, F. R. (1996). Molecular genetics of fungal photobiology. *J. Genet.* 75: 375–386.
1142. Lauter, F. R., U. Marchfelder, V. E. Russo, C. T. Yamashiro, E. Yatzkan, and O. Yarden (1998). Photoregulation of *cot-1*, a kinase-encoding gene involved in hyphal growth in *Neurospora crassa. Fungal Genet. Biol.* 23: 300–310.
1143. Lauter, F. R., V. E. Russo, and C. Yanofsky (1992). Developmental and light regulation of *eas*, the structural gene for the rodlet protein of *Neurospora. Genes Dev.* 6: 2373–2381.
1144. Lauter, F. R., and C. Yanofsky (1993). Day/night and circadian rhythm control of *con* gene expression in *Neurospora. Proc. Natl. Acad. Sci. USA* 90: 8249–8253.
1145. Lauter, F. R., and V. E. A. Russo (1991). Blue light induction of conidiation-specific genes in *Neurospora crassa. Nucleic Acids Res.* 19: 6883–6886.
1146. Le Page, R. W. F., Personal communication.
1147. Le Page, R. W. F. (1975). Altered cyclic AMP metabolism and pleiotropic catabolic deficiencies in mutants of the *crisp* gene complex of *Neurospora crassa. Heredity* 34: 293 (Abstr.).
1148. Leary, J. V., and A. M. Srb (1969). *Giant spore*, a new developmental mutant of *N. crassa. Neurospora Newsl.* 15: 22–23.
1149. Leckie, B. J., and J. R. Fincham (1971). A structural gene for *Neurospora crassa* isocitrate lyase. *J. Gen. Microbiol.* 65: 35–43.
1150. Lee, D. B., and S. J. Free (1984). Isolation and characterization of *Neurospora* mutants affected in invertase synthesis. *Genetics* 106: 591–599.
1151. Lee, H., Y. H. Fu, and G. A. Marzluf (1990). Nucleotide sequence and DNA recognition elements of *alc*, the structural gene which encodes allantoicase, a purine catabolic enzyme of *Neurospora crassa. Biochemistry* 29: 8779–8787.
1152. Lee, I. H., R. G. Walline, and M. Plamann (1998). Apolar growth of *Neurospora crassa* leads to increased secretion of extracellular proteins. *Mol. Microbiol.* 29: 209–218.
1153. Lee, K., and D. J. Ebbole (1998). Analysis of two transcription activation elements in the promoter of the developmentally regulated *con-10* gene of *Neurospora crassa. Fungal Genet Biol.* 23: 259–268.
1154. Lee, K., and D. J. Ebbole (1998). Tissue-specific repression of starvation and stress responses of the *Neurospora crassa con-10* gene is mediated by RCO1. *Fungal Genet. Biol.* 23: 269–278.
1155. Lee, K. Y., S. S. Pan, R. Erickson, and A. Nason (1974). Involvement of molybdenum and iron in the *in vitro* assembly of assimilatory nitrate reductase utilizing *Neurospora* mutant *nit-1. J. Biol. Chem.* 249: 3941–3952.
1156. Legerton, T. L., and R. L. Weiss (1979). Mobilization of sequestered metabolites into degradative reactions by nutritional stress in *Neurospora. J. Bacteriol.* 138: 909–914.
1157. Lehman, J. F., M. K. Gleason, S. K. Ahlgren, and R. L. Metzenberg. (1973). Regulation of phosphate metabolism in *Neurospora crassa*. Characterization of regulatory mutants. *Genetics* 75: 61–73.
1158. Lehman, J. F., and R. L. Metzenberg (1976). Regulation of phosphate metabolism in *Neurospora crassa*: Identification of the structural gene for repressible alkaline phosphatase. *Genetics* 84: 175–182.
1159. Lein, J., H. K. Mitchell, and M. B. Houlahan (1948). A method for selection of biochemical mutants of *Neurospora. Proc. Natl. Acad. Sci. USA* 34: 435–442.
1160. Leinweber, F.-J., and K. J. Monty (1965). Cysteine biosynthesis in *Neurospora crassa*. I. The metabolism of sulfite, sulfide, and cysteinesulfinic acid. *J. Biol. Chem.* 240: 782–787.
1161. Leiter, E. H., D. A. LaBrie, A. Bergquist, and R. P. Wagner (1971). *In vitro* mitochondrial complementation in *Neurospora crassa. Biochem. Genet.* 5: 549–561.
1162. Leong, S. A., and G. Winkelmann (1998). Molecular biology of iron transport in fungi. *Metal Ions Biol. Sys.* 35: 147–186.

1163. Lerch, K. (1978). Amino acid sequence of tyrosinase from *Neurospora crassa*. *Proc. Natl. Acad. Sci. USA* **75**: 3605–3609.

1164. Lerch, K. (1980). Copper metallothionein, a copper-binding protein from *Neurospora crassa*. *Nature* **284**: 368–370.

1165. Lerch, K. (1982). Primary structure of tyrosinase from *Neurospora crassa*. II. Complete amino acid sequence and chemical structure of a tripeptide containing an unusual thioether. *J. Biol. Chem.* **257**: 6414–6419.

1166. Lerch, K. (1983). *Neurospora* tyrosinase: Structural, spectroscopic and catalytic properties. *Mol. Cell. Biochem.* **52**: 125–138.

1167. Lerch, K. (1984). S-β-(2-Histidyl)cysteine: Properties, assay, and occurrence. *Methods Enzymol.* **106**: 355–359.

1168. Lerch, K. (1991). Purification of *Neurospora crassa* copper metallothionein. *Methods Enzymol.* **205**: 278–283.

1169. Lerch, K., C. Longoni, and E. Jordi (1982). Primary structure of tyrosinase from *Neurospora crassa*. I. Purification and amino acid sequence of the cyanogen bromide fragments. *J. Biol. Chem.* **257**: 6408–6413.

1170. Lerch, K., W. B. Mims, and J. Peisach (1981). Pulsed EPR studies of peroxide-activated cytochrome *c* peroxidase and of the mercaptoethanol derivative of *Neurospora* tyrosinase. *J. Biol. Chem.* **256**: 10088–10091.

1171. Lerch, K., and E. Schenk (1985). Primary structure of copper–zinc superoxide dismutase from *Neurospora crassa*. *J. Biol. Chem.* **260**: 9559–9566.

1172. Leslie, J. F., Personal communication.

1173. Leslie, J. F. (1982). Reciprocal translocation AR30 has a breakpoint distal to all known IIL markers. *Neurospora Newsl.* **29**: 18.

1174. Leslie, J. F., and N. B. Raju (1985). Recessive mutations from natural populations of *Neurospora crassa* that are expressed in the sexual diplophase. *Genetics* **111**: 759–777.

1175. Leslie, J. F., and C. T. Yamashiro (1997). Effects of the *tol* mutation on allelic interactions at het loci in *Neurospora crassa*. *Genome* **40**: 834–840.

1176. Leslie, J. F., and K. A. Zeller (1996). Heterokaryon incompatibility in fungi—More than just another way to die. *J. Genet.* **75**: 415–424.

1177. Lester, G. (1966). Genetic control of amino acid permeability in *Neurospora crassa*. *J. Bacteriol.* **107**: 193–202.

1178. Lester, G. (1971). End-product regulation of the tryptophan–nicotinic acid pathway in *Neurospora crassa*. *J. Bacteriol.* **107**: 448–455.

1179. Lester, G., and A. Byers (1965). Properties of two β-galactosidases of *Neurospora crassa*. *Biochem. Biophys. Res. Commun.* **18**: 725–734.

1180. Lester, G., and P. J. Russell, Personal communication.

1181. Lester, H. E., and S. R. Gross (1959). Efficient method for selection of auxotrophic mutants of *Neurospora*. *Science* **129**: 572.

1182. Levine, W. B., and G. A. Marzluf (1989). Isolation and characterization of cadmium-resistant mutants of *Neurospora crassa*. *Can. J. Microbiol.* **35**: 359–365.

1183. Lewis, C. M., and R. Holliday (1970). Mistranslation and aging in *Neurospora*. *Nature* **228**: 877–880.

1184. Lewis, M. T. (1995). Molecular genetic analysis of circadian clock genes in *Neurospora crassa*. Ph.D. Thesis, University of California, Santa Cruz, CA.

1185. Lewis, M. T., and J. F. Feldman (1998). Genetic mapping of the *bd* locus. *Fungal Genet. Newsl.* **45**: 21.

1186. Lewis, M. T., V. Lam, R. Williams, and J. F. Feldman, Personal communication.

1187. Lewis, R. W. (1948). Mutants of *Neurospora* requiring succinic acid or a biochemically related acid for growth. *Am. J. Bot.* **35**: 292–295.

1188. Li, C., M. S. Sachs, and T. J. Schmidhauser (1997). Developmental and photoregulation of three *Neurospora crassa* carotenogenic genes during conidiation induced by desiccation. *Fungal Genet. Biol.* **21**: 101–108.

1189. Li, C. G., and T. J. Schmidhauser (1995). Developmental and photoregulation of *al-1* and *al-2*, structural genes for two enzymes essential for carotenoid biosynthesis in *Neurospora*. *Dev. Biol.* **169**: 90–95.

1190. Li, Q., G. Jarai, B. Yaghmai, and G. A. Marzluf (1993). The *leu-1* gene of *Neurospora crassa*: Nucleotide and deduced amino acid sequence comparisons. *Gene* **136**: 301–305.

1191. Li, Q., and G. A. Marzluf (1996). Determination of the *Neurospora crassa* CYS3 sulfur regulatory protein consensus DNA-binding site: Amino acid substitutions in the CYS3 bZIP domain that alter DNA-binding specificity. *Curr. Genet.* **30**: 298–304.

1192. Li, Q., R. G. Ritzel, L. L. McLean, L. McIntosh, T. Ko, H. Bertrand, and F. E. Nargang (1996). Cloning and analysis of the alternative oxidase gene of *Neurospora crassa*. *Genetics* **142**: 129–140.

1193. Li, Q., L. Zhou, and G. A. Marzluf (1996). Functional *in vivo* studies of the *Neurospora crassa cys-14* gene upstream region: Importance of CYS3-binding sites for regulated expression. *Mol. Microbiol.* **22**: 109–117.

1194. Lill, R., F. E. Nargang, and W. Neupert (1996). Biogenesis of mitochondrial proteins. *Curr. Opin. Cell. Biol.* **8**: 505–512.

1195. Lin, J., and R. Addison (1994). Topology of the *Neurospora* plasma membrane H$^+$-ATPase. Localization of a transmembrane segment. *J. Biol. Chem.* **269**: 3887–3990.

1196. Lin, J., and R. Addison (1995). The membrane topology of the carboxyl-terminal third of the *Neurospora* plasma membrane H$^+$-ATPase. *J. Biol. Chem.* **270**: 6942–6948.

1197. Lin, J., and R. Addison (1995). A novel integration signal that is composed of two transmembrane segments is required to integrate the *Neurospora* plasma membrane H$^+$-ATPase into microsomes. *J. Biol. Chem.* **270**: 6935–6941.

1198. Lindberg, R. A., and H. Drucker (1984). Regulation of a *Neurospora crassa* extracellular RNase by phosphorus, nitrogen, and carbon derepressions. *J. Bacteriol.* **157**: 380–384.

1199. Lindegren, C. C. (1933). The genetics of *Neurospora*. III. Pure bred stocks and crossing over in *N. crassa*. *Bull. Torrey Bot. Club* **60**: 133–154.

1200. Lindegren, C. C. (1936). A six point map of the sex chromosome of *Neurospora crassa*. *J. Genet.* **32**: 243–256.

1201. Lindegren, C. C. (1936). The structure of the sex chromosomes of *Neurospora crassa* suggested by genetical analysis. *J. Hered* 27: 251–259.

1202. Lindegren, C. C., V. Beanfield, and R. Barber (1939). Increasing the fertility of *Neurospora* by selective inbreeding. *Bot. Gaz.* 100: 592–599.

1203. Lindegren, C. C., and G. Lindegren (1939). Nonrandom crossing over in the second chromosome of *Neurospora crassa*. *Genetics* 24: 1–7.

1204. Lindegren, C. C., and G. Lindegren (1941). X-ray and ultraviolet induced mutations in *Neurospora*. I. X-ray mutations. *J. Hered.* 32: 404–412.

1205. Linden, H., P. Ballario, G. Arpaia, and G. Macino (1999). Seeing the light: News in *Neurospora* blue light signal transduction. *Adv. Genet.* 41: 35–54.

1206. Linden, H., P. Ballario, and G. Macino (1997). Blue light regulation in *Neurospora crassa*. *Fungal Genet. Biol.* 22: 141–150.

1207. Linden, H., and G. Macino (1997). White collar 2, a partner in bluelight signal transduction, controlling expression of light-regulated genes in *Neurospora crassa*. *EMBO J.* 16: 98–109.

1208. Linden, H., M. Rodriguez-Franco, and G. Macino (1997). Mutants of *Neurospora crassa* defective in regulation of blue light perception. *Mol. Gen. Genet.* 254: 111–118.

1209. Linden, R. M., B. C. Schilling, U. A. Germann, and K. Lerch (1991). Regulation of laccase synthesis in induced *Neurospora crassa* cultures. *Curr. Genet.* 19: 375–381.

1210. Littlewood, R. K., and K. D. Munkres (1972). Simple and reliable method for replica plating *Neurospora crassa*. *J. Bacteriol.* 110: 1017–1021.

1211. Liu, C.-K., C.-A. Hsu, and M. T. Abbott (1973). Catalysis of three sequential dioxygenase reactions by thymine 7-hydroxylase. *Arch. Biochem. Biophys.* 159: 180–187.

1212. Liu, L., and J. C. Dunlap (1989). Molecular cloning and analysis of *arg-13*, *os-1* and *prd-4*. *Fungal Genet. Newsl.* 36: 15.

1213. Liu, Q., and J. C. Dunlap (1996). Isolation and analysis of the *arg-13* gene of *Neurospora crassa*. *Genetics* 143: 1163–1174.

1214. Liu, Y., N. Garceau, J. J. Loros, and J. C. Dunlap (1997). Thermally regulated translational control of FRQ mediates aspects of temperature responses in the *Neurospora* circadian clock. *Cell* 89: 477–486.

1215. Liu, Y., M. Merrow, J. J. Loros, and J. C. Dunlap (1998). How temperature changes reset a circadian oscillator. *Science* 281: 825–829.

1216. Llédias, F., P. Rangel, and W. Hansberg (1998). Oxidation of catalase by singlet oxygen. *J. Biol. Chem.* 273: 10630–10637.

1217. Lobo-Hajdu, G., H. P. Braun, N. Romp, L. A. Grivell, J. A. Berden, and U. K. Schmitz (1996). Subunit VII of ubiquinol:cytochrome *c* oxidoreductase from *Neurospora crassa* is functional in yeast and has an N-terminal extension that is not essential for mitochondrial targeting. *Biochem. J.* 320: 769–75.

1218. Logan, J. B. (1969). Biochemistry and genetics of canavanine resistance in *Neurospora*. Ph.D. Thesis, California Institute of Technology, Pasadena, CA.

1219. Loo, M. (1974). Temperature-sensitive mutants of *Neurospora* and the role of some cell functions in germination. Ph.D. Thesis, University of Washington, Seattle, WA.

1220. Loo, M. (1975). *Neurospora crassa* temperature-sensitive mutant apparently defective in protein synthesis. *J. Bacteriol.* 121: 286–295.

1221. Loo, M. W. (1975). A temperature-sensitive mutant of *Neurospora* defective in ribosome processing (*rip-1*). *Neurospora Newsl.* 22: 10–11.

1222. Loo, M. W., N. S. Schricker, and P. J. Russell (1981). Heat-sensitive mutant strain of *Neurospora crassa*, 4M(t), conditionally defective in 25S ribosomal ribonucleic acid production. *Mol. Cell. Biol.* 1: 199–207.

1223. Loring, H. S., and J. G. Pierce (1944). Pyrimidine nucleosides and nucleotides as growth factors for mutant strains of *Neurospora*. *J. Biol. Chem.* 153: 61–69.

1224. Loros, J. (1995). The molecular basis of the *Neurospora* clock. *Semin. Neurosci.* 7: 3–13.

1225. Loros, J. J. (1998). Time at the end of the millennium: The *Neurospora* clock. *Curr. Opin. Microbiol.* 1: 698–706.

1226. Loros, J. J., S. A. Denome, and J. C. Dunlap (1989). Molecular cloning of genes under control of the circadian clock in *Neurospora*. *Science* 243: 385–388.

1227. Loros, J. J., and J. C. Dunlap (1991). *Neurospora crassa* clock-controlled genes are regulated at the level of transcription. *Mol. Cell. Biol.* 11: 558–563.

1228. Loros, J. J., A. Richman, and J. F. Feldman (1986). A recessive circadian clock mutation at the *frq* locus of *Neurospora crassa*. *Genetics* 114: 1095–1110.

1229. Lowendorf, H. S., and C. W. Slayman (1975). Genetic regulation of phosphate transport system II in *Neurospora*. *Biochim. Biophys. Acta* 413: 95–103.

1230. Lowry, R. J., T. L. Durkee, and A. S. Sussman (1967). Ultrastructural studies of microconidium formation in *Neurospora crassa*. *J. Bacteriol.* 94: 1757–1763.

1231. Lu, B. C. (1993). Spreading the synaptonemal complex of *Neurospora crassa*. *Chromosoma* 102: 464–472.

1232. Lu, B. C., and D. R. Galeazzi (1978). Light and electron microscope observations of a meiotic mutant of *Neurospora crassa*. *Can. J. Bot.* 56: 2694–2706.

1233. Luck, D. J. L. (1963). Genesis of mitochondria in *Neurospora crassa*. *Proc. Natl. Acad. Sci. USA* 49: 233–240.

1234. Luker, M. A., and B. J. Kilbey (1982). A simplified method for the simultaneous detection of intragenic and intergenic mutations (deletions) in *Neurospora crassa*. *Mutat. Res.* 92: 63–68.

1235. Lumsden, J., and J. R. Coggins (1978). The subunit structure of the arom multienzyme complex of *Neurospora crassa*. Evidence from peptide "maps" for the identity of the subunits. *Biochem J.* 169: 441–414.

1236. Luo, C., J. J. Loros, and J. C. Dunlap (1998). Nuclear localization is required for function of the essential clock protein FRQ. *EMBO J.* 17: 1228–1235.

1237. Luo, Z., M. Freitag, and M. S. Sachs (1995). Translational regulation in response to changes in amino acid availability in *Neurospora crassa*. *Mol. Cell. Biol.* 15: 5235–5245.

1238. Luo, Z., and M. S. Sachs (1996). Role of an upstream open reading frame in mediating arginine-specific translational control in *Neurospora crassa*. *J. Bacteriol.* 178: 2172–2177.

1239. Macleod, H., and D. Stadler (1986). Excision of pyrimidine dimers from the DNA of *Neurospora*. *Mol. Gen. Genet.* 202: 321–326.

1240. MacPhee, K. G., R. E. Nelson, and S. M. Schuster, Personal communication.

1241. Madi, L., D. J. Ebbole, B. T. White, and C. Yanofsky (1994). Mutants of *Neurospora crassa* that alter gene expression and conidia development. *Proc. Natl. Acad. Sci. USA* 91: 6226–6230.

1242. Madi, L., S. A. McBride, L. A. Bailey, and D. J. Ebbole (1997). *rco-3*, a gene involved in glucose transport and conidiation in *Neurospora crassa*. *Genetics* 146: 499–508.

1243. Magill, C. W., H. Sweeney, and V. W. Woodward (1972). Histidine uptake in strains of *Neurospora crassa* with normal and mutant transport systems. *J. Bacteriol.* 110: 313–320.

1244. Magill, J. M., P. Dalke, T. S. Lyda, and C. W. Magill (1982). Adenosine kinase-deficient mutant of *Neurospora crassa*. *J. Bacteriol.* 152: 1292–1294.

1245. Magill, J. M., E. S. Edwards, R. L. Sabina, and C. W. Magill (1976). Depression of uracil uptake by ammonium in *Neurospora crassa*. *J. Bacteriol.* 127: 1265–1269.

1246. Mahadevan, P. R., and B. M. Eberhart (1962). A dominant regulatory gene for aryl-β-glucosidase in *Neurospora crassa*. *J. Cell. Comp. Physiol.* 60: 281–283.

1247. Mahadevan, P. R., and U. R. Mahadkar (1970). Role of enzymes in growth and morphology of *Neurospora crassa*: Cell-wall-bound enzymes and their possible role in branching. *J. Bacteriol.* 101: 941–947.

1248. Mahanty, S. K., U. S. Rao, R. A. Nicholas, and G. A. Scarborough (1994). High-yield expression of the *Neurospora crassa* plasma membrane H(+)-ATPase in *Saccharomyces cerevisiae*. *J. Biol. Chem.* 269: 17705–17712.

1249. Mahanty, S. K., and G. A. Scarborough (1996). Site-directed mutagenesis of the cysteine residues in the *Neurospora crassa* plasma membrane H(+)-ATPase. *J. Biol. Chem.* 271: 367–371.

1250. Maheshwari, R. (1991). Microcycle conidiation and its genetic basis in *Neurospora crassa*. *J. Gen. Microbiol.* 137: 2103–2115.

1251. Maheshwari, R. (1991). A new genotype of *Neurospora crassa* that selectively produces abundant microconidia in submerged shake culture. *Exp. Mycol.* 15: 346–350.

1252. Maheshwari, R. (1999). Microconidia of *Neurospora crassa*. *Fungal Genet. Biol.* 26: 1–18.

1253. Maier, J., K. Witter, M. Gutlich, I. Ziegler, T. Werner, and H. Ninnemann (1995). Homology cloning of GTP-cyclohydrolase I from various unrelated eukaryotes by reverse-transcription polymerase chain reaction using a general set of degenerate primers. *Biochem. Biophys. Res. Commun.* 212: 705–711.

1254. Maling, B. (1959). The effect of environmental factors on crossing over in *Neurospora crassa*. Ph.D. Thesis, Stanford University, Stanford, CA.

1255. Maling, B. (1959). Linkage data for group IV markers in *Neurospora*. *Genetics* 44: 1212–1220.

1256. Maling, B. (1960). Replica plating and rapid ascus collection of *Neurospora*. *J. Gen. Microbiol.* 23: 257–260.

1257. Mandala, S. M., and C. W. Slayman (1988). Identification of tryptic cleavage sites for two conformational states of the *Neurospora* plasma membrane H$^+$-ATPase. *J. Biol. Chem.* 263: 15122–15128.

1258. Mandala, S. M., and C. W. Slayman (1989). The amino and carboxyl termini of the *Neurospora* plasma membrane H$^+$-ATPase are cytoplasmically located. *J. Biol. Chem.* 264: 16276–16281.

1259. Mann, B. J., B. J. Bowman, J. Grotelueschen, and R. L. Metzenberg (1989). Nucleotide sequence of *pho-4$^+$*, encoding a phosphate-repressible phosphate permease of *Neurospora crassa*. *Gene* 83: 281–289.

1260. Mannella, C. A. (1997). Minireview: On the structure and gating mechanism of the mitochondrial channel, VDAC. *J. Bioenerg. Biomembr.* 29: 525–531.

1261. Mannella, C. A. (1998). Conformational changes in the mitochondrial channel protein, VDAC, and their functional implications. *J. Struct. Biol.* 121: 207–218.

1262. Mannella, C. A., R. A. Collins, M. R. Green, and A. M. Lambowitz (1979). Defective splicing of mitochondrial rRNA in cytochrome-deficient nuclear mutants of *Neurospora crassa*. *Proc. Natl. Acad. Sci. USA* 76: 2635–2639.

1263. Marathe, S., I. F. Connerton, and J. R. Fincham (1990). Duplication-induced mutation of a new *Neurospora* gene required for acetate utilization: Properties of the mutant and predicted amino acid sequence of the protein product. *Mol. Cell. Biol.* 10: 2638–2644.

1264. Marathe, S., Y. G. Yu, G. E. Turner, C. Palmier, and R. L. Weiss (1998). Multiple forms of arginase are differentially expressed from a single locus in *Neurospora crassa*. *J. Biol. Chem.* 273: 29776–29785.

1265. Margolin, B. S., M. Freitag, and E. U. Selker (1997). Improved plasmids for gene targeting at the *his-3* locus of *Neurospora crassa* by electroporation. *Fungal Genet. Newsl.* 44: 34–36.

1266. Margolin, B. S., P. W. Garrett-Engele, J. N. Stevens, D. Y. Fritz, C. Garrett-Engele, R. L. Metzenberg, and E. U. Selker (1998). A methylated *Neurospora* 5S rRNA pseudogene contains a transposable element inactivated by repeat-induced point mutation. *Genetics* 149: 1787–1997.

1267. Margolis, P., Personal communication.

1268. Margolis, P., and K. Black, Personal communication.

1269. Margolis, P., and C. Yanofsky (1998). Reverse genetic analysis of signal transduction loci of *N. crassa*. *Neurospora 1998*, Asilomar, CA, 24 (Abstr.).

1270. Margolles-Clark, E., and B. J. Bowman, Personal communication.

1271. Margolles-Clark, E., K. Tenney, E. J. Bowman, and B. J. Bowman (1999). The structure of the vacuolar ATPase in *Neurospora crassa*. *J. Bioenerg. Biomembr.* 31: 29–37.

1272. Margolles-Clark, E. C., S. Abreu, and B. J. Bowman (1999). Characterization of a vacuolar Ca^{2+}/H$^+$ exchanger (CAX) of *Neurospora crassa*. *Fungal. Genet. Newsl.* 46 (Suppl.): 137.

1273. Martin, C. E., and A. G. DeBusk (1975). Temperature-sensitive, osmotic-remedial mutants of *Neurospora crassa*: Osmotic-pressure-induced alterations of enzyme stability. *Mol. Gen. Genet.* 136: 31–40.

1274. Martin, P. G. (1967). A gene modifying pigment production by *pdx-1* (44602) of *Neurospora crassa*. *Microb. Genet. Bull.* 27: 9–10.

1275. Martinoia, E., U. Heck, T. Boller, A. Wiemken, and P. Matile (1979). Some properties of vacuoles isolated from *Neurospora crassa* slime variant. *Arch. Microbiol.* **120**: 31–34.

1276. Mary, A., G. L. Kidd, and S. R. Gross (1989). The response time of transcription and translation of the *leu-2* gene of *Neurospora* to its inducer, α-isopropylmalate, approaches the permissible minimum. *Biochem. Biophys. Res. Commun.* **161**: 1286–1290.

1277. Marzluf, G. A, Personal communication.

1278. Marzluf, G. A. (1970). Genetic and biochemical studies of distinct sulfate permease species in different developmental stages of *Neurospora crassa*. *Arch. Biochem. Biophys.* **138**: 254–263.

1279. Marzluf, G. A. (1970). Genetic and metabolic controls for sulfate metabolism in *Neurospora crassa*: Isolation and study of chromate-resistant and sulfate transport-negative mutants. *J. Bacteriol.* **102**: 716–721.

1280. Marzluf, G. A. (1973). Regulation of sulfate transport in *Neurospora* by transinhibition and by inositol depletion. *Arch. Biochem. Biophys.* **156**: 244–254.

1281. Marzluf, G. A. (1977). Regulation of gene expression in fungi. *In* "Regulatory Biology" (J. Copeland, and G. Marzluf, eds.), pp. 196–242. Ohio State University Press, Columbus, OH.

1282. Marzluf, G. A. (1981). Regulation of nitrogen metabolism and gene expression in fungi. *Microbiol. Rev.* **45**: 437–461.

1283. Marzluf, G. A. (1993). Regulation of sulfur and nitrogen metabolism in filamentous fungi. *Annu. Rev. Microbiol.* **47**: 31–55.

1284. Marzluf, G. A. (1994). Genetics and molecular genetics of sulfur assimilation in the fungi. *Adv. Genet.* **31**: 187–206.

1285. Marzluf, G. A. (1996). Regulation of nitrogen metabolism in mycelial fungi. *In* "The Mycota: Biochemistry and Molecular Biology" (R. Brambl, and G. A Marzluf, eds.), Vol. III, 357–368. Springer-Verlag, Heidelberg.

1286. Marzluf, G. A. (1997). Genetic regulation of nitrogen metabolism in the fungi. *Microbiol. Mol. Biol. Rev.* **61**: 17–32.

1287. Marzluf, G. A. (1997). Molecular genetics of sulfur assimilation in filamentous fungi and yeast. *Annu. Rev. Microbiol.* **51**: 73–96.

1288. Matchett, W. H. (1974). Indole channeling by tryptophan synthase of *Neurospora*. *J. Biol. Chem.* **249**: 4041–4049.

1289. Matchett, W. H., and J. A. DeMoss (1975). The subunit structure of tryptophan synthase from *Neurospora crassa*. *J. Biol. Chem.* **250**: 2941–2946.

1290. Matchett, W. H., A. M. Lacy, and J. A. DeMoss (1987). Orientation of enzymic domains in tryptophan synthase of *Neurospora crassa*: An immunoblot analysis of TRP3 mutant products. *Mol. Gen. Genet.* **208**: 398–407.

1291. Mathieson, M. J., and D. G. Catcheside (1955). Inhibition of histidine uptake in *Neurospora crassa*. *J. Gen. Microbiol.* **13**: 72–83.

1292. Matile, P. (1966). Inositol deficiency resulting in death: An explanation of its occurrence in *Neurospora crassa*. *Science* **151**: 86–88.

1293. Matsuyama, S. S., R. E. Nelson, and R. W. Siegel (1974). Mutations specifically blocking differentiation of macroconidia *Neurospora crassa*. *Dev. Biol.* **41**: 278–287.

1294. Mattern, D., and S. Brody (1979). Circadian rhythms in *Neurospora crassa*: Effects of saturated fatty acids. *J. Bacteriol.* **139**: 977–983.

1295. Mattern, D. L., L. R. Forman, and S. Brody (1982). Circadian rhythms in *Neurospora crassa*: A mutation affecting temperature compensation. *Proc. Natl. Acad. Sci. USA* **79**: 825–829.

1296. Matzanke, B. F., E. Bill, G. I. Muller, A. X. Trautwein, and G. Winkelmann (1987). Metabolic utilization of ^{57}Fe-labeled coprogen in *Neurospora crassa*. An *in vivo* Mossbauer study. *Eur. J. Biochem.* **162**: 643–650.

1297. Matzanke, B. F., E. Bill, A. X. Trautwein, and G. Winkelmann (1988). Ferricrocin functions as the main intracellular iron-storage compound in mycelia of *Neurospora crassa*. *Biol. Metab.* **1**: 18–25.

1298. Mautino, M. R., J. L. Barra, and A. L. Rosa (1996). *eth-1*, the *Neurospora crassa* locus encoding S-adenosylmethionine synthetase: Molecular cloning, sequence analysis and *in vivo* overexpression. *Genetics* **142**: 789–800.

1299. Mautino, M. R., S. D. Haedo, and A. L. Rosa (1993). Physical mapping of meiotic crossover events in a 200-kb region of *Neurospora crassa* linkage group I. *Genetics* **134**: 1077–1083.

1300. Maxwell, J., R. Bleek, S. Growther, M. Neal, T. Parker, and L. Winikur (1980). Location of *ser-4* near *arg-2* on linkage group IV. *Neurospora Newsl.* **27**: 35.

1301. Maxwell, J. B. (1970). Part I. Synthesis of L-amino acid oxidase by a serine- or glycine-requiring strain of *Neurospora*. Part II. Studies concerning multiple electrophoretic forms of tyrosinase in *Neurospora*. Ph.D. Thesis, California Institute of Technology, Pasadena, CA.

1302. Maxwell, J. B., F. Kline, and R. S. Bengston (1974). Linkage data on two new serine-requiring mutants, one of which is a new locus, serine-5 (*ser-5*). *Neurospora Newsl.* **21**: 23.

1303. Mayer, A., F. E. Nargang, W. Neupert, and R. Lill (1995). MOM22 is a receptor for mitochondrial targeting sequences and cooperates with MOM19. *EMBO J.* **14**: 4204–4211.

1304. Mays, L. L. (1969). Isolation, characterization, and genetic analysis of osmotic mutants of *Neurospora crassa*. *Genetics* **63**: 781–794.

1305. Mays, L. L., and R. W. Barratt (1974). Note on Yale osmotic stocks. *Neurospora Newsl.* **21**: 25.

1306. Mazur, P., W. J. Henzel, S. Mattoo, and J. W. Kozarich (1994). 3-Carboxy-*cis,cis*-muconate lactonizing enzyme from *Neurospora crassa*: An alternate cycloisomerase motif. *J. Bacteriol.* **176**: 1718–1728.

1307. McClintock, B. (1945). *Neurospora*. I. Preliminary observations of the chromosomes of *Neurospora crassa*. *Am. J. Bot.* **32**: 671–678.

1308. McClung, C. R., C. R. Davis, K. M. Page, and S. A. Denome (1992). Characterization of the *formate* (*for*) locus, which encodes the cytosolic serine hydroxymethyltransferase of *Neurospora crassa*. *Mol. Cell. Biol.* **12**: 1412–1421.

1309. McClung, C. R., B. A. Fox, and J. C. Dunlap (1989). The *Neurospora* clock gene *frequency* shares a sequence element with the *Drosophila* clock gene *period*. *Nature* **339**: 558–562.

1310. McDougall, K. J., J. Deters, and J. Miskimen (1977). Isolation of putrescine-requiring mutants of *Neurospora crassa*. *Antonie van Leeuwenhoek* **43**: 143–151.

1311. McDougall, K. J., R. Ostman, and V. W. Woodward (1969). The isolation and analysis of one functional type of *pyr-3* mutant in *Neurospora*. *Genetica* **40**: 527–531.

1312. McDougall, K. J., and V. W. Woodward (1964). Suppression of *pyr-3* mutants of *Neurospora*. *Neurospora Newsl.* **6**: 14.

1313. McDougall, K. J., and V. W. Woodward (1965). Suppression of arginine- and pyrimidine-requiring mutants of *Neurospora crassa*. *Genetics* **52**: 397–406.

1314. McElroy, W. D., and H. K. Mitchell (1946). Enzyme studies on a temperature-sensitive mutant of *Neurospora*. *Fed. Proc.* **5**: 376–379.

1315. McNally, M. T., and S. J. Free (1988). Isolation and characterization of a *Neurospora* glucose-repressible gene. *Curr. Genet.* **14**: 545–551.

1316. Mehta, R. D., and J. Weijer (1971). UV mutability in γ-ray-sensitive mutants of *Neurospora crassa*. Symposium on use of radiation and radioisotopes for genetic improvement of industrial microorganisms, pp. 63–71. International Atomic Energy Agency, Vienna.

1317. Melnik, V. I., and B. J. Bowman (1996). Isolation of the *vma-6* gene encoding a 41-kDa subunit of the *Neurospora crassa* vacuolar ATPase, and an adjoining gene encoding a ribosome-associated protein. *Biochim. Biophys. Acta* **1273**: 77–83.

1318. Melo, A. M., M. Duarte, and A. Videira (1999). Primary structure and characterisation of a 64-kDa NADH dehydrogenase from the inner membrane of *Neurospora crassa* mitochondria. *Biochim. Biophys. Acta* **1412**: 282–287.

1319. Merino, S. T., and M. A. Nelson, Personal communication.

1320. Merrow, M., M. Brunner, and T. Roenneberg (1999). Assignment of circadian function for the *Neurospora* clock gene *frequency*. *Nature* **399**: 584–586.

1321. Merrow, M. W., N. Y. Garceau, and J. C. Dunlap (1997). Dissection of a circadian oscillation into discrete domains. *Proc. Natl. Acad. Sci. USA* **94**: 3877–3882.

1322. Metzenberg, R. L., Personal communication.

1323. Metzenberg, R. L. (1962). A gene affecting the repression of invertase and trehalase in *Neurospora*. *Arch. Biochem. Biophys.* **96**: 468–474.

1324. Metzenberg, R. L. (1968). Repair of multiple defects of a regulatory mutant of *Neurospora* by high osmotic pressure and by reversion. *Arch. Biochem. Biophys.* **125**: 532–541.

1325. Metzenberg, R. L. (1979). Implications of some genetic control mechanisms in *Neurospora*. *Microbiol. Rev.* **43**: 361–383.

1326. Metzenberg, R. L. (1998). How *Neurospora crassa* gets its phosphorus. *In* "Phosphorus in Plant Biology: Regulatory Roles in Molecular, Cellular, Organismic, and Ecosystem Processes" (J. P. Lynch and J. Deikman, eds.), pp. 181–191. American Society of Plant Physiologists, Rockville, MD.

1327. Metzenberg, R. L. (1999). Mutation affecting the housekeeping (low affinity) phosphate permease of *Neurospora crassa*. *Fungal Genet. Newsl.* **46**: 21.

1328. Metzenberg, R. L., and S. K. Ahlgren (1970). Mutants of *Neurospora* deficient in aryl sulfatase. *Genetics* **64**: 409–422.

1329. Metzenberg, R. L., and S. K. Ahlgren (1971). Structural and regulatory control of aryl sulfatase in *Neurospora*: The use of interspecific differences in structural genes. *Genetics* **68**: 369–381.

1330. Metzenberg, R. L., G. S. Chen, and S. K. Ahlgren (1971). Reversion of aryl sulfataseless mutants of *Neurospora*. *Genetics* **68**: 359–368.

1331. Metzenberg, R. L., and W. Chia (1979). Genetic control of phosphorus assimilation in *Neurospora crassa*: Dose-dependent dominance and recessiveness in constitutive mutants. *Genetics* **93**: 625–643.

1332. Metzenberg, R. L., and N. L. Glass (1990). Mating type and mating strategies in *Neurospora*. *Bioessays* **12**: 53–59.

1333. Metzenberg, R. L., M. K. Gleason, and B. S. Littlewood (1974). Genetic control of alkaline phosphatase synthesis in *Neurospora*: The use of partial diploids in dominance studies. *Genetics* **77**: 25–43.

1334. Metzenberg, R. L., and J. Grotelueschen (1995). Restriction polymorphism maps of *Neurospora crassa*: Update. *Fungal Genet. Newsl.* **42**: 82–90.

1335. Metzenberg, R. L., and J. S. Grotelueschen (1992). Disruption of essential genes in *Neurospora* by RIP. *Fungal Genet. Newsl.* **39**: 37–49.

1336. Metzenberg, R. L., M. S. Kappy, and J. W. Parson (1964). Irreparable mutations and ethionine resistance in *Neurospora*. *Science* **145**: 1435–1435.

1337. Metzenberg, R. L., and T. A. Randall (1995). Mating type in *Neurospora* and closely related ascomycetes. Some current problems. *Can. J. Bot.* **73**: S251–S257.

1338. Metzenberg, R. L., and M. S. Sachs, Unpublished results.

1339. Metzenberg, R. L., J. N. Stevens, E. U. Selker, and E. Morzycka-Wrobleska (1985). Identification and chromosomal distribution of 5S RNA genes in *Neurospora crassa*. *Proc. Natl. Acad. Sci. USA* **82**: 2067–2071.

1340. Metzenberg, R. L., J. N. Stevens, E. U. Selker, and E. Morzycka-Wroblewska (1984). A method for finding the genetic map position of cloned DNA fragments. *Neurospora Newsl.* **31**: 35–39.

1341. Meyer, M. M., D. Techel, and L. Rensing (1997). Direct database submission.

1342. Meyer, U., Personal communication.

1343. Miao, V. P. W., M. J. Singer, M. R. Rountree, and E. U. Selker (1994). A targeted-replacement system for identification of signals for *de novo* methylation in *Neurospora crassa*. *Mol. Cell. Biol.* **14**: 7059–7067.

1344. Mich'ea-Hamzehpour, M., and G. Turian (1987). GMP stimulation of the cyanide-insensitive mitochondrial respiration in heat-shocked conidia of *Neurospora crassa*. *Experientia* **43**: 439–440.

1345. Min, J. (1995). Studies on heterokaryon incompatibility in *Neurospora*. Ph.D. Thesis, Wayne State University, Detroit, MI.

1346. Min, J., C. Xu, M. P. Lamberti, M. T. Arganoza, and R. A. Akins (1994). Cloning and mapping of heterokaryon incompatibility genes in *Neurospora crassa* by using ectopically integrating plasmids. *Fifth Int. Congr. Mycol.* Vancouver, BC, Canada, 142 (Abstr.).

1347. Minke, P., J. Tinsley, H. Lee, and M. Plamann (1997). Analysis of *Neurospora ropy* genes encoding novel proteins required for normal hyphal growth and nuclear distribution. *19th Fungal Genet. Conf.*, Asilomar, CA, 45 (Abstr.).

1348. Minke, P. F., I. H. Lee, and M. Plamann (1999). Microscopic analysis of *Neurospora ropy* mutants defective in nuclear distribution. *Fungal Genet. Biol.* 28: 55–67.

1349. Minke, P. F., I. H. Lee, J. H. Tinsley, K. S. Bruno, and M. Plamann (1999). *Neurospora crassa ro-10* and *ro-11* genes encode novel proteins required for nuclear distribution. *Mol. Microbiol.* 32: 1065–1076.

1350. Minson, A. C., and E. H. Creaser (1969). Purification of a trifunctional enzyme, catalysing three steps of the histidine pathway, from *Neurospora crassa. Biochem. J.* 114: 49–56.

1351. Mir-Rashed, N., D. J. Jacobson, R. M. Dehghany, O. C. Micali, and M. L. Smith (2000). Molecular and functional analysis of incompatibility genes at *het-6* in a population of *Neurospora crassa. Fungal Genet. Biol.* (In press).

1352. Mishra, N. C. (1977). Characterization of the new osmotic mutants (*os*) which originated during genetic transformation in *Neurospora crassa. Genet. Res.* 29: 9–19.

1353. Mishra, N. C. (1977). Genetics and biochemistry of morphogenesis in *Neurospora. Adv. Genet.* 19: 341–405.

1354. Mishra, N. C., A. Alamasan, and M. Cooley (1990). Characterization of eukaryotic DNA polymerases: Aphidicolin-resistant mutants of *Neurospora* with altered DNA polymerase. *In* "Isozymes: Structure, Function, and Use in Biology and Medicine" (Z.-I. Ogita, and C. L. Markert, eds.), pp. 295–313. Wiley-Liss, New York.

1355. Mishra, N. C., and A. M. Forsthoefel (1983). Biochemical genetics of *Neurospora* nuclease II: Mutagen sensitivity and other characteristics of the nuclease mutants. *Genet. Res.* 41: 287–297.

1356. Mishra, N. C., G. Szabó, and E. L. Tatum (1973). Nucleic acid-induced genetic changes in *Neurospora. In* "The Role of RNA in Reproduction and Development" (M. C. Niu and S. J. Segal, eds.), pp. 259–268. North Holland, Amsterdam.

1357. Mishra, N. C., and E. L. Tatum (1970). Phosphoglucomutase mutants of *Neurospora sitophila* and their relation to morphology. *Proc. Natl. Acad. Sci. USA* 66: 638–645.

1358. Mishra, N. C., and E. L. Tatum (1973). Non-Mendelian inheritance of DNA-induced inositol independence in *Neurospora. Proc. Natl. Acad. Sci. USA* 70: 3875–3879.

1359. Mishra, N. C., and S. F. Threlkeld (1967). Variation in the expression of the ragged mutant in *Neurospora. Genetics* 55: 113–121.

1360. Mitchell, H. K., and M. B. Houlahan (1946). Adenine-requiring mutants of *Neurospora crassa. Fed. Proc.* 5: 370–375.

1361. Mitchell, H. K., and M. B. Houlahan (1946). *Neurospora.* IV. A temperature-sensitive riboflavinless mutant. *Am. J. Bot.* 33: 31–35.

1362. Mitchell, H. K., and J. Lein (1948). A *Neurospora* mutant deficient in the enzymatic synthesis of tryptophan. *J. Biol. Chem.* 175: 481–482.

1363. Mitchell, K. J., and M. B. Houlahan (1947). Investigations on the biosynthesis of pyrimidine nucleosides in *Neurospora. Fed. Proc.* 6: 506–509.

1364. Mitchell, M. B. (1955). Aberrant recombination of pyridoxine mutants of *Neurospora. Proc. Natl. Acad. Sci. USA* 41: 215–220.

1365. Mitchell, M. B. (1959). Detailed analysis of a *Neurospora* cross. *Genetics* 44: 847–856.

1366. Mitchell, M. B. (1963). Indications of pre-ascus recombination in *Neurospora crassa. Genetics* 48: 553–559.

1367. Mitchell, M. B. (1966). A round spore character in *N. crassa. Neurospora Newsl.* 10: 6.

1368. Mitchell, M. B., and H. K. Mitchell (1950). The selective advantage of an adenineless double mutant over one of the single mutants involved. *Proc. Natl. Acad. Sci. USA* 36: 115–119.

1369. Mitchell, M. B., and H. K. Mitchell (1954). A partial map of linkage group D in *Neurospora crassa. Proc. Natl. Acad. Sci. USA* 40: 436–440.

1370. Mitchell, M. B., and H. K. Mitchell (1956). A nuclear gene suppressor of a cytoplasmically inherited character in *Neurospora crassa. J. Gen. Microbiol.* 14: 84–89.

1371. Mitchell, M. B., H. K. Mitchell, and A. Tissières (1953). Mendelian and non-Mendelian factors affecting the cytochrome system in *Neurospora crassa. Proc. Natl. Acad. Sci. USA* 39: 606–613.

1372. Mitchell, M. B., T. H. Pittenger, and H. K. Mitchell (1952). Pseudo- wild types in *Neurospora crassa. Proc. Natl. Acad. Sci. USA* 38: 569–580.

1373. Moczko, M., K. Dietmeier, T. Söllner, B. Segui, H. F. Steger, W. Neupert, and N. Pfanner (1992). Identification of the mitochondrial receptor complex in *Saccharomyces cerevisiae. FEBS Lett.* 310: 265–268.

1374. Mohan, P. M., and K. S. Sastry (1983). Interrelationships in trace-element metabolism in metal toxicities in nickel-resistant strains of *Neurospora crassa. Biochem. J.* 212: 205–210.

1375. Monnat, J., R. Ortega-Perez, and G. Turian (1997). Molecular cloning and expression studies of two divergent α-tubulin genes in *Neurospora crassa. FEMS Microbiol. Lett.* 150: 33–41.

1376. Mora, J. (1990). Glutamine metabolism and cycling in *Neurospora crassa. Microbiol. Rev.* 54: 293–304.

1377. More, S., Personal communication.

1378. Morelli, G., M. A. Nelson, P. Ballario, and G. Macino (1993). Photoregulated carotenoid biosynthetic genes of *Neurospora crassa. Methods Enzymol.* 214: 412–424.

1379. Morgan, D. H. (1965). Acetylornithine transaminase in *Neurospora. Neurospora Newsl.* 8: 8.

1380. Morgan, D. H. (1970). Selection and characterisation of mutants lacking arginase in *Neurospora crassa. Mol. Gen. Genet.* 108: 291–302.

1381. Morgan, J. L., S. Brown, and S. N. Bennett (1985). Isolation and characterization of an osmotic-sensitive conidial mutant of *Neurospora crassa. Genetics* 110: s79 (Abstr.).

1382. Morgan, L. W., and J. F. Feldman (1997). Isolation and characterization of a temperature-sensitive circadian clock mutant of *Neurospora crassa. Genetics* 146: 525–530.

1383. Morgan, M. P., L. Garnjobst, and E. L. Tatum (1967). Linkage relations of new morphological mutants in linkage group V of *Neurospora crassa. Genetics* 57: 605–612.

1384. Morjana, N. A., and G. A. Scarborough (1989). Evidence for an essential histidine residue in the *Neurospora crassa* plasma membrane H$^+$-ATPase. *Biochim. Biophys. Acta* **985**: 19–25.

1385. Morris, D. C., S. Safe, and R. E. Subden (1974). Detection of the ergosterol and episterol isomers lichesterol and fecosterol in nystatin-resistant mutants of *Neurospora crassa. Biochem. Genet.* **12**: 459–466.

1386. Müller, F., D. Krüger, E. Sattlegger, B. Hoffmann, P. Ballario, M. Kanaan, and I. B. Barthelmess (1995). The *cpc-2* gene of *Neurospora crassa* encodes a protein entirely composed of WD-repeat segments that is involved in general amino acid control and female fertility. *Mol. Gen. Genet.* **248**: 162–173.

1387. Münger, K., U. A. Germann, and K. Lerch (1985). Isolation and structural organization of the *Neurospora crassa* copper metallothionein gene. *EMBO J.* **4**: 2665–2668.

1388. Münger, K., U. A. Germann, and K. Lerch (1987). The *Neurospora crassa* metallothionein gene: Regulation of expression and chromosomal location. *J. Biol. Chem.* **262**: 7363–7367.

1389. Munkres, K. D. (1977). Selection of improved microconidial strains of *Neurospora crassa. Neurospora Newsl.* **24**: 9–10.

1390. Munkres, K. D. (1979). A novel class of biochemical mutants in *Neurospora crassa*: Nutritionally irreparable, temperature extremity sensitive. *J. Gen. Appl. Microbiol.* **25**: 137–144.

1391. Munkres, K. D. (1981). Biochemical genetics of aging of *Neurospora crassa* and *Podospora anserina*: A review. *In* "Age Pigments" (R. Sohal, ed.), pp. 83–100. North Holland Publishing Co., Amsterdam.

1392. Munkres, K. D. (1992). Selection and analysis of *superoxide dismutase* mutants of *Neurospora. Free Radical. Biol. Med.* **13**: 305–318.

1393. Munkres, K. D., K. Benveniste, J. Gorski, and C. A. Zuiches (1970). Genetically induced subcellular mislocation of *Neurospora* mitochondrial malate dehydrogenase. *Proc. Natl. Acad. Sci. USA* **67**: 263–270.

1394. Munkres, K. D., and C. A. Furtek (1984). Linkage of conidial longevity determinant genes in *Neurospora crassa. Mech. Ageing Dev.* **25**: 63–77.

1395. Munkres, K. D., and C. A. Furtek (1984). Selection of conidial longevity mutants of *Neurospora crassa. Mech. Ageing Dev.* **25**: 47–62.

1396. Munkres, K. D., N. H. Giles, and M. E. Case (1965). Genetic control of *Neurospora* malate dehydrogenase and aspartate aminotransferase. I. Mutant selection, linkage, and complementation studies. *Arch. Biochem. Biophys.* **109**: 397–403.

1397. Munkres, K. D., and M. Minssen (1976). Ageing of *Neurospora crassa*. I. Evidence for the free radical theory of aging from studies of a natural-death mutant. *Mech. Ageing Dev.* **5**: 79–98.

1398. Munkres, K. D., and F. M. Richards (1965). Genetic alteration of *Neurospora* malate dehydrogenase. *Arch. Biochem. Biophys.* **109**: 457–465.

1399. Murayama, T., Y. Fujisawa, and Y. Okano (1995). A suppressor mutation which suppresses adenylyl cyclase mutations in *Neurospora crassa. Exp. Mycol.* **19**: 320–323.

1400. Murayama, T., and T. Ishikawa (1973). Mutation in *Neurospora crassa* affecting some of the extracellular

enzymes and several growth characteristics. *J. Bacteriol.* **115**: 796–804.

1401. Murayama, T., and T. Ishikawa (1975). Characterization of *Neurospora crassa* mutants deficient in glucosephosphate isomerase. *J. Bacteriol.* **122**: 54–58.

1402. Murayama, T., and T. Ishikawa (1977). Temperature sensitivity of a sorbose-resistant mutant. *Neurospora Newsl.* **24**: 3.

1403. Murayama, T., I. Uno, K. Hamamoto, and T. Ishikawa (1985). A cyclic adenosine 3',5'-monophosphate-dependent protein kinase mutant of *Neurospora crassa. Arch. Microbiol.* **142**: 109–112.

1404. Murayama, T., C. T. Yamashiro, and S. Kore-eda (1993). Genes related to regulation of conidial formation in *Neurospora crassa. Recent Adv. Mycol. Res., Proc. 1st Int. Symp. Mycol. Soc. Jpn.,* Chiba, Japan, 34–37.

1405. Murray, J. C. (1959). Studies of morphology, genetics, and culture of wild-type and morphological mutant strains of *Neurospora crassa.* Ph.D. Thesis, Cornell University, Ithaca, NY.

1406. Murray, J. C., and A. M. Srb (1959). A recessive gene determining colonial growth of the mycelium and aberrant asci in *Neurospora crassa. Proc. 9th Int. Bot. Congr.,* Montreal, Canada, 276 (Abstr.).

1407. Murray, J. C., and A. M. Srb (1960). Physiological and morphological studies of a morphological mutant in *Neurospora crassa* stimulated by sorbose. *Bot. Gaz.* **122**: 72–76.

1408. Murray, J. C., and A. M. Srb (1961). A mutant locus determining abnormal morphology and ascospore lethality in *Neurospora. J. Hered.* **52**: 149–153.

1409. Murray, J. C., and A. M. Srb (1962). The morphology and genetics of wild-type and seven morphological mutant strains of *Neurospora crassa. Can. J. Bot.* **40**: 337–350.

1410. Murray, N. E. (1960). Complementation and recombination between *methionine-2* alleles in *Neurospora crassa. Heredity* **15**: 207–217.

1411. Murray, N. E. (1960). The distribution of *methionine* loci in *Neurospora crassa. Heredity* **15**: 199–206.

1412. Murray, N. E., and M. Glassey (1962). A second "leaky" histidine mutant in linkage group IV. *Neurospora Newsl.* **1**: 15–16.

1413. Murray, N. E. (1963). Polarized recombination and fine structure within the *me-2* gene of *Neurospora crassa. Genetics* **48**: 1163–1183.

1414. Murray, N. E. (1965). Cysteine mutant strains of *Neurospora. Genetics* **52**: 801–808.

1415. Murray, N. E. (1968). Linkage information for cysteine and methionine mutants. *Neurospora Newsl.* **13**: 19.

1416. Murray, N. E. (1968). Polarized intragenic recombination in chromosome rearrangements of *Neurospora. Genetics* **58**: 181–191.

1417. Murray, N. E. (1969). Reversal of polarized recombination of alleles in *Neurospora* as a function of their position. *Genetics* **61**: 67–77.

1418. Murray, N. E. (1970). Recombination events that span sites within neighbouring gene loci of *Neurospora. Genet. Res.* **15**: 109–121.

1419. Myers, J. W., and E. A. Adelberg (1954). The biosynthesis of isoleucine and valine. I. Enzymatic transformation of the dihydroxy acid precursors to the keto acid precursors. *Proc. Natl. Acad. Sci. USA* 40: 493–499.

1420. Myers, M., and B. Eberhart (1966). Regulation of cellulase and cellobiase in *Neurospora crassa*. *Biochem. Biophys. Res. Commun.* 22: 782–785.

1421. Mylyk, O. (1975). Heterokaryon incompatibility genes in *Neurospora crassa* detected using duplication-producing chromosome rearrangements. *Genetics* 80: 107–124.

1422. Mylyk, O., E. G. Barry, and D. R. Galeazzi (1974). New isogenic wild types in *N. crassa*. *Neurospora Newsl.* 21: 24.

1423. Mylyk, O., and S. Threlkeld (1974). A genetic study of female sterility in *Neurospora crassa*. *Genet. Res.* 24: 91–102.

1424. Mylyk, O. M., Personal communication.

1425. Mylyk, O. M. (1976). Heteromorphism for *heterokaryon incompatibility* genes in natural populations of *Neurospora crassa*. *Genetics* 83: 275–284.

1426. Nagai, S., and M. Flavin (1966). Acetylhomoserine and methionine biosynthesis in *Neurospora*. *J. Biol. Chem.* 241: 3861–3863.

1427. Nakamoto, R. K., and C. W. Slayman (1989). Molecular properties of the fungal plasma membrane H^+-ATPase. *J. Bioenerg. Biomembr.* 21: 621–632.

1428. Nakamura, K. (1961). An ascospore color mutant of *Neurospora crassa*. *Bot. Mag. Tokyo* 74: 104–109.

1429. Nakamura, K. (1961). Preferential segregation in linkage group V of *Neurospora crassa*. *Genetics* 46: 887 (Abstr.).

1430. Nakamura, K. (1966). Heterogeneity in crossing-over frequency in *Neurospora*. *Genetica* 37: 235–246.

1431. Nakamura, K., and T. Egashira (1961). Genetically mixed perithecia in *Neurospora*. *Nature* 190: 1129–1130.

1432. Nakashima, H. (1986). Phase shifting of the circadian conidiation rhythm in *Neurospora crassa* by calmodulin antagonists. *J. Biol. Rhythms* 1: 163–169.

1433. Nakashima, H., and J. F. Feldman (1980). Temperature-sensitivity of light-induced phase shifting of the circadian clock of *Neurospora*. *Photochem. Photobiol.* 32: 247–251.

1434. Nakashima, H., and K. Onai (1996). The circadian conidiation rhythm in *Neurospora crassa*. *Semin. Cell Dev. Biol.* 7: 765–774.

1435. Nakashima, H., J. Perlman, and J. F. Feldman (1981). Genetic evidence that protein synthesis is required for the circadian clock of *Neurospora*. *Science* 212: 361–362.

1436. Nargang, F. E., Personal communication.

1437. Nargang, F. E. (1978). The isolation and characterization of temperature sensitive and nonsensitive mutants affected in the production of cytochrome aa_3 or cytochromes aa_3 and *b* in *Neurospora crassa*. Ph.D. Thesis, University of Regina, Regina, Saskatchewan, Canada.

1438. Nargang, F. E., H. Bertrand, and S. Werner (1978). A nuclear mutant of *Neurospora crassa* lacking subunit 1 of cytochrome *c* oxidase. *J. Biol. Chem.* 253: 6364–6369.

1439. Nargang, F. E., K. P. Künkele, A. Mayer, R. G. Ritzel, W. Neupert, and R. Lill (1995). Sheltered disruption of *Neurospora crassa* MOM22, an essential component of the mitochondrial protein import complex. *EMBO J.* 14: 1099–1108.

1440. Nargang, F. E., D. Rapaport, R. G. Ritzel, W. Neupert, and R. Lill (1998). Role of the negative charges in the cytosolic domain of TOM22 in the import of precursor proteins into mitochondria. *Mol. Cell. Biol.* 18: 3173–3181.

1441. Nason, A., K. Y. Lee, S. S. Pan, P. A. Ketchum, A. Lamberti, and J. DeVries (1971). *In vitro* formation of assimilatory reduced nicotinamide adenine dinucleotide phosphate: Nitrate reductase from a *Neurospora* mutant and a component of molybdenum enzymes. *Proc. Natl. Acad. Sci. USA* 68: 3242–3246.

1441a. Natvig, D. O., Personal communication.

1441b. Navaraj, A., A. Pandit, and R. Maheshwari (2000). *senescent*: A new *Neurospora crassa* nuclear gene mutant derived from nature exhibits mitochondrial abnormalities and a "death" phenotype. *Fungal Genet. Biol.* 29: 165–173.

1442. Nazario, M., J. A. Kinsey, and M. Ahmad (1971). *Neurospora* mutant deficient in tryptophanyl-tRNA ribonucleic acid synthetase activity. *J. Bacteriol.* 105: 121–126.

1443. Nehls, U., T. Friedrich, A. Schmiede, T. Ohnishi, and H. Weiss (1992). Characterization of assembly intermediates of NADH:ubiquinone oxidoreductase (complex I) accumulated in *Neurospora* mitochondria by gene disruption. *J. Mol. Biol.* 227: 1032–1042.

1444. Nehls, U., S. Hemmer, D. A. Rohlen, J. C. Van der Pas, D. Preis, U. Sackmann, and H. Weiss (1991). cDNA and genomic DNA sequence of the 21.3-kDa subunit of NADH:ubiquinone reductase (complex I) from *Neurospora crassa*. *Biochim. Biophys. Acta* 1088: 325–326.

1445. Nehls, U., C. Kruell, and H. Weiss (1992). Direct database submission.

1446. Nelson, M. A., Personal communication.

1447. Nelson, M. A., M. E. Crawford, and D. O. Natvig (1998). Restriction polymorphism maps of *Neurospora crassa*: 1998 update. *Fungal Genet. Newsl.* 45: 44–54.

1448. Nelson, M. A., S. Kang, E. L. Braun, M. E. Crawford, P. L. Dolan, P. M. Leonard, J. Mitchell, A. M. Armijo, L. Bean, E. Blueyes, T. Cushing, A. Errett, M. Fleharty, M. Gorman, K. Judson, R. Miller, J. Ortega, I. Pavlova, J. Perea, S. Todisco, R. Trujillo, J. Valentine, A. Wells, M. Werner-Washburne, and D. O. Natvig (1997). Expressed sequences from conidial, mycelial, and sexual stages of *Neurospora crassa*. *Fungal Genet. Biol.* 21: 348–363.

1449. Nelson, M. A., S. T. Merino, and R. L. Metzenberg (1997). A putative rhamnogalacturonase required for sexual development of *Neurospora crassa*. *Genetics* 146: 531–540.

1450. Nelson, M. A., and R. L. Metzenberg (1992). Sexual development genes of *Neurospora crassa*. *Genetics* 132: 149–162.

1451. Nelson, M. A., G. Morelli, A. Carattoli, N. Romano, and G. Macino (1989). Molecular cloning of a *Neurospora crassa* carotenoid biosynthetic gene (*albino-3*) regulated by blue light and the products of the *white collar* genes. *Mol. Cell. Biol.* 9: 1271–1276.

1452. Nelson, M. A., and D. D. Perkins (2000). Restriction polymorphism maps of *Neurospora crassa*: 2000 update. *Fungal Genet. Newsl.* 47.

1453. Nelson, R. E. (1976). Report, Ad hoc session: *Slime*. *Neurospora Newsl.* 23: 14.

1454. Nelson, R. E., T. Chandler, and C. P. Selitrennikoff (1973). *sn cr*: The significance of macroconidiation for mutant hunts. *Neurospora Newsl.* 20: 33–34.

1455. Nelson, R. E., J. F. Lehman, and R. L. Metzenberg (1976). Regulation of phosphate metabolism in *Neurospora crassa*: Identification of the structural gene for repressible acid phosphatase. *Genetics* **84**: 183–192.

1456. Nelson, R. E., B. S. Littlewood, and R. L. Metzenberg (1975). Toward the domestication of *slime*. *Neurospora Newsl.* **22**: 15–16.

1457. Nelson, R. E., C. P. Selitrennikoff, and R. W. Siegel (1975). Mutants of *Neurospora* deficient in nicotinamide adenine dinucleotide (phosphate) glycohydrolase. *J. Bacteriol.* **122**: 695–709.

1458. Neuhaüser, A., W. Klingmüller, and F. Kaudewitz (1970). Selektion Actidion-resistenter Mutanten bei *Neurospora crassa* sowie ihre genetische und biochemische Analyse. *Mol. Gen. Genet.* **106**: 180–194.

1459. Neupert, W. (1997). Protein import into mitochondria. *Annu. Rev. Biochem.* **66**: 863–917.

1460. Neves, M. J., and H. F. Terenzi (1989). *In vivo* control of gluconeogenesis in wild-type *Neurospora crassa* and in the adenylate cyclase-deficient *cr-1 (crisp)* mutant. *J. Bacteriol.* **171**: 1767–1771.

1461. Newcombe, K. D., and A. J. Griffiths (1973). The suppression of *ad-3B* mutants by supersuppressors in *Neurospora crassa*. *Genetics* **75**: 615–622.

1462. Newmeyer, D. (1957). Arginine synthesis in *Neurospora*: Genetic studies. *J. Gen. Microbiol.* **16**: 449–462.

1463. Newmeyer, D. (1962). Genes influencing the conversion of citrulline to arginosuccinate in *Neurospora crassa*. *J. Gen. Microbiol.* **28**: 215–230.

1464. Newmeyer, D. (1963). Altered phenotype of *phen*. *Neurospora Newsl.* **4**: 10.

1465. Newmeyer, D. (1964). Growth and allelism of *arg-11* and *adg*. *Neurospora Newsl.* **6**: 14–15.

1466. Newmeyer, D. (1970). A suppressor of the heterokaryon incompatibility associated with mating type in *Neurospora crassa*. *Can. J. Genet. Cytol.* **12**: 914–926.

1467. Newmeyer, D. (1984). *Neurospora* mutants sensitive both to mutagens and to histidine. *Curr. Genet.* **9**: 65–74.

1468. Newmeyer, D., and D. R. Galeazzi (1977). The instability of *Neurospora* duplication Dp(IL → IR)H4250, and its genetic control. *Genetics* **85**: 461–487.

1469. Newmeyer, D., and D. R. Galeazzi (1978). A meiotic UV-sensitive mutant which causes deletion of duplications in *Neurospora*. *Genetics* **89**: 245–269.

1470. Newmeyer, D., H. B. Howe, Jr., and D. R. Galeazzi (1973). A search for complexity at the *mating-type* locus of *Neurospora crassa*. *Can. J. Genet. Cytol.* **15**: 577–585.

1471. Newmeyer, D., D. D. Perkins, and E. G. Barry (1987). An annotated pedigree of *Neurospora crassa* laboratory wild types, showing the probable origin of the nucleolus satellite and showing that certain stocks are not authentic. *Fungal Genet. Newsl.* **34**: 46–51.

1472. Newmeyer, D., V. C. Pollard, and D. D. Perkins (1988). New supplements for the formate mutant of *Neurospora*. A possible role for ascorbic acid? *Fungal Genet. Newsl.* **35**: 36–37.

1473. Newmeyer, D., A. L. Schroeder, and D. R. Galeazzi (1978). An apparent connection between histidine, recombination and repair in *Neurospora*. *Genetics* **89**: 271–279.

1474. Newmeyer, D., and E. L. Tatum (1953). Gene expression in *Neurospora* mutants requiring nicotinic acid or tryptophan. *Am. J. Bot.* **40**: 393–400.

1475. Newmeyer, D., and C. W. Taylor (1967). A pericentric inversion in *Neurospora*, with unstable duplication progeny. *Genetics* **56**: 771–791.

1476. Newmeyer, D. L., Personal communication.

1477. Newmeyer, D. L. (1951). Biochemical genetics of tryptophan mutants of *Neurospora crassa*. Ph.D. Thesis, Stanford University, Stanford, CA.

1478. Nichoalds, G. E., M. A. Bromley, and H. P. Broquist (1971). Glutarate and the lysine requirement of *Neurospora crassa* STL-7. *Fed. Proc.* **30**: 463 (Abstr.).

1479. Nichoalds, G. E., M. A. Bromley, and H. P. Broquist (1971). Glutarate and the lysine requirement of *Neurospora crassa* STL-7. *Fed. Proc.* **30**: 463 (Abstr.).

1480. Niedermann, D. M., and K. Lerch (1990). Molecular cloning of the L-amino acid oxidase gene from *Neurospora crassa*. *J. Biol. Chem.* **265**: 17246–17251.

1481. Nilheden, E., K. Holm, and H. G. Kølmark (1975). Glycerol-non-utilizing mutants in *Neurospora crassa*. Isolation by net replication. *Hereditas* **79**: 239–250.

1482. Ninnemann, H. (1991). Photostimulation of conidiation in mutants of *Neurospora crassa*. *J. Photochem. Photobiol. B* **9**: 189–199.

1483. Noguchi, Y., N. Handa, P. Ballario, G. Macino, and H. Inoue (1995). Analysis of the *Neurospora crassa* gene *mus-25* for DNA repair. *Jpn. J. Genet.* **70**: 767 (Abstr.).

1483a. Novak, D. (1971). A study of developmental mutants of *Neurospora*. Ph.D. Thesis, Cornell University, Ithaca, NY.

1484. Nyc, J. F., and S. Brody (1971). Effects of mutations and growth conditions on lipid synthesis in *Neurospora crassa*. *J. Bacteriol.* **108**: 1310–1317.

1485. Oda, K., and K. Hasunuma (1997). Genetic analysis of signal transduction through light-induced protein phosphorylation in *Neurospora crassa* perithecia. *Mol. Gen. Genet.* **256**: 593–601.

1486. Oda, K., K. Ichimura, and K. Hasunuma (1995). Genetic analysis of a newly identified light insensitive mutant *ps15-1* in *Neurospora crassa*. *Jpn. J. Genet.* **70**: 768 (Abstr.).

1487. O'Donovan, G. A., and J. Neuhard (1970). Pyrimidine metabolism in microorganisms. *Bacteriol. Rev.* **34**: 278–345.

1488. Ogilvie-Villa, S., R. M. DeBusk, and A. G. DeBusk (1981). Characterization of 2-aminoisobutyric acid transport in *Neurospora crassa*: A general amino acid permease-specific substrate. *J. Bacteriol.* **147**: 944–948.

1489. Ogura, Y., Y. Yoshida, K. Ichimura, C. Aoyagi, N. Yabe, and K. Hasanuma (1999). Isolation and characterization of *Neurospora crassa* nucleoside diphosphate kinase NNK-1. *Eur. J. Biochem.* **266**: 709–714.

1489a. Ogura, Y., Y. Yoshida, N. Yabe, and K. Hasanuma (2000). Point mutation of nucleoside diphosphate kinase results in deficiency of the light response of perithecial polarity in *Neurospora*. In preparation.

1490. Ohnishi, E., H. Macleod, and N. H. Horowitz (1962). Mutants of *Neurospora* deficient in D-amino acid oxidase. *J. Biol. Chem.* **237**: 138–142.

1491. Ohnishi, T., G. Cornelius, and H. Nakashima (1992). A mutant of *Neurospora crassa* that has a long lag phase in low-calcium medium. *J. Gen. Microbiol.* **138**: 1573–1578.

1492. Ohrenberger, J., M. T. Arganosa, and R. Akins (1994). Cloning a new *het* gene, *het-12*, in *Neurospora crassa*. *Abstr. 5th Int. Mycol. Congr.*, Vancouver, BC, Canada, 161.

1493. Ohrenberger, J. M. (1995). Cloning and characterization of a heterokaryon incompatibility suppressor locus. Ph.D. Thesis, Wayne State University, Detroit, MI.

1494. Okamoto, P. M., R. H. Garrett, and G. A. Marzluf (1993). Molecular characterization of conventional and new repeat-induced mutants of *nit-3*, the structural gene that encodes nitrate reductase in *Neurospora crassa*. *Mol. Gen. Genet.* **238**: 81–90.

1495. Okamoto, P. M., and G. A. Marzluf (1993). Nitrate reductase of *Neurospora crassa*: The functional role of individual amino acids in the heme domain as examined by site-directed mutagenesis. *Mol. Gen. Genet.* **240**: 221–230.

1496. Okamura, R., and H. Kuwana (1979). Gene–protein relationships in acetate-requiring mutants of *Neurospora crassa* deficient in activity of pyruvate dehydrogenase complex. *Jpn. J. Genet.* **54**: 245–257.

1497. Olshan, A. R., and S. R. Gross (1974). Role of the *leu-3* cistron in the regulation of the synthesis of isoleucine and valine biosynthetic enzymes in *Neurospora*. *J. Bacteriol.* **118**: 374–384.

1498. Onai, K., S. Katagiri, M. Akiyama, and H. Nakashima (1998). Mutation of the gene for the second largest subunit of RNA polymerase I prolongs the period length of the circadian conidiation rhythm in *Neurospora crassa*. *Mol. Gen. Genet.* **259**: 264–271.

1499. Onai, K., and H. Nakashima (1997). Mutation of the *cys-9* gene, which encodes thioredoxin reductase, affects the circadian conidiation rhythm in *Neurospora crassa*. *Genetics* **146**: 101–110.

1500. Ong, T. M. (1978). Use of the spot, plate and suspension test systems for the detection of the mutagenicity of environmental agents and chemical carcinogens in *Neurospora crassa*. *Mutat. Res.* **53**: 297–308.

1501. Orbach, M. (1992). Untitled. *Fungal Genet. Newsl.* **39**: 92.

1502. Orbach, M. J. (1994). A cosmid with a HyR marker for fungal library construction and screening. *Gene* **150**: 159–162.

1503. Orbach, M. J., E. B. Porro, and C. Yanofsky (1986). Cloning and characterization of the gene for β-tubulin from a benomyl-resistant mutant of *Neurospora crassa* and its use as a dominant selectable marker. *Mol. Cell. Biol.* **6**: 2452–2461.

1504. Orbach, M. J., and M. S. Sachs (1991). The Orbach-Sachs cosmid library of *N. crassa* DNA sequences (pMOcosX). *Fungal Genet. Newsl.* **38**: 97.

1505. Orbach, M. J., D. Vollrath, R. W. Davis, and C. Yanofsky (1988). An electrophoretic karyotype of *Neurospora crassa*. *Mol. Cell. Biol.* **8**: 1469–1473.

1506. Ortega Perez, R., I. Irminger Finger, J. F. Arrighi, N. Capelli, D. Van Tuinen, and G. Turian (1994). Identification and partial purification of calmodulin-binding microtubule-associated proteins from *Neurospora crassa*. *Eur. J. Biochem.* **226**: 303–310.

1507. Ortega Perez, R., D. Van Tuinen, D. Marme, J. A. Cox, and G. Turian (1981). Purification and identification of calmodulin from *Neurospora crassa*. *FEBS Lett.* **133**: 205–208.

1508. Ortega Perez, R., D. Van Tuinen, D. Marme, and G. Turian (1983). Calmodulin-stimulated cyclic nucleotide phosphodiesterase from *Neurospora crassa*. *Biochim. Biophys. Acta* **758**: 84–87.

1509. Oulevey, N., J. W. Dicker, and G. Turian (1978). Striated inclusions and defective mitochondria in the restricted form of the "amycelial" mutant of *Neurospora crassa*. *Experientia* **34**: 840–841.

1510. Oulevey-Matikian, N., and G. Turian (1968). Contrôle metabolique et aspects ultrastructuraux de la conidiation (macro-microconidies) de *Neurospora crassa*. *Arch. Mikrobiol.* **60**: 35–58.

1511. Overton, L. K., J. S. Dubins, and F. J. de Serres (1989). Molecular and classical genetic analyses of *his-3* mutants of *Neurospora crassa*: I. Tests for allelic complementation and specific revertibility. *Mutat. Res.* **214**: 267–283.

1512. Owen, N. E., P. T. Chaure, and I. F. Connerton (1992). Isolation and characterization of new fluoroacetate-resistant/acetate-non-utilizing mutants of *Neurospora crassa*. *J. Gen. Microbiol.* **138**: 2599–2608.

1513. Paietta, J., and M. L. Sargent (1981). Photoreception in *Neurospora crassa*: Correlation of reduced light sensitivity with flavin deficiency. *Proc. Natl. Acad. Sci. USA* **78**: 5573–5577.

1514. Paietta, J., and M. L. Sargent (1982). Isolation and characterization of light-insensitive mutants of *Neurospora crassa*. *Genetics* **100**: s52.

1515. Paietta, J., and M. L. Sargent (1983). Isolation and characterization of light-insensitive mutants of *Neurospora crassa*. *Genetics* **104**: 11–21.

1516. Paietta, J. V. (1990). Molecular cloning and analysis of the *scon-2* negative regulatory gene of *Neurospora crassa*. *Mol. Cell. Biol.* **10**: 5207–5214.

1517. Paietta, J. V. (1992). Production of the CYS3 regulator, a bZIP DNA-binding protein, is sufficient to induce sulfur gene expression in *Neurospora crassa*. *Mol. Cell. Biol.* **12**: 1568–1577.

1518. Paietta, J. V. (1995). Analysis of CYS3 regulator function in *Neurospora crassa* by modification of leucine zipper dimerization specificity. *Nucleic Acids Res.* **23**: 1044–1049.

1519. Paietta, J. V., R. A. Akins, A. M. Lambowitz, and G. A. Marzluf (1987). Molecular cloning and characterization of the *cys-3* regulatory gene of *Neurospora crassa*. *Mol. Cell. Biol.* **7**: 2506–2511.

1520. Pall, M. L. (1968). Kinetic and genetic studies of amino acid transport in *Neurospora*. *Genetics* **60**: 209.

1521. Pall, M. L. (1969). Amino acid transport in *Neurospora crassa*. I. Properties of two amino acid transport systems. *Biochim. Biophys. Acta* **173**: 113–127.

1522. Pall, M. L. (1970). Amino acid transport in *Neurospora crassa*. II. Properties of a basic amino acid transport system. *Biochim. Biophys. Acta* **203**: 139–149.

1523. Pall, M. L., and C. K. Robertson (1988). Growth regulation by GTP. Regulation of nucleotide pools in *Neurospora* by nitrogen and sulfur control systems. *J. Biol. Chem.* **263**: 11168–11174.

1524. Pall, M. L., J. M. Trevillyan, and N. Hinman (1981). Deficient cyclic adenosine 3′,5′-monophosphate control in mutants of two genes of *Neurospora crassa*. *Mol. Cell. Biol.* **1**: 1–8.

1525. Paluh, J. L., M. J. Orbach, T. L. Legerton, and C. Yanofsky (1988). The cross-pathway control gene of *Neurospora crassa*, *cpc-1*, encodes a protein similar to GCN4 of yeast and the DNA-binding domain of the oncogene *v-jun*-encoded protein. *Proc. Natl. Acad. Sci. USA* **85**: 3728–3732.

1526. Paluh, J. L., M. Plamann, D. Krüger, I. B. Barthelmess, C. Yanofsky, and D. D. Perkins (1990). Determination of the inactivating alterations in two mutant alleles of the *Neurospora crassa* cross-pathway control gene *cpc-1*. *Genetics* **124**: 599–606.

1527. Pan, H., B. Feng, and G. A. Marzluf (1997). Two distinct protein–protein interactions between the NIT2 and NMR regulatory proteins are required to establish nitrogen metabolite repression in *Neurospora crassa*. *Mol. Microbiol.* **26**: 721–729.

1528. Pandit, A., and R. Maheshwari (1994). A simple method of obtaining pure microconidia in *Neurospora crassa*. *Fungal Genet. Newsl.* **41**: 64–65.

1529. Pandit, N. N., and V. E. Russo (1992). Reversible inactivation of a foreign gene, *hph*, during the asexual cycle in *Neurospora crassa* transformants. *Mol. Gen. Genet.* **234**: 412–422.

1530. Pandit, N. N., and V. E. A. Russo (1991). The *bli* regulon—A network of blue light inducible gene of *N. crassa*. *Fungal Genet. Newsl.* **38**: 93–94.

1531. Pao, W. K. (1950). Investigations of the thermophobic character in *Neurospora crassa*, especially of the relationships between temperature and carbohydrate utilization. Ph.D. Thesis, California Institute of Technology, Pasadena, CA.

1532. Papavinasasundaram, K. G., and D. P. Kasbekar (1993). Pisatin resistance in *Dictyostelium discoideum* and *Neurospora crassa*: Comparison of mutant phenotypes. *J. Gen. Microbiol.* **139**: 3035–3041.

1533. Papavinasasundaram, K. G., and D. P. Kasbekar (1994). The *Neurospora crassa erg-3* gene encodes a protein with sequence homology to both yeast sterol C-14 reductase and chicken lamin B receptor. *J. Genet.* **73**: 33–41.

1534. Pardo, J. P., and C. W. Slayman (1989). Cysteine 532 and cysteine 545 are the N-ethylmaleimide-reactive residues of the *Neurospora* plasma membrane H$^+$-ATPase. *J. Biol. Chem.* **264**: 9373–9379.

1535. Parra-Gessert, L., K. Koo, J. Fajardo, and R. L. Weiss (1998). Processing and function of a polyprotein precursor of two mitochondrial proteins in *Neurospora crassa*. *J. Biol. Chem.* **273**: 7972–7980.

1536. Pateman, J. A. (1956). The stability of an adaptive system in *Neurospora*. *Microb. Genet. Bull.* **14**: 22–25.

1537. Pateman, J. A. (1957). Back-mutation studies at the *am* locus in *Neurospora crassa*. *J. Genet.* **55**: 444–455.

1538. Pazirandeh, M. (1996). Development of a metallothionein-based heavy metal biosorbent. *Biochem. Mol. Biol. Int.* **39**: 789–95.

1539. Peduzzi, R., and G. Turian (1972). Recherches sur la différenciation conidienne de *Neurospora crassa* III. Activité malico-déshydrogénasique de structures antigéniques et ses relations avec la compétence conidienne. *Ann. Inst. Pasteur (Paris)* **122**: 1081–1097.

1540. Peleg, Y., R. Aramayo, S. Kang, J. G. Hall, and R. L. Metzenberg (1996). NUC-2, a component of the phosphate-regulated signal transduction pathway in *Neurospora crassa*, is an ankyrin repeat protein. *Mol. Gen. Genet.* **252**: 709–716.

1541. Peleg, Y., and R. L. Metzenberg (1994). Analysis of the DNA-binding and dimerization activities of *Neurospora crassa* transcription factor NUC-1. *Mol. Cell. Biol.* **14**: 7816–7826.

1542. Pendyala, L., J. Smyth, and A. M. Wellman (1979). Nature of 6-methylpurine inhibition and characterization of two 6-methylpurine-resistant mutants of *Neurospora crassa*. *J. Bacteriol.* **137**: 248–255.

1543. Pendyala, L., and A. M. Wellman (1975). Effect of histidine on purine nucleotide synthesis and utilization in *Neurospora crassa*. *J. Bacteriol.* **124**: 78–85.

1544. Pendyala, L., and A. M. Wellman (1977). Developmental-stage-dependent adenine transport in *Neurospora crassa*. *J. Bacteriol.* **131**: 453–462.

1545. Pendyala, L., and A. M. Wellman (1980). Purine biosynthesis and its regulation in *Neurospora crassa*. *Biochim. Biophys. Acta* **607**: 350–360.

1546. Perkins, D. D. Unpublished results.

1547. Perkins, D. D. (1953). The detection of linkage in tetrad analysis. *Genetics* **38**: 187–197.

1548. Perkins, D. D. (1959). New markers and multiple point linkage data in *Neurospora*. *Genetics* **44**: 1185–1208.

1549. Perkins, D. D. (1960). Incidence of canavanine sensitivity in *Neurospora*. *Microb. Genet. Bull.* **17**: 17.

1550. Perkins, D. D. (1962). Asci of *bis* × *bis* crosses for chromosome cytology. *Neurospora Newsl.* **2**: 14–15.

1551. Perkins, D. D. (1962). Crossing over and interference in a multiply marked chromosome arm of *Neurospora*. *Genetics* **47**: 1253–1274.

1552. Perkins, D. D. (1968). Heterozygosity for the *C/c* heterokaryon compatibility alleles in duplications generated by a translocation in *Neurospora*. *Proc. 12th Int. Congr. Genet. (Tokyo)* **1**: 67 (Abstr.).

1553. Perkins, D. D. (1971). Conidiating colonial strains that are homozygous-fertile and suitable for replication. *Neurospora Newsl.* **18**: 12.

1554. Perkins, D. D. (1971). Gene order in the *albino* region of linkage group I. *Neurospora Newsl.* **18**: 14–15.

1555. Perkins, D. D. (1972). An insertional translocation in *Neurospora* that generates duplications heterozygous for mating type. *Genetics* **71**: 25–51.

1556. Perkins, D. D. (1972). Linkage testers having markers near the centromere. *Neurospora Newsl.* **19**: 33.

1557. Perkins, D. D. (1972). Response of *thi-5* and *thi-1* to vitamin pyrimidine. *Neurospora Newsl.* **19**: 16.

1558. Perkins, D. D. (1974). The manifestation of chromosome rearrangements in unordered asci of *Neurospora*. *Genetics* **77**: 459–489.

1559. Perkins, D. D. (1974). Osmotic mutants. *Neurospora Newsl.* **21**: 25–26.

1560. Perkins, D. D. (1975). The use of duplication-generating rearrangements for studying heterokaryon incompatibility genes in *Neurospora*. *Genetics* **80**: 87–105.

1561. Perkins, D. D. (1977). Behavior of *Neurospora sitophila* mating-type alleles in heterozygous duplications after introgression into *Neurospora crassa*. *Exp. Mycol.* **1**: 166–172.

1562. Perkins, D. D. (1977). Evidence confirming location of *het-d* in linkage group IIR. *Neurospora Newsl.* **24**: 11–12.

1563. Perkins, D. D. (1979). A new, highly fertile microconidiating combination, *dingy, fluffy. Neurospora Newsl.* **26**: 9.

1564. Perkins, D. D. (1984). Advantages of using the inactive-mating-type *a^{m1}* strain as a helper component in heterokaryons. *Neurospora Newsl.* **31**: 41–42.

1565. Perkins, D. D. (1986). *col-3: colonial-3* is an allele of *bn: button* in Neurospora. *Fungal Genet. Newsl.* **33**: 33–34.

1566. Perkins, D. D. (1986). Determining the order of chromosomal loci in Neurospora by tests of duplication coverage. *J. Genet.* **65**: 121–144.

1567. Perkins, D. D. (1986). *ro-9: ropy-9* (R2526) is apparently an allele of *da: dapple* (R2375) in Neurospora. *Fungal Genet. Newsl.* **33**: 34.

1568. Perkins, D. D. (1988). Main features of vegetative incompatibility in Neurospora. *Fungal Genet. Newsl.* **35**: 44–46.

1569. Perkins, D. D. (1988). Photoinduced carotenoid synthesis in perithecial-wall tissue of Neurospora crassa. *Fungal Genet. Newsl.* **35**: 38–39.

1570. Perkins, D. D. (1989). Visibly distinguishable *albino* alleles in Neurospora crassa. *Fungal Genet. Newsl.* **36**: 63.

1571. Perkins, D. D. (1990). New *multicent* linkage testers for centromere-linked genes and rearrangements in Neurospora. *Fungal Genet. Newsl.* **37**: 31–32.

1572. Perkins, D. D. (1991). Neurospora *alcoy* tester strains with linkage group VII marked, and their use for mapping translocations. *Fungal Genet. Newsl.* **38**: 83.

1573. Perkins, D. D. (1992). Neurospora chromosomes. *In* "The Dynamic Genome: Barbara McClintock's Ideas in the Century of Genetics" (N. Federoff and D. Botstein, eds.), pp. 33–44. Cold Spring Harbor Laboratory Press, Cold Spring Harbor, NY.

1574. Perkins, D. D. (1992). *Neurospora crassa* genetic maps. *Fungal Genet. Newsl.* **39**: 61–70.

1575. Perkins, D. D. (1992). *Neurospora*: The organism behind the molecular revolution. *Genetics* **130**: 687–701.

1576. Perkins, D. D. (1993). Use of a helper strain in Neurospora *crassa to maintain stocks of uvs-4 and uvs-5*, which deteriorate unless sheltered in heterokaryons. *Fungal Genet. Newsl.* **40**: 66.

1577. Perkins, D. D. (1994). How should the infertility of interspecies crosses be designated? *Mycologia* **86**: 758–761.

1578. Perkins, D. D. (1997). Chromosome rearrangements in Neurospora and other filamentous fungi. *Adv. Genet.* **36**: 239–398.

1579. Perkins, D. D., and R. W. Barratt (1973). A modest proposal regarding gene symbols. *Neurospora Newsl.* **20**: 38.

1580. Perkins, D. D., and E. G. Barry (1977). The cytogenetics of Neurospora. *Adv. Genet.* **19**: 133–285.

1581. Perkins, D. D., and E. G. Barry (1977). Information on chromosome rearrangements. *Neurospora Newsl.* **24**: 12–13.

1582. Perkins, D. D., and M. Björkman, Unpublished results.

1583. Perkins, D. D., and M. Björkman (1978). A temperature-sensitive morphological mutant present in Beadle–Tatum and Rockefeller–Lindegren "wild-type" stocks and their derivatives. *Neurospora Newsl.* **25**: 24–25.

1584. Perkins, D. D., and M. Björkman (1979). Additional special purpose stocks. *Neurospora Newsl.* **26**: 9–10.

1585. Perkins, D. D., M. Glassey, and B. A. Bloom (1962). New data on markers and rearrangements in Neurospora. *Can J. Genet. Cytol.* **4**: 187–205.

1586. Perkins, D. D., and T. Ishikawa (1972). Locus designations for irreparable temperature-sensitive mutants. *Neurospora Newsl.* **19**: 24.

1587. Perkins, D. D., and C. Ishitani (1959). Linkage data for group III markers in Neurospora. *Genetics* **44**: 1209–1213.

1588. Perkins, D. D., J. F. Leslie, and D. J. Jacobson (1993). Strains for identifying and studying individual vegetative (heterokaryon) incompatibility loci in Neurospora crassa. *Fungal Genet. Newsl.* **40**: 69–73.

1589. Perkins, D. D., B. S. Margolin, E. U. Selker, and S. D. Haedo (1997). Occurrence of repeat-induced point mutation in long segmental duplications of Neurospora. *Genetics* **141**: 125–136.

1590. Perkins, D. D., R. L. Metzenberg, N. B. Raju, E. U. Selker, and E. G. Barry (1986). Reversal of a Neurospora translocation by crossing over involving displaced rDNA, and methylation of the rDNA segments that result from recombination. *Genetics* **114**: 791–817.

1591. Perkins, D. D., and N. E. Murray (1963). New markers and linkage data. *Neurospora Newsl.* **4**: 26–27.

1592. Perkins, D. D., D. Newmeyer, C. W. Taylor, and D. C. Bennett (1969). New markers and map sequences in Neurospora crassa, with a description of mapping by duplication coverage, and of multiple translocations stocks for testing linkage. *Genetica* **40**: 247–278.

1593. Perkins, D. D., and V. C. Pollard (1987). Newly mapped chromosomal loci of Neurospora crassa. *Fungal Genet. Newsl.* **34**: 53.

1594. Perkins, D. D., and V. C. Pollard (1988). A highly fertile *fluffy* allele *fl^Y*, which produces macroconidia. *Fungal Genet. Newsl.* **35**: 40–41.

1595. Perkins, D. D., and V. C. Pollard (1989). Alternative fluffy testers for detecting and diagnosing chromosome rearrangements in Neurospora crassa. *Fungal Genet. Newsl.* **36**: 63–64.

1596. Perkins, D. D., A. Radford, D. Newmeyer, and M. Björkman (1982). Chromosomal loci of Neurospora crassa. *Microbiol. Rev.* **46**: 426–570.

1597. Perkins, D. D., and N. B. Raju (1986). *Neurospora discreta*, a new heterothallic species defined by its crossing behavior. *Exp. Mycol.* **10**: 323–338.

1598. Perkins, D. D., N. B. Raju, and E. G. Barry (1980). A chromosome rearrangement of Neurospora that produces viable progeny containing two nucleolus organizers. *Chromosoma* **76**: 255–275.

1599. Perkins, D. D., N. B. Raju, and E. G. Barry (1984). A chromosome rearrangement in Neurospora that produces segmental aneuploid progeny containing only part of the nucleolus organizers. *Chromosoma* **89**: 8–17.

1600. Perkins, D. D., N. B. Raju, E. G. Barry, and D. K. Butler (1995). Chromosome rearrangements that involve the nucleolus organizer region in Neurospora. *Genetics* **141**: 909–923.

1601. Perkins, D. D., N. B. Raju, E. G. Barry, and D. K. Butler (1995). Chromosome rearrangements that involve the nucleolus organizer region in *Neurospora*. *Genetics* **141**: 909–923.

1602. Perkins, D. D., N. B. Raju, V. C. Pollard, J. L. Campbell, and A. M. Richman (1986). Use of *Neurospora* Spore killer strains to obtain centromere linkage data without dissecting asci. *Can. J. Genet. Cytol.* **28**: 971–981.

1603. Perkins, D. D., M. Smith, and D. R. Galeazzi (1973). New markers and linkage data. *Neurospora Newsl.* **20**: 45–49.

1604. Perkins, D. D., C. W. Taylor, D. C. Bennett, and B. C. Turner (1969). New morphological mutants that have been localized to linkage group. *Neurospora Newsl.* **14**: 13–18.

1605. Perkins, D. D., B. C. Turner, E. G. Barry, and V. C. Pollard (1995). Cytogenetics of an intrachromosomal transposition in *Neurospora*. *Chromosoma* **104**: 260–273.

1606. Perkins, D. D., B. C. Turner, V. C. Pollard, and A. Fairfield (1989). *Neurospora* strains incorporating *fluffy*, and their use as testers. *Fungal Genet. Newsl.* **36**: 64–66.

1607. Perlin, D. S., K. Kasamo, R. J. Brooker, and C. W. Slayman (1984). Electrogenic H⁺ translocation by the plasma membrane ATPase of *Neurospora*. Studies on plasma membrane vesicles and reconstituted enzyme. *J. Biol. Chem.* **259**: 7884–7892.

1608. Perlin, D. S., M. J. San Francisco, C. W. Slayman, and B. P. Rosen. (1986). H⁺-ATP stoichiometry of proton pumps from *Neurospora crassa* and *Escherichia coli*. *Arch. Biochem. Biophys.* **248**: 53–61.

1609. Perlman, J., and J. F. Feldman (1982). Cycloheximide and heat shock induce new polypeptide synthesis in *Neurospora crassa*. *Mol. Cell. Biol.* **2**: 1167–1173.

1610. Perrine, K. G., and G. A. Marzluf (1986). Amber nonsense mutations in regulatory and structural genes of the nitrogen control circuit of *Neurospora crassa*. *Curr. Genet.* **10**: 677–684.

1611. Pfanner, N., E. A. Craig, and M. Meijer (1994). The protein import machinery of the mitochondrial inner membrane. *Trends Biochem. Sci.* **19**: 368–372.

1612. Pfanner, N., M. G. Douglas, T. Endo, N. J. Hoogenraad, R. E. Jensen, M. Meijer, W. Neupert, G. Schatz, U. K. Schmitz, and G. C. Shore (1996). Uniform nomenclature for the protein transport machinery of the mitochondrial membranes. *Trends Biochem. Sci.* **21**: 51–52.

1613. Pfiffner, E., and K. Lerch (1981). Histidine at the active site of *Neurospora* tyrosinase. *Biochemistry* **20**: 6029–6035.

1614. Philley, M. L., and C. Staben (1994). Functional analyses of the *Neurospora crassa* MT *a-1* mating-type polypeptide. *Genetics* **137**: 715–722.

1615. Philley, M. L., and C. S. Staben, Personal communication.

1616. Phillips, R. L., Personal communication.

1617. Phillips, R. L. (1967). The association of linkage group V with chromosome 2 in *Neurospora crassa*. *J. Hered.* **58**: 263–265.

1618. Phillips, R. L., and R. M. Srb (1967). A new white ascospore mutant of *Neurospora crassa*. *Can. J. Genet. Cytol.* **9**: 766–775.

1619. Phinney, B. O. (1948). Cysteine mutants in *Neurospora*. *Genetics* **33**: 624.

1620. Pierce, J. G., and H. S. Loring (1945). Growth requirements of a purine-deficient strain of *Neurospora*. *J. Biol. Chem.* **160**: 409–415.

1621. Pierce, J. G., and H. S. Loring (1948). Purine and pyrimidine antagonism in a pyrimidine-deficient mutant of *Neurospora*. *J. Biol. Chem.* **176**: 1131–1140.

1622. Pietro, R. C. L. R., J. A. Jorge, and H. F. Terenzi (1989). Pleiotropic deficiency in the control of carbon-regulated catabolic enzymes in the "Slime" variant of *Neurospora crassa*. *J. Gen. Microbiol.* **135**: 1374–1382.

1623. Pincheira, G. (1967). Studies of ascus differentiation in wild types and mutants of *Neurospora*. Ph.D. Thesis, Cornell University, Ithaca, NY.

1624. Pincheira, G., and A. M. Srb (1969). Cytology and genetics of two abnormal ascus mutants of *Neurospora*. *Can. J. Genet. Cytol.* **11**: 281–286.

1625. Pincheira, G., and A. M. Srb (1969). Genetic variation in the orientation of nuclear spindles during the development of asci in *Neurospora*. *Am. J. Bot.* **56**: 846–852.

1626. Pitkin, J., and R. H. Davis (1990). The genetics of polyamine synthesis in *Neurospora crassa*. *Arch. Biochem. Biophys.* **278**: 386–391.

1627. Pitkin, J., M. Perriere, A. Kanehl, J. L. Ristow, and R. H. Davis (1994). Polyamine metabolism and growth of *Neurospora* strains lacking *cis*-acting control sites in the ornithine decarboxylase gene. *Arch. Biochem. Biophys.* **315**: 153–160.

1628. Pittendrigh, C. S., V. G. Bruce, N. S. Rosensweig, and M. L. Rubin (1959). A biological clock in *Neurospora*. *Nature* **184**: 169–170.

1629. Pittenger, T. H. (1957). The mating-type alleles and heterokaryon formation in *Neurospora crassa*. *Microb. Genet. Bull.* **15**: 21–22.

1630. Pittenger, T. H., and T. G. Brawner (1961). Genetic control of nuclear selection in *Neurospora* heterokaryons. *Genetics* **46**: 1645–1663.

1631. Pittenger, T. H., and D. J. West (1979). Isolation and characterization of temperature-sensitive respiratory mutants of *Neurospora crassa*. *Genetics* **93**: 539–555.

1632. Plamann, M., Personal communication.

1633. Plamann, M., and D. J. Ebbole, Personal communication.

1634. Plamann, M., P. E. Minke, J. H. Tinsley, and K. S. Bruno (1994). Cytoplasmic dynein and actin-related protein Arp1 are required for normal nuclear distribution in filamentous fungi. *J. Cell Biol.* **127**: 139–149.

1634a. Platero, H., Personal communication.

1635. Plesofsky, N., and R. Brambl (1999). Glucose metabolism in *Neurospora* is altered by heat shock and by disruption of HSP30. *Biochim. Biophys. Acta* **1449**: 73–82.

1636. Plesofsky-Vig, N., and R. Brambl (1987). Two developmental stages of *Neurospora crassa* utilize similar mechanisms for responding to heat shock but contrasting mechanisms for recovery. *Mol. Cell. Biol.* **7**: 3041–3048.

1637. Plesofsky-Vig, N., and R. Brambl (1995). Disruption of the gene for Hsp30, an α-crystallin-related heat shock protein of *Neurospora crassa*, causes defects in thermotolerance. *Proc. Natl. Acad. Sci. USA* **92**: 5032–5036.

1638. Plesofsky-Vig, N., and R. Brambl (1998). Characterization of an 88-kDa heat shock protein of *Neurospora crassa* that interacts with Hsp30. *J. Biol. Chem.* **273**: 11335–11341.

1639. Pöggeler, S. (1999). Phylogenetic relationships between *mating-type* sequences from homothallic and heterothallic ascomycetes. *Curr. Genet.* **36**: 222–231.

1640. Polacco, J., and S. R. Gross (1973). The product of the *leu-3* cistron as a regulatory element for the production of the leucine biosynthetic enzymes of *Neurospora*. *Genetics* **74**: 443–459.

1641. Polizeli, M. d. L., J. A. Jorge, and H. F. Terenzi (1991). Pectinase production by *Neurospora crassa*: Purification and biochemical characterization of extracellular polygalacturonase activity. *J. Gen. Microbiol.* **137**: 1815–1823.

1642. Pollard, V. C., and D. D. Perkins (1994). Convenient scoring of deoxyglucose resistance in *Neurospora*. *Fungal Genet. Newsl.* **41**: 86.

1643. Pongratz, M., and W. Klingmüller (1973). Role of ribosomes in cycloheximide resistance of *Neurospora* mutants. *Mol. Gen. Genet.* **124**: 359–363.

1644. Popp, B., D. A. Court, R. Benz, W. Neupert, and R. Lill (1996). The role of the N- and C-termini of recombinant *Neurospora* mitochondrial porin in channel formation and voltage-dependent gating. *J. Biol. Chem.* **271**: 13593–13599.

1645. Pounder, J. I., and B. J. Bowman, Personal communication.

1646. Prade, R. A., and H. F. Terenzi (1982). Role of sulfhydryl compounds in the control of tyrosinase activity in *Neurospora crassa*. *Biochem. Genet.* **20**: 1235–1243.

1647. Prakash, A., S. Sengupta, K. Aparna, and D. P. Kasbekar (1999). The *erg-3* (sterol Δ14,15-reductase) gene of *Neurospora crassa*: Generation of null mutants by repeat-induced point mutation and complementation by proteins chimeric for human lamin B receptor sequences. *Microbiology* **145**: 1443–1451.

1648. Premakumar, R., G. J. Sorger, and D. Gooden (1979). Nitrogen metabolite repression of nitrate reductase in *Neurospora crassa*. *J. Bacteriol.* **137**: 1119–1126.

1649. Premakumar, R., G. J. Sorger, and D. Gooden (1980). Physiological characterization of a *Neurospora crassa* mutant with impaired regulation of nitrate reductase. *J. Bacteriol.* **144**: 542–551.

1650. Premakumar, R., G. J. Sorger, and D. Gooden (1980). Repression of nitrate reductase in *Neurospora* studied by using L-methionine-DL-sulfoximine and glutamine auxotroph *gln-1b*. *J. Bacteriol.* **143**: 411–415.

1651. Printz, D. B., and S. R. Gross (1967). An apparent relationship between mistranslation and an altered leucyl-tRNA synthetase in a conditional lethal mutant of *Neurospora crassa*. *Genetics* **55**: 451–467.

1652. Prokisch, H., O. Yarden, M. Dieminger, M. Tropschug, and I. B. Barthelmess (1997). Impairment of calcineurin function in *Neurospora crassa* reveals its essential role in hyphal growth, morphology and maintenance of the apical Ca^{2+} gradient. *Mol. Gen. Genet.* **256**: 104–114.

1653. Pu, R. T., G. Xu, L. Wu, J. Vierula, K. O'Donnell, X. S. Ye, and S. A. Osmani (1995). Isolation of a functional homolog of the cell-cycle-specific NIMA protein kinase of *Aspergillus nidulans* and functional analysis of conserved residues. *J. Biol. Chem.* **270**: 18110–18116.

1654. Pyle, J. E. (1980). Genetic and biochemical studies of glycerol uptake and dissimilation in *Neurospora crassa*. Ph.D. Thesis, University of Georgia, Athens, GA.

1655. Pynadath, T. I., and R. M. Fink (1967). Studies of orotidine 5'-phosphate decarboxylase in *Neurospora crassa*. *Arch. Biochem. Biophys.* **118**: 185–189.

1656. Quigley, D. R., and C. P. Selitrennikoff (1987). Sorbose-resistant mutants of *Neurospora crassa* do not have altered (1–3)glucan synthase activity. *Curr. Microbiol.* **15**: 185–192.

1657. Rabinowitz, J. C., and E. E. Snell (1953). Vitamin B$_6$ antagonists and growth of microorganisms. I. 4-Desoxypyridoxine. *Arch. Biochem. Biophys.* **43**: 399–407.

1658. Rachmeler, M. (1967). Altered protein formation as a result of suppression in *Neurospora crassa*. *J. Bacteriol.* **93**: 1863–1868.

1659. Radford, A., Unpublished results.

1660. Radford, A. (1965). Heterokaryon complementation among the pyridoxine auxotrophs of *Neurospora crassa*. *Can. J. Genet. Cytol.* **7**: 472–477.

1661. Radford, A. (1966). Further studies on the complementation at the *pdx-1* locus of *Neurospora crassa*. *Can. J. Genet. Cytol.* **8**: 672–676.

1662. Radford, A. (1967). The *en* gene of *Neurospora* in single strains and heterokaryons. *Microb. Genet. Bull.* **27**: 10–11.

1663. Radford, A. (1968). High resolution recombination analysis of the pyridoxine-1 locus of *Neurospora*. *Can. J. Genet. Cytol.* **10**: 893–897.

1664. Radford, A. (1969). Information from ICR-170-induced mutations on the structure of the pyrimidine-3 locus in *Neurospora*. *Mutat. Res.* **8**: 537–544.

1665. Radford, A. (1969). Polarised complementation at the pyrimidine-3 locus of *Neurospora*. *Mol. Gen. Genet.* **104**: 288–294.

1666. Radford, A. (1970). Intragenic mapping of the *Neurospora* pyrimidine-3 locus by functional deletions. *Mol. Gen. Genet.* **109**: 241–245.

1667. Radford, A. (1972). Restoration of a single enzyme function in bifunctionally defective nonpolar pyrimidine-3 mutants of *Neurospora*. *Mutat. Res.* **15**: 23–29.

1668. Radford, A. (1974). Change in nomenclature of sorbose-resistant mutants. *Neurospora Newsl.* **21**: 23.

1669. Radford, A., and J. H. Parish (1997). The genome and genes of *Neurospora crassa*. *Fungal Genet. Biol.* **21**: 258–266.

1670. Radhakrishnan, A. N., R. P. Wagner, and E. E. Snell (1960). Biosynthesis of valine and isoleucine. III. ketohydroxy acid reductase and hydroxy-keto acid reductoisomerase. *J. Biol. Chem.* **235**: 2322–2331.

1671. Raju, N. (1986). Postmeiotic mitoses without chromosome replication in a mutagen-sensitive *Neurospora* mutant. *Exp. Mycol.* **10**: 243–251.

1672. Raju, N. (1986). A simple fluorescent staining method for meiotic chromosomes of *Neurospora*. *Mycologia* **78**: 901–906.

1673. Raju, N. B., Personal communication.

1674. Raju, N. B. (1979). Cytogenetic behavior of *Spore killer* genes in *Neurospora*. *Genetics* **93**: 607–623.

1675. Raju, N. B. (1979). A four-spored mutant of *Neurospora crassa*. *Exp. Mycol.* **3**: 270–280.

1676. Raju, N. B. (1980). Meiosis and ascospore genesis in *Neurospora*. *Eur. J. Cell. Biol.* **23**: 208–223.

1677. Raju, N. B. (1986). Ascus development in two tempera-ture-sensitive Four-spore mutants of *Neurospora crassa*. *Can. J. Genet. Cytol.* **28:** 982–990.

1678. Raju, N. B. (1987). A *Neurospora* mutant with abnormal croziers, giant ascospores, and asci having multiple apical pores. *Mycologia* **79:** 696–706.

1679. Raju, N. B. (1988). Nonlinear asci without apical pores in the *peak* mutant of *Neurospora*. *Mycologia* **80:** 825–831.

1680. Raju, N. B. (1992). Genetic control of the sexual cycle in *Neurospora*. *Mycol. Res.* **96:** 241–262.

1681. Raju, N. B. (1994). Ascomycete spore killers: Chromoso-mal elements that distort genetic ratios among the products of meiosis. *Mycologia* **86:** 461–473.

1682. Raju, N. B. (1996). Meiotic drive in fungi: Chromosomal elements that cause fratricide and distort genetic ratios. *J. Genet.* **75:** 287–296.

1683. Raju, N. B., and E. G. Barry, Personal communication.

1684. Raju, N. B., and D. Newmeyer (1977). Giant ascospores and abnormal croziers in a mutant of *Neurospora crassa*. *Exp. Mycol.* **1:** 152–165.

1685. Raju, N. B., and D. D. Perkins (1978). Barren perithecia in *Neurospora crassa*. *Can. J. Genet. Cytol.* **20:** 41–59.

1686. Raju, N. B., and D. D. Perkins (1991). Expression of meiotic drive elements Spore killer-2 and Spore killer-3 in asci of *Neurospora tetrasperma*. *Genetics* **129:** 25–37.

1687. Raju, N. B., D. D. Perkins, and D. Newmeyer (1987). Genetically determined nonselective abortion of asci in *Neurospora crassa*. *Genetics* **141:** 119–136.

1688. Rambosek, J. A., and J. A. Kinsey (1984). An unstable mutant gene of the *am* locus of *Neurospora* results from a small duplication. *Gene* **27:** 101–107.

1689. Rand, J. B. (1975). The regulation of sugar transport and galactose metabolism in *Neurospora crassa*. Ph.D. Thesis, Rockefeller University, New York.

1690. Randall, T. A., and R. L. Metzenberg (1995). Species-specific and mating-type-specific DNA regions adjacent to mating-type idiomorphs in the genus *Neurospora*. *Genet-ics* **141:** 119–136.

1691. Randall, T. A., and R. L. Metzenberg (1998). The mating-type locus of *Neurospora crassa*: Identification of an ad-jacent gene and characterization of transcripts surrounding the idiomorphs. *Mol. Gen. Genet.* **259:** 615–621.

1692. Randall, T. A., and D. A. Stadler, Personal communica-tion.

1693. Rao, E., and A. G. DeBusk (1975). A mutant of *Neurospora* deficient in the general (Pm G) amino acid transport system. *Neurospora Newsl.* **22:** 12–13.

1694. Rao, E. Y., T. K. Rao, and A. G. Debusk (1975). Isolation and characterization of a mutant of *Neurospora crassa* deficient in general amino acid permease activity. *Biochim. Biophys. Acta* **413:** 45–51.

1695. Rao, T. K., and A. G. DeBusk (1973). Characteristics of a transport-deficient mutant (*nap*) of *Neurospora crassa*. *Biochim. Biophys. Acta* **323:** 619–626.

1696. Rao, T. K., and A. G. DeBusk (1977). An inducible acetate transport system in *Neurospora crassa* conidia. *Biochim. Biophys. Acta* **470:** 475–483.

1697. Rao, U. S., D. D. Bauzon, and G. A. Scarborough (1992). Cytoplasmic location of amino acids 359–440 of the *Neurospora crassa* plasma membrane H(+)-ATPase. *Biochim. Biophys. Acta* **1108:** 153–158.

1698. Rao, U. S., J. P. Hennessey, Jr., and G. A. Scarborough (1991). Identification of the membrane-embedded regions of the *Neurospora crassa* plasma membrane H(+)-ATPase. *J. Biol. Chem.* **266:** 14740–14746.

1699. Rao, V. S. K. V. R., C. H. Wilson, and P. M. Mohan (1997). Zinc resistance in *Neurospora crassa*. *Biometals* **10:** 147–166.

1700. Rapaport, D., K. P. Künkele, M. Dembowski, U. Ahting, F. E. Nargang, W. Neupert, and R. Lill (1998). Dynamics of the TOM complex of mitochondria during binding and translocation of preproteins. *Mol. Cell. Biol.* **18:** 5256–5262.

1701. Rapaport, D., W. Neupert, and R. Lill (1997). Mitochon-drial protein import. Tom40 plays a major role in targeting and translocation of preproteins by forming a specific binding site for the presequence. *J. Biol. Chem.* **272:** 18725–18731.

1702. Rassow, J., K. Mohrs, S. Koidl, I. B. Barthelmess, N. Pfanner, and M. Tropschug (1995). Cyclophilin 20 is involved in mitochondrial protein folding in cooperation with molecular chaperones Hsp70 and Hsp60. *Mol. Cell. Biol.* **15:** 2654–2662.

1703. Rau, W., and U. Mitzka-Schnabel (1985). Carotenoid synthesis in *Neurospora crassa*. *Methods Enzymol.* **110:** 253–267.

1704. Regnery, D. C., Personal communication.

1705. Regnery, D. C. (1944). A leucineless mutant strain of *Neurospora crassa*. *J. Biol. Chem.* **154:** 151–160.

1706. Regnery, D. C. (1947). A study of the leucineless mutants of *Neurospora crassa*. Ph.D. Thesis, California Institute of Technology, Pasadena, CA.

1707. Reich, E., and S. Silagi (1963). Glutamine metabolism and glutamine-requiring mutants of *Neurospora crassa*. *Proc. Eleventh Int. Congr. Genet.* (The Hague), 49–50. Perga-mon Press, Oxford.

1708. Reichenbecher, V. E., Jr., M. Fischer, and S. R. Gross (1978). Regulation of isopropylmalate isomerase synthesis in *Neurospora crassa*. *J. Bacteriol.* **133:** 794–801.

1709. Reichenbecher, V. E., Jr., and S. R. Gross (1978). Structural features of normal and complemented forms of the *Neurospora* isopropylmalate isomerase. *J. Bacteriol.* **133:** 802–810.

1710. Reig, J. A., M. T. Tellez-Inon, M. M. Flawia, and H. N. Torres (1984). Activation of *Neurospora crassa* soluble adenylate cyclase by calmodulin. *Biochem. J.* **221:** 541–543.

1711. Reimer, H., and H. Ninneman (1992). Effect of light on conidiation of *nit* and *lis* mutants of *Neurospora crassa* grown on different nitrogen sources. *Fungal Genet. Newsl.* **39:** 71–72.

1712. Reinert, W. R., and G. A. Marzluf (1975). Genetic and metabolic control of the purine catabolic enzymes of *Neurospora crassa*. *Mol. Gen. Genet.* **139:** 39–55.

1713. Reisner, A., R. W. Barratt, and D. Newmeyer (1953). Confirmation of the seventh linkage group of *Neurospora crassa*. *Genetics* **38:** 685 (Abstr.).

1714. Reissig, J. (1958). A marker for chromosome V of *Neurospora*. *Microb. Genet. Bull.* **16:** 21.

1715. Reissig, J. L. (1960). Forward and back mutation in the *pyr-3* region of *Neurospora*. *Genet. Res.* **1:** 356–374.

1716. Reissig, J. L. (1960). Forward and back mutation in the *pyr-3* region of *Neurospora*. I. Mutations from arginine dependence to prototrophy. *Genet. Res.* **1:** 356–374.

1717. Reissig, J. L. (1963). Spectrum of forward mutants in the *pyr-3* region of *Neurospora*. *J. Gen. Microbiol.* **30:** 327–337.

1718. Reissig, J. L. (1974). Decoding of regulatory signals at the microbial surface. *Curr. Topics Microbiol. Immunol.* **67:** 43–96.

1719. Reissig, J. L., and J. E. Glasgow (1971). Mucopolysaccharide which regulates growth in *Neurospora*. *J. Bacteriol.* **106:** 882–889.

1720. Reissig, J. L., A. S. Issaly, and I. M. De Issaly (1967). Arginine–pyrimidine pathways in microorganisms. *Natl. Cancer Inst. Monogr.* **27:** 259–271.

1721. Reissig, J. L., A. S. Issaly, and I. M. Issaly (1967). Arginine–pyrimidine pathways in microorganisms. *Natl. Cancer Inst. Monogr.* **27:** 259–271.

1722. Reissig, J. L., A. S. Issaly, M. Nazario, and A. J. Jobbagy (1965). *Neurospora* mutants deficient in aspartate transcarbamylase. *Natl. Cancer Inst. Monogr.* **18:** 21–23.

1723. Rigby, D. J., and A. Radford (1981). The involvement of proteolysis in conformational stability of the carbamoyl phosphate synthetase/aspartate carbamoyltransferase enzyme of *Neurospora crassa*. *Biochim. Biophys. Acta* **661:** 315–322.

1724. Rines, H. W. (1968). The recovery of mutants in the inducible quinic acid catabolic pathway in *Neurospora crassa*. *Genetics* **60:** 215.

1725. Rines, H. W. (1969). Genetical and biochemical studies on the inducible quinic acid catabolic pathway in *Neurospora crassa*. Ph.D. Thesis, Yale University, New Haven, CT.

1726. Rines, H. W., M. E. Case, and N. H. Giles (1969). Mutants in the *arom* gene cluster of *Neurospora crassa* specific for biosynthetic dehydroquinase. *Genetics* **61:** 789–800.

1727. Riquelme, M., and S. Bartnicki-Garcia (1998). Morphometric and cytologic analysis of hyphae in *Neurospora crassa ropy* mutants. *Inoculum (Mycologia* 49 Suppl.): 44 (Abstr.).

1728. Ritari, S. J., W. Sakami, and C. W. Black (1973). Identification of two genes specifying folylpolyglutamic synthases. *Neurospora Newsl.* **20:** 27.

1729. Robb, M. J., M. A. Wilson, and P. J. Vierula (1995). A fungal actin-related protein involved in nuclear migration. *Mol. Gen. Genet.* **247:** 583–590.

1730. Roberts, A. N., V. Berlin, K. M. Hager, and C. Yanofsky (1988). Molecular analysis of a *Neurospora crassa* gene expressed during conidiation. *Mol. Cell. Biol.* **8:** 2411–2418.

1731. Roberts, A. N., and C. Yanofsky (1989). Genes expressed during conidiation in *Neurospora crassa*: Characterization of *con-8*. *Nucleic Acids Res.* **17:** 197–214.

1732. Roberts, C. J., and E. U. Selker (1995). Mutations affecting the biosynthesis of *S*-adenosylmethionine cause reduction of DNA methylation in *Neurospora crassa*. *Nucleic Acids Res.* **23:** 4818–4826.

1733. Rodland, K. D., and P. J. Russell (1982). Regulation of ribosomal RNA cistron number in a double nucleolar organizer strain of *Neurospora crassa* possessing heterogeneous rDNA. *Fed. Proc.* **41:** 1037.

1734. Rodland, K. D., and P. J. Russell (1982). Regulation of ribosomal RNA cistron number in a strain of *Neurospora crassa* with a duplication of the nucleolus organizer region. *Biochim. Biophys. Acta* **697:** 162–169.

1735. Rodriguez-Cousino, N., F. E. Nargang, R. Baardman, W. Neupert, R. Lill, and D. A. Court (1998). An import signal in the cytosolic domain of the *Neurospora* mitochondrial outer membrane protein TOM22. *J. Biol. Chem.* **273:** 11527–11532.

1736. Roess, W. B., and A. G. DeBusk (1968). Properties of a basic amino acid permease in *Neurospora crassa*. *J. Gen. Microbiol.* **52:** 421–432.

1737. Romano, N., and G. Macino (1992). Quelling: transient inactivation of gene expression in *Neurospora crassa* by transformation with homologous sequences. *Mol. Microbiol.* **6:** 3343–3353.

1738. Romero, D., and G. Davilá (1986). Genetic and biochemical identification of the glutamate synthase structural gene in *Neurospora crassa*. *J. Bacteriol.* **167:** 1043–1047.

1738a. Rosa, A. L., Personal communication.

1739. Rosa, A. L., M. E. Alvarez, and C. Maldonado (1990). Abnormal cytoplasmic bundles of filaments in the *Neurospora crassa snowflake* colonial mutant contain P59Nc. *Exp. Mycol.* **14:** 372–380.

1740. Rosa, A. L., S. D. Haedo, E. D. Temporini, G. A. Barioli, and M. R. Mautino (1997). Mapping chromosome landmarks in the centromere I region of *Neurospora crassa*. *Fungal Genet. Biol.* **21:** 315–322.

1741. Rosenberg, E., C. Mora, and D. L. Edwards (1976). Selection of extranuclear mutants of *Neurospora crassa*. *Genetics* **83:** 11–24.

1742. Rosenberg, G., and M. L. Pall (1978). Cyclic AMP and cyclic GMP in germinating conidia of *Neurospora crassa*. *Arch. Microbiol.* **118:** 87–90.

1743. Rosenberg, G., and M. L. Pall (1979). Properties of two cyclic nucleotide-deficient mutants of *Neurospora crassa*. *J. Bacteriol.* **137:** 1140–1144.

1744. Rossier, C., T. C. Ton-That, and G. Turian (1977). Microcyclic microconidiation in *Neurospora crassa*. *Exp. Mycol.* **1:** 52–62.

1745. Rothschild, H., J. Germershausen, and S. R. Suskind (1975). Biochemical genetic studies of cycloheximide resistance in *Neurospora crassa*. *Biochem. Genet.* **13:** 283–300.

1746. Rountree, M. R., and E. U. Selker (1997). DNA methylation inhibits elongation but not initiation of transcription in *Neurospora crassa*. *Genes Dev.* **11:** 2383–2395.

1747. Roychowdhury, H. S., D. Wong, and M. Kapoor (1992). *hsp80* of *Neurospora crassa*: cDNA cloning, gene mapping, and studies of mRNA accumulation under stress. *Biochem. Cell Biol.* **70:** 1356–1367.

1748. Ruegg, C., D. Ammer, and K. Lerch (1982). Comparison of amino acid sequence and thermostability of tyrosinase from three wild-type strains of *Neurospora crassa*. *J. Biol. Chem.* **257:** 6420–6426.

1749. Ruegg, C., and K. Lerch (1981). Cobalt tyrosinase: Replacement of the binuclear copper of *Neurospora* tyrosinase by cobalt. *Biochemistry* **20:** 1256–1262.

1750. Russell, G. C., and J. R. Guest (1991). Sequence similarities within the family of dihydrolipoamide acyltransferases and discovery of a previously unidentified fungal enzyme. *Biochim. Biophys. Acta* **1076:** 225–232.

1751. Russell, P. J., Personal communication.

1752. Russell, P. J. (1977). Cold-sensitive growth phenotype of some supersuppressor strains of *Neurospora crassa. Neurospora Newsl.* 24: 6–7.

1753. Russell, P. J., R. R. Granville, and N. J. Tublitz (1980). A cold-sensitive mutant of *Neurospora crassa* obtained using tritium-suicide enrichment that is conditionally defective in the biosynthesis of cytoplasmic ribosomes. *Exp. Mycol.* 4: 23–32.

1754. Russell, P. J., J. R. Hammett, and E. U. Selker (1976). *Neurospora crassa* cytoplasmic ribosomes: Cold-sensitive mutant defective in ribosomal ribonucleic acid synthesis. *J. Bacteriol.* 125: 1112–1119.

1755. Russell, P. J., M. W. Loo, and N. S. Schricker (1985). Growth and macromolecular synthesis phenotypes of a heat-sensitive mutant strain, *rip-1*, of *Neurospora crassa. Mol. Gen. Genet.* 200: 247–251.

1756. Russell, P. J., E. U. Selker, and J. A. Jackson (1981). Cold-sensitive mutation in *Neurospora crassa* affecting the production of 17S ribosomal RNA from ribosomal precursor RNA. *Curr. Genet.* 4: 1–5.

1757. Russell, P. J., and A. M. Srb (1972). Dominance modifiers in *Neurospora crassa*: Phenocopy selection and influence of certain ascus mutants. *Genetics* 71: 233–245.

1758. Russell, P. J., and A. M. Srb (1974). A study of L-glutamine: D-fructose 6-phosphate amidotransferase in certain developmental mutants of *Neurospora crassa. Mol. Gen. Genet.* 129: 77–86.

1759. Russell, P. J., and K. Talbot (1980). S-Adenosyl methionine synthetase in a ribosome biosynthesis mutant of *Neurospora crassa. Genetics* 94: s90.

1760. Russo, V. E. A. (1986). Are carotenoids the blue light photoreceptor in the photoinduction of protoperithecia in *Neurospora crassa? Planta* 168: 56–60.

1761. Rutledge, B. J. (1984). Molecular characterization of the *qa-4* gene of *Neurospora crassa. Gene* 32: 275–287.

1762. Ryan, F. J., and J. Lederberg (1946). Reverse mutation and adaptation in leucineless *Neurospora. Proc. Natl. Acad. Sci. USA* 32: 163–173.

1763. Sabina, R. L. (1979). Regulation of nucleotide metabolism during conidial germination in *Neurospora crassa.* Ph.D. Thesis, Texas A&M University, College Station, TX.

1764. Sabina, R. L., J. M. Magill, and C. M. Magill (1980). Identification of the biochemical blocks in the *ad-1* and *ad-5* strains of *Neurospora crassa. Genetics* 94: s91 (Abstr.).

1765. Sabina, R. L., J. M. Magill, and C. W. Magill (1976). Regulation of hypoxanthine transport in *Neurospora crassa. J. Bacteriol.* 128: 598–603.

1766. Sachs, M. S., Unpublished results.

1767. Sachs, M. S. (1986). The nuclear genes for cytochrome *c* oxidase subunits of *Neurospora crassa.* Ph.D. Thesis, Massachusetts Institute of Technology, Cambridge, MA.

1768. Sachs, M. S. (1996). General and cross-pathway controls of amino acid biosynthesis. *In* "The Mycota: Biochemistry and Molecular Biology" (R. Brambl and G. A. Murzluf, eds.), Vol. III, pp. 315–345. Springer-Verlag, Heidelberg.

1769. Sachs, M. S., H. Bertrand, R. L. Metzenberg, and U. L. RajBhandary (1989). Cytochrome oxidase subunit V gene of *Neurospora crassa*: DNA sequences, chromosomal mapping, and evidence that the *cya-4* locus specifies the structural gene for subunit V. *Mol. Cell. Biol.* 9: 566–577.

1770. Sachs, M. S., M. David, S. Werner, and U. L. RajBhandary (1986). Nuclear genes for cytochrome *c* oxidase subunits of *Neurospora crassa*: Isolation and characterization of cDNA clones for subunits IV, V, VI, and possibly VII. *J. Biol. Chem.* 261: 869–873.

1771. Sachs, M. S., and D. Ebbole (1990). The use of *lacZ* gene fusions in *Neurospora crassa. Fungal Genet. Newsl.* 37: 35–36.

1772. Sachs, M. S., E. U. Selker, B. Lin, C. J. Roberts, Z. Luo, D. Vaught-Alexander, and B. Margolin (1997). Expression of herpes virus thymidine kinase in *Neurospora crassa. Nucleic. Acids. Res.* 25: 2389–2395.

1773. Sachs, M. S., and C. Yanofsky (1991). Developmental expression of genes involved in conidiation and amino acid biosynthesis in *Neurospora crassa. Dev. Biol.* 148: 117–128.

1774. Sackmann, U., R. Zensen, D. Rohlen, U. Jahnke, and H. Weiss (1991). The acyl carrier protein in *Neurospora crassa* mitochondria is a subunit of NADH:ubiquinone reductase (complex I). *Eur. J. Biochem.* 200: 463–469.

1775. Sadakane, Y., and H. Nakashima (1996). Light-induced phase shifting of the circadian conidiation rhythm is inhibited by calmodulin antagonists in *Neurospora crassa. J. Biol. Rhythms* 11: 234–240.

1776. Sadler, R., and S. Ogilvie-Villa, Personal communication.

1777. Sahni, M., and J. A. Kinsey (1997). Identification and cloning of the *Neurospora crassa* glyceraldehyde-3-phosphate dehydrogenase gene, *gpd-1. Fungal Genet. Newsl.* 44: 47.

1778. Said, S., and H. F. Terenzi (1979). Cyclic AMP deficiency, modifier mutations, and instability of the *cr-1* phenotype. *Neurospora Newsl.* 26: 15–16.

1779. Said, S., and H. F. Terenzi (1981). Enzymatic alterations secondary to adenyl cyclase deficiency in the *cr-1 (crisp)* mutant of *Neurospora crassa*: Anomalies of the glucosidase system. *Braz. J. Med. Biol. Res.* 14: 3–10.

1780. Sajani, L. S., and P. M. Mohan (1998). Cobalt resistance in *Neurospora crassa*: Overproduction of a cobaltoprotein in a resistant strain. *Biometals* 11: 33–40.

1781. Sanchez, F., G. Dávila, J. Mora, and R. Palacios (1979). Immunochemical characterization of glutamine synthetase from *Neurospora crassa* glutamine auxotrophs. *J. Bacteriol.* 139: 537–543.

1782. Sanchez, S., L. Martinez, and J. Mora (1972). Interactions between amino acid transport systems in *Neurospora crassa. J. Bacteriol.* 112: 276–284.

1783. Sandal, N. N., and K. A. Marcker (1994). Similarities between a soybean nodulin, *Neurospora crassa* sulphate permease II and a putative human tumour suppressor. *Trends Biochem. Sci.* 19: 19.

1784. Sandeman, R. A., M. J. Hynes, J. R. S. Fincham, and I. F. Connerton (1991). Molecular organisation of the malate synthase genes of *Aspergillus nidulans* and *Neurospora crassa. Mol. Gen. Genet.* 228: 445–452.

1785. Sandmann, G. (1993). Photoregulation of carotenoid biosynthesis in mutants of *Neurospora crassa*: Activities of enzymes involved in the synthesis and conversion of phytoene. *Z. Naturforsch.* 48c: 570–574.

1786. Sansome, E. R. (1946). Heterokaryosis, *mating type* factors, and sexual reproduction in *Neurospora. Bull. Torrey Bot. Club* 73: 397–409.

1787. Sansome, E. R., M. Demerec, and A. Hollaender (1945). Quantitative irradiation experiments with *Neurospora crassa*. I. Experiments with X rays. *Am. J. Bot.* **32:** 218–226.

1788. Santosa, S., and H. Kuwana (1990). Isolation and characterization of a new acetate-requiring mutant strain, *ace-9*, of *Neurospora crassa*. *Fungal Genet. Newsl.* **37:** 37 (Abstr.).

1789. Santosa, S., and H. Kuwana (1992). Isolation and characterization of a new acetate-requiring mutant strain, *ace-9*, of *Neurospora crassa*. *Jpn. J. Genet.* **67:** 85–95.

1790. Sargent, M. L. (1985). *eas* strains of *Neurospora* for the classroom. *Neurospora Newsl.* **32:** 12–13.

1791. Sargent, M. L., W. R. Briggs, and D. O. Woodward (1966). Circadian nature of a rhythm expressed by an invertaseless strain of *Neurospora crassa*. *Plant Physiol.* **41:** 1343–1349.

1792. Sargent, M. L., and S. H. Kaltenborn (1972). Effects of medium composition and carbon dioxide on circadian conidiation in *Neurospora*. *Plant Physiol.* **50:** 171–175.

1793. Sargent, M. L., and D. O. Woodward (1969). Gene-enzyme relationships in *Neurospora* invertase. *J. Bacteriol.* **97:** 867–872.

1794. Sargent, M. L., and D. O. Woodward (1969). Genetic determinants of circadian rhythmicity in *Neurospora*. *J. Bacteriol.* **97:** 861–866.

1795. Sattlegger, E., A. G. Hinnebusch, and I. B. Barthelmess (1998). *cpc-3*, the *Neurospora crassa* homologue of yeast GCN2, encodes a polypeptide with juxtaposed eIF2α kinase and histidyl-tRNA synthetase-related domains required for general amino acid control. *J. Biol. Chem.* **273:** 20404–20416.

1796. Saupe, S., L. Stenberg, K. T. Shiu, A. J. Griffiths, and N. L. Glass (1996). The molecular nature of mutations in the *mt A-1* gene of the *Neurospora crassa A* idiomorph and their relation to mating-type function. *Mol. Gen. Genet.* **250:** 115–122.

1797. Saupe, S. J., and N. L. Glass (1997). Allelic specificity at the *het-c* heterokaryon incompatibility locus of *Neurospora crassa* is determined by a highly variable domain. *Genetics* **146:** 1299–1309.

1798. Savtchenko, E. S., H. F. Terenzi, and J. A. Jorge (1986). Alkaline protease deficiency in the *cr-1 (crisp)* mutant of *Neurospora crassa*. *Braz. J. Med. Biol. Res.* **19:** 27–32.

1799. Scarborough, G. A. (1973). Transport in *Neurospora*. *Int. Rev. Cytol.* **34:** 103–122.

1800. Scarborough, G. A. (1978). The *Neurospora* plasma membrane: A new experimental system for investigating eukaryote surface membrane structure and function. *Methods Cell Biol.* **20:** 117–133.

1801. Scarborough, G. A. (1985). A variant of the cell wall-less *fz;sg;os-1* strain of *Neurospora crassa* with altered morphology and improved growth. *Exp. Mycol.* **9:** 275–278.

1802. Scarborough, G. A. (1988). Large-scale purification of plasma membrane H⁺-ATPase from a cell-wall-less mutant of *Neurospora crassa*. *Methods. Enzymol.* **157:** 574–579.

1803. Scarborough, G. A. (1992). Probing the structure of the *Neurospora crassa* plasma membrane H⁺-ATPase. *Mol. Cell. Biochem.* **114:** 49–56.

1804. Scarborough, G. A., and R. Addison (1984). On the subunit composition of the *Neurospora* plasma membrane H⁺-ATPase. *J. Biol. Chem.* **259:** 9109–9114.

1805. Scarborough, G. A., and J. P. Hennessey, Jr. (1990). Identification of the major cytoplasmic regions of the *Neurospora crassa* plasma membrane H⁺-ATPase using protein chemical techniques. *J. Biol. Chem.* **265:** 16145–16149.

1806. Scarborough, G. A., and J. F. Nyc (1967). Methylation of ethanolamine phosphatides by microsomes from normal and mutant strains of *Neurospora crassa*. *J. Biol. Chem.* **242:** 238–242.

1807. Schablik, M., A. Kiss, A. Zsindely, and G. Szabó (1988). Characterization of a mutation that causes overproduction of inositol in *Neurospora crassa*. *Mol. Gen. Genet.* **213:** 140–143.

1808. Schafer, M. P., and G. E. Dean (1993). Cloning and sequence analysis of an H⁺-ATPase-encoding gene from the human dimorphic pathogen *Histoplasma capsulatum*. *Gene* **136:** 295–300.

1809. Schechtman, M. G. (1987). Isolation of telomere DNA from *Neurospora crassa*. *Mol. Cell. Biol.* **7:** 3168–3177.

1810. Schechtman, M. G. (1989). Segregation patterns of *Neurospora* chromosome ends: Mapping chromosome tips. *Fungal Genet. Newsl.* **36:** 71–73.

1811. Schechtman, M. G. (1990). Characterization of telomere DNA from *Neurospora crassa*. *Gene* **88:** 159–165.

1812. Schilling, B., R. M. Linden, U. Kupper, and K. Lerch (1992). Expression of *Neurospora crassa* laccase under the control of the copper-inducible metallothionein promoter. *Curr. Genet.* **22:** 197–203.

1813. Schlitt, S. C., G. Lester, and P. J. Russell (1974). Isolation and characterization of low-kynureninase mutants of *Neurospora crassa*. *J. Bacteriol.* **117:** 1117–1120.

1814. Schlitt, S. C., and P. J. Russell (1974). *Neurospora crassa* cytoplasmic ribosomes: Isolation and characterization of a cold-sensitive mutant defective in ribosome biosynthesis. *J. Bacteriol.* **120:** 666–671.

1815. Schmidhauser, T. J., Personal communication.

1816. Schmidhauser, T. J., F. R. Lauter, V. E. A. Russo, and C. Yanofsky (1990). Cloning, sequence, and photoregulation of *al-1*, a carotenoidbiosynthetic gene of *Neurospora crassa*. *Mol. Cell. Biol.* **10:** 5064–5070.

1817. Schmidhauser, T. J., F. R. Lauter, M. Schumacher, W. Zhou, V. E. Russo, and C. Yanofsky (1994). Characterization of *al-2*, the phytoene synthase gene of *Neurospora crassa*. Cloning, sequence analysis, and photoregulation. *J. Biol. Chem.* **269:** 12060–12066.

1818. Schmidhauser, T. J., Y.-Z. Liu, H. Liu, and S. Zhou (1997). Genome analysis in *Neurospora crassa*; cloning of four loci arginine-1 (*arg-1*), methionine-6 (*met-6*), unknown-7 (*un-7*), and ribosome production-1 (*rip-1*) and associated chromosome walking. *Fungal Genet. Biol.* **21:** 323–328.

1819. Schmidhauser, T. L., and D. J. Ebbole, Personal communication.

1820. Schmit, J. C., and S. Brody (1975). Developmental control of glucosamine and galactosamine levels during conidiation in *Neurospora crassa*. *J. Bacteriol.* **122:** 1071–1075.

1821. Schmit, J. C., and S. Brody (1982). Temperature-sensitive mutant of *Neurospora crassa* that affects mycelial growth and morphology. *J. Gen. Microbiol.* **128**: 1147–1158.

1822. Schmit, J. C., M. Cohen, and S. Brody (1974). Conidial germination in *scon^c^*. *Neurospora Newsl.* **21**: 17–18.

1823. Schneider, H., M. Arretz, E. Wachter, and W. Neupert (1990). Matrix processing peptidase of mitochondria. Structure–function relationships. *J. Biol. Chem.* **265**: 9881–9887.

1824. Schneider, R., B. Brors, M. Massow, and H. Weiss (1997). Mitochondrial fatty acid synthesis: A relic of endosymbiontic origin and a specialized means for respiration. *FEBS Lett.* **407**: 249–252.

1825. Schneider, R., M. Massow, T. Lisowsky, and H. Weiss (1995). Different respiratory-defective phenotypes of *Neurospora crassa* and *Saccharomyces cerevisiae* after inactivation of the gene encoding the mitochondrial acyl carrier protein. *Curr. Genet.* **29**: 10–17.

1826. Schönbrunner, E. R., S. Mayer, M. Tropschug, G. Fischer, N. Takahashi, and F. X. Schmid (1991). Catalysis of protein folding by cyclophilins from different species. *J. Biol. Chem.* **266**: 3630–3635.

1827. Schopfer, W. H., and T. Posternak (1958). Action d'anti-inositols sur *Neurospora crassa* "inositol-less" cultivé en milieu hautement purifié. *Arch. Mikrobiol.* **31**: 240–243.

1828. Schroeder, A. L., Personal communication.

1829. Schroeder, A. L. (1968). Ultraviolet-sensitive mutants in *Neurospora crassa*. *Genetics* **60**: 223 (Abstr.).

1830. Schroeder, A. L. (1970). Ultraviolet-sensitive mutants of *Neurospora*. I. Genetic basis and effect on recombination. *Mol. Gen. Genet.* **107**: 291–304.

1831. Schroeder, A. L. (1970). Ultraviolet-sensitive mutants of *Neurospora*. II. Radiation studies. *Mol. Gen. Genet.* **107**: 305–320.

1832. Schroeder, A. L. (1972). Photoreactivating enzyme in a UV-sensitive *Neurospora* mutant with abnormal photoreactivation. *Genetics* **71**: s56.

1833. Schroeder, A. L. (1974). Properties of a UV-sensitive *Neurospora* strain defective in pyrimidine dimer excision. *Mutat. Res.* **24**: 9–16.

1834. Schroeder, A. L. (1975). Genetic control of radiation sensitivity and DNA repair in *Neurospora*. *In* "Molecular Mechanisms for Repair of DNA" (P. C. Hanawalt and R. B. Setlow, eds.), Vol. Part B. Plenum Press, New York.

1835. Schroeder, A. L. (1986). Chromosome instability in mutagen-sensitive mutants of *Neurospora*. *Curr. Genet.* **10**: 381–387.

1836. Schroeder, A. L. (1988). Use of *Neurospora* to study DNA repair. *In* "DNA Repair: A Laboratory Manual of Research Procedures" (E. C. Friedberg and P. C. Hanawalt, eds.), Vol. 3, pp. 77–98. Marcel Dekker, Inc., New York.

1837. Schroeder, A. L., F. J. de Serres, and M. E. Schüpbach (1972). A new ultraviolet-light-sensitive mutant in *Neurospora, uvs-6*. *Neurospora Newsl.* **19**: 17.

1838. Schroeder, A. L., H. Inoue, and M. S. Sachs (1998). DNA repair in *Neurospora*. *In* "DNA Damage and Repair, Vol. 1: DNA Repair in Prokaryotes and Lower Eukaryotes" (M. F. Hoekstra and J. A. Nickoloff, eds.), 503–538. Humana Press, Totowa, NJ.

1839. Schroeder, A. L., and L. D. Olson (1980). Mutagen-sensitive mutants in *Neurospora*. *In* "Conference on DNA Repair and Mutagenesis in Eukaryotes" (F. J. de Serres, W. M. Generoso, and M. D. Shelby, eds.), pp. 55–62. Plenum Publishing Corp., New York.

1840. Schroeder, A. L., and L. D. Olson (1983). Mutagen sensitivity of *Neurospora* meiotic mutants. *Can. J. Genet. Cytol.* **25**: 16–25.

1841. Schroeder, A. L., and N. B. Raju (1991). *mei-2*, a mutagen-sensitive mutant of *Neurospora* defective in chromosome pairing and meiotic recombination. *Mol. Gen. Genet.* **231**: 41–48.

1842. Schulte, U., V. Haupt, A. Abelmann, W. Fecke, B. Brors, T. Rasmussen, T. Friedrich, and H. Weiss (1999). A reductase/isomerase subunit of mitochondrial NADH:ubiquinone oxidoreductase (complex I) carries an NADPH and is involved in the biogenesis of the complex. *J. Mol. Biol.* **292**: 569–580.

1843. Schulte, U., and H. Weiss (1995). Generation and characterization of NADH:ubiquinone oxidoreductase mutants in *Neurospora crassa*. *Methods Enzymol* **260**: 3–14.

1844. Schumacher, M. M., C. S. Enderlin, and C. P. Selitrennikoff (1997). The *osmotic-1* locus of *Neurospora crassa* encodes a putative histidine kinase similar to osmosensors of bacteria and yeast. *Curr. Microbiol.* **34**: 340–347.

1845. Schumacker, M., Y.-Z. Liu, H. Liu, and T. J. Schmidhauser (1995). Characterization of *Neurospora crassa* albino mutants that were previously unassigned to locus. *Fungal Genet. Newsl.* **42**: 69–70.

1846. Schüpbach, M. E., and F. J. de Serres (1981). Mutagenesis at the *ad-3A* and *ad-3B* loci in haploid UV-sensitive strains of *Neurospora crassa*. III. Comparison of dose–response curves for inactivation. *Mutat. Res.* **81**: 49–58.

1847. Scott, W. A. (1976). Adenosine 3′,5′-cyclic monophosphate deficiency in *Neurospora crassa*. *Proc. Natl. Acad. Sci. USA* **73**: 2995–2999.

1848. Scott, W. A. (1976). Biochemical genetics of morphogenesis in *Neurospora*. *Annu. Rev. Microbiol.* **30**: 85–104.

1849. Scott, W. A. (1977). Mutations resulting in an unsaturated fatty acid requirement in *Neurospora*: Evidence for Δ9-desaturase defects. *Biochemistry* **16**: 5274–5280.

1850. Scott, W. A. (1977). Unsaturated fatty acid mutants of *Neurospora crassa*. *J. Bacteriol.* **130**: 1144–1148.

1851. Scott, W. A. (1979). Biochemical–genetic control of morphogenesis. *In* "Physiological Genetics" (J. G. Scandalios, ed.), pp. 141–170. Academic Press, New York.

1852. Scott, W. A., and T. Abramsky (1973). *Neurospora* 6-phosphogluconate dehydrogenase. II. Properties of two purified mutant enzymes. *J. Biol. Chem.* **248**: 3542–3545.

1853. Scott, W. A., and S. Brody (1973). Effects of suppressor mutations on nonallelic glucose 6-phosphate dehydrogenase mutants of *Neurospora crassa*. *Biochem. Genet.* **10**: 285–295.

1854. Scott, W. A., and E. Mahoney (1976). Defects of glucose-6-phosphate and 6-phosphogluconate dehydrogenases in *Neurospora* and their pleiotropic effects. *Curr. Topics Cell Regul.* **10**: 205–236.

1855. Scott, W. A., N. C. Mishra, and E. L. Tatum (1974). Biochemical genetics of morphogenesis in *Neurospora*. *Brookhaven Symp. Biol.* **25**: 1–18.

1856. Scott, W. A., and B. Solomon (1975). Adenosine 3',5'-cyclic monophosphate and morphology in *Neurospora crassa*: Drug-induced alterations. *J. Bacteriol.* **122:** 454–463.

1857. Scott, W. A., and E. L. Tatum (1970). Glucose-6-phosphate dehydrogenase and *Neurospora* morphology. *Proc. Natl. Acad. Sci. USA* **66:** 515–522.

1858. Seale, T. (1968). Reversion of the *am* locus in *Neurospora*: Evidence for nonsense suppression. *Genetics* **58:** 85–99.

1859. Seale, T. W. (1972). Supersuppressors in *Neurospora crassa*. I. Induction, genetic localization and relationship to a missense suppressor. *Genetics* **70:** 385–396.

1860. Seale, T. W. (1976). Supersuppressor action spectrum in *Neurospora*. *Mol. Gen. Genet.* **148:** 105–108.

1861. Seale, T. W., M. Brett, A. J. Baron, and J. R. Fincham (1977). Amino acid replacements resulting from suppression and missense reversion of a chain-terminator mutation in *Neurospora*. *Genetics* **86:** 261–274.

1862. Seale, T. W., M. Case, and R. W. Barratt (1969). Supersuppressors in *Neurospora crassa*. *Neurospora Newsl.* **15:** 5.

1863. Sebald, W., J. Hoppe, and E. Wachter (1979). Amino acid sequence of the ATPase proteolipid from mitochondria, chloroplasts and bacteria (wild type and mutants). *In* "Symposium on Function and Molecular Aspects of Biomembrane Transport" (E. Quagliariello, F. Palmieri, S. Papa, and M. Kingenberg, eds.), pp. 63–73. North-Holland Publishing Co., Amsterdam.

1864. Sebald, W., and B. Kruse (1984). Nucleotide sequences of the nuclear genes for the proteolipid and δ-subunit of the mitochondrial ATP synthase from *Neurospora crassa*. *In* "H⁺-ATPase (ATP Synthase): Structure, Function, Biogenesis. The F_0F_1 Complex of Coupling Membranes" (S. Papa, K. Altendorf, L. Ernster, and L. Packer, eds.), pp. 67–75. Adriatica Editrice, Bari, Italy.

1865. Sebald, W., M. Sebald-Althaus, and E. Wachter (1977). Altered amino acid sequence of the DCCD-binding protein of the nuclear oligomycin-resistant mutant AP-2 from *Neurospora crassa*. *In* "Mitochondria 1977: Genetics and Biogenesis of Mitochondria" (W. Bandlow, R. J. Schweyen, K. Wolf, and F. Kaudewitz, eds.), pp. 433–440. Walter de Gruyter, Berlin.

1866. Sebald, W., and E. Wachter (1978). Amino acid sequence of the putative protonophore of the energy-transducing ATPase complex. *In* "Energy Conservation in Biological Membranes" (G. Schäfer, and M. Klingenberg, eds.), pp. 228–236. Springer-Verlag, Berlin.

1867. Sebo, T. J., and J. C. Schmit (1980). A convenient method for the storage and cultivation of slime. *Neurospora Newsl.* **27:** 30.

1868. Seidel, B. L. (1980). Characterization of *Neurospora crassa* adenylated messenger RNA: Structure of the 5'-terminus and metabolism of the polyadenylate region. Ph.D. Thesis, Rutgers University, New Brunswick, NJ.

1869. Seidel-Rogel, B. L., J. King, and H. Bertrand (1989). Unstable mitochondrial DNA in natural-death nuclear mutants of *Neurospora crassa*. *Mol. Cell. Biol.* **9:** 4259–4264.

1870. Seiler, S., F. E. Nargang, G. Steinberg, and M. Schliwa (1997). Kinesin is essential for cell morphogenesis and polarized secretion in *Neurospora crassa*. *EMBO J.* **16:** 3025–3034.

1871. Seiler, S., M. Plamann, and M. Schliwa (1999). Kinesin and dynein mutants provide novel insights into the roles of vesicle traffic during cell morphogenesis in *Neurospora*. *Curr. Biol.* **12:** 779–785.

1872. Selhub, J. (1970). Defects of the four types of *Neurospora crassa* mutants that require methionine and are unable to utilize homocyseine. Ph.D. Thesis, Case Western Reserve University, Cleveland, OH.

1873. Selhub, J., E. Burton, and W. Sakami (1969). Identification of three enzymes specifically involved in the *de novo* methionine methyl biosynthesis of *N. crassa*. *Fed. Proc.* **28:** 352.

1874. Selhub, J., M. A. Savin, W. Sakami, and M. Flavin (1971). Synchronization of converging metabolic pathways: Activation of the cystathionine γ-synthase of *Neurospora crassa* by methyltetrahydrofolate. *Proc. Natl. Acad. Sci. USA* **68:** 312–314.

1875. Selitrenikoff, C. P. (1976). Easily-wettable, a new mutant. *Neurospora Newsl.* **23:** 23.

1876. Selitrennikoff, C. P. (1974). Use of conidial separation-defective strains. *Neurospora Newsl.* **21:** 22.

1877. Selitrennikoff, C. P. (1978). Storage of slime strains. *Neurospora Newsl.* **25:** 16.

1878. Selitrennikoff, C. P. (1981). A new allele of *csp-2* which does not complement *csp-1*. *Neurospora Newsl.* **21:** 22.

1879. Selitrennikoff, C. P. (1982). Temperature-sensitive, protoplast-forming *os-1* variant of *Neurospora*—A few tricks. *Neurospora Newsl.* **29:** 27.

1880. Selitrennikoff, C. P., and M. Bailey (1974). A simple classroom complementation experiment. *Neurospora Newsl.* **21:** 22.

1881. Selitrennikoff, C. P., B. L. Lilley, and R. Zuker (1981). Formation and regeneration of protoplasts derived from a temperature-sensitive osmotic strain of *Neurospora crassa*. *Exp. Mycol.* **5:** 155–161.

1882. Selitrennikoff, C. P., R. E. Nelson, and R. W. Siegel (1974). Phase-specific genes for macroconidiation in *Neurospora crassa*. *Genetics* **78:** 679–690.

1883. Selitrennikoff, C. P., S. E. Slemmer, and R. E. Nelson (1979). Cell surface changes associated with the loss of cell–cell recognition in *Neurospora*. *Exp. Mycol.* **3:** 363–373.

1884. Selker, E., and C. Yanofsky (1980). A phenylalanine tRNA gene from *Neurospora crassa*: Conservation of secondary structure involving an intervening sequence. *Nucleic Acids Res* **8:** 1033–1042.

1885. Selker, E. U., Personal communication.

1886. Selker, E. U. (1990). Premeiotic instability of repeated sequences in *Neurospora crassa*. *Annu. Rev. Genet.* **24:** 579–613.

1887. Selker, E. U., S. J. Free, R. L. Metzenberg, and C. Yanofsky (1981). An isolated pseudogene related to the 5S RNA genes in *Neurospora crassa*. *Nature* **294:** 576–578.

1888. Selker, E. U., and P. W. Garrett (1988). DNA sequence duplications trigger gene inactivation in *Neurospora crassa*. *Proc. Natl. Acad. Sci. USA* **85:** 6870–6874.

1889. Selker, E. U., E. Morzycka-Wroblewska, J. N. Stevens, and R. L. Metzenberg (1987). An upstream signal is required for *in vitro* transcription of *Neurospora* 5S RNA genes. *Mol. Gen. Genet.* **205:** 189–1192.

1890. Selker, E. U., and J. N. Stevens (1985). DNA methylation at asymmetric sites is associated with numerous transition mutations. *Proc. Natl. Acad. Sci. USA* **82:** 8114–8118.

1891. Selker, E. U., and J. N. Stevens (1987). Signal for DNA methylation associated with tandem duplication in *Neurospora crassa*. *Mol. Cell. Biol.* **7**: 1032–1038.

1892. Selker, E. U., J. N. Stevens, and R. L. Metzenberg (1985). Heterogeneity of 5S RNA in fungal ribosomes. *Science* **227**: 1340–1343.

1893. Selker, E. U., C. Yanofsky, K. Driftmier, R. L. Metzenberg, B. Alzner-DeWeerd, and U. L. RajBhandary (1981). Dispersed 5S RNA genes in *N. crassa*: Structure, expression and evolution. *Cell* **24**: 819–828.

1894. Sengupta, S., T. B. Prasanna, and D. P. Kaskebar (1995). Sterol-14,15-reductase (*erg-3*) mutations switch the phenotype of *Neurospora crassa* from sensitivity to the tomato saponin α-tomatine to sensitivity to the pea phytoalexin pisatin. *Fungal Genet. Newsl.* **42**: 71–72.

1895. Serna, L., and D. Stadler (1978). Nuclear division cycle in germinating conidia of *Neurospora crassa*. *J. Bacteriol.* **136**: 341–351.

1896. de Serres, F. J. (1960). Studies with purple adenine mutants in *Neurospora crassa*. IV. Lack of complementation between different *ad-3A* mutants in heterokaryons and pseudowild types. *Genetics* **45**: 555–566.

1897. Shaffer, P. M., C.-A. Hsu, and M. T. Abbott (1975). Metabolism of pyrimidine deoxyribonucleotides in *Neurospora crassa*. *J. Bacteriol.* **121**: 648–655.

1898. Shaw, N. M., and R. W. Harding (1983). Calcium inhibition of a heat-stable cyclic nucleotide phosphodiesterase from *Neurospora crassa*. *FEBS Lett.* **152**: 295–299.

1899. Shear, C. L., and B. O. Dodge (1927). Life histories and heterothallism of the red bread mold fungi of the *Monilia sitophila* group. *J. Agric. Res.* **34**: 1019–1042.

1900. Shelby, D. M., F. J. de Serres, and G. J. Stine (1975). Ultraviolet-inactivation of conidia from heterokaryons of *Neurospora crassa* containing UV-sensitive mutations. *Mutat. Res.* **27**: 45–58.

1901. Shen, S.-C (1950). Genetics and biochemistry of the cysteine–tyrosine relationship in *Neurospora crassa*. Ph.D. Thesis, California Institute of Technology, Pasadena, CA.

1902. Shen, W. C., J. Wieser, T. H. Adams, and D. J. Ebbole (1998). The *Neurospora rca-1* gene complements an *Aspergillus flbD* sporulation mutant but has no identifiable role in *Neurospora* sporulation. *Genetics* **148**: 1031–1041.

1903. Sheng, T. C. (1951). A gene that causes natural death in *Neurospora crassa*. *Genetics* **36**: 199–212.

1904. Sheng, T. C., and G. Sheng (1952). Genetic and nongenetic factors in pigmentation of *Neurospora crassa*. *Genetics* **37**: 264–269.

1905. Shi, Y. G., and B. M. Tyler (1991). Coordinate expression of ribosomal protein genes in *Neurospora crassa* and identification of conserved upstream sequences. *Nucleic Acids. Res.* **19**: 6511–6517.

1906. Shields, M. (1968). A mutation which prevents adaptation of *am* mutants of *Neurospora crassa*. M.S. Thesis, University of Utah, Salt Lake City, UT.

1907. Shimizu-Takahama, M., T. Egashira, and U. Takahama (1981). Inhibition of respiration and loss of membrane integrity by singlet oxygen generated by a photosensitized reaction in *Neurospora crassa* conidia. *Photochem. Photobiol.* **33**: 689–694.

1908. Shimura, M., Y. Ito, C. Ishii, H. Yajima, H. Linden, T. Harashima, A. Yasui, and H. Inoue (1999). Characterization of a *Neurospora crassa* photolyase-deficient mutant generated by repeat-induced point mutation (RIP) of the *phr* gene. *Fungal Genet. Biol.* **28**: 12–20.

1909. Shinohara, M., D. Bell-Pederson, J. Loros, and J. Dunlap, Personal communication.

1910. Shinohara, M. L., J. J. Loros, and J. C. Dunlap (1998). Glyceraldehyde-3-phosphate dehydrogenase is regulated on a daily basis by the circadian clock. *J. Biol. Chem.* **273**: 446–452.

1911. Shiu, P. K. T., and N. L. Glass (1999). Molecular characterization of *tol*, a mediator of mating-type-associated vegetative incompatibility in *Neurospora crassa*. *Genetics* **151**: 545–555.

1912. Siddig, M. A., J. A. Kinsey, J. R. Fincham, and M. Keighren (1980). Frameshift mutations affecting the N-terminal sequence of *Neurospora* NADP-specific glutamate dehydrogenase. *J. Mol. Biol.* **137**: 125–135.

1913. Sigmund, R. D., M. T. McNally, D. B. Lee, and S. J. Free (1985). *Neurospora* glucamylase and a mutant affected in its regulation. *Biochem Genet* **23**: 89–103.

1914. Sikora, L., and G. A. Marzluf (1982). Regulation of L-amino acid oxidase and of D-amino acid oxidase in *Neurospora crassa*. *Mol. Gen. Genet.* **186**: 33–39.

1915. Silagi, S. (1965). Interactions between an extrachromosomal factor, *poky*, and nuclear genes in *Neurospora crassa*. *Genetics* **52**: 341–347.

1916. Silva, M., and G. Pincheira (1984). Genetic control of a structural polymer of the *Neurospora crassa* cell wall. *J. Bacteriol.* **158**: 1104–1108.

1917. Simmons, J., P. Chary, and D. O. Natvig (1987). Linkage group assignments for two *Neurospora crassa* catalase genes: The Metzenberg RFLP mapping kit applied to an enzyme polymorphism. *Fungal Genet. Newsl.* **34**: 55–56.

1918. Singleton, J. R. (1948). Cytogenic studies on *Neurospora crassa*. Ph.D. Thesis, California Institute of Technology, Pasadena, CA.

1919. Sista, H., and B. Bowman (1992). Characterization of the *ilv-2* gene from *Neurospora crassa* encoding α-keto-β-hydroxylacyl reductoisomerase. *Gene* **120**: 115–118.

1920. Sista, H., M. A. Wechser, and B. J. Bowman (1994). The proteolipid subunit of the *Neurospora crassa* vacuolar ATPase: Isolation of the protein and the *vma-3* gene. *Mol. Gen. Genet.* **243**: 82–90.

1921. Slayman, C. L. (1965). Electrical properties of *Neurospora crassa*. Respiration and the intracellular potential. *J. Gen. Physiol.* **49**: 93–116.

1922. Slayman, C. L. (1987). The plasma membrane ATPase of *Neurospora*: A proton-pumping electroenzyme. *J. Bioenerg. Biomembr.* **19**: 1–20.

1923. Slayman, C. L., W. S. Long, and C. Y. Lu (1973). The relationship between ATP and an electrogenic pump in the plasma membrane of *Neurospora crassa*. *J. Membr. Biol.* **14**: 305–338.

1924. Slayman, C. L., C. Y. Lu, and L. Shane (1970). Correlated changes in membrane potential and ATP concentrations in *Neurospora*. *Nature* **226**: 274–276.

1925. Slayman, C. W., Personal communication.

1926. Slayman, C. W., and E. L. Tatum (1965). Potassium transport in *Neurospora*. 3. Isolation of a transport mutant. *Biochim. Biophys. Acta* **109**: 184–193.

1927. Smiley, J. A., J. M. Angelot, R. C. Cannon, E. M. Marshall, and D. K. Asch (1999). Radioactivity-based and spectrophotometric assays for iso-orotate decarboxylase: Identification of the thymidine salvage pathway in lower eukaryotes. *Anal. Biochem.* **266**: 85–92.

1928. Smith, B. R. (1962). Linkage data for group IV. *Neurospora Newsl.* **1**: 16–17.

1929. Smith, B. R. (1962). Mating-type tests using plating technique. *Neurospora Newsl.* **1**: 14.

1930. Smith, B. R. (1966). Genetic controls of recombination I. The *recombination-2* gene of *Neurospora crassa*. *Heredity* **21**: 481–498.

1931. Smith, B. R. (1968). A genetic control of recombination in *Neurospora crassa*. *Heredity* **23**: 162–163 (Abstr.).

1932. Smith, B. V., and E. G. Barry (1966). A modifier of the *arg-8* mutant in *Neurospora*. *Genetics* **54**: 363.

1933. Smith, D. A. (1975). A mutant affecting meiosis in *Neurospora*. *Genetics* **80**: 125–133.

1934. Smith, M. L., and N. L. Glass (1994). Molecular analysis of *het-6*, a locus controlling vegetative incompatibility in *Neurospora crassa*. *Fifth Int. Congr. Mycol.* Vancouver, BC, Canada, 203 (Abstr.).

1935. Smith, M. L., and N. L. Glass (1996). Mapping translocation breakpoints by orthogonal field agarose gel electrophoresis. *Curr. Genet.* **29**: 301–305.

1936. Smith, M. L., S. P. Hubbard, D. J. Jacobson, O. C. Micali, and N. L. Glass (2000). An osmotic-remedial, temperature-sensitive mutation in the allosteric activity site of ribonucleotide reductase in *Neurospora crassa*. *Mol. Gen. Genet.* **262**: 1022–1035.

1937. Smith, M. L., O. C. Micali, S. P. Hubbard, N. Mir-Rashed, D. J. Jacobson, and N. L. Glass (2000). Vegetative incompatibility in the *het-6* region of *Neurospora crassa* is mediated by two linked genes. *Genetics* **155**: 1095–1104.

1938. Smith, M. L., C. J. Yang, R. L. Metzenberg, and N. L. Glass (1996). Escape from *het-6* incompatibility in *Neurospora crassa* partial diploids involves preferential deletion within the ectopic segment. *Genetics* **144**: 523–531.

1939. Smyth, D. R. (1970). Genetic control of recombination in the *amination-1* region of *Neurospora crassa*. Ph.D. Thesis, Australian National University, Canberra, Australia.

1940. Smyth, D. R. (1971). Effect of *rec-3* on polarity of recombination in the *amination-1* locus of *Neurospora crassa*. *Aust. J. Biol. Sci.* **24**: 97–106.

1941. Smyth, D. R. (1973). Action of *rec-3* on recombination near the *amination-1* locus of *Neurospora crassa*. *Aust. J. Biol. Sci.* **26**: 439–444.

1942. Sokolove, P. M., and K. W. Kinnally (1996). A mitochondrial signal peptide from *Neurospora crassa* increases the permeability of isolated rat liver mitochondria. *Arch. Biochem. Biophys.* **336**: 69–76.

1943. Söllner, T., G. Griffiths, R. Pfaller, N. Pfanner, and W. Neupert (1989). MOM19, an import receptor for mitochondrial precursor proteins. *Cell* **59**: 1061–1070.

1944. Söllner, T., R. Pfaller, G. Griffiths, N. Pfanner, and W. Neupert (1990). A mitochondrial import receptor for the ADP/ATP carrier. *Cell* **62**: 107–115.

1945. Söllner, T., J. Rassow, M. Wiedmann, J. Schlossmann, P. Keil, W. Neupert, and N. Pfanner (1992). Mapping of the protein import machinery in the mitochondrial outer membrane by cross-linking of translocation intermediates. *Nature* **355**: 84–87.

1946. Sommer, T., J. A. A. Chambers, J. Eberle, F. R. Lauter, and V. E. A. Russo (1989). Fast light-regulated genes of *Neurospora crassa*. *Nucleic Acids Res.* **17**: 5713–5723.

1946a. Sone, T., and Griffiths, A. J. F. 2000. The *frost* gene of *Neurospora crassa* is a homolog of yeast *cde-1*, and affects hyphal branching via manganese homeostasis. *Fungal Genet. Biol.* **28**: 227–237.

1947. Sorger, G. J., Personal communication.

1948. Sorger, G. J. (1966). Nitrate reductase electron transport systems in mutant and in wild-type strains of *Neurospora*. *Biochim. Biophys. Acta* **118**: 484–494.

1949. Sorger, G. J., D. Brown, M. Farzannejad, A. Guerra, M. Jonathan, S. Knight, and R. Sharda (1989). Isolation of a gene that down-regulates nitrate assimilation and influences another regulatory gene in the same system. *Mol. Cell. Biol.* **9**: 4113–4117.

1950. Sorger, G. J., and N. H. Giles (1965). Genetic control of nitrate reductase in *Neurospora crassa*. *Genetics* **52**: 777–788.

1951. Sorger, G. J., R. Premakumar, and D. Gooden (1978). Demonstration *in vitro* of two intracellular inactivators of nitrate reductase from *Neurospora*. *Biochim. Biophys. Acta* **540**: 33–47.

1952. Soshi, T., Y. Sakuraba, E. Kafer, and H. Inoue (1996). The *mus-8* gene of *Neurospora crassa* encodes a structural and functional homolog of the Rad6 protein of *Saccharomyces cerevisiae*. *Curr. Genet.* **30**: 224–231.

1953. Sousa, R., B. Barquera, M. Duarte, M. Finel, and A. Videira (1999). Characterisation of the last Fe-cluster-binding subunit of *Neurospora crassa* complex I. *Biochim. Biophys. Acta* **1411**: 142–146.

1954. Spinelli, L. B. B., M. L. Polizeli, H. F. Terenzi, and J. A. Jorge (1996). Biochemical characterisation of glucoamylase from the hyperproducer *exo-1* mutant strain of *Neurospora crassa*. *FEMS Microbiol. Lett.* **138**: 173–177.

1955. Springer, M. L. (1991). Genetic control of morphological differentiation in *Neurospora crassa*. Ph.D. Thesis, Stanford University, Stanford, CA.

1956. Springer, M. L. (1993). Genetic control of fungal differentiation: The three sporulation pathways of *Neurospora crassa*. *Bioessays* **15**: 365–374.

1957. Springer, M. L., K. M. Hager, C. Garrett-Engele, and C. Yanofsky (1992). Timing of synthesis and cellular localization of two conidiation-specific proteins of *Neurospora crassa*. *Dev. Biol.* **152**: 255–262.

1958. Springer, M. L., and C. Yanofsky (1989). A morphological and genetic analysis of conidiophore development in *Neurospora crassa*. *Genes Dev.* **3**: 559–571.

1959. Springer, M. L., and C. Yanofsky (1992). Expression of *con* genes along the three sporulation pathways of *Neurospora crassa*. *Genes Dev.* **6**: 1052–1057.

1960. Springer, W. D., and A. M. Srb (1978). Molecular alteration in a *Neurospora crassa* morphological mutant and its phenocopy. *Proc. Natl. Acad. Sci. USA* 75: 1461–1465.

1961. Spurgeon, S. L., and W. H. Matchett (1977). Inhibition of aminoacyl-transfer ribonucleic acid synthetases and the regulation of amino acid biosynthetic enzymes in *Neurospora crassa*. *J. Bacteriol.* 129: 1303–1312.

1962. Srb, A. M. Personal communication.

1963. Srb, A. M. (1946). Ornithine–arginine metabolism in *Neurospora* and its genetic control. Ph.D. Thesis, Stanford University, Stanford, CA.

1964. Srb, A. M. (1950). Complex growth requirement of an arginine-less mutant of *Neurospora. Bot. Gaz.* 111: 470–476.

1965. Srb, A. M., and M. Basl. (1969). Abnormal ascus mutants. *Neurospora Newsl.* 15: 22.

1966. Srb, A. M., M. Basl, M. Bobst, and J. V. Leary (1973). Mutations in *Neurospora crassa* affecting ascus and ascospore development. *J. Hered.* 64: 242–246.

1967. Srb, A. M., J. R. S. Fincham, and D. Bonner (1950). Evidence from gene mutations in *Neurospora* for close metabolic relationships among ornithine, proline and α-amino-δ-hydroxyvaleric acid. *Am. J. Bot.* 37: 533–538.

1968. Srb, A. M., and N. H. Horowitz (1944). The ornithine cycle in *Neurospora* and its genetic control. *J. Biol. Chem.* 154: 129–139.

1969. Srb, A. M., J. B. Nasrallah, and M. Basl (1974). Genetic control of the development of the sexual reproductive apparatus of *Neurospora. Brookhaven Symp. Biol.* 25: 40–50.

1970. St. Lawrence, P., Personal communication.

1971. St. Lawrence, P (1956). The *q* locus of *Neurospora crassa. Proc. Natl. Acad. Sci. USA.* 42: 189–194.

1972. St. Lawrence, P. (1959). Gene conversion at the *nic-2* locus of *Neurospora crassa* in crosses between strains with normal chromosomes and a strain carrying a translocation at the locus. *Genetics* 44: 532 (Abstr.).

1973. St. Lawrence, P., B. D. Maling, L. Altwerger, and M. Rachmeler (1964). Mutational alteration of permeability in *Neurospora*: Effects on growth and the uptake of certain amino acids and related compounds. *Genetics* 50: 1383–1402.

1974. St. Lawrence, P., R. Naish, and B. Burr (1965). The action of suppressors of a tryptophan synthetase mutant of *Neurospora* in heterokaryons. *Biochem. Biophys. Res. Commun.* 18: 868–876.

1975. St. Lawrence, P., and J. R. Singleton (1963). The cytogenetic effects of a paracentric inversion in *Neurospora crassa. Eleventh Int. Congr. Genet.* (The Hague), 119. Pergamon Press, Oxford.

1976. Staben, C. (1996). The mating-type locus of *Neurospora crassa. J. Genet.* 75: 341–350.

1977. Staben, C., B. Jensen, M. Singer, J. Pollock, M. Schechtman, J. Kinsey, and E. Selker (1989). Use of a bacterial hygromycin B resistance gene as a dominant selectable marker in *Neurospora crassa* transformation. *Fungal Genet. Newsl.* 36: 79–81.

1978. Staben, C., and C. Yanofsky (1990). *Neurospora crassa* mating-type region. *Proc. Natl. Acad. Sci. USA* 87: 4917–4921.

1979. Stadler, D., and E. Crane (1979). Analysis of lethal events induced by ultraviolet in a heterokaryon of *Neurospora. Mol. Gen. Genet.* 171: 59–68.

1980. Stadler, D., and H. Macleod (1984). A dose–rate effect in UV mutagenesis in *Neurospora. Mutat. Res.* 127: 39–47.

1981. Stadler, D., H. Macleod, and D. Dillon (1991). Spontaneous mutation at the *mtr* locus of *Neurospora*: The spectrum of mutant types. *Genetics* 129: 39–45.

1982. Stadler, D., H. Macleod, and M. Loo (1987). Repair-resistant mutation in *Neurospora. Genetics* 116: 207–214.

1983. Stadler, D., and R. Moyer (1981). Induced repair of genetic damage in *Neurospora. Genetics* 98: 763–774.

1984. Stadler, D. R., Personal communication.

1985. Stadler, D. R. (1956). Double crossing over in *Neurospora. Genetics* 41: 623–630.

1986. Stadler, D. R. (1956). A map of linkage group VI of *Neurospora crassa. Genetics* 41: 528–543.

1987. Stadler, D. R. (1959). Gene conversion of cysteine mutants in *Neurospora. Genetics* 44: 647–655.

1988. Stadler, D. R. (1959). Genetic control of a cyclic growth pattern in *Neurospora. Nature (London)* 184: 170–171.

1989. Stadler, D. R. (1959). The relationship of gene conversion to crossing over in *Neurospora. Proc. Natl. Acad. Sci. USA* 45: 1625–1629.

1990. Stadler, D. R. (1966). Genetic control of the uptake of amino acids in *Neurospora. Genetics* 54: 677–685.

1991. Stadler, D. R (1967). Suppressors of amino acid uptake mutants of *Neurospora. Genetics* 57: 935–942.

1992. Stadler, D. R. (1981). Temperature-sensitive mutants. *Neurospora Newsl.* 28: 18–19.

1993. Stadler, D. R., and D. A. Smith (1968). A new mutation in *Neurospora* for sensitivity to ultraviolet. *Can. J. Genet. Cytol.* 10: 916–919.

1994. Stadler, D. R., and A. M. Towe (1963). Recombination of allelic cysteine mutants in *Neurospora. Genetics* 48: 1323–1344.

1995. Stadler, D. R., and A. M. Towe (1968). A test of coincident recombination in closely linked genes of *Neurospora. Genetics* 58: 327–336.

1996. Stadler, D. R., A. M. Towe, and N. Murray (1965). Intragenic and intergenic recombination in *Neurospora. Genetics* 52: 477 (Abstr.).

1997. Stafford, A. E., T. A. McKeon, and M. Goodrich-Tanrikulu (1998). Conversion of palmitate to unsaturated fatty acids differs in a *Neurospora crassa* mutant with impaired fatty acid synthase activity. *Lipids* 33: 303–306.

1998. Steger, H. F., T. Söllner, M. Kiebler, K. A. Dietmeier, R. Pfaller, K. S. Trulzsch, M. Tropschug, W. Neupert, and N. Pfanner (1990). Import of ADP/ATP carrier into mitochondria: Two receptors act in parallel. *J. Cell Biol.* 111: 2353–2363.

1999. Steinberg, G., and M. Schliwa (1995). The *Neurospora* organelle motor: A distant relative of conventional kinesin with unconventional properties. *Mol. Biol. Cell* 6: 1605–1618.

2000. Stephens, S. P., W. A. Krissinger, and S. N. Bennett (1993). Characterization of SS-1018, a newly isolated osmotic-sensitive mutant of *Neurospora crassa. Fungal Genet. Newsl.* 40A: 67 (Abstr.).

2001. Stephens, S. P., W. A. Krissinger, and S. N. Bennett (1994). Genetic analysis and morphological characterization of an osmotic-sensitive mutant of *Neurospora crassa*. *Georgia J. Sci.* **52**: 32 (Abstr.).

2002. Sternberg, D., and A. S. Sussman (1974). Hyperproduction of some glycosidases in *Neurospora crassa*. *Arch. Microbiol.* **101**: 303–320.

2003. Stevens, C. M., and A. Mylroie (1953). Inhibition effects in back-mutation tests with mutants of *Neurospora*. *Nature* **171**: 179–180.

2004. Stokes, D. L., M. Auer, P. Zhang, and W. Kühlbrandt (1999). Comparison of H^+-ATPase and Ca^{2+}-ATPase suggests that a large conformational change initiates P-type ion pump reaction cycles. *Curr. Biol.* **9**: 672–679.

2005. Stone, P. J., A. J. Makoff, J. H. Parish, and A. Radford (1993). Cloning and sequence analysis of the glucoamylase gene of *Neurospora crassa*. *Curr. Genet.* **24**: 205–211.

2006. Strauss, B. S. (1951). Studies on the vitamin B_6-requiring, pH-sensitive mutants of *Neurospora crassa*. *Arch. Biochem.* **30**: 292–305.

2007. Strauss, B. S. (1952). Aspects of the carbohydrate metabolism of a mutant of *Neurospora crassa* requiring acetate for growth. *Arch. Biochem. Biophys.* **36**: 33–47.

2008. Strauss, B. S. (1956). The nature of the lesion in the succinate-requiring mutants of *Neurospora crassa*: Interaction between carbohydrate and nitrogen metabolism. *J. Gen. Microbiol.* **14**: 494–511.

2009. Strauss, B. S. (1957). Oxaloacetic acid carboxylase deficiency of the succinate-requiring mutants of *Neurospora crassa*. *J. Biol. Chem.* **225**: 535–544.

2010. Strauss, B. S. (1958). Cell death and "unbalanced growth" in *Neurospora*. *J. Gen. Microbiol.* **18**: 658–669.

2011. Strauss, B. S., and S. Pierog (1954). Gene interactions: The mode of action of the suppressor of acetate-requiring mutants of *Neurospora crassa*. *J. Gen. Microbiol.* **10**: 221–235.

2012. Strauss, B. S., and S. Tokuno, Personal communication.

2013. Strickland, W. N. (1961). Tetrad analysis of short chromosome regions of *Neurospora crassa*. *Genetics* **46**: 1125–1141.

2014. Strickland, W. N., D. D. Perkins, and C. C. Veach (1959). Linkage data for group V markers in *Neurospora*. *Genetics* **44**: 1221–1226.

2015. Stuart, R. A., W. Neupert, and M. Tropschug (1987). Deficiency in mRNA splicing in a cytochrome *c* mutant of *Neurospora crassa*: Importance of carboxy terminus for import of apocytochrome *c* into mitochondria. *EMBO J.* **6**: 2131–2137.

2016. Stuart, W. D., and A. G. DeBusk (1975). Genetic alterations of ribonuclease-sensitive glycoprotein subunits of amino acid transport systems in *Neurospora crassa* conidia. *Arch. Biochem. Biophys.* **166**: 213–222.

2017. Stuart, W. D., K. Koo, and S. J. Vollmer (1988). Cloning of *mtr*, an amino acid transport gene of *Neurospora crassa*. *Genome* **30**: 198–203.

2018. Subden, R. E., and A. H. Goldie (1973). Biochemical analysis of isoallelic series of the *al-1* locus of *Neurospora crassa*. *Genetica* **44**: 615–620.

2019. Subden, R. E., and S. F. Threlkeld (1969). Some aspects of complementation with carotenogenic *al* loci in *Neurospora crassa*. *Experientia* **25**: 1106–1107.

2020. Subden, R. E., and S. F. Threlkeld (1970). Genetic fine structure of the *albino* (*al*) region of *Neurospora crassa*. *Genet. Res.* **15**: 139–146.

2021. Sugisaki, M., C. Ito, A. Toh-e, Y. Matsui, and H. Inoue (1999). Mutational analyses of the *NC-ras* gene in *Neurospora crassa*. *Fungal Genet. Newsl.* **46** (**Suppl.**): 147 (Abstr.).

2022. Sullivan, J. L., and A. G. DeBusk (1971). Method for specific selection of temperature-sensitive mutants. *Neurospora Newsl.* **18**: 13.

2023. Sundaram, T. K., and J. R. S. Fincham (1964). A mutant enzyme in *Neurospora crassa* interconvertible between electrophoretically distinct active and inactive forms. *J. Mol. Biol.* **10**: 423–437.

2024. Suresh, K., and C. Subramanyam (1997). A putative role for calmodulin in the activation of *Neurospora crassa* chitin synthase. *FEMS Microbiol. Lett.* **150**: 95–100.

2025. Sussman, A. S., M. K. Garrett, M. Sargent, and S. A. Yu (1971). Isolation, mapping, and characterization of trehalaseless mutants of *Neurospora crassa*. *J. Bacteriol.* **108**: 59–68.

2026. Sussman, A. S., R. J. Lowry, and T. Durkee (1964). Morphology and genetics of a periodic colonial mutant of *Neurospora crassa*. *Am. J. Bot.* **51**: 243–252.

2027. Sussman, M. R., J. E. Strickler, K. M. Hager, and C. W. Slayman (1987). Location of a dicyclohexylcarbodiimide-reactive glutamate residue in the *Neurospora crassa* plasma membrane H^+-ATPase. *J. Biol. Chem.* **262**: 4569–4573.

2028. Suyama, Y., and D. M. Bonner (1964). Complementation between tryptophan synthetase mutants of *Neurospora crassa*. *Biochim. Biophys. Acta* **81**: 565–575.

2029. Suyama, Y., A. M. Lacy, and D. M. Bonner (1964). A genetic map of the *td* locus of *Neurospora crassa*. *Genetics* **49**: 135–144.

2030. Suyama, Y., K. D. Munkres, and V. W. Woodward (1959). Genetic analyses of the *pyr-3* locus of *Neurospora crassa*: The bearing of recombination and gene conversion upon interallelic linearity. *Genetica* **30**: 293–311.

2031. Suzuki, S., S. Katagiri, and H. Nakashima (1996). Mutants with altered sensitivity to a calmodulin antagonist affect the circadian clock in *Neurospora crassa*. *Genetics* **143**: 1175–1180.

2032. Swanson, J., and E. U. Selker, Personal communication.

2033. Szakacs, N. A. (1978). The isolation and characterization of alternate oxidase deficient mutants of *Neurospora crassa*. Ph.D. Thesis, University of Regina, Regina, Saskatchewan, Canada.

2034. Szczesna-Skorupa, E., W. Filipowicz, and A. Paszewski (1981). The cell-free protein synthesis system from the "slime" mutant of *Neurospora crassa*: Preparation and characterisation of importance of 7-methylguanosine for translation of viral and cellular mRNAs. *Eur. J. Biochem.* **121**: 163–168.

2035. Szöör, B., Z. Fehér, T. Zeke, P. Gergely, E. Yatzkan, O. Yarden, and V. Dombrádi (1998). *pzl-1* encodes a novel protein phosphatase-Z-like Ser/Thr protein phosphatase in *Neurospora crassa*. *Biochim. Biophys. Acta* **1388**: 260–266.

2036. Taccioli, G. E., E. Grotewold, G. O. Aisemberg, and N. D. Judewicz (1989). Ubiquitin expression in *Neurospora crassa*: Cloning and sequencing of a polyubiquitin gene. *Nucleic Acids Res.* **17**: 6153–6165.

2037. Taccioli, G. E., E. Grotewold, G. O. Aisemberg, and N. D. Judewicz (1991). The cDNA sequence and expression of an ubiquitin–tail gene fusion in *Neurospora crassa. Gene* **102**: 133–137.

2038. Takahashi, K. (1988). The amino acid sequence of ribonuclease-N1, a guanine-specific ribonuclease from the fungus *Neurospora crassa. J. Biochem.* **104**: 375–382.

2039. Takao, M., R. Yonemasu, K. Yamamoto, and A. Yasui (1996). Characterization of a UV endonuclease gene from the fission yeast *Schizosaccharomyces pombe* and its bacterial homolog. *Nucleic. Acids Res.* **24**: 1267–1271.

2040. Talora, C., L. Franchi, H. Linden, P. Ballario, and G. Macino (1999). Role of a white collar-1–white collar-2 complex in blue light signal transduction. *EMBO J.* **18**: 4961–4968.

2041. Tamaru, H., and H. Inoue (1989). Isolation and characterization of a laccase-derepressed mutant of *Neurospora crassa. J. Bacteriol.* **171**: 6288–6293.

2042. Tamaru, H., and H. Inoue (1991). *cpc-1*, the general regulatory gene for amino acid biosynthetic genes is required for transcriptional activation of the *lacc* gene in *Neurospora crassa. Jpn. J. Genet.* **66**: 767 (Abstr.).

2043. Tamaru, H., T. Nishida, T. Harashima, and H. Inoue (1994). Transcriptional activation of a cycloheximide-inducible gene encoding laccase is mediated by *cpc1*, the cross-pathway control gene, in *Neurospora crassa. Mol. Gen. Genet.* **243**: 548–554.

2044. Tan, S. T. (1972). Genes controlling the development of protoperithecium in *Neurospora*. M.S. Thesis, University of Malaya, Kuala Lumpur, Malaysia.

2045. Tan, S. T., and C. C. Ho (1970). A gene controlling the early development of protoperithecium in *Neurospora crassa. Mol. Gen. Genet.* **107**: 158–161.

2046. Tan, W.-L., and F. E. Nargang (1999). Cloning of the *Neurospora crassa tim17. Bull. Genet. Soc. Can.* **30**: 51 (Abstr.).

2047. Tanaka, S., C. Ishii, and H. Inoue (1989). Effects of heat shock on the induction of mutations by chemical mutagens in *Neurospora crassa. Mutat. Res.* **223**: 233–242.

2048. Tanenbaum, S. W., L. Garnjobst, and E. Tatum (1954). A mutant of *Neurospora* requiring asparagine for growth. *Am. J. Bot.* **41**: 484–488.

2049. Tanizawa, K., and K. Soda (1979). Inducible and constitutive kynureninases (EC 3.7.1.3) control of the inducible enzyme activity by transamination and inhibition of the constitutive enzyme by 3-hydroxyanthranilate. *J. Biochem. (Tokyo)* **86**: 499–508.

2050. Tao, Y., and K. Y. Chen (1994). PCR-based cloning of the full-length *Neurospora* eukaryotic initiation factor 5A cDNA: Polyhistidine tagging and overexpression for protein affinity binding. *Biochem. J.* **302**: 517–525.

2051. Tao, Y., and K. Y. Chen (1995). Molecular cloning and functional expression of *Neurospora* deoxyhypusine synthase cDNA and identification of yeast deoxyhypusine synthase cDNA. *J. Biol. Chem.* **270**: 23984–23987.

2052. Tao, Y., and K. Y. Chen (1995). Purification of deoxyhypusine synthase from *Neurospora crassa* to homogeneity by substrate elution affinity chromatography. *J. Biol. Chem.* **270**: 383–386.

2053. Tao, Y., and G. A. Marzluf (1998). Analysis of a distal cluster of binding elements and other unusual features of the promoter of the highly regulated *nit-3* gene of *Neurospora crassa. Biochemistry* **37**: 11136–11142.

2054. Tao, Y., and G. A. Marzluf (1998). Synthesis and differential turnover of the CYS3 regulatory protein of *Neurospora crassa* are subject to sulfur control. *J. Bacteriol.* **180**: 478–482.

2055. Tarawneh, K. A. (1992). Identification of a *cis*-acting regulatory element involved in the transcriptional regulation of a *Neurospora crassa* glucose-repressible gene. Ph.D. Thesis, State University of New York at Buffalo, Buffalo, NY.

2056. Tarawneh, K. A., K. R. Anumula, and S. J. Free (1994). The isolation and characterization of a *Neurospora crassa* gene *(ubi::crp-6)* encoding a ubiquitin-40S ribosomal protein fusion protein. *Gene* **147**: 137–140.

2057. Tatum, E. L. (1951). Genetic aspects of growth responses in fungi. *In* "Plant Growth Substances" (F. Skoog, ed.), pp. 447–461. University of Wisconsin Press, Madison, WI.

2058. Tatum, E. L., and G. W. Beadle (1942). Genetic control of biochemical reactions in *Neurospora*: An "aminobenzoicless" mutant. *Proc. Natl. Acad. Sci. USA.* **28**: 234–243.

2059. Tatum, E. L., and G. W. Beadle (1945). Biochemical genetics of *Neurospora. Ann. Missouri Bot. Garden.* **32**: 125–219.

2060. Tatum, E. L., and T. T. Bell (1946). *Neurospora*. III. Biosynthesis of thiamin. *Am. J. Bot.* **33**: 15–20.

2061. Tatum, E. L., and D. M. Bonner (1944). Indole and serine in the biosynthesis and breakdown of tryptophane. *Proc. Natl. acad. Sci. USA* **30**: 30–37.

2062. Tazuya, K., M. Morisaki, K. Yamada, H. Kumaoka, and K. Saiki (1987). Biosynthesis of thiamin. Different biosynthetic routes of the thiazole moiety of thiamin in aerobic organisms and anaerobic organisms. *Biochem. Int.* **14**: 153–160.

2063. Tchorzewski, M., T. Kurihara, N. Gorlatova, N. Esaki, and K. Soda (1995). Direct database submission.

2064. Teas, H. J. (1947). The biochemistry and genetics of threonine-requiring mutants of *Neurospora crassa*. Ph.D. Thesis, California Institute of Technology, Pasadena, CA.

2065. Teas, H. J. (1951). Effects of canavanine on mutants of *Neurospora* and *Bacillus subtilis. J. Biol. Chem.* **190**: 369–375.

2066. Teas, H. J., N. H. Horowitz, and M. Fling (1948). Homoserine as a precursor of threonine and methionine in *Neurospora. J. Biol. Chem.* **172**: 651–658.

2067. Techel, D., G. Gebauer, W. Kohler, T. Braumann, B. Jastorff, and L. Rensing (1990). On the role of Ca^{2+}-calmodulin-dependent and cAMP-dependent protein phosphorylation in the circadian rhythm of *Neurospora crassa. J. Comp. Physiol. B* **159**: 695–706.

2068. Teles-Grilo, M. L., and W. Klingmüller (1975). Influence of edeine on intergenic and interallelic recombination in *Neurospora crassa. Mol. Gen. Genet.* **136**: 309–316.

2069. Teles-Grilo, M. L., and W. Klingmüller (1975). Influence of light on the inhibition of *Neurospora crassa* growth by the antibiotic edeine. *Naturwissenschaften* **62**: 444–445.

2070. Tellez-Inon, M. T., R. M. Ulloa, G. C. Glikin, and H. N. Torres (1985). Characterization of *Neurospora crassa* cyclic AMP phosphodiesterase activated by calmodulin. *Biochem. J.* **232**: 425–430.

2071. Templeton, M. D., D. R. Greenwood, and R. E. Beever (1995). Solubilization of *Neurospora crassa* rodlet proteins and identification of the predominant protein as the proteolytically processed *eas* (*ccg-2*) gene product. *Exp. Mycol.* **19**: 166–169.

2072. Temporini, E. D., and A. L. Rosa (1992). Pleiotropic and differential phenotypic expression of two *sn* (*snowflake*) mutant alleles of *Neurospora crassa*. *Curr. Genet.* **23**: 129–133.

2073. Tenney, K., and B. J. Bowman, Personal communication.

2074. Tentler, S., J. Palas, C. Enderlin, J. Campbell, C. Taft, T. K. Miller, R. L. Wood, and C. P. Selitrennikoff (1997). Inhibition of *Neurospora crassa* growth by a glucan synthase-1 antisense construct. *Curr. Microbiol.* **34**: 303–308.

2075. Terenzi, H. F., Personal communication.

2076. Terenzi, H. F., M. M. Flawia, M. T. Tellez-Inon, and H. N. W. Torres (1976). Control of *Neurospora crassa* morphology by cyclic adenosine 3′,5′-monophosphate and dibutyryl cyclic adenosine 3′,5′-monophosphate. *J. Bacteriol.* **126**: 91–99.

2077. Terenzi, H. F., M. M. Flawia, and H. N. Torres (1974). A *Neurospora crassa* morphological mutant showing reduced adenylate cyclase activity. *Biochem. Biophys. Res. Commun.* **58**: 990–996.

2078. Terenzi, H. F., J. A. Jorge, J. E. Roselino, and R. H. Migliorini (1979). Adenylyl cyclase-deficient *cr-1* (*crisp*) mutant of *Neurospora crassa*: Cyclic AMP-dependent nutritional deficiencies. *Arch. Microbiol.* **123**: 251–258.

2079. Terenzi, H. F., and J. L. Reissig (1967). Modifiers of the *cot* gene in *Neurospora*: The gulliver mutants. *Genetics* **56**: 321–329.

2080. Thayer, P. S., J. F. Ames, N. Ditullio, and W. Wallace (1960). A strain of *Neurospora crassa* requiring 2-acetylaminofluorene: Genetics and alternative growth requirements. *Fed. Proc.* **19**: 392 (Abstr.).

2081. Thedei, G., Jr., S. R. Nozawa, A. L. Simoes, and A. Rossi (1997). Gene *pho-2* codes for the multiple active forms of P_i-repressible alkaline phosphatase in the mould *Neurospora crassa*. *World J. Microbiol. Biotechnol* **13**: 609–611.

2082. Thomas, G. H., and R. L. Baxter (1987). Analysis of mutational lesions of acetate metabolism in *Neurospora crassa* by ^{13}C nuclear magnetic resonance. *J. Bacteriol.* **169**: 359–366.

2083. Thomas, P. L., and D. G. Catcheside (1969). Genetic control of flanking marker behaviour in an allelic cross of *Neurospora crassa*. *Can. J. Genet. Cytol.* **11**: 558–566.

2084. Thomas, S. A., M. L. Sargent, and R. W. Tuveson (1981). Inactivation of normal and mutant *Neurospora crassa* conidia by visible light and near-UV: Role of $^1O_2$, carotenoid composition and sensitizer location. *Photochem. Photobiol.* **33**: 349–54.

2085. Thompson-Coffe, C., G. Borioli, D. Zickler, and A. L. Rosa (1999). Pyruvate decarboxylase filaments are associated with the cortical cytoskeleton of asci and spores over the sexual cycle of filamentous ascomycetes. *Fungal Genet. Biol.* **26**: 71–80.

2086. Threlkeld, S. F. H. (1965). Pantothenic acid requirement for spore color in *Neurospora crassa*. *Can. J. Genet. Cytol.* **7**: 171–173.

2087. Threlkeld, S. F. H., K. Newcombe, and O. M. Mylyk (1969). Nonreciprocal recombination at the *pan-2* locus in *Neurospora crassa*. *Can. J. Genet. Cytol.* **11**: 54–59.

2088. Thwaites, W. M. (1967). A mutation reducing feedback regulation by arginine in suppressed *pyr-3* mutants in *Neurospora*. *Genetics* **55**: 769–781.

2089. Thwaites, W. M., F. K. Knauert, Jr., and S. S. Carney (1973). Complementation analysis of metabolite-resistant mutations with forced heterokaryons of *Neurospora crassa*. *Genetics* **74**: 581–593.

2090. Thwaites, W. M., and L. Pendyala (1969). Regulation of amino acid assimilation in a strain of *Neurospora crassa* lacking basic amino acid transport activity. *Biochim. Biophys. Acta* **192**: 455–461.

2091. Tifford, A. H. (1977). The regulation of the nitrate assimilatory pathway in *Neurospora crassa*. Ph.D. Thesis, Johns Hopkins University, Baltimore, MD.

2092. Tinsley, J. H., I. H. Lee, P. F. Minke, and M. Plamann (1998). Analysis of actin and actin-related protein 3 (ARP3) gene expression following induction of hyphal tip formation and apolar growth in *Neurospora*. *Mol. Gen. Genet.* **259**: 601–609.

2093. Tinsley, J. H., P. F. Minke, K. S. Bruno, and M. Plamann (1996). p150^{glued}, the largest subunit of the dynactin complex, is nonessential in *Neurospora* but required for nuclear distribution. *Mol. Biol. Cell* **7**: 731–742.

2094. Todd, R. B., and A. Andrianopoulos (1997). Evolution of a fungal regulatory gene family: the Zn(II)2Cys6 binuclear cluster DNA-binding motif. *Fungal Genet. Biol.* **21**: 388–405.

2095. Tognolli, M., A. Utz-Pugin, G. Turian, and C. Rossier (1998). Developmental abnormalities in benomyl-resistant strains of *Neurospora crassa*. *Mycol. Res* **102**: 869–875.

2096. Toh-e, A., and T. Ishikawa (1971). Genetic control of the synthesis of repressible phosphatases in *Neurospora crassa*. *Genetics* **69**: 339–351.

2097. Toledo, I., J. Aguirre, and W. Hansberg (1994). Enzyme inactivation related to a hyperoxidant state during conidiation of *Neurospora crassa*. *Microbiology* **140**: 2391–2397.

2098. Tom, G. D., M. Viswanath-Reddy, and J. Howe (1978). Effect of carbon source on enzymes involved in glycerol metabolism in *Neurospora crassa*. *Arch. Microbiol.* **117**: 259–263.

2099. Tomita, H., T. Soshi, and H. Inoue (1993). The *Neurospora uvs-2* gene encodes a protein which has homology to yeast Rad18 with unique zinc finger motifs. *Mol. Gen. Genet.* **238**: 225–233.

2100. Tomsett, A. B., N. S. Dunn-Coleman, and R. H. Garrett (1981). The regulation of nitrate assimilation in *Neurospora crassa*: The isolation and genetic analysis of *nmr-1* mutants. *Mol. Gen. Genet.* **182**: 229–233.

2101. Tomsett, A. B., and R. H. Garrett (1980). The isolation and characterization of mutants defective in nitrate assimilation in *Neurospora crassa*. *Genetics* **95**: 649–660.

2102. Tomsett, A. B., and R. H. Garrett (1981). Biochemical analysis of mutants defective in nitrate assimilation in *Neurospora crassa*: Evidence for autogenous control by nitrate reductase. *Mol. Gen. Genet.* **184**: 183–190.

2103. Toussaint, O., and K. Lerch (1987). Catalytic oxidation of 2-aminophenols and o-hydroxylation of aromatic amines by tyrosinase. *Biochemistry* **26**: 8567–8571.

2104. Trevillyan, J. M., M. L. Pall, and J. L. Paznokas (1981). Purification and characterization of a cyclic AMP binding protein from *Neurospora crassa*. *Abstr. Annu. Meeting Am. Soc. Microbiol* **81**: 148 (Abstr.).

2105. Trevithick, J. R., and R. L. Metzenberg (1966). Genetic alteration of pore size and other properties of the *Neurospora* cell wall. *J. Bacteriol.* **92**: 1016–1020.

2106. Trevithick, J. R., and R. L. Metzenberg (1966). Molecular sieving by *Neurospora* cell walls during secretion of invertase isozymes. *J. Bacteriol.* **92**: 1010–1015.

2107. Trinci, A. P. (1973). Growth of wild type and spreading colonial mutants of *Neurospora crassa* in batch culture and on agar medium. *Arch. Mikrobiol.* **91**: 113–126.

2108. Trinci, A. P. (1973). The hyphal growth unit of wild-type and spreading colonial mutants of *Neurospora crassa*. *Arch. Mikrobiol.* **91**: 127–136.

2109. Trinci, A. P., and A. J. Collinge (1973). Structure and plugging of septa of wild type and spreading colonial mutants of *Neurospora crassa*. *Arch. Mikrobiol.* **91**: 355–364.

2110. Trinci, A. P., and A. J. Collinge (1974). Occlusion of the septal pores of damaged hyphae of *Neurospora crassa* by hexagonal crystals. *Protoplasma* **80**: 57–67.

2111. Tropschug, M., I. B. Barthelmess, and W. Neupert (1989). Sensitivity to cyclosporin A is mediated by cyclophilin in *Neurospora crassa* and *Saccharomyces cerevisiae*. *Nature* **342**: 953–955.

2112. Tropschug, M., D. W. Nicholson, F. U. Hartl, H. Kohler, N. Pfanner, E. Wachter, and W. Neupert (1988). Cyclosporin A-binding protein (cyclophilin) of *Neurospora crassa*. One gene codes for both the cytosolic and mitochondrial forms. *J. Biol. Chem.* **263**: 14433–14440.

2113. Trupin, J. S., and H. P. Broquist (1965). Saccharopine, an intermediate of the aminoadipic acid pathway of lysine biosynthesis. I. Studies in *Neurospora crassa*. *J. Biol. Chem.* **240**: 2524–2530.

2114. Tudzynski, B., V. Homann, B. Feng, and G. A. Marzluf (1999). Isolation, characterization and disruption of the *areA* nitrogen regulatory gene of *Gibberella fujikuroi*. *Mol. Gen. Genet.* **261**: 106–114.

2115. Turcq, B., K. F. Dobinson, N. Serizawa, and A. M. Lambowitz (1992). A protein required for RNA processing and splicing in *Neurospora* mitochondria is related to gene products involved in cell-cycle protein phosphatase functions. *Proc. Natl. Acad. Sci. USA* **89**: 1676–1680.

2116. Turian, G. (1988). Polarité de croissance-différenciation des hyphes fongiques (modèles). *Cryptogam. Mycol.* **9**: 239–249.

2117. Turian, G., and D. E. Bianchi (1972). Conidiation in *Neurospora*. *Bot. Rev.* **38**: 119–154.

2118. Turian, G., and T. C. Caeser (1987). Multipolar germination of conditionally produced conidia in an "amycelial" mutant of *Neurospora crassa*. *J. Gen. Appl. Microbiol.* **33**: 543–545.

2119. Turian, G., N. Oulevey, and N. Coniordos (1971). Recherches sur la différenciation conidienne de *Neurospora crassa*. I. Organisation chimio-structurale de la conidiation conditionelle d'un mutant amycélien. *Ann. Inst. Pasteur (Paris)* **121**: 325–335.

2120. Turian, G., N. Oulevey, and F. Tissot (1967). Preliminary studies on pigmentation and ultrastructure of microconidia of *Neurospora crassa*. *Neurospora Newsl.* **11**: 17.

2121. Turner, B. C., Personal communication.

2122. Turner, B. C. (1976). Dominance of the wild-type (sensitive) allele of *cyh-1*. *Neurospora Newsl.* **23**: 24.

2123. Turner, B. C. (1977). Euploid derivatives of duplications from a translocation in *Neurospora*. *Genetics* **85**: 439–460.

2124. Turner, B. C. (2000). Two new genes that modify the action of Spore killer factors in *Neurospora*. In preparation.

2125. Turner, B. C., and D. D. Perkins (1979). *Spore killer*, a chromosomal factor in *Neurospora* that kills meiotic products not containing it. *Genetics* **93**: 587–606.

2126. Turner, B. C., and D. D. Perkins (1991). Meiotic drive in *Neurospora* and other fungi. *Am. Nat.* **137**: 416–429.

2127. Turner, B. C., and D. D. Perkins (1993). Strains for studying *Spore killer* elements in four *Neurospora* species. *Fungal Genet. Newsl.* **40**: 76–78 (corrected in Turner and Perkins, 1994).

2128. Turner, B. C., and D. D. Perkins (1994). Strains for studying *Spore killer* elements in four *Neurospora* species (correction). *Fungal Genet. Newsl.* **41**: 14.

2129. Turner, B. C., C. W. Taylor, D. D. Perkins, and D. Newmeyer (1969). New duplication-generating inversions in *Neurospora*. *Can. J. Genet. Cytol.* **11**: 622–638.

2130. Turner, G. E., and K. A. Borkovich (1993). Identification of a G protein α subunit from *Neurospora crassa* that is a member of the G_i family. *J. Biol. Chem.* **268**: 14805–14811.

2131. Turner, G. E., T. J. Jimenez, S. K. Chae, R. A. Baasiri, and K. A. Borkovich (1997). Utilization of the *Aspergillus nidulans pyrG* gene as a selectable marker for transformation and electroporation of *Neurospora crassa*. *Fungal Genet. Newsl.* **44**: 57–59.

2132. Tuveson, R. W. (1972). Genetic and enzymatic analysis of a gene controlling UV sensitivity in *Neurospora crassa*. *Mutat. Res.* **15**: 411–424.

2133. Tuveson, R. W., and J. Mangan (1970). A UV-sensitive mutant of *Neurospora* defective for photoreactivation. *Mutat. Res.* **9**: 455–466.

2134. Tyler, B. M., Personal communication.

2135. Ueki, K., and R. L. Kincaid (1993). Interchangeable associations of calcineurin regulatory subunit isoforms with mammalian and fungal catalytic subunits. *J. Biol. Chem.* **268**: 6554–6559.

2136. Ulloa, R. M., G. C. Glikin, M. T. Tellez-Inon, H. N. Torres, and N. D. Judewicz (1987). A novel stimulator of protein phosphorylation in *Neurospora crassa*. *Mol. Cell. Biochem.* **77**: 11–17.

2137. Ulloa, R. M., H. N. Torres, C. M. Ochatt, and M. T. Tellez-Inon (1991). Ca^{2+} calmodulin-dependent protein kinase activity in the ascomycete *Neurospora crassa*. *Mol. Cell. Biochem.* **102**: 155–163.

2138. Umbarger, H. E., and E. A. Adelberg (1951). The role of α-keto-β-ethylbutyric acid in the biosynthesis of isoleucine. *J. Biol. Chem.* **192**: 883–889.

2139. Urey, J. C. (1966). Enzyme induction in *Neurospora crassa*. Ph.D. Thesis, California Institute of Technology, Pasadena, CA.

2140. Van der Pas, J. C., D. A. Rohlen, U. Weidner, and H. Weiss (1991). Primary structure of the nuclear-encoded 29.9-kDa subunit of NADH:ubiquinone reductase from *Neurospora crassa* mitochondria. *Biochim. Biophys. Acta* **1089:** 389–390.

2141. van Heemst, D., K. Swart, E. F. Holub, R. van Dijk, H. H. Offenberg, T. Goosen, H. W. van den Broek, and C. Heyting (1997). Cloning, sequencing, disruption and phenotypic analysis of *uvsC*, an *Aspergillus nidulans* homologue of yeast *RAD51*. *Mol. Gen. Genet.* **254:** 654–664.

2142. Vanderleyden, J., C. Peeters, H. Verachtert, and H. Bertrand (1980). Stimulation of the alternative oxidase of *Neurospora crassa* by nucleoside phosphates. *Biochem. J.* **188:** 141–144.

2143. Vassilev, A. O., N. Plesofsky-Vig, and R. Brambl (1992). Isolation, partial amino acid sequence, and cellular distribution of heat-shock protein hsp98 from *Neurospora crassa*. *Biochim. Biophys. Acta* **1156:** 1–6.

2144. Vázquez-Laslop, N., and B. J. Bowman, Personal communication.

2145. Vázquez-Laslop, N., K. Tenney, and B. J. Bowman (1996). Characterization of a vacuolar protease in *Neurospora crassa* and the use of gene RIPing to generate protease-deficient strains. *J. Biol. Chem.* **271:** 21944–21949.

2146. Vellani, T. S., A. J. Griffiths, and N. L. Glass (1994). New mutations that suppress mating-type vegetative incompatibility in *Neurospora crassa*. *Genome* **37:** 249–255.

2146a. Verdoes, J. C., P. Krubasik, G. Sandmann, and A. J. J. van Ooyen (1999). Isolation and functional characterization of a novel type of carotenoid biosynthetic gene from *Xanthophyllomyces dendrorhous*. *Mol. Gen. Genet.* **262:** 453–461.

2147. Versaw, W. K., and R. L. Metzenberg (1995). Genetic mapping of the *N. crassa pho-5* gene. *Fungal Genet. Newsl.* **42:** 78.

2148. Versaw, W. K., and R. L. Metzenberg (1996). Activator-independent gene expression in *Neurospora crassa*. *Genetics* **142:** 417–423.

2149. Videira, A. (1998). Complex I from the fungus *Neurospora crassa*. *Biochim. Biophys. Acta* **1364:** 89–100.

2150. Videira, A., J. E. Azevedo, S. Werner, and P. Cabral (1993). The 12.3-kDa subunit of complex I (respiratory-chain NADH dehydrogenase) from *Neurospora crassa*: cDNA cloning and chromosomal mapping of the gene. *Biochem. J.* **291 (Pt. 3):** 729–732.

2151. Videira, A., M. L. T. Grilo, S. Werner, and H. Bertrand (1988). Mitochondrial gene expression in a nuclear mutant of *Neurospora* deficient in large subunits of mitochondrial ribosomes. *Genome* **30:** 802–807.

2152. Viebrock, A., A. Perz, and W. Sebald (1982). The imported preprotein of the proteolipid subunit of the mitochondrial ATP synthase from *Neurospora crassa*. Molecular cloning and sequencing of the mRNA. *EMBO J.* **1:** 565–571.

2153. Vierula, K., K. Campsall, P. Sallmen, and Y. Yan (1999). Cloning and characterization of the colonial, *sbr* mutant of *Neurospora crassa*. *Fungal Genet. Newsl.* **46 (Suppl.):** 91 (Abstr.).

2154. Vierula, P. J. (1997). Cloning and characterization of a *Neurospora crassa* ribosomal protein gene, *crps-7*. *Curr. Genet.* **31:** 139–143.

2155. Vierula, P. J., and J. M. Mais (1997). A gene required for nuclear migration in *Neurospora crassa* codes for a protein with cysteine-rich, LIM/RING-like domains. *Mol. Microbiol.* **24:** 331–340.

2156. Vigfusson, N. V., D. G. Walker, M. S. Islam, and J. Weijer (1971). The genetics and biochemical characterization of sterility mutants in *Neurospora crassa*. *Folia Microbiol. (Praha)* **16:** 166–196.

2157. Vigfusson, N. V., and J. Weijer (1972). Sexuality in *Neurospora crassa*. II. Genes affecting the sexual development cycle. *Genet. Res.* **19:** 205–211.

2158. Vigneron, L., J. M. Ruysschaert, and E. Goormaghtigh (1995). Fourier transform infrared spectroscopy study of the secondary structure of the reconstituted *Neurospora crassa* plasma membrane H⁺-ATPase and of its membrane-associated proteolytic peptides. *J. Biol. Chem.* **270:** 17685–17696.

2159. Vigneron, L., G. A. Scarborough, J. M. Ruysschaert, and E. Goormaghtigh (1995). Reconstitution of the *Neurospora crassa* plasma membrane H⁺-adenosine triphosphatase. *Biochim. Biophys. Acta* **1236:** 95–104.

2160. Viswanath-Reddy, M., S. N. Bennett, and H. B. Howe, Jr. (1977). Characterization of glycerol-non-utilizing and protoperithecial mutants of *Neurospora*. *Mol. Gen. Genet.* **153:** 29–38.

2161. Vittorioso, P., A. Carattoli, P. Londei, and G. Macino (1994). Internal translational initiation in the mRNA from the *Neurospora crassa albino-3* gene. *J. Biol. Chem.* **269:** 26650–26654.

2162. Vogel, H. J. (1956). A convenient growth medium for *Neurospora* (Medium N). *Microbiol. Genet. Bull.* **13:** 42–43.

2163. Vogel, H. J. (1964). Distribution of lysine pathways among fungi: Evolutionary implications. *Am. Nat.* **98:** 435–446.

2164. Vogel, H. J., and D. M. Bonner (1954). On the glutamine-proline–ornithine interrelation in *Neurospora crassa*. *Proc. Natl. Acad. Sci. USA* **40:** 688–694.

2165. Vogel, R. H., and M. J. Kopac (1959). Glutamic-γ-semialdehyde in arginine and proline synthesis of *Neurospora*: A mutant-tracer analysis. *Biochim. Biophys. Acta* **36:** 505–510.

2166. Vogel, R. H., and H. J. Vogel (1965). Repression of ornithine–glutamate transacetylase in *Neurospora*. *Genetics* **52:** 482.

2167. Vollmer, S. J., and C. Yanofsky (1986). Efficient cloning of genes of *Neurospora crassa*. *Proc. Natl. Acad. Sci. USA* **83:** 4869–4873.

2168. Vomvoyanni, V. (1974). Multigenic control of ribosomal properties associated with cycloheximide sensitivity in *Neurospora crassa*. *Nature* **248:** 508–510.

2169. Vomvoyanni, V. E., and M. P. Argyrakis (1979). Pleiotropic effects of ribosomal mutations for cycloheximide resistance in a double-resistant homocaryon of *Neurospora crassa*. *J. Bacteriol.* **139:** 620–624.

2170. von Ahsen, O., M. Tropschug, N. Pfanner, and J. Rassow (1997). The chaperonin cycle cannot substitute for prolyl isomerase activity, but GroEL alone promotes productive folding of a cyclophilin-sensitive substrate to a cyclophilin-resistant form. *EMBO J.* **16**: 4568–4578.

2171. Wachter, E., W. Sebald, and A. Tzagoloff (1977). The altered amino acid sequence of the DCCD-binding protein in the *oli-1* resistant mutant D273-10B/A21 of *Saccharomyces cerevisiae*. In " Mitochondria 1977: Genetics and Biogenesis of Mitochondria" (W. Bandlow, R. J. Schweyen, K. Wolf, and F. Kaudewitz, eds.), pp. 441–449. Walter de Gruyter, Berlin.

2172. Wagenmann, M., W. Klingmüller, and W. Neupert (1974). Edeine inhibition and resistance in *Neurospora*. *Arch. Microbiol.* **100**: 105–114.

2173. Wagner, R. P. (1949). The *in vitro* synthesis of pantothenic acid by pantothenicless and wild-type *Neurospora*. *Proc. Natl. Acad. Sci. USA.* **35**: 185–189.

2174. Wagner, R. P., A. Bergquist, T. Barbee, and K. Kiritani (1964). Genetic blocks in the isoleucine–valine pathway of *Neurospora crassa*. *Genetics* **49**: 865–882.

2175. Wagner, R. P., and B. M. Guirard (1948). A gene-controlled reaction in *Neurospora* involving the synthesis of pantothenic acid. *Proc. Natl. Acad. Sci. USA.* **34**: 398–402.

2176. Wagner, R. P., and C. H. Haddox (1951). A further analysis of the pantothenicless mutants of *Neurospora*. *Am. Nat.* **85**: 319–330.

2177. Wagner, R. P., and H. K. Mitchell (1955). "Genetics and Metabolism," 2nd ed. Wiley, New York.

2178. Walker, G. E., B. Dunbar, I. S. Hunter, H. G. Nimmo, and J. R. Coggins (1996). Evidence for a novel class of microbial 3-deoxy-D-arabino-heptulosonate-7-phosphate synthase in *Streptomyces coelicolor* A3(2), *Streptomyces rimosus* and *Neurospora crassa*. *Microbiology* **142**: 1973–1982.

2179. Walker, P. A. (1963). Mapping a tyrosinaseless mutant of *Neurospora*. *Neurospora Newsl.* **3**: 15.

2180. Wallace, D. G. (1970). 2,3,5-Triphenyltetrazolium chloride: Effect on *Neurospora crassa*. M.A. Thesis, University of North Carolina, Greensboro, NC.

2181. Wallace, D. G., and J. F. Wilson (1971). Nuclear and cytoplasmic inheritance of resistance to 2,3,4-triphenyltetrazolium chloride in *Neurospora crassa*. *Genetics* **68**: s72–s73 (Abstr.).

2182. Wan, Y., H. Liu, C. Li, and T. J. Schmidhauser (1997). Genome analysis on linkage group VI of *Neurospora crassa*. *Fungal Genet. Biol.* **21**: 329–336.

2183. Wang, D. C., S. W. Meinhardt, U. Sackmann, H. Weiss, and T. Ohnishi (1991). The iron–sulfur clusters in the two related forms of mitochondrial NADH:ubiquinone oxidoreductase made by *Neurospora crassa*. *Eur. J. Biochem.* **197**: 257–264.

2184. Wang, L.-W. C., and G. A. Marzluf (1979). Nitrogen regulation of uricase synthesis in *Neurospora crassa*. *Mol. Gen. Genet.* **176**: 385–392.

2185. Wang, S. S., J. M. Magill, and R. W. Phillips (1971). Auxotrophic and visible mutations in white spore (*ws-1*). *Neurospora Newsl.* **18**: 16–17.

2186. Wang, Z., M. Deak, and S. J. Free (1994). A *cis*-acting region required for the regulated expression of *grg-1*, a *Neurospora* glucose-repressible gene. Two regulatory sites (CRE and NRS) are required to repress *grg-1* expression. *J. Mol. Biol.* **237**: 65–74.

2187. Wang, Z., and M. S. Sachs (1997). Ribosome stalling is responsible for arginine-specific translational attenuation in *Neurospora crassa*. *Mol. Cell. Biol.* **17**: 4904–4913.

2188. Wang, Z., K. A. Tarawneh, and S. J. Free (1993). Isolation, sequencing, and characterization of *crp-5*, a gene encoding a *Neurospora* ribosomal protein. *Curr. Genet.* **23**: 330–333.

2189. Warncke, J., and C. L. Slayman (1980). Metabolic modulation of stoichiometry in a proton pump. *Biochim. Biophys. Acta* **591**: 224–233.

2190. Watanabe, K., Y. Sakuraba, and H. Inoue (1997). Genetic and molecular characterization of *Neurospora crassa mus-23*: A gene involved in recombinational repair. *Mol. Gen. Genet.* **256**: 436–445.

2191. Watson, D. H., and J. C. Wootton (1977). Affinity chromatography of the *Neurospora* NADP-specific glutamate dehydrogenase, its mutational variants and hybrid hexamers. *Biochem. J.* **167**: 95–108.

2192. Watters, M. K., and D. R. Stadler (1995). Spontaneous mutation during the sexual cycle of *Neurospora crassa*. *Genetics* **139**: 137–145.

2193. Weaver, W. M. (1988). Nucleotide sequence and initial protein characterization of a *Neurospora crassa* nuclear gene encoding a mitochondrial small subunit protein with homology to eubacterial ribosomal proteins S4 and S5. Ph.D. Thesis, Indiana University, Bloomington, IN.

2194. Webber, B. B. (1965). Genetical and biochemical studies of histidine-requiring mutants of *Neurospora crassa*. III. Correspondence between biochemical characteristics and complementation map position of *his-3* mutants. *Genetics* **51**: 263–273.

2195. Webber, B. B. (1965). Genetical and biochemical studies of histidine-requiring mutants of *Neurospora crassa*. IV. Linkage relationships of *his-3* mutants. *Genetics* **51**: 275–283.

2196. Webber, B. B., and M. E. Case (1960). Genetical and biochemical studies of histidine-requiring mutants of *Neurospora crassa*. I. Classification of mutants and characterization of mutant groups. *Genetics* **45**: 1605–1615.

2197. Webber, B. B., and H. V. Malling (1967). Relation between complementation pattern and genetic alteration in *his-3* mutants of *Neurospora*. *Genetics* **56**: 595.

2198. Webster, R. E., C. A. Nelson, and S. R. Gross (1965). The α-isopropylmalate synthetase of *Neurospora*. II. The relation between structure and complementation interactions. *Biochemistry* **4**: 2319–2327.

2199. Wegman, J., and J. A. DeMoss (1965). The enzymatic conversion of anthranilate to indolylglycerol phosphate in *Neurospora crassa*. *J. Biol. Chem.* **240**: 3781–3788.

2200. Weidner, U., U. Sackmann, U. Nehls, and H. Weiss (1991). Primary structure of the nuclear-encoded 18.3-kDa subunit of NADH:ubiquinone reductase (complex I) from *Neurospora crassa* mitochondria. *Biochim. Biophys. Acta* **1089**: 391–392.

2201. Weijer, J. (1954). A genetical investigation into the *td* locus of *Neurospora crassa*. *Genetica* **28**: 173–252.

2202. Weijer, J., and N. V. Vigfusson (1972). Sexuality in *Neurospora crassa*. I. Mutations to male sterility. *Genet. Res.* **19**: 191–204.

2203. Weiss, H., T. Friedrich, G. Hofhaus, and D. Preis (1991). The respiratory-chain NADH dehydrogenase (complex I) of mitochondria. *Eur. J. Biochem.* **197:** 563–576.

2204. Weiss, R. L., and C. A. Lee (1980). Isolation and characterization of *Neurospora crassa* mutants impaired in feedback control of ornithine synthesis. *J. Bacteriol.* **141:** 1305–1311.

2205. Welch, G. R., and F. H. Gaertner (1980). Enzyme organization in the polyaromatic–biosynthetic pathway: The *arom* conjugate and other multienzyme systems. *Curr. Topics Cell. Regul.* **16:** 113–162.

2206. West, D. J. (1975). Effect of pH and biotin on a circadian rhythm of conidiation in *Neurospora crassa*. *J. Bacteriol.* **123:** 387–389.

2207. West, D. J., and T. H. Pittenger (1977). A temperature-sensitive mutant of *Neurospora crassa* deficient in cytochrome *b*. *Mol. Gen. Genet.* **152:** 77–82.

2208. Westergaard, M., and H. K. Mitchell (1947). *Neurospora* V. A synthetic medium favoring sexual reproduction. *Am. J. Bot.* **34:** 573–577.

2209. White, B. T., and C. Yanofsky (1993). Structural characterization and expression analysis of the *Neurospora* conidiation gene *con-6*. *Dev. Biol.* **160:** 254–264.

2210. White, C., D. B. Lee, and S. J. Free (1985). *Neurospora* trehalase and its structural gene. *Genetics* **110:** 217–227.

2211. Wiebers, J. L., and H. R. Garner (1964). Use of S-methylcysteine and cystathionine by methionineless *Neurospora* mutants. *J. Bacteriol.* **88:** 1798–1804.

2212. Wieland, C. R., and K. J. McDougall (1969). Suppression of *pyr-3* mutants by *am* mutants in *Neurospora*. *Genetics* **61:** s62–s63 (Abstr.).

2213. Williams, C. A., and J. F. Wilson (1966). Cytoplasmic incompatibility reactions in *Neurospora crassa*. *Ann. NY Acad. Sci.* **129:** 853–863.

2214. Williams, J. G. K., A. R. Kubelik, J. A. Rafalski, and S. V. Tingey (1991). Genetic analysis with RAPD markers. *In* "More Gene Manipulations in Fungi" (J. W. Bennet and L. L. Lasure, eds.), pp. 431–439. Academic Press, San Diego, CA.

2215. Williams, L. G. (1968). Thymidine metabolism in *Neurospora crassa*. Ph.D. Thesis, California Institute of Technology, Pasadena, CA.

2216. Williams, L. G., and R. H. Davis (1970). Pyrimidine-specific carbamyl phosphate synthetase in *Neurospora crassa*. *J. Bacteriol.* **103:** 335–341.

2217. Williams, L. G., and H. K. Mitchell (1969). Mutants affecting thymidine metabolism in *Neurospora crassa*. *J. Bacteriol.* **100:** 383–389.

2218. Williams, L. J., G. R. Barnett, J. L. Ristow, J. Pitkin, M. Perriere, and R. H. Davis (1992). Ornithine decarboxylase gene of *Neurospora crassa*: Isolation, sequence, and polyamine-mediated regulation of its mRNA. *Mol. Cell. Biol.* **12:** 347–359.

2219. Wilson, C. H. (1985). Production of microconidia by several *fl* strains. *Neurospora Newsl.* **32:** 18.

2220. Wilson, C. H. (1990). Location of *mea-1* in LG IL of *Neurospora crassa*. *Fungal Genet. Newsl.* **37:** 46.

2221. Wilson, C. H., L. S. Sajani, and P. M. Mohan (1992). Location of a mutant resistant to cobalt and nickel in LG IIIR of *Neurospora crassa*. *Fungal Genet. Newsl.* **39:** 89.

2222. Wilson, J. F. (1998). Heterokaryon incompatibility revisited. *Neurospora 1998*, Asilomar, CA, 8 (Abstr.).

2223. Wilson, J. F., C. P. Calligan, and J. A. Dempsey (1999). A heterokaryon instability gene in the Rockefeller–Lindegren strains of *Neurospora crassa* and its possible relation to the *het i* gene in Oak Ridge–St. Lawrence strains. *Fungal Genet. Newsl.* **46:** 25–30.

2224. Wilson, J. F., and J. A. Dempsey (1999). A hyphal fusion mutant in *Neurospora crassa*. *Fungal Genet. Newsl.* **46:** 31.

2225. Wilson, J. F., and L. Garnjobst (1966). A new incompatibility locus in *Neurospora crassa*. *Genetics* **53:** 621–631.

2226. Wilson, J. F., L. Garnjobst, and E. L. Tatum (1961). Heterokaryon incompatibility in *Neurospora crassa*—Microinjection studies. *Am. J. Bot.* **48:** 299–305.

2227. Winkelmann, G. (1979). Surface iron polymers and hydroxy acids. A model of iron supply in sideramine-free fungi. *Arch. Mikrobiol.* **121:** 43–51.

2228. Winkelmann, G. (1991). Importance of siderophores in fungal growth, sporulation and spore germination. *In* "Frontiers in Mycology" (D. L. Hawksworth, ed.), pp. 49–65. C. A. B. International, Wallingford, UK.

2229. Winkelmann, G., and H. Zälmer (1973). Stoffwechselproducte von Mikroorganismen. I. Zur Spezifität des Eisentransportes. *Arch. Mikrobiol.* **88:** 49–60.

2230. Wolfinbarger, L., Jr. (1976). Mutations in *Neurospora crassa* which affect multiple amino acid transport systems. *Biochim. Biophys. Acta* **436:** 774–788.

2231. Wolfinbarger, L., Jr. (1980). Transport and utilization of amino acids by fungi. *In* "Microorganisms and Nitrogen Sources" (J. Payne, ed.), pp. 63–87. John Wiley & Sons, New York.

2232. Wolfinbarger, L., Jr. (1980). Transport and utilization of peptides by fungi. *In* "Microorganisms and Nitrogen Sources" (J. Payne, ed.), pp. 281–300. John Wiley & Sons, New York.

2233. Wolfinbarger, L., Jr., and A. G. DeBusk (1971). Molecular transport. I. *In vivo* studies of transport mutants of *Neurospora crassa* with altered amino acid competition patterns. *Arch. Biochem. Biophys.* **144:** 503–511.

2234. Wolfinbarger, L., Jr., and G. Marzluf (1976). Characterization of a mutant of *Neurospora crassa* sensitive to L-tyrosine. *J. Gen. Microbiol.* **93:** 189–193.

2235. Wolfinbarger, L., Jr., and G. A. Marzluf (1974). Peptide utilization by amino acid auxotrophs of *Neurospora crassa*. *J. Bacteriol.* **119:** 371–378.

2236. Wolfinbarger, L., Jr., and G. A. Marzluf (1975). Size restriction on utilization of peptides by amino acid auxotrophs of *Neurospora crassa*. *J. Bacteriol.* **122:** 949–956.

2237. Wolfinbarger, L., Jr., and G. A. Marzluf (1975). Specificity and regulation of peptide transport on *Neurospora crassa*. *Arch. Biochem. Biophys.* **171:** 637–644.

2238. Wolfinbarger, L., Jr., F. Snyder, and J. Castellano (1983). Peptide utilization by nitrogen-starved *Neurospora crassa*. *J. Bacteriol.* **153:** 1567–1569.

2239. Woodward, D. O. (1959). Enzyme complementation *in vitro* between adenylosuccinaseless mutants of *Neurospora crassa*. *Proc. Natl. Acad. Sci. USA.* **45:** 846–850.

2240. Woodward, D. O., C. W. H. Partridge, and N. H. Giles (1960). Studies of adenylosuccinase in mutants and revertants of *Neurospora crassa*. *Genetics* **45:** 535–554.

2241. Woodward, V. W. (1962). Complementation and recombination among *pyr-3* heteroalleles of *Neurospora crassa*. *Proc. Natl. Acad. Sci. USA.* **48**: 348–356.

2242. Woodward, V. W., and P. Schwarz (1964). *Neurospora* mutants lacking ornithine transcarbamylase. *Genetics* **49**: 845–853.

2243. Woodward, V. W., and C. K. Woodward (1968). The care and feeding of slime. *Neurospora Newsl.* **13**: 18.

2244. Woolery, G. L., L. Powers, M. Winkler, E. I. Solomon, K. Lerch, and T. G. Spiro (1984). Extended X-ray absorption fine structure study of the coupled binuclear copper active site of tyrosinase from *Neurospora crassa*. *Biochim. Biophys. Acta* **788**: 155–161.

2245. Wootton, J. C., M. J. Fraser, and A. J. Baron (1980). Efficient transformation of germinating *Neurospora* conidia using total nuclear DNA fragments. *Neurospora Newsl.* **27**: 33.

2246. Worthy, T. E., and J. L. Epler (1972). Repair of ultraviolet light induced damage to the deoxyribonucleic acid of *Neurospora crassa*. *J. Bacteriol.* **110**: 1010–1016.

2247. Worthy, T. E., and J. L. Epler (1973). Biochemical basis of radiation sensitivity in mutants of *Neurospora crassa*. *Mutat. Res.* **19**: 167–173.

2248. Woudt, L. P., A. Pastink, A. E. Kempers-Veenstra, A. E. M. Jansen, W. H. Mager, and R. J. Planta (1983). The genes coding for histones H3 and H4 in *Neurospora crassa* are unique and contain intervening sequences. *Nucleic Acids Res.* **11**: 5347–5360.

2249. Wrathall, C. R., and E. L. Tatum (1974). Hyphal wall peptides and colonial morphology in *Neurospora crassa*. *Biochem. Genet.* **12**: 59–68.

2250. Xiao, X., Y. H. Fu, and G. A. Marzluf (1995). The negative-acting NMR regulatory protein of *Neurospora crassa* binds to and inhibits the DNA-binding activity of the positive-acting nitrogen regulatory protein NIT2. *Biochemistry* **34**: 8861–8868.

2251. Xiao, X., and G. A. Marzluf (1996). Identification of the native NIT2 major nitrogen regulatory protein in nuclear extracts of *Neurospora crassa*. *Genetica* **97**: 153–163.

2252. Yajima, H., H. Inoue, A. Oikawa, and A. Yasui (1991). Cloning and functional characterization of a eucaryotic DNA photolyase from *Neurospora crassa*. *Nucleic Acids Res.* **19**: 5359–5362.

2253. Yajima, H., M. Takao, S. Yasuhira, J. H. Zhao, C. Ishii, H. Inoue, and A. Yasui (1995). A eukaryotic gene encoding an endonuclease that specifically repairs DNA damaged by ultraviolet light. *EMBO J.* **14**: 2393–2399.

2254. Yamamoto, K., D. H. Huber, H. Bertrand, C. Ishii, and H. Inoue (1998). Isolation and characterization of *Neurospora crassa* genes encoding homologs of the bacterial MutS and MutL mismatch repair proteins. *Neurospora 1998*, Asilomar, CA, 28 (Abstr.).

2255. Yamashiro, C., Personal communication.

2256. Yamashiro, C. T., D. J. Ebbole, B. U. Lee, R. E. Brown, C. Bourland, L. Madi, and C. Yanofsky (1996). Characterization of *rco-1* of *Neurospora crassa*, a pleiotropic gene affecting growth and development that encodes a homolog of Tup1 of *Saccharomyces cerevisiae*. *Mol. Cell. Biol.* **16**: 6218–6228.

2257. Yamashiro, C. T., O. Yarden, and C. Yanofsky (1992). A dominant selectable marker that is meiotically stable in *Neurospora crassa*: The *amdS* gene of *Aspergillus nidulans*. *Mol. Gen. Genet.* **236**: 121–124.

2258. Yamashita, R. A., and G. S. May (1998). Motoring along the hyphae: Molecular motors and the fungal cytoskeleton. *Curr. Opin. Cell. Biol.* **10**: 74–79.

2259. Yamashita, R. A., and W. D. Stuart (1996). Identification of several novel *Neurospora crassa* genes in the λ-ZAP I and λ-ZAP II libraries. *Fungal Genet. Newsl.* **43**: 66–67.

2260. Yang, Q., and K. A. Borkovich (1999). Mutational activation of a Gα$_i$ causes uncontrolled proliferation of aerial hyphae and increased sensitivity to heat and oxidative stress in *Neurospora crassa*. *Genetics* **151**: 107–117.

2261. Yang, Y. C., K. Y. Chen, M. Seyfzadeh, and R. H. Davis (1990). Deoxyhypusine/hypusine formation on a 21,000-Da cellular protein in a *Neurospora crassa* mutant *in vivo* and *in vitro*. *Biochim. Biophys. Acta* **1033**: 133–138.

2262. Yanofsky, C. (1952). The effects of gene change on tryptophan desmolase formation. *Proc. Natl. Acad. Sci. USA* **38**: 215–226.

2263. Yanofsky, C. (1955). Nonallelic suppressor genes affecting a single *td* locus. *Genetics* **40**: 602–603.

2264. Yanofsky, C. (1960). The tryptophan synthetase system. *Bacteriol. Rev.* **24**: 221–245.

2265. Yanofsky, C., and D. M. Bonner (1950). Evidence for the participation of kynurenine as a normal intermediate in the biosynthesis of niacin in *Neurospora*. *Proc. Natl. Acad. Sci. USA* **36**: 167–176.

2266. Yanofsky, C., and D. M. Bonner (1955). Gene interaction in tryptophan synthetase formation. *Genetics* **40**: 761–769.

2267. Yarden, O., Personal communication.

2268. Yarden, O., M. Plamann, D. J. Ebbole, and C. Yanofsky (1992). *cot-1*, a gene required for hyphal elongation in *Neurospora crassa*, encodes a protein kinase. *EMBO J.* **11**: 2159–2166.

2269. Yarden, O., and C. Yanofsky (1991). Chitin synthase 1 plays a major role in cell wall biogenesis in *Neurospora crassa*. *Genes Dev.* **5**: 2420–2430.

2270. Yatzkan, E., B. Szoor, Z. Feher, V. Dombradi, and O. Yarden (1998). Protein phosphatase 2A is involved in hyphal growth of *Neurospora crassa*. *Mol. Gen. Genet.* **259**: 523–531.

2271. Yatzkan, E., and O. Yarden (1995). Inactivation of a single type-2A phosphoprotein phosphatase is lethal in *Neurospora crassa*. *Curr. Genet.* **28**: 458–466.

2272. Yatzkan, E., and O. Yarden (1997). *ppt-1*, a *Neurospora crassa* PPT/PP5 subfamily serine/threonine protein phosphatase. *Biochim. Biophys. Acta* **1353**: 18–22.

2273. Yatzkan, E., and O. Yarden (1999). The B regulatory subunit of protein phosphatase 2A is required for completion of macroconidiation and other developmental processes in *Neurospora crassa*. *Mol. Microbiol.* **31**: 197–209.

2274. Yeadon, P. J., and D. E. Catcheside (1995). Guest: A 98-bp inverted repeat transposable element in *Neurospora crassa*. *Mol. Gen. Genet.* **247**: 105–109.

2275. Yeadon, P. J., and D. E. A. Catcheside (1995). The chromosomal region which includes the recombinator *cog* in *Neurospora crassa* is highly polymorphic. *Curr. Genet.* **28**: 155–163.

2276. Yeadon, P. J., and D. E. A. Catcheside (1999). Polymorphism around the *cog* recombinator of *Neurospora crassa* extends into adjacent structural genes. *Curr. Genet.* **35**: 631–637.
2277. Yoder, O. C., Personal communication.
2278. Yoder, O. C. (1979). Guanine-requiring mutants. *Neurospora Newsl.* **26**: 11.
2279. Yoder, O. C. (1979). Mutants with increased sensitivity to caffeine. *Neurospora Newsl.* **26**: 11.
2280. Yonemasu, R., S. J. McCready, J. M. Murray, F. Osman, M. Takao, K. Yamamoto, A. R. Lehmann, and A. Yasui (1997). Characterization of the alternative excision repair pathway of UV-damaged DNA in *Schizosaccharomyces pombe*. *Nucleic Acids Res.* **25**: 1553–1558.
2281. Yoon, J. H., B. J. Lee, and H. S. Kang (1995). The *Aspergillus uvsH* gene encodes a product homologous to yeast RAD18 and *Neurospora* UVS-2. *Mol. Gen. Genet.* **248**: 174–181.
2282. Young, J. L., G. Jarai, Y.-H. Fu, and G. A. Marzluf (1990). Nucleotide sequence and analysis of *NMR*, a negative-acting regulatory gene in the nitrogen circuit of *Neurospora crassa*. *Mol. Gen. Genet.* **222**: 120–128.
2283. Yourno, J. D., C. Juhala, and S. R. Suskind (1966). Nonallelic suppressors of a *Neurospora* tryptophan synthetase mutant. *Genetics* **53**: 437–444.
2284. Yourno, J. D., and S. R. Suskind (1964). Suppressor gene action in the tryptophan synthetase system of *Neurospora crassa*. I. Genetic studies. *Genetics* **50**: 803–816.
2285. Yourno, J. D., and S. R. Suskind (1964). Suppressor gene action in the tryptophan synthetase system of *Neurospora crassa*. II. Biochemical studies. *Genetics* **50**: 817–828.
2286. Yu, S. A., K. Garrett, and A. S. Sussman (1971). Genetic control of multiple forms of trehalase in *Neurospora crassa*. *Genetics* **68**: 473–481.
2287. Yu, Y. G., G. E. Turner, and R. L. Weiss (1996). Acetylglutamate synthase from *Neurospora crassa*: Structure and regulation of expression. *Mol. Microbiol.* **22**: 545–554.
2288. Yuan, G. F., Y. H. Fu, and G. A. Marzluf (1991). *nit-4*, a pathway-specific regulatory gene of *Neurospora crassa*, encodes a protein with a putative binuclear zinc DNA-binding domain. *Mol. Cell. Biol.* **11**: 5735–5745.
2289. Yuan, G. F., and G. A. Marzluf (1992). Molecular characterization of mutations of *nit-4*, the pathway-specific regulatory gene which controls nitrate assimilation in *Neurospora crassa*. *Mol. Microbiol.* **6**: 67–73.
2290. Yuan, G. F., and G. A. Marzluf (1992). Transformants of *Neurospora crassa* with the *nit-4* nitrogen regulatory gene: Copy number, growth rate and enzyme activity. *Curr. Genet.* **22**: 205–211.
2291. Yura, T. (1959). Genetic alteration of pyrroline-5-carboxylate reductase in *Neurospora crassa*. *Proc. Natl. Acad. Sci. USA* **45**: 197–204.
2292. Yura, T., and H. J. Vogel (1959). An ω-hydroxy-α-amino acid dehydrogenase of *Neurospora crassa*: Partial purification and some properties. *J. Biol. Chem.* **234**: 339–342.
2293. Zalokar, M. (1948). The *p*-aminobenzoic acid requirement of the "sulfonamide-requiring" mutant strain of *Neurospora*. *Proc. Natl. Acad. Sci. USA* **34**: 32–36.
2294. Zalokar, M. (1950). The sulfonamide-requiring mutant of *Neurospora*: Threonine–methionine antagonism. *J. Bacteriol.* **60**: 191–203.
2295. Zalokar, M. (1955). Biosynthesis of carotenoids in *Neurospora*. Action spectrum of photoactivation. *Arch. Biochem. Biophys.* **56**: 318–325.
2296. Zamma, A., H. Tamaru, T. Harashima, and H. Inoue (1993). Isolation and characterization of mutants defective in production of laccase in *Neurospora crassa*. *Mol. Gen. Genet.* **240**: 231–237.
2297. Zapella, P. D., A. M. da Silva, J. C. da Costa Maia, and H. F. Terenzi (1996). Serine/threonine protein phosphatases and a protein phosphatase 1 inhibitor from *Neurospora crassa*. *Braz. J. Med. Biol. Res.* **29**: 599–604.
2298. Zapella, P. D. A., H. F. Terenzi, and A. M. da Silva (1998). Direct database submission.
2299. Zensen, R., H. Husmann, R. Schneider, T. Peine, and H. Weiss (1992). *De novo* synthesis and desaturation of fatty acids at the mitochondrial acyl-carrier protein, a subunit of NADH:ubiquinone oxidoreductase in *Neurospora crassa*. *FEBS Lett.* **310**: 179–181.
2300. Zhang, X. P., A. Elofsson, D. Andreu, and E. Glaser (1999). Interaction of mitochondrial presequences with DnaK and mitochondrial hsp70. *J. Mol. Biol.* **288**: 177–190.
2301. Zhang, Y., R. Lamm, C. Pillonel, J.-R. Xu, and S. Lam (1999). The hyper-osmotic stress response pathway of *Neurospora crassa* is the target of phenylpyrrole fungicides. *Fungal Genet. Newsl.* **46 (Suppl.)**: 72 (Abstr.).
2302. Zhou, L. W., H. Haas, and G. A. Marzluf (1998). Isolation and characterization of a new gene, *sre*, which encodes a GATA-type regulatory protein that controls iron transport in *Neurospora crassa*. *Mol. Gen. Genet.* **259**: 532–540.
2303. Ziegenhagen, R., and H. P. Jennissen (1990). Plant and fungus calmodulins are polyubiquitinated at a single site in a Ca^{2+}-dependent manner. *FEBS Lett.* **273**: 253–256.
2304. Zimmer, E. (1946). Mutant strains of *Neurospora* deficient in *p*-aminobenzoic acid. M.A. Thesis, Stanford University, Stanford, CA.
2305. Zsindely, A., M. Szabolcs, M. Kavai, M. Schablik, J. Aradi, and G. Szabo (1979). Demonstration of myoinositol-1-phosphate synthase and its assumed defective variant in various *Neurospora crassa* strains by immunological methods. *Acta Biol. Acad. Sci. Hung.* **30**: 141–149.

Printed and bound by CPI Group (UK) Ltd, Croydon, CR0 4YY

03/10/2024

01040323-0020